Nano-Structures for Optics and Photonics

NATO Science for Peace and Security Series

This Series presents the results of scientific meetings supported under the NATO Programme: Science for Peace and Security (SPS).

The NATO SPS Programme supports meetings in the following Key Priority areas: (1) Defence Against Terrorism; (2) Countering other Threats to Security and (3) NATO, Partner and Mediterranean Dialogue Country Priorities. The types of meeting supported are generally "Advanced Study Institutes" and "Advanced Research Workshops". The NATO SPS Series collects together the results of these meetings. The meetings are co-organized by scientists from NATO countries and scientists from NATO's "Partner" or "Mediterranean Dialogue" countries. The observations and recommendations made at the meetings, as well as the contents of the volumes in the Series, reflect those of participants and contributors only; they should not necessarily be regarded as reflecting NATO views or policy.

Advanced Study Institutes (ASI) are high-level tutorial courses to convey the latest developments in a subject to an advanced-level audience

Advanced Research Workshops (ARW) are expert meetings where an intense but informal exchange of views at the frontiers of a subject aims at identifying directions for future action

Following a transformation of the programme in 2006 the Series has been re-named and re-organised. Recent volumes on topics not related to security, which result from meetings supported under the programme earlier, may be found in the NATO Science Series.

The Series is published by IOS Press, Amsterdam, and Springer, Dordrecht, in conjunction with the NATO Emerging Security Challenges Division.

Sub-Series

A.	Chemistry and Biology	Springer
B.	Physics and Biophysics	Springer
C.	Environmental Security	Springer
D.	Information and Communication Security	IOS Press
E.	Human and Societal Dynamics	IOS Press

http://www.nato.int/science
http://www.springer.com
http://www.iospress.nl

Series B: Physics and Biophysics

Nano-Structures for Optics and Photonics

Optical Strategies for Enhancing Sensing, Imaging, Communication and Energy Conversion

edited by

Baldassare Di Bartolo
Boston College
Chestnut Hill, MA, USA

John Collins
Wheaton College
Norton, MA, USA

and

Luciano Silvestri
Boston College
Chestnut Hill, USA

Springer
Published in Cooperation with NATO Emerging Security Challenges Division

Proceedings of the NATO Advanced Study Institute on
Nano-Structures for Optics and Photonics: Optical Strategies for Enhancing Sensing, Imaging, Communication, and Energy Conversion
Erice, Sicily, Italy
4–19 July 2013

Library of Congress Control Number: 2014950450

ISBN 978-94-017-9142-7 (PB)
ISBN 978-94-017-9132-8 (HB)
ISBN 978-94-017-9133-5 (e-book)
DOI 10.1007/978-94-017-9133-5

Published by Springer,
P.O. Box 17, 3300 AA Dordrecht, The Netherlands.

www.springer.com

Printed on acid-free paper

Italien ohne Sizilien macht gar kein Bild in der Seele: Hier ist erst der Schlüssel zu allen.

Johann W. Goethe

Preface

This book presents an account of the NATO Advanced Study Institute on "Nano-structures for Optics and Photonics: Optical Strategies for Enhancing Sensing, Imaging, Communication, and Energy Conversion," held in Erice, Sicily, Italy, from 4–19 July 2013. This meeting was organized by the International School of Atomic and Molecular Spectroscopy of the "Ettore Majorana" Center for Scientific Culture.

Nanotechnology, by taking advantage of the special properties of nano-scale matter, imitates nature that designs its structures at the molecular level. As pointed out by one of our lecturers (Novotny), "the power of optics is based on the simple fact that the energy of light quanta lies in the energy range of electronic and vibrational transitions in matter. Pushing the optical interaction to the nanometer scale opens new perspectives, properties and phenomena." The growing connection between optics and electronics, due to the increasing important role played by semiconductor materials and devices, find their expression in the term Photonics, which also reflects the importance of the photon aspect of light in the description of the performance of several optical systems. Nano-structures have unique capabilities that allow the enhanced performance of processes of interest in optical and photonic devices. In particular, these structures permit the nano-scale manipulation of photons, electrons and atoms; they represent a very hot topic of research and are relevant to many devices and applications.

Nanotechnology, optics and photonics are key technologies of the twenty first century. The rapidly increasing possibilities of nano-science enable a completely new level of molding the flow of light and controlling light-matter interaction, nearly on the atomic scale. This has for instance led to nano-antennas for light that can modify spontaneous emission of nearby molecules or that can act as high-figure-of-merit sensors. Antennas combined with emitters can also be used as novel nano-lasers (spasers). Other interesting areas are artificial optical materials called metamaterials and nano-plasmonic structures. Here, applications include but are not limited to ultra-compact and ultra-fast optical telecommunication devices, efficient sustainable solar energy conversion, and bio-photonics. Transformation optics expands the concept of metamaterials towards intentionally spatially inhomogeneous structures, e.g., invisibility cloaks. Conversely, optics and photonics also fuel nanotechnology, e.g., by novel super-resolution approaches in optical microscopy and lithography.

This Institute introduced the students to the field of nano-structures and provided a comprehensive overview on experiments and theory, basic physics and applications as well as on nano-fabrication and optical characterization. It brought together physicists, chemists, and engineers. In the best tradition of our past Institutes, it started from the consideration of fundamental principles, and reached the frontier of research in a systematic and didactic fashion.

The participants came from 25 different countries: Italy, United States, Germany, Belgium, Sweden, Belarus, Switzerland, France, Ukraine, Czech Republic, Estonia, Russia, United Kingdom, Canada, The Netherlands, Slovak Republic, Spain, Finland, Greece, Hungary, Israel, Morocco, New Zealand, Poland, and Turkey. Over the 2 weeks of the course, participants were given numerous opportunities to interact with one another, at both formal (poster sessions, seminars) and informal (e.g. dinners, excursions) events. The goal was to allow the participants to learn from one another about their scientific work and to expose them to other researchers from various cultures.

Two roundtable discussions were conducted during the course. The first discussion, conducted early in the course, allowed for the organizers and lectures to get immediate feedback from the participants regarding the organizational aspects of the course. The second roundtable meeting, held on the last day of the course, assessed the overall effectiveness of the course from the points of view of the participants. All participants filled out an evaluation form of the course and were given the opportunity to express their views regarding the meeting. The discussion and the evaluation forms indicated that the participants overwhelmingly felt that the course was a success. They appreciated the didactic nature of the course, and found some of the lecturers very inspiring. They felt that the scientific level of the course was very high, and that both the breadth and balance of the subjects covered were appropriate. They believed the atmosphere of the course helped to promote interaction between all participants, especially between students and lecturers, and that these interactions often led to creative discussions. They also appreciated that the lectures were made available to all the participants online. The evaluations provided many helpful suggestions that we will implement in the next course. Several participants suggested that additional information be made available online regarding some practical aspects of the coming to Erice (accommodations, food, climate, etc.). Another frequent suggestion was for more and/or longer poster sessions, something that can be easily accommodated in the next course. They expressed a great appreciation for those lecturers who included interactive exercises to their presentations, and expressed a desire for more such activities. This was a very useful comment, and we will seek to encourage all of our lecturers of the next course to include such activities, especially to help students understand better the related fundamental physical principles. Finally, some students requested lectures on methods of computer modelling of physical systems and on experimental techniques.

The students were enthusiastic about the meeting and felt there were ample opportunities to discuss their work not only with the lecturers, but also with one another. They indicated that their discussions with graduate students, post-doctoral researchers, and professors from other countries working in the same field, or in related fields, would likely enhance their own work. Finally, we anticipate that many of the friendships and contacts that were established at the school will lead to new collaborations and the enhancement of their research work in nanostructured materials. The following quotes from the evaluation forms give insights into how the students viewed their experience at the Institute. "Gathering high-level scientists and students in such an 'open mind way of being' is really great and stimulating for our future research life." "In general I enjoyed the lectures and especially the interaction with the other students it led to new knowledge and good ideas for my project." I wish to thank Academician Aleksandr Voitovich for his support and collaboration, Prof. John Collins, Dr. Maura Cesaria, the secretary of the Course Luciano Silvestri, and Ms. Elena Buscemi for the help they gave me during the Course. I wish to acknowledge the sponsorship of the meeting by the NATO Organization, Boston College, the Italian Ministry of Scientific Research and Technology, and the Sicilian Regional Government.

I am looking forward to our activities at the Ettore Majorana Center in years to come, including the next 2015 meeting of the International School of Atomic and Molecular Spectroscopy.

Boston, USA
December, 2013

Baldassare (Rino) Di Bartolo
Director of the International School
of Atomic and Molecular Spectroscopy
of the "Ettore Majorana" Center

Fig. 1 The co-directors

Fig. 2 The team

Fig. 3 NATO ASI Summer School Erice, July 4–19, 2013

Contents

Part III Short Seminars

Contributors

Adulfas Abrutis Department of General and Inorganic Chemistry, University of Vilnius, Vilnius, Lithuania

M. Aghaeipour Solid State Physics, Lund University, Lund, Sweden

V. Ahmadi Department of Electrical and Computer Engineering, Tarbiat Modares University, Tehran, Iran

Ismail Mekkaoui Alaoui Faculty of Sciences Semlalia, Physics Department, Cadi Ayyad University, Marrakech, Morocco

Marco Allione PSE and BESE Divisions, King Abdullah University of Science and Technology (KAUST), Thuwal, Kingdom of Saudi Arabia

Filippo Alpeggiani Department of Physics, University of Pavia, Pavia, Italy

José Alvarez LGEP, CNRS UMR 8507, Supelec, University of Paris-Sud, UPMC Univ Paris 06, Gif-sur-Yvette Cedex, France

L.C. Andreani Department of Physics, University of Pavia, Pavia, Italy

M. Andrulevicius Institute of Materials Science, Kaunas University of Technology, Kaunas, Lithuania

Nicola Armaroli Istituto per la Sintesi Organica e la Fotoreattività, Bologna, Italy

C.L. Arnold Department of Physics, Lund University, Lund, Sweden

Stephen Arnold MicroParticle PhotoPhysics Laboratory (MP^{3}L), New York University Polytechnic School of Engineering, Brooklyn, NY, USA

D. Asoli Sol Voltaics AB, Ideon Science Park, Lund, Sweden

B. Auguié The MacDiarmid Institute for Advanced Materials and Nanotechnology, School of Chemical and Physical Sciences, Victoria University of Wellington, Wellington, New Zealand

M. Aziz School of Engineering and Applied Sciences, Harvard University, Cambridge, MA, USA

Ausrine Bartasyte Institute Jean Lamour (UMR 7189) CNRS – Lorraine University, Nancy, France

E.K. Bartosiewicz Faculty of Chemistry, Chemistry Department, University of Wroclaw, Wroclaw, Poland

M. Benkovicova Department of Multilayers and Nanostructures, Institute of Physics, Slovak Academy of Sciences, Bratislava, Slovak Republic

Romeo Beccherelli Consiglio Nazionale delle Ricerche, Istituto per la Microelettronica e Microsistemi (CNR-IMM), Roma, Italy

Torsten Beck Institute of Applied Physics, Karlsruhe Institute of Technology, Karlsruhe, Germany

Edward H. Bernhardi Integrated Optical Microsystems Group, MESA+ Institute for Nanotechnology, University of Twente, Enschede, The Netherlands

Marco Bettinelli Luminescent Materials Laboratory, Universitd́i Verona, Verona, Italy

Palash Bharadwaj ETH Zürich, Zürich, Switzerland

M. Bieza Faculty of Chemistry, Chemistry Department, University of Wroclaw, Wroclaw, Poland

Gokhan Bilir Department of Physics, Istanbul Technical University, Istanbul, Turkey

Department of Physics, Boston College, Chestnut Hill, MA, USA

Sean Blakley Department of Physics and Astronomy, Texas A&M University, College Station, TX, USA

I. Bolesta Department of Electronics, Ivan Franko National University of Lviv, Lviv, Ukraine

F. Bonfigli ENEA C.R. Frascati, UTAPRAD-MNF, Rome, Italy

A. Bonifacio Center of Excellence for Nanostructured Materials, Graduate School of Nanotechnology, University of Trieste, Trieste, Italy

F. Bordi Department of Physics, Sapienza University of Rome, Rome, Italy

CNR – Istituto dei Processi Chimico-Fisici, Rome, Italy

M.T. Borgström Solid State Physics and the Nanometer Structure Consortium, Lund University, Lund, Sweden

D. Bouchier IEF, CNRS UMR 8622, University of Paris-Sud, Orsay, France

George Boulon Institute Light Matter (ILM), UMR 5306 UCB Lyon 1-CNRS, University of Lyon, Villeurbanne, France

John W. Bowen School of Systems Engineering, University of Reading, Reading, UK

J.D.B. Bradley School of Engineering and Applied Sciences, Harvard University, Cambridge, MA, USA

Laura Brigo Industrial Engineering Department and INSTM, University of Padova, Padova, Italy

G. Brusatin Industrial Engineering Department and INSTM, University of Padova, Padova, Italy

L. Calvez Equipe Verres et Céramiques, UMR-CNRS 6226, Institute des Sciences chimiques de Rennes, Université de Rennes 1, Rennes Cedex, France

Patrizio Candeloro Bionem Lab, Department of Experimental Medicine, University Magna Graecia, Germaneto-Catanzaro, Italy

F. Capasso School of Engineering and Applied Sciences, Harvard University, Cambridge, MA, USA

Rossella Capasso Institute of Cybernetics "E. Caianello" of CNR, Pozzuoli, Italy

S. Carlström Department of Physics, Lund University, Lund, Sweden

Maura Cesaria Department of Physics, Universita del Salento, Lecce, Italy

Department of Mathematics and Physics "E. De Giorgi", Universitá del Salento, Lecce, Italy

Y. Caudano Research Centre in Physics of Matter and Radiation (PMR), University of Namur, Namur, Belgium

F. Cecchet Research Centre in Physics of Matter and Radiation (PMR), University of Namur, Namur, Belgium

Laboratoire Lasers et Spectroscopies – Centre PMR – NAmur Research Institute for LIfe Sciences, University of Namur, Namur, Belgium

N. Cherkasin CEMES, UPR CNRS 8011, 29 rue Jeanne Marvig, 31055 Toulouse, France

M. Cittadini Industrial Engineering Department and INSTM, University of Padova, Padova, Italy

M. Clerici INRS-EMT, Varennes, QC, Canada

School of Engineering and Physical Sciences, Edinburgh, UK

John Collins Department of Physics and Astronomy, Wheaton College, Norton, MA, USA

Jean-François Colomer Department of Physics, University of Namur, Namur, Belgium

Maria Laura Coluccio Bionem Lab, Department of Experimental Medicine, University Magna Graecia, Germaneto-Catanzaro, Italy

Giovanni Costa Department of Physics, University of Padova, Padova, Italy

S. D'Agostino Department of Physics, University of Pavia, Pavia, Italy

Fausto D'Apuzzo Department of Physics, Sapienza University of Rome, Rome, Italy

Istituto Italiano di Tecnologia, Rome, Italy

M. Daneau INRS-EMT, Varennes, QC, Canada

University of Ottawa, Ottawa, ON, Canada

Gobind Das PSE and BESE Divisions, King Abdullah University of Science and Technology (KAUST), Thuwal, Kingdom of Saudi Arabia

F. Javier García de Abajo ICFO-Institut de Ciencies Fotoniques, Mediterranean Technology Park, Spain

Bartolomeo Della Ventura Universitá degli Studi di Roma "La Sapienza", Roma, Italy

S. Demoustier-Champagne Bio and Soft Matter (BSMA), Institute of Condensed Matter and Nanosciences (IMCN), Université Catholique de Louvain (UCL), Louvain-la-Neuve, Belgium

Olivier Deparis Department of Physics, University of Namur, Namur, Belgium

Laboratoire de Physique du Solide, University of Namur, Namur, Belgium

Baldassare Di Bartolo Department of Physics, Boston College, Chestnut Hill, MA, USA

Enzo Di Fabrizio PSE and BESE Divisions, King Abdullah University of Science and Technology (KAUST), Thuwal, Kingdom of Saudi Arabia

Bionem Lab, Department of Experimental Clinics, University Magna Graecia, Germaneto-Catanzaro, Italy

A. Di Gaspare CNR – Institute of Photonics and nanotechnologies, Rome, Italy

Shawn Divitt ETH Zürich, Zürich, Switzerland

F. Domenici Department of Physics, Sapienza University of Rome, Rome, Italy

L.V. Doronina-Amitonova Department of Neuroscience, Kurchatov Institute National Research Center, Moscow, Russia

Physics Department, International Laser Center, M.V. Lomonosov Moscow State University, Moscow, Russia

M.K. Ekmekçi Chemistry, Marmara University, Istanbul, Turkey

Murat Erdem Physics, Marmara University, Istanbul, Turkey

Thierry Epicier MATEIS, UMR 5510 CNRS, INSA of Lyon, University of Lyon, Villeurbanne, France

C.C. Evans School of Engineering and Applied Sciences, Harvard University, Cambridge, MA, USA

V.V. Fadeev Department of Physics, M.V. Lomonosov Moscow State University, Moscow, Russia

Andrea Falqui Dipartimento di Scienze Chimiche e Geologiche, Università di Cagliari, Monserrato (CA), Italy

Xudong Fan Department of Biomedical Engineering, University of Michigan, Ann Arbor, MI, USA

Andrei B. Fedotov Physics Department, International Laser Center, M.V. Lomonosov Moscow State University, Moscow, Russia

Russian Quantum Center, Moscow, Russia

Department of Neuroscience, Kurchatov Institute National Research Center, Moscow, Russia

F. Fossard IEF, CNRS UMR 8622, University of Paris-Sud, Orsay, France

Marco Francardi PSE and BESE Divisions, King Abdullah University of Science and Technology (KAUST), Thuwal, Kingdom of Saudi Arabia

Yan Francescato The Blackett Laboratory, Imperial College London, London, UK

Benjamin Franta School of Engineering and Applied Sciences, Harvard University, Cambridge, MA, USA

Riccardo Funari Universitá degli Studi di Napoli "Federico II", Napoli, Italy

Tibor Gál Department of Atomic Physics, Budapest University of Technology and Economics, Budapest, Hungary

Lucie Gaouyat Department of Physics, University of Namur, Namur, Belgium

Sergei Gaponenko Institute of Physics of NAS of Belarus, Minsk, Belarus

E. Gazzola Physics and Astronomy Department, University of Padova, Padova, Italy

Francesco Gentile Bionem Lab, Department of Experimental Medicine, University Magna Graecia, Germaneto-Catanzaro, Italy

Giorgos Georgiou Center for Nanophotonics, FOM Institute AMOLF, Amsterdam, The Netherlands

S. Ghafouri Department of Physics, Istanbul Technical University, Istanbul, Turkey

V. Giannini The Blackett Laboratory, Imperial College London, London, UK

Andrea Giugni PSE and BESE Divisions, King Abdullah University of Science and Technology (KAUST), Thuwal, Kingdom of Saudi Arabia

V. Giliberti Department of Physics, Sapienza University of Rome, Rome, Italy

Pawel Gluchowski Institute of Low Temperature and Structure Research, Polish Academy of Sciences, Wroclaw, Poland

S. Griesse-Nascimento School of Engineering and Applied Sciences, Harvard University, Cambridge, MA, USA

Tobias Grossmann Institute of Applied Physics, Karlsruhe Institute of Technology, Karlsruhe, Germany

M. Guglielmi Industrial Engineering Department and INSTM, University of Padova, Padova, Italy

Stéphane-Olivier Guillaume Department of Physics, University of Namur, Namur, Belgium

C. Guo Department of Physics, Lund University, Lund, Sweden

Yannick Guyot Institute Light Matter (ILM), UMR 5306 UCB Lyon 1-CNRS, University of Lyon, Villeurbanne, France

Malgorzata Guzik Faculty of Chemistry, Chemistry Department, University of Wroclaw, Wroclaw, Poland

Y. Halahovets Department of Multilayers and Nanostructures, Institute of Physics, Slovak Academy of Sciences, Bratislava, Slovak Republic

R. Hamouche IEF, CNRS UMR 8622, University of Paris-Sud, Orsay, France

Muhammad Hanif Department of Physics and Nanotechnology, Aalborg University, Aalborg, Denmark

G. Hawkins School of Systems Engineering, University of Reading, Reading, UK

Z. He Electron Microscopy for Materials Research (EMAT), University of Antwerp, Antwerp, Belgium

Luc Henrard Laboratoire de Physique du Solide, University of Namur, Namur, Belgium

M. Henry

Aline Herman Department of Physics, University of Namur, Namur, Belgium

M. Heurlin Solid State Physics and the Nanometer Structure Consortium, Lund, University, Lund, Sweden

C.M. Heyl Department of Physics, Lund University, Lund, Sweden

Stephen Holler Department of Physics, Fordham University, Bronx, NY, USA

A. Hospodková Institute of Physics of the Academy of Sciences of the Czech Republic, Prague 6, Czech Republic

Dariusz Hreniak Institute of Low Temperature and Structure Research, Polish Academy of Sciences, Wroclaw, Poland

E. Hulicius Institute of Physics of the Academy of Sciences of the Czech Republic, Prague 6, Czech Republic

S. Ibrahim LGEP, CNRS UMR 8507, Supelec, University of Paris-Sud, UPMC Univ Paris 06, Gif-sur-Yvette Cedex, France

A. Ingram Opole University of Technology, Opole, Poland

J. Iwanska Faculty of Chemistry, Chemistry Department, University of Wroclaw, Wroclaw, Poland

A. Jaffré LGEP, CNRS UMR 8507, Supelec, University of Paris-Sud, UPMC Univ Paris 06, Gif-sur-Yvette Cedex, France

Vishal Jain Solid State Physics and the Nanometer Structure Consortium, Lund University, Lund, Sweden

Laboratory of Mathematics, Physics and Electrical Engineering, Halmstad University, Halmstad, Sweden

M. Jergel Department of Multilayers and Nanostructures, Institute of Physics, Slovak Academy of Sciences, Bratislava, Slovak Republic

M. Kaiser Department of Multilayers and Nanostructures, Institute of Physics, Slovak Academy of Sciences, Bratislava, Slovak Republic

Gergely Kajtár Department of Physics, INPE FEI STU, Bratislava, Slovak Republic

V.S. Kalinov B.I. Stepanov Institute of Physics, National Academy of Sciences of Belarus, Minsk, Belarus

Heinz Kalt Institute of Applied Physics, Karlsruhe Institute of Technology, Karlsruhe, Germany

Ivan Karbovnyk Department of Electronics, Ivan Franko National University of Lviv, Lviv, Ukraine

J.P. Kleider LGEP, CNRS UMR 8507, Supelec, University of Paris-Sud, UPMC Univ Paris 06, Gif-sur-Yvette Cedex, France

Halyna Klym Lviv Polytechnic National University, Lviv, Ukraine

Lyudmyla M. Kokhtych Institute of Physics, NAS of Ukraine, Kyiv, Ukraine

P. Koppa Department of Atomic Physics, Budapest University of Technology and Economics, Budapest, Hungary

M. Kotlar Department of Multilayers and Nanostructures, Institute of Physics, Slovak Academy of Sciences, Bratislava, Slovak Republic

Institute of Electronics and Photonics, Slovak University of Technology, Bratislava, Slovak Republic

M. Kozák Faculty of Mathematics and Physics, Charles University in Prague, Praha 2, Czech Republic

Jana Kubiŝtová Faculty of Nuclear Sciences and Physical Engineering, Czech Technical University in Prague, Prague 6, Czech Republic

Institute of Physics of the Academy of Sciences of the Czech Republic, Prague 6, Czech Republic

K. Kuldová Institute of Physics of the Academy of Sciences of the Czech Republic, Prague 6, Czech Republic

S. Kumar Department of Chemistry and Biomedical Sciences, Linnaeus University, Kalmar, Sweden

Department of Biotechnology, Delhi Technological University, Delhi, India

A.S. Kutsenko L.V. Pisarzhevsky Institute for Physical Chemistry, Kyiv, Ukraine

P. La Manna Institute of Chemistry and Technology of Polymers of CNR, Pozzuoli, Italy

Aleksandr A. Lanin Physics Department, International Laser Center, M.V. Lomonosov Moscow State University, Moscow, Russia

Russian Quantum Center, Moscow, Russia

Zachary J. Lapin ETH Zürich, Zürich, Switzerland

Mercy Lard The Nanometer Structure Consortium (nmC@LU), Division of Solid State Physics, Lund University, Lund, Sweden

E.W. Larsen Department of Physics, Lund University, Lund, Sweden

P.-L. Lavertu INRS-EMT, Varennes, QC, Canada

E.C. Le Ru The MacDiarmid Institute for Advanced Materials and Nanotechnology, School of Chemical and Physical Sciences, Victoria University of Wellington, Wellington, New Zealand

Maksim S. Leanenia Stepanov Institute of Physics of NAS of Belarus, Minsk, Belarus

J. Legendziewicz Faculty of Chemistry, Chemistry Department, University of Wroclaw, Wroclaw, Poland

A. L'Huillier Department of Physics, Lund University, Lund, Sweden

O. Limaj Department of Physics, Sapienza University of Rome, Rome, Italy

INFN Laboratori Nazionali di Frascati, Rome, Italy

Tania Limongi PSE and BESE Divisions, King Abdullah University of Science and Technology (KAUST), Thuwal, Kingdom of Saudi Arabia

H. Linke The Nanometer Structure Consortium (nmC@LU), Division of Solid State Physics, Lund University, Lund, Sweden

D. Lis Research Centre in Physics of Matter and Radiation (PMR), University of Namur, Namur, Belgium

Laboratoire Lasers et Spectroscopies – Centre PMR – NAmur Research Institute for LIfe Sciences, University of Namur, Namur, Belgium

Michaël Lobet Laboratoire de Physique du Solide, University of Namur, Namur, Belgium

Eleonora Lorek Department of Physics, Lund University, Lund, Sweden

V. Lughi Center of Excellence for Nanostructured Materials, Graduate School of Nanotechnology, University of Trieste, Trieste, Italy

S. Lupi Department of Physics, Sapienza University of Rome, Rome, Italy

INFN Laboratori Nazionali di Frascati, Rome, Italy

CNR – Istituto Officina dei Materiali, Rome, Italy

E.V. Lutsenko Stepanov Institute of Physics of NAS of Belarus, Minsk, Belarus

Martin Mai Institute of Applied Physics, Karlsruhe Institute of Technology, Karlsruhe, Germany

S.A. Maier The Blackett Laboratory, Imperial College London, London, UK

E. Majkova Department of Multilayers and Nanostructures, Institute of Physics, Slovak Academy of Sciences, Bratislava, Slovak Republic

A. Månsson Department of Chemistry and Biomedical Sciences, Linnaeus University, Kalmar, Sweden

Timo Mappes Institute of Microstructure Technology, Karlsruhe Institute of Technology, Karlsruhe, Germany

Carl Zeiss AG, Corporate Research and Technology, Jena, Germany

J. Marchand-Brynaert Laboratoire de Chimie Organique et Médicale, Institute of Condensed Matter and Nanosciences (IMCN), Université Catholique de Louvain (UCL), Bâtiment Lavoisier, Louvain-la-Neuve, Belgium

Samuel Margueron Laboratoire Matériaux Optiques, Photonique et Systèmes, EA 4423, Lorraine University and Supelec, Metz, France

Monica Marini PSE and BESE Divisions, King Abdullah University of Science and Technology (KAUST), Thuwal, Kingdom of Saudi Arabia

P. Markoš Department of Physics, INPE FEI STU, Bratislava, Slovak Republic

E. Mårsell Department of Physics, Lund University, Lund, Sweden

A. Martucci Industrial Engineering Department and INSTM, University of Padova, Padova, Italy

A.A. Maskevich Department of Physics, Y. Kupala State University of Grodno, Grodno, Belarus

J. Mauritsson Department of Physics, Lund University, Lund, Sweden

Anna Mazhorova INRS-EMT, Varennes, QC, Canada

E. Mazur School of Engineering and Applied Sciences, Harvard University, Cambridge, MA, USA

D. Mencaraglia LGEP, CNRS UMR 8507, Supelec, University of Paris-Sud, UPMC Univ Paris 06, Gif-sur-Yvette Cedex, France

A. Mergen Metalurgy & Material Engineering, Marmara University, Istanbul, Turkey

G. Messina ENEA C.R. Frascati, UTAPRAD-MNF, Rome, Italy

A. Mikkelsen Department of Physics, Lund University, Lund, Sweden

F. Mirabella CRM Group AC&CS, Boulevard de Colonster 57B, Liege B-4000, Belgium

M. Miranda Department of Physics, Lund University, Lund, Sweden

Aleksandr V. Mitrofanov Institute of Laser and Information Technologies, Russian Academy of Sciences, Moscow, Russia

Physics Department, International Laser Center, M.V. Lomonosov Moscow State University, Moscow, Russia

Russian Quantum Center, Moscow, Russia

Michael Moebius School of Engineering and Applied Sciences, Harvard University, Cambridge, MA, USA

T. Molière LGEP, CNRS UMR 8507, Supelec, University of Paris-Sud, UPMC Univ Paris 06, Gif-sur-Yvette Cedex, France

R.M. Montereali ENEA C.R. Frascati, UTAPRAD-MNF, Rome, Italy

Filippo Monti Istituto per la Sintesi Organica e la Fotoreattività, Bologna, Italy

Marienette Morales Vega Center of Excellence for Nanostructured Materials, Graduate School of Nanotechnology, University of Trieste, Trieste, Italy

R. Morandotti INRS-EMT, Varennes, QC, Canada

J. Moreau Laboratoire de Chimie Organique et Médicale, Institute of Condensed Matter and Nanosciences (IMCN), Université Catholique de Louvain (UCL), Bâtiment Lavoisier, Louvain-la-Neuve, Belgium

P. Mormile Institute of Cybernetics "E. Caianello" of CNR, Pozzuoli, Italy

Alexander Moroz Gitschiner Strasse, Berlin, Germany

M.K. Mridha INRS-EMT, Varennes, QC, Canada

Sébastien Mouchet Department of Physics, University of Namur, Namur, Belgium

Philip Munoz School of Engineering and Applied Sciences, Harvard University, Cambridge, MA, USA

P. Musto Institute of Chemistry and Technology of Polymers of CNR, Pozzuoli, Italy

V. Nadazdy Department of Multilayers and Nanostructures, Institute of Physics, Slovak Academy of Sciences, Bratislava, Slovak Republic

E. Nichelatti ENEA C.R. Casaccia, UTTMAT-OTT, Rome, Italy

Aleksei. N. Novikov B.I. Stepanov Institute of Physics, National Academy of Sciences of Belarus, Minsk, Belarus

Lukas Novotny ETH Zürich, Zürich, Switzerland

A. Nowzari Solid State Physics and the Nanometer Structure Consortium, Lund University, Lund, Sweden

S. Örs Department of Atomic Physics, Budapest University of Technology and Economics, Budapest, Hungary

M. Ortolani Department of Physics, Sapienza University of Rome, Rome, Italy

CNR – Institute of Photonics and nanotechnologies, Rome, Italy

J. Oswald Institute of Physics of the Academy of Sciences of the Czech Republic, Prague 6, Czech Republic

Gönül Özen Department of Physics, Istanbul Technical University, Istanbul, Turkey

D. Paleček Department of Chemistry, Lund University, Lund, Sweden

I.M. Panchishin Department of Physics, M.V. Lomonosov Moscow State University, Moscow, Russia

J. Pangrác Institute of Physics of the Academy of Sciences of the Czech Republic, Prague 6, Czech Republic

M. Pannico Institute of Chemistry and Technology of Polymers of CNR, Pozzuoli, Italy

D. Pastor School of Engineering and Applied Sciences, Harvard University, Cambridge, MA, USA

Eleonora Pavoni Istituto per la Sintesi Organica e la Fotoreattività, Bologna, Italy

M. Peccianti Institute for Complex Systems-CNR, Rome, Italy

Nadia Peerboom Research Centre in Physics of Matter and Radiation (PMR), University of Namur, Namur, Belgium

Wolfram H.P. Pernice Institute of Nanotechnology (INT), Karlsruhe Institute of Technology (KIT), Karlsruhe, Eggenstein-Leopoldshafen, Germany

Gerardo Perozziello Bionem Lab, Department of Experimental Medicine, University Magna Graecia, Germaneto-Catanzaro, Italy

M. Persson Department of Chemistry and Biomedical Sciences, Linnaeus University, Kalmar, Sweden

H. Pettersson Solid State Physics and the Nanometer Structure Consortium, Lund University, Lund, Sweden

Laboratory of Mathematics, Physics and Electrical Engineering, Halmstad University, Halmstad, Sweden

L. Petti Institute of Cybernetics "E. Caianello" of CNR, Pozzuoli, Italy

Fabio Piccinelli Luminescent Materials Laboratory, Universitdi Verona, Verona, Italy

Valentina Plausinaitiene Department of General and Inorganic Chemistry, University of Vilnius, Vilnius, Lithuania

Martin Pokorný Faculty of Mathematics and Physics, Charles University in Prague, Praha 2, Czech Republic

Markus Pollnau Integrated Optical Microsystems Group, MESA+ Institute for Nanotechnology, University of Twente, Enschede, The Netherlands

Vladimir Popok Department of Physics and Nanotechnology, Aalborg University, Aalborg, Denmark

Remo Proietti Zaccaria Italian Institute of Technology, Genova, Italy

V. Qaradaghi Department of Electrical and Computer Engineering, Tarbiat Modares University, Tehran, Iran

Raffaella Raimondo PSE and BESE Divisions, King Abdullah University of Science and Technology (KAUST), Thuwal, Kingdom of Saudi Arabia

Bionem Lab, Department of Experimental Medicine, University Magna Graecia, Germaneto-Catanzaro, Italy

C. Renard IEF, CNRS UMR 8622, University of Paris-Sud, Orsay, France

Indrek Renge Institute of Physics, University of Tartu, Tartu, Estonia

Orad Reshef School of Engineering and Applied Sciences, Harvard University, Cambridge, MA, USA

M. Rippa Institute of Cybernetics "E. Caianello" of CNR, Pozzuoli, Italy

J.G. Rivas Center for Nanophotonics, FOM Institute AMOLF, Amsterdam, The Netherlands

F. Romanato Physics and Astronomy Department, University of Padova, Padova, Italy

X. Ropagnol INRS-EMT, Varennes, QC, Canada

I. Rovetskii Department of Electronics, Ivan Franko National University of Lviv, Lviv, Ukraine

P. Rudawski Department of Physics, Lund University, Lund, Sweden

A. Rufoloni ENEA C.R. Frascati, UTFUS-COND, Rome, Italy

L. Samuelson Solid State Physics and the Nanometer Structure Consortium, Lund University, Lund, Sweden

M. Sarrazin Department of Physics, University of Namur, Namur, Belgium

M.C. Schaafsma Center for Nanophotonics, FOM Institute AMOLF, Amsterdam, The Netherlands

D. Schryvers Electron Microscopy for Materials Research (EMAT), University of Antwerp, Antwerp, Belgium

S. Sennato Department of Physics, Sapienza University of Rome, Rome, Italy

CNR – Istituto dei Processi Chimico-Fisici, Rome, Italy

V. Sergo Center of Excellence for Nanostructured Materials, Graduate School of Nanotechnology, University of Trieste, Trieste, Italy

Elena Shabunya-Klyachkovskaya Institute of Physics of NAS of Belarus, Minsk, Belarus

E.A. Shirshin Department of Physics, M.V. Lomonosov Moscow State University, Moscow, Russia

O. Shpotyuk Lviv Institute of Materials of SRC Carat, Lviv, Ukraine

Dmitrii A. Sidorov-Biryukov Physics Department, International Laser Center, M.V. Lomonosov Moscow State University, Moscow, Russia

Russian Quantum Center, Moscow, Russia

P. Siffalovic Department of Multilayers and Nanostructures, Institute of Physics, Slovak Academy of Sciences, Bratislava, Slovak Republic

K. Šlapikas Institute of Materials Science, Kaunas University of Technology, Kaunas, Lithuania

T.N. Smirnova Institute of Physics, NAS of Ukraine, Kyiv, Ukraine

Walter R.C. Somerville The MacDiarmid Institute for Advanced Materials and Nanotechnology, School of Chemical and Physical Sciences, Victoria University of Wellington, Wellington, New Zealand

Shammai Speiser Schulich Faculty of Chemistry, Technion – Israel Institute of Technology, Haifa, Israel

V. Stankevich Institute of Physics of NAS of Belarus, Minsk, Belarus

Mark I. Stockman Center for Nano-Optics (CeNO) and Department of Physics and Astronomy, Georgia State University, Atlanta, GA, USA

Wieslaw Strek Institute of Low Temperature and Structure Research, Polish Academy of Sciences, Wroclaw, Poland

A.P. Stupak B.I. Stepanov Institute of Physics, National Academy of Sciences of Belarus, Minsk, Belarus

G. te Kronnie Department of Women's and Children's Health, University of Padova, 35131 Padova, Italy

L. ten Siethoff Department of Chemistry and Biomedical Sciences, Linnaeus University, Kalmar, Sweden

A. Tamulevičiené Institute of Materials Science, Kaunas University of Technology, Kaunas, Lithuania

S. Tamulevičius Institute of Materials Science, Kaunas University of Technology, Kaunas, Lithuania

T. Tamulevičius Institute of Materials Science, Kaunas University of Technology, Kaunas, Lithuania

Timothy Threadgold School of Systems Engineering, University of Reading, Reading, UK

Luca Tirinato PSE and BESE Divisions, King Abdullah University of Science and Technology (KAUST), Thuwal, Kingdom of Saudi Arabia

Xavier Toledo-Fuentes Laboratoire Lasers et Spectroscopies – Centre PMR – NAmur Research Institute for LIfe Sciences, University of Namur, Namur, Belgium

E. Tomaszewicz Department of Inorganic and Analytical Chemistry, West Pomeranian University of Technology, Szczecin, Poland

P. Tomaszewicz Institute of Low Temperature and Structure Research, PAS, Wroclaw, Poland

Bruno Torre PSE and BESE Divisions, King Abdullah University of Science and Technology (KAUST), Thuwal, Kingdom of Saudi Arabia

F. Trojánek Faculty of Mathematics and Physics, Charles University in Prague, Praha 2, Czech Republic

S. Vaschenko Institute of Physics of NAS of Belarus, Minsk, Belarus

F. Vidal INRS-EMT, Varennes, QC, Canada

Jean-Pol Vigneron Department of Physics, University of Namur, Namur, Belgium

L. Vincent IEF, CNRS UMR 8622, University of Paris-Sud, Orsay, France

Maria Aurora Vincenti ENEA C.R. Frascati, UTAPRAD-MNF, Rome, Italy

Andrej Vojtko Department of Multilayers and Nanostructures, Institute of Physics, Slovak Academy of Sciences, Bratislava, Slovak Republic

J. Wallentin Solid State Physics and the Nanometer Structure Consortium, Lund University, Lund, Sweden

S. Wautier Laboratoire de Chimie Organique et Médicale, Institute of Condensed Matter and Nanosciences (IMCN), Université Catholique de Louvain (UCL), Bâtiment Lavoisier, Louvain-la-Neuve, Belgium

Martin Wegener Institute of Applied Physics, Institute of Nanotechnology, and DFG-Center for Functional Nanostructures (CFN), Karlsruhe Institute of Technology (KIT), Karlsruhe, Germany

Sarah Wiegele Institute of Applied Physics, Karlsruhe Institute of Technology, Karlsruhe, Germany

Jean-Pierre Wolf GAP-Biophotonics, University of Geneva, Geneva 4, Switzerland

G.P. Yablonskii Stepanov Institute of Physics of NAS of Belarus, Minsk, Belarus

V. Yam IEF, CNRS UMR 8622, University of Paris-Sud, Orsay, France

Iryna Yaremchuk Department of Photonics, Lviv Polytechnic National University, Lviv, Ukraine

G. Zacco Physics and Astronomy Department, University of Padova, Padova, Italy

Nadezda G. Zhdanova Department of Physics, M.V. Lomonosov Moscow State University, Moscow, Russia

Aleksei M. Zheltikov Physics Department, International Laser Center, M.V. Lomonosov Moscow State University, Moscow, Russia

Russian Quantum Center, Moscow, Russia

Department of Physics and Astronomy, Texas A&M University, College Station, TX, USA

Department of Neuroscience, Kurchatov Institute National Research Center, Moscow, Russia

D. Zigmantas Department of Chemistry, Lund University, Lund, Sweden

M. Zíková Institute of Physics of the Academy of Sciences of the Czech Republic, Prague 6, Czech Republic

Dimitrios C. Zografopoulos Consiglio Nazionale delle Ricerche, Istituto per la Microelettronica e Microsistemi (CNR-IMM), Roma, Italy

E. Zych Faculty of Chemistry, Chemistry Department, University of Wroclaw, Wroclaw, Poland

Part I
Lectures

1 Nanoplasmonics: Fundamentals and Applications

Mark I. Stockman

Abstract A review of nanoplasmonics is given. This includes fundamentals, nanolocalization of optical energy and hot spots, ultrafast nanoplasmonics and control of the spatiotemporal nanolocalization of optical fields, and quantum nanoplasmonics (spaser and gain plasmonics). This chapter reviews both fundamental theoretical ideas in nanoplasmonics and selected experimental developments. It is designed both for specialists in the field and general physics readership.

Keywords Plasmonics • Nanoconcentration of optical energy • Plasmonic eigenmodes • Hot spots • Nanoscale localization • Ultrafast nanoplasmonics • Spaser • Nanolasers • Amplification and loss compensation

1.1 Introduction

1.1.1 Preamble

This is a review chapter on fundamentals of nanoplasmonics. Admittedly, the selection of the included subjects reflects the interests and expertise of the author.

We have made a conscious decision not to include such important and highly developed subject as SERS (Surface Enhanced Raman Scattering). The reason is that this subject is too large and too specialized for this chapter. There is an extensive literature devoted to SERS. This includes both reviews and original publications – see, e.g., Refs. [1–5] and a representative collective monograph [6]. Another important subject that we do not include in this review is the extraordinary transmission

---

M.I. Stockman (✉)
Center for Nano-Optics (CeNO) and Department of Physics and Astronomy,
Georgia State University, Atlanta, GA 30340, USA
e-mail: mstockman@gsu.edu

B. Di Bartolo et al. (eds.), *Nano-Structures for Optics and Photonics*,
NATO Science for Peace and Security Series B: Physics and Biophysics,
DOI 10.1007/978-94-017-9133-5_1,

of light through subwavelength holes – there are extensive literature and excellent reviews on this subject – see, e.g., [7–11]. Also, due to limitations of time and space we do not cover systematically a subject of particular interest to us: the adiabatic nanoconcentration of optical energy [12]. There are many important experimental developments and promising applications of this phenomenon [12–22]. This field by itself is large enough to warrant a dedicated review. We only briefly touch this subject in Sect. 1.4.5.

Another important class of questions that we leave mostly outside of this review chapter are concerned with applications of nanoplasmonics. Among this applications are sensing, biomedical diagnostics, labels for biomedical research, nanoantennas for light-emitting diodes, etc. There exist a significant number of reviews on the applications of nanopalsmonics, of which we mention just a few below, see also a short feature article [23]. Especially promising and important are applications to cancer treatment [24, 25], sensing and solar energy conversion to electricity [26], and photo-splitting of hydrogen [27] and water [28] ("artificial photosynthesis" for solar production of clean fuels).

Presently, nanoplasmonics became a highly developed and advanced science. It would have been an impossible task to review even a significant part of it. We select some fundamental subjects in plasmonics of high and general interest. We hope that our selection reflects the past, shows the modern state, and provides an attempt of a glimpse into the future. Specifically, our anticipation is that the ultrafast nanoplasmonics, nanoplasmonics in strong field, and the spaser as a necessary active element will be prominently presented in this future. On the other hand, it is still just a glimpse into it.

1.1.2 Composition of the Chapter

In Sect. 1.2, we present an extended introduction to nanoplasmonics. Then we consider selected subfields of nanoplasmonics in more detail. Nanoplasmonics is presently a rather developed science with a number of effects and rich applications [23]. In the center of our interest and, in our opinion, the central problem of nanoplasmonics is control and monitoring of the localization of optical energy in space on the nanometer scale and in time on the femtosecond or even attosecond scale.

In Sect. 1.3, we consider ultimately small nanoplasmonic systems with size less or on the order of skin depth l_s where we employ the so-called quasistatic approximation to describe in an analytical form the nanolocalized optical fields, their eigenmodes and hot spots, and introduce the corresponding Green's functions and solutions. This section is focused on the spatial nanoconcentration of the local optical fields.

In Sect. 1.4 we present ideas and results of ultrafast nanoplasmonics and coherent control of nanoscale localization of the optical fields, including control in time with femtosecond resolution. We will describe both theoretical ideas and some experimental results.

One of the most important problems of the nanoplasmonics, where only recently solutions and first experimental results have been obtained, is the active and gain nanoplasmonics. Its major goal is to create nanoscale quantum generators and amplifiers of optical energy. In Sect. 1.5, we present theory and a significant number of experimental results available to date regarding the spaser and related polaritonic spasers (nanolasers or plasmonic lasers). We also consider a related problem of loss compensation in metamaterials.

1.2 Basics of Nanoplasmonics

1.2.1 Fundamentals

Nanoplasmonics is a branch of optical condensed matter science devoted to optical phenomena on the nanoscale in nanostructured metal systems. A remarkable property of such systems is their ability to keep the optical energy concentrated on the nanoscale due to modes called surface plasmons (SPs). It is well known [29] and reviewed below in this chapter that the existence of SPs depends entirely on the fact that dielectric function ε_m has a negative real part, $\mathrm{Re}\,\varepsilon_m < 0$. The SPs are well pronounced as resonances when the losses are small enough, i.e., $\mathrm{Im}\,\varepsilon_m \ll -\mathrm{Re}\,\varepsilon_m$. This is a known property of a good plasmonic metal, valid, e.g., for silver in the most of the visible region. We will call a substance a good plasmonic metal if these two properties

$$\mathrm{Re}\,\varepsilon_m < 0\,, \quad \mathrm{Im}\,\varepsilon_m \ll -\mathrm{Re}\,\varepsilon_m \tag{1.1}$$

are satisfied simultaneously.

There is a limit to which an electromagnetic wave can be concentrated. We immediately note that, as we explain below, nanoplasmonics is about concentration of *electromechanical* energy at optical frequencies (in contrast to electromagnetic energy) on the nanoscale.

The scale of the concentration of electromagnetic energy is determined by the wavelength and can be understood from Fig. 1.1a. Naively, let us try to achieve the strongest light localization using two parallel perfect mirrors forming an ideal Fabry-Perot resonator. A confined wave (resonator mode) should propagate normally to the surface of the mirrors. In this case, its electric field $\mathbf{E}$ is parallel to the surface of the mirror. The ideal mirror can be thought of as a metal with a zero skin depth that does not allow the electric field of the wave $\mathbf{E}$ to penetrate inside. Therefore the field is zero inside the mirror and, due to the Maxwell boundary conditions, must be zero on the surface of the mirror. The same condition should be satisfied at the surface of the second mirror. Thus, the length L of this Fabry-Perot cavity should be equal an integer number n of the half-wavelengths of light in the inner dielectric, $L = n\lambda/2$. The minimum length of this resonator is, obviously $\lambda/2$. This implies that light cannot be confined tighter than to a length of $\lambda/2$ in each direction, with the minimum modal volume of $\lambda^3/8$.

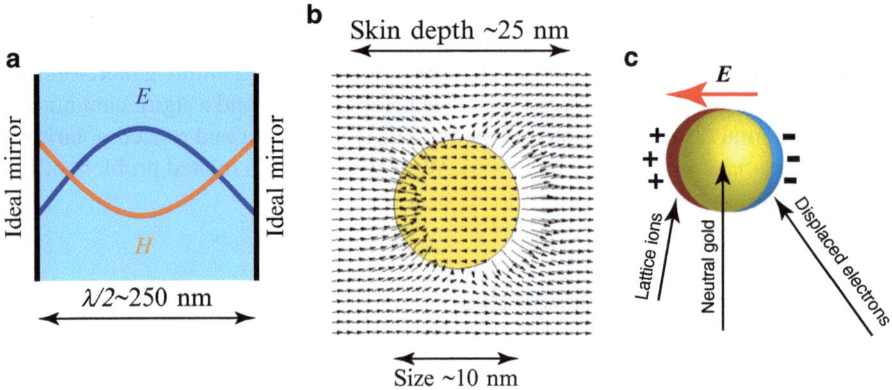

Fig. 1.1 (**a**) Localization of optical fields by ideal mirrors and (**b**) by a gold nanoparticle. (**c**) Schematic of charge separation is shown in panel

One may think that it is impossible to achieve a localization of the optical energy to smaller volume than $\lambda^3/8$ by any means, because the ideal mirrors provide the best confinement of electromagnetic waves. There are two implied assumptions: (i) The optical energy is electromagnetic energy, and (ii) The best confinement is provided by ideal mirrors. Both these assumptions must be abandoned to achieve nanolocalization of optical energy.

Consider a nanoplasmonic system whose size is less than or comparable to the skin depth

$$l_s = \lambdabar \left[\mathrm{Re} \left(\frac{-\varepsilon_m^2}{\varepsilon_m + \varepsilon_d} \right)^{1/2} \right]^{-1} , \tag{1.2}$$

where $\lambdabar = \lambda/(2\pi) = \omega/c$ is the reduced vacuum wavelength. For for single-valence plasmonic metals (silver, gold, copper, alkaline metals) $l_s \approx 25\,\mathrm{nm}$ in the entire optical region.

For such a plasmonic nanosystem with $R \lesssim l_s$, the optical electric field penetrates the entire system and drives oscillations of the metal electrons. The total energy of the system in this case is a sum of the potential energy of the electrons in the electric field and their mechanical kinetic energy. While the magnetic field is present, non-relativistic electrons' interaction with it is weak proportional to a small parameter $v_F/c{\sim}\alpha{\sim}10^{-2}$, where v_F is the electron speed at the Fermi surface, c is speed of light, and $\alpha = e^2/\hbar c$ is the fine structure constant. Thus in this limit, which is conventionally called quasistatic, the effects of the magnetic component of the total energy is relatively small. Hence, this total energy is mostly *electromechanical* (and not electromagnetic) energy. (At this point, it may be useful to refer to Eq. (1.107), which expresses the Brillouin formula for the total energy $\mathscr{E}$ of a system in such a quasistatic case.) This is why the wavelength, which determines the length scale of the energy exchange between the electric and magnetic components of an electromagnetic wave does not define the limit of the spatial localization of energy. Because the size of the system R is smaller than any electromagnetic length scale, of which smallest is l_s, it is R that defines the spatial scale of the optical energy localization. Thus the optical fields are confined on the nanoscale, and their spatial distribution scales with the system's size. This physical picture is at the heart of the nanoplasmonics.

Consider as an example a gold nanosphere of radius $R < l_s$, e.g., $R \sim 10\,\mathrm{nm}$, subjected to a plane electromagnetic wave, as shown in Fig. 1.1b. The field penetrates the metal and causes displacement of electrons with respect to the lattice resulting in the opposite charges appearing at the opposing surfaces, as illustrated in Fig. 1.1c. The attraction of these charges causes a restoring force that along with the (effective) mass of the electrons defines an electromechanical oscillator called a SP. When the frequency ω_{sp} of this SP is close to the frequency of the excitation light wave, a resonance occurs leading to the enhanced local field at the surface, as illustrated in Fig. 1.1b.

This resonant enhancement has also an adverse side: loss of energy always associated with a resonance. The rate of this loss is proportional to $\mathrm{Im}\,\varepsilon_m$ [30]. This leads to a finite lifetime of SPs. The decay rate of the plasmonic field γ is $\propto (\mathrm{Im}\,\varepsilon_m)^{-1}$. In fact, it is given below in this chapter as Eq. (1.49) in Sect. 1.3.4. This expression has originally been obtained in Ref. [31] and is also reproduced below for convenience,

$$\gamma = \frac{\mathrm{Im}\, s(\omega)}{\frac{\partial \mathrm{Re}\, s(\omega)}{\partial \omega}} \approx \frac{\mathrm{Im}\, \varepsilon_m(\omega)}{\frac{\partial \mathrm{Re} \varepsilon_m(\omega)}{\partial \omega}}, \tag{1.3}$$

where

$$s(\omega) = \frac{\varepsilon_d}{\varepsilon_d - \varepsilon_m(\omega)} \tag{1.4}$$

is Bergman's spectral parameter [29]. Note that γ does not explicitly depend on the system geometry but only on the optical frequency ω and the permittivities. However, the system's geometry determines the SP frequency ω and, thus, implicitly enters these equations. The approximate equality in Eq. (1.3) is valid for relatively small relaxation rates, $\gamma \ll \omega$. Apart from γ, an important parameter is the so-called quality factor

$$Q = \frac{\omega}{2\gamma} \approx \frac{\omega \frac{\partial \mathrm{Re} \varepsilon_m(\omega)}{\partial \omega}}{2 \mathrm{Im}\, \varepsilon_m(\omega)} \tag{1.5}$$

The quality factor determines how many optical periods free SP oscillations occur before field decays. It also shows how many times the local optical field at the surface of a plasmonic nanoparticle exceeds the external field.

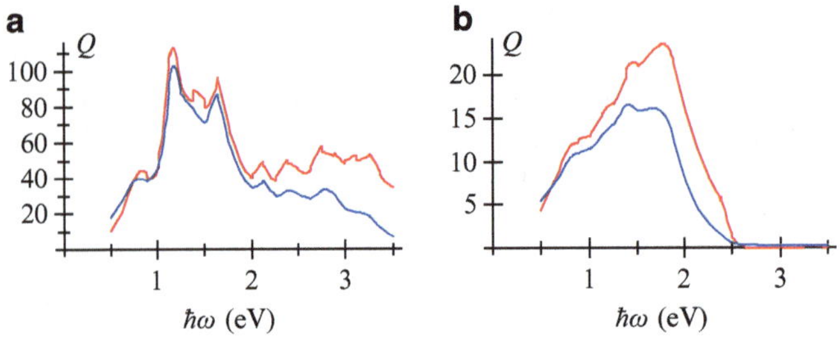

Fig. 1.2 (**a**) Quality factor Q for silver and (**b**) for gold calculated according to Eq. (1.5) (*red*) and Eq. (1.6) (*blue*) as a function of frequency ω

Note that another definition of the quality factor, which is often used, is

$$Q = \frac{-\mathrm{Re}\, \varepsilon_m(\omega)}{\mathrm{Im}\, \varepsilon_m(\omega)} . \tag{1.6}$$

The SP quality factors Q calculated according to Eqs. (1.5) and (1.6) for gold and silver using the permittivity data of Ref. [32] are shown in Fig. 1.2. The Q-factors found from these two definitions agree reasonably well in the red to near-infrared (near-ir) region but not in the yellow to blue region of the visible spectrum. The reason is that these two definitions would be equivalent if metals' permittivity were precisely described by a Drude-type formula $\mathrm{Re}\, \varepsilon_m(\omega) = -\omega_p^2/\omega^2$, where ω_p is the bulk plasma frequency; $\hbar\omega_p \approx 9\,\mathrm{eV}$ for one-electron metals such as silver, copper, gold, and alkaline metals. This formula is reasonably well applicable in the red and longer wavelength part of the spectrum, but not in the yellow to blue part where the d-band transitions are important. Note that silver is a much better plasmonic metal than gold: its Q-factor is several-fold of that of gold.

The finite skin depth of real metals leads to an effect related to nanoplasmonic confinement: a phase shift $\Delta\varphi$ for light reflected from a metal mirror deviates from a value of $\Delta\varphi = \pi$ characteristic of an ideal metal. As suggested in Ref. [33], this allows for ultrasmall cavities whose length $L \ll \lambda$. While generally this is a valid idea, there two problems with Ref. [33] that affect the validity of its specific results. First, the Fresnel reflection formulas used in this article to calculate $\Delta\varphi$ are only valid for infinite surfaces but not for the "nanomirrors" in a nanocavity. Second, Eq. (1.1) of this article expressing Q is incorrect: it contains in the denominator a quantity $\partial\left[\omega \mathrm{Im} \varepsilon_m(\omega)\right]/\partial\omega$ instead of $2\mathrm{Im}\, \varepsilon_m(\omega)$ as in Eq. (1.5). The correct expression [30] for Ohmic losses defining the Q-factor, which we reproduce as Eq. (1.108), is proportional to $\mathrm{Im}\, \varepsilon_m(\omega)$ as in Eq. (1.5) and not to $\partial\left[\omega \mathrm{Im} \varepsilon_m(\omega)\right]/\partial\omega$, which constitutes a significant difference.

The lifetime τ of the SPs is related to the spectral width as

$$\tau = \frac{1}{2\gamma} . \tag{1.7}$$

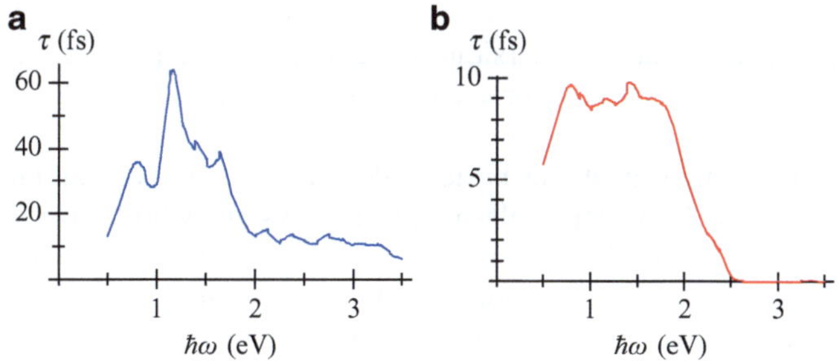

Fig. 1.3 (**a**) Lifetime τ of SPs for silver and (**b**) for gold calculated according to Eq. (1.7) as a function of frequency ω

Note that the SP spectral width γ, quality factor Q, and lifetime τ depend explicitly only on frequency ω and the type of the metal (permittivity ε_m) but not on the nanosystem's geometry or surrounding dielectric. However, this geometry and the ambient-dielectric permittivity ε_d do affect the modal frequency and enter the corresponding Eqs. (1.3), (1.5), and (1.7) implicitly via ω.

The dependence of the SP lifetime τ on frequency ω calculated for gold and silver using permittivity [32] is illustrated in Fig. 1.3. This lifetime is in the range 10–60 fs for silver and 1–10 fs for gold in the plasmonic region. These data show that nanoplasmonic phenomena are ultrafast (femtosecond).

However, the fastest linear response time τ_c of SPs, as any other linear response system, depends not on the relaxation time but solely on the bandwidth. In fact, it can be calculated as a quarter period (i.e., a time interval between zero and the maximum field) of the beating between the extreme spectral components of the plasmonic oscillations,

$$\tau_c = \frac{1}{4}\frac{2\pi}{\Delta\omega}, \tag{1.8}$$

where $\Delta\omega$ is the spectral bandwidth of the plasmonic spectrum. For gold and silver, this bandwidth is the entire optical spectrum, i.e., $\hbar\Delta\omega \approx 3.5\,\text{eV}$. If aluminum is included among system's plasmonic metals, this bandwidth is increased to $\hbar\Delta\omega \approx 9\,\text{eV}$. This yields this coherent reaction time $\tau_c \sim 100\,\text{as}$. Thus nanoplasmonics is potentially attosecond science.

While the characteristic size of a nanoplasmonic system should be limited from the top by the skin depth, $R \ll l_s$, it is also limited from the bottom by the so called nonlocality length l_{nl} – see, e.g., [34,35]. This nonlocality length is the distance that an electron with the Fermi velocity v_F moves in space during a characteristic period of the optical field,

$$l_{nl} \sim v_F/\omega \sim 1\,\text{nm}, \tag{1.9}$$

where an estimate is shown for the optical spectral region. For metal nanoparticles smaller than l_{nl}, the spatial dispersion of the dielectric response function and the related Landau damping cause broadening and disappearance of SP resonances [34, 35].

Thus, we have arrived at the basic understanding of the qualitative features of nanoplasmonics. Consider a plasmonic nanosystem whose size R satisfies a condition $l_{nl} \ll R \ll l_s$. This nanosystem is excited by an external field in resonance. In this case, the local optical field in the vicinity of such a nanosystem is enhanced by a factor $\sim Q$, which does not depend on R. The spatial extension of the local field scales with the size of the nanosystem $\propto R$. This is because $R \ll l_s$, and l_s is the smallest electromagnetic length; thus there is no length in the system that R can be comparable to. When the external field changes, the local field relaxes with the relaxation time Q/ω that does not depend on R; the lifetimes of the SP are in the femtosecond range.

In many cases of fundamental and applied significance, the size of a nanosystem can be compoarable to or even greater than l_s but still subwavelength, $\lambda \gg R \gtrsim l_s$. In such a case, the coupling to far-field radiation and radiative losses may greatly increase as we will discuss below in Sects. 1.2.2 and 1.2.3. Another important subfield of nanoplasmonics that is related to extended systems is the surface plasmon polaritons – see, e.g., a collective monograph [36]. We consider some polaritonic phenomena relevant to coherent control below in Sect. 1.4.5.

1.2.2 Nanoantennas

Consider a molecule situated in the near-field of a metal plasmonic nanosystem. Such a molecule interacts not with the external field but with the local optical field $\mathbf{E}(\mathbf{r})$ at its location $\mathbf{r}$. The interactions Hamiltonian of such a molecule with the optical field is $H' = -\mathbf{E}(\mathbf{r})\mathbf{d}$, where $\mathbf{d}$ is the dipole operator of this molecule. Note that a modal expansion of the quantized local field operator is given below in this chapter by Eq. (1.64).

Consequently, the enhanced local fields cause enhancement of radiative and nonradiative processes in which such a molecule participate. In particular, the rates of both the excitation and emission are enhanced proportionality to the local field intensity, i.e., by a factor of $\sim Q^2$. This effect is often referred to as nanoantenna effect [37–64] in analogy with the common radio-frequency antennas For the recent review of the concept and applications of optical nanoantennas see Ref. [65]. Currently, the term nanoantenna or optical antenna is used so widely that it has actually became synonymous with the entire field of nanoplasmonics: any enhancement in nanoplasmonic systems is called a nanoantenna effect.

General remarks about the terms "nanoantenna" or "optical antenna" are due. The term "antenna" has originated in the conventional radio-frequency technology where it is used in application to receivers for devices that convert the wave energy of far-field radio waves into local (near-field) electric power used to drive the input circuitry. For transmitters, antennas perform the inverse transformation: from the local field electric power to that of the emitted radio waves. Due to the general properties of time reversal symmetry there is no principal difference between the receiving and transmitting antennas: any receiving antenna can work as a transmitting one and vice versa. The mechanism of the efficiency enhancement in the radio frequency range is a combination of spatial focusing (e.g., for parabolic antennas) and resonant enhancement (e.g., for a dipole antenna). In all cases, the size of the radio antenna is comparable to or greater than the wavelength. Thus one may think that a receiving antenna collects energy from a large geometric cross section and concentrates it in a small, subwalength area.

The receiving antennas in radio and microwave technology are loaded by matched impedance loads that effectively withdraw the energy from them. This suppresses the radiation by such antennas but simultaneously dampens their resonances and makes them poor resonators.

In majority of cases, the optical antennas are not matched-loaded because they are designed not to transduce energy efficiently but to create high local fields interacting with molecules or atoms, which do not load these antennas significantly. (There are exceptions though: for instance, the nanoantenna in Ref. [66] is loaded with an adiabatic nanofocusing waveguide.) The unloaded antennas efficiently loose energy to radiation (scattering), which dampens their resonances.

A question is whether this concept of collecting energy form a large geometric cross section is a necessary paradigm also in nanoplasmonics. The answer is no, which is clear already from the fact that the enhancement of the rates of both the excitation and emission of a small chromophore (molecule, rear earth ion, etc.) in the near field of a small ($R \lesssim l_s$) plasmonic nanoparticle is $\sim Q^2$ and *does not depend* on the nanoparticle size R. This enhancement is due to the coherent resonant accumulation of the energy of the SPs during $\sim Q$ plasmonic oscillations and has nothing to do with the size of the nanoparticle. Thus such an enhancement does not quite fit into the concept of antennas as established in the radio or microwave technology.

Another test of the nanoantenna concept is whether the efficiency of a nanoantenna is necessarily increased with its size. The answer to this question is generally no. This is because for plasmonic nanoparticles, with the increase of size there is also an increased radiative loss – see below Sect. 1.2.3. In contrast, for many types of radio-frequency antennas (dish antennas or microwave-horn antennas, for instance), the efficiency does increase with the size.

1.2.3 Radiative Loss

As we described above in conjunction with Fig. 1.1c, the interaction of optical radiation with a nanoplasmonic system occurs predominantly via the dipole oscillations. The radiative decay of SPs occur via spontaneous emission of photons, which is a process that does not exist in classical physics and requires a quantum-mechanical treatment. To find the radiative life time of a SP state quantum-mechanically, we need to determine the transitional dipole matrix element $\mathbf{d}_{0p}$ between the ground state $|0\rangle$ and a single-plasmon excited state $|p\rangle$. To carry out such a computation consistently, one needs to quantize the SPs, which we have originally done in Ref. [31] and present below in Sect. 1.5.4.1.

However, there is a general way to do it without the explicit SP quantization, which we present below in this section. We start with the general expression for the polarizability α of a nanosystem obtained using quantum mechanics – see e.g., Ref. [67], which near the plasmon frequency has a singular form,

$$\alpha = \frac{1}{\hbar} \frac{\left|\mathbf{d}_{0p}\right|^2}{\omega - \omega_{sp}} , \tag{1.10}$$

where ω_{sp} is the frequency of the resonant SP mode. This can compared with the corresponding pole expression of the polarizability of a nanoplasmonic system, which is given below as Eq. (1.55), to find absolute value of the matrix element $\left|\mathbf{d}_{0p}\right|$.

Here, for the sake of simplicity, we will limit ourselves to a particular case of a nanosphere whose polarizability is given by a well-known expression

$$\alpha = R^3 \frac{\varepsilon_m(\omega) - \varepsilon_d}{\varepsilon_m(\omega) + 2\varepsilon_d} , \tag{1.11}$$

where R is the radius of the nanosphere. The SP frequency $\omega = \omega_{sp}$ corresponds to the pole of α, i.e., it satisfies an equation

$$\mathrm{Re}\, \varepsilon_m(\omega_{sp}) = -2\varepsilon_d , \tag{1.12}$$

where we neglect $\mathrm{Im}\, \varepsilon_m$. In the same approximation, near $\omega = \omega_{sp}$, we obtain from Eq. (1.11),

$$\alpha = -3R^3 \varepsilon_d \left[\left(\omega - \omega_{sp}\right) \frac{\partial \mathrm{Re}\, \varepsilon_m(\omega_{sp})}{\partial \omega_{sp}} \right]^{-1} . \tag{1.13}$$

Comparing the two pole approximations of Eqs. (1.10) and (1.13), we obtain the required expression for the dipole moment of a quantum transition between the ground state and the SP state,

$$\left|\mathbf{d}_{0p}\right|^2 = \hbar 3 R^3 \varepsilon_d \left[\frac{\partial \mathrm{Re}\, \varepsilon_m(\omega_{sp})}{\partial \omega_{sp}} \right]^{-1} . \tag{1.14}$$

Consider the well-known quantum-mechanical expression for the dipole-radiation rate (see, e.g., Ref. [67]),

$$\gamma^{(r)} = \frac{4}{3} \frac{\omega^3 \sqrt{\varepsilon_d}}{\hbar c^3} \left|\mathbf{d}_{0p}\right|^2 . \tag{1.15}$$

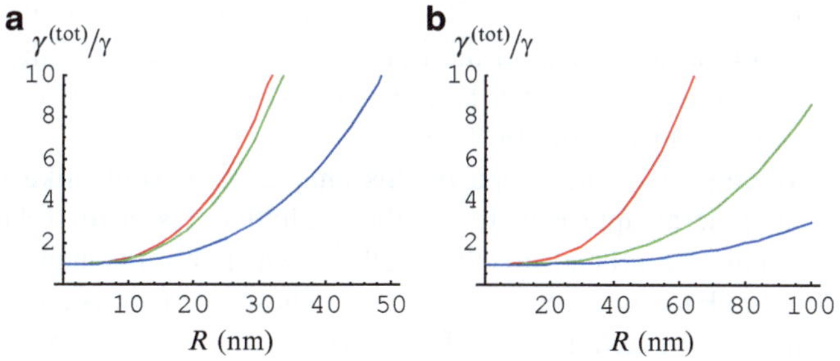

Fig. 1.4 Ratio of the rates of the total to internal loss, $\gamma^{(tot)}/\gamma$, for a nanosphere as a function of its radius R for (**a**) silver and (**b**) gold. The *blue*, *green*, and *red lines* correspond to the embedding dielectric with $\varepsilon_d = 1, 2$, and 5, respectively. The computations are made at the SP frequency ω_{sp}, which for these value of ε_d is for silver $\hbar\omega_{sp} = 3.5, 3.2, 2.5\,\text{eV}$, and for gold $\hbar\omega_{sp} = 2.6, 2.4, 2.0\,\text{eV}$, correspondingly

Substituting Eq. (1.14) into (1.15), we obtain the desired expression for the quantum-mechanical rate of the radiative decay of the SP state as

$$\gamma^{(r)} = 4\varepsilon_d^{3/2}\left(\frac{\omega_{sp}R}{c}\right)^3\left[\frac{\partial \text{Re}\,\varepsilon_m(\omega_{sp})}{\partial\omega_{sp}}\right]^{-1} . \tag{1.16}$$

Note that for losses not very large (which is the case in the entire plasmonic region for noble metals), the Kramers-Kronig relations for $\varepsilon_m(\omega)$ predict [30] that

$$\frac{\partial \text{Re}\,\varepsilon_m(\omega_{sp})}{\partial\omega_{sp}} > 0 , \tag{1.17}$$

which guarantees that $\gamma^{(r)} > 0$ in Eq. (1.16).

Comparing this expression to Eq. (1.3) (see also Eq. (1.49)), we immediately conclude that, in contrast to the internal (radiationless) loss rate γ, the radiative rate is proportional to the volume of the system (i.e., the number of the conduction electrons in it), which is understandable. Thus for systems small enough, the radiative rate can be neglected. The quality factor of the SP resonance is actually defined by the total decay rate $\gamma^{(tot)}$ (cf. Eq. (1.5)),

$$Q = \frac{\omega_{sp}}{2\gamma^{(tot)}} , \quad \gamma^{(tot)} = \gamma + \gamma^{(r)} . \tag{1.18}$$

Therefore, Q is lower for larger nanoparticles, tending to a constant for small R. To quantify it, we find a ratio

$$\frac{\gamma^{(tot)}}{\gamma} = 1 + \frac{4}{\text{Im}\,\varepsilon_m(\omega_{sp})}\left(\frac{\sqrt{\varepsilon_d}\,\omega_{sp}R}{c}\right)^3 . \tag{1.19}$$

We illustrate behavior of this rate ratio of the total to internal loss, $\gamma^{(tot)}/\gamma$, in Fig. 1.4. General conclusion is that the radiative loss for silver is not very important for nanospheres in the true quasistatic regime, i.e., for $R < l_s \approx 25\,\text{nm}$ but is a dominant mechanism of loss for $R > 30\,\mathrm{nm}$, especially in high-permittivity environments. In contrast, for gold the radiative loss is not very important in the quasistatic regime due to the much higher intrinsic losses, except for a case of a relatively high ambient permittivity, $\varepsilon_d = 5$.

Though it is outside of the scope of this chapter, we would like to point out that there is a general approach to combat radiative losses in relatively large nanoparticles. This is related to the well-known Fano resonances originally discovered by Ugo Fano in atomic spectra [68]. These resonances can be described in the following way. In certain cases of optical excitation, when two quantum paths lead to the same final quantum state of the system, the resonance peaks have specific asymmetric line shapes due to the interference of these quantum paths.

An analogous phenomenon is also known in nanoplasmonics and metamaterials [69–77]. They can be explained in the following way [77]. Apart from bright plasmonic resonances with high transitional dipole moment, there are also dark ones [78], which by themselves are not very prominent in optical spectra. However, if a bright resonance and a dark resonance coexist in a certain spectral range – which is not unlikely, because the bright resonances are wide spanning relatively wide wavelength ranges – then their optical fields interfere. This interference significantly enhances the manifestation of the dark resonance: it acquires strength from the bright resonance and shows itself as an asymmetric peak-and-dip profile characteristic of a Fano resonance. An important, albeit counterintuitive, property of the Fano resonances is that, exactly at the frequency of the Fano dip, the hot spots of the nanolocalized optical fields in the nanosystem are strongest. This is because at this frequency the nanosystem emits minimal light intensity and, consequently, it does not wastefully deplete the energy of the plasmon oscillations. This leads to a decreased radiative loss and a high quality resonance quality factor.

Thus at the frequency of a Fano resonance, the radiative loss is significantly suppressed. The width of the Fano resonances is ultimately determined by the internal (Ohmic) losses described by $\mathrm{Im}\,\varepsilon_m$. Summarizing, the Fano resonances enable one using relatively large nanoplasmonic particles or plasmonic metamaterials to achieve narrow spectral features with high local fields. These can be applied to plasmonic sensing and to produce spasers and nanolasers – see Sect. 1.5.

1.2.4 Other Important Issues of Plasmonics in Brief

There are other very important issues and directions of investigation in plasmonics that we will not be able to review in any details in this chapter due to the limitations of time and space. Below we will briefly list some of them.

1.2.4.1 Enhanced Mechanical Forces in Nanoplasmonic Systems

The resonantly enhanced local fields in the vicinity of plasmonic nanoparticles lead to enhanced nanolocalized forces acting between the nanoparticles, see, e.g.,

Refs. [79–85]. A perspective application of plasmonically-enhanced forces is optical manipulation (tweezing) of micro- and nanoparticles [86–92].

Another direction of research is opened up by the recently introduced theoretically surface-plasmon-induced drag-effect rectification (SPIDER) [93], which is based on transfer of the linear momentum from decaying surface-plasmon polaritons (SPPs) to the conduction electrons of a metal nanowire. The SPIDER effect bears a promise to generate very high terahertz fields in the vicinity of the metal nanowire.

1.2.4.2 Interaction Between Electrons and Surface Plasmons

The surface plasmonics, as it is called today, originated by a prediction of electron energy losses for an electron beam in thin metal films below the energy of the bulk plasmons [94]. This is how coherent electronic excitations called SPPs today were predicted. Soon after this prediction, the SPP-related energy losses were experimentally confirmed [95, 96]. Presently, the electron energy loss spectroscopy (EELS) in nanopalsmonics is a thriving field of research. We refer to a recent review [97] for further detail.

A distinct and original direction of research is control of mechanical motion of metal nanoparticles using electron beams [98]. It is based on the same principles as optically-induced forces. The difference in this case is that the SP oscillations in nanoparticles are excited locally, with an Angstrom precision, by a beam of fast electrons – see also Sect. 1.2.4.1 above.

There are other important phenomena in plasmonics based on electron-SP interaction called nonlocality [99]. One of them is dephasing of plasmons causing their decay into electron-hole pairs, which is called Landau damping, contributing to $\mathrm{Im}\,\varepsilon_m$. There is necessarily a related phenomenon of spatial dispersion contributing to $\mathrm{Re}\,\varepsilon_m$. These become important for plasmonics when the size of the nanosystem become too small, $R \lesssim l_{nl}$ – see Eq. (1.9). The nonlocality and Landau damping degrade plasmonic effects. The nonlocal effects lead to an increased decay rate of dipolar emitters at metal surfaces [34] and limits resolution of plasmonic imaging, making the so-called "perfect" lens [100] rather imperfect [35]. In aggregates, the nonlocality of dielectric responses causes reduction of local fields and widening of plasmonic resonances [101]. These broadening effects have initially been taken into account purely phenomenologically by adding an additional contribution to the width of plasmonic resonances $\sim A/\tau_{nl}$, where $A = const$ [102]. Practically, if the size of a nanoparticle is less then 3 nm, the non-local broadening of the SP resonances is very significant; otherwise, it can be neglected in a reasonable approximation.

The above-mentioned publications [34, 35, 99, 101] on the nonlocality phenomena are based on a semi-phenomenological approach where the nonlocality is treated via applying additional boundary conditions stemming from the electron scattering by the boundaries of the plasmonic system. A more advanced approach to nonlocality in nanoplasmonics, albeit treatable only for very small, $R \lesssim 1$ nm, nanoparticles, is based on an ab initio quantum-chemical approach of time-dependent density functional theory (usually abbreviated as TD-DFT) [103–109].

It shows that while for larger particles and relatively large spacing between them ($\gtrsim$1 nm), the semi-phenomenological models work quite well, for smaller nanoparticles and gaps the predicted local fields are significantly smaller. This is understandable because in ab initio theories there are phenomena that are important in the extremely small nanosystem such as a significant dephasing due to the stronger coupling between the collective plasmon and one-particle electron degrees of freedom, discreetness of the one-electron spectrum, spill-out of the conduction-band electrons (extension of their wave function outside of the lattice region) and the corresponding undescreening of the d-band electrons, and simply the discreetness of the lattice.

In the latest set of publications, e.g., [108, 109], this approach is called quantum nanoplasmonics. We would argue that this approach is traditionally called quantum chemistry because what is found from the TD-DFT quantum-mechanically is the dielectric response (susceptibility or polarizability) of the nanosystems. However, even to calculate theoretically the permittivity of a bulk method, one has to employ quantum-mechanical many-body approaches such as the random-phase approximation, self-consistent random-phase approximation (or GW-aproximation), or TD-DFT, etc. The only difference from the above-sited works is that for bulk metals the size effects are absent. Therefore permittivities can be adopted from experimental measurements such as Ref. [32, 110].

Based on the arguments of the preceding paragraph, we would reserve the therm "quantum plasmonics" for the subfields of nanoplasmonics studying phenomena related to quantum nature and behavior SPs and SPPs. This term has been proposed in our 2003 paper [31] introducing the spaser as a quantum generator of nanolocalized optical fields – see Sect. 1.5 and references sited therein. A related field of studies devoted to quantum behavior of single SPPs also can reasonably be called quantum plasmonics as proposed later in Refs. [111, 112].

While the decay of SP excitations is usually a parasitic phenomenon, there are some effects that completely depend on it. One of them is the SPIDER [93] mentioned above in Sect. 1.2.4.1. It is based on the transfer of the energy and momentum from SPPs to the conduction electrons, which microscopically occurs through the decay of the SPPs into electron-hole pairs leading to production of hot electrons.

Yet another range of phenomena associated with a plasmon-dephasing decay into incoherent electron-hole pairs (Landau damping) has come to the forefront lately. This is the plasmon-assisted and enhanced generation of a dc electric current due to rectification in Schottky diodes involving hot electrons [61, 113–115]. This phenomenon is promising for applications to photodetection and solar energy conversion. Note that the use of the Schottky contacts between the plasmonic metal and a semiconductor permits one to eliminate a requirement that the photon energy $\hbar\omega$ is greater that the band gap. This is replaced by a much weaker requirement that $\hbar\omega$ is greater than a significantly lower Schottky-barrier potential [116].

1.2.4.3 Nonlinear Photoprocesses in Nanoplasmonics

As became evident from the first steps of what now is called nanoplasmonics, the enhanced local fields in resonant metal nanosystems bring about strongly enhanced nonlinear responses [117–120].

Nonlinear nanoplasmonics is presently a very large and developed field. Some of its phenomena related to coherent control and spasing are discussed in Sects. 1.4, and 1.5. Here we will give a classification of the nonlinear nanoplasmonic phenomena and provide some examples, not attempting at being comprehensive.

Nonlinearities in nanoplasmonics can occur in the nanostructured plasmonic metal, in the embedding medium (dielectric), or in both. Correspondingly, we classify them as intrinsic, extrinsic, or combined. As an independent classification, these nonlinearities can be classified as weak (perturbative) or strong (nonperturbative). The perturbative nonlinearities can be coherent (or parametric), characterized by nonlinear polirizabilities [121] and incoherent such as nonlinear absorption, two-photon fluorescence, surface-enhanced hyper-Raman scattering (SEHRS) [122], nonlinear photo-modification, two-photon electron emission [123], etc.

Let us give some examples illustrating a variety of nonlinear photoprocesses in nanoplasmonics.

- Second-harmonic generation from nanostructured metal surfaces and metal nanoparticles [57, 124–132] is a coherent, perturbative (second-order or three-wave mixing), intrinsic nonlinearity.
- Enhanced four wave mixing (sum- or difference frequency generation) at metal surfaces [133] is a coherent, perturbative (third-order or four-wave), intrinsic nonlinearity.
- Another four-wave mixing process in a hybrid plasmonic-photonic waveguide involves nonlinearities in both metal and dielectric [134] and, therefore, is classified as a coherent, combined, perturbative third-order nonlinear process.
- An all-optical modulator consisting of a plasmonic waveguide covered with CdSe quantum dots [135] is based on a perturbative third-order, combined nonlinearity. To the same class belongs a nanoscale-thickness metamaterial modulator [136].
- An ultrafast all-optical modulator using polaritons in an aluminum plasmonic waveguide is based on perturbative third-order, intrinsic nonlinearity [137]. There are arguments that this nonlinearity is incoherent, based on interband population transfer of carriers [137].
- Nonperturbative (strong-field), coherent, extrinsic nonlinearity is plasmon-enhanced generation of high harmonics [138] where the enhanced nanoplasmonic fields excite argon atoms in the surrounding medium. Spaser [31] belongs to the same class where the nonlinearity is the saturation of the gain medium by the coherent plasmonic field [139]. The same is true for the loss compensation by gain [140, 141].
- Intrinsic perturbative nonlinearities in nanoplasmonics stemming from a redistribution of the electron density caused by the ponderomotive forces of nanoplasmonic fields have been predicted for surface plasmon polaritons [93, 142].

An intrinsic nonperturbative nonlinear process is the predicted plasmon soliton [143] where strong local optical fields in a plasmonic waveguide cause a significant redistribution of the conduction-electron density.

- There are also relevant strongly-nonlinear processes in non-plasmonic materials that are based on nanolocalized fields and are very similar to those in plasmonics. Among them are near-field enhanced electron acceleration from dielectric nanospheres with intense few-cycle laser fields [144]. Another such a process is a strong optical-field electron emission from tungsten nanotips controlled with an attosecond precision [145].
- Finally, a recently predicted phenomenon of metallization of dielectrics by strong optical fields [146, 147] belongs to a new class of highly-nonlinear phenomena where strong optical fields bring a dielectric nanofilm into a plasmonic metal-like state.

1.3 Nanolocalized Surface Plasmons (SPs) and Their Hot Spots

1.3.1 SPs as Eigenmodes

Assuming that a nanoplasmonic system is small enough, $R \ll \lambda\!\!\!^{-}, R \lesssim l_s$, we employ the so-called quasistatic approximation where the Maxwell equations reduce to the continuity equation for the electrostatic potential $\varphi(\mathbf{r})$,

$$\frac{\partial}{\partial \mathbf{r}}\varepsilon(\mathbf{r})\frac{\partial}{\partial \mathbf{r}}\varphi(\mathbf{r}) = 0\,. \tag{1.20}$$

The systems permittivity (dielectric function) varying in space is expressed as

$$\varepsilon(\mathbf{r}) = \varepsilon_m(\omega)\Theta(\mathbf{r}) + \varepsilon_d[1 - \Theta(\mathbf{r})]\,. \tag{1.21}$$

Here $\Theta(\mathbf{r})$ is the so-called characteristic function of the nanosystem, which is equal to 1 when $\mathbf{r}$ belongs to the metal and 0 otherwise. We solve this equation following the spectral theory developed in Refs. [78, 148, 149].

Consider a nanosystem excited by an external field with potential $\varphi_0(\mathbf{r})$ at an optical frequency ω. This potential is created by external charges and, therefore, satisfies the Laplace equation within the system,

$$\frac{\partial^2}{\partial \mathbf{r}^2}\varphi_0(\mathbf{r}) = 0\,. \tag{1.22}$$

We present the field potential as

$$\varphi(\mathbf{r}) = \varphi_0(\mathbf{r}) + \varphi_1(\mathbf{r})\,, \tag{1.23}$$

where $\varphi_1(\mathbf{r})$ is the local field.

Substituting Eq. (1.23) into Eq. (1.20) and taking Eqs. (1.21) and (1.22) into account, we obtain a second-order elliptic equation with the right-hand side that describes the external excitation source,

$$\frac{\partial}{\partial \mathbf{r}}\Theta(\mathbf{r})\frac{\partial}{\partial \mathbf{r}}\varphi_1(\mathbf{r}) - s(\omega)\frac{\partial^2}{\partial \mathbf{r}^2}\varphi_1(\mathbf{r}) = -\frac{\partial}{\partial \mathbf{r}}\Theta(\mathbf{r})\frac{\partial}{\partial \mathbf{r}}\varphi_0(\mathbf{r}) , \tag{1.24}$$

where $s(\omega)$ is Bergman's spectral parameter [148] defined by Eq. (1.4).

As a convenient basis to solve this field equation we introduce eigenmodes (SPs) with eigenfunctions $\varphi_n(\mathbf{r})$ and the corresponding eigenvalues s_n, where n is the full set of indices that identify the eigenmodes. These eigenmodes are defined by the following generalized eigenproblem,

$$\frac{\partial}{\partial \mathbf{r}}\Theta(\mathbf{r})\frac{\partial}{\partial \mathbf{r}}\varphi_n(\mathbf{r}) - s_n\frac{\partial^2}{\partial \mathbf{r}^2}\varphi_n(\mathbf{r}) = 0 , \tag{1.25}$$

where the eigenfunctions $\varphi_n(\mathbf{r})$ satisfy the homogeneous Dirichlet-Neumann boundary conditions on a surface S surrounding the system. These we set as

$$\varphi_1(\mathbf{r})|_{\mathbf{r}\in S} = 0 , \quad \text{or} \quad \mathbf{n}(\mathbf{r})\frac{\partial}{\partial \mathbf{r}}\varphi_1(\mathbf{r})\bigg|_{\mathbf{r}\in S} = 0 , \tag{1.26}$$

with $\mathbf{n}(\mathbf{r})$ denoting a normal to the surface S at a point of $\mathbf{r}$. These boundary conditions (1.26) are essential and necessary to define the eigenproblem.

From Eqs. (1.25), (1.26) applying the Gauss theorem, we find

$$s_n = \frac{\int_V \Theta(\mathbf{r}) \left|\frac{\partial}{\partial \mathbf{r}}\varphi_n(\mathbf{r})\right|^2 \mathrm{d}^3 r}{\int_V \left|\frac{\partial}{\partial \mathbf{r}}\varphi_n(\mathbf{r})\right|^2 \mathrm{d}^3 r} . \tag{1.27}$$

From this equation, it immediately follows that all the eigenvalues are real numbers and

$$1 \geq s_n \geq 0 . \tag{1.28}$$

Physically, as one can judge from Eq. (1.27), an eigenvalue of s_n is the integral fraction of the eigenmode (surface plasmon) intensity $|\partial\varphi_n(\mathbf{r})/\partial\mathbf{r}|^2$ that is localized within the metal.

Because the SP eigenproblem is real, and all the eigenvalues s_n are all real, the eigenfunctions φ_n can also be chosen real, though are not required to be chosen in such a way. Physically, it means that the quasistatic nanoplasmonic eigenproblem is time-reversible.

For the eigenproblem (1.25) and (1.26), we can introduce a scalar product of any two functions ψ_1 and ψ_2 as

$$(\psi_1|\,\psi_2) = \int_V \left[\frac{\partial}{\partial \mathbf{r}}\psi_2^*(\mathbf{r})\right]\left[\frac{\partial}{\partial \mathbf{r}}\psi_1(\mathbf{r})\right] d^3 r , \tag{1.29}$$

This construct possesses all the necessary and sufficient properties of a scalar product: it is a binary, Hermitian self-adjoined, and positive-defined operation. It is easy to show that the eigenfunctions of Eqs. (1.25) and (1.26) are orthogonal. They can be normalized as

$$(\varphi_n | \varphi_m) = \delta_{nm} , \tag{1.30}$$

1.3.2 Inhomogeneous Localization of SPs and Hot Spots of Local Fields

One of the most fundamental properties of eigenmodes is their localization. By nature, the SP eigenmodes of small nanoplasmonic systems are localized and non-propagating. This generally follows from the fact that the eigenproblem (1.25) is real and has real eigenvalues, implying time-reversal invariance and, consequently, zero current carried by any eigenmode.

From the early days of nanoplasmonics, there has been keen attention paid to the localization of SP eigenmodes, because it was immediately clear that absence of any characteristic wavelength of the localized SPs leads to the possibility of their concentration in nanoscopic volumes of the space [117, 120, 150]. Many early publications claimed that the SPs in disordered nanoplasmonics systems, e.g., fractal clusters, experience Anderson localization [151–157].

However, a different picture of the SP localization, named inhomogeneous localization, has been introduced [78, 158–161]. In this picture of inhomogeneous localization, eigenmodes of very close frequencies with varying degree of localization, from strongly localized at the minimum scale of the system to delocalized over the entire nanosystem coexist. This phenomenon of inhomogeneous localization has been experimentally confirmed recently [162]. The eigenmodes experiencing the Anderson localization are dark, corresponding to dipole-forbidden transitions, and thus can only be excited from the near field [78].

A related phenomenon is the formation of hot spots in local fields of nanoplasmonic system that we introduced in Refs. [158, 159, 163, 164]. As characteristic of the inhomogeneous localization, the energy is localized by different SP eigenmodes at vastly different scales. However, it is the localization at the minimum scale that gives the highest local fields and energy density; these tightly-localized modes are the most conspicuous in the near-field intensity distributions as the hot spots. The hot spots exist in all kind of nanoplasmonic system but they are especially strongly pronounced in disordered and aperiodic systems [165].

We will illustrate the hot spots and the inhomogeneous localization of the SP eigenmodes using the results of the original works that established the phenomena [158,159] using plasmonic-metal fractal clusters as objects. The model of these fractals were the so-called cluster-cluster aggregates (CCA) [166, 167]. In Fig. 1.5, we show two representative eigenmodes with Bergman's eigenvalues of $s_n = 0.3202$

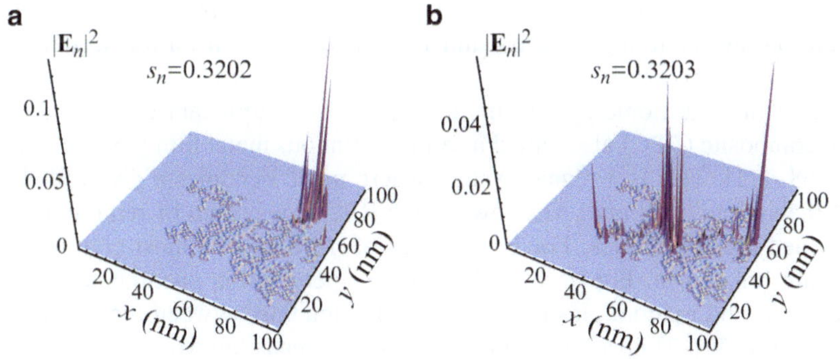

Fig. 1.5 Near-field intensity of eigenmodes computed for cluster-cluster aggregate (CCA) cluster. Square of the eigenmode electric field $|\mathbf{E}_n|^2$ is displayed against the projection of the cluster for two eigenmodes with close eigenvalues: (**a**) $s_n = 0.3202$ and (**b**) $s_n = 0.3203$. For silver embedding medium with a permittivity $\varepsilon_d \approx 2.0$, which is an approximate value for water, these modes correspond to a *blue* spectral range with $\hbar\omega \approx 3.13$ eV (Adapted from Ref. [158])

and $s_n = 0.3203$, which are very close in frequency (the blue spectral range for the case of silver in water). Both the eigenmodes are highly singular and are represented by sharp peaks – hot spots – that may be separated by the distances from the minimum scale of the system to the maximum scale that is on the order of the total size of the entire system. These eigenmodes possess very different topologies but very close eigenvalues and, consequently, have almost the same frequency $\hbar\omega \approx 3.13$ eV corresponding to the blue spectral range. This coexistence of the very different eigenmodes at the same frequency was called the inghomogeneous localization [158, 159].

The formation of host spots by the SP eigenmodes and the inhomogeneous localization of the eigenmodes are very pronounced for the fractal clusters. However, the same phenomena also take place in all dense random plasmonic systems. Physically, this phenomena is related to the absence of the characteristic length scale for SPs: the smallest electromagnetic scale is the skin depth $l_s \approx 25$ nm, which is too large on the scale of the system to affect the SP localization. The inhomogeneous localization implies that eigenmodes can be localized on all scales but this localization is always singular. The hot spots are the concentration regions of the optical energy: sharp peaks on the minimum scale ("fine grain" size) of the system are most visible.

Note that there is a fundamental difference between the plasmonic hot spots and their counterpart in the wave optics: speckles produced by scattering of laser light from a random medium. In the speckle case, there is a characteristic size of the speckles on the order of a character distance L_s between them that is determined by diffraction:

$$L_s \sim \lambda D / A \,, \tag{1.31}$$

where λ is wavelength of light, A is an aperture (cross-size of the coherent spot of light on the scattering system), and D is the distance from the scatterer to the observation screen.

One of the plasmonic system models studied in significant detail is a random metal composite (RPC) also called a semi-continuous metal film [78, 128, 149, 156, 162, 168–171]. This is a planar system where metal occupies a given fill fraction f of the system's volume. At a low f, the RPC is a system of remote, randomly positioned metal particles. For high values of f, it is an almost continuous film with rare holes in it. For $f \approx 0.5$, there are percolation phenomena: there is a large connected random cluster of the metal extending between the boundaries of the system [172]. This connected percolation cluster is known to possess a fractal geometry.

To consider statistical measures of the SP localization, we introduce the localization radius L_n of an eigenmode, which is defined as the gyration radius of its electric field intensity $|\mathbf{E}_n(\mathbf{r})|^2$, where

$$E_n(\mathbf{r}) = -\frac{\partial}{\partial \mathbf{r}}\varphi_n(\mathbf{r}) \tag{1.32}$$

is the eigenmode electric field, as

$$L_n^2 = \int_V \mathbf{r}^2|\mathbf{E}_n(\mathbf{r})|^2 \mathrm{d}^3 r - \left(\int_V \mathbf{r}|\mathbf{E}_n(\mathbf{r})|^2 \mathrm{d}^3 r\right)^2 . \tag{1.33}$$

We remind that due to Eq. (1.30), the eigenmode fields are normalized

$$\int_V |\mathbf{E}_n(\mathbf{r})|^2 \mathrm{d}^3 r = 1 , \tag{1.34}$$

so Eq. (1.33) is a standard definition of the gyration radius.

In Fig. 1.6a, we show the smoothed, discretized nanostructure of one particular sample of a RPC. This system is generated in the following way. We consider a volume of size, in our case, $32 \times 32 \times 32$ grid steps. In the central xz plane of this cube we randomly fill a cell of size 2×2 grid steps with metal with some probability f (fill factor or filling factor). Then we repeat this procedure with other 2×2 cells in that central xz plane. As a result, we arrive at a thin planar layer of thickness 2 grid steps in the y direction and fill factor of f in the central xz plane.

In Fig. 1.6b, we display all of the eigenmodes (SPs) of the above-described RPC in a plot of oscillator strength F_n versus localization length L_n. These eigenmodes are strikingly unusual.

First, there is a large number of eigenmodes with negligible oscillator strengths $F_n \lesssim 10^{-5}$. Note that the rounding-up relative error in our computations is $\sim 10^{-6}$, so these eigenmodes' oscillator strengths do not significantly differ from zero. Such eigenmodes do not couple to the far-field electromagnetic waves, and they can be neither observed nor excited from the far-field (wave) zone. We call them

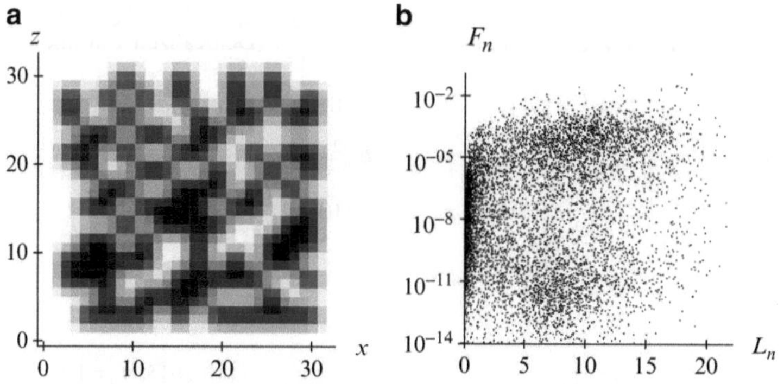

Fig. 1.6 For a planar random composite (in the xz-plane), the density of the metal component (panel (**a**)) and all eigenmodes plotted in the coordinates of oscillator strength F_n versus localization radius L_n (panel (**b**))

dark modes. They can, however, be excited and observed by NSOM (near-field scanning optical microscope) type probes in the near-field region. Such eigenmodes are also important from the computational-mathematical point of view because they are necessary for the completeness of the eigenmode set.

Second, in Fig. 1.6, there also are many eigenmodes with relatively large oscillator strengths, $F_n \gtrsim 10^{-4}$, which we call *luminous* or *bright* modes. These do couple efficiently to the far-zone fields.

Third, both the luminous and the dark modes have localization radii L_n with all possible values, from zero to one half of the diagonal system size, and with very little correlation between F_n and L_n, except for the superlocalized (zero-size) eigenmodes that are all dark. This wide range of L_n shows that *the Anderson localization does not occur for most of the modes, including all the luminous modes*. Similar to these findings in certain respects, deviations from the simple Anderson localization have been seen in some studies of the spatial structure of vibrational modes [173, 174], dephasing rates [175] in disordered solids induced by long-range (dipole-type) interactions. A direct confirmation of this picture of the inhomogeneous localization has been obtained in experiments studying fluctuations of the local density of states of localized SPs on disordered metal films [162].

To gain more insight, we show in Fig. 1.7 the local electric field intensities $|\mathbf{E}_n(\mathbf{r})|^2$ for particular eigenmodes of four extreme types, all with eigenvalues very close to $s_n = 0.2$. As a measure of the eigenmode oscillator strength, we show a normalized oscillator strength F_n. The data of Fig. 1.7 confirm the above-discussed absence of correlation between the localization length and oscillator strength, and also show that there is no correlation between the topology of the local field intensity and the oscillator strength – compare the pairs of eigenmodes: $s_n = 0.1996$ with $s_n = 0.2015$, and $s_n = 0.2$ with $s_n = 0.2011$. Note that the large and random changes of the intensities between the close eigenmodes evident in Fig. 1.7 is an underlying cause of the giant fluctuations [176] and chaos [177–179] of local fields.

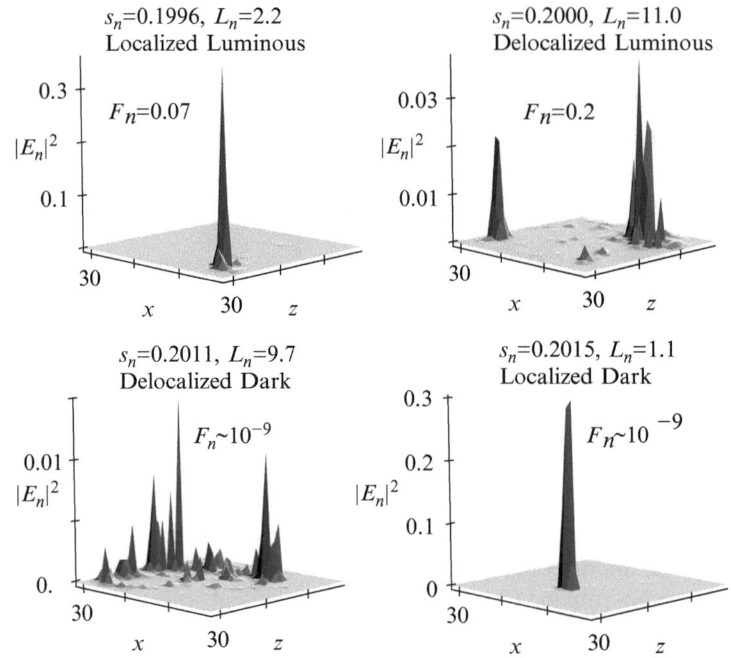

Fig. 1.7 Hot spots: Local field intensities $|\mathbf{E}_n(\mathbf{r})|^2$ of eigenmodes at the surface of the system shown in Fig. 1.6, versus spatial coordinates in the xz plane

A fundamental property of the SP eigenmodes, whether localized or delocalized, is that they may be thought of as consisting of hot spots. While the localized eigenmodes possess a single tight hot spots, the delocalized ones consist of several or many host spots. Note that the fields in the hot spots constituting a single eigenmode are coherent. In a sense, the hot spots are somewhat analogous to speckles produced by laser light scattered from a random system. However, such speckles are limited by the half-wavelength of light and cannot be smaller than that. In contrast, there is no wavelength limitations for the SP hot spots. They are limited only by the minimum scale of the underlying plasmonic system.

1.3.3 Retarded Green's Function and Field Equation Solution

Retarded Green's function $G^r(\mathbf{r},\mathbf{r}';\omega)$ of field equation (1.24), by definition, satisfies the same equation with the Dirac δ-function on the right-hand side,

$$\left[\frac{\partial}{\partial \mathbf{r}}\Theta(\mathbf{r})\frac{\partial}{\partial \mathbf{r}} - s(\omega)\frac{\partial^2}{\partial \mathbf{r}^2}\right] G^r(\mathbf{r},\mathbf{r}';\omega) = \delta(\mathbf{r}-\mathbf{r}')\,, \tag{1.35}$$

We expand this Green's function over the eigenfunctions φ_n using the orthonormality Eq. (1.30), obtaining

$$G^r(\mathbf{r},\mathbf{r}';\omega) = \sum_n \frac{\varphi_n(\mathbf{r})\,\varphi_n(\mathbf{r}')^*}{s(\omega)-s_n}\,. \tag{1.36}$$

This expression for Green's function is exact (within the quasistatic approximation) and contains the maximum information on the linear responses of a nanosystem to an arbitrary excitation field at any frequency. It satisfies all the general properties of Green's functions due to the analytical form of Eq. (1.36) as an expansion over the eigenmodes (surface plasmons). This result demonstrates separation of geometry of a nanosystem from its material properties and the excitation field. The eigenfunctions $\varphi_n(\mathbf{r})$ and eigenvalues s_n in Eq. (1.36) depend only on geometry of the nanosystem, but not on its material composition or the optical excitation frequency. In contrast, the spectral parameter $s(\omega)$ depends only on the material composition and the excitation frequency, but not on the system's geometry. One of the advantages of this approach is in its applications to numerical computations: the eigenproblem has to be solved only once, and then the optical responses of the nanosystem are determined by Green's function that can be found by a simple summation in Eq. (1.36).

This Green's function is called retarded because it describes responses that occur necessarily at later time moments with respect to the forces that cause them. (Note that this name and property have nothing to do with the electromagnetic retardation, which is due to the finite speed of light and is absent in the quasistatic approximation.) This property, also called Kramers-Kronig causality, is mathematically equivalent to all singularities of $G^r(\mathbf{r},\mathbf{r}';\omega)$ as a function of complex ω being situated in the lower half-plane. Consequently, $G^r(\mathbf{r},\mathbf{r}';\omega)$ as a function of ω satisfies the Kramers-Kronig dispersion relations [30]. By the mere form of the spectral expansion (1.36), this Green's function satisfies all other exact analytical properties. This guarantees that in numerical simulations it will possess these properties irrespectively of the numerical precision with which the eigenproblem is solved. This insures an exceptional numerical stability of computational Green's function approaches.

Once the Green's function is found from Eq. (1.36), the local optical field potential is found as contraction of this Green's function with the excitation potential $\varphi_0(\mathbf{r})$ as

$$\varphi_1(\mathbf{r}) = -\int_V G^r(\mathbf{r},\mathbf{r}';\omega)\frac{\partial}{\partial\mathbf{r}'}\Theta(\mathbf{r}')\frac{\partial}{\partial\mathbf{r}'}\varphi_0(\mathbf{r}')\,\mathrm{d}^3r'\,. \tag{1.37}$$

From Eqs. (1.23) and (1.37) using the Gauss theorem, we obtain an expression for the field potential $\varphi(\mathbf{r})$ as a functional of the external (excitation) potential $\varphi_0(\mathbf{r})$,

$$\varphi(\mathbf{r}) = \varphi_0(\mathbf{r}) - \int_V \varphi_0(\mathbf{r}')\frac{\partial}{\partial\mathbf{r}'}\Theta(\mathbf{r}')\frac{\partial}{\partial\mathbf{r}'}G^r(\mathbf{r},\mathbf{r}';\omega)\,\mathrm{d}^3r'\,. \tag{1.38}$$

Finally, differentiating this, we obtain a closed expression for the optical electric field $\mathbf{E}(\mathbf{r})$ as a functional of the excitation (external) field $\mathbf{E}^{(0)}(\mathbf{r})$ as

$$E_\alpha(\mathbf{r}) = E_\alpha^{(0)}(\mathbf{r}) + \int_V G^r_{\alpha\beta}(\mathbf{r},\mathbf{r}';\omega)\Theta(\mathbf{r}')E_\beta^{(0)}(\mathbf{r}')\,\mathrm{d}^3r' \,, \tag{1.39}$$

where $\alpha, \beta, \ldots$ are Euclidean vector indices ($\alpha, \beta, \cdots = x, y, z$) with summation over repeated indices implied; the fields are

$$\mathbf{E}(\mathbf{r}) = -\frac{\partial\varphi(\mathbf{r})}{\partial\mathbf{r}} \,, \quad \mathbf{E}^{(0)}(\mathbf{r}) = -\frac{\partial\varphi_0(\mathbf{r})}{\partial\mathbf{r}} \,, \tag{1.40}$$

and the tensor (dyadic) retarded Green's function is defined as

$$G^r_{\alpha\beta}(\mathbf{r},\mathbf{r}';\omega) = \frac{\partial^2}{\partial r_\alpha \partial r'_\beta} G^r(\mathbf{r},\mathbf{r}';\omega) \,. \tag{1.41}$$

One of the exact properties of this Green's function is its Hermitian symmetry,

$$G^r_{\alpha\beta}(\mathbf{r},\mathbf{r}';\omega) = G^r_{\beta\alpha}(\mathbf{r}',\mathbf{r};-\omega)^* \,. \tag{1.42}$$

If the excitation is an optical field, its wave front is flat on the scale of the nanosystem, i.e., $\mathbf{E}^{(0)} = \text{const}$. Then from Eq. (1.39) we get

$$E_\alpha(\mathbf{r}) = \left[\delta_{\alpha\beta} + g_{\alpha\beta}(\mathbf{r},\omega)\right] E_\beta^{(0)} \,, \tag{1.43}$$

where the local field enhancement (tensorial) factor is a contraction of the retarded dyadic Green's function,

$$g_{\alpha\beta}(\mathbf{r},\omega) = \int_V G^r_{\alpha\beta}(\mathbf{r},\mathbf{r}';\omega)\Theta(\mathbf{r}')\,\mathrm{d}^3r' \,. \tag{1.44}$$

1.3.4 SP Modes as Resonances

Each physical eigenmode is described by the corresponding pole of Green's function (1.36). Close to such a pole, Green's function and, consequently, local fields (1.43) become large, which describes the surface plasmon resonance of the nanosystem. A complex frequency of such a resonance can be found from the position of the corresponding pole in the complex plane of frequency,

$$s(\omega_n - i\gamma_n) = s_n \,, \tag{1.45}$$

where ω_n is the real frequency of the surface plasmon, and γ_n is its spectral width (relaxation rate).

Note that we presume $\gamma_n > 0$, i.e., a negative sign of the imaginary part of the physical surface frequency. This a presumption, which is confirmed by the solution presented below in this section, is based on the standard convention of the sign of an exponential in the field temporal evolution,

$$\mathbf{E}_n(\mathbf{r}, t) \propto \exp\left[-i(\omega_n - i\gamma_n)t\right] \propto \exp(-\gamma_n t) , \tag{1.46}$$

which decays exponentially for $t \rightarrow +\infty$, as should be. The wave functions of physical surface plasmons are the familiar eigenfunctions $\varphi_n(\mathbf{r})$, i.e., those of the geometric eigenmodes. However, their physical frequencies, of course, depend on the material composition of the system.

For weak relaxation, $\gamma_n \ll \omega_n$, one finds that this real surface plasmon frequency satisfies an equation

$$\mathrm{Re}[s(\omega_n)] = s_n , \tag{1.47}$$

and that the surface plasmon spectral width is expressed as

$$\gamma_n = \frac{\mathrm{Im}[s(\omega_n)]}{s_n'}, \qquad s_n' \equiv \left.\frac{\partial \mathrm{Re}[s(\omega)]}{\partial \omega}\right|_{\omega=\omega_n} . \tag{1.48}$$

In terms of the dielectric permittivity as functions of frequency

$$s'(\omega) = \frac{\varepsilon_d}{|\varepsilon_d - \varepsilon(\omega)|^2} \mathrm{Re} \frac{\partial \varepsilon_m(\omega)}{\partial \omega} , \quad \gamma(\omega) = \frac{\mathrm{Im}\varepsilon_m(\omega)}{\mathrm{Re}\frac{\partial \varepsilon_m(\omega)}{\partial \omega}} . \tag{1.49}$$

This expression has been given in Sect. 1.2.1 as Eq. (1.3). Importantly, the spectral width γ is a universal function of frequency ω and does not explicitly depend on the eigenmode wave function $\varphi_n(\mathbf{r})$ or system's geometry. However, the system's geometry does, of course, define the plasmon eigenfrequencies ω_n. This property has been successfully used in Ref. [180] where a method of designing nanoplasmonic systems with desired spectra has been developed. Note also that the classical SPs have been quantized in Ref. [31] in connection with the prediction of spaser, a nanoscale counterpart of laser (see Sect. 1.5).

As follows from Eq. (1.28), external frequency ω is within the range of the physical surface plasmon frequencies and, therefore, can be close to a surface plasmon resonance (pole of Green's function (1.36) as given by Eq. (1.45)) under the following conditions

$$0 \leq \mathrm{Re}\, s(\omega) \leq 1 , \quad \mathrm{Im}\, s(\omega) \ll \mathrm{Re}\, s(\omega) . \tag{1.50}$$

These conditions are equivalent to

$$\varepsilon_d > 0 , \quad 0 \leq \mathrm{Re}\, \varepsilon_m(\omega) < 0 , \quad \mathrm{Im}\, \varepsilon_m(\omega) \ll |\mathrm{Re}\, \varepsilon_m(\omega)| . \tag{1.51}$$

These conditions, in fact, constitute a definition of a plasmonic system, i.e., a system where a position of surface plasmon resonance can be physically approached: the dielectric permittivity of the metal component should be negative and almost real, while the permittivity of the second constituent (dielectric) should be positive, as assumed.

It is useful to write down an expression for Green's function (1.36) that is asymptotically valid near its poles, which can be obtained from Eqs. (1.47) and (1.48) as

$$G^r(\mathbf{r}, \mathbf{r}'; \omega) = \frac{1}{s'(\omega)} \sum_n \frac{\varphi_n(\mathbf{r})\, \varphi_n(\mathbf{r}')^*}{\omega - \omega_n + i\gamma_n}\,, \tag{1.52}$$

where γ_n is given above by Eqs. (1.48) or (1.49). This expression constitutes what is called the singular approximation or pole approximation of the Green's function. When an excitation frequency is in resonance with an SP frequency, i.e., $\omega = \omega_n$, the Green's function (1.52) increases in magnitude by $\sim\omega_n/\gamma_n\sim Q$ times, where the quality factor Q is given by Eq. (1.5).

Below, for the sake of reference, we give a modal expansion for the polarizability α of a nanoplasmonic system as a tensor,

$$\alpha_{\alpha\beta} = -\frac{\varepsilon_d}{4\pi} \sum_n \frac{1}{s_n(s - s_n)} M_{n\alpha} M^*_{n\beta}\,, \tag{1.53}$$

where the indexes α, β denote Cartesian components, and $\mathbf{M}_n$ is a coupling vector defined as

$$\mathbf{M}_n = -\int_V \Theta(\mathbf{r}) \frac{\partial \varphi_n(\mathbf{r})}{\partial \mathbf{r}}\, \mathrm{d}^3 r\,. \tag{1.54}$$

Near a SP frequency, $\omega \approx \omega_n$, a singular part of the polarizability (1.53) acquires a form

$$\alpha_{\alpha\beta} = -\frac{\varepsilon_d}{4\pi s'_n s_n} \frac{M_{n\alpha} M^*_{n\beta}}{\omega - \omega_n + i\gamma_n}\,. \tag{1.55}$$

Also, for the reference sake, we give a general expression for the SP radiative decay rate, $\gamma_n^{(r)}$. This can be obtained from Eq. (1.55) taking into account Eqs. (1.10) and (1.15) as

$$\gamma_n^{(r)} = \frac{\varepsilon_d^{3/2} \omega^3 \left|\mathbf{M}_n\right|^2}{9\pi c^3 s'_n s_n}\,. \tag{1.56}$$

Note that $|\mathbf{M}_n|^2 \sim V_n$, where V_n is the modal volume of the n-th eigenmode. Thus Eq. (1.56) is consistent with Eq. (1.16) obtained earlier in this chapter.

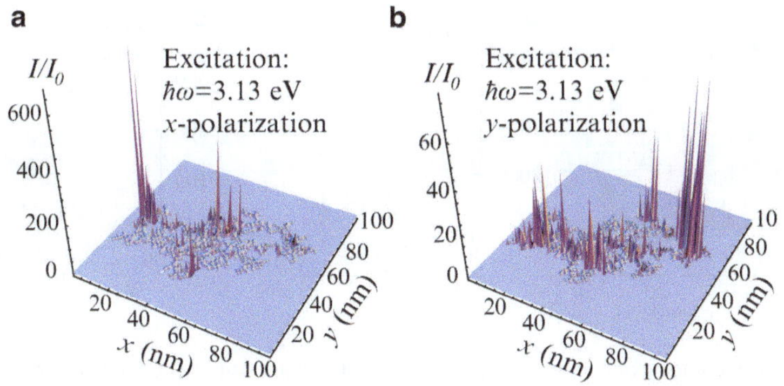

Fig. 1.8 Spatial distributions of local field intensity I relative to the external intensity I_0 for an individual CCA cluster of $N = 1{,}500$ silver nanospheres in water ($\varepsilon_d = 2.0$) for the frequency $\hbar\omega = 3.13\,\text{eV}$. The polarizations of the excitation radiation is x (**a**) and y (**b**), as indicated in the panels. The projection of the cluster nanospheres to the xy plane is also shown (Adapted from Ref. [158])

1.3.5 *Examples of Local Fields and their Hot Spots*

Let us give an example of local fields computed using Eq. (1.39). We start with the results of the original publications Ref. [158, 159] where the hot spots of the plasmonic local fields have been predicted. This prediction was made for fractal clusters because the fractals were expected to possess highly inhomogeneous and fluctuating local optical fields as was shown in pioneering papers in a subfield of physical optics that today is called nanoplasmonics [117, 150, 181].

In Fig. 1.8 adapted from Ref. [158], we illustrate the local-field hot spots for a silver CCA cluster of $N = 1{,}500$ identical nanospheres embedded in water. We show local field intensity $I = |\mathbf{E}(\mathbf{r}, \omega)|^2$ relative to the excitation field intensity I_0 at the surface of the silver nanospheres at a relatively high frequency $\hbar\omega = 3.13\,\text{eV}$ corresponding to vacuum wavelength $\lambda = 390\,\text{nm}$ in the far blue end of the visible spectrum. We can clearly see that the local intensity is highly non-uniform, exhibiting pronounced singular hot spots. These hot spots are localized at the minimum scale of the system (on the order of the radius of the nanospheres). The local intensity in the hot spots is greatly enhanced (by a factor of up to $\sim$600) as one would expect from an estimate $I/I_0 \sim Q^2$ – cf. Fig. 1.2.

This hot spotting is nothing else as random nanofocusing. It is similar in this respect to the formation of speckles in the wave optics, as we have discussed above in conjunction with Fig. 1.5. However, reflecting the properties of the corresponding SP eigenmodes, there is no characteristic wavelength that limits this hot spot singularity by defining the characteristic size L_s of the speckles, which is also a characteristic separation between them – see Eq. (1.31).

Another property of the local fields of a great significance is the dramatic dependence of the intensity distribution on the polarization: the local distributions or

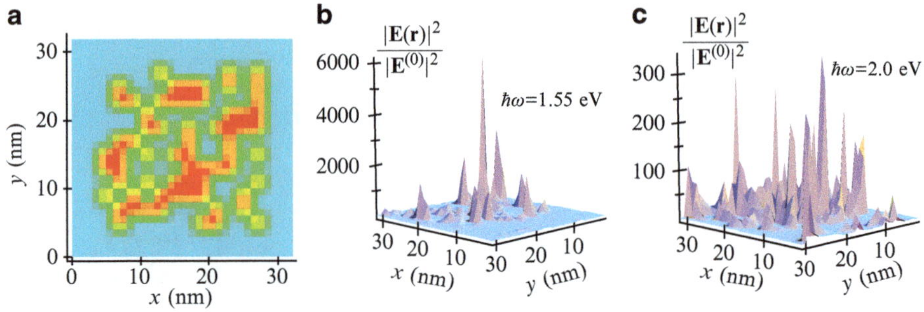

Fig. 1.9 (**a**) Geometry of nanostructured random planar composite (RPC): characteristic function $\Theta(\mathbf{r})$ is displayed in the xz plane of the RPC. Axes unit is nm; thickness of the system in the y direction (normal to its plane) is 2 nm. The fill factor is $p = 0.5$. Characteristic function $\Theta(\mathbf{r})$ is smoothed by a Gaussian filter with a radius of 1 nm to improve numerical accuracy (shown in the panel by the halftone density). (**b**) Local field intensity $|\mathbf{E}(\mathbf{r})|^2$ in the plane of the nanostructure displayed relative to the excitation field intensity $|\mathbf{E}^{(0)}|^2$; excitation frequency $\hbar\omega = 1.55\,\mathrm{eV}$; computed using Eq. (1.38). The metal is silver embedded in the dielectric with $\varepsilon_d = 2$. (**c**) Same as (**b**) but for $\hbar\omega = 2.0\,\mathrm{eV}$ (Adapted from data computed for Ref. [182])

the x-polarization (Fig. 1.5a) and y-polarization (panel b) are completely different. An experimental observation of this effect has been obtained in Ref. [118] already at a very early stage of the development of nanoplasmonics.

Note that the SP eigenmode geometry is also strongly dependent on its frequency – see Fig. 1.5. However, in externally-excited local fields, this frequency dependence is obscured by the resonance broadening due to the losses, as is evident from the expression for the resonant part of the Green's function

We will present below spectral and statistical properties of the local fields using a model of random planar composite (RPC). A specific RPC system used in the computation is shown in Fig. 1.9a. To improve numerical accuracy, we smooth the unit-step characteristic function $\Theta(\mathbf{r})$ with a Gaussian filter with a radius of 1 grid step: this dramatically improves numerical accuracy of a grid method that we use to solve the eigenproblem. Such a smoothing is clearly seen in Fig. 1.9a.

In Fig. 1.9b, c, we display the spatial distribution of the local field intensity $|\mathbf{E}(\mathbf{r})|^2$in the plane of the nanostructure at the surface of the metal. These computations are described in Ref. [182]. They are done for silver whose dielectric function is adopted from Ref. [32]; the embedding dielectric has permittivity is set as $\varepsilon_d = 2.0$. This intensity is plotted relative to the excitation field intensity $|\mathbf{E}_0|^2$; thus the quantity displayed is the enhancement factor of the local field intensity. Panel (b) shows the intensity computed from Eq. (1.38). The maximum of the local intensity enhancement of $\approx$6,000 is in a reasonable agreement with the estimate $\sim Q^2 \sim 10^4$, where Q is displayed in Fig. 1.2.

Dependence of the local fields on frequency is dramatic: cf. Fig. 1.9b, c. As frequency increases from the near-IR (1.55 eV) to visible (2.0 eV), the distribution becomes much more delocalized and its magnitude dramatically decreases, which cannot be explained by some decrease of quality factor Q alone. Most importantly, at all frequencies these near-field intensity distributions are dominated by the pronounced hots spots. These are manifestation of the hot spots of the SP eigenmodes – see Fig. 1.7.

Generally, the intensity distribution of local field intensity in Fig. 1.9b, c is highly singular: it consists of relatively narrow peaks (hot spots [159, 164]) separated by regions of a low intensity. This is a typical distribution of intensity in plasmonic nanosystems, which is a reflection of the inhomogeneous localization of the SP eigenmodes.

1.3.6 Experimental Examples of Nanoplasmonic Hot Spots

There has been a significant number of experimental studies of near-field distributions of optical fields in plasmonic nanostructures. In all cases, a pronounced picture of the hots spots [158, 159] has been exhibited, see, e.g., Refs. [123, 156, 169]. The inhomogeneous localization of the SP eigenmodes (see Sect. 1.3.2), which is inherently related to hot spots, has recently been confirmed experimentally [162].

The photoemission electron microscope (PEEM) is a powerful tool of analyzing the distribution of the local field intensity without perturbing it in any way. In the PEEM approach, the plasmonic nanosystem to be analyzed serves as a cathode and an object of an electron microscope. The electron emission is caused by the local field $\mathbf{E}(\mathbf{r}, \omega)$ of the plasmonic system. The photoelectrons are analyzed by the electron optics of the PEEM that creates a magnified image of the system in "light" of the photo-emitted electrons.

For silver, the work function W_f (i.e., the minimum energy needed to excite an electron from the Fermi surface to the zero energy that is the energy in vacuum far away from the metal) is approximately 4.2 eV. The highest energy of an optical quantum (at the vacuum wavelength of 390 nm) is 3.2 eV, i.e., it is significantly less than W_f. Thus, a single optical photon cannot emit an electron from a silver surface. Such an emission can, however, occur through two-photon absorption, leaving for the emitted electron the kinetic energy at infinity of $E_\infty \leq 2\hbar\omega - W_f$. Such a two-photon electron photoemission is in the foundation of the so-called two-photon photoemission PEEM (or, 2PP-PEEM). On the other hand, for ultraviolet radiation (say, from a Hg lamp), the energy of a photon is sufficient for the one-photon photoemission PEEM (1PP-PEEM). The 2PP-PEEM electron intensity mirrors the distribution of $I_2 = |\mathbf{E}(\mathbf{r}, \omega)|^4$.

A model system to illustrate the hot spots used in a 2PP-PEEM experiment of Ref. [123] is shown in Fig. 1.10a. This is a diffraction grating covered with silver layer with roughness of a $<$10 nm RMS grain size, as the scanning electron micrograph (SEM) shows in the insert. The Hg lamp illumination (the energy of the quantum $\hbar\omega = 4.89$ eV exceeds $W_f = 4.2$ eV, thus allowing one-photon photoemission, 1PP-PEEM) shows a smooth image of the underlying diffraction grating with the resolution of the PEEM ($\lesssim$100 nm).

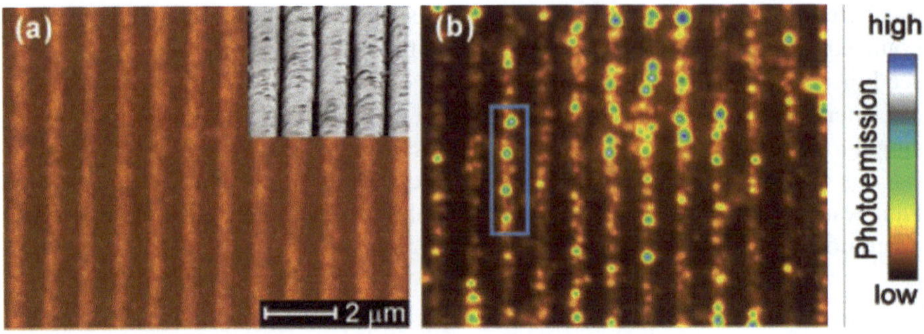

Fig. 1.10 PEEM micrographs of the same region on the silver grating obtained with (**a**) 254-nm line of a Hg lamp (1PP-PEEM) and (**b**) p-polarized 400-nm femtosecond laser excitation (2PP-PEEM). A scanning electron micrograph (SEM) of the silver grating in (**a**) is superimposed with the 1PP-PEEM image to show correspondence in the >100 nm scale topographical contrast. The surface roughness with <10 nm RMS distribution in the SEM image, which is too fine to resolve with the PEEM, gives rise to excitation of the localized SP modes seen as the hot spots in the 2PP-PEEM image of (**b**). The *blue rectangle* locates the four hot spots that were used for a coherent control experiment (Adapted from Ref. [123])

A dramatically different picture is observed in Fig. 1.10b. In this case, the irradiation is with femtosecond laser pulses of $\lambda = 400$ nm vacuum wavelength. The corresponding energy of the quantum is below the work function, $\hbar\omega = 3$ eV $< W_f = 4.2$ eV. Thus the electron photoemission is two-photon. The corresponding 2PP-PEEM image in Fig. 1.10b exhibits a pronounced picture of the hot spots due to the fact that in this case the optical frequency is in the plasmonic range. These hot spots are localized SPs that are excited by the p-polarized radiation with a significantly greater efficiency than by an s-polarized one. This suggests that SPPs excitation may play a role as an intermediate process for the localized SP excitation. In a full qualitative agreement with theory (see Sect. 1.3.2), these hot spots are singular, highly localized, and randomly distributed in space. The local fields in these hot spots are highly enhanced as witnessed by their dominance in the 2PP process.

Formation of the hot spots for random nanostructured plasmonic systems is a universal phenomenon whose physics is defined by the absence of the characteristic wavelength of the localized SPs, which localize at all available scales and whose fields are highly singular and highest at the minimum scale [78, 158, 159, 184].

One of the most convincing and comprehensive studies of geometry and statistics of the plasmonic hots spots is recently published Ref. [183] performed using PEEM and semicontinuous gold film whose model is RPC. Adapted from this, in Fig. 1.11, we show spatial distributions of the hot spots for a semicontinuous film with a fill factor (percentage of the area occupied by metal) $f = 0.53$. At this f, the film is close to the percolation threshold for static conductivity. The connected clusters in such a film have a fractal nature where we expect giant fluctuations and inhomogeneous localization of the SP fields [158, 159]. In fact, the distributions in Fig. 1.11 do demonstrate pronounced hot-spot behavior with inhomogeneous

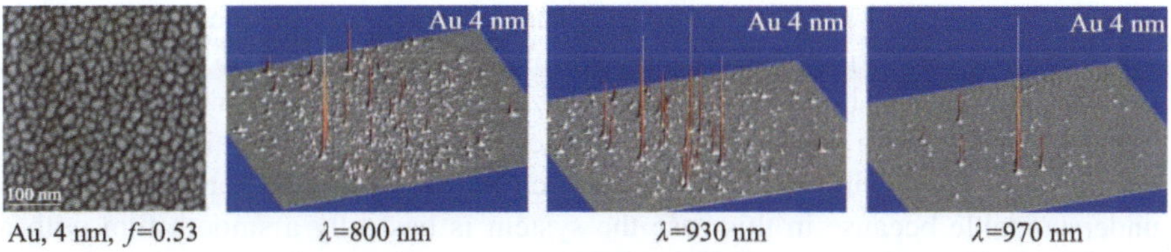

Fig. 1.11 *Left column*, scanning electron microscope images of the gold/glass films for the 4 nm grain size (filling factor $f = 0.53$). *Right*, PEEM distributions corresponding to gold/glass films for three different wavelengths. For each PEEM image, excitation wavelength λ is indicated (Adapted from Ref. [183])

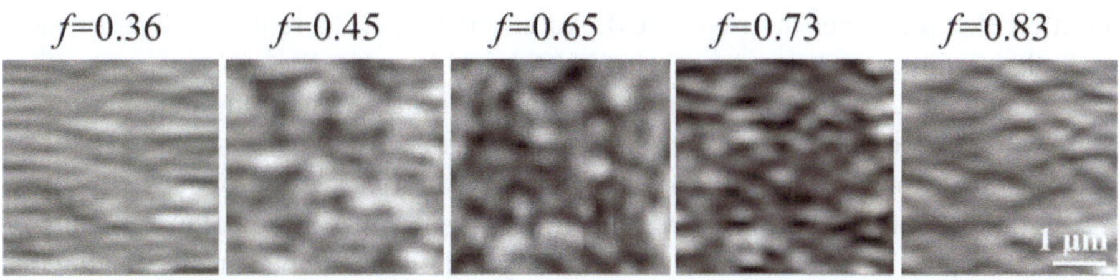

Fig. 1.12 NSOM images of $4 \times 4\,\mu m^2$ semi-continuous silver films with different metal filling fractions f as indicated above the graphs. Local intensity distribution is displayed as a function of the spatial coordinates in the plane of the film. The *white areas* correspond to higher intensities (Adapted from Ref. [169])

localization, giant fluctuations in space, where the distributions and intensities of individual hot spots strongly and randomly change with frequency. These distributions are in a full qualitative agreement with the theoretical predictions for the hot spots of local nanoplasmonic fields [158, 159] – cf. above Figs. 1.8 and 1.9.

We emphasize again that the PEEM-based observation of the plasmonic hot spots is completely non-perturbing. The photo-emitted electrons that are used in the PEEM fly away from the metal surface naturally, no matter whether they are used for imaging or not.

There has also been a series of research dealing with the observation of the plasmonic hot spots using the scanning near-field optical microscope (NSOM or SNOM) [156, 163, 169]. In fact, the first experimental evidence of the nanoplasmonic hot spots has been obtained [163] using an aperture-type NSOM, which is a based on a tapered optical fiber with the tip covered by a metal. A general concern about such observation is that they are perturbative: the tip of NSOM (or nanoscope, as it is often called) is typically much larger than a hot spot. Made of metal, it can, in principle, modify the host spot by both shifting its resonant frequency and decreasing the quality factor.

As an example, we present Fig. 1.12 adapted from Ref. [169]. This study is done on the semicontinuous metal film (random planar composite, or RPC). At relatively low values of the fill factor, $f = 0.36$ and $f = 0.45$, the local intensity distribution $I(\mathbf{r})$ shows relatively delocalized regions elongated normally to the direction of propagation (vertical axis in the figure). These are analogous to the caustics of the usual 3d optics. Relatively close to the percolation point, $f = 0.66$ and $f = 0.73$, the distribution $I(\mathbf{r})$ becomes highly localized exhibiting singular hot spots. The behavior of $I(\mathbf{r})$ at a relatively high fill factor of $f = 0.83$ again reminds that for the low f showing delocalized caustics but not singular hot spots. This is understandable because in this case the system is basically a smooth film with a few defects. This film supports SPPs that are weakly scattered by the relatively few defects.

As we have discussed above in this section, NSOM measurements of hot spots are inherently perturbative. While PEEM is nonperturbative, the spatial resolution so far has been insufficient (due to aberrations in the electron optics and large spread of the emitted electrons over their energies). Additionally, PEEM requires clean surfaces in high vacuum.

A fundamentally different non-perturbing approach to studying nanoplasmonic hot spots has been pioneered in Refs. [185, 186]. It is based on the so-called photon-localization super-resolution far-field microscopy. This method of far-field super-resolution has originally been developed in application to biological imaging [187].

This method's fundamentals can be very briefly described as the following. Assume that there is a *single* radiating chromophore (say, fluorescing molecule) in the view field of an optical microscope. Alternatively, there may be a number of such chromophores but their concentration should be low enough so they are resolved separately by the microscope (i.e., the distance between these molecules are greater than the microscope's resolution). The center of the emission of such a single (or separately resolved) emitter can be found with any precision that is only limited by statistical fluctuations of the number of the recorded photons but not by the resolution of the microscope provided that this microscope or the system under study do not change in the course of the observation.

After the position and brightness of a given single molecule are recorded, this molecule is naturally bleached. Then another molecule comes into the hot spot and its position and brightness are recorded until it is bleached. The process is repeated until the distribution of the brightness of emitters is built with a sufficient statistical precision.

It is assumed that the emission brightness of a single chromophore is proportional to the local field intensity of the hot spot at its position and that this chromophore exerts a negligibly weak perturbation on the local field of the hot spot. Thus this photon-localization nanoscopy is a non-perturbative method allowing one to find the intensity distribution at the hot spot on the nanoscale limited only by the statistical fluctuations (inversely proportional to the accumulation time) and the size of the chromophore itself, which is negligible in realistic situations.

The results of the hot spot local intensity-distribution measurements for an aluminum surface are shown in Fig. 1.13a. This distribution is a narrow peak with

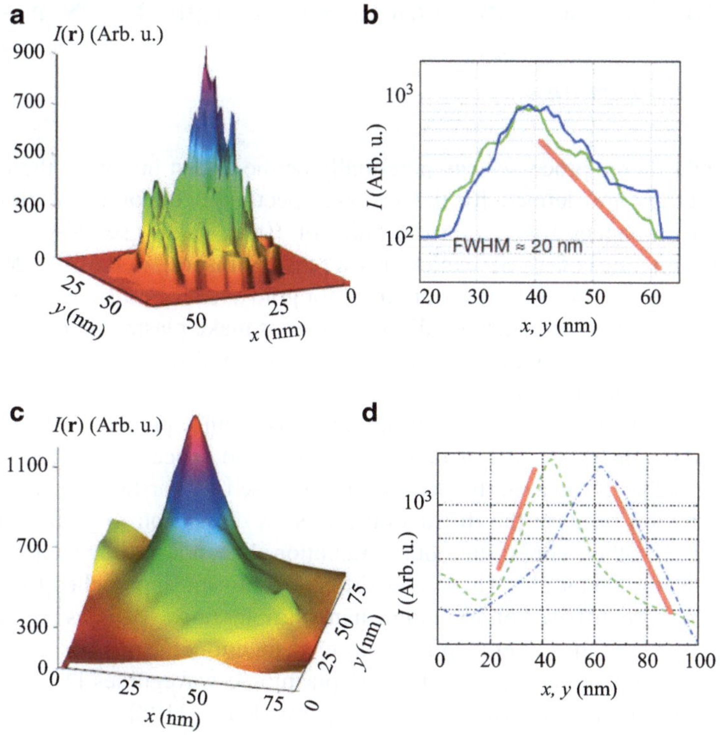

Fig. 1.13 Hot spots at the surfaces of metals measured by the photon localization method (see the text). (**a**) and (**b**) Distribution of the local intensity for a hot spot at the surface of aluminum. The kernel window size is 2.1 nm; this small window size makes the image appear noisy. The dye is Chromeo-542 with excitation at 532 nm and the emission centered around 580 nm. (**b**) An exponential decay field profile is visible, and is more evident on a log scale, shown as almost a decade of straight line (*red solid line*). The *blue* and *green curves* are two cross sections of the hot spot along x and y directions through the peak. The FWHM of the spot is ~20 nm. (**c**) and (**d**) is the same as (**a**) and (**b**), respectively, but for the case of a silver metal colloid cluster precipitated on a surface. A Chromeo-642 dye (Active Motif) – whose emission centers around 660 nm – is used (Adapted from Ref. [185])

the width of ≈20 nm. The observed fine structure of this distribution is attributed to statistical fluctuations [185]. The cross section through this distribution displayed in Fig. 1.13b suggests an exponential decay of this distribution function in space with the FWHM = 20 nm.

Very similar results are obtained for the silver colloid clusters as shown in Fig. 1.13c, d. Note that the aluminum surface studied is nominally smooth and contains only random roughness while the silver colloid clusters are fractals whose density fundamentally possesses large and correlated fluctuations.

1.4 Ultrafast Plasmonics and Coherent Control on Nanoscale

1.4.1 Introduction

The nanoplasmonic processes can potentially be the fastest in optics: their shortest evolution times are defined by the inverse spectral width of the region of the plasmonic resonances and are on the order of 100 as [188], see also Sect. 1.2.1. The relaxation times of the SP excitations are also ultrashort, in the 10–100 fs range [189–193]. See also the SP relaxation times for gold and silver displayed in Fig. 1.3. The nanolocalization and such an ultrafast kinetics make plasmonic nanostructures promising for various applications, especially for the ultrafast computations, data control and storage on the nanoscale.

These and potentially many other applications require precise control over the optical excitations of the nanostructures in time and space on the femtosecond-nanometer scale. Such a control cannot be imposed by far-field focusing of the optical radiation because the diffraction limits its dimension to greater than half wavelength. In other words, the optical radiation does not have spatial degrees of freedom on the nanoscale. There is a different class of approaches to control a system on nanoscale based on plasmonic nanoparticles or waveguides brought to the *near-field* region of the system. Among these we mention: the tips of scanning near-field optical microscopes [194], adiabatic plasmonic waveguides [12], nanowires [195, 196], plasmonic superlenses [197] or hyperlenses [198]. In all these cases, massive amount of metal is brought to the vicinity of the plasmonic nanosystem, which will produce strong perturbations of its spectrum and SP eigenmodes, cause additional optical losses, and adversely affect the ultrafast dynamics and energy nanolocalization in the system. This nanowaveguide approach also may not work because of the excitation delocalization due to the strong interaction (capacitive coupling) at the nanoscale distances for optical frequencies.

We have proposed [199] a principally different approach to ultrafast optical control on the nanoscale based on the general idea of coherent control. The coherent control of the quantum state of atom and molecules is based on the directed interference of the different quantum pathways of the optical excitation [200–209], which is carried out by properly defining the phases of the corresponding excitation waves. This coherent control can also be imposed by an appropriate phase modulation of the excitation ultrashort (femtosecond) pulse [206,210–212]. Shaping the polarization of a femtosecond pulse has proven to be a useful tool in controlling quantum systems [213].

Our idea of the coherent control on the nanoscale by the phase modulation of the excitation pulse can be explained with a schematic shown in Fig. 1.14. Phase modulation of the excitation pulse can be thought of as changing the frequency (color) light as the pulse progresses in time. For the sake of argument, let us assume that, as shown in Fig. 1.14, that initially the pulse contains blue colors that gradually change to red with the time progression. At earlier times, the dominating blue component of the pulse will excite the SP eigenmodes with corresponding high

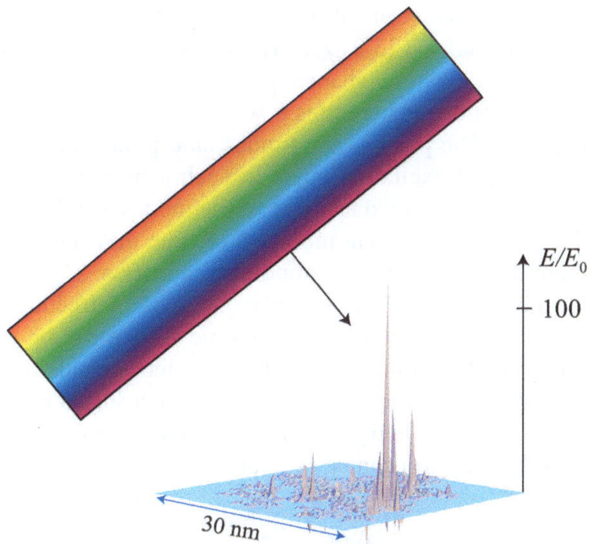

Fig. 1.14 Schematic of the fundamentals of the coherent control of nanoscale optical energy distribution. An excitation pulse is phase-modulated (shown by different colors changing with the progression of the pulse), which may be qualitatively thought of as different frequencies (colors) are incident on the nanosystem at different times, in a certain sequence. The system (a fractal cluster) is indicated by its projection on the horizontal coordinate plane. In response to this pulse, different SP eigenmodes are excited in a sequence. As time progresses, these eigenmodes interfere between themselves leading to a hot spot appearing at a required position at a given time. This leads to a large enhancement of the local field E relative to the excitation field E_0

optical frequencies. As the pulse progresses, the lower-frequency eigenmodes are excited. It is assumed that the total duration τ_p of the pulse is less than the decay (decoherence) time $\tau = \gamma^{-1}$ of the SPs, i.e., $\tau_p \lesssim \tau$ (for the decay rates and life times of the SPs see Eq. (1.3) or (1.49) and Fig. 1.3). In such a case, the SPs of different frequencies will coexist simultaneously, and their fields will interfere. This interference depends on the relative phases and amplitudes of the SPs of different frequencies that, in turn, are determined by the relative phases of different spectral components of the excitation pulse. The ultimate goal of the spatio-temporal coherent control on the nanoscale is to have a hot spot of the local fields at a given nanosite at a given femtosecond temporal interval. Below in this chapter we show that this problem is solved both theoretically and experimentally.

Another approach that we have proposed [214] invokes spatial modulation of the excitation field on the microscale in a polaritonic system. This field excites SPPs whose phases are determined by those of the original field. This determines the wave fronts of the SPP waves that focus on the nanoscale at the targeted nanofoci at the required times with femtosecond temporal resolution. The spatial-phase coherent control of the SPPs has been demonstrated experimentally by different groups [215, 216].

Our initial idea [199] has been subsequently developed theoretically [149, 213, 217, 218] and experimentally [123, 219–221]. In this coherent control approach, one sends from the far-field zone a shaped pulse (generally, modulated by phase, amplitude, and polarization) that excites a wide-band packet of SP excitations in the entire nanosystem. The phases, amplitudes, and polarizations of these modes are forced by this shaped excitation pulse in such a manner that at the required moment of time and at the targeted nanosite, these modes' oscillations add in phase while at the other sites and different moments of time they interfere destructively, which brings about the desired spatio-temporal localization.

Theoretically, the number of the effective degrees of freedom that a shaped femtosecond pulse may apply to a nanoplasmonic system can be estimated in the following way. The number of the independent frequency bands is $\sim\Delta\omega/\gamma$, where $\Delta\omega$ is the bandwidth of the plasmonic system. For each such a band, there are two degrees of freedom: amplitude and phase. Thus, the total number N_{DF} of the degrees of freedom for coherent control can be estimated as

$$N_{DF} \sim 2\frac{\Delta\omega}{\gamma} . \tag{1.57}$$

For a plasmonic system with the maximum bandwidth $\Delta\omega \sim \omega$, and Eq. (1.57) becomes

$$N_{DF} \sim 4Q , \tag{1.58}$$

where we took into account Eq. (1.5). In the optical region for noble metals $Q \sim 100$ (see Fig. 1.2), providing a rich, $\sim$100-dimensional space of controlling parameters. The coherent control approach is non-invasive: in principle, it does not perturb or change the nanosystem's material structure in any way.

However, how to actually determine a shaped femtosecond pulse that compels the optical fields in the nanosystem to localize at a targeted nanosite at the required femtosecond time interval is a formidable problem to which until now there has been no general and effective approach. To compare, our original chirped pulses possessed only two effective degrees of freedom (carrier frequency ω_0 and chirp) which allowed one to concentrate optical energy at the tip of a V-shape structure vs. its opening [149, 199]. Similarly, the two unmodulated pulses with the regulated delay τ between them used in the interferometric coherent control [123, 217, 220] also possess only two degrees of freedom (τ and ω_0) and can only select one of any two local-field hot spots against the other; it is impossible, in particular, to select one desired hot spot against *several* others.

There exists another method based on the adaptive genetic algorithms [206]. However, its application to the spatial-temporal localization in nanosystem is difficult due to the complexity of the problem. To date, the only example is the spatial concentration of the excitation on one arm of the three-pronged metal nanostar [219] where the obtained controlling pulses are very complicated and difficult to interpret though the nanosystem itself is rather simple. A general problem with this method is that the adaptive genetic algorithms are actually refined trial-and-error methods; they do not allow one to obtain the required controlling pulses as a result of the solution of a set of deterministic equations or an application of any regular deterministic procedure such as Green's function integration.

1.4.2 Time-Reversal Solution for Coherent Control

Our solution of this major problem of the coherent control, which is proposed and theoretically developed in Ref. [222], is based on an idea of time-reversal that has originally been proposed and used to control the focusing of acoustic waves and microwave radiation [223–225]. Some of these studies required use of a reverberating chamber to cause multiple interactions of the waves with the system needed to transfer the information to the far field. The electromagnetic subwavelength focusing also required a subwavelength-scale metal structure (a metal wire brush) to be positioned in the vicinity of the target system as a focusing antenna. In contrast, in nanoplasmonics there is no need for the reverberating chamber or the metal brush antenna, because the plasmonic nanosystem plays the roles of both of them. It confines the plasmonic modes for long times relative to their oscillation periods and also nano-localizes these modes.

1.4.3 Qualitative Description of Time-Reversal Coherent Control

The idea of the time-reversal solution of the nanoscale coherent control can be described using a schematic of Fig. 1.15. Consider a metal plasmonic nanosystem, indicated by blue in Fig. 1.15a, which may be embedded in a host dielectric (or be in vacuum). The nanosystem is excited by an external ultrafast (femtosecond) nanosource of radiation at its surface. As such we choose an oscillating dipole indicated by a double red arrow. This dipole generates a local optical electric field shown by a bold red waveform. This field excites SP oscillations of the system in its vicinity. In turn these oscillations excite other, more distant regions, and so forth until the excitation spreads out over the entire system. The relatively long relaxation time of these SP modes leads to the long "reverberations" of the plasmonic fields and the corresponding far-zone optical electric field. The latter is shown in Fig. 1.15b where one can see that a complicated vector waveform is predicted. This waveform is time reversed, as shown in panel (c), and send back to the system as an excitation plane wave from the *far-field* zone. If the entire field, in the whole space including the near-field (evanescent) zone, were time reversed and the system would have been completely time-reversible, which would imply the absence of any dielectric losses, then the system would have been compelled by this field exactly to back-trace its

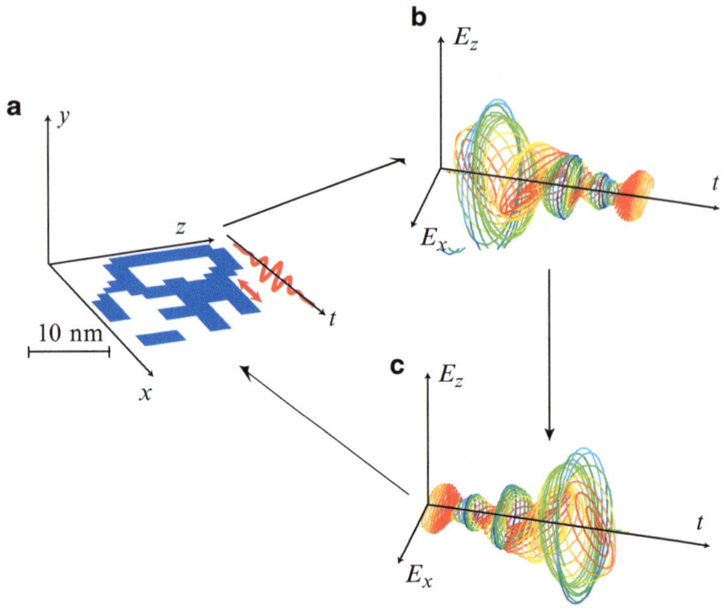

Fig. 1.15 (**a**) Geometry of nanosystem, initial seed oscillating dipole and its oscillation waveform. The nanosystem as a thin nanostructured silver film is depicted in *blue*. A position of the oscillating dipole that initially excites the system is indicated by a *double red arrow*, and its oscillation in time is shown by a *bold red waveform*. (**b**) Field in the far-field zone that is generated by the system following the excitation by the local oscillating dipole: vector $\{E_x(t), E_z(t)\}$ is shown as a function of the observation time t. The color corresponds to the instantaneous ellipticity as explained in the text. (**c**) Same as in panel (**b**) but for a time-reversed pulse in the far zone that is used as an excitation pulse to drive the optical energy nanolocalization at the position of the initial dipole

own evolution in time. This would have lead to the concentration of the local optical energy exactly at the position of the initial dipole at a time corresponding to the end of the excitation pulse.

Indeed, the system is somewhat lossy, which means that it is not exactly time reversible. Nevertheless, these losses are small, and one may expect that they will not fundamentally change the behavior of the system. Another problem appear to be more significant: the evanescent fields contain the main information of the nano-distribution of the local fields in the system, and they cannot be time reversed from the far zone because they are exponentially small, practically lost there. However, our idea is that the nanostructured metal system itself plays the role of the metal brush of Ref. [225] continuously coupling the evanescent fields to the far zone. Therefore the fields in the far zone actually contain, in their reverberations, most information about the evanescent fields that will be regenerated in the process of the time reversal.

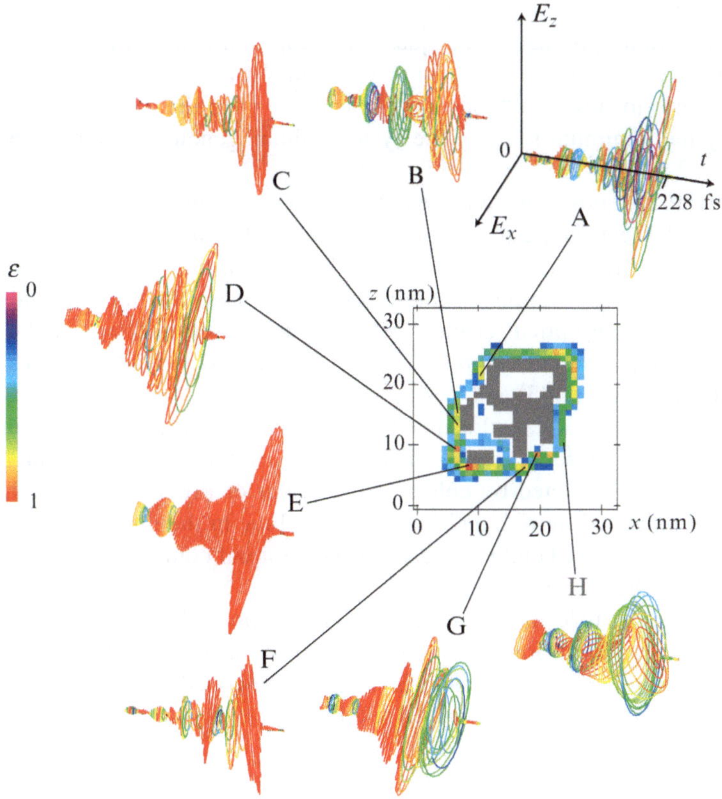

Fig. 1.16 Schematic of plasmonic-nanosystem geometry, local fields, and pulses generated in the far field. *Central insert*: The geometry of a nanosystem is shown by *dark gray*, and the local fields in the region surrounding it are shown by colors. The highest local field intensity is depicted by *red* and the lowest intensity is indicated by *blue* (in the rainbow sequence of colors). Panels *A–H*: The excitation waveforms in the far fields obtained as described in the text by positioning the initial excitation dipole at the metal surface at the locations indicated by the corresponding lines. Coordinate vectors $\boldsymbol{\rho}$ of points A–H in the xz plane are (in nm): $\boldsymbol{\rho}_A = (11, 22)$, $\boldsymbol{\rho}_B = (7, 16)$, $\boldsymbol{\rho}_C = (7, 14)$, $\boldsymbol{\rho}_D = (7, 10)$, $\boldsymbol{\rho}_E = (9, 7)$, $\boldsymbol{\rho}_F = (18, 7)$, $\boldsymbol{\rho}_G = (20, 9)$, and $\boldsymbol{\rho}_H = (24, 11)$. The instantaneous degree of linear polarization ε is calculated as the eccentricity of an instantaneous ellipse found from an fit to a curve formed by vector $\{E_x(t), E_y(t)\}$ during an instantaneous optical period. The pure circular polarization corresponds to $\varepsilon = 0$ and is denoted by *blue-violet color*; the pure linear polarization is for $\varepsilon = 1$ indicated by *red*. The corresponding polarization color-coding bar is shown at the left edge of the figure

We will illustrate this idea by considering a random planar composite (RPC) whose geometry is shown in gray in the center of Fig. 1.16. In specific computations, as the plasmonic metal, we consider silver whose dielectric permittivity ε_m we adopt from bulk data [32]. This system has been generated by randomly positioning $2 \times 2 \times 2\,\mathrm{nm}^3$ metal cubes on a plane, which for certainty we will consider as the xz coordinate plane. The random system shown in the center of Fig. 1.16 has filling factor of $f = 0.5$.

The interaction of a nanosystem with electromagnetic pulses is described in Green's function approach using quasistatic approximation [149, 199, 226] – see Sect. 1.3.3. It is known that the optical excitation energy in random plasmonic nanostructures localizes in "hot spots" whose size is on the nanoscale and is determined by the minimum scale of the system inhomogeneities [78, 159, 160, 227] – see Sect. 1.3.5.

Initially, to find positions of these hot spots in our system, we apply an ultrashort near-infrared (near-ir) pulse whose spectral width was very large, covering a frequency band from 1.1 to 1.7 eV. The pulse polarization is along the z axis (the incidence direction is normal to the plane of the nanostructure, i.e. along the y axis). The resulting optical electric field $\mathbf{E}$ is expressed in terms of the external electric field of the excitation optical wave $\mathbf{E}_0$ and retarded dyadic Green's function $\mathbf{G}^r$, as given by Eqs. (1.43) and (1.44).

The hot spots are always localized at the surface of the metal, predominantly at the periphery of the system. Their intensities found as the result of these computations are depicted by colors in the center of Fig. 1.16. The highest local intensity is indicated by red, and the lowest by blue in the region surrounding the metal. We have selected eight of these hot spots for our computations as denoted by letters A–H in the figure.

To generate the field in the far zone, we take a point dipole and position it at a surface of the metal at point $\mathbf{r}_0$ at such a hot spot, as described in the discussion of Fig. 1.15 above. The near-zone field $\mathbf{E}^L(\mathbf{r}, t)$ generated in response to this point dipole is found from Green's function relation

$$\mathbf{E}^L(\mathbf{r}, t) = \frac{4\pi}{\varepsilon_d} \int dt' \mathbf{G}^r(\mathbf{r}, \mathbf{r}_0; t - t')\, \mathbf{d}(\mathbf{r}_0, t') \,. \tag{1.59}$$

Knowing this local electric field, we calculate the total radiating optical dipole moment of the nanosystem in the frequency domain as

$$\mathbf{D}(\omega) = \frac{1}{4\pi} \int d^3r\, [\varepsilon_m(\omega) - \varepsilon_d]\, \Theta(\mathbf{r}) \mathbf{E}^L(\mathbf{r}, \omega) \,. \tag{1.60}$$

Here and below, the frequency- and time-domain quantities, as indicated by their arguments ω and t, are Fourier transforms of each other. The field in the far zone produced by this radiating dipole is given by standard electrodynamic formula – see, e.g. §67 in Ref. [228]. The time-reversed field is generated by time-reversed dipole $\mathbf{D}^T(t)$ that is complex-conjugated in the frequency domain, $\mathbf{D}^T(\omega) = \mathbf{D}(\omega)^*$.

The dependence on time of the initial excitation dipole, $\mathbf{d}(\mathbf{r}_0, t)$ is set as an ultrashort Gaussian-shaped pulse of 12 fs duration with the carrier frequency $\hbar\omega_0 =$ 1.2 eV. Following the procedure described above, the fields shown in Figs. 1.15 and 1.16 have been calculated for the radiation propagating in the y direction (normal to the plane of the nanostructure). These fields simply copy the retarded time evolution of the emitting dipole.

At the completing stage of our calculations, the time-reversed excitation pulse is sent back to the system as a plane wave propagating along the y direction (normal to the nanosystem plane). To calculate the resulting local fields, we again use Green's function Eq. (1.43) where the shaped excitation pulse substitutes for field $\mathbf{E}_0$.

1.4.4 Numerical Results for Time-Reversal Coherent Control

The electric field of the excitation wave is chosen as a modulated waveform (including amplitude, phase, and polarization modulation) that has been computed as described above in the previous subsection. The optical excitation energy can only be concentrated at sites where SP eigenmodes localize. For the present system, these are the hot spots shown by color in the central insert of Fig. 1.16, labeled A–H. The corresponding calculated excitation waveforms are displayed in panels as vector plots shown as functions of time $\{E_x(t), E_z(t)\}$.

There are several important features of these waveforms deserving our attention and discussion. First, these waveforms are rather long in duration: much longer than the excitation-dipole 12 fs pulses. This confirms our understanding that the initial dipole field excites local SP fields that, in a cascade manner, excite a sequence of the system SPs, which ring down relatively long time (over 200 fs, as shown in the figure). This long ring-down process is exactly what is required for the nanostructure to transfer to the far-field zone the information on the near-zone local (evanescent) fields as is suggested by our idea presented above in the introduction. The obtained fields are by shape resembling the controlling pulses for the microwave radiation [225]. However, a fundamental difference is that in the microwave case the long ringing-down is due to the external reverberation chamber, while for the nanoplasmonic systems it is due to the intrinsic evolution of the highly resonant SP eigenmodes that possess high Q-factors (setting a reverberation chamber around a nanosystem would have been, indeed, unrealistic).

Second, one can see that the pulses in Fig. 1.16 have a very nontrivial polarization properties ranging from the pure linear polarization (indicated by red as explained in the caption to Fig. 1.16) to the circular polarization indicated by blue, including all intermediate degrees of circularity. The temporal-polarization structure of pulses A–H in Fig. 1.16 is very complicated, somewhat reminding that of Ref. [219], which was obtained by a genetic adaptive algorithm. However, in our case these pulses are obtained in a straightforward manner, by applying the well-known, deterministic Green's function of the system, which is a highly efficient and fast method.

Third, and most important, feature of the waveforms in Fig. 1.16 is that they are highly site-specific: pulses generated by the initial dipole in different positions are completely different. This is a very strong indication that they do transfer to the far-field zone the information about the complicated spatio-temporal structure of the local, near-zone fields. This creates a pre-requisite for studying a possibility to use these pulses for the coherently-controlled nano-targeting.

Now we turn to the crucial test of the nanofocusing induced by the excitation pulses discussed above in conjunction with Fig. 1.16. Because of the finite time window ($T = 228$ fs) used for the time reversal, all these excitation pulses end and should cause the concentration of the optical energy (at the corresponding sites) at the same time, $t = T = 228$ fs (counted from the moment the excitation pulse starts impinging on the system). After this concentration instant, the nanofocused fields can, in principle, disappear (dephase) during a very short period on the order of the initial dipole pulse length, i.e. $\sim$12 fs. Thus this nanofocusing is a dynamic, transient phenomenon.

Note that averaging (or, integration) of the local-field intensity $I(\mathbf{r},t) = |\mathbf{E}(\mathbf{r},t)|^2$ over time t would lead to the loss of the effects of the phase modulation. This is due to a mathematical equality $\int_{-\infty}^{\infty} I(\mathbf{r},t)dt = \int_{-\infty}^{\infty} |\mathbf{E}(\mathbf{r},\omega)|^2 d\omega/(2\pi)$, where the spectral-phase modulation of the field certainly eliminates from the expression in the right-hand side. Thus the averaged intensity of the local fields is determined only by the local power spectrum of the excitation $|\mathbf{E}(\mathbf{r},\omega)|^2$ and, consequently, is not coherently controllable. Very importantly, such a cancellation does not take place for nonlinear phenomena. In particular, two-photon processes such as two-photon fluorescence or two-photon electron emission that can be considered as proportional to the squared intensity $I^2(\mathbf{r},t) = |\mathbf{E}(\mathbf{r},t)|^4$ are coherently controllable even after time averaging (integration), as we have argued earlier [149, 217]. Note the distributions measured in nonlinear optical experiments with the detection by the PEEM [123, 219, 220, 229] and in the fluorescence upconversion experiments [230] can be modeled as such nonlinear processes that yield distributions $\langle I^n(\mathbf{r})\rangle = \int_{-\infty}^{\infty} I^n(\mathbf{r},t)dt/T$, where $n \geq 2$. Inspired by this, we will consider below, in particular, the coherent control of the two-photon process averaged intensity $\left\langle I^2(\mathbf{r})\right\rangle$.

Let us investigate how precisely one can achieve the spatio-temporal focusing of the optical excitation at a given nanosite of a plasmonic nanostructure using the full shaping (amplitude, phase, and polarization) of the excitation pulses found from the time-reversal method. The results for the present nanostructure, targeting sites A–H, are shown in Fig. 1.17. For each excitation pulse, the spatial distribution of the local field intensity is displayed for the moment of time when this local intensity acquires its global (highest) maximum. The most important conclusion that one can draw from comparing panels (a)–(h) is that for each pulse A–H this global maximum corresponds to the maximum concentration of the optical energy at the corresponding targeted nanosite A–H. This obtained spatial resolution is as good as 4 nm, which is determined by the spatial size of inhomogeneities of the underlying plasmonic metal nanosystem. It is very important that this localization occurs not only at the desired nanometer-scale location but also very close to the targeted time that in our case is $t = 228$ fs. Thus the full shaping of femtosecond pulses by the time reversal is an efficient method of controlling the spatio-temporal localization of energy at the femtosecond-nanometer scale.

Let us turn to the temporal dynamics of intensity of the nanoscale local fields at the targeted sites A–H, which is shown in Fig. 1.18a–h. As we can see, in each of the panels there is a sharp spike of the local fields very close to the target time of

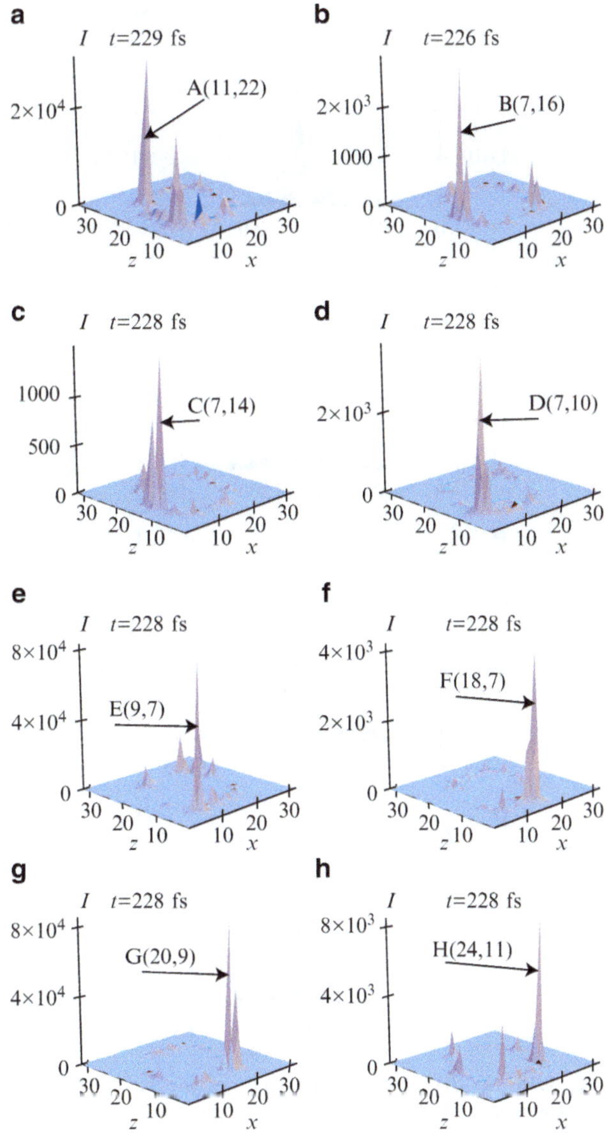

Fig. 1.17 Spatial distributions of the local optical field intensities at the surface of the metal nanostructure. Panels (**a**)–(**h**) correspond to the excitation with pulses A–H. Each such a distribution is displayed for the instance t at which the intensity for a given panel reaches its global maximum in space and time. This time t is displayed at the top of the corresponding panels. The corresponding targeted sites are indicated by *arrows* and labeled by the corresponding letters A–H and the coordinates (x, z). No special normalization has been applied so the distribution within any given panel is informative but not necessarily the magnitudes of the intensities between the panels

$t = 228$. The duration of this spike in most panels (a–f) is close to that of the initial dipole, i.e., 12 fs. This shows a trend to the reproduction of the initial excitation state due to the evolution of the time-reversed SP packet induced by the shaped pulses. There is also a pedestal that shows that this reproduction is not precise, which is expected due to the fact that the time reversal is incomplete: only the far-zone field propagating in one direction (along the y axis) is reversed. Nevertheless, as the discussion of Fig. 1.17 shows, this initial excitation-state reproduction is sufficient

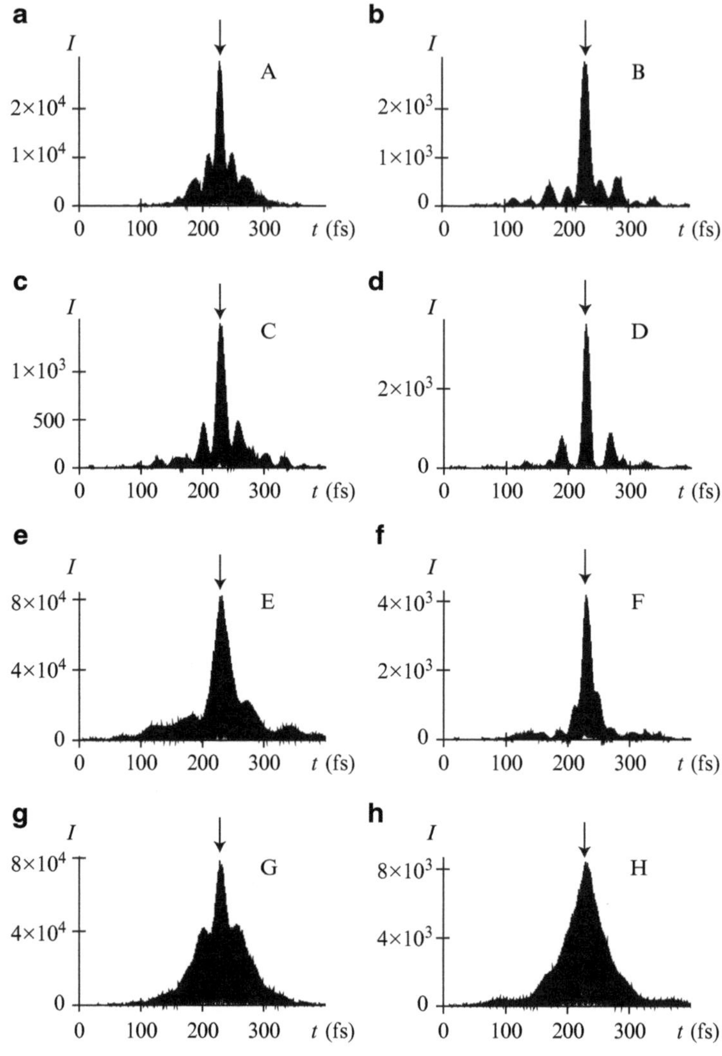

Fig. 1.18 (**a**)–(**h**) Temporal dynamics of the local field Intensity $I(\mathbf{r}, t) = \mathbf{E}^2(\mathbf{r}, t)$ at the corresponding hot spots A–H. The *down-arrows mark* the target time $t = 228$ fs where the local energy concentration is expected to occur

to guarantee that the targeted (initial excitation) site develops the global maximum (in time and space) of the local-field intensity. Interesting enough, the trend to reproduce the initial excitation state is also witnessed by almost symmetric (with respect to the maximum points $t = 228$ fs) shapes of all waveforms, which occurs in spite of the very asymmetric shapes of the excitation waveforms (cf. Fig. 1.16).

Apart from the ultrafast (femtosecond) dynamics of the nanolocalized optical fields discussed above in conjunction with Figs. 1.17 and 1.18, there is a considerable interest in its the time-integrated or averaged distributions, in particular, the mean squared intensity $\langle I^2(\mathbf{r})\rangle$. This quantity defines the nanoscale spatial distribution of the incoherent two-photon processes such as two-photon electron emission or two-photon luminescence. For example, in some approximation, the spatial distribution of the two-photon electron emission recorded by PEEM [123, 219, 220, 229] is determined by $\langle I^2(\mathbf{r})\rangle$.

Now we test the spatial concentration of time-averaged mean-squared intensity $\langle I^2(\mathbf{r})\rangle$ for all sites, which is displayed in Fig. 1.19. As clearly follows from this figure, in all cases, there are leading peaks at the targeted sites. Thus the two-photon excitation, even after the time averaging, can be concentrated at desired sites using the coherent-control by the time-reversed shaped pulses.

We point out that there has recently been an experimental demonstration of a coherent spatiotemporal control on the nanoscale by polarization and phase pulse shaping [221]. The optical energy concentration at a given site on a ~50 nm spatial scale at a given time on a ~100 fs temporal scale has been demonstrated. Since this time scale is comparable to or longer than the SP dephasing time, the time-reversal method could not be employed.

1.4.5 Coherent Control by Spatiotemporal Pulse Shaping

For coherent control on the nanoscale, as we have described above in Sect. 1.4, the phase of the excitation waveform along with its polarization provide functional degrees of freedom to control the nanoscale distribution of energy [123, 149, 199, 217–219, 221, 229, 231]. Spatiotemporal pulse shaping permits one to generate dynamically predefined waveforms modulated both in frequency and in space to focus ultrafast pulses in the required microscopic spatial and femtosecond temporal domains [232, 233].

Here we follow Ref. [214] that has introduced a method of full coherent control on the nanoscale where a temporally and *spatially* modulated waveform is launched in a graded nanostructured system, specifically a wedge – see schematic of Fig. 1.20. Its propagation from the thick (macroscopic) to the thin (nanoscopic) edge of the wedge and the concurrent adiabatic concentration provide a possibility to focus the optical energy in nanoscale spatial and femtosecond temporal regions.

This method unifies three components that individually have been developed and experimentally tested. The coupling of the external radiation to the surface plasmon polaritons (SPPs) propagating along the wedge occurs through an array of nanoobjects (nanoparticles or nanoholes) that is situated at the thick edge of the wedge. The phases of the SPPs emitted (scattered) by individual nanoobjects are determined by a spatio-temporal modulator. The nanofocusing of the SPPs occurs due to their propagation toward the nanofocus and the concurrent adiabatic concentration [12, 234, 235].

The coupling of the external radiation to SPPs and their nanofocusing have been observed – see, e.g., Refs. [236, 237]. The second component of our approach,

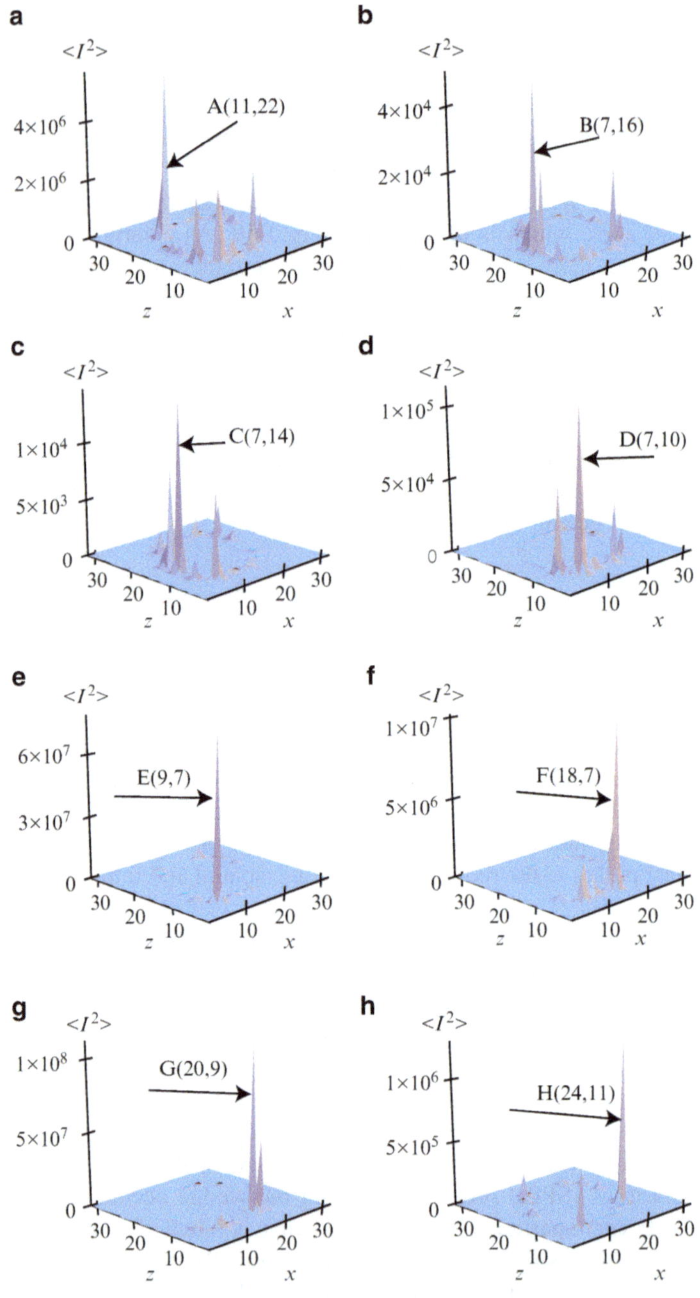

Fig. 1.19 Spatial distributions of the time-averaged mean-squared intensity $\langle I^2(\mathbf{r})\rangle$. This represents, in particular, the spatial distribution of the two-photon excited photocurrent density. Panels (**a**)–(**h**) correspond to the excitation with pulses A–H. The corresponding targeted sites are indicated by *arrows* and labeled by the corresponding letters A–H and coordinates (x, z). No special normalization has been applied so the distribution within any given panel is informative on spatial distribution but not necessarily the magnitudes of the intensities between the panels

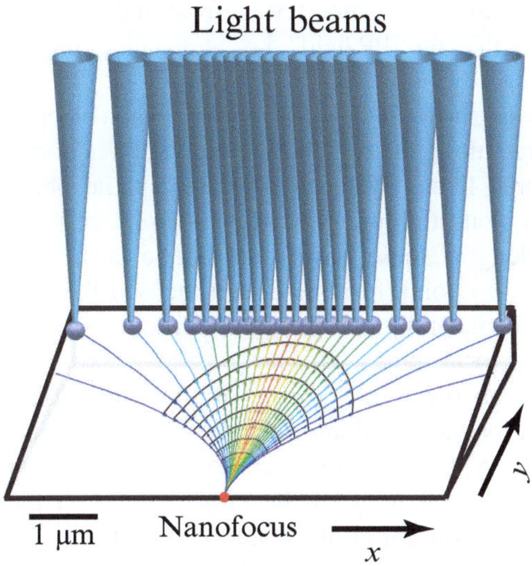

Fig. 1.20 Schematic of spatiotemporal coherent control on nanoscale (Adapted from Ref. [214]). Independently controlled light beams (shown by *blue cones*) are focused on launch pads depicted as silver spheres that are positioned on a thick edge of a wedge. SPP wavelets generated by the launchpads are shown by *black arcs*. Normal to them are rays (SPP trajectories) that are displayed by color lines coded accordingly to their origination points. These wavefronts and trajectories converge at the nanofocus indicated by the *red dot*

the spatio-temporal coherent control of such nanofocusing has been developed [232, 233]. The third component, the adiabatic concentration of SPPs also has been observed and extensively studied experimentally [13–16, 18, 19, 22].

The adiabatic concentration (nanofocusing) is based on adiabatic following by a propagating SPP wave of a graded plasmonic waveguide, where the phase and group velocities decrease while the propagating SPP wave is adiabatically transformed into a standing, localized SP mode. A new quality that is present in this approach is a possibility to arbitrary move the nanofocus along the nanoedge of the wedge. Moreover, it is possible to superimpose any number of such nanofoci simultaneously and, consequently, create any distribution of the nanolocalized fields at the thin edge of the wedge.

To illustrate this idea of the full spatiotemporal coherent control, now let us turn to a wedge that contains a line of nanosize scatterers (say, nanoparticles or nanoholes) located at the thick edge and parallel to it, i.e. in the x direction in Fig. 1.20. Consider first monochromatic light incident on these nanoparticles or nanoholes that scatter and couple it into SPP wavelets. Every such a scatterer emits SPPs in all directions; there is, of course, no favored directionality of the scattering.

At this point, we assume that the excitation radiation and, correspondingly, the scattered wavelets of the SPP are coherent, and their phases smoothly vary in space along the thick edge, i.e., in the x direction. Then the SPP wavelets emitted by different scatterers will interfere, which in accord with the Huygens-Fresnel principle leads to formation of a smooth wavefront of the SPP wave at some distance from the scatterers in the "far SPP field", i.e., at distances much greater than the SPP wavelength $2\pi/k_{SPP}$.

Such wavefronts are shown in Fig. 1.20 with concave black curves. The energy of the SPP is transferred along the rays, which are the lines normal to the wavefronts, shown by the colored lines. By the appropriate spatial phase modulation of the excitation radiation along the line of scatterers (in the x direction) over distances of many SPP wavelengths, these wavefronts can be formed in such a way that the rays intersect at a given point, forming a nanofocus at the thin (sharp) edge of the wedge, as shown schematically in Fig. 1.20. Diffraction of the SPP waves will lead to a finite size of this focal spot.

By changing the spatial phase profile of the excitation radiation, this focal spot can be arbitrarily moved along the thin edge. This focusing and adiabatic concentration, as the SPPs slow down approaching the sharp edge, will lead to the enhancement of the intensity of the optical fields in the focal region. This dynamically-controlled concentration of energy is a plasmonic counterpart of a large phased antenna array (also known as an aperture synthesis antenna), widely used in radar technology (synthetic aperture radar or SAR) and radio astronomy [238].

Now we can consider excitation by spatiotemporally shaped ultrashort pulses independently in space. Such pulses are produced by spatio-temporal modulators [232, 233]. The field produced by them is a coherent superposition of waves with different frequencies whose amplitudes and phases can arbitrarily vary in space and with frequency. This modulation can be chosen so that all the frequency components converge at the same focal spot at the same time forming an ultrashort pulse of the nanolocalized optical fields.

As an example we consider a silver [32] nanowedge illustrated in Fig. 1.20 whose maximum thickness is $d_m = 30\,\text{nm}$, the minimum thickness is $d_f = 4\,\text{nm}$, and whose length (in the y direction) is $L = 5\,\mu\text{m}$. Trajectories calculated by the Wentzel-Kramers-Brillouin (WKB) method in Ref. [214] for $\hbar\omega = 2.5\,\text{eV}$ are shown by lines (color used only to guide eye); the nanofocus is indicated by a bold red dot. In contrast to focusing by a conventional lens, the SPP rays are progressively bent toward the wedge slope direction.

Now consider the problem of coherent control. The goal is to excite a spatiotemporal waveform at the thick edge of the wedge in such a way that the propagating SPP rays converge at an arbitrary nanofocus at the sharp edge where an ultrashort pulse is formed. To solve this problem, we use the idea of back-propagation or time-reversal [224, 225, 239]. We generate rays at the nanofocus as an ultrashort pulse containing just several oscillations of the optical field. Propagating these rays, we find amplitudes and phases of the fields at the thick edge at each frequency as given by the complex propagation phase (eikonal) $\Phi(\boldsymbol{\rho})$, where $\boldsymbol{\rho}$ is a 2-d coordinate vector in the plane of the wedge. Then we complex conjugate the amplitudes of frequency components, which corresponds to the time reversal. We also multiply these amplitudes by $\exp(2\,\text{Im}\,\Phi)$, which pre-compensates for the Ohmic losses. This provides the required phase and amplitude modulation at the thick edge of the wedge.

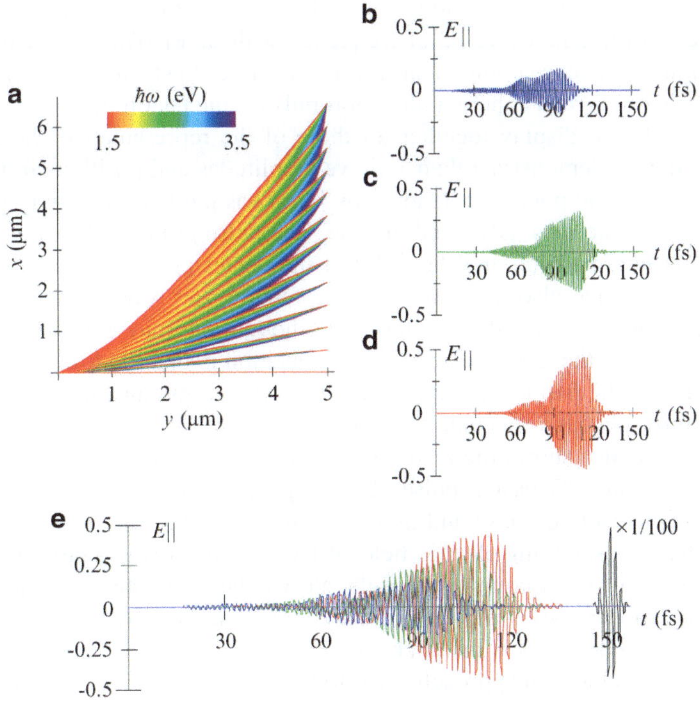

Fig. 1.21 (**a**) Trajectories (rays) of SPP packets propagating from the thick edge to the nanofocus displayed in the xy plane of the wedge. The frequencies of the individual rays in a packet are indicated by color as coded by the bar at the top. (**b**)–(**d**) Spatiotemporal modulation of the excitation pulses at the thick edge of the wedge required for nanofocusing. The temporal dependencies (waveforms) of the electric field for the phase-modulated pulses for three points at the thick edge boundary: two extreme points and one at the center, as indicated, aligned with the corresponding x points at panel (**a**). (**e**) The three excitation pulses of panels (**b**)–(**d**) (as shown by their colors), superimposed to elucidate the phase shifts, delays, and shape changes between these pulses. The resulting ultrashort pulse at the nanofocus is shown by the *black line*. The scale of the electric fields is arbitrary but consistent throughout the figure

We show an example of such calculations in Fig. 1.21. Panel (a) displays the trajectories of SPPs calculated [214] by the WKB method. The trajectories for different frequencies are displayed by colors corresponding to their visual perception. There is a very significant spectral dispersion: trajectories with higher frequencies are much more curved. The spatial-frequency modulation that we have found succeeds in bringing all these rays (with different frequencies and emitted at different x points) to the same nanofocus at the sharp edge.

The required waveforms at different x points of the thick edge of the wedge are shown in Fig. 1.21b–d where the corresponding longitudinal electric fields are shown. The waves emitted at large x, i.e., at points more distant from the nanofocus, should be emitted significantly earlier to pre-compensate for the longer propagation times. They should also have different amplitudes due to the differences in the adiabatic compression along the different rays. Finally, there is clearly a negative chirp (gradual decrease of frequency with time). This is due to the fact that the higher frequency components propagate more slowly and therefore must be emitted earlier to form a coherent ultrashort pulse at the nanofocus.

In Fig. 1.21e we display together all three of the representative waveforms at the thick edge to demonstrate their relative amplitudes and positions in time. The pulse at the extreme point in x (shown by blue) has the longest way to propagate and therefore is the most advanced in time. The pulse in the middle point (shown by green) is intermediate, and the pulse at the center ($x = 0$, shown by red) is last. One can notice also a counterintuitive feature: the waves propagating over longer trajectories are smaller in amplitude though one may expect the opposite to compensate for the larger losses. The explanation is that the losses are actually insignificant for the frequencies present in these waveforms, and the magnitudes are determined by adiabatic concentration factor.

Figure 1.21e also shows the resulting ultrashort pulse in the nanofocus. This is a transform-limited, Gaussian pulse. The propagation along the rays completely compensates the initial phase and amplitude modulation, exactly as intended. As a result, the corresponding electric field of the waveform is increased by a factor of 100. Taking the other component of the electric field and the magnetic field into account, the corresponding increase of the energy density is by a factor $\sim 10^4$ with respect to that of the SPPs at the thick edge.

To briefly conclude, an approach [214] to full coherent control of spatiotemporal energy localization on the nanoscale has been presented. From the thick edge of a plasmonic metal nanowedge, SPPs are launched, whose phases and amplitudes are independently modulated for each constituent frequency of the spectrum and at each spatial point of the excitation. This pre-modulates the departing SPP wave packets in such a way that they reach the required point at the sharp edge of the nanowedge in phase, with equal amplitudes forming a nanofocus where an ultrashort pulse with required temporal shape is generated. This system constitutes a "nanoplasmonic portal" connecting the incident light field, whose features are shaped on the microscale, with the required point or features at the nanoscale.

1.4.6 Experimental Demonstrations of Coherent Control on the Nanoscale

The ideas of the coherent control of the nanoscale distribution of ultrafast optical fields both space and in time, which have been introduced theoretically in Refs. [149, 199, 214, 218, 222, 240, 241], have been investigated and confirmed experimentally. Using the full phase and amplitude modulation of the excitation-pulse wavefront in both polarizations (the so-called polarization pulse shaping), the experiments have achieved both spatial control [123, 219] and spatiotemporal control [221] on nanometer-femtosecond scale.

Recently spatiotemporal nanofocusing via the adiabatic concentration along the lines of ideas presented above in Sect. 1.4.5 has been successfully demonstrated experimentally [21]. In this work, a shaped femtosecond pulse has been coupled by a grating to a TM_0 SPP mode on the surface of an adiabatically-tapered nanocone. The spatiotemporal concentration of optical energy in space to a ~10 nm region and in time to a 15 fs duration (Fourier-transform limited, i.e., the shortest possible at a given bandwidth). Indeed the position of the nanofocus in Ref. [21] is always the tip of the nanocone; so the possibility of moving the nanofocus in space is not available.

The ideas of employing the spatial modulation of the excitation wavefront [214] described above in Sect. 1.4.5 have been experimentally tested and confirmed for continuous wave (CW) excitation [215, 216]. We will present some of these experimental results below in this section.

We start with experiments on polarization-shaping coherent control that we adapt from Ref. [219]. The corresponding experimental approach is schematically illustrated in Fig. 1.22. Polarization-shaped ultrashort laser pulses illuminate a planar nanostructure, with two-photon photoemission electron microscopy (PEEM) [242] providing the feedback signal from the nanoscale field distribution that is essential for adaptive near-field control.

The spatial resolution of two-photon PEEM (~50 nm) is determined by its electron optics and is, thus, independent of the electromagnetic light-field diffraction limit. The sensitivity of the two-photon PEEM patterns to the optical field intensities arises from the nonlinear two-photon photoemission process whose intensity is proportional to the time-integrated fourth power of the local electric-field amplitude. With these elements in place, a user-specified nanoscopic optical field distribution is realized by processing recorded photoemission patterns in an evolutionary algorithm that directs the iterative optimization of the irradiating laser pulse shape.

The basic idea of the experiment is that the measured PEEM pattern identifies the origin of ejected photoelectrons and hence the regions of high local field intensity. A controlled variation of the PEEM pattern then proves the spatial control over the nanoscopic field distribution. We have already discussed such an approach above – see Fig. 1.10 [123] and the corresponding discussion in Sect. 1.3.6.

The nanostructure used consists of circular Ag disks with 180 nm diameter and 30 nm height, fabricated by electron-beam lithography on a conductive, 40-nm-thick indium-tin oxide (ITO) film grown on a quartz substrate. The disks are arranged into three dimers that form the arms of a star-like shape (Fig. 1.22a, lower right). The whole nanostructure is about 800 nm across, while the gap between two of the dimer disks is ~10 nm wide. After inspection by scanning-electron microscopy (SEM), the sample is mounted in the ultrahigh-vacuum PEEM set-up. The deposition of a small amount of caesium (~0.1 monolayers) reduces the work function of the Ag nanostructure to about 3.1 eV, that is, just below the threshold for two-photon photoemission with 790 nm photons.

The PEEM pattern obtained after maximization of the photoemission from the upper two arms of the Ag nanostructure in shown in Fig. 1.22c. It shows strong emission from these two upper arms and almost no emission from the bottom arm. Analogously, the photoemission after minimization of the upper part PEEM

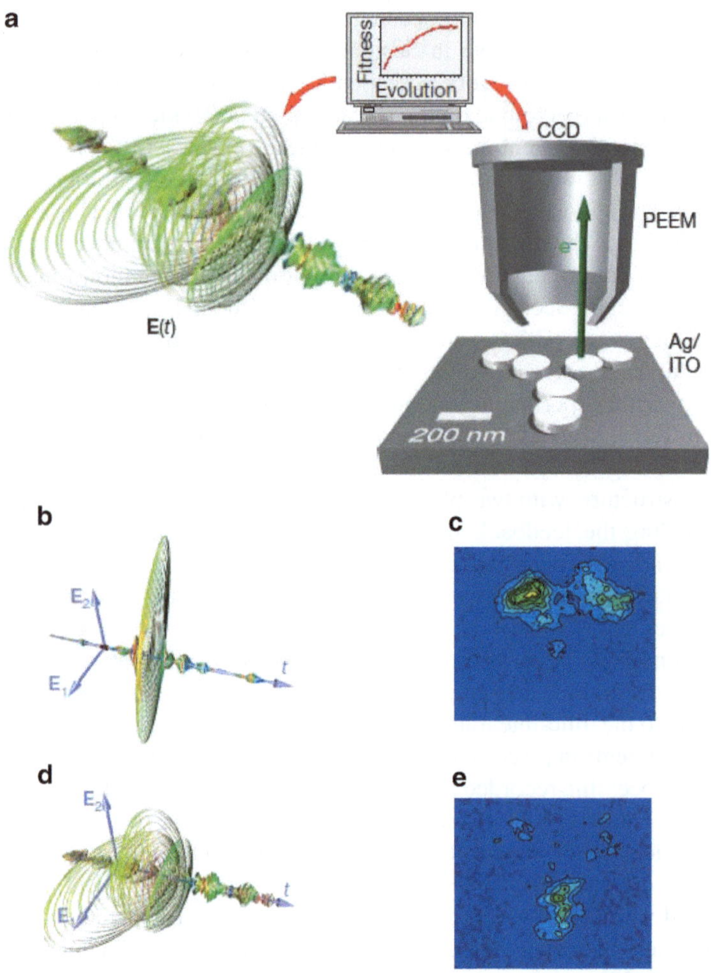

Fig. 1.22 Schematic and experimental results of coherent control with polarization shaping (Adapted from Ref. [219]). (**a**) Schematic of the experiment. A polarization shaper for ultrashort laser pulses controls the temporal evolution of the vectorial electric field $E(t)$ on a femtosecond timescale. These pulses illuminate a planar nanostructure in an ultrahigh-vacuum chamber that is equipped with a photoemission electron microscope (PEEM). The nanostructure consists of six circular Ag islands on an indium-tin oxide (ITO) film and a quartz substrate. A computer-controlled charge-coupled device (CCD) camera records the photoemission image and provides a feedback signal for an evolutionary learning algorithm. Iterative optimization of the pulse-shaper settings leads to an increase in the fitness value and correspondingly allows control over the nanooptical fields. (**b**) and (**c**) The optimal laser pulses, as experimentally characterized, display complex temporal electric-field evolution for the objectives of (**b**) minimizing and (**d**) maximizing the concentration of the excitation on the lower branch. E_1 and E_2 indicate the two field components that are phase-modulated in the polarization pulse shaper in the first and second LCD layer, respectively. They are at 45° angles with respect to the p-polarization. The overall time window shown is 2 ps. (**c**) The experimental PEEM image after adaptive maximization of the upper region intensity using complex polarization-shaped laser pulses (fittest individual of the final generation) shows predominant emission from the upper region. (**e**) Photoemission after minimization of the intensity in the upper region is concentrated in the lower region

brightness (Fig. 1.22e) occurs mainly in the lower area while the contribution from the upper two arms is extremely weak. The adaptively determined solution to each optimization problem has been proven to be robust with respect to slight imperfections in the experimental nanostructures. These successful optimizations demonstrate that polarization pulse shaping allows adaptive control of the spatial distribution of photoelectrons on a subwavelength scale, and thus of the nanoscopic optical fields that induce photoemission.

The optimally polarization-shaped laser pulses after adaptive maximization and minimization described above are shown in Fig. 1.22b, d, respectively, as determined by dual-channel spectral interferometry [243, 244]. In this representation, the shape of the quasi-three-dimensional figure indicates the temporal evolution of the polarization state of the electric field, with the color representing the instantaneous oscillation frequency. Contributions from both transverse polarization components are visible in each of the two cases. Whereas the upper-region photoemission maximization is achieved with a comparatively simple time evolution, the corresponding minimization requires a more complex field with varying degrees of ellipticity, orientation and temporal amplitudes.

Our idea [214] of the coherent control on the nanoscale by spatial modulation (shaping) of the excitation waveform has been developed theoretically [241] and experimentally [215, 216]. The coherent control of nanoscale distribution of local optical fields based on CW excitation aimed at achieving a deterministic control of plasmonic fields by using the spatial shaping of high order beams such as Hermite-Gaussian (HG) and Laguerre-Gaussian (LG) beams has been carried out in Ref. [215]. It has been shown experimentally that the spatial phase shaping of the excitation field provides an additional degree of freedom to drive optical nanoantennas and consequently control their near field response.

An example of such a deterministic coherent control is illustrated in Fig. 1.23. It shows a double gap antenna formed by three 500 nm aligned gold bars forming two identical 50 nm air gaps separated by 500 nm. For reference, in panel (a) it displays a measured two-photon luminescence (TPL) map when driving the whole antenna with a Gaussian beam linearly polarized along the x-axis. Note that similar to what has been discussed above in Sect. 1.4.4, in particular, in conjunction with Fig. 1.19, the TPL reflects the time-averaged distribution of the local field intensity $\langle I^2(\mathbf{r})\rangle$. As we see from Fig. 1.23a and as expected, a field concentration is observed in both gaps. Figure 1.23b, c show TPL maps recorded when the π-phase shift of a HG10 beam coincides, respectively, with the right and left gaps. These data demonstrate how a suitable positioning of the phase jump over the double antenna enables us to selectively switch on and off one of the two hot-spot sites.

Even closer to the original idea [214] that a plasmonic wavefront can be shaped and focused at a predetermined spot by a spatial phase modulation of the excitation waveform incident on optically-addressable launch pads is a recent publication [216]. This article achieves controlled launching and propagation of SPPs by spatially designing the amplitude and phase of the incident light. The chosen amplitude profile, consisting of four bright ("on") SPP launching platforms and one central dark ("off") arena, fully separates plasmonic effects from photonic

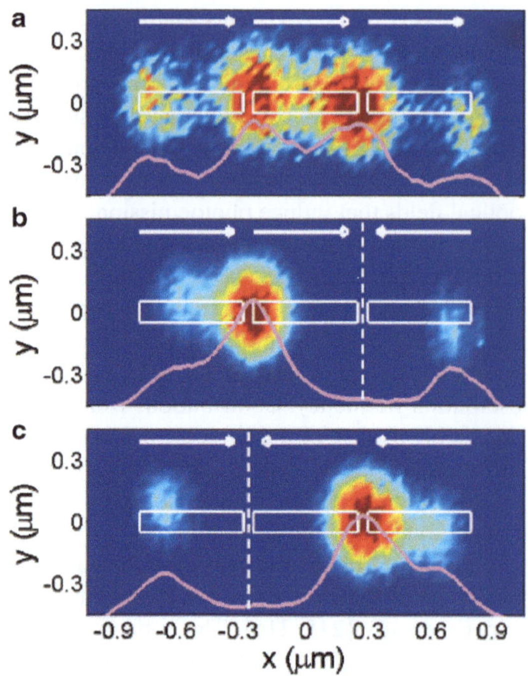

Fig. 1.23 Experimental results on spatial coherent control of nanoantennas (Adapted from Ref. [215]). Experimental two-photon luminescence (TPL) maps recorded for (**a**) a Gaussian beam and (**b**, **c**) a Hermite-Gaussian (HG10) beam whom phase shift (indicated by the *vertical dashed line*) coincides with (**b**) the right gap and (**c**) the left gap

effects and in addition is the necessary starting point for later focusing and scanning experiments. Any intensity detected inside the arena is purely plasmonic.

Adapting from Ref. [216], we present the achieved SPP focusing in Fig. 1.24. A phase optimization loop is used to focus SPPs at a pre-chosen target. This loop yields the optimal phase for each launching pad ("superpixel") as well as the relative intensity to focus. The amplitude profile is the same in all cases including the bare gold case, with four launching areas and a central dark arena where only SPPs can propagate. The incident polarization is diagonal in relation to the grating lines so as to have all available angles (2π range) contributing to the focus, thereby maximizing the numerical aperture and resolution.

Successful focusing at the center of the SPP arena is shown in Fig. 1.24a. The structured SPP wavefront produces an intensity in the designated target that is at least 20 times higher than the average SPP background of an unstructured wavefront. The measured size of the plasmonic focus is 420 nm, consistent with the diffraction limit of the SPPs. The flexibility of the method (scanning the focus) is demonstrated in Fig. 1.24b, c, which shows the SPP focus relocated without mechanical motion to controlled positions in the plasmonic arena.

The work of Ref. [216] has fully implemented the idea of Ref. [214] on the spatial-phase-modulation control of the SPP wavefronts to position a SPP nanofocus at a desired location at the surface. However, it employs only CW excitation and does not exploit a potential femtosecond temporal degree of freedom to achieve such a nanofocusing at a predetermined moment of time as in Ref. [214].

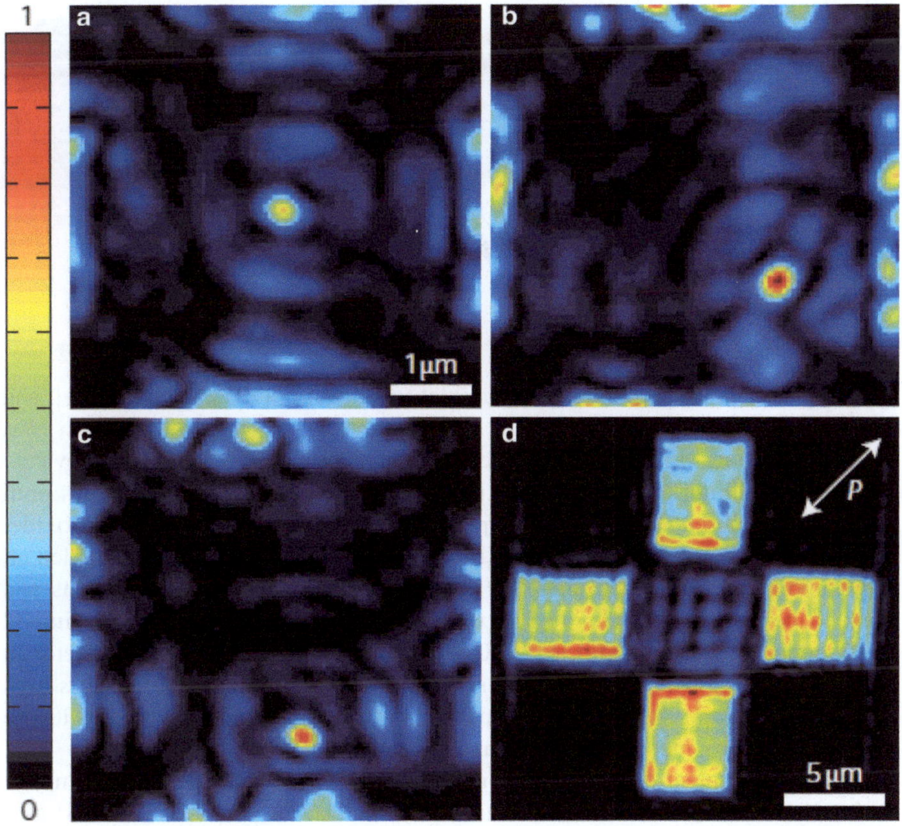

Fig. 1.24 Experiment on coherent control (dynamic focusing) of SPPs (Adapted from Ref. [216]). (**a**) Relative phases of the superpixels are optimized to focus SPPs at the center of the SPP arena. The intensity in the target spot is purely plasmonic and 20 times higher than the average background of an unstructured plasmonic wavefront. The focus size is diffraction limited by the detecting optics. (**b**) and (**c**), Demonstration of SPP focusing on freely chosen targets in the SPP arena. (**d**) Background reference of an unstructured SPP wavefront (uniform phase profile)

1.5 Quantum Nanoplasmonics: Spaser and Nanoplasmonics with Gain

1.5.1 Introduction to Spasers and Spasing

Not just a promise anymore [245], nanoplasmonics has delivered a number of important applications: ultrasensing [246], scanning near-field optical microscopy [194, 247], SP-enhanced photodetectors [53], thermally assisted magnetic recording [248], generation of extreme uv [138], biomedical tests [246, 249], SP-assisted thermal cancer treatment [250], plasmonic enhanced generation of extreme ultraviolet

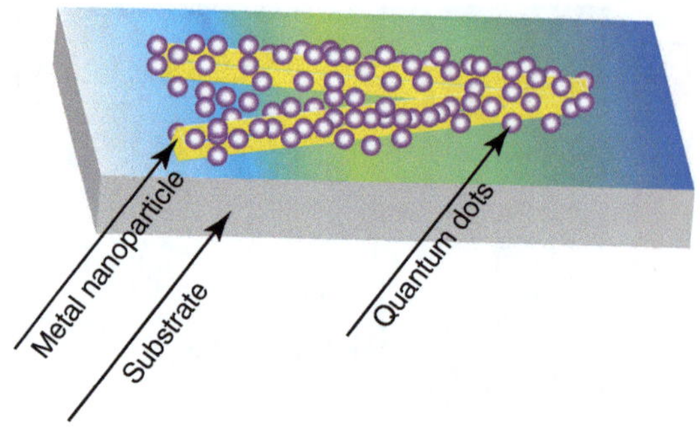

Fig. 1.25 Schematic of the spaser as originally proposed in Ref. [31]. The resonator of the spaser is a metal nanoparticle shown as a gold V-shape. It is covered by the gain medium depicted as nanocrystal quantum dots. This active medium is supported by a neutral substrate

(EUV) pulses [138] and extreme ultraviolet to soft x-ray (XUV) pulses [251], and many others – see also Ref. [23].

To continue its vigorous development, nanoplasmonics needs an active device – near-field generator and amplifier of nanolocalized optical fields, which has until recently been absent. A nanoscale amplifier in microelectronics is the metal-oxide-semiconductor field effect transistor (MOSFET) [252, 253], which has enabled all contemporary digital electronics, including computers and communications and enabled the present day technology as we know it. However, the MOSFET is limited by frequency and bandwidth to $\lesssim$100 GHz, which is already a limiting factor in further technological development. Another limitation of the MOSFET is its high sensitivity to temperature, electric fields, and ionizing radiation, which limits its use in extreme environmental conditions and nuclear technology and warfare.

An active element of nanoplasmonics is the spaser (Surface Plasmon Amplification by Stimulated Emission of Radiation), which was proposed [31, 254] as a nanoscale quantum generator of nanolocalized coherent and intense optical fields. The idea of spaser has been further developed theoretically [139–141, 255]. Spaser effect has recently been observed experimentally [256]. Also a number of SPP spasers (also called nanolasers) have been experimentally observed [257–260].

Spaser is a nanoplasmonic counterpart of laser: it is a quantum generator and nanoamplifier where photons as the generated quanta are replaced by SPs. Spaser consists of a metal nanoparticle, which plays a role of the laser cavity (resonator), and the gain medium. Figure 1.25 schematically illustrates geometry of a spaser introduced in the original article [31], which contains a V-shaped metal nanoparticle surrounded by a layer of semiconductor nanocrystal quantum dots.

1.5.2 Spaser Fundamentals

As we have already mentioned, the spaser is a nanoplasmonic counterpart of the laser [31, 255]. The laser has two principal elements: resonator (or cavity) that supports photonic mode(s) and the gain (or active) medium that is population-inverted and supplies energy to the lasing mode(s). An inherent limitation of the laser is that the size of the laser cavity in the propagation direction is at least half wavelength and practically more than that even for the smallest lasers developed [257, 258, 261]. In the spaser [31] this limitation is overcome. The spasing modes are surface plasmons (SPs) whose localization length is on the nanoscale [78] and is only limited by the minimum inhomogeneity scale of the plasmonic metal and the nonlocality radius [35] $l_{nl} \sim 1$ nm. So, the spaser is truly nanoscopic – its minimum total size can be just a few nanometers.

The resonator of a spaser can be any plasmonic metal nanoparticle whose total size R is much less than the wavelength λ and whose metal thickness is between l_{nl} and l_s, which supports a SP mode with required frequency ω_n. This metal nanoparticle should be surrounded by the gain medium that overlaps with the spasing SP eigenmode spatially and whose emission line overlaps with this eigenmode spectrally [31]. As an example, we consider a model of a nanoshell spaser [139, 255, 262], which is illustrated in Fig. 1.26. Panel (a) shows a silver nanoshell carrying a single SP (plasmon population number $N_n = 1$) in the dipole eigenmode. It is characterized by a uniform field inside the core and hot spots at the poles outside the shell with the maximum field reaching $\sim 10^6$ V/cm. Similarly, Fig. 1.26b shows the quadrupole mode in the same nanoshell. In this case, the mode electric field is non-uniform, exhibiting hot spots of $\sim 1.5 \times 10^6$ V/cm of the modal electric field at the poles. These high values of the modal fields is the underlying physical reason for a very strong feedback in the spaser. Under our conditions, the electromagnetic retardation within the spaser volume can be safely neglected. Also, the radiation of such a spaser is a weak effect: the decay rate of plasmonic eigenmodes is dominated by the internal loss in the metal. Therefore, it is sufficient to consider only quasistatic eigenmodes [29, 78] and not their full electrodynamic counterparts [263].

For the sake of numerical illustrations of our theory, we will use the dipole eigenmode (Fig. 1.26a). There are two basic ways to place the gain medium: (i) outside the nanoshell, as shown in panel (c), and (ii) in the core, as in panel (d), which was originally proposed in Ref. [262]. As we have verified, these two designs lead to comparable characteristics of the spaser. However, the placement of the gain medium inside the core illustrated in Fig. 1.26d has a significant advantage because the hot spots of the local field are not covered by the gain medium and are sterically available for applications.

Note that any l-multipole mode of a spherical particle is, indeed, $2l + 1$-times degenerate. This may make the spasing mode to be polarization unstable, like in lasers without polarizing elements. In reality, the polarization may be clamped and become stable due to deviations from the perfect spherical symmetry, which exist naturally or can be introduced deliberately. More practical shape for a spaser may be a nanorod, which has a mode with the stable polarization along the major axis. However, a nanorod is a more complicated geometry for theoretical treatment, and we will consider it elsewhere.

The level diagram of the spaser gain medium and the plasmonic metal nanoparticle is displayed in Fig. 1.26e along with a schematic of the relevant energy transitions in the system. The gain medium chromophores may be semiconductor

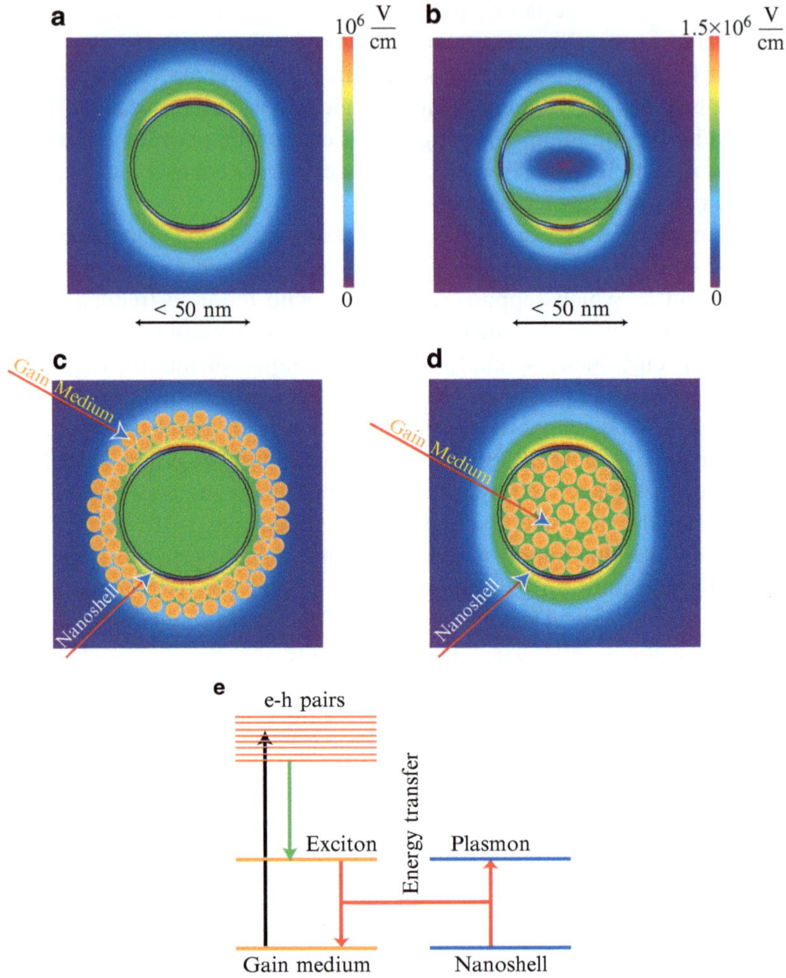

Fig. 1.26 Schematic of spaser geometry, local fields, and fundamental processes leading to spasing (Adapted from Ref. [139]). (**a**) Nanoshell geometry and the local optical field distribution for one SP in an axially-symmetric dipole mode. The nanoshell has aspect ratio $\eta = 0.95$. The local field magnitude is color-coded by the scale bar in the *right-hand side* of the panel. (**b**) The same as (**a**) but for a quadrupole mode. (**c**) Schematic of a nanoshell spaser where the gain medium is outside of the shell, on the background of the dipole-mode field. (**d**) The same as (**c**) but for the gain medium inside the shell. (**e**) Schematic of the spasing process. The gain medium is excited and population-inverted by an external source, as depicted by the *black arrow*, which produces electron-hole pairs in it. These pairs relax, as shown by the *green arrow*, to form the excitons. The excitons undergo decay to the ground state emitting SPs into the nanoshell. The plasmonic oscillations of the nanoshell stimulates this emission, supplying the feedback for the spaser action

nanocrystal quantum dots [31, 264], dye molecules [265, 266], rare-earth ions [262], or electron-hole excitations of an unstructured semiconductor [257, 261]. For certainty, we will use a semiconductor-science language of electrons and holes in quantum dots.

The pump excites electron-hole pairs in the chromophores (Fig. 1.26e), as indicated by the vertical black arrow, which relax to form excitons. The excitons constitute the two-level systems that are the donors of energy for the SP emission into the spasing mode. In vacuum, the excitons would recombine emitting photons. However, in the spaser geometry, the photoemission is strongly quenched due to the resonance energy transfer to the SP modes, as indicated by the red arrows in the panel. The probability of the radiativeless energy transfer to the SPs relative to that of the radiative decay (photon emission) is given by the so-called Purcell factor

$$\sim \frac{\lambda\!\!\!^{-3} Q}{R^3} \gg 1 , \tag{1.61}$$

where R is a characteristic size of the spaser metal core. Thus this radiativeless energy transfer to the spaser mode is the dominant process whose probability is by orders of magnitude greater than that of the free-space (far-field) emission.

The plasmons already in the spaser mode create the high local fields that excite the gain medium and stimulate more emission to this mode, which is the feedback mechanism. If this feedback is strong enough, and the life time of the spaser SP mode is long enough, then an instability develops leading to the avalanche of the SP emission in the spasing mode and spontaneous symmetry breaking, establishing the phase coherence of the spasing state. Thus the establishment of spasing is a non-equilibrium phase transition, as in the physics of lasers.

1.5.3 Brief Overview of Latest Progress in Spasers

After the original theoretical proposal and prediction of the spaser [31], there has been an active development in this field, both theoretical and experimental. There has also been a US patent issued on spaser [254].

Among theoretical developments, a nanolens spaser has been proposed [267], which possesses a nanofocus ("the hottest spot") of the local fields. In Refs. [31, 267], the necessary condition of spasing has been established on the basis of the perturbation theory.

There have been theories published describing the SPP spasers (or, "nanolasers" as sometimes they are called) phenomenologically, on the basis of classic linear electrodynamics by considering the gain medium as a dielectric with a negative imaginary part of the permittivity, e.g., [262]. Very close fundamentally and technically are works on the loss compensation in metamaterials [268–271]. Such linear-response approaches do not take into account the nature of the spasing as a non-equilibrium phase transition, at the foundation of which is spontaneous symmetry breaking: establishing coherence with an arbitrary but sustained phase of the SP quanta in the system [139]. Spaser is necessarily a deeply-nonlinear (nonperturbative) phenomenon where the coherent SP field always saturates the gain medium, which eventually brings about establishment of the stationary (or, continuous wave, CW) regime of the spasing [139]. This leads to principal differences of the linear-response results from the microscopic quantum-mechanical theory in the region of spasing, as we discuss below in conjunction with Fig. 1.29.

There has also been a theoretical publication on a bowtie spaser (nanolaser) with electrical pumping [272]. It is based on balance equations and only the CW spasing generation intensity is described. Yet another theoretical development has been a proposal of the lasing spaser [273], which is made of a plane array of spasers.

There have also been a theoretical proposal of a spaser ("nanolaser") consisting of a metal nanoparticle coupled to a single chromophore [274]. In this paper, a dipole-dipole interaction is illegitimately used at very small distances r where it has a singularity (diverging for $r \to 0$), leading to a dramatically overestimated coupling with the SP mode. As a result, a completely unphysical prediction of CW spasing due to single chromophore has been obtained [274]. In contrast, our theory [139] is based on the full (exact) field of the spasing SP mode without the dipole (or, any multipole) approximation. As our results of Sect. 1.5.5 below show, hundreds of chromophores per metal nanoparticle are realistically requited for the spasing even under the most favorable conditions.

There has been a vigorous experimental investigation of the spaser and the concepts of spaser. Stimulated emission of SPPs has been observed in a proof-of-principle experiment using pumped dye molecules as an active (gain) medium [265]. There have also been later experiments that demonstrated strong stimulated emission compensating a significant part of the SPP loss [266, 275–278]. As a step toward the lasing spaser, the first experimental demonstration has been reported of a partial compensation of the Joule losses in a metallic photonic metamaterial using optically pumped PbS semiconductor quantum dots [264]. There have also been experimental investigations reporting the stimulated emission effects of SPs in plasmonic metal nanoparticles surrounded by gain media with dye molecules [279, 280].

The full loss compensation and amplification of the long-range SPPs at $\lambda = 882$ nm in a gold nanostrip waveguide with a dyes solution as a gain medium has been observed [281]. Another example of full loss compensation has recently been obtained for thin ($\sim$20 nm thickness) gold stripes (width $\sim$1 μm) surrounded by a gain medium containing donor-acceptor with a Förster energy transfer to increase the Stokes shift and decrease absorption at the probe frequency.

At the present time, there have been a number of the successful experimental observations of the spaser and SPP spasers (the so-called nanolasers). An electrically-pumped nanolaser with semiconductor gain medium have been demonstrated [257] where the lasing modes are SPPs with a one-dimensional confinement to a $\sim$50 nm size. Other electrically-pumped nanolasers (SPP spasers) have recently been fabricated and their lasing observed based on a diode with an intrinsic InGaAs gain media and silver nanocavities as plasmonic cores [282–284]. The latest of these nanolasers [284] operates at a room temperature and has a relatively small cavity volume $V_c \approx 0.67\lambda^3$, where vacuum wavelength $\lambda = 1{,}591$ nm. This volume is still much larger than the modal volumes of the spasers with tighter confinement, especially SP-mode spasers – see below.

A nanolaser with an optically-pumped semiconductor gain medium and a hybrid semiconductor/metal (CdS/Ag) SPP waveguide has been demonstrated with an extremely tight transverse (two-dimensional) mode confinement to $\sim$10 nm size [258]. This has been followed by the development of CdS/Ag nanolasers generating a visible single mode at a room temperature with a tight one-dimensional confinement ($\sim$20 nm) and a two-dimensional confinement in the plane of the structure to an area $\sim 1\,\mu\text{m}^2$ [259]. A highly efficient SPP spaser in the communication range ($\lambda = 1.46\,\mu$m) with an optical pumping based on a gold film and an InGaAs semiconductor quantum-well gain medium has recently been reported [260].

Another class of spasers observed are random spasers comprised of a rough metal nanofilm as a plasmonic component and a dye-doped polymeric film as a gain medium [285]. The spasing in such systems competes with loss compensation for SPPs propagating at the interface – see also Sect. 1.5.7.

Hitorically, the first spaser observed was a nanoparticle spaser [256]. This spaser is a chemically synthesized gold nanosphere of radius 7 nm surrounded by a dielectric shell of a 21 nm outer radius containing immobilized dye molecules. Under nanosecond optical pumping in the absorption band of the dye, this spaser develops a relatively narrow-spectrum and intense visible emission that exhibits a pronounced threshold in pumping intensity. The observed characteristics of this spaser are in an excellent qualitative agreement and can be fully understood on the basis of the corresponding theoretical results described below in Sect. 1.5.5.

1.5.3.1 Nanospaser with Semiconductor Gain Media

It is of both fundamental and applied importance to develop nanoscale-size spasers (nanospasers) with semiconductor gain media. The photochemical and electrochemical stability of the semiconductor gain media is the main attraction of such a design. Belonging to this class, spasers have recently been fabricated and their operation observed, comprised of a InGaN-core/InN-shell semiconductor-nanorod gain medium and silver film as a plasmonic component [286, 287]. They generate on localized SP modes. One of these [287] is a nanospaser with a deeply subwavelength mode size based on an epitaxial silver nanofilm [287]. Such a design bears a promise of practical applications due to its stability and small modal volume leading to high operational speed – see below Sect. 1.5.6.

In Fig. 1.27, we display geometry of this InGaN-core/InN-shell nanorod spaser and properties of its spasing mode. The active region of the spaser (Fig. 1.27a, left panel) is a core-shell nanocylinder with a 30-nm diameter core of InGaN surrounded by think shell if GaN. The latter is a wide band-gap semiconductor that plays a role of insulator. The active nanorod is separated by the metal by a 5-nm layer of silica.

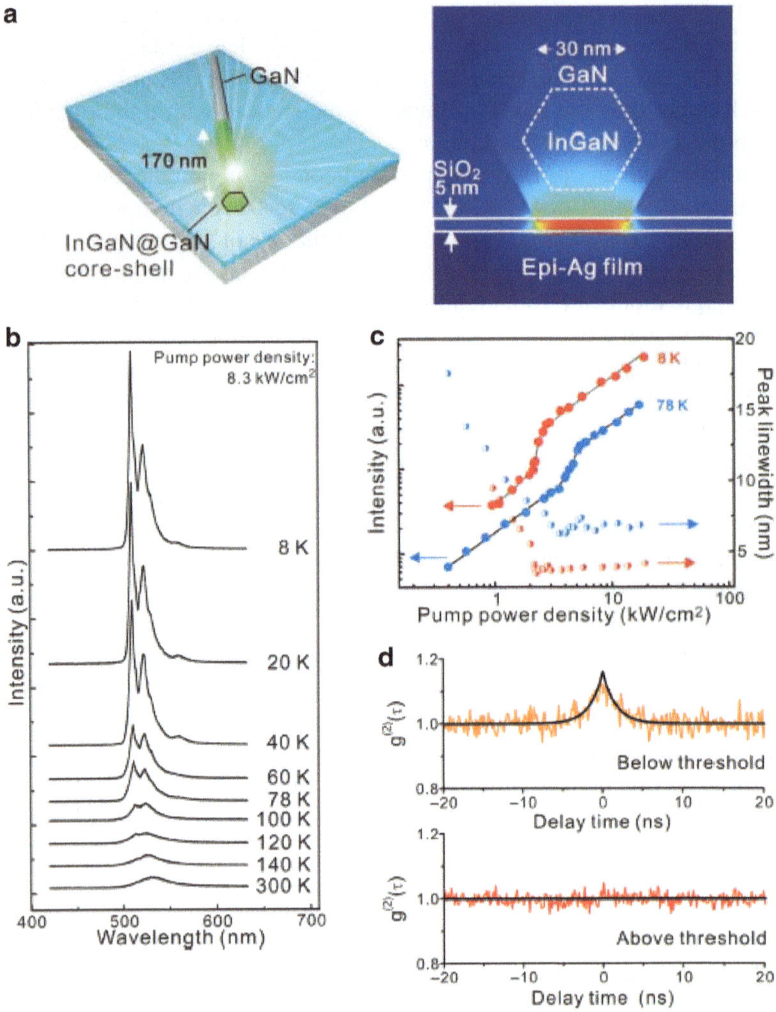

Fig. 1.27 InGaN nanospaser and its properties. (**a**) Schematics of geometry of InGaN/GaN core-shell nanospaser (*left*) and theoretical intensity of its spasing eigenmode. (**b**) Series of emission spectra: Temperature-dependent spasing behavior from 8 to 300 K. The spasing threshold at 140 K is clearly visible. (**c**) The L-L (light-light) plots at the main lasing peak (510 nm) are shown with the corresponding linewidth-narrowing behavior when the spaser is measured at 8 K (*red*) and 78 K (*blue*), with lasing thresholds of 2.1 and 3.7 kW/cm^2, respectively. (**d**) Second-order photon correlation function $g^{(2)}(\tau)$ measured at 8 K. The *upper curve* is recorded below the spasing threshold, and the *lower* above the threshold (Adapted from Ref. [287])

The plasmonic component of this spaser is a flat layer of epitaxial silver. The high monocrystalline quality of the silver film is instrumental in reducing the threshold of the spaser and increasing its output. The calculated intensity for the spasing eigenmode is shown in the right panel of Fig. 1.27a. Similar to the gap modes introduced in Ref. [288], this eigenmode is concentrated in the thin layer of a low-permittivity dielectric (silica) between the two high-permittivity media: GaN and silver. The modal fields do penetrate sufficiently into the gain medium providing the feedback necessary for the spaser functioning.

Under $8.3\,\mathrm{kW/cm^2}$ optical pumping with frequency above the band gap of InGaN, a series of the emission spectra of a single spaser is displayed in Fig. 1.27b, At a room temperature, $T = 300\,\mathrm{K}$, the emission is a spontaneous fluorescence in a wide yellow-green spectral band near the band gap of InGaN. The first evidence of the spasing appears at $T = 120\,\mathrm{K}$ as a small notch at the green side of the spectrum. As the temperature decreases to $T = 8\,\mathrm{K}$, the narrow line at $\lambda \approx 500\,\mathrm{nm}$ becomes dominant and narrow. This change of the spectrum over the threshold is in a qualitative agreement with theory – see below Sect. 1.5.5 and, in particular, Fig. 1.29d–f.

The light-line (L-L) line is the dependence of the light intensity out (the intensity of the radiation emitted by the spaser within the linewidth spectral range) versus the intensity of the pumping radiation. The theoretical prediction for the spaser is that after reaching the spasing threshold, the L-L line becomes linear with universally unit slope – see Fig. 1.29a and its discussion in Sect. 1.5.5.

The experimentally obtained L-L line of the nanorod spaser shown in Fig. 1.27c is in an excellent agreement with this prediction. Note that this figure is presented in the double-logarithmic scale. There are two curves in this figure taken at different temperatures, which are similar though at a lower temperature the intensity out is higher and the threshold is lower. The parts of the curves at lower pumping intensities are also unit-slope straight lines corresponding to spontaneous fluorescence. With the increased intensity, the curves enter a transitional regime of amplified spontaneous emission where the slopes are greater than one. The regime of developed spasing takes place at high intensities where the L-L curves become unit-slope straight lines without a saturation. As have already been mentioned above, this is a universal behavior.

This universal unsaturable behavior can be very simply understood qualitatively – cf. Ref. [289]. The excitation rate $\dot{N}_e$ of the upper spasing level is linearly proportional to pumping intensity I_p, $\dot{N}_e = \sigma_e I_p$, where σ_e is the total excitation cross section into the conduction band of the semiconductor gain medium. In the developed spasing regime, plasmon population N_n of the spasing eigenmode becomes large, asymptotically $N_n \to \infty$. Correspondingly, the stimulated decay rate, which is $\propto N_n$, becomes large and dominates over any spontaneous decay rate. Thus, all the excitation events to the conduction band end up with the emission of a SP into the spasing mode whose SP population becomes $N_n = \dot{N}_e / \gamma_n$, where γ_n is the SP decay rate – see above Eq. (1.48). Finally, radiation rate $\dot{N}_r$ for a spaser becomes

$$\dot{N}_r = \sigma_e \gamma^{(r)} / \gamma_n \,, \tag{1.62}$$

where $\gamma^{(r)}$ is the SP radiative decay rate, which for a plasmonic metal sphere is given by Eq. (1.16) and in, general case, by Eq. (1.56). Of course, in reality the straight-line, unsaturable L-L curves will end when the pumping intensities become so high that the nonlinearity in the spaser metal develops (including, but not limited to, thermal nonlinearity), or optical breakdown occurs, or heat production will physically damage the spaser.

As theory shows (see below Sect. 1.5.6.1 and Fig. 1.30a), under steady pumping, the generating spaser reaches its stationary regime within ~100 fs. Correspondingly, we expect that any fluctuation in the emission radiated by the generating spaser relaxes back to the mean level within the same time. A measure of the fluctuations of the spaser-radiation intensity $I(t)$ with time t is the second-order autocorrelation function

$$g^{(2)}(\tau) = \frac{\langle I(t+\tau)I(t)\rangle}{\langle I(t)\rangle^2}, \tag{1.63}$$

where τ is the delay time, and $\langle\cdots\rangle$ denotes quantum-mechanical (theory) or temporal (experiment) averaging.

Experimentally, $g^{(2)}(\tau)$ has been measured for a single spaser in Ref. [287]. The result is reproduced in Fig. 1.27d. The upper curve is recorded below the spasing threshold; at the zero delay, it shows a peak, which is characteristic of incoherent radiation. If such radiation is produced by many independent emitters, it has Gaussian statistics, and the peak value should be $g^{(2)}(0) = 2$ – this effect was introduced by Hanbury Brown and Twiss and used by them for stellar interferometry [290]. For the upper curve of Fig. 1.27d, $g^{(2)}(0)$ is significantly less. This may be due to various reasons, in particular, insufficient temporal resolution of the photodetection or partial coherence between the individual emitters of the gain medium induced by their interaction via plasmonic fields.

In sharp contrast, above the spasing threshold, the autocorrelation function in Fig. 1.27d is a constant at all delays. As we have already pointed out this is due to the fact that after an emission of a photon, the number of plasmons in the spaser is restored within ~100 fs, while the temporal resolution of the photodetection in Ref. [287] is $\Delta\tau \gtrsim 100$ ps, i.e., three orders of magnitude coarser. The physical reason for $g^{(2)}(\tau) = \text{const}$ is that the spaser under steady-state pumping tends to keep a constant plasmon population. After the emission of a photon, this population is decreased by one. However, very rapidly, within ~100 fs, it restores to the pre-emission level. This transitional restoration process is too fast and the photodetectors of Ref. [287] miss it, producing $g^{(2)}(\tau) = \text{const}$.

1.5.4 Equations of Spaser

1.5.4.1 Quantum Density Matrix Equations (Optical Bloch Equations) for Spaser

The SP eigenmodes $\varphi_n(\mathbf{r})$ are described by a wave equation (1.25) [31, 78]. The electric field operator of the quantized SPs is an operator [31]

$$\hat{\mathbf{E}}(\mathbf{r}) = -\sum_n A_n \nabla\varphi_n(\mathbf{r})(\hat{a}_n + \hat{a}_n^\dagger)\,, \quad A_n = \left(\frac{4\pi\hbar s_n}{\varepsilon_d s_n'}\right)^{1/2}, \tag{1.64}$$

where $\hat{a}_n^\dagger$ and $\hat{a}_n$ are the SP creation and annihilation operators, $-\nabla\varphi_n(\mathbf{r}) = \mathbf{E}_n(\mathbf{r})$ is the modal field of an nth mode, and $s_n' = \mathrm{Re}\,[ds(\omega_n)/d\omega_n]$. Note that we have corrected a misprint in Ref. [31] by replacing the coefficient 2π by 4π.

The spaser Hamiltonian has the form

$$\hat{H} = \hat{H}_g + \hbar\sum_n \omega_n \hat{a}_n^\dagger \hat{a}_n - \sum_p \hat{\mathbf{E}}(\mathbf{r}_p)\hat{\mathbf{d}}^{(p)}\,, \tag{1.65}$$

where $\hat{H}_g$ is the Hamiltonian of the gain medium, p is a number (label) of a gain medium chromophore, $\mathbf{r}_p$ is its coordinate vector, and $\hat{\mathbf{d}}^{(p)}$ is its dipole moment operator. In this theory, we treat the gain medium quantum mechanically but the SPs quasiclassically, considering $\hat{a}_n$ as a classical quantity (c-number) a_n with time dependence as $a_n = a_{0n}\exp(-i\omega t)$, where a_{0n} is a slowly-varying amplitude. The number of coherent SPs per spasing mode is then given by $N_p = |a_{0n}|^2$. This approximation neglects the quantum fluctuations of the SP amplitudes. However, when necessary, we will take into account these quantum fluctuations, in particular, to describe the spectrum of the spaser.

Introducing $\rho^{(p)}$ as the density matrix of a pth chromophore, we can find its equation of motion in a conventional way by commutating it with the Hamiltonian (1.65) as

$$i\hbar\dot{\rho}^{(p)} = [\rho^{(p)}, \hat{H}]\,, \tag{1.66}$$

where the dot denotes temporal derivative. We use the standard rotating wave approximation (RWA), which only takes into account the resonant interaction between the optical field and chromophores. We denote $|1\rangle$ and $|2\rangle$ as the ground and excited states of a chromophore, with the transition $|2\rangle \rightleftharpoons |1\rangle$ resonant to the spasing plasmon mode n. In this approximation, the time dependence of the nondiagonal elements of the density matrix is $\left(\rho^{(p)}\right)_{12} = \bar{\rho}_{12}^{(p)}\exp(i\omega t)$, and $\left(\rho^{(p)}\right)_{21} = \bar{\rho}_{12}^{(p)*}\exp(-i\omega t)$, where $\bar{\rho}_{12}^{(p)}$ is an amplitude slowly varying in time, which defines the coherence (polarization) for the $|2\rangle \rightleftharpoons |1\rangle$ spasing transition in a pth chromophore of the gain medium.

Introducing a rate constant Γ_{12} to describe the polarization relaxation and a difference $n_{21}^{(p)} = \rho_{22}^{(p)} - \rho_{11}^{(p)}$ as the population inversion for this spasing transition, we derive an equation of motion for the non-diagonal element of the density matrix as

$$\dot{\bar{\rho}}_{12}^{(p)} = -\left[i\left(\omega - \omega_{12}\right) + \Gamma_{12}\right]\bar{\rho}_{12}^{(p)} + i a_{0n} n_{21}^{(p)} \tilde{\Omega}_{12}^{(p)*}\,, \tag{1.67}$$

where

$$\tilde{\Omega}_{12}^{(p)} = -A_n \mathbf{d}_{12}^{(p)} \nabla \varphi_n(\mathbf{r}_p)/\hbar \tag{1.68}$$

is the one-plasmon Rabi frequency for the spasing transition in a pth chromophore, and $\mathbf{d}_{12}^{(p)}$ is the corresponding transitional dipole element. Note that always $\mathbf{d}_{12}^{(p)}$ is either real or can be made real by a proper choice of the quantum state phases, making the Rabi frequency $\tilde{\Omega}_{12}^{(p)}$ also a real quantity.

An equation of motion for n_{21}^{p} can be found in a standard way by commutating it with $\hat{H}$. To provide conditions for the population inversion ($n_{21}^{p} > 0$), we imply existence of a third level. For simplicity, we assume that it very rapidly decays into the excited state $|2\rangle$ of the chromophore, so its own populations is negligible. It is pumped by an external source from the ground state (optically or electrically) with some rate that we will denote g. In this way, we obtain the following equation of motion:

$$\dot{n}_{21}^{(p)} = -4\mathrm{Im}\left[a_{0n}\bar{\rho}_{12}^{(p)}\tilde{\Omega}_{21}^{(p)}\right] - \gamma_2\left(1 + n_{21}^{(p)}\right) + g\left(1 - n_{21}^{(p)}\right), \tag{1.69}$$

where γ_2 is the decay rate $|2\rangle \rightarrow |1\rangle$.

The stimulated emission of the SPs is described as their excitation by the coherent polarization of the gain medium. The corresponding equation of motion can be obtained using Hamiltonian (1.65) and adding the SP relaxation with a rate of γ_n as

$$\dot{a}_{0n} = \left[i\left(\omega - \omega_n\right) - \gamma_n\right] a_{0n} + i a_{0n} \sum_p \rho_{12}^{(p)*} \tilde{\Omega}_{12}^{(p)} . \tag{1.70}$$

As an important general remark, the system of Eqs. (1.67), (1.69), and (1.70) is highly nonlinear: each of these equations contains a quadratic nonlinearity: a product of the plasmon-field amplitude a_{0n} by the density matrix element ρ_{12} or population inversion n_{21}. Altogether, this is a six-order nonlinearity. This nonlinearity is a fundamental property of the spaser equations, which makes the spaser generation always an essentially nonlinear process that involves a noneqilibrium phase transition and a spontaneous symmetry breaking: establishment of an arbitrary but sustained phase of the coherent SP oscillations.

A relevant process is spontaneous emission of SPs by a chromophore into a spasing SP mode. The corresponding rate $\gamma_2^{(p)}$ for a chromophore at a point $\mathbf{r}_p$ can be found in a standard way using the quantized field (1.64) as

$$\gamma_2^{(p)} = 2\frac{A_n^2}{\hbar\gamma_n}\left|\mathbf{d}_{12}\nabla\varphi_n(\mathbf{r}_p)\right|^2 \frac{\left(\Gamma_{12} + \gamma_n\right)^2}{\left(\omega_{12} - \omega_n\right)^2 + \left(\Gamma_{12} + \gamma_n\right)^2} . \tag{1.71}$$

As in Schawlow-Towns theory of laser-line width [291], this spontaneous emission of SPs leads to the diffusion of the phase of the spasing state. This defines width γ_s of the spasing line as

$$\gamma_s = \frac{\sum_p \left(1 + n_{21}^{(p)}\right) \gamma_2^{(p)}}{2(2N_p + 1)} . \tag{1.72}$$

This width is small for a case of developed spasing when $N_p \gg 1$. However, for $N_p \sim 1$, the predicted width may be too high because the spectral diffusion theory assumes that $\gamma_s \lesssim \gamma_n$. To take into account this limitation in a simplified way, we will interpolate to find the resulting spectral width Γ_s of the spasing line as $\Gamma_s = \left(\gamma_n^{-2} + \gamma_s^{-2}\right)^{-1/2}$.

We will also examine the spaser as a bistable (logical) amplifier. One of the ways to set the spaser in such a mode is to add a saturable absorber. This is described by the same Eqs. (1.67)–(1.70) where the chromophores belonging to the absorber are not pumped by the external source directly, i.e., for them in Eq. (1.69) one has to set $g = 0$.

Numerical examples are given for a silver nanoshell where the core and the external dielectric have the same permittivity of $\varepsilon_d = 2$; the permittivity of silver is adopted from Ref. [32]. The following realistic parameters of the gain medium are used (unless indicated otherwise): $d_{12} = 1.5 \times 10^{-17}$ esu, $\hbar\Gamma_{12} = 10\,\text{meV}$, $\gamma_2 = 4 \times 10^{12}\,\text{s}^{-1}$ (this value takes into account the spontaneous decay into SPs), and density of the gain medium chromophores is $n_c = 2.4 \times 10^{20}\,\text{cm}^{-3}$, which is realistic for dye molecules but may be somewhat high for semiconductor quantum dots that were proposed as the chromophores [31] and used in experiments [264]. We will assume a dipole SP mode and chromophores situated in the core of the nanoshell as shown in Fig. 1.26d. This configuration are of advantage both functionally (because the region of the high local fields outside the shell is accessible for various applications) and computationally (the uniformity of the modal fields makes the summation of the chromophores trivial, thus greatly facilitating numerical procedures).

1.5.4.2 Equations for CW Regime

Physically, the spaser action is a result of spontaneous symmetry breaking when the phase of the coherent SP field is established from the spontaneous noise. Mathematically, the spaser is described by homogeneous differential Eqs. (1.67)–(1.70). These equations become homogeneous algebraic equations for the CW case. They always have a trivial, zero solution. However, they may also possess a nontrivial solution describing spasing. An existence condition of such a nontrivial solution is

$$(\omega_s - \omega_n + i\gamma_n)^{-1} \times (\omega_s - \omega_{21} + i\Gamma_{12})^{-1} \sum_p \left|\tilde{\Omega}_{12}^{(p)}\right|^2 n_{21}^{(p)} = -1 . \tag{1.73}$$

The population inversion of a pth chromophore $n_{21}^{(p)}$ is explicitly expressed as

$$n_{21}^{(p)} = (g - \gamma_2) \times \left\{ g + \gamma_2 + 4N_n \left| \tilde{\Omega}_{12}^{(p)} \right|^2 \Big/ \left[(\omega_s - \omega_{21})^2 + \Gamma_{12}^2 \right] \right\}^{-1} . \quad (1.74)$$

From the imaginary part of Eq. (1.74) we immediately find the spasing frequency ω_s,

$$\omega_s = (\gamma_n \omega_{21} + \Gamma_{12} \omega_n) / (\gamma_n + \Gamma_{12}) , \quad (1.75)$$

which generally does not coincide with either the gain transition frequency ω_{21} or the SP frequency ω_n, but is between them (this is a frequency walk-off phenomenon similar to that of laser physics). Substituting Eq. (1.75) back into Eqs. (1.74) and (1.75), we obtain a system of equations

$$\frac{(\gamma_n + \Gamma_{12})^2}{\gamma_n \Gamma_{12} \left[(\omega_{21} - \omega_n)^2 + (\Gamma_{12} + \gamma_n)^2 \right]} \times \sum_p \left| \tilde{\Omega}_{12}^{(p)} \right|^2 n_{21}^{(p)} = 1 , \quad (1.76)$$

$$n_{21}^{(p)} = (g - \gamma_2) \times \left[g + \gamma_2 + \frac{4N_n \left| \tilde{\Omega}_{12}^{(p)} \right|^2 (\Gamma_{12} + \gamma_n)}{(\omega_{12} - \omega_n)^2 + (\Gamma_{12} + \gamma_n)^2} \right]^{-1} . \quad (1.77)$$

This system defines the stationary (CW-generation) number of SPs per spasing mode, N_n.

Since $n_{21}^{(p)} \leq 1$, from Eqs. (1.76), (1.77) we immediately obtain a necessary condition of the existence of spasing,

$$\frac{(\gamma_n + \Gamma_{12})^2}{\gamma_n \Gamma_{12} \left[(\omega_{21} - \omega_n)^2 + (\Gamma_{12} + \gamma_n)^2 \right]} \sum_p \left| \tilde{\Omega}_{12}^{(p)} \right|^2 \geq 1 . \quad (1.78)$$

This expression is fully consistent with Ref. [31]. The following order of magnitude estimate of this spasing condition has a transparent physical meaning and is of heuristic value,

$$\frac{d_{12}^2 Q N_c}{\hbar \Gamma_{12} V_n} \gtrsim 1 , \quad (1.79)$$

where $Q = \omega/\gamma_n$ is the quality factor of SPs, V_n is the volume of the spasing SP mode, and N_c is the of number of the gain medium chromophores within this volume. Deriving this estimate, we have neglected the detuning, i.e., set $\omega_{21} - \omega_n = 0$. We also used the definitions of A_n of Eq. (1.64) and $\tilde{\Omega}_{12}^{(p)}$ given by Eq. (1.68), and the estimate $|\nabla\varphi_n(\mathbf{r})|^2 \sim 1/V$ following from the normalization of the SP eigenmodes $\int |\nabla\varphi_n(\mathbf{r})|^2 d^3r = 1$ of Ref. [78]. The result of Eq. (1.79) is, indeed, in agreement with Ref. [31] where it was obtained in different notations.

It follows from Eq. (1.79) that for the existence of spasing it is beneficial to have a high quality factor Q, a high density of the chromophores, and a large transition dipole (oscillator strength) of the chromophore transition. The small modal volume V_n (at a given number of the chromophores N_c) is beneficial for this spasing condition: physically, it implies strong feedback in the spaser. Note that for the given density of the chromophores $n_c = N_c/V_n$, this spasing condition does not explicitly depend on the spaser size, which opens up a possibility of spasers of a very small size limited from the bottom by only the nonlocality radius $l_{nl} \sim 1$ nm. Another important property of Eq. (1.79) is that it implies the quantum-mechanical nature of spasing and spaser amplification: this condition essentially contains the Planck constant $\hbar$ and, thus, does not have a classical counterpart. Note that in contrast to lasers, the spaser theory and Eqs. (1.78), (1.79) in particular do not contain speed of light, i.e., they are quasistatic.

Now we will examine the spasing condition and reduce it to a requirement for the gain medium. First, we substitute all the definitions and assume the perfect resonance between the generating SP mode and the gain medium, i.e., $\omega_n = \omega_{21}$. As a result, we obtain from Eq. (1.78),

$$\frac{4\pi}{3}\frac{s_n\,|\mathbf{d}_{12}|^2}{\hbar\gamma_n\Gamma_{12}\varepsilon_d s_n'}\int_V [1-\Theta(\mathbf{r})]\,|\mathbf{E}_n(\mathbf{r})|^2\,d^3r \geq 1\,, \tag{1.80}$$

where the integral is extended over the volume V of the system, and the Θ-function takes into account a simplifying realistic assumption that the gain medium occupies the entire space free from the core's metal. We also assume that the orientations of the transition dipoles $\mathbf{d}_{12}^{(p)}$ are random and average over them, which results in the factor of 3 in the denominator in Eq. (1.80).

From Eqs. (1.27) and (1.34), it follows that

$$\int_V [1-\Theta(\mathbf{r})]\,|\mathbf{E}_n(\mathbf{r})|^2\,d^3r = 1 - s_n\,. \tag{1.81}$$

Next, we give approximate expressions for the spectral parameter (1.4), which are very accurate for the realistic case of $Q \gg 1$,

$$\mathrm{Im}\,s(\omega) = \frac{s_n^2}{\varepsilon_d}\mathrm{Im}\,\varepsilon_m(\omega) = \frac{1}{Q}s_n\,(1-s_n)\;, \tag{1.82}$$

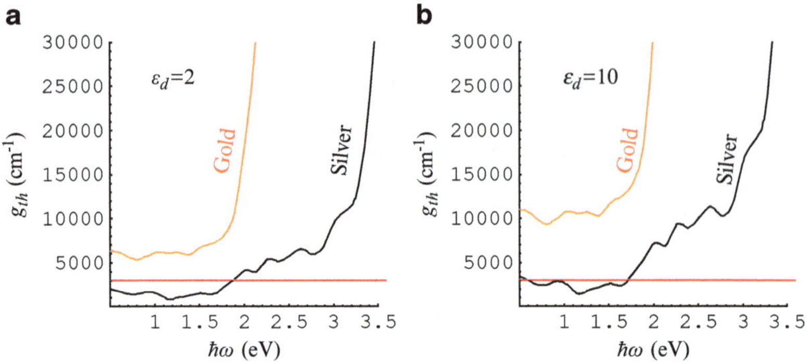

Fig. 1.28 Threshold gain for spasing g_{th} for silver and gold, as indicated in the graphs, as a function of the spasing frequency ω. The *red line* separates the area $g_{th} < 3 \times 10^3\,\mathrm{cm}^{-1}$, which can relatively easily be achieved with direct band-gap semiconductors (DBGSs). The real part of the gain medium permittivity is denoted in the corresponding panels as ε_d

where definition (1.6) is used. Taking into account Eqs. (1.47), (1.48) and (1.81), (1.82), we obtain from Eq. (1.80) a necessary condition of spasing at a frequency ω as

$$\frac{4\pi}{3}\frac{|\mathbf{d}_{12}|^2 n_c\,[1-\mathrm{Re}\,s(\omega)]}{\hbar\Gamma_{12}\mathrm{Re}\,s(\omega)\mathrm{Im}\,\varepsilon_m(\omega)} \geq 1\,, \tag{1.83}$$

For the sake of comparison, consider a continuous gain medium comprised of the same chromophores as the gain shell of the spaser. Its gain g (whose dimensionality is cm^{-1}) is given by a standard expression

$$g = \frac{4\pi}{3}\frac{\omega}{c}\frac{\sqrt{\varepsilon_d}\,|\mathbf{d}_{12}|^2 n_c}{\hbar\Gamma_{12}}\,. \tag{1.84}$$

Substituting it into Eq. (1.83), we obtain the spasing criterion in terms of the gain as

$$g \geq g_{th}\,, \quad g_{th} = \frac{\omega}{c\sqrt{\varepsilon_d}}\frac{\mathrm{Re}\,s(\omega)}{1-\mathrm{Re}\,s(\omega)}\mathrm{Im}\,\varepsilon_m(\omega)\,, \tag{1.85}$$

where g_{th} has a meaning of the threshold gain needed for spasing. Importantly, this gain depends only on the dielectric properties of the system and spasing frequency but not on the geometry of the system or the distribution of the local fields of the spasing mode (hot spots, etc.) explicitly. However note that the system's geometry (along with the permittivities) does define the spasing frequencies.

In Fig. 1.28a, b, correspondingly, we illustrate the analytical expression (1.85) for gold and silver embedded in a dielectric with $\varepsilon_d = 2$ (simulating a light glass) and $\varepsilon_d = 10$ (simulating a semiconductor), correspondingly. These are computed from Eq. (1.85) assuming that the metal core is embedded into the gain medium with the real part of the dielectric function equal to ε_d. As we see from Fig. 1.28, the spasing is possible for silver in the near-ir communication range and the adjacent red portion of the visible spectrum for a gain $g < 3{,}000\,\text{cm}^{-1}$ (regions below the red line in Fig. 1.28), which is realistically achievable with direct band-gap semiconductors (DBDSs).

1.5.5 Spaser in CW Mode

The "spasing curve" (a counterpart of the light-light curve, or L-L curve, for lasers), which is the dependence of the coherent SP population N_n on the excitation rate g, obtained by solving Eqs. (1.76) and (1.77), is shown in Fig. 1.29a for four types of the silver nanoshells with the frequencies of the spasing dipole modes as indicated, which are in the range from near-ir ($\hbar\omega_s = 1.2\,\text{eV}$) to mid-visible ($\hbar\omega_s = 2.2\,\text{eV}$). In all cases, there is a pronounced threshold of the spasing at an excitation rate $g_{th} \sim 10^{12}\,\text{s}^{-1}$. Soon after the threshold, the dependence $N_n(g)$ becomes linear, which means that every quantum of excitation added to the active medium with a high probability is stimulated to be emitted as a SP, adding to the coherent SP population.

While this is similar to conventional lasers, there is a dramatic difference for the spaser. In lasers, a similar relative rate of the stimulated emission is achieved at a photon population of $\sim 10^{18}$–10^{20}, while in the spaser the SP population is $N_n \lesssim 100$. This is due to the much stronger feedback in spasers because of the much smaller modal volume V_n – see discussion of Eq. (1.79). The shape of the spasing curves of Fig. 1.29a (the well-pronounced threshold with the linear dependence almost immediately above the threshold) is in a qualitative agreement with the experiment [256].

The population inversion number n_{21} as a function of the excitation rate g is displayed in Fig. 1.29b for the same set of frequencies (and with the same color coding) as in panel (a). Before the spasing threshold, n_{21} increases with g to become positive with the onset of the population inversion just before the spasing threshold. For higher g, after the spasing threshold is exceeded, the inversion n_{21} becomes constant (the inversion clamping). The clamped levels of the inversion are very low, $n_{21} \sim 0.01$, which again is due to the very strong feedback in the spaser.

The spectral width Γ_s of the spaser generation is due to the phase diffusion of the quantum SP state caused by the noise of the spontaneous emission of the SPs into the spasing mode, as described by Eq. (1.72). This width is displayed in Fig. 1.29c as a function of the pumping rate g. At the threshold, Γ_s is that of the SP line γ_n but for stronger pumping, as the SPs accumulate in the spasing mode, it decreases $\propto N_n^{-1}$, as given by Eq. (1.72). This decrease of Γ_s reflects the higher coherence of the spasing state with the increased number of SP quanta and, correspondingly, lower quantum fluctuations. As we have already mentioned, this is similar to the lasers as described by the Schawlow-Townes theory [291].

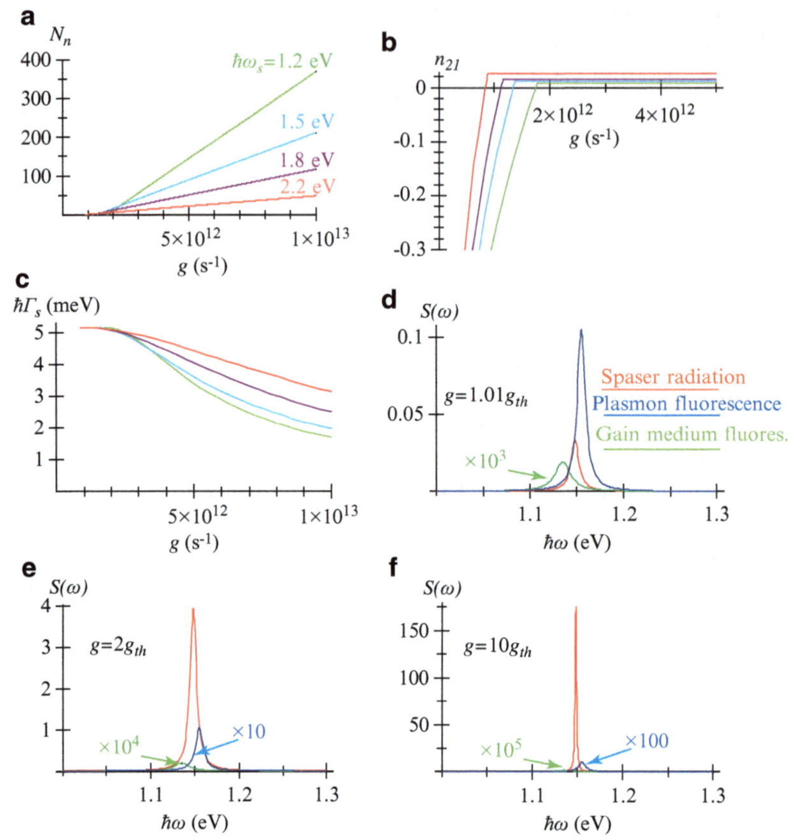

Fig. 1.29 Spaser SP population and spectral characteristics in the stationary state. The computations are done for a silver nanoshell with the external radius $R_2 = 12$ nm; the detuning of the gain medium from the spasing SP mode is $\hbar(\omega_{21} - \omega_n) = -0.02$ eV. The other parameters are indicated in Sect. 1.5.4. (**a**) Number N_n of plasmons per spasing mode as a function of the excitation rate g (per one chromophore of the gain medium). Computations are done for the dipole eigenmode with the spasing frequencies ω_s as indicated, which were chosen by the corresponding adjustment of the nanoshell aspect ratio. (**b**) Population inversion n_{12} as a function of the pumping rate g. The color coding of the lines is the same as in panel (**a**). (**c**) The spectral width Γ_s of the spasing line (expressed as $\hbar\Gamma_s$ in meV) as a function of the pumping rate g. The color coding of the lines is the same as in panel (**a**). (**d**)–(**f**) Spectra of the spaser for the pumping rates g expressed in the units of the threshold rate g_{th}, as indicated in the panels. The curves are color coded and scaled as indicated

The developed spasing in a dipole SP mode will show itself in the far field as an anomalously narrow and intense radiation line. The shape and intensity of this line in relation to the lines of the spontaneous fluorescence of the isolated gain medium and its SP-enhanced fluorescence line in the spaser is illustrated in Fig. 1.29d–f. Note that for the system under consideration, there is a 20 meV red shift of the gain medium fluorescence with respect to the SP line center. It is chosen so to illustrate the spectral walk-off of the spaser line. For one percent in the excitation rate above the threshold of the spasing (panel d), a broad spasing line (red color) appears comparable in intensity to the SP-enhanced spontaneous fluorescence line (blue color). The width of this spasing line is approximately the same as of the fluorescence, but its position is shifted appreciably (spectral walk-off) toward the isolated gain medium line (green color). For the pumping twice more intense (panel e), the spaser-line radiation dominates, but its width is still close to that of the SP line due to significant quantum fluctuations of the spasing state phase. Only when the pumping rate is an order of magnitude above the threshold, the spaser line strongly narrows (panel f), and it also completely dominates the spectrum of the radiation. This is a regime of small quantum fluctuations, which is desired in applications.

These results in the spasing region are different in the most dramatic way from previous phenomenological models, which are based on linear electrodynamics where the gain medium that has negative imaginary part of its permittivity plus lossy metal nanosystem, described purely electrodynamically [262, 269]. For instance, in a "toy model" [269], the width of the resonance line tends to zero at the threshold of spasing and then broadens up again. This distinction of the present theory is due the nature of the spasing as a spontaneous symmetry breaking (nonequilibrium phase transition with a randomly established but sustained phase) leading to the establishment of a coherent SP state. This non-equilibrium phase transition to spasing and the spasing itself are contained in the present theory due to the fact that the fundamental equations of the spasing (1.67), (1.69), and (1.70) are nonlinear, as we have already discussed above in conjunction with these equations – see the text after Eq. (1.70). The previous publications on gain compensation by loss [262, 269, 271] based on linear electrodynamic equations do not contain spasing. Therefore, they are not applicable in the region of the complete loss compensation and spasing, though their results are presented for that region.

1.5.6 Spaser as Ultrafast Quantum Nanoamplifier

1.5.6.1 Problem of Setting Spaser as an Amplifier

As we have already mentioned in Sect. 1.5.1, a fundamental and formidable problem is that, in contrast to the conventional lasers and amplifiers in quantum electronics, the spaser has an inherent feedback that typically cannot be removed. Such a spaser will develop generation and accumulation of the macroscopic number of coherent SPs in the spasing mode. This leads to the population inversion clamping in the CW regime at a very low level – cf. Fig. 1.29b. This CW regime corresponds to the net amplification equal zero, which means that the gain exactly compensates the loss, which condition is expressed by Eq. (1.76). This is a consequence of the nonlinear gain saturation. This holds for any stable CW generator (including any spaser or laser) and precludes using them as amplifiers.

There are several ways to set a spaser as a quantum amplifier. One of them is to reduce the feedback, i.e., to allow some or most of the SP energy in the spaser to escape from the active region, so the spaser will not generate in the region of amplification. Such a root has successfully been employed to build a SPP plasmonic amplifier on the long-range plasmon polaritons [281]. A similar root for the SP spasers would be to allow some optical energy to escape either by a near-field coupling or by a radiative coupling to far-field radiation. The near-field coupling approach is promising for building integrated active circuits out of the spasers. Another root has been used in Ref. [292], which employed symmetric SPP modes in a thin gold strip. Such modes have much lower loss that the antisymmetric modes at the expense of much weaker confinement (transverse modal area $\sim\lambda^2$). The lower loss allows one to use the correspondingly lower gain and, therefore, avoid both spasing at localized SP modes and random lasing due to back-scattering from gold imperfections.

Following Ref. [139], we consider here two distinct approaches for setting the spasers as quantum nanoamplifiers. The first is a transient regime based on the fact that the establishment of the CW regime and the consequent inversion clamping and the total gain vanishing require some time that is determined mainly by the rate of the quantum feedback and depends also on the relaxation rates of the SPs and the gain medium. After the population inversion is created by the onset of pumping and before the spasing spontaneously develops, as we show below in this section, there is a time interval of approximately 250 fs, during which the spaser provides usable (and as predicted, quite high) amplification – see Sect. 1.5.6.2 below.

The second approach to set the spaser as a logical quantum nanoamplifier is a bistable regime that is achieved by introducing a saturable absorber into the active region, which prevents the spontaneous spasing. Then injection of a certain above-threshold amount of SP quanta will saturate the absorber and initiate the spasing. Such a bistable quantum amplifier will be considered in Sect. 1.5.6.3.

The temporal behavior of the spaser has been found by direct numerical solution of Eqs. (1.67)–(1.70). This solution is facilitated by the fact that in the model under consideration all the chromophores experience the same local field inside the nanoshell, and there are only two types of such chromophores: belonging to the gain medium and the saturable absorber, if it is present.

1.5.6.2 Monostable spaser as a Nanoamplifier in Transient Regime

Here we consider a monostable spaser in a transient regime. This implies that no saturable absorber is present. We will consider two pumping regimes: stationary and pulse.

Starting with the stationary regime, we assume that the pumping at a rate (per one chromophore) of $g = 5 \times 10^{12}\,\text{s}^{-1}$ starts at a moment of time $t = 0$ and stays constant after that. Immediately at $t = 0$, a certain number of SPs are injected into the spaser. We are interested in its temporal dynamics from this moment on.

The dynamical behavior of the spaser under this pumping regime is illustrated in Figs. 1.30a, b. As we see, the spaser, which starts from an arbitrary initial population N_n, rather rapidly, within a few hundred femtoseconds approaches the same stationary ("logical") level. At this level, an SP population of $N_n = 67$ is established, while the inversion is clamped at a low level of $n_{21} = 0.02$. On the way to this stationary state, the spaser experiences relaxation oscillations in both the SP numbers and inversion, which have a trend to oscillate out of phase (compare panels (a) and (b)). This temporal dynamics of the spaser is quite complicated and highly nonlinear (unharmonic). It is controlled not by a single relaxation time but by a set of the relaxation rates. Clearly, among these are the energy transfer rate from the gain medium to the SPs and the relaxation rates of the SPs and the chromophores.

In this mode, the main effect of the initial injection of the SPs (described theoretically as different initial values of N_n) is in the interval of time it is required for the spaser to reach the final (CW) state. For very small N_n, which in practice can be supplied by the noise of the spontaneous SP emission into the mode, this time is approximately 250 fs (cf.: the corresponding SP relaxation time is less then 50 fs). In contrast, for the initial values of $N_n = 1$–5, this time shortens to 150 fs.

Now consider the second regime: pulse pumping. The gain-medium population of the spaser is inverted at $t = 0$ to saturation with a short (much shorter than 100 fs) pump pulse. Simultaneously, at $t = 0$, some number of plasmons are injected (say, by an external nanoplasmonic circuitry). In response, the spaser should produce an amplified pulse of the SP excitation. Such a function of the spaser is illustrated in Fig. 1.30c, d.

As we see from panel (c), independently from the initial number of SPs, the spaser always generates a series of SP pulses, of which only the first pulse is large (at or above the logical level of $N_n \sim 100$). (An exception is a case of little practical importance when the initial $N_n = 120$ exceeds this logical level, when two large pulses are produced.) The underlying mechanism of such a response is the rapid depletion of the inversion seen in panel (d), where energy is dissipated in the metal of the spaser. The characteristic duration of the SP pulse $\sim$100 fs is defined by this depletion, controlled by the energy transfer and SP relaxation rates. This time is much shorter than the spontaneous decay time of the gain medium. This acceleration is due to the stimulated emission of the SPs into the spasing mode (which can be called a "stimulated Purcell effect"). There is also a pronounced trend: the lower is initial SP population N_n, the later the spaser produces the amplified pulse. In a sense, this spaser functions as a pulse-amplitude to time-delay converter.

1.5.6.3 Bistable Spaser with Saturable Absorber as an Ultrafast Nanoamplifier

Now let us consider a bistable spaser as a quantum threshold (or, logical) nanoamplifier. Such a spaser contains a saturable absorber mixed with the gain medium with parameters indicated at the end of Sect. 1.5.4.1 and the concentration of the saturable absorber $n_a = 0.66 n_c$. This case of a bistable spaser amplifier is of

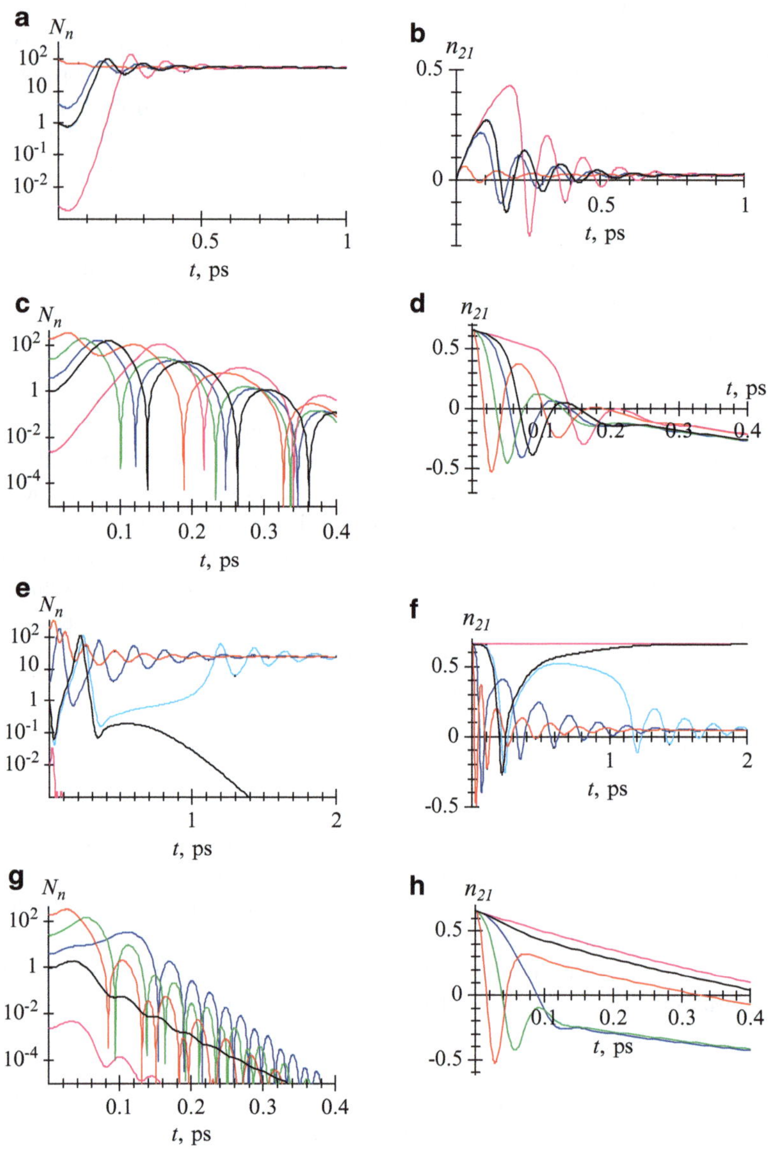

Fig. 1.30 Ultrafast dynamics of spaser. (**a**) For monostable spaser (without a saturable absorber), dependence of SP population in the spasing mode N_n on time t. The spaser is stationary pumped at a rate of $g = 5 \times 10^{12}\ \mathrm{s}^{-1}$. The color-coded curves correspond to the initial conditions with the different initial SP populations, as shown in the graphs. (**b**) The same as (**a**) but for the temporal behavior of the population inversion n_{21}. (**c**) Dynamics of a monostable spaser (no saturable absorber) with the pulse pumping described as the initial inversion $n_{21} = 0.65$. Coherent SP population N_n is displayed as a function of time t. Different initial populations are indicated by color-coded curves. (**d**) The same as (**c**) but for the corresponding population inversion n_{21}. (**e**) The same as (**a**) but for bistable spaser with the saturable absorber in concentration $n_a = 0.66 n_c$. (**f**) The same as (**b**) but for the bistable spaser. (**g**) The same as (**e**) but for the pulse pumping with the initial inversion $n_{21} = 0.65$. (**h**) The same as (**g**) but for the corresponding population inversion n_{21}

a particular interest because in this regime the spaser comes as close as possible in its functioning to the semiconductor-based (mostly, MOSFET-based) digital nanoamplifiers. As in the previous Subsection, we will consider two cases: the stationary and short-pulse pumping.

We again start with the case of the stationary pumping at a rate of $g = 5 \times 10^{12}\,\mathrm{s}^{-1}$. We show in Fig. 1.30e, f the dynamics of such a spaser. For a small initial population $N_n = 5 \times 10^{-3}$ simulating the spontaneous noise, the spaser is rapidly (faster than in 50 fs) relaxing to the zero population (panel e), while its gain-medium population is equally rapidly approaching a high level (panel f) $n_{21} = 0.65$ that is defined by the competition of the pumping and the enhanced decay into the SP mode (the purple curves). This level is so high because the spasing SP mode population vanishes and the stimulated emission is absent. After reaching this stable state (which one can call, say, "logical zero"), the spaser stays in it indefinitely long despite the continuing pumping.

In contrast, for initial values N_n of the SP population large enough (for instance, for $N_n = 5$, as shown by the blue curves in Figs. 1.30e, f), the spaser tends to the "logical one" state where the stationary SP population reaches the value of $N_n \approx 60$. Due to the relaxation oscillations, it actually exceeds this level within a short time of $\lesssim$100 fs after the seeding with the initial SPs. As the SP population N_n reaches its stationary (CW) level, the gain medium inversion n_{21} is clamped down at a low level of a few percent, as typical for the CW regime of the spaser. This "logical one" state salso persists indefinitely, as long as the inversion is supported by the pumping.

There is a critical curve (separatrix) that divide the two stable dynamics types (leading to the logical levels of zero and one). For the present set of parameters this separatrix starts with the initial population of $N_n \approx 1$. For a value of the initial N_n slightly below 1, the SP population N_n experiences a slow (hundreds fs in time) relaxation oscillation but eventually relaxes to zero (Fig. 1.30e, black curve), while the corresponding chromophore population inversion n_{21} relaxes to the high value $n_{21} = 0.65$ (panel (f), black curve). In contrast, for a value of N_n slightly higher than 1 (light blue curves in panels (e) and (f)), the dynamics is initially close to the separaratrix but eventually the initial slow dynamics tends to the high SP population and low chromophore inversion through a series of the relaxation oscillations. The dynamics close to the separatrix is characterized by a wide range of oscillation times due to its highly nonlinear character. The initial dynamics is slowest (the "decision stage" of the bistable spaser that lasts $\gtrsim$1 ps). The "decision time" is diverging infinitesimally close to the separatrix, as is characteristic of any threshold (logical) amplifier.

The gain (amplification coefficient) of the spaser as a logical amplifier is the ratio of the high CW level to the threshold level of the SP population N_n. For this specific spaser with the chosen set of parameters, this gain is $\approx$60, which is more than sufficient for the digital information processing. Thus this spaser can make a high-gain, $\sim$10 THz-bandwidth logical amplifier or dynamical memory cell with excellent prospects of applications.

The last but not the least regime to consider is that of the pulse pumping in the bistable spaser. In this case, the population inversion ($n_{21} = 0.65$) is created by a short pulse at $t = 0$ and simultaneously initial SP population N_n is created. Both are simulated as the initial conditions in Eqs. (1.67)–(1.70). The corresponding results are displayed in Fig. 1.30g, h.

When the initial SP population exceeds the critical one of $N_n = 1$ (the blue, green, and red curves), the spaser responds with generating a short (duration less than 100 fs) pulse of the SP population (and the corresponding local fields) within a time $\lesssim$100 fs (panel g). Simultaneously, the inversion is rapidly (within $\sim$100 fs) exhausted (panel h).

In contrast, when the initial SP population N_n is less than the critical one (i.e., $N_n < 1$ in this specific case), the spaser rapidly (within a time $\lesssim$100 fs) relaxes as $N_n \to 0$ through a series of realaxation oscillations – see the black and magenta curves in Fig. 1.30g. The corresponding inversion decays in this case almost exponentially with a characteristic time $\sim$1 ps determined by the enhanced energy transfer to the SP mode in the metal – see the corresponding curves in panel (h). Note that the SP population decays faster when the spaser is above the generation threshold due to the stimulated SP emission leading to the higher local fields and enhanced relaxation.

1.5.7 Compensation of Loss by Gain and Spasing

1.5.7.1 Introduction to Loss Compensation by Gain

A problem for many applications of plasmonics and metamaterials is posed by losses inherent in the interaction of light with metals. There are several ways to bypass, mitigate, or overcome the detrimental effects of these losses, which we briefly discuss below.

(i) The most common approach consists in employing effects where the losses are not fundamentally important such as surface plasmon polariton (SPP) propagation used in sensing [23], ultramicroscopy [16, 19], and solar energy conversion [26]. For realistic losses, there are other effects and applications that are not prohibitively suppressed by the losses and useful, in particular, sensing based on SP resonances and surface enhanced Raman scattering (SERS) [23, 182, 246, 293, 294].

(ii) Another promising idea is to use superconducting plasmonics to dramatically reduce losses [74, 295–297]. However, this is only applicable for frequencies below the superconducting gaps, i.e., in the terahertz region.

(iii) Yet another proposed direction is using highly doped semiconductors where the Ohmic losses can be significantly lower due to much lower free carrier concentrations [298]. However, a problem with this approach may lie in the fact that the usefulness of plasmonic modes depends not on the loss per se but on the quality factor Q, which for doped semiconductors may not be higher than for the plasmonic metals.

(iv) One of the alternative approaches to low-loss plasmonic metamaterials is based on our idea of the spaser: it is using a gain to compensate the dielectric (Ohmic) losses [299, 300]. In this case the gain medium is included into the metamaterials. It surrounds the metal plasmonic component in the same manner as in the spasers. The idea is that the gain will provide quantum amplification compensating the loss in the metamaterials quite analogously to the spasers.

We will consider theory of the loss compensation in the plasmonic metamaterials using gain [140, 141]. Below we show that the full compensation or overcompensation of the optical loss in a dense resonant gain metamaterial leads to an instability that is resolved by its spasing (i.e., by becoming a generating spaser). We further show analytically that the conditions of the complete loss compensation by gain and the threshold condition of spasing – see Eqs. (1.83) and (1.85) – are identical. Thus the full compensation (overcompensation) of the loss by gain in such a metamaterial will cause spasing. This spasing limits (clamps) the gain – see Sect. 1.5.5 – and, consequently, inhibits the complete loss compensation (overcompensation) at any frequency.

1.5.7.2 Permittivity of Nanoplasmonic Metamaterial

We will consider, for certainty, an isotropic and uniform metamaterial that, by definition, in a range of frequencies ω can be described by the effective permittivity $\bar{\varepsilon}(\omega)$ and permeability $\bar{\mu}(\omega)$. We will concentrate below on the loss compensation for the optical electric responses; similar consideration with identical conclusions for the optical magnetic responses is straightforward. Our theory is applicable for the true three-dimensional (3d) metamaterials whose size is much greater than the wavelength λ (ideally, an infinite metamaterial).

Consider a small piece of such a metamaterial with sizes much greater that the unit cell but much smaller than λ. Such a piece is a metamaterial itself. Let us subject this metamaterial to a uniform electric field $\mathbf{E}(\omega) = -\nabla\phi(\mathbf{r}, \omega)$ oscillating with frequency ω. Note that $\mathbf{E}(\omega)$ is the amplitude of the macroscopic electric field inside the metamaterial. We will denote the local field at a point $\mathbf{r}$ inside this metamaterial as $\mathbf{e}(\mathbf{r}, \omega) = -\nabla\varphi(\mathbf{r}, \omega)$. We assume standard boundary conditions

$$\varphi(\mathbf{r}, \omega) = \phi(\mathbf{r}, \omega), \tag{1.86}$$

for $\mathbf{r}$ belonging to the surface S of the volume under consideration.

To present our results in a closed form, we first derive a homogenization formula used in Ref. [301] (see also references cited therein). By definition, the electric displacement in the volume V of the metamaterial is given by a formula

$$\mathbf{D}(\mathbf{r}, \omega) = \frac{1}{V}\int_V \varepsilon(\mathbf{r}, \omega)\mathbf{e}(\mathbf{r}, \omega)dV\ , \tag{1.87}$$

where $\varepsilon(\mathbf{r},\omega)$ is a position-dependent permittivity. This can be identically expressed (by multiplying and dividing by the conjugate of the macroscopic field E^*) and, using the Gauss theorem, transformed to a surface integral as

$$D = \frac{1}{VE^*(\omega)}\int_V \mathbf{E}^*(\omega)\varepsilon(\mathbf{r},\omega)\mathbf{e}(\mathbf{r},\omega)dV$$
$$= \frac{1}{VE^*(\omega)}\int_S \phi^*(\mathbf{r},\omega)\varepsilon(\mathbf{r},\omega)\mathbf{e}(\mathbf{r},\omega)d\mathbf{S}\,, \tag{1.88}$$

where we took into account the Maxwell continuity equation $\nabla\left[\boldsymbol{\varepsilon}(\mathbf{r},\omega)\mathbf{e}(\mathbf{r},\omega)\right] = 0$. Now, using the boundary conditions of Eq. (1.86), we can transform it back to the volume integral as

$$D = \frac{1}{VE^*(\omega)}\int_S \varphi^*(\mathbf{r})\varepsilon(\mathbf{r},\omega)\mathbf{e}(\mathbf{r},\omega)d\mathbf{S}$$
$$= \frac{1}{VE^*(\omega)}\int_V \varepsilon(\mathbf{r},\omega)\left|\mathbf{e}(\mathbf{r},\omega)\right|^2 dV\,. \tag{1.89}$$

From the last equality, we obtain the required homogenization formula as an expression for the effective permittivity of the metamaterial:

$$\bar{\varepsilon}(\omega) = \frac{1}{V\left|E(\omega)\right|^2}\int_V \varepsilon(\mathbf{r},\omega)\left|\mathbf{e}(\mathbf{r},\omega)\right|^2 dV\,. \tag{1.90}$$

1.5.7.3 Plasmonic Eigenmodes and Effective Resonant Permittivity of Metamaterials

This piece of the metamaterial with the total size $R \ll \lambda$ can be treated in the quasistatic approximation. The local field inside the nanostructured volume V of the metamaterial is given by the eigenmode expansion [78, 149, 222]

$$\mathbf{e}(\mathbf{r},\omega) = \mathbf{E}(\omega) - \sum_n \frac{a_n}{s(\omega) - s_n}\mathbf{E}_n(\mathbf{r})\,, \tag{1.91}$$
$$a_n = \mathbf{E}(\omega)\int_V \theta(\mathbf{r})\mathbf{E}_n(\mathbf{r})dV,$$

where we remind that $\mathbf{E}(\omega)$ is the macroscopic field. In the resonance, $\omega = \omega_n$, only one term at the pole of in Eq. (1.91) dominates, and it becomes

$$\mathbf{e}(\mathbf{r},\omega) = \mathbf{E}(\omega) + i\frac{a_n}{\operatorname{Im} s(\omega_n)}\mathbf{E}_n(\mathbf{r})\,. \tag{1.92}$$

The first term in this equation corresponds to the mean (macroscopic) field and the second one describes the deviations of the local field from the mean field containing contributions of the hot spots [159]. The mean root square ratio of the second term (local field) to the first (mean field) is estimated as

$$\sim \frac{f}{\operatorname{Im} s(\omega_n)} = \frac{fQ}{s_n(1-s_n)} , \tag{1.93}$$

where we took into account that, in accord with Eq. (1.34), $E_n \sim V^{-1/2}$, and

$$f = \frac{1}{V}\int_V \theta(\mathrm{r}) dV , \tag{1.94}$$

where f is the metal fill factor of the system, and Q is the plasmonic quality factor. Deriving expression (1.93), we have also taken into account an equality $\operatorname{Im} s(\omega_n) = s_n(1-s_n)/Q$, which is valid in the assumed limit of the high quality factor, $Q \gg 1$ (see the next paragraph).

For a good plasmonic metal $Q \gg 1$ – see Fig. 1.2. For most metal-containing metamaterials, the metal fill factor is not small, typically $f \gtrsim 0.5$. Thus, keeping Eq. (1.28) in mind, it is very realistic to assume the following condition

$$\frac{fQ}{s_n(1-s_n)} \gg 1 . \tag{1.95}$$

If so, the second (local) term of the field (1.92) dominates and, with a good precision, the local field is approximately the eigenmode's field:

$$\mathbf{e}(\mathbf{r},\omega) = i\frac{a_n}{\operatorname{Im} s(\omega_n)}\mathbf{E}_n(\mathbf{r}) . \tag{1.96}$$

Substituting this into Eq. (1.90), we obtain a homogenization formula

$$\bar{\varepsilon}(\omega) = b_n \int_V \varepsilon(\mathbf{r},\omega) \left[\mathbf{E}_n(\mathbf{r})\right]^2 dV , \tag{1.97}$$

where $b_n > 0$ is a real positive coefficient whose specific value is

$$b_n = \frac{1}{3V}\left(\frac{Q\int_V \theta(\mathbf{r})\mathbf{E}_n(\mathbf{r})dV}{s_n\,(1-s_n)}\right)^2 \tag{1.98}$$

Using Eqs. (1.97) and (1.27), (1.34), it is straightforward to show that the effective permittivity (1.97) simplifies exactly to

$$\bar{\varepsilon}(\omega) = b_n\left[s_n\varepsilon_m(\omega) + (1-s_n)\varepsilon_h(\omega)\right] . \tag{1.99}$$

1.5.8 Conditions of Loss Compensation by Gain and Spasing

In the case of the full inversion (maximum gain) and in the exact resonance, the host medium permittivity acquires the imaginary part describing the stimulated emission as given by the standard expression

$$\varepsilon_h(\omega) = \varepsilon_d - i\frac{4\pi}{3}\frac{|\mathbf{d}_{12}|^2 n_c}{\hbar\Gamma_{12}}\,, \tag{1.100}$$

where $\varepsilon_d = \mathrm{Re}\,\varepsilon_h$, $\mathbf{d}_{12}$ is a dipole matrix element of the gain transition in a chromophore center of the gain medium, Γ_{12} is a spectral width of this transition, and n_c is the concentration of these centers (these notations are consistent with those used above in Sects. 1.5.4.1–1.5.6.3). Note that if the inversion is not maximum, then this and subsequent equations are still applicable if one sets as the chromophore concentration n_c the inversion density: $n_c = n_2 - n_1$, where n_2 and n_1 are the concentrations of the chromophore centers of the gain medium in the upper and lower states of the gain transition, respectively.

The condition for the full electric loss compensation in the metamaterial and amplification (overcompensation) at the resonant frequency $\omega = \omega_n$ is

$$\mathrm{Im}\,\bar{\varepsilon}(\omega) \le 0 \tag{1.101}$$

Taking Eq. (1.99) into account, this reduces to

$$s_n \mathrm{Im}\,\varepsilon_m(\omega) - \frac{4\pi}{3}\frac{|\mathbf{d}_{12}|^2 n_c(1-s_n)}{\hbar\Gamma_{12}} \le 0\,. \tag{1.102}$$

Finally, taking into account Eqs. (1.28), (1.47) and that $\mathrm{Im}\,\varepsilon_m(\omega) > 0$, we obtain from Eq. (1.102) the condition of the loss (over)compensation as

$$\frac{4\pi}{3}\frac{|\mathbf{d}_{12}|^2 n_c\,[1-\mathrm{Re}\,s(\omega)]}{\hbar\Gamma_{12}\mathrm{Re}\,s(\omega)\mathrm{Im}\,\varepsilon_m(\omega)} \ge 1\,, \tag{1.103}$$

where the strict inequality corresponds to the overcompensation and net amplification. In Eq. (1.100) we have assumed non-polarized gain transitions. If these transitions are all polarized along the excitation electric field, the concentration n_c should be multiplied by a factor of 3.

Equation (1.103) is a fundamental condition, which is precise (assuming that the requirement (1.95) is satisfied, which is very realistic for metamaterials) and general. Moreover, it is fully analytical and, actually, very simple. Remarkably, it depends only on the material characteristics and does not contain any geometric properties of the metamaterial system or the local fields. (Note that the system's geometry does affect the eigenmode frequencies and thus enters the problem implicitly.) In particular, the hot spots, which are prominent in the local fields of nanostructures [78, 159], are completely averaged out due to the integrations in Eqs. (1.90) and (1.97).

The condition (1.103) is completely non-relativistic (quasistatic) – it does not contain speed of light c, which is characteristic of also of the spaser. It is useful to express this condition also in terms of the total stimulated emission cross section $\sigma_e(\omega)$ (where ω is the central resonance frequency) of a chromophore of the gain medium as

$$\frac{c\sigma_e(\omega)\sqrt{\varepsilon_d}n_c\left[1-\mathrm{Re}\,s(\omega)\right]}{\omega\mathrm{Re}\,s(\omega)\mathrm{Im}\,\varepsilon_m(\omega)} \geq 1\,. \tag{1.104}$$

We see that Eq. (1.103) *exactly* coincides with a spasing condition expressed by Eq. (1.83). This brings us to an important conclusion: the full compensation (overcompensation) of the optical losses in a metamaterial (which is resonant and dense enough to satisfy condition (1.95)) and the spasing occur under precisely the same conditions.

We have considered above in Sect. 1.5.4.2 the conditions of spasing, which are equivalent to (1.104). These are given by one of equivalent conditions of Eqs. (1.83), (1.85), (1.103). It is also illustrated in Fig. 1.28. We stress that exactly the same conditions are for the full loss compensation (overcompensation) of a dense resonant plasmonic metamaterial with gain.

We would like also to point out that the criterion given by the equivalent conditions of Eqs. (1.83), (1.85), (1.103), or (1.104) is derived for localized SPs, which are describable in the quasistatic approximation, and is not directly applicable to the propagating plasmonic modes (SPPs). However, we expect that very localized SPPs, whose wave vector $k \lesssim l_s$, can be described by these conditions because they are, basically, quasistatic. For instance, the SPPs on a thin metal wire of a radius $R \lesssim l_s$ are described by a dispersion relation [12]

$$k \approx \frac{1}{R}\left[-\frac{\varepsilon_m}{2\varepsilon_d}\left(\ln\sqrt{-\frac{4\varepsilon_m}{\varepsilon_d}}-\gamma\right)\right]^{-1/2}, \tag{1.105}$$

where $\gamma \approx 0.57721$ is the Euler constant. This relation is obviously quasistatic because it does not contain speed of light c.

1.5.8.1 Discussion of Spasing and Loss Compensation by Gain

This fact of the equivalence of the full loss compensation and spasing is intimately related to the general criteria of the thermodynamic stability with respect to small fluctuations of electric and magnetic fields – see Chap. IX of Ref. [30],

$$\mathrm{Im}\,\bar{\varepsilon}(\omega) > 0\,, \quad \mathrm{Im}\,\bar{\mu}(\omega) > 0\,, \tag{1.106}$$

which must be *strict* inequalities for all frequencies for electromagnetically stable systems. For systems in thermodynamic equilibrium, these conditions are automatically satisfied.

However, for the systems with gain, the conditions (1.106) can be violated, which means that such systems can be electromagnetically unstable. The first of conditions (1.106) is opposite to Eqs. (1.101) and (1.103). This has a transparent meaning: the electrical instability of the system is resolved by its spasing.

The significance of these stability conditions for gain systems can be elucidated by the following *gedanken* experiment. Take a small isolated piece of such a metamaterial (which is a metamaterial itself). Consider that it is excited at an optical frequency ω either by a weak external optical field $\mathbf{E}$ or acquires such a field due to fluctuations (thermal or quantum). The energy density $\mathscr{E}$ of such a system is given by the Brillouin formula [30]

$$\mathscr{E} = \frac{1}{16\pi} \frac{\partial \omega \mathrm{Re}\, \bar{\varepsilon}}{\partial \omega} |\mathbf{E}|^2 \ . \tag{1.107}$$

Note that for the energy of the system to be definite, it is necessary to assume that the loss is not too large, $|\mathrm{Re}\, \bar{\varepsilon}| \gg \mathrm{Im}\, \bar{\varepsilon}$. This condition is realistic for many metamaterials, including all potentially useful ones.

The internal optical energy-density loss per unit time Q (i.e., the rate of the heat-density production in the system) is [30]

$$Q = \frac{\omega}{8\pi} \mathrm{Im}\, \bar{\varepsilon}\, |\mathbf{E}|^2 \ . \tag{1.108}$$

Assume that the internal (Ohmic) loss dominates over other loss mechanisms such as the radiative loss, which is also a realistic assumption since the Ohmic loss is very large for the experimentally studied systems and the system itself is very small (the radiative loss rate is proportional to the volume of the system). In such a case of the dominating Ohmic losses, we have $d\mathscr{E}/dt = Q$. Then Eqs. (1.107) and (1.108) can be resolved together yielding the energy $\mathscr{E}$ and electric field $|\mathbf{E}|$ of this system to evolve with time t exponentially as

$$|\mathbf{E}| \propto \sqrt{\mathscr{E}} \propto e^{-\Gamma t} \ , \quad \Gamma = \omega \mathrm{Im}\, \bar{\varepsilon} \Bigg/ \frac{\partial(\omega \mathrm{Re}\, \bar{\varepsilon})}{\partial \omega} \ . \tag{1.109}$$

We are interested in a resonant case when the metamaterial possesses a resonance at some eigenfrequency $\omega_n \approx \omega$. For this to be true, the system's behavior must be plasmonic, i.e., $\mathrm{Re}\, \bar{\varepsilon}(\omega) < 0$. Then the dominating contribution to $\bar{\varepsilon}$ comes from a resonant SP eigenmode n with a frequency $\omega_n \approx \omega$. In such a case, the dielectric function [78] $\bar{\varepsilon}(\omega)$ has a simple pole at $\omega = \omega_n$. As a result, $\partial\left(\omega \mathrm{Re}\, \bar{\varepsilon}\right)/\partial \omega \approx \omega \partial \mathrm{Re}\, \bar{\varepsilon}/\partial \omega$ and, consequently, $\Gamma = \gamma_n$, where γ_n is the SP decay rate given by Eqs. (1.3) or (1.48), and the metal dielectric function ε_m is replaced by the effective permittivity $\bar{\varepsilon}$ of the metamaterial. Thus, Eq. (1.109) is fully consistent with the spectral theory of SPs – see Sect. 1.3.4.

If the losses are not very large so that energy of the system is meaningful, the Kramers-Kronig causality requires [30] that $\partial(\omega \mathrm{Re}\,\bar{\varepsilon})/\partial\omega > 0$. Thus, $\mathrm{Im}\,\bar{\varepsilon} < 0$ in Eq. (1.109) would lead to a negative decrement,

$$\Gamma < 0 , \qquad (1.110)$$

implying that the initial small fluctuation starts exponentially grow in time in its field and energy, which is an instability. Such an instability is indeed not impossible: it will result in spasing that will eventually stabilize $|\mathbf{E}|$ and $\mathscr{E}$ at finite stationary (CW) levels of the spaser generation.

Note that the spasing limits (clamps) the gain and population inversion making *the net gain to be precisely zero* [139] in the stationary (continuous wave or CW) regime see Sect. 1.5.6 and Fig. 1.29b. Above the threshold of the spasing, the population inversion of the gain medium is clamped at a rather low level $n_{21} \sim 1\,\%$. The corresponding net amplification in the CW spasing regime is exactly zero, which is a condition for the CW regime. This makes the complete loss compensation and its overcompensation impossible in a dense resonant metamaterial where the feedback is created by the internal inhomogeneities (including its periodic structure) and the facets of the system.

1.5.8.2 Discussion of Published Research on Spasing and Loss Compensations

In an experimental study of the lasing spaser [264], a nanofilm of PbS quantum dots (QDs) was positioned over a two-dimensional metamaterial consisting of an array of negative split ring resonators. When the QDs were optically pumped, the system exhibited an increase of the transmitted light intensity on the background of a strong luminescence of the QDs but apparently did not reach the lasing threshold. The polarization-dependent loss compensation was only $\sim$1 %. Similarly, for an array of split ring resonators over a resonant quantum well, where the inverted electron-hole population was excited optically [302], the loss compensation did not exceed $\sim$8 %. The relatively low loss compensation in these papers may be due either to random spasing and/or spontaneous or amplified spontaneous emission enhanced by this plasmonic array, which reduces the population inversion.

A dramatic example of possible random spasing is presented in Ref. [266]. The system studied was a Kretschmann-geometry SPP setup [303] with an added $\sim$1 μm polymer film containing Rodamine 6G dye in the $n_c = 1.2 \times 10^{19}\,\mathrm{cm}^{-3}$ concentration. When the dye was pumped, there was outcoupling of radiation in a range of angles. This was a threshold phenomenon with the threshold increasing with the Kretschmann angle. At the maximum of the pumping intensity, the widest range of the outcoupling angles was observed, and the frequency spectrum at every angle narrowed to a peak near a single frequency $\hbar\omega \approx 2.1$ eV.

These observations of Ref. [266] can be explained by the spasing where the feedback is provided by roughness of the metal. At the high pumping, the localized

SPs (hots spots), which possess the highest threshold, start to spase in a narrow frequency range around the maximum of the spasing criterion – the left-hand side of Eq. (1.103). Because of the sub-wavelength size of these hot spots, the Kretschmann phase-matching condition is relaxed, and the radiation is outcoupled into a wide range of angles.

The SPPs of Ref. [266] excited by the Kretschmann coupling are short-range SPPs, very close to the antisymmetric SPPs. They are localized at subwavelength distances from the surface, and their wave length in the plane is much shorter the ω/c. Thus they can be well described by the quasistatic approximation and the present theory is applicable to them. Substituting the above-given parameters of the dye and the extinction cross section $\sigma_e = 4 \times 10^{-16}$ cm^2 into Eq. (1.104), we find that the conditions of Ref. [266] are above the threshold, supporting our assertion of the spasing. Likewise, the amplified spontaneous emission and, possibly spasing, appear to have prevented the full loss compensation in a SPP system of Ref. [278]. Note that recently, random spasing for rough surfaces surrounded by dye gain media was shown experimentally in two independent observations [285, 304].

Note that the long-range SPPs of Ref. [281] are localized significantly weaker (at distances $\sim\lambda$) than those excited in Kretschmann geometry. Thus the long-range SPPs experience a much weaker feedback, and the amplification instead of the spasing can be achieved. Generally, the long-range SPPs are fully electromagnetic (non-quasistatic) and are not describable in the present theory. Similarly, relatively weakly confined, full electromagnetic are symmetric SPP modes on thin gold strips in Ref. [292] where the amplification has been demonstrated.

As we have already discussed in conjunction with Fig. 1.28, the spasing is readily achievable with the gain medium containing common DBGSs or dyes. There have been numerous experimental observations of the spaser. Among them is a report of a SP spaser with a 7-nm gold nanosphere as its core and a laser dye in the gain medium [256], observations of the SPP spasers (also known as nanolasers) with silver as a plasmonic-core metal and DBGS as the gain medium with a 1d confinement [257, 260], a tight 2d confinement [258], and a 3d confinement [259]. There also has been a report on observation of a SPP microcylinder spaser [305]. A high efficiency room-temperature semiconductor spaser with a DBGS InGaAS gain medium operating near 1.5 μm (i.e., in the communication near-ir range) has been reported [260].

The research and development in the area of spasers (nanolasers) as quantum nano-generators is very active and will undoubtedly lead to further rapid advances. The next in line is the spaser as an ultrafast nanoamplifier, which is one of the most important tasks in nanotechnology.

In periodic metamaterials, plasmonic modes generally are propagating waves (SPPs) that satisfy Bloch theorem [306] and are characterized by quasi-wavevector **k**. These are propagating waves except for the band edges where $\mathbf{ka} = \pm\pi$, where **a** is the lattice vector. At the band edges, the group velocity v_g of these modes is zero, and these modes are localized, i.e., they are SPs. Their wave function is periodic with period $2a$, which may be understood as a result of the Bragg reflection from the crystallographic planes. Within this $2a$ period, these band-edge modes can, indeed, be treated quasistatically because $2a \ll l_s, \lambdabar$. If any of the band-edge frequencies is within the range of compensation (where the condition (1.83) [or, (1.85)] is satisfied), the system will spase. In fact, at the band edge, this metamaterial with gain is similar to a distributed feedback (DFB) laser [307]. It actually is a DFB spaser, which, as all the DFB lasers, generates in a band-edge mode.

In fact, there have recently been two observations of lasing spasers with optical pumping generating on the band–edge modes [308, 309], see also our Research Highlight in Ref. [310]. In Ref. [308], the metal component of the lasing spaser was a periodic nanohole array in a silver nanofilm, and the gain component was semiconductor diode of InGaAs/InP. In Ref. [309], the plasmonic metal component was a periodic planar array of eithet gold or silver nanoparticle while the gain medium was a polymer nanolayer composed of polyurethane and IR-140 dye.

Moreover, not only the SPPs, which are exactly at the band edge, will be localized. Due to unavoidable disorder caused by fabrication defects in metamaterials, there will be scattering of the SPPs from these defects. Close to the band edge, the group velocity becomes small, $v_g \to 0$. Because the scattering cross section of any wave is $\propto v_g^{-2}$, the corresponding SPPs experience Anderson localization [311]. Also, there always will be SPs nanolocalized at the defects of the metamaterial, whose local fields are hot spots – see Fig. 1.10 and, generally, Sect. 1.3.5 and the publications referenced therein. Each of such hot spots within the bandwidth of conditions (1.83) or (1.85) will be a generating spaser, which clamps the inversion and precludes the full loss compensation.

Acknowledgements The primary support for this work was provided by MURI Grant No. N00014-13-1-0649 from the U.S. Office of Navy Research. Additional support was provided by Grant No. DE-FG02-11ER46789 from the Materials Sciences and Engineering Division, Office of the Basic Energy Sciences, Office of Science, U.S. Department of Energy, and Grant No. DE-FG02-01ER15213 from the Chemical Sciences, Biosciences and Geosciences Division, Office of the Basic Energy Sciences, Office of Science, U.S. Department of Energy.

References

1. Moskovits M (1985) Surface-enhanced spectroscopy. Rev Mod Phys 57:783–826
2. Stockman MI, Shalaev VM, Moskovits M, Botet R, George TF (1992) Enhanced Raman scattering by fractal clusters: scale invariant theory. Phys Rev B 46:2821–2830
3. Gunnarsson L, Petronis S, Kasemo B, Xu H, Bjerneld J, Kall M (1999) Optimizing nanofabricated substrates for surface enhanced Raman scattering. Nanostruct Mater 12:783–788
4. Xu HX, Bjerneld EJ, Kall M, Borjesson L (1999) Spectroscopy of single hemoglobin molecules by surface enhanced Raman scattering. Phys Rev Lett 83:4357–4360
5. Xu H, Aizpurua J, Kall M, Apell P (2000) Electromagnetic contributions to single-molecule sensitivity in surface-enhanced Raman scattering. Phys Rev E 62:4318–4324
6. Kneipp K, Moskovits M, Kneipp H (eds) (2006) Electromagnetic theory of SERS, vol 103. Springer, Heidelberg
7. Ebbesen TW, Lezec HJ, Ghaemi HF, Thio T, Wolff PA (1998) Extraordinary optical transmission through sub-wavelength hole arrays. Nature 391:667–669
8. Lezec HJ, Degiron A, Devaux E, Linke RA, Martin-Moreno L, Garcia-Vidal FJ, Ebbesen TW (2002) Beaming light from a subwavelength aperture. Science 297:820–822

9. Martin-Moreno L, Garcia-Vidal FJ, Lezec HJ, Degiron A, Ebbesen TW (2003) Theory of highly directional emission from a single subwavelength aperture surrounded by surface corrugations. Phys Rev Lett 90:167401–1–4
10. Genet C, Ebbesen TW (2007) Light in tiny holes. Nature 445:39–46
11. Garcia-Vidal FJ, Martin-Moreno L, Ebbesen TW, Kuipers L (2010) Light passing through subwavelength apertures. Rev Mod Phys 82:729–787
12. Stockman MI (2004) Nanofocusing of optical energy in tapered plasmonic waveguides. Phys Rev Lett 93:137404–1–4
13. Verhagen E, Polman A, Kuipers L (2008) Nanofocusing in laterally tapered plasmonic waveguides. Opt Express 16:45–57
14. Verhagen E, Spasenovic M, Polman A, Kuipers L (2009) Nanowire plasmon excitation by adiabatic mode transformation. Phys Rev Lett 102:203904–1–4
15. De Angelis F, Patrini M, Das G, Maksymov I, Galli M, Businaro L, Andreani LC, Di Fabrizio E (2008) A hybrid plasmonic-photonic nanodevice for label-free detection of a few molecules. Nano Lett 8:2321–2327
16. De Angelis F, Das G, Candeloro P, Patrini M, Galli M, Bek A, Lazzarino M, Maksymov I, Liberale C, Andreani LC, Di Fabrizio E (2009) Nanoscale chemical mapping using three-dimensional adiabatic compression of surface plasmon polaritons. Nat Nanotechnol 5:67–72
17. Angelis FD, Gentile F, Das FMG, Moretti M, Candeloro P, Coluccio ML, Cojoc G, Accardo A, Liberale C, Zaccaria RP, Perozziello G, Tirinato L, Toma A, Cuda G, Cingolani R, Di Fabrizio E (2011) Breaking the diffusion limit with super-hydrophobic delivery of molecules to plasmonic nanofocusing SERS structures. Nat Photonics 5:682–687
18. Ropers C, Neacsu CC, Elsaesser T, Albrecht M, Raschke MB, Lienau C (2007) Grating-coupling of surface plasmons onto metallic tips: a nano-confined light source. Nano Lett 7:2784–2788
19. Neacsu CC, Berweger S, Olmon RL, Saraf LV, Ropers C, Raschke MB (2010) Near-field localization in plasmonic superfocusing: a nanoemitter on a tip. Nano Lett 10:592–596
20. Raschke MB, Berweger S, Atkin JM, Olmon RL (2010) Adiabatic tip-plasmon focusing for nano-Raman spectroscopy. J Phys Chem Lett 1:3427–3432
21. Berweger S, Atkin JM, Xu XG, Olmon RL, Raschke MB (2011) Femtosecond nanofocusing with full optical waveform control. Nano Lett 11:4309–4313
22. Sadiq D, Shirdel J, Lee JS, Selishcheva E, Park N, Lienau C (2011) Adiabatic nanofocusing scattering-type optical nanoscopy of individual gold nanoparticles. Nano Lett 11:1609–1613
23. Stockman MI (2011) Nanoplasmonics: the physics behind the applications. Phys Today 64:39–44
24. Lal S, Clare SE, Halas NJ (2008) Nanoshell-enabled photothermal cancer therapy: impending clinical impact. Acc Chem Res 41:1842–1851
25. Huang XH, Neretina S, El-Sayed MA (2009) Gold nanorods: from synthesis and properties to biological and biomedical applications. Adv Mater 21:4880–4910
26. Atwater HA, Polman A (2010) Plasmonics for improved photovoltaic devices. Nat Mater 9:205–213
27. Mukherjee S, Libisch F, Large N, Neumann O, Brown LV, Cheng J, Lassiter JB, Carter EA, Nordlander P, Halas NJ (2012) Hot electrons do the impossible: plasmon-induced dissociation of H_2 on Au. Nano Lett 13:240–247
28. Mubeen S, Lee J, Singh N, Kramer S, Stucky GD, Moskovits M (2013) An autonomous photosynthetic device in which all charge carriers derive from surface plasmons. Nat Nano 8:247–251
29. Bergman DJ, Stroud D (1992) Properties of macroscopically inhomogeneous media. In: Ehrenreich H, Turnbull D (eds) Solid state physics, vol 46. Academic, Boston, pp 148–270
30. Landau LD, Lifshitz EM (1984) Electrodynamics of continuous media. Pergamon, Oxford/New York
31. Bergman DJ, Stockman MI (2003) Surface plasmon amplification by stimulated emission of radiation: quantum generation of coherent surface plasmons in nanosystems. Phys Rev Lett 90:027402–1–4

32. Johnson PB, Christy RW (1972) Optical constants of noble metals. Phys Rev B 6:4370–4379
33. Feigenbaum E, Orenstein M (2008) Ultrasmall volume plasmons, yet with complete retardation effects. Phys Rev Lett 101:163902–1–4
34. Larkin IA, Stockman MI, Achermann M, Klimov VI (2004) Dipolar emitters at nanoscale proximity of metal surfaces: giant enhancement of relaxation in microscopic theory. Phys Rev B 69:121403(R)–1–4
35. Larkin IA, Stockman MI (2005) Imperfect perfect lens. Nano Lett 5:339–343
36. Bozhevolny SI (ed) (2008) Plasmonic nanoguides and circuits. World Scientific, Singapore
37. Kramer A, Keilmann F, Knoll B, Guckenberger R (1996) The coaxial tip as a nano-antenna for scanning near-field microwave transmission microscopy. Micron 27:413–417
38. Oldenburg SJ, Hale GD, Jackson JB, Halas NJ (1999) Light scattering from dipole and quadrupole nanoshell antennas. Appl Phys Lett 75:1063–1065
39. Kalkbrenner T, Hkanson U, Schadle A, Burger S, Henkel C, Sandoghdar V (2005) Optical microscopy via spectral modifications of a nanoantenna. Phys Rev Lett 95:200801–1–4
40. Muhlschlegel P, Eisler HJ, Martin OJF, Hecht B, Pohl DW (2005) Resonant optical antennas. Science 308:1607–1609
41. Schuck PJ, Fromm DP, Sundaramurthy A, Kino GS, Moerner WE (2005) Improving the mismatch between light and nanoscale objects with gold bowtie nanoantennas. Phys Rev Lett 94:017402–1–4
42. Kuhn S, Hakanson U, Rogobete L, Sandoghdar V (2006) Enhancement of single-molecule fluorescence using a gold nanoparticle as an optical nanoantenna. Phys Rev Lett 97:017402–1–4
43. Novotny L (2007) Effective wavelength scaling for optical antennas. Phys Rev Lett 98:266802–1–4
44. Taminiau TH, Segerink FB, Moerland RJ, Kuipers L, van Hulst NF (2007) Near-field driving of a optical monopole antenna. J Opt A 9:S315–S321
45. Taminiau TH, Segerink FB, van Hulst NF (2007) A monopole antenna at optical frequencies: single-molecule near-field measurements. IEEE Trans Antennas Propag 55:3010–3017
46. Behr N, Raschke MB (2008) Optical antenna properties of scanning probe tips: plasmonic light scattering, tip-sample coupling, near-field enhancement. J Phys Chem C 112:3766–3773
47. Bryant GW, de Abajo FJG, Aizpurua J (2008) Mapping the plasmon resonances of metallic nanoantennas. Nano Lett 8:631–636
48. Ghenuche P, Cherukulappurath S, Taminiau TH, van Hulst NF, Quidant R (2008) Spectroscopic mode mapping of resonant plasmon nanoantennas. Phys Rev Lett 101:116805–1–4
49. Guo HC, Meyrath TP, Zentgraf T, Liu N, Fu LW, Schweizer H, Giessen H (2008) Optical resonances of bowtie slot antennas and their geometry and material dependence. Opt Express 16:7756–7766
50. Bakker RM, Yuan HK, Liu ZT, Drachev VP, Kildishev AV, Shalaev VM, Pedersen RH, Gresillon S, Boltasseva A (2008) Enhanced localized fluorescence in plasmonic nanoantennae. Appl Phys Lett 92:043101–1–3
51. Olmon RL, Krenz PM, Jones AC, Boreman GD, Raschke MB (2008) Near-field imaging of optical antenna modes in the mid-infrared. Opt Express 16:20295–20305
52. Taminiau TH, Stefani FD, Segerink FB, Hulst NFV (2008) Optical antennas direct single-molecule emission. Nat Photonics 2:234–237
53. Tang L, Kocabas SE, Latif S, Okyay AK, Ly-Gagnon DS, Saraswat KC, Miller DAB (2008) Nanometre-scale germanium photodetector enhanced by a near-infrared dipole antenna. Nat Photonics 2:226–229
54. Bharadwaj P, Deutsch B, Novotny L (2009) Optical antennas. Adv Opt Photonics 1:438–483
55. Eghlidi H, Lee KG, Chen XW, Gotzinger S, Sandoghdar V (2009) Resolution and enhancement in nanoantenna-based fluorescence microscopy. Nano Lett 9:4007–4011
56. Hanke T, Krauss G, Trauetlein D, Wild B, Bratschitsch R, Leitenstorfer A (2009) Efficient nonlinear light emission of single gold optical antennas driven by few-cycle near-infrared pulses. Phys Rev Lett 103:257404–1–4

57. Palomba S, Danckwerts M, Novotny L (2009) Nonlinear plasmonics with gold nanoparticle antennas. J Opt A 11:114030
58. Cao LY, Park JS, Fan PY, Clemens B, Brongersma ML (2010) Resonant germanium nanoantenna photodetectors. Nano Lett 10:1229–1233
59. Giannini V, Vecchi G, Rivas JG (2010) Lighting up multipolar surface plasmon polaritons by collective resonances in arrays of nanoantennas. Phys Rev Lett 105:266801–1–4
60. Weber-Bargioni A, Schwartzberg A, Schmidt M, Harteneck B, Ogletree DF, Schuck PJ, Cabrini S (2010) Functional plasmonic antenna scanning probes fabricated by induced-deposition mask lithography. Nanotechnology 21:065306–1–6
61. Knight MW, Sobhani H, Nordlander P, Halas NJ (2011) Photodetection with active optical antennas. Science 332:702–704
62. Li W-D, Ding F, Hu J, Chou SY (2011) Three-dimensional cavity nanoantenna coupled plasmonic nanodots for ultrahigh and uniform surface-enhanced Raman scattering over large area. Opt Express 19:3925–3936
63. Liu N, Tang ML, Hentschel M, Giessen H, Alivisatos AP (2011) Nanoantenna-enhanced gas sensing in a single tailored nanofocus. Nat Mater 10:631–636
64. Maksymov IS, Miroshnichenko AE (2011) Active control over nanofocusing with nanorod plasmonic antennas. Opt Express 19:5888–5894
65. Novotny L, van Hulst N (2011) Antennas for light. Nat Photonics 5:83–90
66. Schnell M, Gonzalez PA, Arzubiaga L, Casanova F, Hueso LE, Chuvilin A, Hillenbrand R (2011) Nanofocusing of mid-infrared energy with tapered transmission lines. Nat Photonics 5:283–287
67. Berestetskii VB, Lifshits EM, Pitaevskii LP (1982) Quantum electrodynamics. Pergamon, Oxford/New York
68. Fano U (1935) On the absorption spectrum of noble gases at the arc spectrum limit. Nuovo Cimento 12:154–161
69. Fedotov VA, Rose M, Prosvirnin SL, Papasimakis N, Zheludev NI (2007) Sharp trapped-mode resonances in planar metamaterials with a broken structural symmetry. Phys Rev Lett 99:147401–1–4
70. Hao F, Sonnefraud Y, Dorpe PV, Maier SA, Halas NJ, Nordlander P (2008) Symmetry breaking in plasmonic nanocavities: subradiant LSPR sensing and a tunable Fano resonance. Nano Lett 8:3983–3988
71. Mirin NA, Bao K, Nordlander P (2009) Fano resonances in plasmonic nanoparticle aggregates. J Phys Chem A 113:4028–4034
72. Brown LV, Sobhani H, Lassiter JB, Nordlander P, Halas NJ (2010) Heterodimers: plasmonic properties of mismatched nanoparticle pairs. ACS Nano 4:819–832
73. Fan JA, Wu CH, Bao K, Bao JM, Bardhan R, Halas NJ, Manoharan VN, Nordlander P, Shvets G, Capasso F (2010) Self-assembled plasmonic nanoparticle clusters. Science 328:1135–1138
74. Fedotov VA, Tsiatmas A, Shi JH, Buckingham R, de Groot P, Chen Y, Wang S, Zheludev NI (2010) Temperature control of Fano resonances and transmission in superconducting metamaterials. Opt Express 18:9015–9019
75. Hentschel M, Saliba M, Vogelgesang R, Giessen H, Alivisatos AP, Liu N (2010) Transition from isolated to collective modes in plasmonic oligomers. Nano Lett 10:2721–2726
76. Luk'yanchuk B, Zheludev NI, Maier SA, Halas NJ, Nordlander P, Giessen H, Chong CT (2010) The Fano resonance in plasmonic nanostructures and metamaterials. Nat Mater 9:707–715
77. Stockman MI (2010) Dark-hot resonances. Nature 467:541–542
78. Stockman MI, Faleev SV, Bergman DJ (2001) Localization versus delocalization of surface plasmons in nanosystems: can one state have both characteristics? Phys Rev Lett 87:167401–1–4
79. Novotny L (2001) Forces in optical near-fields. In: Kawata S, Ohtsu M, Irie M (eds) Near-field optics and surface plasmon polaritons, vol 81. Springer, Berlin, pp 123–141

80. Ignatovich FV, Novotny L (2003) Experimental study of nanoparticle detection by optical gradient forces. Rev Sci Instrum 74:5231–5235
81. Joulain K, Mulet J-P, Marquier F, Carminati R, Greffet J-J (2005) Surface electromagnetic waves thermally excited: radiative heat transfer, coherence properties and Casimir forces revisited in the near field. Surf Sci Rep 57:59–112
82. Li XT, Bergman DJ, Stroud D (2005) Electric forces among nanospheres in a dielectric host. Europhys Lett 69:1010–1016
83. Volpe G, Quidant R, Badenes G, Petrov D (2006) Surface plasmon radiation forces. Phys Rev Lett 96:238101
84. Zelenina AS, Quidant R, Nieto-Vesperinas M (2007) Enhanced optical forces between coupled resonant metal nanoparticles. Opt Lett 32:1156–1158
85. Takuya I, Hajime I (2008) Theory of resonant radiation force exerted on nanostructures by optical excitation of their quantum states: from microscopic to macroscopic descriptions. Phys Rev B 77:245319–1–16
86. Quidant R, Zelenina S, Nieto-Vesperinas M (2007) Optical manipulation of plasmonic nanoparticles. Appl Phys A 89:233–239
87. Righini M, Zelenina AS, Girard C, Quidant R (2007) Parallel and selective trapping in a patterned plasmonic landscape. Nat Phys 3:477–480
88. Quidant R, Girard C (2008) Surface-plasmon-based optical manipulation. Laser Photonics Rev 2:47–57
89. Righini M, Volpe G, Girard C, Petrov D, Quidant R (2008) Surface plasmon optical tweezers: tunable optical manipulation in the femtonewton range. Phys Rev Lett 100:186804–1–4
90. Juan ML, Gordon R, Pang YJ, Eftekhari F, Quidant R (2009) Self-induced back-action optical trapping of dielectric nanoparticles. Nat Phys 5:915–919
91. Righini M, Ghenuche P, Cherukulappurath S, Myroshnychenko V, de Abajo FJG, Quidant R (2009) Nano-optical trapping of Rayleigh particles and Escherichia coli bacteria with resonant optical antennas. Nano Lett 9:3387–3391
92. Tong LM, Righini M, Gonzalez MU, Quidant R, Kall M (2009) Optical aggregation of metal nanoparticles in a microfluidic channel for surface-enhanced Raman scattering analysis. Lab Chip 9:193–195
93. Durach M, Rusina A, Stockman MI (2009) Giant surface-plasmon-induced drag effect in metal nanowires. Phys Rev Lett 103:186801–1–4
94. Ritchie RH (1957) Plasma losses by fast electrons in thin films. Phys Rev 106:874–881
95. Blackstock AW, Ritchie RH, Birkhoff RD (1955) Mean free path for discrete electron energy losses in metallic foils. Phys Rev 100:1078
96. Swanson N, Powell CJ (1966) Inelastic scattering cross sections for 20-keV electrons in Al, Be, polystyrene. Phys Rev 145:195
97. de Abajo FJG (2010) Optical excitations in electron microscopy. Rev Mod Phys 82:209
98. Reyes-Coronado A, Barrera RG, Batson PE, Echenique PM, Rivacoba A, Aizpurua J (2010) Electromagnetic forces on plasmonic nanoparticles induced by fast electron beams. Phys Rev B 82:235429–1–19
99. Dasgupta BB, Fuchs R (1981) Polarizability of a small sphere including nonlocal effects. Phys Rev B 24:554–561
100. Pendry JB (2000) Negative refraction makes a perfect lens. Phys Rev Lett 85:3966–3969
101. de Abajo FJG (2008) Nonlocal effects in the plasmons of strongly interacting nanoparticles, dimers, and waveguides. J Phys Chem C 112:17983–17987
102. Kreibig U, Vollmer M (1995) Optical properties of metal clusters. Springer, New York
103. Pustovit VN, Shahbazyan TV (2005) Quantum-size effects in SERS from noble-metal nanoparticles. Microelectron J 36:559–563
104. Pustovit VN, Shahbazyan TV (2006) Finite-size effects in surface-enhanced Raman scattering in noble-metal nanoparticles: a semiclassical approach. J Opt Soc Am A 23:1369–1374
105. Pustovit VN, Shahbazyan TV (2006) Surface-enhanced Raman scattering on the nanoscale: a microscopic approach. J Opt A 8:S208–S212

106. Pustovit VN, Shahbazyan TV (2006) SERS from molecules adsorbed on small Ag nanoparticles: a microscopic model. Chem Phys Lett 420:469–473
107. Pustovit VN, Shahbazyan TV (2006) Microscopic theory of surface-enhanced Raman scattering in noble-metal nanoparticles. Phys Rev B 73:085408–1–7
108. Zuloaga J, Prodan E, Nordlander P (2009) Quantum description of the plasmon resonances of a nanoparticle dimer. Nano Lett 9:887–891
109. Nordlander P, Zuloaga J, Prodan E (2010) Quantum plasmonics: optical properties and tunability of metallic nanorods. ACS Nano 4:5269–5276
110. Palik ED (1998) Handbook of optical constants of solids. Academic, San Diego
111. Chang DE, Sorensen AS, Hemmer PR, Lukin MD (2006) Quantum optics with surface plasmons. Phys Rev Lett 97:053002–1–4
112. Akimov AV, Mukherjee A, Yu CL, Chang DE, Zibrov AS, Hemmer PR, Park H, Lukin MD (2007) Generation of single optical plasmons in metallic nanowires coupled to quantum dots. Nature 450:402–406
113. Berini P, Akbari A, Tait RN (2010) Surface plasmon waveguide Schottky detector. Opt Express 18:8505–8514
114. Scales C, Breukelaar I, Berini P (2010) Surface-plasmon Schottky contact detector based on a symmetric metal stripe in silicon. Opt Lett 35:529–531
115. Levy U, Goykhman I, Desiatov B, Khurgin J, Shappir J (2011) Locally oxidized silicon surface-plasmon Schottky detector for telecom regime. Nano Lett 11:2219–2224
116. Sze SM (2007) Physics of semiconductor devices. Wiley-Interscience, Hoboken
117. Butenko AV, Shalaev VM, Stockman MI (1988) Giant impurity nonlinearities in optics of fractal clusters. Sov Phys JETP 67:60–69
118. Karpov AV, Popov AK, Rautian SG, Safonov VP, Slabko VV, Shalaev VM, Stockman MI (1988) Observation of a wavelength- and polarization-selective photomodification of silver clusters. JETP Lett 48:571–573
119. Rautian SG, Safonov VP, Chubakov PA, Shalaev VM, Stockman MI (1988) Surface-enhanced parametric scattering of light by silver clusters. JETP Lett 47:243–246
120. Shalaev VM, Stockman MI, Botet R (1992) Resonant excitations and nonlinear optics of fractals. Physica A 185:181–186
121. Boyd RW (2003) Nonlinear optics. Academic, London/San Diego
122. Kneipp J, Kneipp H, Kneipp K (2006) Two-photon vibrational spectroscopy for biosciences based on surface-enhanced hyper-Raman scattering. Proc Natl Acad Sci U S A 103:17149–17153
123. Kubo A, Onda K, Petek H, Sun Z, Jung YS, Kim HK (2005) Femtosecond imaging of surface plasmon dynamics in a nanostructured silver film. Nano Lett 5:1123–1127
124. Zayats AV, Smolyaninov II, Davis CC (1999) Observation of localized plasmonic excitations in thin metal films with near-field second-harmonic microscopy. Opt Commun 169:93–96
125. Bouhelier A, Beversluis M, Hartschuh A, Novotny L (2003) Near-field second-harmonic generation induced by local field enhancement. Phys Rev Lett 90:13903–1–4
126. Bozhevolnyi SI, Beermann J, Coello V (2003) Direct observation of localized second-harmonic enhancement in random metal nanostructures. Phys Rev Lett 90:197403–1–4
127. Labardi M, Allegrini M, Zavelani-Rossi M, Polli D, Cerullo G, Silvestri SD, Svelto O (2004) Highly efficient second-harmonic nanosource for near-field optics and microscopy. Opt Lett 29:62–64
128. Stockman MI, Bergman DJ, Anceau C, Brasselet S, Zyss J (2004) Enhanced second-harmonic generation by metal surfaces with nanoscale roughness: nanoscale dephasing, depolarization, and correlations. Phys Rev Lett 92:057402–1–4
129. Zheludev NI, Emelyanov VI (2004) Phase matched second harmonic generation from nanostructured metal surfaces. J Opt A 6:26–28
130. Jin RC, Jureller JE, Kim HY, Scherer NF (2005) Correlating second harmonic optical responses of single Ag nanoparticles with morphology. J Am Chem Soc 127:12482–12483

131. Canfield BK, Husu H, Laukkanen J, Bai BF, Kuittinen M, Turunen J, Kauranen M (2007) Local field asymmetry drives second-harmonic generation in noncentrosymmetric nanodimers. Nano Lett 7:1251–1255
132. Zdanowicz M, Kujala S, Husu H, Kauranen M (2011) Effective medium multipolar tensor analysis of second-harmonic generation from metal nanoparticles. New J Phys 13:023025–1–12
133. Renger J, Quidant R, van Hulst N, Novotny L (2010) Surface-enhanced nonlinear four-wave mixing. Phys Rev Lett 104:046803–1–4
134. Utikal T, Stockman MI, Heberle AP, Lippitz M, Giessen H (2010) All-optical control of the ultrafast dynamics of a hybrid plasmonic system. Phys Rev Lett 104:113903–1–4
135. Pacifici D, Lezec HJ, Atwater HA (2007) All-optical modulation by plasmonic excitation of CdSe quantum dots. Nat Photonics 1:402–406
136. Samson ZL, MacDonald KF, De Angelis F, Gholipour B, Knight K, Huang CC, Di Fabrizio E, Hewak DW, Zheludev NI (2010) Metamaterial electro-optic switch of nanoscale thickness. Appl Phys Lett 96:143105–1–3
137. MacDonald KF, Samson ZL, Stockman MI, Zheludev NI (2009) Ultrafast active plasmonics. Nat Photonics 3:55–58
138. Kim S, Jin JH, Kim YJ, Park IY, Kim Y, Kim SW (2008) High-harmonic generation by resonant plasmon field enhancement. Nature 453:757–760
139. Stockman MI (2010) The spaser as a nanoscale quantum generator and ultrafast amplifier. J Opt 12:024004–1–13
140. Stockman MI (2011) Spaser action, loss compensation, stability in plasmonic systems with gain. Phys Rev Lett 106:156802–1–4
141. Stockman MI (2011) Loss compensation by gain and spasing. Philos Trans R Soc A 369:3510–3524
142. Ginzburg P, Hayat A, Berkovitch N, Orenstein M (2010) Nonlocal ponderomotive nonlinearity in plasmonics. Opt Lett 35:1551–1553
143. Feigenbaum E, Orenstein M (2007) Plasmon-soliton. Opt Lett 32:674–676
144. Zherebtsov S, Fennel T, Plenge J, Antonsson E, Znakovskaya I, Wirth A, Herrwerth O, Suessmann F, Peltz C, Ahmad I, Trushin SA, Pervak V, Karsch S, Vrakking MJJ, Langer B, Graf C, Stockman MI, Krausz F, Ruehl E, Kling MF (2011) Controlled near-field enhanced electron acceleration from dielectric nanospheres with intense few-cycle laser fields. Nat Phys 7:656–662
145. Kruger M, Schenk M, Hommelhoff P (2011) Attosecond control of electrons emitted from a nanoscale metal tip. Nature 475:78–81
146. Durach M, Rusina A, Kling MF, Stockman MI (2010) Metallization of nanofilms in strong adiabatic electric fields. Phys Rev Lett 105:086803–1–4
147. Durach M, Rusina A, Kling MF, Stockman MI (2011) Predicted ultrafast dynamic metallization of dielectric nanofilms by strong single-cycle optical fields. Phys Rev Lett 107:086602–1–5
148. Bergman DJ, Stroud D (1992) Properties of macroscopically inhomogeneous media. In: Ehrenreich H, Turnbull D (eds) Solid state physics, vol 46. Academic, Boston, pp 148–270
149. Stockman MI, Bergman DJ, Kobayashi T (2004) Coherent control of nanoscale localization of ultrafast optical excitation in nanosystems. Phys Rev B 69:054202–1–10
150. Shalaev VM, Stockman MI (1987) Optical properties of fractal clusters (susceptibility, surface enhanced Raman scattering by impurities). Sov Phys JETP 65:287–294
151. Shalaev VM, Botet R, Butenko AV (1993) Localization of collective dipole excitations on fractals. Phys Rev B 48:6662–6664
152. Shalaev VM, Botet R, Tsai DP, Kovacs J, Moskovits M (1994) Fractals – localization of dipole excitations and giant optical polarizabilities. Physica A 207:197–207
153. Sarychev AK, Shubin VA, Shalaev VM (2000) Anderson localization of surface plasmons and Kerr nonlinearity in semicontinuous metal films. Physica B 279:87–89

154. Bozhevolnyi SI, Markel VA, Coello V, Kim W, Shalaev VM (1998) Direct observation of localized dipolar excitations on rough nanostructured surfaces. Phys Rev B 58:11441–11448
155. Sarychev AK, Shubin VA, Shalaev VM (1999) Anderson localization of surface plasmons and nonlinear optics of metal-dielectric composites. Phys Rev B 60:16389–16408
156. Gresillon S, Aigouy L, Boccara AC, Rivoal JC, Quelin X, Desmarest C, Gadenne P, Shubin VA, Sarychev AK, Shalaev VM (1999) Experimental observation of localized optical excitations in random metal-dielectric films. Phys Rev Lett 82:4520–4523
157. Shalaev VM (2000) Nonlinear optics of random media: fractal composites and metal-dielectric films. Springer, Berlin/New York
158. Stockman MI (1997) Inhomogeneous eigenmode localization, chaos, and correlations in large disordered clusters. Phys Rev E 56:6494–6507
159. Stockman MI, Pandey LN, George TF (1996) Inhomogeneous localization of polar eigenmodes in fractals. Phys Rev B 53:2183–2186
160. Stockman MI (1997) Chaos and spatial correlations for dipolar eigenproblems. Phys Rev Lett 79:4562–4565
161. Stockman MI (2000) Giant attosecond fluctuations of local optical fields in disordered nanostructured media. Phys Rev B 62:10494–10497
162. Krachmalnicoff V, Castanie E, Wilde YD, Carminati R (2010) Fluctuations of the local density of states probe localized surface plasmons on disordered metal films. Phys Rev Lett 105:183901–1–4
163. Tsai DP, Kovacs J, Wang Z, Moskovits M, Shalaev VM, Suh JS, Botet R (1994) Photon scanning tunneling microscopy images of optical excitations of fractal metal colloid clusters. Phys Rev Lett 72:4149–4152
164. Stockman MI, Pandey LN, Muratov LS, George TF (1995) Photon scanning-tunneling-microscopy images of optical-excitations of fractal metal colloid clusters – comment. Phys Rev Lett 75:2450
165. Negro LD, Boriskina SV (2012) Deterministic aperiodic nanostructures for photonics and plasmonics applications. Laser Photonics Rev 6:178–218
166. Kolb M, Botet R, Julienne J (1983) Scaling of kinetically growing clusters. Phys Rev Lett 51:1123–1126
167. Weitz DA, Oliveria M (1984) Fractal structures formed by kinetic aggregation of aqueous gold colloids. Phys Rev Lett 52:1433–1436
168. Westcott SL, Halas NJ (2002) Electron relaxation dynamics in semicontinuous metal films on nanoparticle surfaces. Chem Phys Lett 356:207–213
169. Seal K, Sarychev AK, Noh H, Genov DA, Yamilov A, Shalaev VM, Ying ZC, Cao H (2005) Near-field intensity correlations in semicontinuous metal-dielectric films. Phys Rev Lett 94:226101–1–4
170. Stockman MI (2005) Giant fluctuations of second harmonic generation on nanostructured surfaces. Chem Phys 318:156–162
171. Fort E, Gresillon S (2008) Surface enhanced fluorescence. J Phys D 41:013001–1–31
172. Efros AL (1986) Physics and geometry of disorder: percolation theory. Mir, Moscow
173. Levitov LS (1990) Delocalization of vibrational modes caused by electric dipole interaction. Phys Rev Lett 64:547–550
174. Parshin DA, Schober HR (1998) Multifractal structure of eigenstates in the Anderson model with long-range off-diagonal disorder. Phys Rev B 57:10232–10235
175. Burin AL, Kagan Y, Maksimov LA, Polischuk IY (1998) Dephasing rate in dielectric glasses at ultralow temperatures. Phys Rev Lett 80:2945–2948
176. Stockman MI, Pandey LN, Muratov LS, George TF (1994) Giant fluctuations of local optical fields in fractal clusters. Phys Rev Lett 72:2486–2489
177. Stockman MI, Pandey LN, George TF (1996) Inhomogeneous localization of polar eigenmodes in fractals. Phys Rev B 53:2183–2186
178. Stockman MI (1997) Chaos and spatial correlations for dipolar eigenproblem. Phys Rev Lett 79:4562–4565

179. Stockman MI (1997) Inhomogeneous eigenmode localization, chaos, and correlations in large disordered clusters. Phys Rev E 56:6494–6507
180. Ginzburg P, Berkovitch N, Nevet A, Shor I, Orenstein M (2011) Resonances on-demand for plasmonic nano-particles. Nano Lett 11:2329–2333
181. Markel VA, Muratov LS, Stockman MI, George TF (1991) Theory and numerical simulation of optical properties of fractal clusters. Phys Rev B 43:8183
182. Stockman MI (2006) Electromagnetic theory of SERS. In: Kneipp MMK, Kneipp H (eds) Surface enhanced Raman scattering, vol 103. Springer, Heidelberg, pp 47–66
183. Awada C, Barbillon G, Charra F, Douillard L, Greffet JJ (2012) Experimental study of hot spots in gold/glass nanocomposite films by photoemission electron microscopy. Phys Rev B 85:045438–1–6
184. Stockman M, George T (1994) Photon tunneling microscope reveals local hot-spots. Phys World 7:27–28
185. Cang H, Labno A, Lu CG, Yin XB, Liu M, Gladden C, Liu YM, Zhang X (2011) Probing the electromagnetic field of a 15-nanometre hotspot by single molecule imaging. Nature 469:385–388
186. McLeod A, Weber-Bargioni A, Zhang Z, Dhuey S, Harteneck B, Neaton JB, Cabrini S, Schuck PJ (2011) Nonperturbative visualization of nanoscale plasmonic field distributions via photon localization microscopy. Phys Rev Lett 106:037402
187. Yildiz A, Forkey JN, McKinney SA, Ha T, Goldman YE, Selvin PR (2003) Myosin V walks hand-over-hand: single fluorophore imaging with 1.5-nm localization. Science 300:2061–2065
188. Stockman MI, Kling MF, Kleineberg U, Krausz F (2007) Attosecond nanoplasmonic field microscope. Nat Photonics 1:539–544
189. Klar T, Perner M, Grosse S, von Plessen G, Spirkl W, Feldman J (1998) Surface-plasmon resonances in single metallic nanoparticles. Phys Rev Lett 80:4249–4252
190. Lehmann J, Merschdorf M, Pfeiffer W, Thon A, Voll S, Gerber G (2000) Surface plasmon dynamics in silver nanoparticles studied by femtosecond time-resolved photoemission. Phys Rev Lett 85:2921–2924
191. Bosbach J, Hendrich C, Stietz F, Vartanyan T, Trager F (2002) Ultrafast dephasing of surface plasmon excitation in silver nanoparticles: influence of particle size, shape, and chemical surrounding. Phys Rev Lett 89:257404–1–4
192. Hendrich C, Bosbach J, Stietz F, Hubenthal F, Vartanyan T, Trager F (2003) Chemical interface damping of surface plasmon excitation in metal nanoparticles: a study by persistent spectral hole burning. Appl Phys B 76:869–875
193. Zentgraf T, Christ A, Kuhl J, Giessen H (2004) Tailoring the ultrafast dephasing of quasiparticles in metallic photonic crystals. Phys Rev Lett 93:243901–1–4
194. Novotny L, Hecht B (2006) Principles of nano-optics. Cambridge University Press, Cambridge/New York
195. Ono A, Kato J, Kawata S (2005) Subwavelength optical imaging through a metallic nanorod array. Phys Rev Lett 95:267407–1–4
196. Shvets G, Trendafilov S, Pendry JB, Sarychev A (2007) Guiding, focusing, and sensing on the subwavelength scale using metallic wire arrays. Phys Rev Lett 99:053903–1–4
197. Pendry JB (2003) Perfect cylindrical lenses. Opt Express 11:755–760
198. Liu Z, Lee H, Xiong Y, Sun C, Zhang X (2007) Far-field optical hyperlens magnifying sub-diffraction-limited objects. Science 315:1686
199. Stockman MI, Faleev SV, Bergman DJ (2002) Coherent control of femtosecond energy localization in nanosystems. Phys Rev Lett 88:067402–1–4
200. Tannor DJ, Rice SA (1985) Control of selectivity of chemical reaction via control of wave packet evolution. J Chem Phys 83:5013–5018
201. Brumer P, Shapiro M (2003) Principles of the quantum control of molecular processes. Wiley, New York
202. Judson RS, Rabitz H (1992) Teaching lasers to control molecules. Phys Rev Lett 68:1500

203. Kurizki G, Shapiro M, Brumer P (1989) Phase-coherent control of photocurrent directionality in semiconductors. Phys Rev B 39:3435–3437
204. Weinacht TC, Ahn J, Bucksbaum PH (1999) Controlling the shape of a quantum wavefunction. Nature 397:233–235
205. Brumer P, Shapiro M (1992) Laser control of molecular processes. Ann Rev Phys Chem 43:257–282
206. Rabitz H, de Vivie-Riedle R, Motzkus M, Kompa K (2000) Chemistry – whither the future of controlling quantum phenomena? Science 288:824–828
207. Geremia JM, Rabitz H (2002) Optimal identification of Hamiltonian information by closed-loop laser control of quantum systems. Phys Rev Lett 89:263902–1–4
208. Nguyen NA, Dey BK, Shapiro M, Brumer P (2004) Coherent control in nanolithography: Rydberg atoms. J Phys Chem A 108:7878–7888
209. Shapiro M, Brumer P (2006) Quantum control of bound and continuum state dynamics. Phys Rep 425:195–264
210. Assion A, Baumert T, Bergt M, Brixner T, Kiefer B, Seyfried V, Strehle M, Gerber G (1998) Control of chemical reactions by feedback-optimized phase-shaped femtosecond laser pulses. Science 282:919–922
211. Bartels R, Backus S, Zeek E, Misoguti L, Vdovin G, Christov IP, Murnane MM, Kapteyn HC (2000) Shaped-pulse optimization of coherent emission of high-harmonic soft X-rays. Nature 406:164–166
212. Dudovich N, Oron D, Silberberg Y (2002) Single-pulse coherently controlled nonlinear Raman spectroscopy and microscopy. Nature 418:512–514
213. Brixner T, Krampert G, Pfeifer T, Selle R, Gerber G, Wollenhaupt M, Graefe O, Horn C, Liese D, Baumert T (2004) Quantum control by ultrafast polarization shaping. Phys Rev Lett 92:208301–1–4
214. Durach M, Rusina A, Nelson K, Stockman MI (2007) Toward full spatio-temporal control on the nanoscale. Nano Lett 7:3145–3149
215. Volpe G, Cherukulappurath S, Parramon RJ, Molina-Terriza G, Quidant R (2009) Controlling the optical near field of nanoantennas with spatial phase-shaped beams. Nano Lett 9:3608–3611
216. Gjonaj B, Aulbach J, Johnson PM, Mosk AP, Kuipers L, Lagendijk A (2011) Active spatial control of plasmonic fields. Nat Photonics 5:360–363
217. Stockman MI, Hewageegana P (2005) Nanolocalized nonlinear electron photoemission under coherent control. Nano Lett 5:2325–2329
218. Sukharev M, Seideman T (2006) Phase and polarization control as a route to plasmonic nanodevices. Nano Lett 6:715–719
219. Aeschlimann M, Bauer M, Bayer D, Brixner T, de Abajo FJG, Pfeiffer W, Rohmer M, Spindler C, Steeb F (2007) Adaptive subwavelength control of nano-optical fields. Nature 446:301–304
220. Bauer M, Wiemann C, Lange J, Bayer D, Rohmer M, Aeschlimann M (2007) Phase propagation of localized surface plasmons probed by time-resolved photoemission electron microscopy. Appl Phys A 88:473–480
221. Aeschlimann M, Bauer M, Bayer D, Brixner T, Cunovic S, Dimler F, Fischer A, Pfeiffer W, Rohmer M, Schneider C, Steeb F, Struber C, Voronine DV (2010) Spatiotemporal control of nanooptical excitations. Proc Natl Acad Sci U S A 107:5329–5333
222. Li X, Stockman MI (2008) Highly efficient spatiotemporal coherent control in nanoplasmonics on a nanometer-femtosecond scale by time reversal. Phys Rev B 77:195109–1–10
223. Derode A, Tourin A, de Rosny J, Tanter M, Yon S, Fink M (2003) Taking advantage of multiple scattering to communicate with time-reversal antennas. Phys Rev Lett 90:014301–1–4
224. Lerosey G, de Rosny J, Tourin A, Derode A, Montaldo G, Fink M (2004) Time reversal of electromagnetic waves. Phys Rev Lett 92:193904–1–3
225. Lerosey G, de Rosny J, Tourin A, Fink M (2007) Focusing beyond the diffraction limit with far-field time reversal. Science 315:1120–1122

226. Stockman MI (2006) Electromagnetic theory of SERS. In: Kneipp K, Moskovits M, Kneipp H (eds) Surface Enhanced Raman scattering – physics and applications. Springer, Heidelberg/New York/Tokyo, pp 47–66
227. Stockman MI, Pandey LN, Muratov LS, George TF (1995) Optical-absorption and localization of eigenmodes in disordered clusters. Phys Rev B 51:185–195
228. Landau LD, Lifshitz EM (1975) The classical theory of fields. Pergamon, Oxford, New York
229. Kubo A, Pontius N, Petek H (2007) Femtosecond microscopy of surface plasmon polariton wave packet evolution at the silver/vacuum interface. Nano Lett 7:470–475
230. Verhagen E, Kuipers L, Polman A (2007) Enhanced nonlinear optical effects with a tapered plasmonic waveguide. Nano Lett 7:334–337
231. Sukharev M, Seideman T (2007) Coherent control of light propagation via nanoparticle arrays. J Phys B 40:S283–S298
232. Wefers MM, Nelson KA (1993) Programmable phase and amplitude femtosecond pulse shaping. Opt Lett 18:2032–2034
233. Feurer T, Vaughan JC, Nelson KA (2003) Spatiotemporal coherent control of lattice vibrational waves. Science 299:374–377
234. Babajanyan AJ, Margaryan NL, Nerkararyan KV (2000) Superfocusing of surface polaritons in the conical structure. J Appl Phys 87:3785–3788
235. Gramotnev DK, Vogel MW, Stockman MI (2008) Optimized nonadiabatic nanofocusing of plasmons by tapered metal rods. J Appl Phys 104:034311–1–8
236. Nomura W, Ohtsu M, Yatsui T (2005) Nanodot coupler with a surface plasmon polariton condenser for optical far/near-field conversion. Appl Phys Lett 86:181108–1–3
237. Yin LL, Vlasko-Vlasov VK, Pearson J, Hiller JM, Hua J, Welp U, Brown DE, Kimball CW (2005) Subwavelength focusing and guiding of surface plasmons. Nano Lett 5:1399–1402
238. Mailloux RJ (2005) Phased array antenna handbook. Artech House, Boston
239. Lerosey G, de Rosny J, Tourin A, Derode A, Fink M (2006) Time reversal of wideband microwaves. Appl Phys Lett 88:154101–1–3
240. Stockman MI (2008) Ultrafast nanoplasmonics under coherent control. New J Phys 10:025031–1–20
241. Kao TS, Jenkins SD, Ruostekoski J, Zheludev NI (2011) Coherent control of nanoscale light localization in metamaterial: creating and positioning isolated subwavelength energy hot spots. Phys Rev Lett 106:085501–1–4
242. Bauer M, Schmidt O, Wiemann C, Porath R, Scharte M, Andreyev O, Schonhense G, Aeschlimann M (2002) Time-resolved two photon photoemission electron microscopy. Appl Phys B 74:223–227
243. Brixner T, Gerber G (2001) Femtosecond polarization pulse shaping. Opt Lett 26:557–559
244. Brixner T, Krampert G, Niklaus P, Gerber G (2002) Generation and characterization of polarization-shaped femtosecond laser pulses. Appl Phys B 74:S133–S144
245. Atwater HA (2007) The promise of plasmonics. Sci Am 296:56–63
246. Anker JN, Hall WP, Lyandres O, Shah NC, Zhao J, Duyne RPV (2008) Biosensing with plasmonic nanosensors. Nat Mater 7:442–453
247. Israel A, Mrejen M, Lovsky Y, Polhan M, Maier S, Lewis A (2007) Near-field imaging probes electromagnetic waves. Laser Focus World 43:99–102
248. Challener WA, Peng C, Itagi AV, Karns D, Peng W, Peng Y, Yang X, Zhu X, Gokemeijer NJ, Hsia YT, Ju G, Rottmayer RE, Seigler MA, Gage EC (2009) Heat-assisted magnetic recording by a near-field transducer with efficient optical energy transfer. Nat Photonics 3:220–224
249. Nagatani N, Tanaka R, Yuhi T, Endo T, Kerman K, Takamura Y, Tamiya E (2006) Gold nanoparticle-based novel enhancement method for the development of highly sensitive immunochromatographic test strips. Sci Technol Adv Mater 7:270–275
250. Hirsch LR, Stafford RJ, Bankson JA, Sershen SR, Rivera B, Price RE, Hazle JD, Halas NJ, West JL (2003) Nanoshell-mediated near-infrared thermal therapy of tumors under magnetic resonance guidance. Proc Natl Acad Sci U S A 100:13549–13554
251. Park I-Y, Kim S, Choi J, Lee D-H, Kim Y-J, Kling MF, Stockman MI, Kim S-W (2011) Plasmonic generation of ultrashort extreme-ultraviolet light pulses. Nat Photonics 5:677–681

252. Kahng D (1963) Electric field controlled semiconductor device. US Patent 3,102,230
253. Tsividis Y (1999) Operation and modeling of the MOS transistor. McGraw-Hill, New York
254. Stockman MI, Bergman DJ (2009) Surface plasmon amplification by stimulated emission of radiation (spaser). US Patent 7,569,188
255. Stockman MI (2008) Spasers explained. Nat Photonics 2:327–329
256. Noginov MA, Zhu G, Belgrave AM, Bakker R, Shalaev VM, Narimanov EE, Stout S, Herz E, Suteewong T, Wiesner U (2009) Demonstration of a spaser-based nanolaser. Nature 460:1110–1112
257. Hill MT, Marell M, Leong ESP, Smalbrugge B, Zhu Y, Sun M, van Veldhoven PJ, Geluk EJ, Karouta F, Oei Y-S, Nötzel R, Ning C-Z, Smit MK (2009) Lasing in metal-insulator-metal sub-wavelength plasmonic waveguides. Opt Express 17:11107–11112
258. Oulton RF, Sorger VJ, Zentgraf T, Ma R-M, Gladden C, Dai L, Bartal G, Zhang X (2009) Plasmon lasers at deep subwavelength scale. Nature 461:629–632
259. Ma R-M, Oulton RF, Sorger VJ, Bartal G, Zhang X (2010) Room-temperature sub-diffraction-limited plasmon laser by total internal reflection. Nat Mater 10:110–113
260. Flynn RA, Kim CS, Vurgaftman I, Kim M, Meyer JR, Mäkinen AJ, Bussmann K, Cheng L, Choa FS, Long JP (2011) A room-temperature semiconductor spaser operating near 1.5 micron. Opt Express 19:8954–8961
261. Hill MT, Oei Y-S, Smalbrugge B, Zhu Y, de Vries T, van Veldhoven PJ, van Otten FWM, Eijkemans TJ, Turkiewicz JP, de Waardt H, Geluk EJ, Kwon S-H, Lee Y-H, Noetzel R, Smit MK (2007) Lasing in metallic-coated nanocavities. Nat Photonics 1:589–594
262. Gordon JA, Ziolkowski RW (2007) The design and simulated performance of a coated nano-particle laser. Opt Express 15:2622–2653
263. Bergman DJ, Stroud D (1980) Theory of resonances in the electromagnetic scattering by macroscopic bodies. Phys Rev B 22:3527–3539
264. Plum E, Fedotov VA, Kuo P, Tsai DP, Zheludev NI (2009) Towards the lasing spaser: controlling metamaterial optical response with semiconductor quantum dots. Opt Express 17:8548–8551
265. Seidel J, Grafstroem S, Eng L (2005) Stimulated emission of surface plasmons at the interface between a silver film and an optically pumped dye solution. Phys Rev Lett 94:177401–1–4
266. Noginov MA, Zhu G, Mayy M, Ritzo BA, Noginova N, Podolskiy VA (2008) Stimulated emission of surface plasmon polaritons. Phys Rev Lett 101:226806–1–4
267. Li K, Li X, Stockman MI, Bergman DJ (2005) Surface plasmon amplification by stimulated emission in nanolenses. Phys Rev B 71:115409–1–4
268. Dong ZG, Liu H, Li T, Zhu ZH, Wang SM, Cao JX, Zhu SN, Zhang X (2008) Resonance amplification of left-handed transmission at optical frequencies by stimulated emission of radiation in active metamaterials. Opt Express 16:20974–20980
269. Wegener M, Garcia-Pomar JL, Soukoulis CM, Meinzer N, Ruther M, Linden S (2008) Toy model for plasmonic metamaterial resonances coupled to two-level system gain. Opt Express 16:19785–19798
270. Fang A, Koschny T, Wegener M, Soukoulis CM (2009) Self-consistent calculation of metamaterials with gain. Phys Rev B (Rapid Commun) 79:241104(R)–1–4
271. Wuestner S, Pusch A, Tsakmakidis KL, Hamm JM, Hess O (2010) Overcoming losses with gain in a negative refractive index metamaterial. Phys Rev Lett 105:127401–1–4
272. Chang SW, Ni CYA, Chuang SL (2008) Theory for bowtie plasmonic nanolasers. Opt Express 16:10580–10595
273. Zheludev NI, Prosvirnin SL, Papasimakis N, Fedotov VA (2008) Lasing spaser. Nat Photonics 2:351–354
274. Protsenko IE, Uskov AV, Zaimidoroga OA, Samoilov VN, O'Reilly EP (2005) Dipole nanolaser. Phys Rev A 71:063812
275. Ambati M, Nam SH, Ulin-Avila E, Genov DA, Bartal G, Zhang X (2008) Observation of stimulated emission of surface plasmon polaritons. Nano Lett 8:3998–4001
276. Zhou ZK, Su XR, Peng XN, Zhou L (2008) Sublinear and superlinear photoluminescence from Nd doped anodic aluminum oxide templates loaded with Ag nanowires. Opt Express 16:18028–18033

277. Noginov MA, Podolskiy VA, Zhu G, Mayy M, Bahoura M, Adegoke JA, Ritzo BA, Reynolds K (2008) Compensation of loss in propagating surface plasmon polariton by gain in adjacent dielectric medium. Opt Express 16:1385–1392
278. Bolger PM, Dickson W, Krasavin AV, Liebscher L, Hickey SG, Skryabin DV, Zayats AV (2010) Amplified spontaneous emission of surface plasmon polaritons and limitations on the increase of their propagation length. Opt Lett 35:1197–1199
279. Noginov MA, Zhu G, Bahoura M, Adegoke J, Small C, Ritzo BA, Drachev VP, Shalaev VM (2007) The effect of gain and absorption on surface plasmons in metal nanoparticles. Appl Phys B 86:455–460
280. Noginov MA (2008) Compensation of surface plasmon loss by gain in dielectric medium. J Nanophotonics 2:021855–1–17
281. Leon ID, Berini P (2010) Amplification of long-range surface plasmons by a dipolar gain medium. Nat Photonics 4:382–387
282. Ding K, Liu ZC, Yin LJ, Hill MT, Marell MJH, van Veldhoven PJ, Noetzel R, Ning CZ (2012) Room-temperature continuous wave lasing in deep-subwavelength metallic cavities under electrical injection. Phys Rev B 85:041301–1–5
283. Ding K, Yin L, Hill MT, Liu Z, van Veldhoven PJ, Ning CZ (2013) An electrical injection metallic cavity nanolaser with azimuthal polarization. Appl Phys Lett 102:041110–1–4
284. Ding K, Hill MT, Liu ZC, Yin LJ, van Veldhoven PJ, Ning CZ (2013) Record performance of electrical injection sub-wavelength metallic-cavity semiconductor lasers at room temperature. Opt Express 21:4728–4733
285. Kitur JK, Zhu G, Yu AB, Noginov MA (2012) Stimulated emission of surface plasmon polaritons on smooth and corrugated silver surfaces. J Opt 14:114015–1–8
286. Wu CY, Kuo CT, Wang CY, He CL, Lin MH, Ahn H, Gwo S (2011) Plasmonic green nanolaser based on a metal-oxide-semiconductor structure. Nano Lett 11:4256–4260
287. Lu Y-J, Kim J, Chen H-Y, Wu C, Dabidian N, Sanders CE, Wang C-Y, Lu M-Y, Li B-H, Qiu X, Chang W-H, Chen L-J, Shvets G, Shih C-K, Gwo S (2012) Plasmonic nanolaser using epitaxially grown silver film. Science 337:450–453
288. Oulton RF, Sorger VJ, Genov DA, Pile DFP, Zhang X (2008) A hybrid plasmonic waveguide for subwavelength confinement and long-range propagation. Nat Photonics 2:496–500
289. Li D, Stockman MI (2013) Electric spaser in the extreme quantum limit. Phys Rev Lett 110:106803–1–5
290. Brown RH, Twiss RQ (1956) A test of a new type of stellar interferometer on Sirius. Nature 178:1046–1048
291. Schawlow AL, Townes CH (1958) Infrared and optical masers. Phys Rev 112:1940
292. Kéna-Cohen S, Stavrinou PN, Bradley DDC, Maier SA (2013) Confined surface plasmon-polariton amplifiers. Nano Lett 13:1323–1329
293. Kneipp K, Moskovits M, Kneipp H (eds) (2006) Surface enhanced Raman scattering: physics and applications. Springer, Heidelberg/New York/Tokyo
294. Kneipp J, Kneipp H, Wittig B, Kneipp K (2010) Novel optical nanosensors for probing and imaging live cells. Nanomed Nanotechnol Biol Med 6:214–226
295. Dunmore FJ, Liu DZ, Drew HD, Dassarma S, Li Q, Fenner DB (1995) Observation of below-gap plasmon excitations in superconducting $YBa_2Cu_3O_7$ films. Phys Rev B 52:R731–R734
296. Schumacher D, Rea C, Heitmann D, Scharnberg K (1998) Surface plasmons and Sommerfeld-Zenneck waves on corrugated surfaces: application to high-T_c superconductors. Surf Sci 408:203–211
297. Tsiatmas A, Buckingham AR, Fedotov VA, Wang S, Chen Y, de Groot PAJ, Zheludev NI (2010) Superconducting plasmonics and extraordinary transmission. Appl Phys Lett 97:111106–1–3
298. Boltasseva A, Atwater HA (2011) Low-loss plasmonic metamaterials. Science 331:290–291
299. Shalaev VM (2007) Optical negative-index metamaterials. Nat Photonics 1:41–48
300. Zheludev NI (2011) A roadmap for metamaterials. Opt Photonics News 22:30–35
301. Stockman MI, Kurlayev KB, George TF (1999) Linear and nonlinear optical susceptibilities of Maxwell Garnett composites: Dipolar spectral theory. Phys Rev B 60:17071–17083

302. Meinzer N, Ruther M, Linden S, Soukoulis CM, Khitrova G, Hendrickson J, Olitzky JD, Gibbs HM, Wegener M (2010) Arrays of Ag split-ring resonators coupled to InGaAs single-quantum-well gain. Opt Express 18:24140–24151
303. Kretschmann E, Raether H (1968) Radiative decay of nonradiative surface plasmons excited by light. Z Naturforsch A 23:2135–2136
304. Heydari E, Flehr R, Stumpe J (2013) Influence of spacer layer on enhancement of nanoplasmon-assisted random lasing. Appl Phys Lett 102:133110–4
305. Kitur JK, Podolskiy VA, Noginov MA (2011) Stimulated emission of surface plasmon polaritons in a microcylinder cavity. Phys Rev Lett 106:183903–1–4
306. Bloch F (1929) Über die Quantenmechanik der Elektronen in Kristallgittern. Z Phys A 52:555–600
307. Ghafouri-Shiraz H (2003) Distributed feedback laser diodes and optical tunable filters. Wiley, West Sussex/Hoboken
308. van Beijnum F, van Veldhoven PJ, Geluk EJ, de Dood MJA, 't Hooft GW, van Exter MP (2013) Surface plasmon lasing observed in metal hole arrays. Phys Rev Lett 110:206802–1–5
309. Zhou W, Dridi M, Suh JY, Kim CH, Co DT, Wasielewski MR, Schatz GC, Odom TW (2013) Lasing action in strongly coupled plasmonic nanocavity arrays. Nat Nanotechnol 8:506–511
310. Stockman MI (2013) Lasing spaser in two-dimensional plasmonic crystals. NPG Asia Mater 5:e71
311. Anderson PW (1958) Absence of diffusion in certain random lattices. Phys Rev 109:1492–1505

2
Transformation Optics

Martin Wegener

Abstract We briefly reviewed the concept of transformation optics for designing functionalities. We also gave recent experimental examples from different areas of physics ranging from optics via mechanics to thermodynamics.

Keywords Transformation optics • Metamaterial • Invisibility cloaking • Conformal mapping • Anisotropy

2.1 Introduction

In physics, we often take a certain material distribution and aim at calculating the way light waves, mechanical waves, or other energy fluxes not even connected with waves emerge from this material distribution. A certain function may result. Let us call this the forward problem. In engineering, the problem is often rather the backward problem: A customer wants a certain function, e.g., he/she wishes that light behaves in a certain manner and the engineer must come up with a material distribution that leads to the requested behavior or function. Usually, such inverse problems are much more difficult to solve than the forward problem. In principle, with computers becoming ever more powerful, one can likely solve each and every such problem numerically in the future. No human brain or intuition will be required at all. Today, this is generally not possible yet, although specific examples have been given. In any case, however, numerical solutions are not entirely satisfactory. One would rather also get some intuitive understanding as to why a certain material distribution solves the problem. The ideas of transformation optics

M. Wegener (✉)
Institute of Applied Physics, Institute of Nanotechnology, and DFG-Center for Functional Nanostructures (CFN), Karlsruhe Institute of Technology (KIT), 76128 Karlsruhe, Germany
e-mail: martin.wegener@kit.edu

B. Di Bartolo et al. (eds.), *Nano-Structures for Optics and Photonics*, NATO Science for Peace and Security Series B: Physics and Biophysics, DOI 10.1007/978-94-017-9133-5_2,

[1–3] to be outlined in the following section help us in solving the inverse problem by connecting wave propagation and/or energy fluxes with fictitious coordinate transformations, i.e., with the geometry of space.

2.2 Principle of Transformation Optics

Suppose you take the rubber sheet shown in Fig. 2.1 with a Cartesian grid drawn onto it [3]. An observer looking normal onto the rubber sheet will see an undistorted rectangular grid. Upon stretching the rubber sheet within the plane or by even pulling and distorting the rubber sheet in the third dimension (mathematically, by performing a coordinate transformation), the observer will see a distorted set of lines. Any of these lines represents the potential path of a light ray [3]. By appropriate pulling on the rubber sheet, essentially any light path can be tailored. For example, if we take a screw driver and punch a hole into the rubber sheet and then open up this hole to macroscopic size in Fig. 2.1, no grid line will pass this hole. Hence light does not enter the hole; we have created an invisibility cloak [1]. Any person within the hole cannot be seen and cannot look outside his/her hole either.

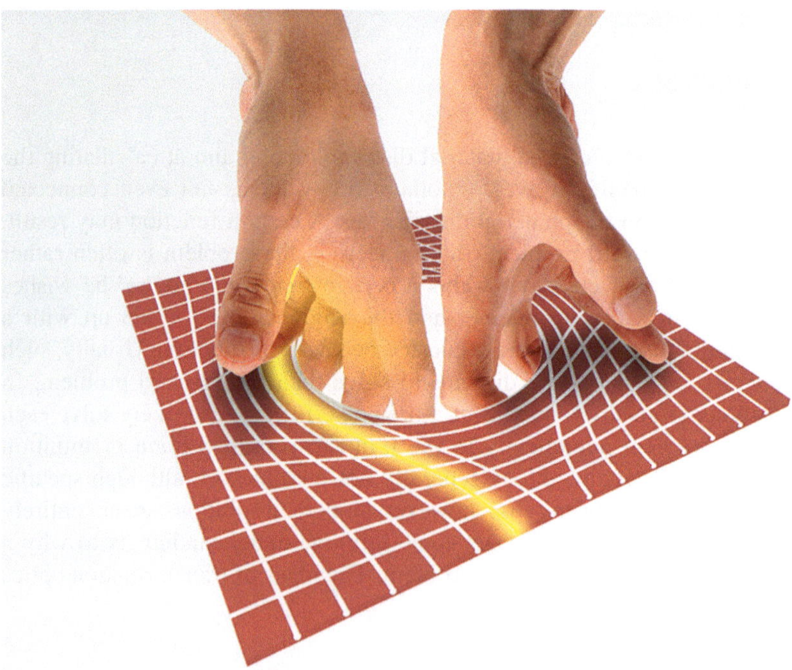

Fig. 2.1 Transformation optics connects the geometry of space with propagation of waves or with energy fluxes in inhomogeneous materials (Taken from Ref. [3])

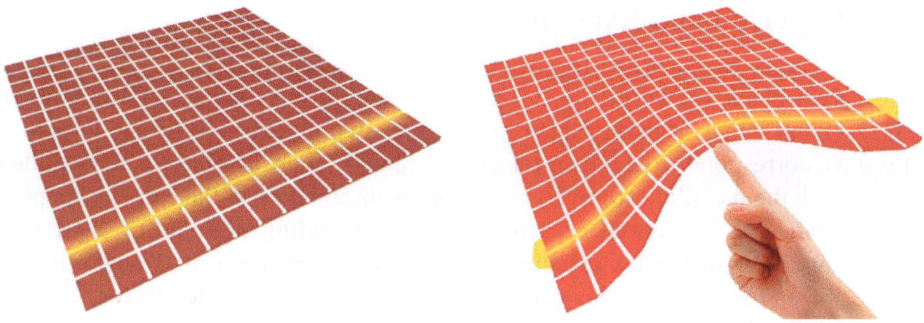

Fig. 2.2 Illustration of the carpet-invisibility-cloak coordinate transformation (Figure provided by Michael S. Rill)

So far, however, the situation is purely fictitious. How can we map it onto an actual material distribution? Suppose the white lines in Fig. 2.1 are roads for cars in a city. All cars drive with the same constant velocity of, e.g., one line per second. For an observer watching the distorted image of the city on a computer screen, it appears as though cars are driving faster in regions where the distance between lines is stretched and slower where the distance between lines are compressed. The basic idea of transformation optics is to make this apparent or fictitious velocity real. This is possible by introducing a (meta-) material distribution leading to exactly that velocity distribution. For the situation depicted in Fig. 2.1, this means that the azimuthal component of the velocity needs to be larger than in the surrounding and that the radial component approaches zero towards the inner radius. The former aspect is somewhat problematic for electromagnetic waves as the azimuthal component needs to be faster than the vacuum speed of light. The phase velocity as well as the group velocity of light may well exceed the vacuum speed. However, the energy velocity cannot exceed the vacuum speed of light c_0 as energy and mass are equivalent. According to special relativity, mass cannot propagate faster than c_0. For a non-dispersive medium, energy and phase velocity are identical. Thus, it is fundamentally impossible to build a non-dispersive invisibility cloak . It can at best operate in certain frequency region. This means, however, that the invisibility cloak can be revealed by a pulsed experiment.

Figure 2.2 shows another example for a coordinate transformation leading to an invisibility cloak , the so-called carpet cloak [4]. Here, the used conformal transformation leads to a locally isotropic distortion of the grid. This means that a locally isotropic material is sufficient to mimic the transformation. Furthermore, the distortions are gentler and, hence, neither zeroes nor infinities occur in the resulting velocity distribution. For moderate bump height to bump width, the resulting refractive indices can stay moderate, e.g., in the interval $[1, 1.5]$.

2.3 Experimental Examples

The carpet cloak shown in Fig. 2.2 does not start from a single point (compare Fig. 2.1), but rather from the edge of a fictitious two-dimensional space. In optics, this edge corresponds to a mirror (or ground plane). Thus, an object can be hidden underneath the metallic carpet. To make the resulting bump in the carpet disappear (which can also be viewed as a particularly demanding example of aberration correction), a graded-index structure needs to be put on top of it. Transformation optics allows for designing this graded-index profile and metamaterials for realizing any refractive index. Corresponding experiments in three-dimensional space have recently been published at telecom [4] and at visible [5, 6] frequencies. A scheme of the fabricated sample is shown in Fig. 2.3, a measured phase image reflecting the time-of-flight in Fig. 2.4. Without cloak, the measured phase image simply resembles the geometrical shape of the bump. With cloak, the phase is nearly flat as expected for a flat metal mirror. The measured cloaking behavior is also broadband [4, 5].

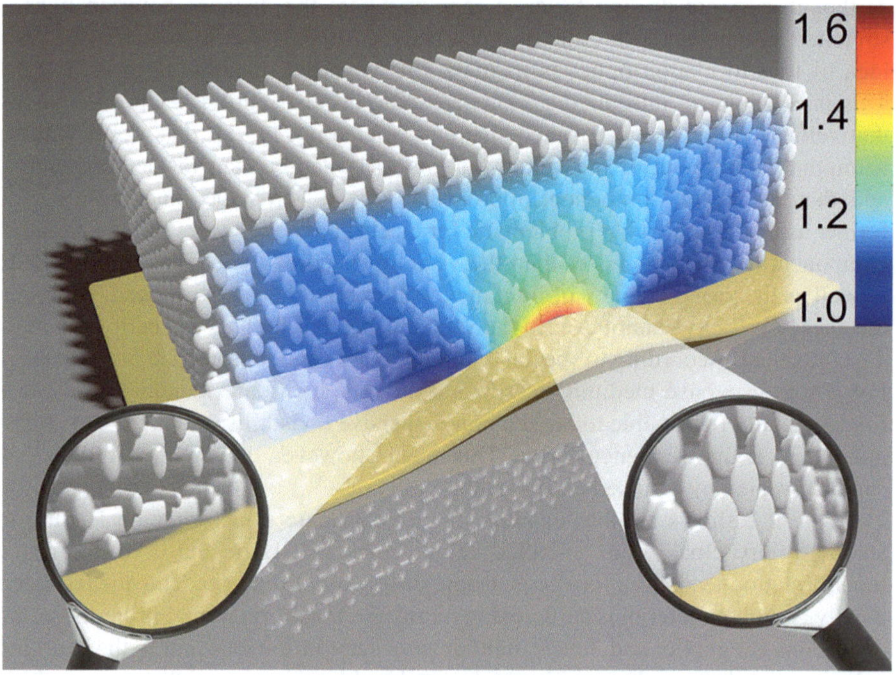

Fig. 2.3 Scheme of a three-dimensional optical carpet invisibility cloak composed of a polymer woodpile photonic crystal used in the long-wavelength limit. Underneath the gold carpet, objects can be hidden (After Refs. [4–6])

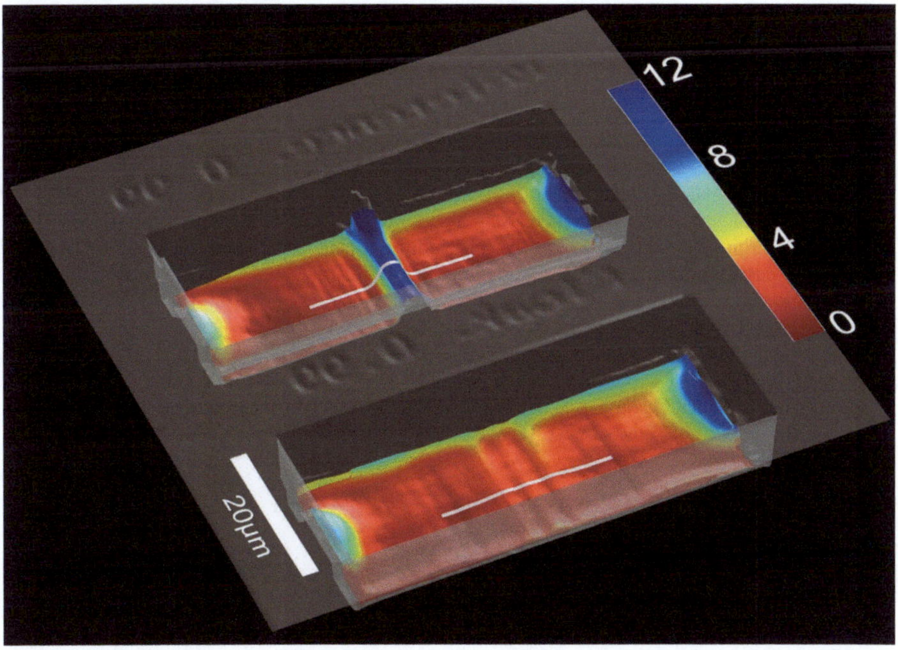

Fig. 2.4 Measured phase images (false-color scale is in units of rad) on a reference structure (top) and on a carpet cloak structure (bottom) like schematically illustrated in Fig. 2.3. Measurements are taken at 700 nm wavelength (Taken from Ref. [6])

In optics, however, as pointed out above, realizing free-space (rather than carpet) broadband cloaks like shown in Fig. 2.1 is not compatible with special relativity, where the vacuum speed of light sets an upper scale. For other situations outside of electromagnetism, for which relevant velocities are far below the vacuum speed of light anyway, no such limitation exists and free-space cloaks can be realized experimentally. For example, mechanical cloaks for vibrations propagating in a thin plate or membrane have been demonstrated [7]. Here, the available off-resonant constituent materials also exhibit a velocity contrast of about 40, which is much larger than in visible optics. This helps to approximate the very low radial and very large azimuthal velocities discussed above.

Measured movies of flexural waves propagating in such structured media have shown excellent broadband free-space cloaking [7] (compare Fig. 2.1). Finally, we mention that the same construction principles can also be applied to heat conduction (or diffusion) in thermodynamics [8] – despite the fact that this type of energy transport is not connected with waves at all. Here, the heat conductivity takes the role of the wave velocity. Material contrast >1,000 enables excellent "free-space" cloaking [8].

References

1. Pendry JB, Schurig D, Smith DR (2006) Controlling electromagnetic fields. Science (New York) 312:1780–1782
2. Leonhardt U (2006) Optical conformal mapping. Science (New York) 312:1777–1780
3. Wegener M, Linden S (2010) Shaping optical space with metamaterials. Phys Today 63(10):32
4. Ergin T, Stenger N, Brenner P, Pendry JB, Wegener M (2010) Three-dimensional invisibility cloak at optical wavelengths. Science (New York) 328:337–339
5. Fischer J, Ergin T, Wegener M (2011) Three-dimensional polarization-independent visible-frequency carpet invisibility cloak. Optics Lett 36:2059–2061
6. Ergin T, Fischer J, Wegener M (2011) Optical phase cloaking of 700 nm light waves in the far field by a three-dimensional carpet cloak. Phys Rev Lett 107:173901
7. Stenger N, Wilhelm M, Wegener M (2012) Experiments on elastic cloaking in thin plates. Phys Rev Lett 108:014301
8. Schittny R, Kadic M, Guenneau S, Wegener M (2013) Experiments on transformation thermodynamics: molding the flow of heat. Phys Rev Lett 110:195901

3
Localized Light-Matter Interactions with Optical Antennas

Zachary J. Lapin, Palash Bharadwaj, Shawn Divitt, and Lukas Novotny

Abstract Standard far-field optical elements, such as lenses and mirrors, are only capable of localizing radiation to about half-a-wavelength—the Abbe criterion. Optical antennas facilitate the further localization of radiation into arbitrarily small spatial volumes. Combining the optical antenna with traditional optical microscopy, a technique termed near-field scanning optical microscopy (NSOM), has enabled the study of biological and solid-state samples at high spatial resolution. Since the development of NSOM in the 1980s, the biggest challenge to researchers has been the design and fabrication of optical antennas functioning as optical near-field probes. We have recently made much progress in the development of widely applicable, and reproducible, optical antennas that provide a high degree of spatial localization. Beyond NSOM, we also explore the electrical excitation, as opposed to photo excitation, of optical antennas with a scanning tunneling microscope (STM). We demonstrate a two-step plasmon-mediated energy conversion from a tunneling current to propagating photons in a smooth gold film as well an extended gold nanowire. We prove that highly localized gap plasmons are first excited in the tunnel gap, which can then couple to propagating plasmons. We elaborate on the role of gap plasmons in explaining the huge variations seen in photon emission yields in the field of STM light emission.

Keywords Optical antenna • Near-field microscopy • NSOM • STM light emission

Z.J. Lapin (✉) • P. Bharadwaj • S. Divitt • L. Novotny
ETH Zürich, Zürich, Switzerland
e-mail: zlapin@ethz.ch; palashb@ethz.ch; sdivitt@ethz.ch; lnovotny@ethz.ch

B. Di Bartolo et al. (eds.), *Nano-Structures for Optics and Photonics*,
NATO Science for Peace and Security Series B: Physics and Biophysics,
DOI 10.1007/978-94-017-9133-5_3,

3.1 Introduction

The need for optical antennas is motivated by the diffraction limit. Standard optical microscopy and spectroscopy rely entirely on far-field optical elements such as lenses and mirrors. These elements limit the obtainable localization of radiation to about half-a-wavelength—the Abbe criterion [1, 2]. The extreme localization of incident radiation afforded by an antenna, as has been well studied at radio and microwave frequencies, is limited only by the antenna geometry. Using our previous definition of an optical antenna as *a device designed to efficiently convert free-propagating optical radiation to localized energy, and vice versa* [3], we find there to be a broad range of applications of the optical antenna in both fundamental research and engineering.

Imaging and spectroscopy at optical frequencies with sub-diffraction spatial resolution, termed near-field scanning optical microscopy (NSOM), has been demonstrated as a robust tool for understanding both physical and biological systems. While there are competing NSOM technologies, the use of the optical antenna has proven to be the most versatile. Applications include single molecule spectroscopy [4], nanoplasmonics [5, 6], biological protein mapping [7–9], fluorescence photophysics [10, 11], Raman spectroscopy [11–14], non-linear optics [15], Green's function measurements [16], interferometry [17, 18], absorption spectroscopy [19], and others.

The broad applicability of the optical antenna is made even more impressive given the infancy of the technology. The history of the optical antenna, specifically the first proposal of accessing optical near-fields, began in 1928 with a prophetic letter to Albert Einstein by a relatively unknown Edward Hutchinson Synge [20]. Synge made two separate suggestions, one relying on localized optical near-fields leaking out of a sub-wavelength aperture (the aperture probe) and the other relying on a single metallic nanoparticle that would be employed to localize and scatter optical near-fields. These ideas, however, were well ahead of the times and are in fact the first proposal of scanning microscopy techniques [21]. Both of Synge's ideas would be realized soon after the development of scanning probe microscopy (SPM) technology.

In general, the extreme localization of radiation afforded by the optical antenna facilitates the study of biological and solid-state systems in two distinct manners—spatial and optical. Spatially, the antenna facilitates resolving features that are separated by sub-diffraction length scales that are therefore simply not resolvable with traditional far-field optical techniques.

3.2 NSOM Imaging with Optical Antennas

The development of near-field optical techniques and the application of optical antennas has a rich history. Many technological advances were required before Synge's 1928 ideas could be experimentally realized. There are several review articles that cover all of the developments of NSOM [20, 22]. Here, advancements that were required to achieve the imaging and spectroscopic technology that is used today will be briefly outlined.

The first necessary advancement came in the 1980s with the development of SPM. First achieved with scanning tunneling microscopy (STM) in 1982 by researchers at IBM in Switzerland [23, 24], SPM techniques allow for sub-nanometer distance control between a probe and a sample of interest. The development of the atomic force microscope (AFM) quickly followed, allowing for the fine distance control between nearly any probe and sample (not just conducting samples and probes as in STM), as well as increasing the probe-sample distance relative to STM (a feature necessary for fluorescence imaging and spectroscopy) [25].

With this technology at hand, the first near-field optical measurements were made on October 22, 1982 by Winfried Denk working with Dieter Pohl at IBM using an aperture probe [20]. Pohl published his first near-field images in 1984 [26]. The technique, termed the "optical stethoscope", a corollary to the medical stethoscope, achieved $\lambda/20$ resolution with a $\lambda = 488$ nm excitation.

Scattering-type NSOM (sometimes referred to as s-NSOM, herein referred to as NSOM) and imaging based on plasmonic resonances was published a few years later by Ulrich Ch. Fischer and Pohl [27]. Their experiment was the first realization of Synge's proposed metallic point scatterer optical antenna. In fact, any near-field probe can be viewed as an antenna since it exploits a near-field interaction for the localization of optical radiation. A further development in NSOM was the use of a sharp tungsten wire as the near-field probe [5]. This type of probe geometry has proved to be a very good optical antenna for many applications.

Today, the biggest remaining challenge for optical antennas as a near-field optical probe is the design and fabrication of the antenna itself. Despite analogies between radio/microwave and optical antennas, which are both described by the same antenna parameters, the realization of the two categories of antennas are quite different. Specifically, standard antenna designs do not always scale well into the optical regime. There are two primary reasons for this: (1) metals behave as strongly coupled plasmas at optical frequencies [28] and (2) the absolute fabrication tolerances of optical antennas is extremely difficult to achieve due to their small physical size. In fact, the motivation to study localized light-matter interactions with optical antennas has helped fuel the development of novel nanofabrication technologies.

3.2.1 Experimental Setup

Most applications of SPM with optical antennas are done in transmission, with an experimental setup representatively shown in Fig. 3.1. In our experiments, an optical excitation is provided by a tightly-focused radially polarized laser source. The excitation is further localized by an optical antenna positioned above the sample and centered in the optical focus. The optical antenna-sample separation

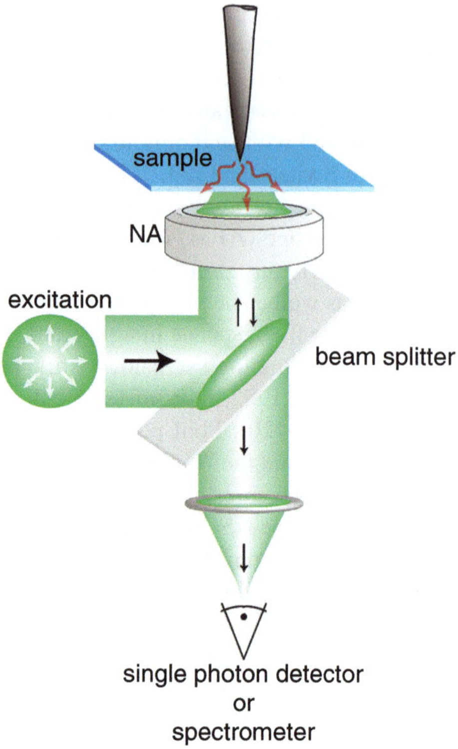

Fig. 3.1 Schematic diagram of the experimental setup used for both NSOM and STM-LE. For NSOM, a radially polarized optical excitation is focused onto the sample by a high-NA objective. Light is further localized by the optical antenna. Spectroscopic emission is collected by the same objective and sent to either a spectrometer or single photon detector. For STM-LE, a bias is placed between the optical antenna and sample and no optical excitation is used. The sample is then raster-scanned between the optical antenna and the objective to form an image

can be maintained with either AFM or STM feedback. Spectroscopic signal from the sample is collected by the objective and sent to either a photodetector or spectrometer. Images are formed by raster-scanning the sample between the tip and focus. The system can easily be modified for STM-based light emission (STM-LE) by simply removing the optical excitation and using an STM feedback loop.

3.2.2 *Optical Antenna Geometries*

Perhaps the most difficult challenge in optical antenna based microscopy is finding an antenna that works well for a single application, let alone a universal antenna geometry. Due to limitations in nanofabrication technologies, it has been a challenge to develop reproducible optical antennas that provide large localized field enhancements and do not quench the signal that is generated by the sample.

In NSOM, as shown in Fig. 3.1, a relatively large area of the sample is illuminated by standard, diffraction-limited, far-field optics. The optical antenna then performs two functions: a local excitation source and a local scatterer. This type of illumination scheme will lead to a detected optical signal that is a superposition of optical signals generated from the far-field excitation (termed background) and the optical

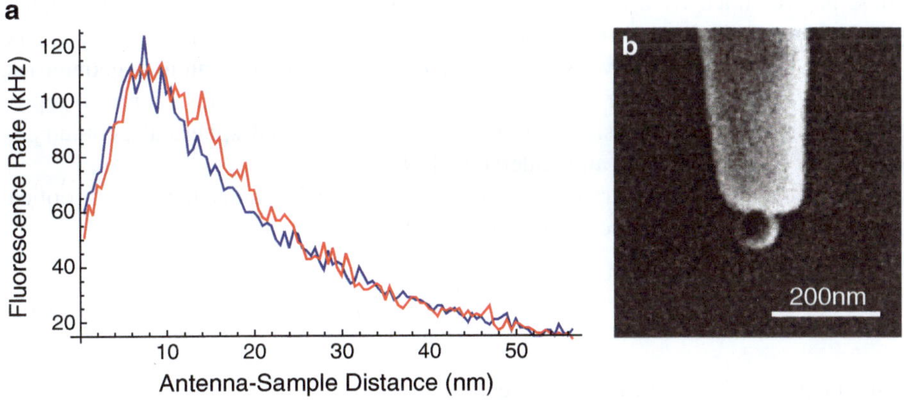

Fig. 3.2 (**a**) An approach curve showing the fluorescence rate enhancement of a single Nile Blue molecule on cover glass as a function of the antenna-sample separation and (**b**) an electron micrograph of a representative particle antenna

antenna. The background is undesirable, however, with the combination of a good antenna, the appropriate sample, and optional real-time background suppression techniques, this methodology is extremely robust.

In general, the optical antenna facilitates the transfer of energy between the near-field and far-field equally well in both directions; however, the proximity of the antenna to the sample can also suppress the signal generated by the sample, for example, it can alter the quantum efficiency (QE) of a fluorescent molecule [11]. In the absence of the antenna (free space), the fluorescence rate of a single molecule excited below saturation, $\Gamma_{\text{fl}}^{\circ}$, is the product of the excitation rate, $\Gamma_{\text{exc}}^{\circ}$, and the intrinsic QE, Q_i, of the molecule,

$$\Gamma_{\text{fl}}^{\circ} = \Gamma_{\text{exc}}^{\circ} \cdot Q_i = \Gamma_{\text{exc}}^{\circ} \cdot \frac{\Gamma_{\text{rad}}^{\circ}}{\Gamma_{\text{rad}}^{\circ} + \Gamma_{\text{nr}}^{\circ}}, \tag{3.1}$$

where, subscripts "rad" and "nr" refer to the radiative and non-radiative decay processes. Coupling a fluorescent molecule, excited below saturation, to an optical antenna will alter both the excitation rate and the QE of the molecule. The increase in the excitation rate will result in an increased radiative rate (enhancement), while the presence of the metallic antenna will also affect the decay rates. Large increases in QE for molecules with a low Q_i have been demonstrated [11].

3.2.2.1 Metallic Particle Antennas

The first optical antenna described as well as the first used was a simple metallic sphere, now referred to as the particle antenna (cf. Fig. 3.2). The metallic nanoparticle is suspended by a dielectric and functions as a dipole resonator to localize an incident optical excitation. Despite the non-optimized spherical geometry, the particle antenna does have several advantages over other antenna geometries. The colloidal synthesis of metallic nanoparticles makes it quite simple to reproducibly fabricate antennas. Additionally, the well defined spherical geometry is relatively easy to simulate and understand [29].

The most important parameters regarding a metallic nanosphere as an optical antenna are the particle's scattering cross-section,

$$\sigma_{scatt} = \frac{k^4}{6\pi\epsilon_0^2}|\alpha(\omega)|^2, \tag{3.2}$$

and the particle's absorption cross-section,

$$\sigma_{abs} = \frac{k}{\epsilon_0}\Im[\alpha(\omega)]. \tag{3.3}$$

Here, $k = \sqrt{\epsilon_d}w/c$ is the wave vector in the surrounding medium and $\alpha(\omega)$ is the quasi-static polarizability of the particle. Since $\alpha(\omega) \propto R^3$, for small particles it can be seen that the particle's absorption cross-section will be larger than the scattering cross-section. While small particles lead to strong absorption, large particles provide poor localization. We have found that an 80 nm diameter gold particle provides a good compromise between absorption and localization.

The 80 nm diameter gold particle antenna provides a 10-fold increase in the fluorescence emission rate of a high-QE single fluorescent emitter excited below saturation—both theoretically and experimentally. Figure 3.2a shows the enhancement of the collected fluorescence emission rate of a single high-QE fluorescent molecule as a function of the particle-molecule separation for a particle antenna shown in Fig. 3.2b. The emission rate is enhanced 10-fold at the optimum antenna-sample separation followed by a net fluorescence quenching as the separation is further decreased. This quenching is due to additional pathways for the excited molecule to non-radiatively transfer energy into the gold particle. The approach curve also shows that the localized field extends ~40 nm below the particle. In addition to the fluorescence-rate enhancement, a localization of radiation in the sample-plane of ~60 nm is measured for an 80 nm gold particle (Fig. 3.3a). This is well below the diffraction-limit for the red light (λ = 632.8 nm) used in this measurement.

We have also extended this imaging system to quantitative biological measurements [9]. Biological measurements impose additional challenges since biological entities are not isolated but instead exist in an optically active matrix. In order to have high quality imaging with a sufficient signal-to-background ratio, a real-time background suppression technique has been employed. This is achieved by modulating the antenna-sample separation, thereby modulating the fluorescence signal, and then demodulating the detected fluorescence signal [8]. In this way, it is possible to isolate when a fluorescent molecule is directly below the antenna

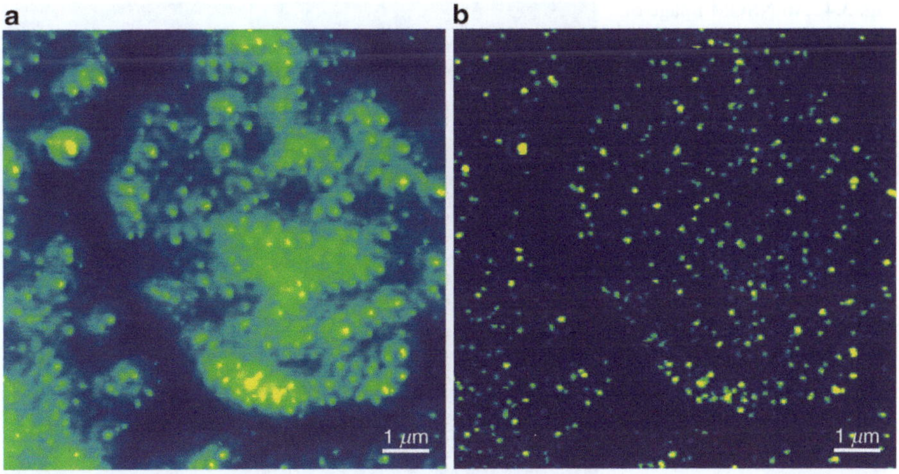

Fig. 3.3 A comparison between an NSOM image (**a**) and a simultaneously acquired background suppressed NSOM image (**b**). The sample is a human erythrocyte ghost on a glass substrate that is labeled against the membrane protein Complement Receptor 1

and successfully suppress all background optical signal arising from the biological matrix. Figure 3.3 shows a comparison between simultaneously acquired NSOM and background suppressed NSOM images of a human erythrocyte membrane ghost labeled against Complement Receptor 1.

The single particle antenna is limited by two competing requirements: large field enhancement and high spatial confinement. These two concerns can be simultaneously satisfied by the self-similar particle antenna. An antenna comprised of a linear chain of particles of decreasing diameter has demonstrated large improvements in both enhancement and resolution for applications in fluorescence imaging and spectroscopy [30].

3.2.2.2 Pyramidal Optical Antennas

The search for a more-ideal optical antenna for fluorescence imaging, meaning improved antenna performance, ease of fabrication, and reproducibility, has lead to the pyramid antenna. The antennas, formed by template-stripping metallic structures from an anisotropically etched silicon wafer, are pyramidal in shape with a tip apex on the order of 10 nm and a cone angle of 70.52°. Such a probe is capable of the localization of optical radiation exceeding that of a gold particle. Additionally, the large cone angle of the pyramid provides a large mode mismatch between radiation emitted by the sample of interest and plasmons propagating along the pyramid, thus limiting fluorescence quenching. Because of this fact, these optical antennas have been shown to have applications in both fluorescence and Raman spectroscopy and imaging [14].

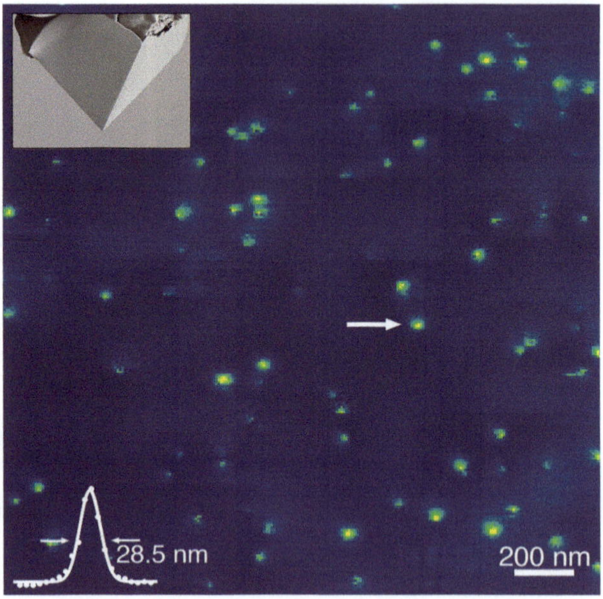

Fig. 3.4 An NSOM image of single Atto 647N dye molecules dispersed on a ~2 nm thick layer of PMMA on coverglass. A cross-section of a single molecule oriented perpendicular to the surface is shown. The *inset* is an electron micrograph of a representative gold pyramid antenna with a 20 μm base

The pyramid antenna has demonstrated single fluorescent molecule imaging with both resolution and enhancement exceeding that of a particle antenna. An approach curve for an isolated molecule oriented along the axis of the antenna maps the localization of radiation of the antenna to ~15 nm and reveals a fluorescence enhancement of 200-fold [14]. Shown in Fig. 3.4, high-resolution imaging of single fluorescent molecules can easily be achieved with the pyramidal antenna geometry. Analogous to the gold particle antenna, the orientation of the molecule's absorption dipole is imaged due to the known excitation polarization [10].

The fluorescence enhancement provided by the pyramid antenna is a complex mixture of the local field enhancement, quenching, and a redirection of fluorescence emission. The Raman enhancement of a single carbon nanotube (CNT) is able to more accurately demonstrate the local field enhancement as Raman is not quenched and the redirection of optical emission can at most enhance the collected spectroscopic signal by a factor of 2. Figure 3.5 shows the Raman spectrum of a CNT. After accounting for the 1-dimensional nature of the CNT and the 0-dimensional localization of optical radiation from the antenna, an electric field enhancement of 7.5 has been demonstrated.

3.2.2.3 Sharp Metallic Optical Antennas

As mentioned previously, the sharp metallic probe has proven to be a very effective optical antenna for many applications. With a relatively simple fabrication process of electrochemically etching a thin metallic wire, high resolution optical antennas with large local electric field enhancements have been realized by many

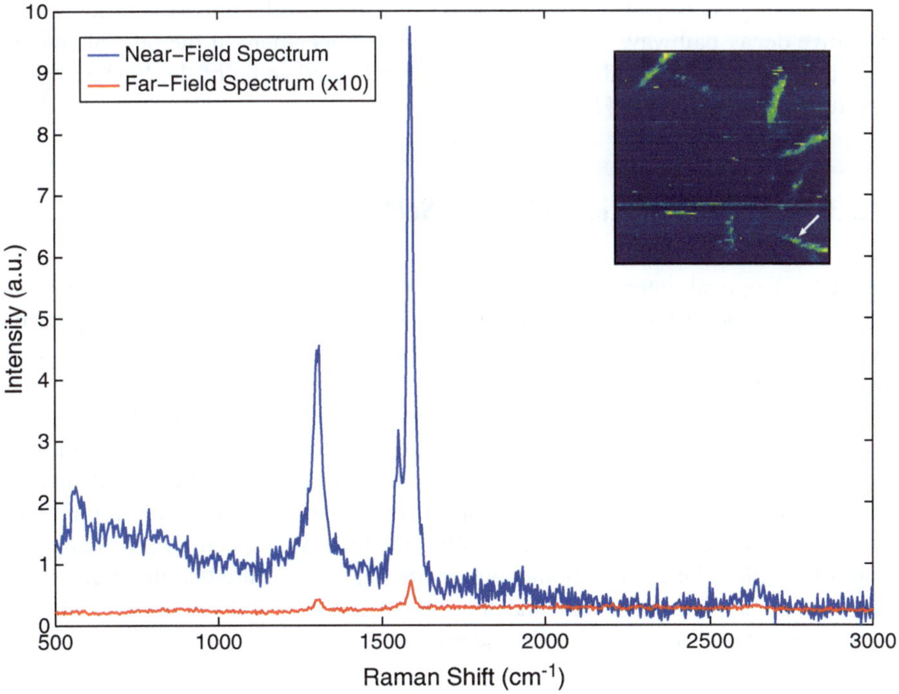

Fig. 3.5 A TERS spectrum of a single CNT acquired with a gold pyramid antenna and the corresponding far-field spectrum. The gold pyramid antenna provides a local electric field enhancement of 7.5. The *inset* shows a $1.5 \times 1.5\,\mu m^2$ area TERS G-band image of the CNTs with the spectrum acquired at the *white arrow*. The far-field spectrum is enhanced 10-fold for visibility

research groups. Little experimental control exists, however, over the exact antenna geometry and crystal structure. This has limited the reproducibility of antennas and additionally makes the exact simulation of antenna behavior impossible.

Despite these challenges, we performed the first simulations of the enhanced electric field at the apex of an ideal sharp metallic probe in 1997 [31]. The results revealed that there can be an electric field enhancement at the tip of a sharp metallic probe if the polarization of the optical illumination is parallel to the axis of the probe. The simulations showed an electric field enhancement of 3,000 for a gold tip surrounded by water. Additionally, in the following year, we also published simulations using a tightly focused HG_{10} laser mode to create a longitudinal polarization at the focus in conjunction with a sharp metallic probe [32].

While these antennas provide a significantly enhanced local electric field at the tip apex, they are generally not well suited for fluorescence measurements. This is due to the extended antenna geometry and small cone angle that allows for a fluorescent molecule in its excited state to efficiently non-radiatively transfer energy into a propagating surface plasmon along the antenna shaft, leading to quenching. For applications in Raman spectroscopy and imaging, however, the inherently low QE and scattering cross-section means that despite the introduction of additional non-radiative decay pathways, the large local field enhancement of the optical antenna results in net emission enhancement and has achieved some of the highest spatial resolution optical images [13].

3.2.3 Concluding Remarks on NSOM

The use of optical antennas for applications in microscopy and spectroscopy has made nanometer spatial resolution imaging possible on many optically active systems. The NSOM community has been rapidly expanding with the commercial development of practical NSOM systems, enabling the study of an increasing variety of samples by experts in their respective fields. Despite this progress, however, it is important to note that challenges still remain. The task of developing the most ideal antenna may end with the template-stripped pyramid antenna but other issues remain. Specifically, the spectroscopic community is concerned with precisely how each individual antenna alters the measured spectroscopic signal of a control sample. The current momentum in the field will insure that these questions, and others, are answered and that the optical antenna will continue to play an increasingly prominent role in microscopy and spectroscopy.

3.3 Towards Electrically Excited Optical Antennas

So far we have considered optical antennas as devices that transduce between propagating and localized fields. On the one hand, a receiving optical antenna takes far-field radiation from a laser and concentrates it into a subwavelength volume. On the other, a transmitting antenna is fed by the near-field of radiating molecules and nanotubes. This 'light-in light-out' modality is different from a typical radio/microwave antenna which is fed by an oscillating electric current at the feedgap. In this section we move away from the theme of antenna-emitter interactions and instead explore the possibility of exciting optical antennas electrically. As will be seen below, it is not necessary to impress an alternating current at optical frequencies to achieve this, rather the discreteness of a direct current across a tunnel junction suffices to create the required optical field oscillations.

3.3.1 Excitation of Surface Plasmons in an STM

Since metallic optical antennas can be seen as leaky resonators for localized surface plasmons (LSPs), a more fundamental question to ask is whether surface plasmons can be excited using electrons. It has been known since the 1960s that high energy electrons impinging on metals can excite both bulk and surface plasmons [33, 34].

With the resurgence of interest in surface plasmons in recent years, there has been a revival of interest in the use of energetic (~30 keV) electron beams to excite propagating [35–37] as well as LSP modes in metal films and nanostructures [38,39]. In a remarkable breakthrough in 1976, Lambe and McCarthy found a fundamentally new way to generate light using inelastic electron tunneling in experiments with metal-insulator-metal (MIM) junctions [40]. The role of surface plasmons as intermediaries in the inelastic process was suspected, if not proven, by these authors. The same phenomenon was rediscovered in a different context with the observation of light emission from the tunnel junction of an STM in the late 1980s [41,42]. It was hypothesized that STM-LE originates from the radiative decay of highly localized gap plasmons from the tip-sample cavity. The gap plasmons, excited by inelastically tunneling electrons, were proposed to behave as a dipole source oriented along the electron tunneling axis [43, 44]. It was further suggested that gap plasmons could also couple to propagating surface plasmon polaritons (SPPs) which could then scatter into photons [45, 46]. At this point, neither the existence of the dipolar gap plasmon nor the coupling to SPPs had been conclusively proven.

We have studied STM-LE in ambient conditions with high-NA light collection (Fig. 3.1) to help answer some of these questions [47]. Initial experiments investigated the electrical excitation of propagating plasmons on a 20 nm thick thermally evaporated gold film on a glass substrate ($n = 1.52$). In a typical experiment, a chemically etched gold tip (diameter $\approx$ 50 nm) was used to tunnel electrons into the film (typical substrate bias $V_t = 2$ V; tunneling current $I_t = 1$ nA; tip-sample distance ~9 Å), and any photons generated were detected with a single photon detector. Consistent with previous studies, it was found that the tunneling electrons did indeed lead to the emission of photons with an emission spectrum peaked around $\lambda = 700$ nm. The origin of the emitted photons remained unclear and could be attributed to either: (1) the direct result of the radiative decay of gap plasmons, or (2) the result of the leakage and subsequent decay of propagating gold-air SPPs into the high-index glass substrate.

In order to identify the origin of the detected photons, we measured their angular radiation pattern by imaging the back focal (Fourier) plane of the objective on to a charge coupled device (CCD) camera. The signature of SPPs is the outcoupling of photons at an angle that conserves the in-plane momentum between SPPs and photons. A free space wavelength of $\lambda = 700$ nm, corresponding to an SPP wavevector of $k_{spp} = 1.06\,k_0$, where $k_0 = 2\pi/\lambda$ is the free space wavevector, gives the Kretchmann angle of $\theta_K = arcsin[1.06/1.52] = 43.8°$. As seen in Fig. 3.6, more than 80 % of the photons are indeed emitted around this so-called Kretchmann angle, and hence arise from the decay of SPPs. When the same measurement was repeated on a ~5 nm thick gold film, which is too lossy to support an SPP, the ring at the Kretchmann angle disappeared, validating our previous conclusions.

While the above experiments on extended films offer indirect evidence of SPP excitation, it would be valuable to observe SPP scattering more directly. Towards this goal, we repeated similar measurements on micron-sized, colloidally synthesized, monocrystalline gold flakes dispersed on an indium tin oxide (ITO) coated glass substrate. The flakes were triangular in shape, ~50 nm thick, with well-

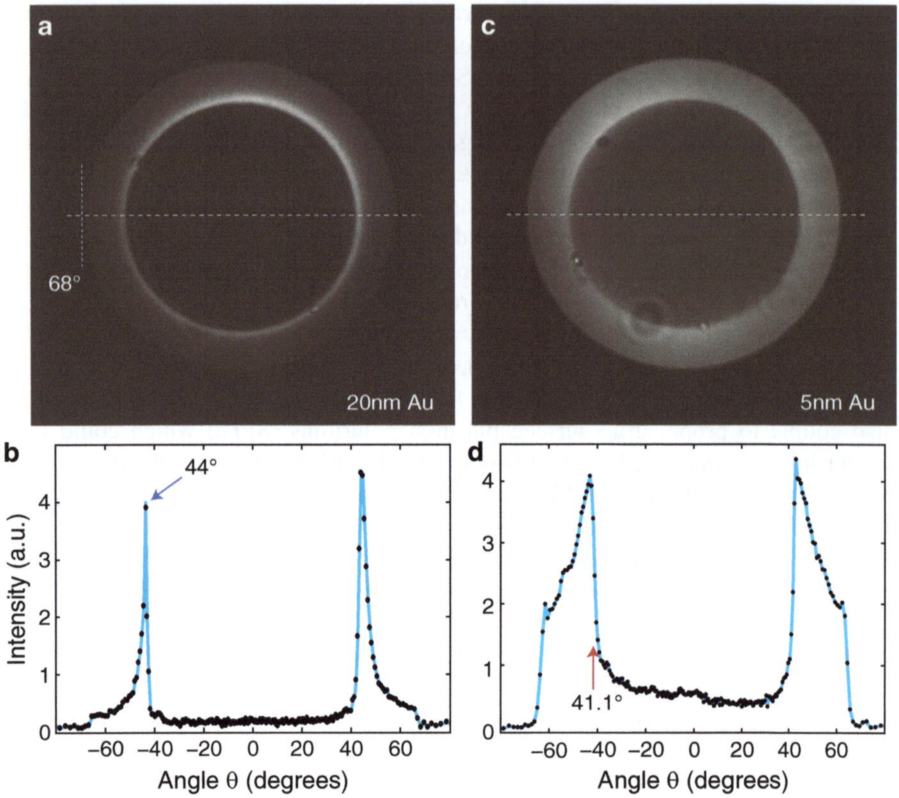

Fig. 3.6 Fourier plane (FP) imaging of STM-LE light from gold films. (**a**) An FP image for a 20 nm gold film showing a predominant ring centered at $\theta \approx 44^\circ$. (**b**) A cross-section taken along the horizontal *dashed line* in (**a**) confirms that ∼80 % of the energy is in the SPP. (**c**) An FP image for a 5 nm gold film featuring a discontinuity at the critical angle $\theta_c = 41^\circ$. (**d**) An intensity cross-section along the *dashed line* in (**c**) shows a random-dipole like emission pattern with a significant fraction of the radiation emitted below the critical angle (denoted by the *arrow*). The maximum angle of collected photons is $\sim 68^\circ$, given by the NA of the objective (1.4)

defined crystal facets on the edges. Figure 3.7 shows a real-space optical image of a flake with the gold STM tip positioned at its center with a steady tunneling current. The image shows both a bright central spot along with light scattering from the edges of the flake. Hence, inelastically tunneling electrons can excite gap plasmons which couple to SPPs propagating along the gold flake. The SPPs then scatter at the edges of the flake and give rise to photon emission.

To demonstrate this two-step excitation pathway unambiguously, we next chose to confine the SPP to 1-dimension using a gold nanowire. Monocrystalline nanowires on a similar ITO coated glass substrate were cut to the desired length using a focused gallium ion beam (FIB). The STM tip is placed on one end of the nanowire to generate gap plasmons (and photons). Part of this energy couples to SPPs that then propagate to the other end of the wire where they scatter into photons

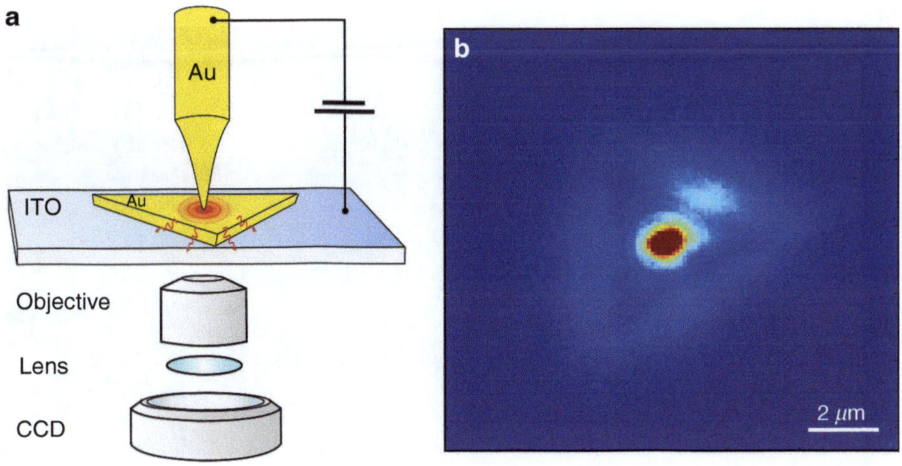

Fig. 3.7 STM-LE from a monocrystalline gold triangle. (**a**) A schematic of the experiment. The tip is positioned near the middle of the gold flake and a real-space image was acquired by refocusing the emitted photons. (**b**) A real-space image where the contrast arises from the intensity of STM-LE photons. In addition to direct light from the where the tip is placed, there is also scattered radiation coming from the three edges of the flake

(Fig. 3.8). Thus, the nanowire functions as a transmission line mediating between the electrical feed point and the outcoupling region, which in the present case is the hard discontinuity at the far end of the wire.

We have demonstrated that low-energy electrons can couple to photons through a cascade of events that begins with the excitation of a gap plasmon. In the next section, we examine more closely the role of the gap plasmon in STM-LE.

3.3.2 Role of the Gap Plasmon in STM-LE

In order to confirm the role of the gap plasmon in STM-LE, it is necessary to uncover the origin of the line-to-line photoemission intensity variations seen in previous 2-dimensional STM-LE studies [48–52]. These line-to-line variations are typically associated with changes in the probe as it interacts with the sample; however, there have also been reports of electroluminescent (EL) metal clusters being created and/or destroyed within tunnel junctions in a planar geometry [53–55]. We have recently used a series of engineered samples to isolate gap plasmon photoemission from other suspected emissive processes [56].

During a physical tip-sample interaction, in which the tip is physically altered, it is expected that the LSP mode structure of the junction will change as it depends sensitively on the tip geometry [57, 58]. There appear to be physical tip-sample interactions in the STM photoemission and topographic maps on samples composed

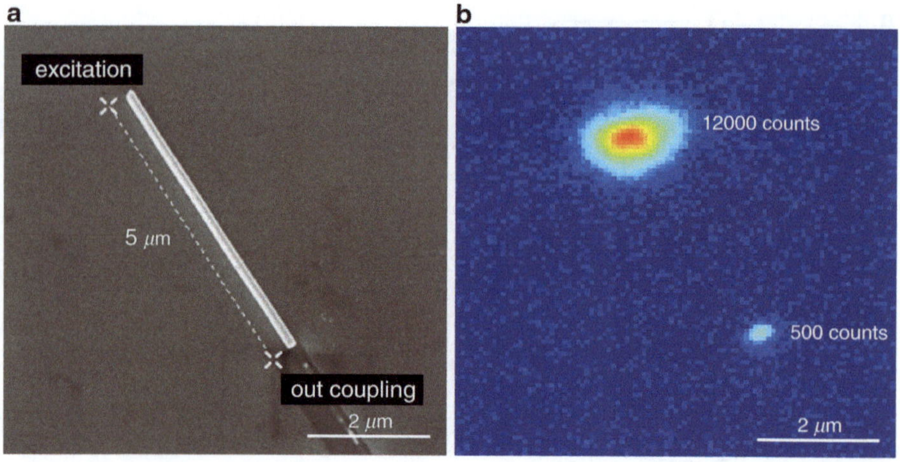

Fig. 3.8 Electrical excitation of SPPs propagating along a monocrystalline gold nanowire. (**a**) An SEM micrograph of a gold nanowire of radius 87 nm and length ∼5 μm. (**b**) A real-space photon emission map obtained by placing the STM tip at one end of the nanowire. SPPs are excited by the STM tip and propagate to the other end of the wire where they are scattered and converted to propagating photons. The wire is immersed in index matching oil to prevent the SPPs from decaying into photons by leakage radiation

of gold particles on a gold film (Fig. 3.9) that could generate or destroy an EL cluster. It is expected that the proximity of a cluster to metal in the tip and sample would lead to significant quenching of electroluminescence, resulting in an emission rate similar to that expected from LSPs alone. Therefore, EL clusters cannot be immediately discounted as a cause of the observed line-to-line emission variations.

One challenge in interpreting the results shown in Fig. 3.9 is that the topographic map obtained with any SPM technique is a convolution between the probe tip geometry and the true sample topography; this convolution can have significant effects on the topographic map [59, 60]. For STM-LE studies, it is important to note that the LSP closely links the photoemission map with the topographic map. In general, larger sample excursions lead to increased sampling of the tip area during the convolution process. Since the mode structure of the LSPs depend on the tip shape, and since different areas of the tip are being sampled by a single topographic feature, the conclusions drawn from STM-LE on high-topography samples can be suspect.

The effects of the tip convolution on the data can be minimized by constructing ultra-flat samples. If the sample has no topographic excursions present to sample the probe then only a specific, and ideally unchanging, part of the probe will take part in the tunneling process. The technique of template-stripping has recently been extended beyond metallic samples to create ultra-smooth samples composed of metals, semiconductors, and/or oxides [61]. We have created specially structured, ultra-flat samples to study STM-LE in a controlled manner [56]. These samples consisted of triangular gold patches, with a side length of 50 nm, surrounded by a 3 nm platinum film.

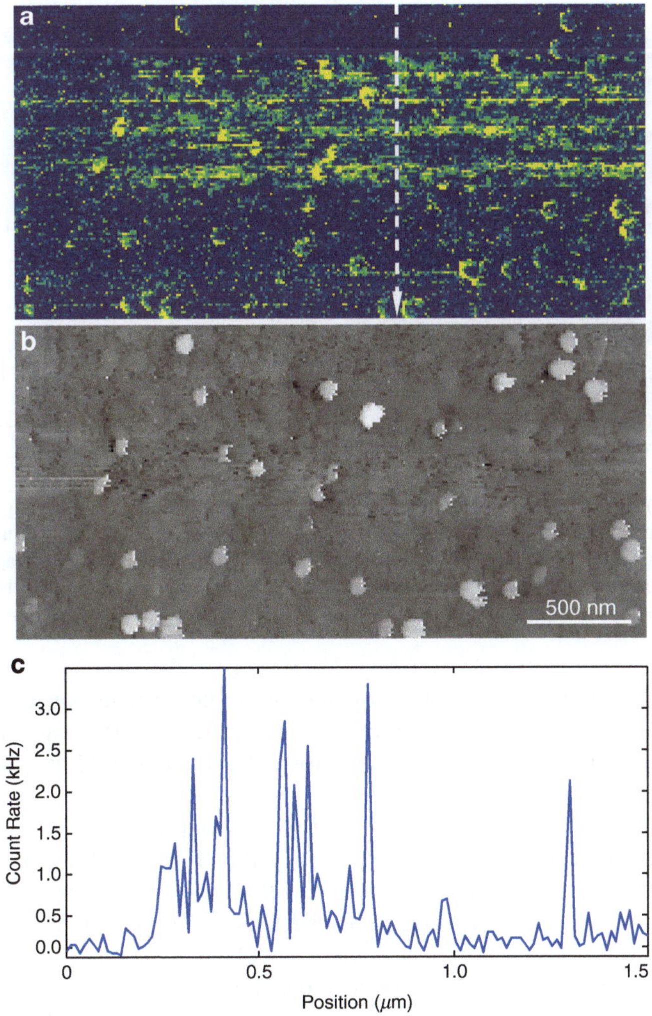

Fig. 3.9 (**a**) An STM-LE map of a sample composed of 30 nm diameter gold particles deposited on a 20 nm thick gold film and (**b**) the corresponding topographic map. (**c**) A cross-section along the *dashed line* in (**a**). Significant line-to-line photoemission variations are seen along with enhanced emission at the particle edges. (STM parameters: $V_s = 2.0\,V$, $I = 1.0\,nA$)

Two specific properties of the metals, as well as their close proximity, allow for the differentiation between emission from an EL cluster and from an LSP. First, gold and platinum have similar work functions and it is therefore expected that the tunneling barrier associated with an STM would have a similar height and width on both materials. A constant tunneling barrier is expected to lead to similar EL output on both materials. Second, the imaginary part of the dielectric function

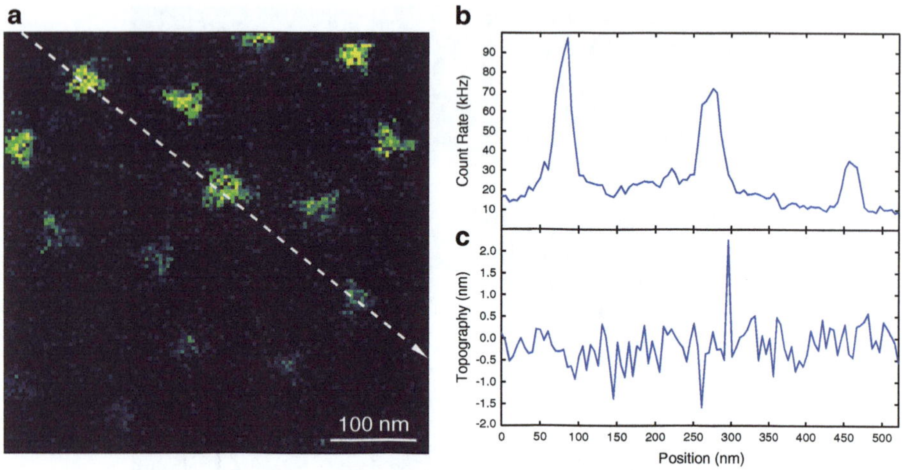

Fig. 3.10 (**a**) An STM-LE map of a low-roughness sample composed of gold triangles embedded in a platinum matrix. (**b**) A smoothed cross-section along the *dashed line* in (**a**) and (**c**) the corresponding raw topographic cross-section. The RMS roughness of the topography shown in (**c**) is ~450 pm. Photoemission is observed to vary line-to-line; however, the ratio of emission rates between gold and platinum remains constant (STM parameters: $V_s = 2.1$ V, I = 0.7 nA)

of platinum is more than an order of magnitude larger than that of gold in the typical STM-LE band near $\lambda = 700$ nm. It is expected that the plasmonic activity on platinum is greatly quenched in comparison to gold due to higher non-radiative plasmonic energy losses in platinum.

A photoemission map generated on such a sample is shown in Fig. 3.10a. Topography confirms that the sample is very flat (Fig. 3.10c) and it is therefore expected to be free of artifacts associated with a tip-sample convolution. In fact, extremely fine (less than 10 nm) features, such as the corners of the gold triangles, are resolvable and this resolution is independent of the tip radius. The photoemission rate varies from line-to-line and greatly depends on sample material. The gold triangles clearly give a larger emission rate than the platinum film. While the emission rate varies between lines, the ratio of emission rates between the gold and platinum remains constant as shown in Fig. 3.10b. This type of sample was studied by systematically varying the sample bias voltage, bias polarity, and set-point current as well as performing the study with multiple tips and samples. The emission behavior remained qualitatively the same in each case. These experiments indicate that EL clusters are rare or non-existent in STM tunnel junctions.

We additionally fabricated a complimentary sample to confirm that EL clusters were not present in STM-LE experiments. In this case, the sample consisted of gold patterns deposited onto ITO. Since LSPs cannot exist on ITO with energies in the 2 eV range, the emission of light on ITO from an STM junction would indicate the presence of an EL cluster. A representative example of a photoemission map on this type of sample is shown in Fig. 3.11a. No light appears to be emitted on the

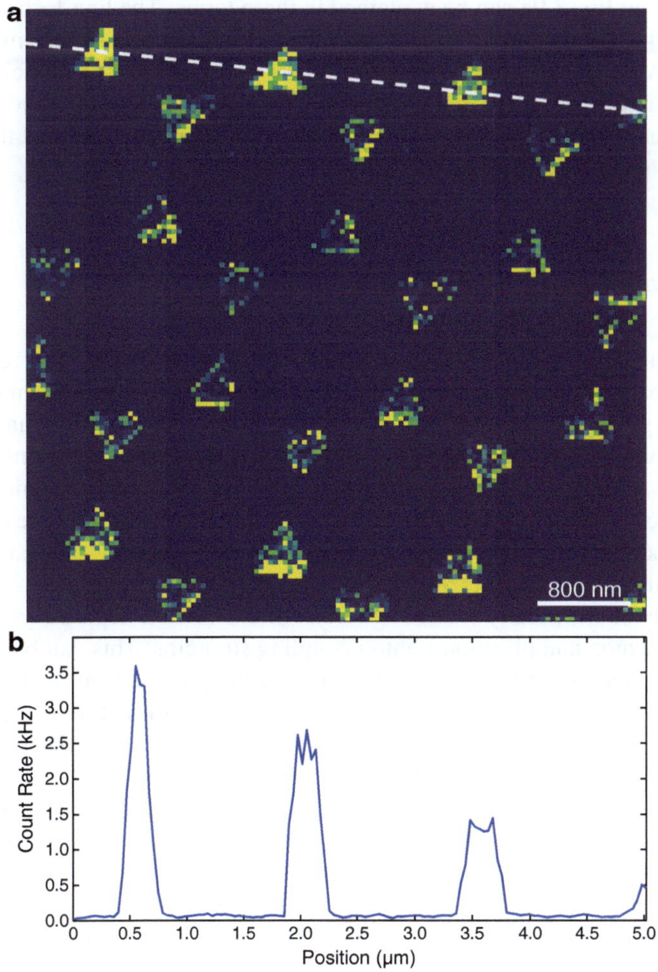

Fig. 3.11 (**a**) An STM-LE map of a sample composed of 20 nm thick gold triangular patterns deposited on a 150 nm thick ITO film on a glass substrate. (**b**) A cross-section along the *dashed line* in (**a**). Photoemission is only observed from the gold surfaces (STM parameters: $V_s = 2.25$ V, $I = 0.2$ nA)

ITO film at any time. The light emission rate varies greatly between lines but the emission regions are strictly confined to the gold triangles as shown in Fig. 3.11b. Again, the emission behavior of this type of sample remained qualitatively the same as experimental parameters were systematically varied.

We can draw two conclusions from these results. First, EL clusters do not appear to be a significant source of light from an STM. Second, the LSP mode structure is greatly affected not only by physical changes in the tip due to interactions with the sample but also by the sample material and topography. The variation in emission rates shown in Fig. 3.9a can be explained in these terms. The line-to-line variations shown are due to tip-sample interactions which lead to a spatially varying LSP mode structure. Also, as the probe approaches close to a particle, the tunnel junction is reoriented and exists between the side of the particle and the side of the probe. This reorientation, from vertical to horizontal, of the LSP modes explains the apparent rings of photoemission around the particles.

3.3.3 Concluding Remarks on STM-LE

Electron tunneling provides a non-optical, voltage-controlled, low energy, and nanoscale pathway for launching SPPs in nanostructures. It is important to point out the advantages of this technique over related schemes. Previous work has been done under vacuum with high energy elections which is not practical for on-chip implementation. Additionally, exciton-mediated approaches using inorganic or organic semiconductors require multi-step fabrication which increases device complexity. The method introduced here only requires metal electrodes that can be easily integrated in a planar geometry.

The study of electrically excited SPPs can be extended to separately optimize the electron-plasmon and plasmon-photon coupling strengths. This can be expected to result in an increase in the overall efficiency of the process from $\sim 10^{-5}$ presently to a level where it can be of technological relevance. In addition to propagating SPPs, low energy electrons can also be used to excite localized plasmon modes in nanostructures, which can then be coupled optical antennas. It is conceivable that the electrical excitation scheme can be applied to highly directional antennas, such as Yagi-Uda antennas, enabling a direct conversion of low energy electrons to a directed beam of far-field photons.

Acknowledgements We are grateful for financial support by the Swiss National Science Foundation (grant 200021_149433).

References

1. Abbe E (1873) Beiträge zur Theorie des Mikroskops und der mikroskopischen Wahrnehmung. Archiv f Mikroskop Anat 9:413
2. Rayleigh L (1896) On the theory of optical images with special reference to the optical microscope. Phil Mag 5(42):167–195
3. Bharadwaj P, Deutsch B, Novotny L (2009) Optical antennas. Adv Opt Phot 1:438–483
4. Xie XS, Trautman JK (1998) Optical studies of single molecules at room temperature. Annu Rev Phys Chem 49:441–480
5. Specht M, Pedarning JD, Heckl WM, Hänsch TW (1992) Scanning plasmon near-field microscope. Phys Rev Lett 68:476–479
6. Xia Y (2005) Shape-controlled synthesis and surface plasmonic properties of metallic nanostructures. MRS Bull 30:338

7. Höppener C, Novotny L (2008) Antenna-based optical imaging of single Ca^{2+} transmembrane proteins in liquids. Nano Lett 8:642–646
8. Höppener C, Beams R, Novotny L (2009) Background suppression in near-field optical imaging. Nano Lett 9:903–908
9. Lapin ZJ, Christiane H, Gelbard HA, Lukas N (2012) Near-field quantification of complement receptor 1 (crl/cd35) protein clustering in human erythrocytes. J Neuroimmune Pharmacol 7(3):539–543
10. Anger P, Bharadwaj P, Novotny L (2006) Enhancement and quenching of single molecule fluorescence. Phys Rev Lett 96:113002
11. Bharadwaj P, Beams R, Novotny L (2011) Nanoscale spectroscopy with optical antennas. Chem Sci 2:136–140
12. Ichimura T, Hayazawa N, Hashimoto M, Inouye Y, Kawata S (2004) Tip-enhanced coherent anti-stokes Raman scattering for vibrational nanoimaging. Phys Rev Lett 92:2004
13. Cançado LG, Hartschuh A, Novotny L (2009) Tip enhanced Raman scattering of carbon nanotubes. J Raman Spectrosc 40:1420–1426
14. Johnson TW, Lapin ZJ, Ryan B, Lindquist NC, Rodrigo SG, Lukas N, Sang-Hyun O (2012) Highly reproducible near-field optical imaging with sub-20-nm resolution based on template-stripped gold pyramids. ACS Nano 6(10):9168–9174
15. Palomba S, Danckwerts M, Novotny L (2009) Nonlinear plasmonics with gold nanoparticle antennas. J Opt A Pure Appl Opt 11:114030
16. Deutsch B, Hillenbrand R, Novotny L (2010) Visualizing the optical interaction tensor of a gold nanoparticle pair. Nano Lett 10:652–656
17. Deutsch B, Hillenbrand R, Novotny L (2008) Near-field amplitude and phase recovery using phase-shifting interferometry. Opt Express 16:494–501
18. Schnell M, Garcia-Etxarri A, Huber AJ, Crozier KB, Borisov A, Aizpurua J, Hillenbrand R (2010) Amplitude- and phase-resolved near-field mapping of infrared antenna modes by transmission-mode scattering-type near-field microscopy. J Phys Chem C 114(16):7341–7345
19. Stiegler JM, Abate Y, Cvitkovic A, Romanyuk YE, Huber AJ, Leone SR, Hillenbrand R (2011) Nanoscale infrared absorption spectroscopy of individual nanoparticles enabled by scattering-type near-field microscopy. ACS Nano 5(8):6494–6499
20. Novotny L (2007) The history of near-field optics. Prog Opt 50:137–180
21. McMullan D (1990) The prehistory of scanned image microscopy. Part I: scanned optical microscopes. Proc Roy Microsc Soc 25(2):127–131
22. Pohl DW (2004) Optics at the nanometre scale. Phil Trans R Soc Lond A 362:701–717
23. Binnig G, Rohrer H, Gerber C, Weibel E (1982) Tunneling through a controllable vacuum gap. Appl Phys Lett 40:178–180
24. Binnig G, Rohrer H (1982) Scanning tunneling microscopy. Helv Phys Acta 55:726
25. Binnig G, Quate CF, Gerber C (1986) Atomic force microscope. Phys Rev Lett 56:930
26. Pohl DW, Denk W, Lanz M (1984) Optical stethoscopy: image recording with resolution l=20. Appl Phys Lett 44:651–653
27. Fischer UC, Pohl DW (1989) Observation on single-particle plasmons by near-field optical microscopy. Phys Rev Lett 62:458–461
28. Novotny L (2007) Effective wavelength scaling for optical antennas. Phys Rev Lett 98:266802
29. Kalkbrenner T, Ramstein M, Mlynek J, Sandoghdar V (2001) A single gold particle as a probe for apertureless scanning near-field optical microscopy. J Microsc 202:72–76
30. Christiane H, Zachary L, Palash B, Lukas N (2012) Self-similar gold-nanoparticle antennas for a cascaded enhancement of the optical field. Phys Rev Lett 109(1):017402
31. Novotny L, Bian RX, Xie XS (1997) Theory of nanometric optical tweezers. Phys Rev Lett 79:645–648
32. Novotny L, Sanchez EJ, Xie XS (1998) Near-field optical imaging using metal tips illuminated by higher-order Hermite-Gaussian beams. Ultramicroscopy 71:21–29
33. Ritchie RH (1957) Plasma losses by fast electrons in thin films. Phys Rev 106:874–881
34. Räther H (1965) Solid state excitations by electrons. In: Springer tracts in modern physics, vol 38. Springer, Berlin, pp 84–157

35. Bashevoy MV, Jonsson F, Krasavin AV, Zheludev NI, Chen Y, Stockman MI (2006) Generation of traveling surface plasmon waves by free-electron impact. Nano Lett 6:1113–1115
36. Vesseur EJR, de Waele R, Kuttge M, Polman A (2007) Direct observation of plasmonic modes in au nanowires using high-resolution cathodoluminescence spectroscopy. Nano Lett 7: 2843–2846
37. Cai W, Sainidou R, Xu J, Polman A, Garcia de Abajo FJ (2009) Efficient generation of propagating plasmons by electron beams. Nano Lett 9:1176–1181
38. Jaysen N, Mathieu K, Stéphan O, García de Abajo FJ, Tencé M, Henrard L, Taverna D, Pastoriza-Santos I, Liz-Marzán LM, Colliex C (2007) Mapping surface plasmons on a single metallic nanoparticle. Nat Phys 3:348–353
39. Kuttge M, Vesseur EJR, Polman A (2009) Fabry-Pérot resonators for surface plasmon polaritons probed by cathodoluminescence. Appl Phys Lett 94:183104
40. Lambe J, McCarthy SL (1976) Light emission from inelastic electron tunneling. Phys Rev Lett 37:923
41. Gimzewski JK, Reihl B, Coombs JH, Schlittler RR (1988) Photon emission with the scanning tunneling microscope. Z Phys B 72:497–501
42. Coombs J, Gimzewski J, Reihl B, Sass J (1988) Photon emission experiments with the scanning tunnelling microscope. J Microsc 152:325–336
43. Johansson P, Monreal R, Peter A (1990) Theory for light emission from a scanning tunneling microscope. Phys Rev B 42:9210–9213
44. Berndt R, Gimzewski JK, Johansson P (1991) Inelastic tunneling excitation of tip-induced plasmon modes on noble-metal surfaces. Phys Rev Lett 67:3796–3799
45. Uehara Y, Kimura Y, Ushioda S, Takeuchi K (1992) Theory of visible light emission from scanning tunneling microscope. Jpn J Appl Phys 31:2465–2469
46. Takeuchi K, Uehara Y, Ushioda S, Morita S (1991) Prism? Coupled light emission from a scanning tunneling microscope. J Vac Sci Technol B 9:557–560
47. Bharadwaj P, Novotny L (2011) Robustness of quantum dot power-law blinking. Nano Lett 11:2137–2141
48. Walmsley DG, Tan T-S, Dawson P (2004) Light emission from gold and silver thin films in a scanning tunneling microscope: role of contamination and interpretation of grain structure in photon maps. Surf Sci 572(23):497–520
49. Silly F, Gusev AO, Taleb A, Pileni M-P, Charra F (2002) Single nanoparticle manipulation with simultaneously recorded stm-induced light emission. Mater Sci Eng C 19(12):193–195
50. Perronet K, Barbier L, Charra F (2004) Influence of the Au(111) reconstruction on the light emission induced by a scanning tunneling microscope. Phys Rev B 70:201405
51. Uemura T, Akai-Kasaya M, Saito A, Aono M, Kuwahara Y (2008) Spatially resolved detection of plasmon-enhanced fluorescence using scanning tunneling microscopy. Surf Interface Anal 40(6-7):1050–1053
52. Fujita D, Onishi K, Niori N (2004) Light emission induced by tunneling electrons from surface nanostructures observed by novel conductive and transparent probes. Microsc Res Tech 64(5–6):403–414
53. Gonzalez JI, Lee T-H, Barnes MD, Antoku Y, Dickson RM (2004) Quantum mechanical single-gold-nanocluster electroluminescent light source at room temperature. Phys Rev Lett 93:147402
54. Tae-Hee L, Gonzalez JI, Dickson RM (2002) Strongly enhanced field-dependent single-molecule electroluminescence. Proc Natl Acad Sci U S A 99(16):10272–10275
55. Tae-Hee L, Dickson RM (2003) Single-molecule leds from nanoscale electroluminescent junctions. J Phys Chem B 107(30):7387–7390
56. Divitt S, Bharadwaj P, Novotny L (2013) The role of gap plasmons in light emission from tunnel junctions. Opt Express 21(22):27452–27459
57. Mitra J, Lei F, Boyle MG, Dawson P (2009) Electromagnetic interaction between a metallic nanoparticle and surface in tunnelling proximitymodelling and experiment. J Phys D Appl Phys 42(21):215101

58. Boyle MG, Mitra J, Dawson P (2009) Infrared emission from tunneling electrons: the end of the rainbow in scanning tunneling microscopy. Appl Phys Lett 94(23):233118
59. Dawson P, Boyle MG (2006) Light emission from scanning tunnelling microscope on polycrystalline Au films? What is happening at the single-grain level? J Opt A Pure Appl Opt 8:S219
60. Branscheid R, Jacobsen V, Kreiter M (2007) STM induced light from nontrivial metal structures: local variations in emission efficiency. Surf Sci 602:176
61. Park JH, Nagpal P, McPeak KM, Lindquist NC, Oh S-H, Norris DJ (2013) Fabrication of smooth patterned structures of refractory metals, semiconductors, and oxides via template stripping. ACS Appl Mater Interfaces 5(19):9701–9708

4
Biosensing Instrumentation

Jean-Pierre Wolf

Abstract Functionalized nanoparticles revolutionized the sensing and imaging of biological samples. A particularly spectacular example was recently achieved with harmonic nanoparticles (HNPs), which allow coherent emission at harmonics of the fundamental wavelength over the whole VIS-NIR spectral range. Deep imaging, nanoscale in situ spatio-temporal focus characterization, on-line monitoring of stem cell differentiation and cancer theranostics are demonstrated. A further ground-breaking coherent approach is based on coherent control of endogenous biomolecules, which provides chemical selectivity at the protein level even "label-free". State of the art and perspectives related to these new biosensing developments are presented

Keywords Bio-imaging • Nano-optics • Cancer • Bacteria detection • Protein identification • Coherent control • Label-free biosensing

4.1 Introduction

Sensing and identifying the composition of biological samples is today mainly based on chemical selectivity, i.e. labeling by functionalized fluorophors. The next generation of bio-markers has to prove to be non toxic, bright, stable against long term excitation, and able to generate a sharp contrast against background fluorescence. In all these respects, Harmonic Nanoparticles (HNPs, "nanodoublers") are receiving an increasing attention as they also open a series of alternative detection possibilities simply not accessible with the present generation of fluorescent dyes and quantum dots. In the first part of the chapter, we report on this novel labelling

J.-P. Wolf (✉)
GAP-Biophotonics, University of Geneva, 22 chemin de Pinchat, 1211 Geneva 4, Switzerland
e-mail: jean-pierre.wolf@unige.ch

B. Di Bartolo et al. (eds.), *Nano-Structures for Optics and Photonics*,
NATO Science for Peace and Security Series B: Physics and Biophysics,
DOI 10.1007/978-94-017-9133-5_4,

method with unprecedented wavelength flexibility, enabled by the non-resonant nature of the second harmonic process. The possibility of employing infrared excitation and the consequent deeper tissue penetration is especially promising for their in vivo applications [1]. The phase-coherent optical response of HNPs can also be exploited to fully characterize the excitation laser pulse in the focal spot of a high-NA objective with nanometric resolution. This proof-of-principle "nano-FROG" experiment [2] sets the ground for further phase-sensitive self-referenced applications, after the recent demonstration of harmonic holography and heterodyne detection with external references.

Although significant progress has been achieved in the quest for better bio-markers, such as the above mentioned HNPs, the ultimate goal would be the use of no external probe for achieving "label-free" imaging. In the last decade coherent control with optimally tailored ultrashort laser pulses revolutionized molecular spectroscopy. It allowed not only observing with unprecedented detail the atomic motion within molecules but also controlling it. In the second part of this chapter, we present how coherent control might significantly contribute towards the goals of label-free imaging using only endogenous molecular processes and quantum control based bioassays.

After an introduction of the basic processes and techniques used in coherent control, we concentrate on recent achievements in the field of coherent manipulation of biological objects. Particular interest will be dedicated to the discrimination of biological molecules that exhibit almost identical linear spectra and the wealth of related applications, such as fluorescent label free microscopy, and specific detection of pathogens in air and water. Recent experiments indeed demonstrated that discrimination between amino acids and similar organic moiety lacking the peptidic backbone was possible using sequences of multicolour pulses. Similar approaches allowed discriminating bacteria from non-biological organic particles such as soot [3, 4]. But the most striking feature of coherent control schemes based on optimally tailored UV pulses was recently the demonstration of its ability to discriminate nearly identical cellular vitamins (riboflavin (RBF) vs. its phosphated counterpart flavin mononucleotide (FMN) [5–7]), which exhibit identical (linear) spectroscopic signatures. This latter demonstration opened the way to peptides and protein discrimination [8, 9] and therefore quantum controlled based assays.

4.2 Harmonic Nanodoublers as Inherent Nonlinear Probes for Multi-photon Imaging

HNPs, based on inorganic noncentrosymmetric crystals, possess a series of attractive properties as compared to fluorescent dyes, including: (i) complete absence of bleaching and blinking, (ii) narrow emission bands, (iii) excitation-wavelength tunability, (iv) orientation retrieval capability, and (v) coherent optical response.

The noncentrosymmetric crystal structure is at the origin of the second-order nonlinear properties of this family of particles. So far, a series of nanomaterials

Table 4.1 Hyperpolarizability, normalized hyperpolarizability, and normalized hyperpolarizability from bulk materials found in the literature of $KNbO_3$, $LiNbO_3$, $BaTiO_3$, KTP, and ZnO, nanocrystals

Nanocrystal	$<\beta_{np}> \pm \delta\beta \times 10^{-24} esu$	$<d_{np}> \pm \delta d pm/V$	$<d_{lit}> pm/V$
$KNbO_3$	7.2 ± 2.4	3.4 ± 1.1	15.3
$LiNbO_3$	21.0 ± 7.0	4.8 ± 1.6	17.3
$BaTiO_3$	11.0 ± 3.5	4.6 ± 0.7	14.1
KTP	25.8 ± 6.0	1.4 ± 0.3	7.6
ZnO	9.6 ± 2.8	1.9 ± 0.6	2.8

including: $LiNbO_3$, $KNbO_3$, $BaTiO_3$, $Fe(IO_3)_3$, ZnO, KTP have been studied [10]. From their common lack of crystal inversion symmetry, it follows that the second order susceptibility tensor, $\chi^{(2)}$, presents not vanishing elements, sustaining a nonlinear polarization at frequency 2ω when excited by an electromagnetic field E^{ω} at frequency ω

$$P_i^{(2\omega)} = \varepsilon_0 \sum_{ij} \chi_{ijk}^{(2)} E_j^{\omega} E_k^{\omega} \tag{4.1}$$

Being function of individual nanoparticles orientation, these responses can be exploited for optical orientation retrieval and electric field probing at microscopic scale (see Refs. [11–13] for details).

Because harmonic generation is a nonresonant process, no absorption occurs. As a consequence, no energy is converted into heat, preventing sample photo-degradation. Moreover signal bleaching and blinking observed with fluorescent dyes and quantum dots are completely absent (Table 4.1).

Interestingly, given that the coherence length of the nonlinear excitation process is longer than the typical dimensions of the nanoparticles (<100 nm), no phase-matching constraints apply and any wavelength can be used for excitation [1, 14]. We have demonstrated such broad excitation tunability under different experimental conditions, with laser pulses ranging from vis-UV (400 nm) to IR (1.5 μm). As an example, Fig. 4.1d shows the responses of Polyethylene Glycol (PEG)-coated $LiNbO_3$ HNPs attached to breast cells membranes stained with FM 1-43FX membrane dye. The response of the nanoparticles can be excited at any wavelength within the Ti:Sapphire spectrum (700–1,100 nm). The optimal contrast obtained with respect to this standard fluorescent dye, widely used for biological assays, testifies the appeal of this approach for in vitro cell imaging. Longer wavelength excitation, due to the reduced tissue scattering, is indeed very promising for extending the penetration depth of imaging techniques and decreasing sample photo-damage. In Fig. 4.1c, as a demonstration, the second and third harmonic spectra generated by a single HNP excited by a 1.55 μm laser show the complete up-conversion of the original laser bandwidth (dotted line).

Finally, Fig. 4.1e, shows the most clearly, the coherent nature of the emission of HNPs as opposed to fluorescence. Two neighbouring KTP nanoparticles lying on

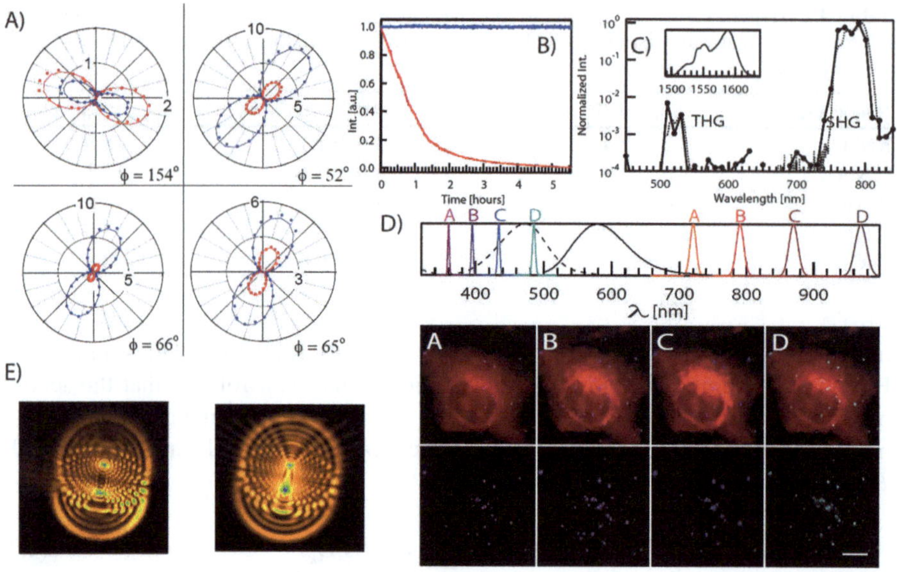

Fig. 4.1 (**a**) The polarized emissions of individual HNPs bear information about their local three-dimensional orientation, in the plots ϕ corresponds to the HNP in-plane angle [12]. (**b**) The excellent photostability upon continuous laser irradiation for several hours of HNPs is compared with the decreasing signal of a fluorescent dye imaged simultaneously. (**c**) The absence of phase-matching constraints allows complete doubling of any input pulse bandwidth. A single HNP up-converts the spectrum of a Erbium ion laser at 1,550 nm (*inset*) generating simultaneously SH and TH. The experimental spectral bandwidths are compared with the calculated ones (*dotted lines*). (**d**) A dye-stained cells treated with HNPs is imaged at different wavelengths within the range of a Ti:Sapphire laser. *Upper plots*: all detection channels overlaid, *lower plots*: single detection channels *A*, *B*, *C*, and *D*. The frequency-doubled signal from the HNPs appear each time uniquely in a narrow spectral detection region. (**e**) *Left image*. Coherent interference pattern between the emissions of two nanoparticles adjacently lying on a waveguide nonlinearly excited by an evanescent wave. *Right image*. Numerical simulations (In collaboration with J. Enderlein, University of Goettingen)

a waveguide are excited at the second harmonic (SH) by the evanescent mode of a laser pulse propagating through the guide. The two emission patterns interfere coherently at the CCD detector [13]. The accompanying image is a numerical simulation of this observation. So far, what we believe is the immense imaging potential of this coherent response has not been fully exploited, just a series of demonstration measurements have been published on especially conceived set-ups: heterodyne detection [15] of nanoparticles for sensitivity increase, harmonic holography [16], digital phase conjugation [17, 18], and our work on spatially resolved pulse phase-retrieval for microscopy (nano-FROG) [2].

The motivations of the nano-FROG procedure reside on the demand for a versatile and reliable pulse phase-sensitive characterization technique, more and more critical as a number of nonlinear approaches have recently been bridged from spectroscopy to imaging. Just to mention a few examples, pulse-shaping has been applied to temporally recompress femtosecond pulses at the measuring site, to improve spatial resolution, and to increase fluorescence excitation and harmonic generation yield [19]. Intra-pulse micro-CARS [20, 21] and coherent control microscopy have demonstrated successful for obtaining higher selectivity in the imaging of structures in very diverse samples [22]. Furthermore, interest has been focused on the effects of pulse polarization, and the related capability to achieve control of optical near field [23]. To fully exploit the potential of these techniques, equally developed pulse diagnostics capabilities are necessary. In particular, considering that femtosecond pulses are prone to undergo modifications during their propagation (temporal stretching by group velocity dispersion, higher order distortions in the spectral-phase induced by microscope objectives [24], spectral amplitudes modulation by sample scattering [25] and absorption) an in situ measurement of the excitation pulse is customary for any advanced application. Since the early work of Muller et al. [26], several techniques, largely based on Frequency Resolved Optical Gating (FROG), have been proposed, but until present none of them incorporates the spatial resolution necessary to resolve features inside the focal region of a high numerical aperture (NA) objective [27]. The starch-based FROG approach proposed by Amat et al. [28] and the scanning SEA TADPOLE technique by the group of Trebino [29] represented a clear progress in this direction, but unfortunately both techniques are limited to a spatial resolution of a few micrometers. We have proposed a technique to achieve unprecedented spatial resolution, without renouncing at the tenability of the FROG technique.

The approach, which can be implemented on a conventional or an inverted laser scanning microscope, is based on the use of individual HNPs as nonlinear medium. An exemplary FROG trace, like the one presented in Fig. 4.2a can be obtained by spectrally resolving the autocorrelation trace of a single HNP and then by filtering out the modulation components in the Fourier space. Successively, the inverse Fourier transform can be applied to return back to the time/wavelength space. The so treated data can successively be fed into a commercial FROG-inversion program [29] to determine the actual pulse characteristics. The retrieved FROG trace is shown in Fig. 4.2b (lower plots series), along with pulse electric field intensity temporal profile and spectrum reported in (c) and (d). The corresponding phase-functions are also shown. The latter indicate the presence of a residual quadratic chirp on the pulse due to group velocity dispersion accumulated during propagation through the microscope optics, which is not fully corrected by the grating compressor.

4.3 3D Movie of Embryonic Stem Cell Differentiation into Cardiomyocites

A very interesting and illustrative application of the unique properties of HNPs is the real time observation of the differentiation process in embryonic stem cells (ESM) [30]. More precisely, in the initial embroid bodies (EB) sample, cells underwent

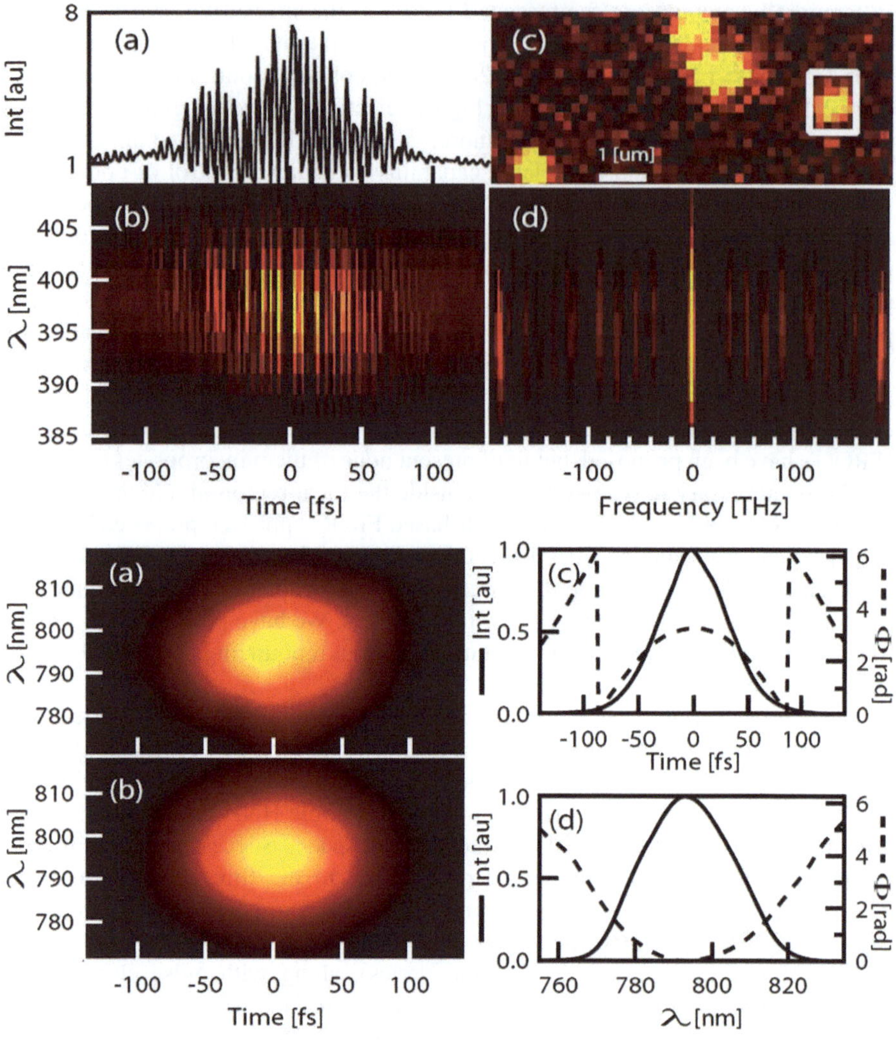

Fig. 4.2 *Upper plots*. (**a**) Interferometric autocorrelation trace. (**b**) FROG trace. (**c**) Image of the nanoparticles dispersed on a microscopic glass slide. The particle highlighted by the 1 um square was used as nonlinear medium for acquiring the autocorrelation and the FROG trace. (**d**) Result of the Fourier transform along the time axis of the FROG trace in (**b**) *Lower plots*. Experimental FROG trace. (**b**) Retrieved FROG trace. Electric field intensity (*solid line*) and phase (*dashed line*) as a function of time (**c**) and of wavelength (**d**)

differentiation into cardiac cells, easily recognizable as clusters displaying rhythmic autonomous contractions. A movie of the rhythmic cluster movements could be recorded over several minutes at high speed over an area of $640 \times 640\ \mu m^2$ detecting SHG radiation from isolated HNPs. It bears 3D information at sub-micrometric resolution about the contraction pattern of the cardiomyocytes. A single frame is

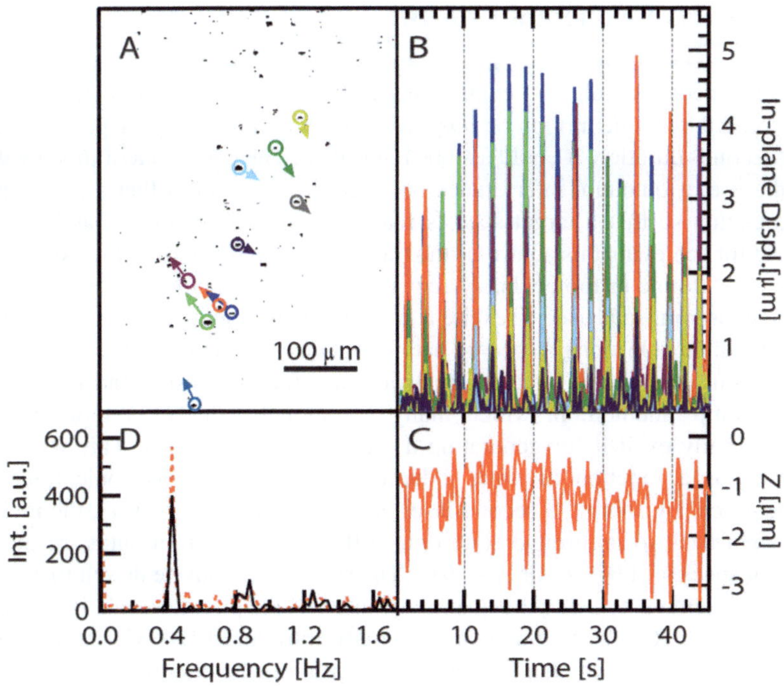

Fig. 4.3 (**a**) Individual frame of the movie of the myocytes contraction. *Arrows* indicate the directions and in-plane displacements of different HNPs. (**b**) Displacements as a function of time of the HNPs highlighted in (**a**). (**c**) Same as B for movement along the optical axis. (**d**) Fourier transform of traces in (**b**) and (**c**). The beating frequency corresponds to 0.43 Hz ([30], with permission)

reported in Fig. 4.3a. The arrows indicate the direction and module of the cells' inplane movements. The absolute distances traveled by several HNPs during the rhythmic EB contractions are plotted as a function of time in Fig. 4.3b, which indicates a clear periodic pattern for all particles. To retrieve the out-of-plane motion, the point spread function of the microscope was determined by performing a 3D nonlinear scan of a single isolated HNP on a substrate. This calibration was then applied to convert the SHG intensity modulations observed in the movie of cardiac contractions into absolute displacements along the z axis. The results of this procedure are reported in Fig. 4.3c. The Fourier transforms of the plots in panels b and c indicate that the observed movements have a constant frequency of 0.43 Hz. Of note, all contractions observed bear the same phase, within the temporal resolution of our measurement. In-plane movements range between 1–5 μm, out-of-plane displacements are slightly shorter (<3 μm).

Notice that the approach described above can be readily applied to monitor the evolution of various stem cell types in a transplanted tissue or their integration in a three dimensional supporting structure [31].

4.4 Basics of Coherent Control

In the last decades, coherent control induced a revolution in photochemistry. Not only could the atomic motion be *observed* within molecules by snapshots of a few femtoseconds duration, it could also be *controlled*. The basic idea of this revolution was the use of quantum interference between different states that are in coherent superposition so that a target state is reached with maximum probability. A key element for achieving this control is the transfer of coherence from the exciting light field to the atomic or molecular system via the electric dipole operator. A tailored coherent laser pulse is used to excite a molecule in a superposition of states, which drives the molecule along a specific path to obtain a targeted result.

A famous example of coherent control is the Tannor, Koslov and Rice [32, 33] "pump-dump" method, in which many vibrational states of a molecule ABC are coherently excited for generating a vibrational wavepacket. This wavepacket explores, as a function of time, a large fraction of the electronically excited hypersurface. By shining a second laser at convenient times t_1 or t_2, the molecule is brought back to its ground state onto different dissociation paths, resulting in different species: $AB + C$ or $A + BC$. The system can thus be driven to a specific daughter fragment by using the quantum control of the photodissociation process.

In 1992, H. Rabitz at Princeton University introduced [34, 35] the concept of "optimal control", in which a feedback loop optimizes the laser pulse characteristics in order to most efficiently reach the desired target. For this, a large number of parameters (corresponding to the amplitude and phase of each spectral component within the exciting laser spectrum) have to be controlled. This "pulse shaping" technique is usually performed by introducing a liquid crystal array in the Fourier plane between 2 gratings (4 f arrangement) [36]. Excellent results in terms of efficiency have been obtained using genetic-type optimization algorithms [37–39].

Besides the above mentioned "pump-dump" scheme, other successful control methods have been extensively applied, including stimulated Raman scattering (STIRAP), Four Wave Mixing like CARS (Coherent AntiStokes Raman Scattering), multiphoton absorption and multiphoton ionization. The controlled targets can be, for instance, specific fragments, specific isomers or specific isotopes, but also the enhancement or reduction of the fluorescence of a specific molecule (by driving it preferentially into other relaxation pathways). In this respect, a pioneering work has been performed by the group of G. Gerber [40, 41], in which the capability of distinguishing two different dyes using 2-photon absorption control was demonstrated, although the dyes had similar linear absorption and fluorescence spectra.

An attractive application is the ability of discriminating different biological systems that are usually undistinguishable with standard linear spectroscopic approaches. This is the approach that is followed in this chapter in order to reach label-free imaging of biological structures or the detection and identification of harmful bacteria.

4.5 Detection of Airborne Bacteria

4.5.1 Multi-photon-excited Fluorescence (MPEF) and Discrimination of Bioaerosols

We recently demonstrated that manipulating the excited states dynamics of biomolecules allows for distinguishing biological microparticles from nonbiological PAH (Polycyclic Aromatic Hydrocarbons)-containing ones. More precisely, we could distinguish amino acids (Tryptophan, Trp) and flavins (riboflavin RBF, FMN and FAD) from PAHs (naphthalene) and diesel fuel in the liquid phase using a "Pump-Probe Deletion" (PPD) technique. We also showed that the non-linear properties of aerosols and droplets produce a strong backward enhancement of the emitted light, which is most favorable for remote sensing applications.

The most prominent feature of non-linear processes in aerosol particles using femtosecond laser pulses is a strong localization of the emitting molecules within the particle, and subsequent backward enhancement of the emitted light [42–45] This unexpected behavior is very attractive for remote detection schemes, such as Lidar applications, where discrimination among the various types of particles in air is particularly difficult [46].

As already mentioned, a major drawback of LIF instruments is the lack of selectivity because UV-Vis fluorescence is incapable of discriminating different molecules with similar absorption and fluorescence signatures. While mineral and carbon black particles do not fluoresce significantly, aromatics and polycyclic aromatic hydrocarbons from organic particles and diesel soot strongly interfere with biological fluorophors such as amino acids. The similarity between the spectral signatures of PAHs and biological molecules under UV excitation lies in the fact that similar π-electrons from carbonic rings are involved. Therefore, PAHs (from Diesel soot for instance) exhibit absorption and emission bands similar to those of amino acids like Tryptophan (Fig. 4.4). Some shifts are present because of differences in specific bonds and the number of aromatic rings, but the broad featureless nature of the bands renders them almost indistinguishable.

In order to discriminate these fluorescing molecules we applied a novel femtosecond pump-probe depletion (PPD) concept. It is based on the time-resolved observation of the competition between excited state absorption (ESA) into higher lying excited states and fluorescence into the ground state. This approach makes use of two physical processes beyond that available in the usual linear fluorescence spectroscopy: (1) the dynamics in the intermediate pumped state (S_1) and (2) the coupling efficiency to higher lying excited states (S_n).

As shown in Fig. 4.5, a first femtosecond pump pulse (at 270 nm for Trp and PAHs, 405 nm for flavins), resonant with the first absorption band of the fluorophores, coherently excites them from the ground state S_0 to a set of vibronic levels $S_1\{\nu'\}$. The vibronic excitation relaxes by internal energy redistribution to lower $\{\nu\}$ modes. Fluorescence relaxation to the ground state occurs within a lifetime of sev-

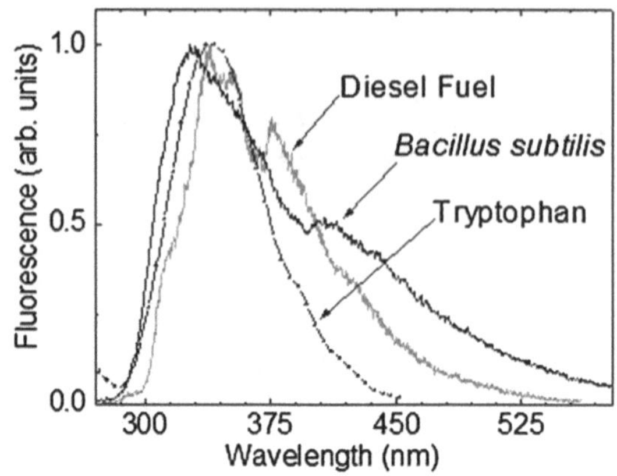

Fig. 4.4 Comparison of the fluorescence spectra of Tryptophan, *B. Subtilis* and Diesel fuel (From Ref. [3])

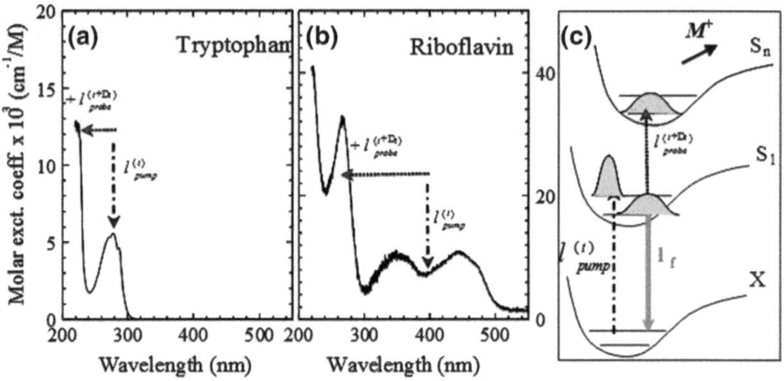

Fig. 4.5 Absorption spectra of Tryptophan (**a**) and Riboflavin (**b**). (**c**) PPD scheme in Trp, flavins and polycyclic aromatics. The pump pulse brings the molecules in their first excited state S_1. The S_1 population (and therefore the fluorescence) is depleted by the second probe pulse [48]

eral nanoseconds. Prior to full relaxation, a second 810 nm femtosecond probe pulse is used to transfer part of the $S_1\{\nu\}$ population to higher lying electronic states S_n.

The depletion of the S_1 population under investigation depends on both the molecular dynamics in this intermediate state and the transition probability to S_n. The relaxation from the intermediate excited state may be associated with different processes, including charge transfer, conformational relaxation and intersystem crossing with repulsive $\pi\sigma^*$ states [47].

S_n states are both autoionizing and relaxing radiationless into S_0. By varying the temporal delay Δt between the UV-Vis and the IR pulses, the dynamics of the internal energy redistribution within the intermediate excited potential hypersurface S_1 is explored. The S_1 population and the fluorescence signal are therefore depleted as a function of Δt. As different species have distinct S_1 hypersurfaces, discriminating signals can be obtained.

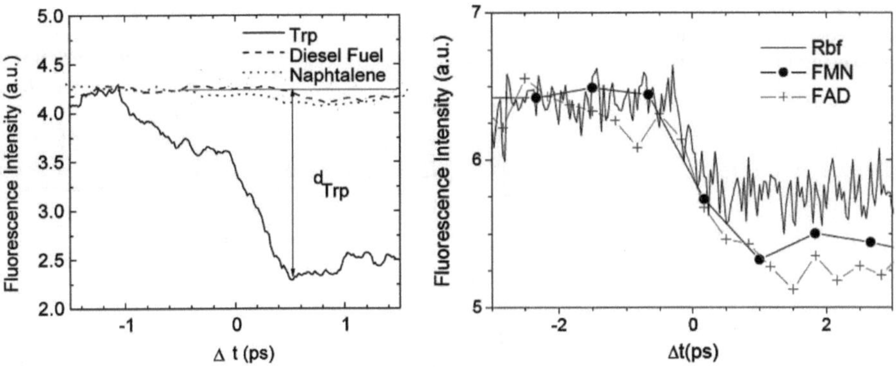

Fig. 4.6 *Left*: PPD experiment on Trp and PAHs, demonstrating discrimination capability between the amino acid and other aromatic molecules. *Right*: Similar results obtained in flavins [48]

Figure 4.6 (left) shows the PPD dynamics of S_1 in Trp as compared to diesel fuel and naphtalene in cyclohexane, one of the most abundant fluorescing PAHs in diesel. While depletion reaches as much as 50 % in Trp, diesel fuel and naphtalene appear almost unaffected (within a few percent), at least on these timescales [3, 4].

The depletion factor δ is defined as $\delta = (P_{undepleted} - P_{depleted})/P_{undepleted}$ (where P is the fluorescence power). This remarkable difference allows for efficient discrimination between Trp and organic species, although they exhibit very similar linear excitation/fluorescence spectra.

Two reasons might be invoked to understand this difference: (1) the intermediate state dynamics are predominantly influenced by the NH- and CO- groups of the amino acid backbone and (2) the ionization efficiency is lower for the PAHs. Further electronic structure calculations are required to better understand the process, especially on the higher lying S_n potential surfaces.

Fluorescence depletion has been obtained as well for RBF, FMN, and FAD (Fig. 4.6, right). However, the depletion in this case is only about 15 % (with a maximum intensity of 5×10^{11} W/cm^2 at 810 nm).

In order to more closely approach the application of detecting and discriminating bioagents from organic particles, we applied PPD spectroscopy to live bacteria ($\lambda_1 = 270$ nm and $\lambda_2 = 810$ nm), such as *Escherichia coli*, *Enterococcus* and *Bacillus subtilis*. Artefacts due to preparation methods have been discarded by using a variety of samples, i.e. lyophilized cells and spores, suspended either in pure or in biologically buffered water (i.e. typically 107–109 bacteria per cc). The bacteria containing solutions replaced the Trp or flavin containing solutions of the formerly described experiment. The observed pump-probe depletion results are remarkably robust (Fig. 4.7), with similar depletion values for all the considered bacteria (results for *Enterococcus*, not shown in the figure, are identical), although the Trp microenvironment within the bacteria proteins is very different from water.

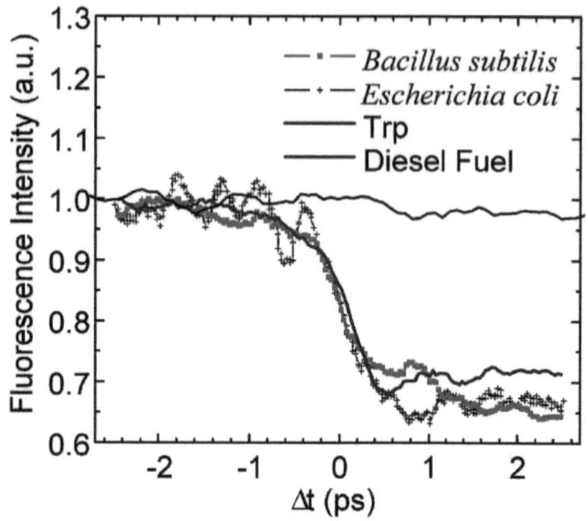

Fig. 4.7 Discrimination between bacteria and diesel fuel using PPD ultrafast spectroscopy (From Ref. [3])

On the other hand the very similar depletion behaviour for all bacteria and Trp also shows the limitations of PPD spectroscopy in the present configuration. Biomolecules can be distinguished from other aromatics but PPD is unable to discriminate two different bacteria in solution.

4.5.2 Remote Discrimination of Bioaerosols

The first multiphoton excited fluorescence Lidar detection of biological aerosols was performed using the "Teramobile" system. The Teramobile (http://www.teramobile.org) is the first femtosecond-terawatt laser based Lidar [49, 50], and was developed by a French-Swiss-German consortium.

The bioaerosol particles, consisting of 1 μm size water droplets containing 0.03 g/l Riboflavin (typical for bacteria), were generated at a distance of 50 m from the Teramobile system. Riboflavin was excited by two photons at 800 nm and emitted a broad fluorescence around 540 nm. This experiment [49, 50] is the first demonstration of the remote detection of bioaerosols using a 2-PEF-femtosecond Lidar. The broad fluorescence signature could clearly observed from the particle cloud (typ. 10^4 p/cm^3), with a range resolution of a few meters. As a comparison, droplets of pure water do not exhibit any parasitic fluorescence in this spectral range. However, a background is observed for both types of particles, arising from the scattering of white light generated by filaments in air.

Primarily, MPEF might be advantageous as compared to linear LIF for the following reasons: (1) MPEF is enhanced in the backward direction as seen above and (2) the transmission of the atmosphere is much higher for longer wavelengths. For example, if we consider the detection of tryptophan (which can be excited with 3 photons of 810 nm), the transmission of the atmosphere is typically 0.6 km^{-1}

at 270 nm, whereas it is $3 \times 10^{-3}\,km^{-1}$ at 810 nm (for a clear atmosphere, depending on the background ozone concentration). This compensates the lower 3-PEF cross-section compared to the 1-PEF cross-section at distances larger than a couple of kilometers [50]. The most attractive feature of MPEF is, however, the possibility of using pump-probe techniques, as described hereafter in order to discriminate bioaerosols from background interferents such as traffic related soot and PAHs.

It is interesting to notice that at these high intensities, competing processes occur such as LIBS (Laser Induced Breakdown Spectroscopy). The ionization potential of water molecules is 6.5 eV, so that 5 photons are required at a laser wavelength of 800 nm to initiate the process of plasma formation. Both localization and backward enhancement strongly increase with the order n of the multiphoton process, exceeding $R_f = 35$ for $n = 5$. As for MPEF, LIBS has the potential of providing information about the aerosols composition, as was demonstrated for bacteria.

Although nanosecond-laser LIBS (nano-LIBS) has already been applied to the study of bacteria [51,52], femtosecond lasers open new perspectives in this respect. The plasma temperature is indeed, much lower in the case of femtosecond excitation, which strongly reduces the blackbody radiation background and interfering lines from excited N_2 and O_2. This allows performing time gated detection with very short delays, and thus observing much richer and cleaner spectra from biological samples [53–55].

4.6 Use of Optimal Control for Discriminating Almost Identical Bio-systems

The identification or discrimination of molecules and quantum systems that are nearly identical (proteins, bacteria, etc.) is normally an exceedingly difficult task. The conventional means of addressing this common need is through various static spectroscopic techniques, which can be especially difficult in complex remote environments. However, recent theoretical work showed that two or more quantum systems, even if they differ only infinitesimally, may be drawn apart to have distinct signatures through tailored control of their dynamics. Ultrafast broad bandwidth lasers with adaptive pulse shaping provide a means to implement this theoretical concept through controlled quantum optimal dynamic discrimination (ODD) [56]. In collaboration with the group of H. Rabitz at Princeton University, we demonstrated the capabilities of ODD by discriminating between two almost identical molecules of riboflavin (RBF) and flavin mononucleotide (FMN) in water solution, which have virtually identical linear absorption and fluorescence spectra (see Fig. 4.8a, b) [7]. The successful implementation of ODD opens up numerous applications including in fluorescence microscopy, protein identification, as well as possible remote discrimination of different bacteria. A key component of ODD

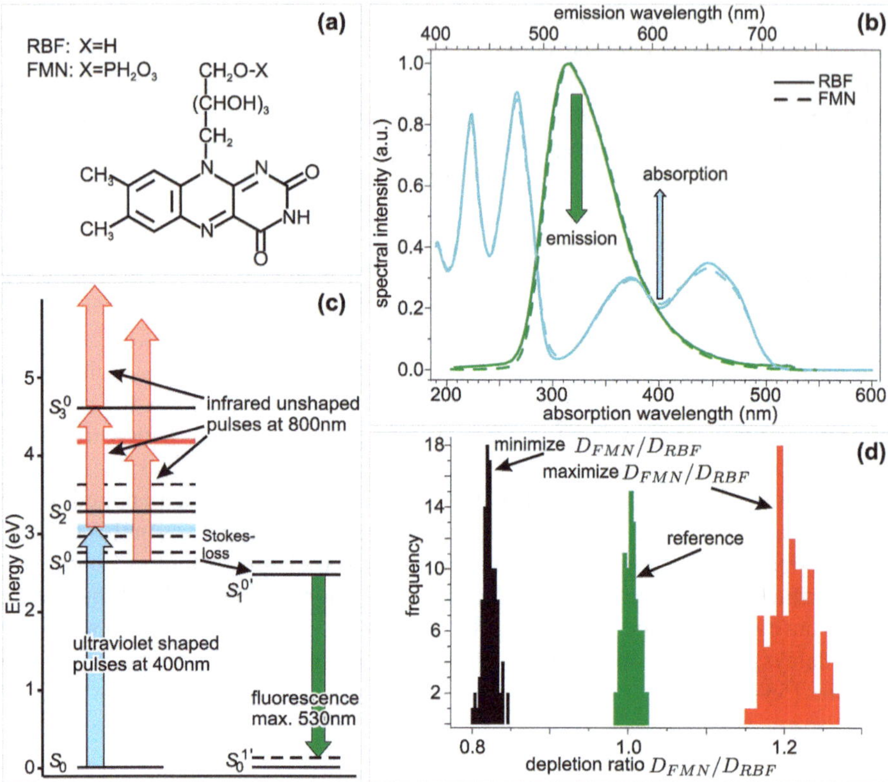

Fig. 4.8 The RBF and FMN molecules in (**a**) are very similar exhibiting nearly identical linear absorption and emission spectra in (**b**). The shaped UV control is centred at 400 nm and the flavin fluorescence signals are recorded over a window around 530 nm indicated by the respective arrows. The controlled optimal dynamic discrimination (ODD) of the flavins operates as indicated in (**c**) with a shaped UV pulse coordinated with a time delayed unshaped IR pulse to draw apart the vibronic dynamics of the flavins in the S_1 or S_2 excited states. The unshaped IR pulse serves to selectively disrupt the subsequent fluorescence signals from the two flavins and the *red arrows* indicate various possible means of action for the IR pulses. A typical outcome is shown in (**d**) where the ratio of flavin depletion signals, either $D(RBF)/D(FMN)$ or $D(FMN)/D(RBF)$, is maximized. The reference case corresponds to the application of transform limited pulses (From Ref. [7])

is operation with shaped laser pulses, which can nonlinearly interact with the molecules to fully exploit their dynamical capabilities and create discriminating signatures. The similar optical spectra of RBF and FMN arise from the common isoalloxazine ring with the only distinguishing feature being changes at the end of the side chain tail (Fig. 4.8a).

The present application of ODD utilizes a control field consisting of a shaped ultraviolet (UV) portion at 400 nm and a near infrared (IR) component at 800 nm, which is Fourier transform limited. The unshaped IR pulse follows the shaped UV pulse by a time delay, and the structure of the shaped UV pulse is deduced optimally under adaptive control to achieve ODD in the fluorescence signals from RBF and FMN. The primary operation of ODD relies on the shaped UV pulse coherently transferring ground state population into the flavin S_1 or S_2 excited states with the IR component likely creating further excitation of the molecules to higher lying S_n states. The IR field disrupts the vibronic excitation in S_1 or S_2 created by the UV pulse in order to make a discriminating difference in the fluorescence spectra of RBF and FMN from S_1 recorded in the window 525–535 nm after a relaxation period of ∼5 ns. Fluorescence is therefore depleted, and this fluorescence depletion value is used as measure of the molecular response to the field. The overall mechanism also may take advantage of any beneficial intra- and inter-molecular dynamical processes, and the ODD discrimination can draw on suitable coherent and incoherent aspects of the molecular dynamics.

In the first series of experiments the RBF and FMN solutions are in separate identical flow cells exposed to the same trial shaped UV pulse and time delayed unshaped IR pulse. An acousto-optic modulator is used to shape the UV pulses. A sequence of experiments was performed to record the relative depletion signal D_n

$$D_n(RBF) = \frac{F_n^{und}(RBF) - F_n^d(RBF)}{F_n^{und}(RBF)}, \tag{4.2}$$

where $F_n^{und}(RBF)$ is the undepleted fluorescence signal from the RBF cell (i.e., without application of the time delayed IR pulse) and $F_n^d(RBF)$ is the depleted signal in the presence of the IR pulse. An analogous expression applies to $D_n(RBF)$ for data collected from the FMN cell exposed to the same laser pulses.

The pulse shaper operated through phase modulation with 50 pixels, each on the range $0{-}2\pi$, over the bandwidth of the UV pulse (only 3–4 nm). A genetic algorithm of 30 individuals was used to optimize the UV pulse phases with convergence typically occurring in $100 \sim 300$ generations.

Figure 4.8d demonstrates the ability of ODD to significantly draw apart the RBF and FMN fluorescence signatures in spite of their nearly identical linear optical spectra in Fig. 4.8b [57]. For a given time delay τ, a specific optimal UV pulse is determined in the adaptive control experiment. The best discrimination was found for $\tau \sim 250{-}500$ fs and the procedure was not able to find significant discrimination for $\tau \geq 1$ ps indicating a loss of coherence and/or insufficient overlap between the UV and IR pulses.

With a delay time τ typically around 250–500 fs, optimal pulse shapes found in different runs were very complex and showed considerable difference. They however produced exactly the same discrimination ratio when applying them, even weeks after the initial experiment. The underlying mechanisms are currently investigated, in order to identify whether this behaviour is generic or particular to this RBF/FMN system.

The assembled set of optimally discriminating laser pulses and their long-term stable performance permits exploiting ODD as a novel means for detection of nearly identical molecules when they are simultaneously present in a sample. In this work the recorded fluorescence depletion signals F_n^d are utilized, as they are proportional to their respective flavin concentrations. Thus, the fluorescence signal $F_n(\text{mix})$ from a mixture of flavins generated with the n-th control UV-IR pulse pair is related to its individual counterparts by

$$F_n(\text{mix}) = c(RBF)F_n^d(RBF) + c(FMN)F_n^d(FMN), \tag{4.3}$$

where $c(RBF)$ is the fraction of RBF present and $c(RBF) + c(FMN) = 1$. The reference signals $F_n^d(RBF)$ and $F_n^d(FMN)$ are normalized to their respective concentrations deduced in learning the n-th control field, and similarly F_n is normalized to the total sum of the two flavins present. Two distinct ODD laser pulses could successfully determine the fractional content of the two flavins (the constrained relation between the fractions was not used in order to test the capabilities of ODD). Increasing the number of interrogating optimal pulses improves the standard deviation of the extracted component concentration fractions and a typical result using six pulses was $c(RBF) = 0.35 \pm 0.04$ and $c(FMN) = 0.68 \pm 0.05$ where the exact values were 0.33 and 0.66, respectively [57].

On a fundamental point of view, the exact molecular mechanisms underlying the selectivity allowed by the application of the optimally shaped pulses could be identified by quantum chemistry calculations [5]. It was shown that the optimal laser field makes use of low-frequency vibrational modes localized on the two biochromophores side chains (terminated by a phosphate group or by hydrogen) and selects the regions of their potential energy landscapes characterized by different transition dipole moments to higher excited states leading to different ionization probabilities. The resulting modulation of the excited state populations eventually controls the emitted fluorescence intensity.

4.7 Label-Free Identification of Proteins

From the experimental standpoint, one of the major difficulties faced for the extension of the ODD scheme to proteins (or DNA) was the lack of deep UV (DUV) laser pulse shapers. This lack of ad hoc instrumentation motivated us to develop a dedicated MEMS based pulse shaper, in collaboration with N. de Rooij at IMT-Neuchatel [58, 59]. This device proved successful for a very broad spectral range from the XUV [60] to the mid-IR.

First DUV-ODD experiments related to the fluorescing building blocks of proteins: Tryptophan and Tyrosine. The success of these experiments [8] enabled to identify tryptophan as ideal molecular target for performing ODD in proteins. In fact this bright emitting fluorophore displays a defined feature at short time-delays (<1 ps) that can be easily acted upon by an UV shaped femtosecond laser pulse and monitored by fluorescence depletion. Given tryptophan sensitivity to its local environment, it was decided to investigate the potential of ODD specifically on tryptophan-containing small peptides. We set-up a discrimination experiment aimed

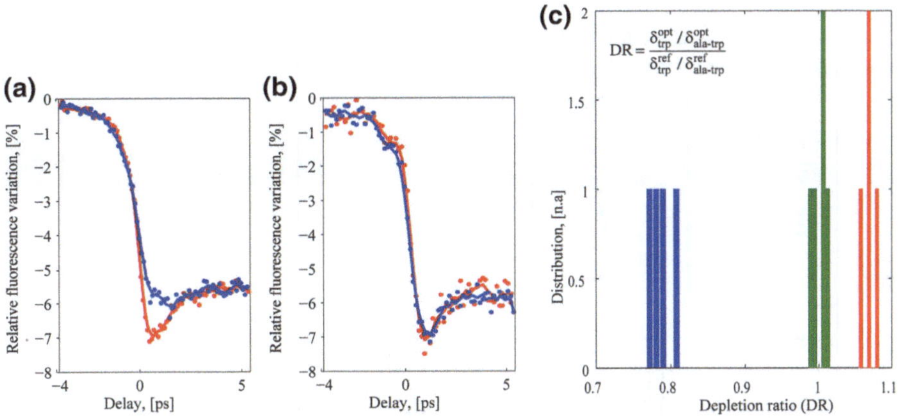

Fig. 4.9 Example of successful optimization for fluorescence depletion modulation in case of ala-trp (**a**) versus trp (**b**) with multi-objective algorithm optimization (**c**). Histogram of fluorescence depletions for ala-trp and trp. *Green (middle column)*: reference obtained with Fourier-transform pulses, *blue (left)*: maximization of fluorescencedepletion forala-trp, *red (right)*: maximization of fluorescence depletion for trp (From Ref. [8])

at the ODD fluorescence-based identification within several peptide pairs, otherwise showing identical fluorescence spectral and time-resolved responses [9].

A first series of feedback driven optimizations aimed at selectivity modifying the depletion amplitude at a given time delay (600 fs) of a defined peptide against another. An exemplary plot summarizing the outcome of an optimization run (ala-trp vs. trp) is shown in Fig. 4.9. In this case, a multi-objective optimization algorithm was launched with two independent goals: (i) increasing ala-trpfluorescence and (ii) decreasing trp fluorescence. One can appreciate the result of the procedure by applying the optimal mask on the two samples and acquiring the timeresolved traces, as reported in Fig. 4.9a, b. In this example, the optimized pulse shape leads to an increase in the ala-trp time-resolved fluorescence variation by 20 %, while trp remains unaffected. Optimization runs with the same objectives were repeated several times to evaluate statistical variations on the results. Note that different pulse shapes were found to lead to similar yields, pointing out the existence of multiple solutions. The closed loop optimization procedure was then performed several times for the opposite set of goals, i.e. (i) decreasing ala-trp-fluorescence and (ii) increasing trp fluorescence. From the comparison of these traces with reference results obtained with Fourier transform pulses, we establish the histogram presented in Fig. 4.9c. The coherent manipulation of the molecular wavepacket leads to a variation of discrimination power that tears apart the trp and alatrp signatures by more than 8σ (where σ is the averaged statistical error on the results measured for a defined pulse shape). We also verified that the discrimination power DR was dependent on the time delay Δt. As expected for a coherent manipulation of the molecular wavepacket, DR decreases to zero for time delays larger than decoherence time (typically 1.5 ps).

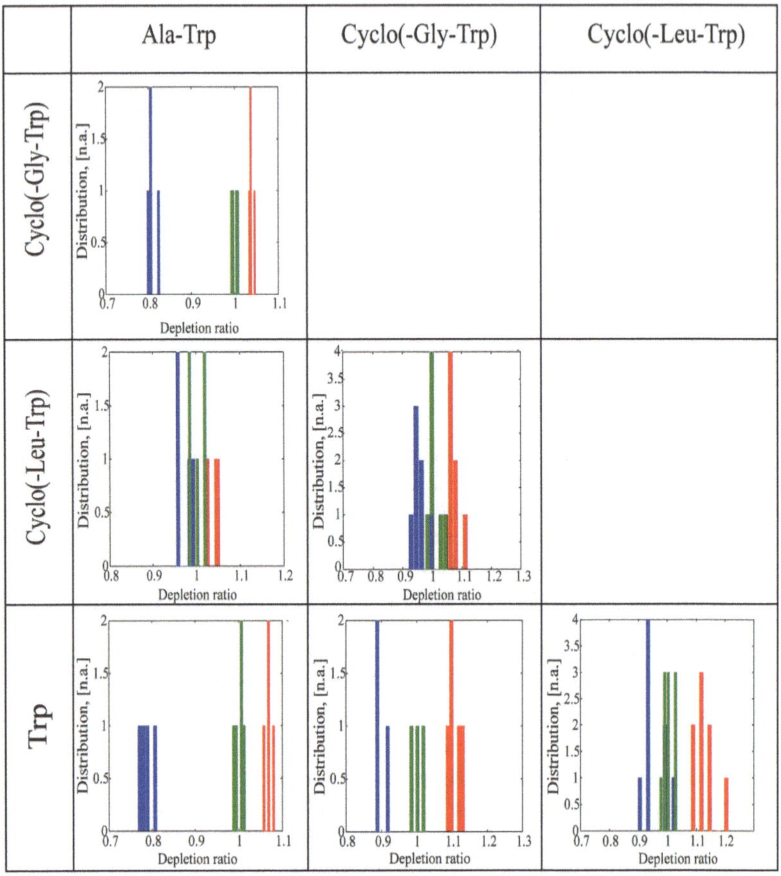

Fig. 4.10 EHistograms of fluorescence depletions for various pairs of molecules. *Green*: reference obtained with Fourier-transform pulses, *blue*: maximization of fluorescence depletion for row molecule, *red*: maximization of fluorescence depletion for column molecule (From Ref. [9])

The same procedure described above was applied to all pairs of dipeptides and free trp, giving rise to the matrix displayed in Fig. 4.10: In the histograms, bars indicate the ratios of the depletion values (row molecules over column molecules) obtained using phase-shaped DUV pulses retrieved by optimizations aimed at minimizing (blue) or maximizing (red) this quantity. Green bars correspond to the ratio of the depletion values measured with unshaped pulses. One can see that for several molecule pairs (trp vs. dipeptides, cyclo(-gly-trp) vs. ala-trp) the discrimination capability of the method is satisfactory, with a fluorescence difference between the optimized pulse shapes ranging from $3s$ (trp vs. cyclo(-leu-trp)) to 8σ (cyclo(-gly-trp) vs. ala-trp), which is remarkably high considering the modest bandwidth available for the shaped DUV pulse. This restricted parameter space does not allow, in opposite, to discriminate other dipeptides like cyclo(-leu-trp) vs. ala-trp.

It is clear from Fig. 4.10 that even for a group of very similar molecules addressed with identical experimental resources (laser spectral bandwidth as compared to absorption bandwidth, shaping spectral resolution), large variations do occur. No obvious reason emerged from our measurements. For instance, no correlation exists between the Stokes shifts in the static spectra and discriminability. Similarly, purely dynamical considerations, based on the time-resolved measurements [9], would suggest that, for instance, ala-trp should be more easily distinguished from cyclo(-leu-trp) (different relaxation time constants) than from free trp, which is not the case. No correlation with far infrared spectra was found either, stressing the importance of more complex quantum interference effects than just impulsive driving of vibrational modes.

As a natural follow up of this benchmark experiment, work is now in progress for transferring the DUV-ODD method to larger proteins with prospective applications in the field of label-free, consumable-free medical sensing. The goal of this undergoing study is currently, for instance, to quantify antibodies (such as IgG, 150 kDa) in blood serum, leading to first "quantum interference based bioassays".

4.8 Conclusion

Quantum control and femtosecond spectroscopy shine a new light onto the detection and identification of biological systems, such as biochromophores but also highly complex systems like cells or bacteria. These applications are only at their beginning, and there is a reasonable hope for widely spread applications in the future, such as label-free microscopy or discrimination of pathogen from non-pathogen bacteria in air. Major technical improvements required for reaching these goals lie in the spectral range and the spectral bandwidth accessible by future femtosecond coherent sources. For instance the access to the deep UV (230–300 nm) will open up applications to proteins and DNA, which constitute major biological targets.

Acknowledgements The authors gratefully acknowledge the collaborators at the Universities of Geneva and Lyon, in particular L. Bonacina, F. Courvoisier, L. Guyon, V. Boutou, E. Salmon, J. Yu, G. Mejean, J. Kasparian, C. Kasparian, A. Rondi, S. Afonina, J. Extermann, P. Bejot, S. Weber, D. Kiselev, and M. Moret, as well as H. Rabitz and his group at Princeton, particularly M. Roth and J. Roslund.

We also acknowledge the financial support of the Swiss National Science Foundation (contracts No. 2000021 111688 and No. 200020 124689), the Swiss NCCR MUST, and European FP7 project NAMDIATREAM (NMP-2009-4.0-3-246479).

References

1. Extermann J, Bonacina L, Cuna E, Kasparian C, Mugnier Y, Feurer T, Wolf JP (2009) Nanodoublers as deep imaging markers for multi-photon microscopy. Optics Express 17:15342–15349
2. Extermann J, Bonacina L, Courvoisier F, Kiselev D, Mugnier Y, Le Dantec R, Galez C, Wolf JP (2008) Nano-FROG: Frequency resolved optical gating by a nanometric object. Opt Express 16:10405–10411

3. Courvoisier F, Boutou V, Guyon L, Roth M, Rabitz H, Wolf JP (2006) Discriminating bacteria from other atmospheric particles using femtosecond molecular dynamics. J Photochem Photobiol A-Chem 180:300–306
4. Courvoisier F, Boutou V, Wood V, Bartelt A, Roth M, Rabitz H, Wolf JP (2005) Femtosecond laser pulses distinguish bacteria from background urban aerosols. Appl Phys Lett 87:063901
5. Petersen J, Mitric R, Bonacic-Koutecky V, Wolf JP, Roslund J, Rabitz H (2010) How shaped light discriminates nearly identical biochromophores. Phys Rev Lett 105:073003
6. Roslund J, Roth M, Guyon L, Boutou V, Courvoisier F, Wolf JP, Rabitz H (2011) Resolution of strongly competitive product channels with optimal dynamic discrimination: application to flavins. J Chem Phys 134:034511
7. Roth M, Guyon L, Roslund J, Boutou V, Courvoisier F, Wolf JP, Rabitz H (2009) Quantum control of tightly competitive product channels. Phys Rev Lett 102:253001
8. Rondi A, Bonacina L, Trisorio A, Hauri C, Wolf J-P (2012) Coherent manipulation of free amino acids fluorescence. Phys Chem Chem Phys 14:9317–9322
9. Afonina S, Nenadl O, Rondi A, Bonacina L, Extermann J, Kiselev D, Dolamic I, Burgi T, Wolf JP (2013) Discriminability of tryptophan containing dipeptides using quantum control. App Phys B 111:541–549
10. Staedler D, Magouroux T, Joulaud C, Extermann J, Hadji R, Kasparian C, Gerber S, Le Dantec R, Mugnier Y, Juillerat L, Rytz D, Ciepielewski D, Bonacina L, Wolf J-P (2012) Harmonic nanocrystals for bio-labeling: a survey of optical properties and biocompatibility. ACS Nano 6(3):2542–2549
11. Le Dantec R, Mugnier Y, Djanta G, Bonacina L, Extermann J, Badie L, Joulaud C, Gerrmann M, Rytz D, Wolf JP, Gale C (2011) Ensemble and individual characterization of the nonlinear optical properties of ZnO and BaTiO3 nanocrystals. J Phys Chem 115(31): 15140–15146
12. Bonacina L, Mugnier Y, Courvoisier F, Le Dantec R, Extermann J, Lambert Y, Boutou V, Galez C, Wolf JP (2007) Polar Fe(IO3)(3) nanocrystals as local probes for nonlinear microscopy. Appl Phys B-Lasers Opt 87:399–403
13. Baumner R, Bonacina L, Enderlein J, Extermann J, Fricke-Begemann T, Marowsky G, Wolf JP (2010) Evanescent-field-induced second harmonic generation by noncentrosymmetric nanoparticles. Opt Express 18:23218–23225
14. Extermann J, Béjot P, Bonacina L, Mugnier Y, Le Dantec R, Mazingue T, Galez C, Wolf JP (2009) An inexpensive nonlinear medium for intense ultrabroadband pulse characterization. Appl Phys B 97:537–540
15. Le Xuan L, Brasselet S, Treussart F, Roch JF, ,Marquier F, Chauvat D, Perruchas S, Tard C, Gacoin, T (2006) Balanced homodyne detection of second-harmonic generation from isolated subwavelength emitters. Appl Phys Lett 89:121118
16. Pu Y, Centurion M, Psaltis D (2008) Harmonic holography: a new holographic principle. Appl Opt 47:A103–A110
17. Hsieh CL, Pu Y, Grange R, Laporte G, Psaltis D (2010) Imaging through turbid layers by scanning the phase conjugated second harmonic radiation from a nanoparticle. Opt Express 18:20723–20731
18. Hsieh CL, Pu Y, Grange R, Psaltis D (2010) Digital phase conjugation of second harmonic radiation emitted by nanoparticles in turbid media. Opt Express 18:12283–12290
19. Squier JA, Muller M, Brakenhoff GJ, Wilson KR (1998) Third harmonic generation microscopy. Opt Express 3:315–324
20. Dudovich N, Oron D, Silberberg Y (2002) Single-pulse coherently controlled nonlinear Raman spectroscopy and microscopy. Nature 418:512–514
21. von Vacano B, Wohlleben W, Motzkus M (2006) Actively shaped supercontinuum from a photonic crystal fiber for nonlinear coherent microspectroscopy. Opt Lett 31:413–415
22. Ogilvie JP, Debarre D, Solinas X, Martin JL, Beaurepaire E, Joffre M (2006) Use of coherent control for selective two-photon fluorescence microscopy in live organisms. Opt Express 14:759–766

23. Aeschlimann M, Bauer M, Bayer D, Brixner T, Cunovic S, Dimler F, Fischer A, Pfeiffer W, Rohmer M, Schneider C, Steeb F, Struber C, Voronine DV (2010) Spatiotemporal control of nanooptical excitations. Proc Natl Acad Sci U S A 107:5329–5333
24. Fuchs U, Zeitner UD, Tunnermann A (2005) Ultra-short pulse propagation in complex optical systems. Opt Express 13:3852–3861
25. Tal E, Oron D, Silberberg Y (2005) Improved depth resolution in video-rate line-scanning multiphoton microscopy using temporal focusing. Opt Lett 30:1686–1688
26. Muller M, Squier J, Brakenhoff GJ (1995) Measurement of femtosecond pulses in the focal point of a high-numerical-aperture lens by 2-photon absorption. Opt Lett 20:1038–1040
27. Brixner T, De Abajo FJG, Spindler C, Pfeiffer W (2006) Adaptive ultrafast nano-optics in a tight focus. Appl Phys B-Lasers Opt 84:89–95
28. Amat-Roldan I, Cormack IG, Loza-Alvarez P, Artigas D (2004) Starch-based second-harmonic-generated collinear frequency-resolved optical gating pulse characterization at the focal plane of a high-numerical-aperture lens. Opt Lett 29:2282–2284
29. Bowlan P, Gabolde P, Trebino R (2007) Directly measuring the spatio-temporal electric field of focusing ultrashort pulses. Opt Express 15:10219–10230
30. Magouroux T, Extermann J, Hoffmann P, Mugnier Y, Le Dantec R, Jaconi M, Kasparian C, Ciepielewski D, Bonacina L, Wolf JP (2012) High-speed tracking of murine cardiacstem cells by harmonic nanodoublers. Small 8(17):2752–2756
31. Vunjak-Novakovic G, Godier-Furnemont AFG, Martens TP, Koeckert MS, Wan L, Parks J, Arai K, Zhang GP, Hudson B, Homma S (2011) Composite scaffold provides a cell delivery platform for cardiovascular repair. Proc Natl Acad Sci U S A 108:7974–7979
32. Tannor DJ, Kosloff R, Rice SA (1986) Coherent pulse sequence induced control of selectivity of reactions – exact quantum-mechanical calculations. J Chem Phys 85:5805–5820
33. Tannor DJ, Rice SA (1985) Control of selectivity of chemical-reaction via control of wave packet evolution. J Chem Phys 83:5013–5018
34. Judson RS, Rabitz H (1992) Teaching lasers to control molecules. Phys Rev Lett 68:1500–1503
35. Warren WS, Rabitz H, Dahleh M (1993) Coherent control of quantum dynamics – the dream is alive. Science 259:1581–1589
36. Weiner AM (2000) Femtosecond pulse shaping using spatial light modulators. Rev Sci Instrum 71:1929–1960
37. Bonacina L, Extermann J, Rondi A, Boutou V, Wolf JP (2007) Multiobjective genetic approach for optimal control of photoinduced processes. Phys Rev A 76:023408
38. Dantus M, Lozovoy VV (2004) Experimental coherent laser control of physicochemical processes. Chem Rev 104:1813–1859
39. Levis RJ, Menkir GM, Rabitz H (2001) Selective bond dissociation and rearrangement with optimally tailored, strong-field laser pulses. Science 292:709–713
40. Brixner T, Damrauer NH, Niklaus P, Gerber G (2001) Photoselective adaptive femtosecond quantum control in the liquid phase. Nature 414:57–60
41. Brixner T, Gerber G (2003) Quantum control of gas-phase and liquid-phase femtochemistry. Chemphyschem 4:418–438
42. Boutou V, Favre C, Hill SC, Pan YL, Chang RK, Wolf JP (2002) Backward enhanced emission from multiphoton processes in aerosols. Appl Phys B-Lasers Opt 75:145–152
43. Favre C, Boutou V, Hill SC, Zimmer W, Krenz M, Lambrecht H, Yu J, Chang RK, Woeste L, Wolf JP (2002) White-light nanosource with directional emission. Phys Rev Lett 89:035002
44. Hill SC, Boutou V, Yu J, Ramstein S, Wolf JP, Pan YL, Holler S, Chang RK (2000) Enhanced backward-directed multiphoton-excited fluorescence from dielectric microcavities. Phys Rev Lett 85:54–57
45. Pan YL, Hill SC, Wolf JP, Holler S, Chang RK, Bottiger JR (2002) Backward-enhanced fluorescence from clusters of microspheres and particles of tryptophan. Appl Opt 41: 2994–2999
46. Kasparian J, Frejafon E, Rambaldi P, Yu J, Ritter P, Viscardi P, Wolf JP (1998) Characterization of urban aerosols using SEM-microscopy, X-ray analysis and lidar measurements. Atmos Env 32(17):2957–2967

47. Iketaki Y, Watanabe T, Ishiuchi S, Sakai M, Omatsu T, Yamamoto K, Fujii M, Watanabe, T (2003) Investigation of the fluorescence depletion process in the condensed phase; application to a tryptophan aqueous solution. Chem Phys Lett 372:773–778
48. Courvoisier F, Bonacina L, Boutou V, Guyon L, Bonnet C, Thuillier B, Extermann J, Roth M, Rabitz H, Wolf JP (2008) Identification of biological microparticles using ultrafast depletion spectroscopy. Faraday Discuss 137:37–49
49. Kasparian J, Rodriguez M, Mejean G, Yu J, Salmon E, Wille H, Bourayou R, Frey S, Andre YB, Mysyrowicz A, Sauerbrey R, Wolf JP, Woste L (2003) White-light filaments for atmospheric analysis. Science 301:61–64
50. Mejean G, Kasparian J, Yu J, Frey S, Salmon E, Wolf JP (2004) Remote detection and identification of biological aerosols using a femtosecond terawatt lidar system. Appl Phys B-Lasers Opt 78:535–537
51. Dixon PB, Hahn DW (2004) Feasibility of detection and identification of individual bioaerosols using laser-induced breakdown spectroscopy. Anal Chem 77:631–638
52. Morel S, Leone N, Adam P, Amouroux J (2003) Detection of bacteria by time-resolved laser-induced breakdown spectroscopy. Appl Opt 42:6184–6191
53. Baudelet M, Guyon L, Yu J, Wolf JP, Amodeo T, Frejafon E, Laloi P (2006) Spectral signature of native CN bonds for bacterium detection and identification using femtosecond laser-induced breakdown spectroscopy. Appl Phys Lett 88:063901
54. Baudelet M, Guyon L, Yu J, Wolf JP, Amodeo T, Frejafon E, Laloi P (2006) Femtosecond time-resolved laser-induced breakdown spectroscopy for detection and identification of bacteria: a comparison to the nanosecond regime. J Appl Phys 99:084701
55. Baudelet M, Yu J, Bossu M, Jovelet J, Wolf JP, Amodeo T, Frejafon E, Laloi P (2006) Discrimination of microbiological samples using femtosecond laser-induced breakdown spectroscopy. Appl Phys Lett 89:163903
56. Li BQ, Rabitz H, Wolf JP (2005) Optimal dynamic discrimination of similar quantum systems with time series data. J Chem Phys 122:154103
57. Roslund J, Roth M, Guyon L, Boutou V, Courvoisier F, Wolf J-P, Rabitz H (2011) Resolution of strongly competitive product channels with optimal dynamic discrimination: application to flavins. J Chem Phys 134:034511
58. Rondi A, Extermann J, Bonacina L, Weber SM, Wolf JP (2009) Characterization of a MEMS-based pulse-shaping device in the deep ultraviolet. Appl Phys B-Lasers Opt 96:757–761
59. Weber S, Barthelemy M, Chatel B (2010) Direct shaping of tunable UV ultra-short pulses. Appl Phys B-Lasers Opt 98:323–326
60. Kiselev D, Kraus PM, Bonacina L, Wörner HJ, Wolf JP (2012) Direct amplitude shaping of high harmonics in the extreme ultraviolet. Opt Express 20(23):25843–25849

5
Ultrafast Photonics with Microstructures Fibers

Aleksandr A. Lanin, Aleksandr V. Mitrofanov, Andrei B. Fedotov, Sean Blakley, Dmitrii A. Sidorov-Biryukov, and Aleksei M. Zheltikov

Abstract Microstructure fibers – new optical fibers with tailored dispersion and nonlinearity – provide a constantly growing platform for the development of advanced fiber-format devices and components for ultrafast photonics. Unique options offered by microstructure fiber technology, such as dispersion management through fiber structure engineering and enhancement of optical nonlinearity due to a strong field confinement in a small-size fiber core, are pushing the frontiers of

---

A.A. Lanin • A.B. Fedotov • D.A. Sidorov-Biryukov
Physics Department, International Laser Center, M.V. Lomonosov Moscow State University, Moscow 119992, Russia

Russian Quantum Center, ul. Novaya 100, Skolkovo, Moscow Region, 1430125, Moscow, Russia

A.V. Mitrofanov
Institute of Laser and Information Technologies, Russian Academy of Sciences, ul. Svyatoozerskaya 1, Shatura, Moscow Region, 140700, Moscow, Russia

Physics Department, International Laser Center, M.V. Lomonosov Moscow State University, Moscow 119992, Russia

Russian Quantum Center, ul. Novaya 100, Skolkovo, Moscow Region, 1430125, Moscow, Russia

S. Blakley
Department of Physics and Astronomy, Texas A&M University, College Station, TX, 77843-4242, USA

A.M. Zheltikov (✉)
Department of Physics and Astronomy, Texas A&M University, College Station, TX, 77843-4242, USA

Physics Department, International Laser Center, M.V. Lomonosov Moscow State University, Moscow 119992, Russia

Russian Quantum Center, ul. Novaya 100, Skolkovo, Moscow Region, 1430125, Moscow, Russia
e-mail: zheltikov@physics.msu.ru

B. Di Bartolo et al. (eds.), *Nano-Structures for Optics and Photonics*, NATO Science for Peace and Security Series B: Physics and Biophysics, DOI 10.1007/978-94-017-9133-5_5, © Springer Science+Business Media Dordrecht 2015

ultrafast photonics, allowing the creation of efficient sources of supercontinuum radiation, novel compact fiber lasers, as well as frequency converters, pulse compressors, and fiber components for microscopy and bioimaging.

5.1 Broadly Wavelength- and Pulse-Width-Tunable High-Repetition-Rate Microstructure-Fiber Light Sources

Following the demonstration of unique properties and excellent performance of photonic-crystal fibers (PCFs, Figs. 5.1 and 5.2) as frequency converters [1–3] and elements of advanced fiber lasers [4, 5], waveguides of this class find growing applications as components of larger-scale laser systems and optical devices for a broad variety of applications. The integration of supercontinuum-generating PCFs with f-to-$2f$ interferometry [6] has resulted in the development of a powerful tool for the control of the carrier-envelope phase of ultrashort optical pulses, paving the ways for revolutionary breakthroughs in frequency-comb metrology [7]

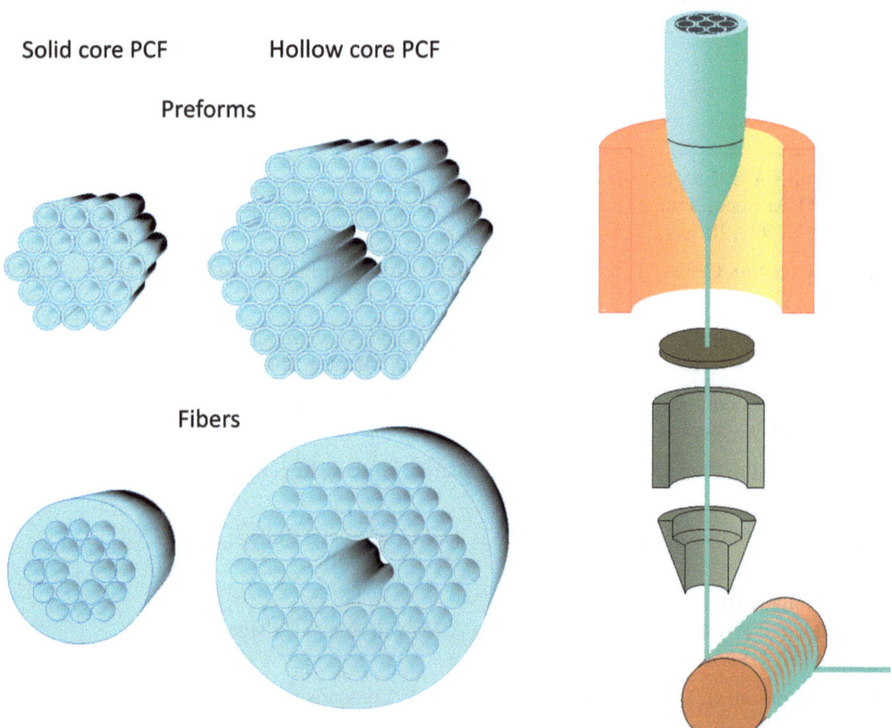

Fig. 5.1 Fabrication of microstructure fibers

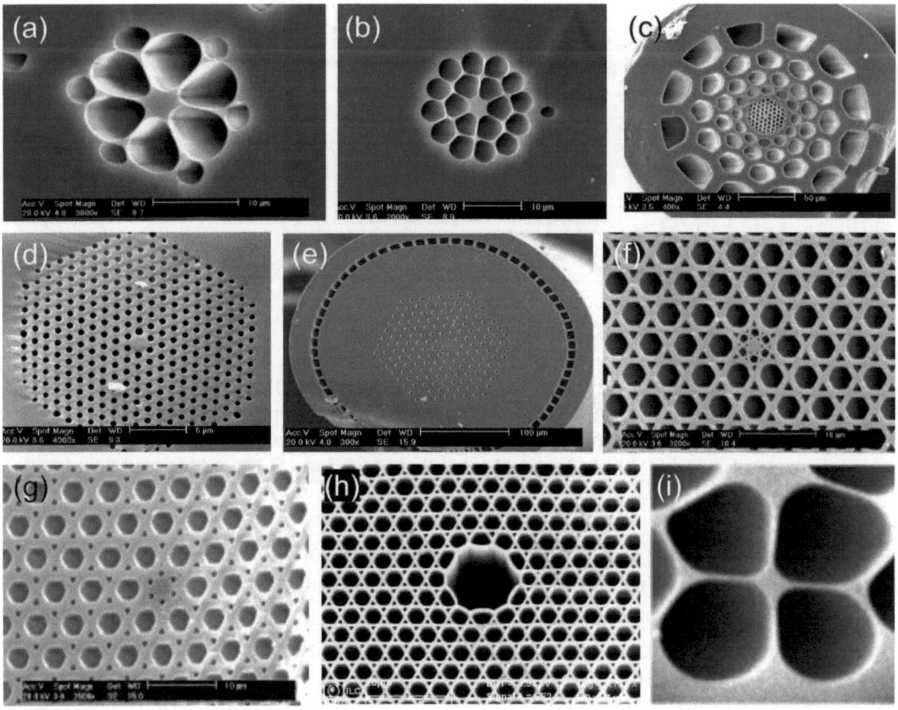

Fig. 5.2 SEM images of photonic-crystal fibers [3]: (**a, b**) solid-core high-index-step silica PCFs, (**c**) solid-core silica PCF with a dual microstructure cladding, (**d**) periodic-cladding PCF, (**e**) YAG-doped PCF, (**f, g**) soft-glass PCF with a nanohole-array-modified core, (**h**) hollow-core PCF, and (**i**) nanowaveguide PCF

and attosecond science [8]. Incorporation of PCF-based soliton frequency shifters into the optical parametric chirped-pulse amplification technology has enabled all-optical synchronization in the amplification of few-cycle field waveforms [9]. In optical imaging, PCFs have been advantageously integrated into the apparatus for coherent Raman microscopy [10, 11], radically simplifying the design of a multicolor optical source required for this method of imaging. PCF-based probes combined with the protocols of optogenetics [12] offer much promise for the development of advanced optical interfaces for neuroscience [13, 14]. Finally, the integration of PCF supercontinuum sources with frequency conversion in nonlinear crystals has been shown to allow the generation of ultrashort pulses within a broad spectral range from the UV to the near-IR [15], as well as to enable the creation of ultracompact sources of terahertz pulses [16].

Here, we demonstrate that soliton self-frequency shift (SSFS) in a PCF pumped by an increased-energy output of a long-cavity mode-locked Cr: forsterite laser can be combined with second-harmonic generation (SHG) in a nonlinear crystal to provide a broad wavelength and pulse-width tunability of ultrashort light pulses in the range of wavelengths from 680 to 1,800 nm at the 20-MHz repetition rate. We

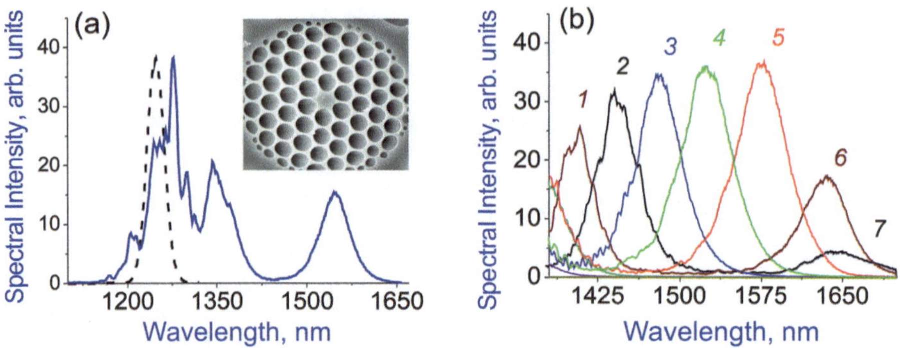

Fig. 5.3 (**a**) The spectrum of Cr: forsterite laser pulses (*dashed line*) and the spectrum at the output of a 20-cm piece of PCF (shown in the *inset*) measured with an input energy of 5 nJ (*solid line*). (**b**) The spectra of wavelength-shifted solitons at the output of a 20-cm piece of PCF measured with an input energy of 3.5 nJ (*1*), 4 nJ (*2*), 4.4 nJ (*3*), 4.9 nJ (*4*), 6.3 nJ (*5*), 7.5. nJ (*6*), and 8.9 nJ (*7*)

show that LBO crystals can support phase matching for short-pulse SHG within the entire tunability range of the nanojoule-level PCF output, stretching from 1.35 to 1.80 μm. Thin LBO crystals are employed in our experiments to generate broadly tunable sub-100-fs second-harmonic pulses in the range of wavelengths from 680 to 900 nm. Thicker nonlinear crystals with properly adapted beam focusing are shown to provide up-to-40 %-efficiency frequency doubling of the PCF output with simultaneous spectral narrowing, yielding ultrashort light pulses with an average power up to 10 mW ideally suited for coherent Raman microspectroscopy.

In experiments, we employ a home-built ytterbium-fiber-laser-pumped mode-locked Cr: forsterite laser oscillator [17], which delivers laser pulses with a central wavelength of 1.25-μm (the dashed line in Fig. 5.3a) and a pulse width of 40–70 fs. The extended-cavity design of the Cr: forsterite laser allows the output laser energy to be increased up to 18 nJ at a pulse repetition rate of 20 MHz. The Cr: forsterite laser output is launched into a silica PCF with a cross-section structure as shown in the inset to Fig. 5.3a. This PCF was designed to provide efficient wavelength conversion of 1.25-μm laser pulses through SSFS, delivering tunable ultrashort light pulses within a wavelength range of 1.35–1.80 μm. The core diameter of the PCF was chosen large enough (4 μm) to allow an increase in the energy of solitons [17] relative to the typical level of soliton energies attainable with highly nonlinear PCFs.

The spectrum of the soliton output of a 20-cm-long PCF pumped by 5-nJ, 68-fs Cr: forsterite laser pulses is shown in Fig. 5.3a. This spectrum features an isolated wavelength-shifted soliton centered at 1550 nm with a bandwidth supporting a transform-limited pulse width of 47 fs. Wavelength tunability of the soliton PCF output is illustrated by Fig. 5.3b, which presents the spectra of frequency-shifted solitons generated in a 20-cm section of PCF by Cr: forsterite laser pulses with different energies.

The wavelength-shifted PCF output was frequency-doubled through SHG in an *xz*-cut LBO crystal. For each wavelength from the tunability range provided by

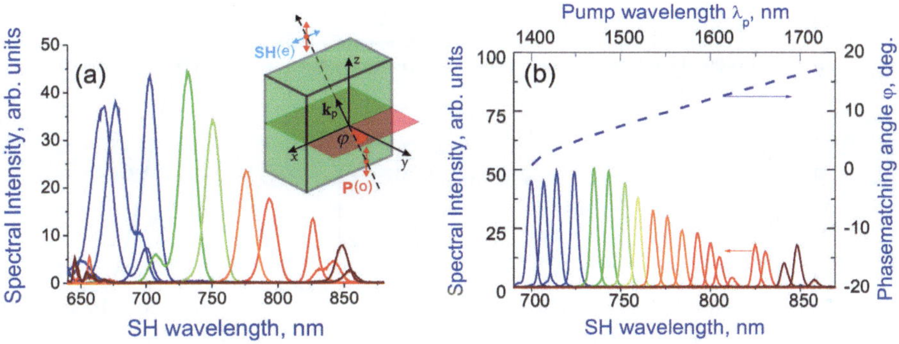

Fig. 5.4 The spectra of the second-harmonic output of (**a**) a 2-mm and (**b**) 20-mm-thick LBO crystal pumped by wavelength-shifted solitons from a 20-cm-long PCF. The *dashed line* in panel (**b**) shows the SHG phase-matching angle ϕ as a function of the soliton pump (the upper abscissa axis) and the second-harmonic (the lower abscissa axis) wavelength for LBO. Diagram of the $o + o \rightarrow e$ SHG process in an xz-cut LBO crystal is shown in the *inset*: P, pump field; SH, second harmonic; ϕ, angle between the wave vector of the pump beam and the y axis in the xy-plane

SSFS in the PCF (Fig. 5.3b), the angle ϕ of the wave vector of the pump field with respect to the y-axis in the xy-plane (see the inset in Fig. 5.2a) was chosen in such a way as to provide phase matching for $o + o \rightarrow e$ SHG at the desired (the dashed line in Fig. 5.4b). The spectra of the second harmonic produced by the wavelength-shifted solitons from the PCF in 2- and 20-mm LBO crystals are presented in Fig. 5.4a, b, respectively. In the range of input wavelengths from 1,400 to 1,500 nm, the 20-mm LBO crystal provided a typical SHG efficiency of about 40 %, yielding second-harmonic pulses in the 700–750 nm wavelength range with an energy up to 0.5 nJ.

The pulse width of the second harmonic generated in the LBO crystal can be tuned by varying the thickness of the nonlinear crystal, beam-focusing geometry, and the wavelength of the soliton PCF output, providing the pump field for the SHG process. Due to the mismatch of the group velocities u_p and u_{SH} of the pump and second-harmonic pulses in an SHG crystal, the pulse width of the second harmonic increases [18, 19] in accordance with $\tau_{SH} \approx (\tau_p^2/2 + \theta^2)^{1/2}$, where τ_p is the pulse width of the pump, $\theta = l_{\text{int}}(u_p^{-1} - u_{SH}^{-1})$, $l_{\text{int}} = \min(L, b)$ is the interaction length, L is the thickness of the nonlinear crystal, $b - 2\pi w_0^2/\lambda_0$ is the confocal parameter, λ_0 is the pump wavelength, and w_0 is the waist radius of the pump beam. The second-harmonic pulse width calculated as a function of the pump and second-harmonic wavelengths for a thin ($L = 2$ mm) and a thick ($L = 20$ cm) nonlinear crystals is shown by the dashed lines in Fig. 5.5a. Dispersion of the nonlinear crystal was included in these calculations through the Sellmeier equation with appropriate coefficients for LBO [20]. In the case of a thin nonlinear crystal, a lens with a focal length $f = 7.5$ cm is used, giving $b = 3.8$ mm for $\lambda_0 = 1{,}500$ nm. In this regime, $L < b$ and τ_p is close to θ, giving rise to a weak dependence ofτ_{SH} on λ_0 (curve 1 in Fig. 5.5a). In the case of a thick crystal, we take $f = 15$ cm, leading to $b = 13.5$ mm

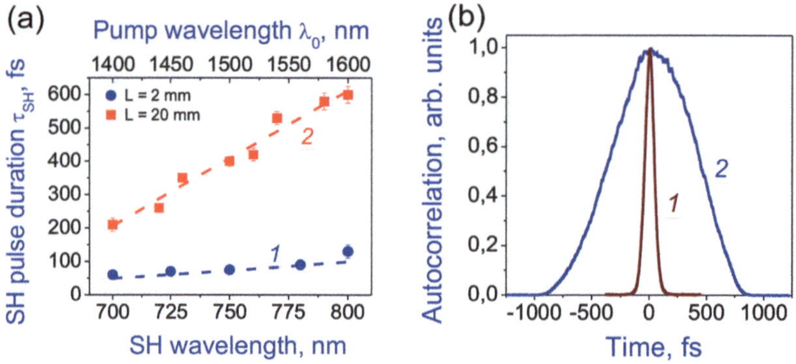

Fig. 5.5 (**a**) The pulse width of the second harmonic as a function of the pump (upper abscissa axis) and second-harmonic (lower abscissa axis) wavelengths: results of experiments using an LBO crystal with $L = 2$ mm (*circles*) and 20 mm (*rectangles*) versus calculations for $L = 2$ mm (*dashed curve 1*) and 20 mm (*dashed curve 2*). (**b**) The autocorrelation traces of (*1*) the 700-nm and (*2*) 790-nm second-harmonic output of (*1*) a 2-mm and (*2*) 20-mm-thick LBO crystal pumped by wavelength-shifted solitons from a 20-cm-long PCF

for $\lambda_0 = 1{,}500$ nm. In this case, $\tau_p \ll \theta$ and $\tau_{SH} \approx b(u_p^{-1} - u_{SH}^{-1})$, leading to a strong dependence of τ_{SH} on λ_0 controlled by the dispersion of the nonlinear crystal (curve 2 in Fig. 5.5a).

Typical autocorrelation traces of second-harmonic pulses produced with thin (2-mm) and thick (20-mm) BBO crystals are presented in Fig. 5.5b. These traces were measured using the SHG process in a 0.5-mm-thick BBO crystal. The second-harmonic output of a 2-mm LBO crystal centered at 700 nm (curve 1 in Fig. 5.5b) can be accurately fitted with a Gaussian envelope with an FWHM pulse width of 73 fs. For the 20-mm crystal, the pulse width of second-harmonic output ranged from 210 to 600 fs within the range of wavelengths from 700 to 800 nm. A typical autocorrelation trace for the second harmonic at 790 nm is shown by curve 2 in Fig. 5.5b, corresponding to an FWHM pulse width of 580 fs. The pulse width of the second-harmonic output measured as a function of λ_0 for 2-mm- and 20-mm-thick nonlinear crystals is shown by the circles and rectangles, respectively, in Fig. 5.5a. Results of these experiments agree well with theoretical predictions (dashed lines in Fig. 5.3a), demonstrating tunability of the second-harmonic pulse width within the range from 70 to 600 fs.

Simultaneous wavelength and pulse-width tunability provided by a combination of the SSFS and SHG technologies is especially helpful for nonlinear Raman spectroscopy and microscopy, where the wavelength tunability is needed for a selective excitation of Raman modes, while the pulse-width tunability is instrumental in reducing the coherent background [21], related to nonresonant four-wave mixing, in the overall coherent Raman response. In Fig. 5.6a, b, we present typical spectra of coherent anti-Stokes Raman scattering (CARS) obtained for synthetic diamond and polystyrene films. The pump field in these experiments had a fixed wavelength, $\lambda_p \approx 623$ nm, and was provided by the Cr: forsterite laser output frequency-doubled

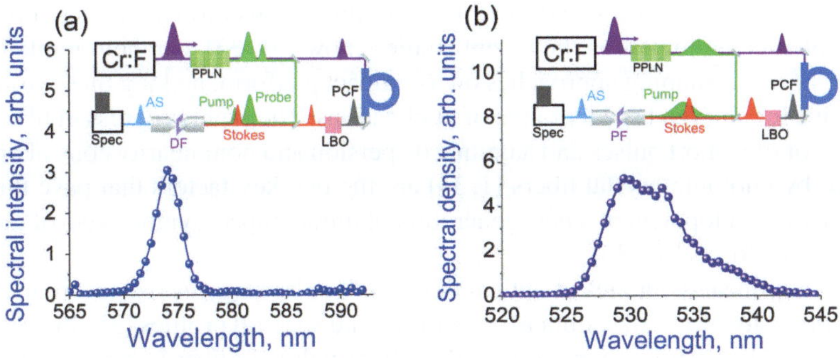

Fig. 5.6 CARS spectra of the 1,332-cm^1 zone-center Γ^{25+} (F_{2g}) symmetry optical phonon in a synthetic diamond film (**a**) and the 2,850-cm^1 CH vibrational mode of polystyrene (**b**): (*dots*) experimental data and (*solid lines*) theoretical fits assuming Lorentzian line profiles for the Raman lines. Diagrams of CARS arrangement are shown in the *insets*: *Cr: F*, Cr: forsterite laser; *PPLN*, periodically poled lithium niobate waveguide; *PCF*, photonic-crystal fiber; *DF*, diamond film; *PF*, polymer film; *AS*, anti-Stokes signal; *Spec*, spectrometer

in a periodically poled lithium niobate (PPLN) waveguide (see the inset in Fig. 5.6a). The Stokes field with a tunable wavelength λ_s was delivered by PCF-based SSFS combined with SHG as described above. In experiments with synthetic diamond, the 1,332-cm^1 zone-center Γ^{25+} (F_{2g}) symmetry optical phonon [22] was excited through the Raman process (sketched in the inset to Fig. 5.6b) by tuning the Stokes wavelength to $\lambda_s \approx 680$ nm. Coherently driven optical-phonon vibrations in diamond were read out with a probe pulse (the inset in Fig. 5.6a), also provided by the second harmonic output of the PPLN waveguide, giving rise to an anti-Stokes signal centered at $\lambda_a \approx 575$ nm (Fig. 5.6a). The probe pulse was applied with a delay time of 400 fs relative to the pump and Stokes pulses in order to reduce the coherent background in the overall signal [23]. In the case of polystyrene, the Stokes wavelength was tuned to $\lambda_s \approx 758$ nm in order to excite the CH vibrational mode at 2,850-cm^1. The probe pulse was applied simultaneously with the pump and Stokes pulses (the inset in Fig. 5.6b), giving rise to an anti-Stokes signal at $\lambda_a \approx 529$ nm observed against an intense coherent background due to the nonresonant four-wave mixing process (Fig. 5.6b).

5.2 Pulse Shaping Using Microstructures Fibers

Ultrafast optical waveforms with accurately controlled temporal shape and pulse width offer a powerful tool to control ultrafast molecular dynamics in physical, chemical, and biological systems [24–26], suggesting new approaches to optical bioimaging [27] and quantum information technologies [28,29]. Spatial light modulators [30] offer a remarkable flexibility in pulse shaping, enabling a broad variety of adaptive-control strategies. Although fiber sources of ultrashort light pulses cannot provide a similar precision and flexibility in optical waveform sculpting, advanced fiber-optic technologies, as the recent studies show [31–33], can help implement a variety of pulse-shaping approaches on a compact platform, making these strategies attractive for an even broader range of applications. The rapid progress in fiber laser sources of ultrashort pulses and a unique dispersion and nonlinearity control options offered by photonic-crystal fibers [1, 34] are the two key factors that pave the way toward the development of a new generation of ultracompact synthesizers of tailored optical waveforms [31–33].

Rich nonlinear dynamics of optical fields in fiber lasers gives rise to a remarkably broad diversity of ultrafast optical waveforms and spectral profiles [35, 36]. Spectral and temporal transformations of ultrashort light pulses in fiber lasers are, however, strongly coupled by the gain, optical nonlinearity, and dispersion, which makes a fine tuning of optical waveforms difficult. Here, we show that an integration of stretched-pulse ytterbium fiber lasers with hollow-core PCFs with a properly designed dispersion can help address this issue. In the earlier work, hollow-core PCFs have been shown to enable efficient pulse compression in fiber laser systems [37–39]. Here, we show that an integration of a stretched-pulse fiber laser with a properly designed hollow PCF enables waveform shaping of short-pulse fiber-laser output beyond the pulse compression option. Experiments presented below in this paper demonstrate such waveform shaping in the regime where a stretched fiber laser is adjusted to deliver a double-pulse output. This regime of mode-locked fiber laser operation is most often treated as unwanted, since it prevents the generation of a well-isolated ultrashort laser pulse. Here, however, we demonstrate that, when combined with a properly designed hollow PCF, a fiber laser operated in the double-pulse regime can serve as a compact source of tunable pairs of ultrashort pulses, offering an attractive solution for coherent control [24,25] and nonlinear microscopy [26, 27, 32, 33].

The fiber source of ultrashort light pulses used in our experiments was based on a home-built ytterbium fiber laser oscillator (Fig. 5.7) with a unidirectional ring cavity [40]. The Yb-doped fiber with a core diameter of 6 μm, numerical aperture of 0.14, and a loss of 250 dB/m at 976 nm was pumped by 980-nm, 600-mW laser-diode radiation, delivered and coupled to the active fiber through a standard wavelength-division multiplexer. Stable mode locking was achieved in the fiber laser through nonlinear polarization evolution with the use of in-line polarizers and optical isolators [35]. With a net positive dispersion provided by this cavity design, stretched laser pulses with a central wavelength of 1.04–1.05 μm were generated at the output of the fiber laser oscillator at a repetition rate of 43 MHz. A standard diffraction-grating pulse compressor was used to dechirp the stretched-pulse fiber laser output and to provide reference measurements for the assessment of the compression ability of the hollow PCF.

The hollow-core PCF used in our experiments guides light due to the photonic band gaps (PBGs) of the periodically structured cladding [34]. The 15-μm-diameter hollow core of these fibers, formed by the omission of 19 canes in the central part of the waveguide, is substantially larger than the core diameter in standard hollow-core PCFs [34] usually employed for pulse compression [38, 39], where

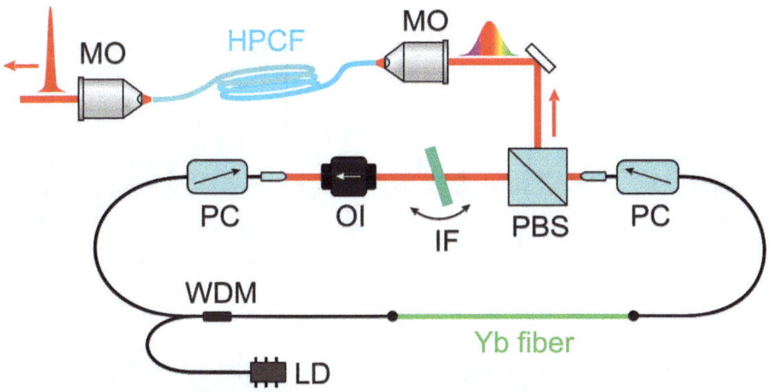

Fig. 5.7 Diagram of the ytterbium fiber laser. Beam coupling into a hollow-core PCF is shown in the *inset*: *MO*, micro-objectives; *HPCF*, hollow-core PCF; *PC*, polarization controllers; *OI*, optical isolator; *PBS*, polarization beam splitter; *IF*, interference filter; *WDM*, wavelength division multiplexer

seven canes are omitted for a hollow core. Due to the larger core diameter, our hollow PCF allowed the delivery of much higher laser energies (up to 8 μJ for 9-ps pulses at 1.04–1.06 μm), suggesting a promising fiber solution for the transportation and transformation of amplified fiber laser output. The large core diameter also helps reduce polarization-mode dispersion. The group delay between orthogonally polarized modes in our PCF was estimated to be on the order of 10 fs/m, having no effect on the performance of the fiber in our experiments. As a penalty, the group-velocity dispersion (GVD) of our large-core hollow PCF was about 2.5 times lower than the GVD of standard, smaller-core hollow PCFs used for pulse compression [38, 39]. The guidance loss of the hollow PCF was estimated from measurements performed on PCF sections of variable lengths yielding a loss below 0.1 dB/m at 1.046 μm.

In Fig. 5.8a, b, we present typical spectral profiles and pulse shapes of the Yb fiber laser output, illustrating two distinctly different regimes of this laser. In accordance with the general tendencies identified in the earlier work [38, 39], larger nonlinear phase shifts of the laser pulses, controlled in our experiments by the in-line polarizers, give rise to broader output spectra with well-pronounced peaks around their edges (cf. Fig. 5.8a). When the bandwidth of the laser output is increased to 18 nm (Fig. 5.8a), these peaks translate into a double-pulse structure in the time domain (Fig. 5.8b), clearly visualized by the FROG trace in Fig. 5.8c. The light pulses delivered by the Yb fiber laser were coupled into the hollow PCF by a micro-objective with a numerical aperture of 0.4. The Yb fiber laser output and light pulses transmitted through the hollow PCF were characterized by means of frequency-resolved optical gating (FROG), implemented through second-harmonic generation in a 0.7-mm-thick KDP crystal. The spectra of laser pulses retrieved from FROG traces (Fig. 5.8) were verified against the spectra measured using a standard Ocean Optics spectrometer.

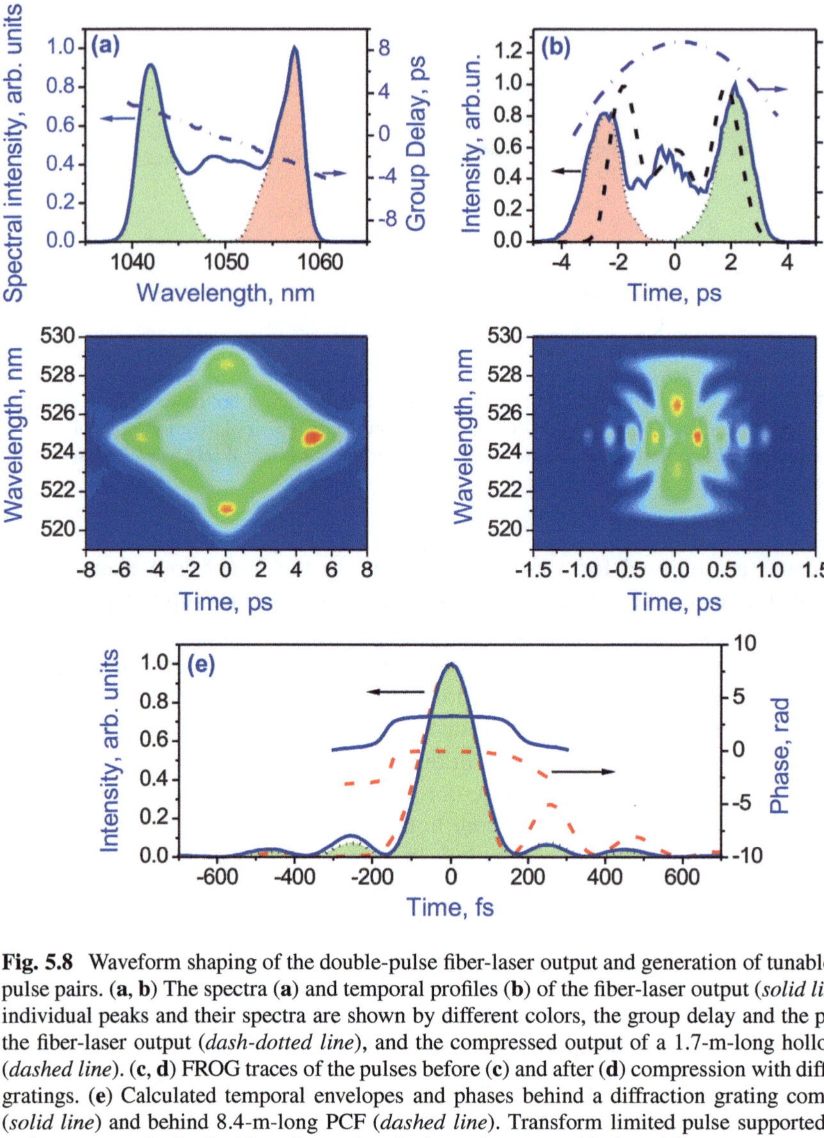

Fig. 5.8 Waveform shaping of the double-pulse fiber-laser output and generation of tunable short-pulse pairs. (**a, b**) The spectra (**a**) and temporal profiles (**b**) of the fiber-laser output (*solid line*), the individual peaks and their spectra are shown by different colors, the group delay and the phase of the fiber-laser output (*dash-dotted line*), and the compressed output of a 1.7-m-long hollow PCF (*dashed line*). (**c, d**) FROG traces of the pulses before (**c**) and after (**d**) compression with diffraction gratings. (**e**) Calculated temporal envelopes and phases behind a diffraction grating compressor (*solid line*) and behind 8.4-m-long PCF (*dashed line*). Transform limited pulse supported by the entire spectrum in the double-pulse regime is shown in *green* with *dotted-line* contours

The individual peaks in the spectrum of the double-pulse fiber laser output (colored in pink and green in Fig. 5.8a) translate, through the Fourier transform, into isolated pulses in the time domain (colored in pink and green, respectively, in Fig. 5.8b). The transform-limited pulse width supported by the spectrum of the double-pulse fiber laser output (the FWHM pulse width is $\tau_0 \approx 150$ fs, the dotted line with green shading in Fig. 5.8e) is substantially shorter than the transform-limited pulse width corresponding to the 10-nm bandwidth of the laser output in the regime of low nonlinear phase.

As can be seen from the GVD profile retrieved from the FROG traces, dispersion of the hollow PCF is anomalous within the entire region covered by the spectrum of the Yb fiber laser. Such a dispersion profile provides compression for the Yb fiber laser output as a whole (cf. solid and dashed curves in Fig. 5.8b), enabling a smooth (along the fiber length) tuning of the time interval between the pulses in the double-pulse output of this laser. Indeed, the first pulse (pink in Fig. 5.8a, b) is red-shifted with respect to the second pulse (green in Fig. 5.8a, b) with virtually no overlap between the spectra of these pulses. The group delay between these pulses at the output of the fiber laser is 4.7 ps. Transmission of this pulse pair through a 170-cm-long piece of anomalously dispersive hollow PCF reduces this time interval between the pulses from 4.7 to 3.6 ps (cf. the two-color pulse profile with the dashed curve in Fig. 5.8b).

We now use the results of these experiments along with the GVD profile of the hollow PCF retrieved, as described above, from our experimental data to simulate the propagation of stretched laser pulses with spectra and envelopes of the Yb fiber laser output as shown in Fig. 5.8a, b in longer sections of the same fiber. This analysis shows that, in the double-pulse regime of fiber laser, longer propagation paths L inside the fiber yield, due to the character of fiber dispersion, pulse pairs with a smoothly decreasing time separation between the pulses until the pulses merge together at $L \approx 8.4$ m. At this point of maximum compression, the FWHM pulse width of this optical waveform ($\tau_0 \approx 159$ fs, dashed line in Fig. 5.8e) is only a factor of 1.06 longer than the FWHM transform-limited pulse width ($\tau_0 \approx 150$ fs, dotted line with green shading in Fig. 5.8e) supported by the entire bandwidth of the double-pulse fiber laser output. The satellites in the pulse compressed by the hollow PCF are due to the residual cubic phase (dashed line in Fig. 5.8e), estimated for this experiment as $\phi^{(3)} \approx 2.2 \times 10^6$ fs^3 at 1.049 μm.

With the fiber dispersion profile of hollow PCF designed to closely match the target dispersion of an ideal pulse compressor, a hollow PCF can outperform a standard diffraction-grating pulse compressor, providing a much more accurate compensation of phase distortions on a stretched-pulse fiber laser output. Pulse compression with a hollow PCF delivers a shorter FWHM pulse width ($\tau_0 \approx 236$ fs) and smaller cubic and quartic phase distortions ($\phi^{(3)} \approx -6.0 \times 10^6$ fs^3, $\phi^{(4)} \approx 8.1 \times 10^6$ fs^4 at 1.045 μm) compared to the diffraction-grating pulse compressor, which yields$\tau_0 \approx 256$ fs, $\phi^{(3)} \approx -9.5 \times 10^6$ fs^3, and $\phi^{(4)} \approx 11.4 \times 10^6$ fs^4.

5.3 Conclusions

In conclusion, SSFS in a PCF pumped by a long-cavity mode-locked Cr: forsterite laser has been integrated with SHG in a nonlinear crystal to generate 20-MHz-repetition-rate ultrashort light pulses wavelength-tunable from 680 to 1,800 nm and pulse-width-tunable from 70 to 600 fs. Nonlinear crystals with properly optimized thickness and beam focusing are shown to enable up-to-40 %-efficiency frequency doubling with simultaneous spectral narrowing of the soliton PCF output, yielding ultrashort light pulses with an average power up to 10 mW ideally suited for coherent Raman microspectroscopy.

To summarize, we have demonstrated that an all-fiber system integrating a stretched-pulse fiber laser and a hollow PCF enables ultrafast optical waveform tailoring, allowing, among other applications, the creation of compact fiber-format sources for coherent control, nonlinear Raman microscopy, quantum information technologies and multicolor brain imaging. Specifically, pairs of ultrashort pulses with a pulse separation tunable on the picosecond scale demonstrated in our experiments can radically enhance the Raman signal relative to the coherent background in CARS microscopy.

References

1. Russell P (2003) Photonic crystal fibers. Science 299(5605):358–362
2. Dudley JM, Coen S (2006) Supercontinuum generation in photonic crystal fiber. Rev Mod Phys 78:1135–1184
3. Zheltikov AM (2004) Nonlinear optics of microstructure fibers. Physics-Uspekhi 47:69–98
4. Limpert J, Roser F, Schreiber T, Tunnermann A (2006) High-power ultrafast fiber laser systems. IEEE J Sel Top Quantum Electron 12:233–244
5. Knight JC (2007) Photonic crystal fibers and fiber lasers (Invited). J Opt Soc Am B 24(8):1661
6. Holzwarth R, Udem T, Hänsch T, Knight J, Wadsworth W, Russell P (2000) Optical frequency synthesizer for precision spectroscopy. Phys Rev Lett 85:2264–2267
7. Udem T, Holzwarth R, Hänsch TW (2002) Optical frequency metrology. Nature 416:233–237
8. Corkum PB, Krausz F (2007) Attosecond science. Nat Phys 3:381–387
9. Teisset CY, Ishii N, Fuji T, Metzger T, Köhler S, Holzwarth R, Baltuska A, Zheltikov AM, Krausz F (2005) Soliton-based pump-seed synchronization for few-cycle OPCPA. Opt Express 13(17):6550
10. Andresen ER, Birkedal V, Thø gersen J, Keiding SRR (2006) Tunable light source for coherent anti-Stokes Raman scattering microspectroscopy based on the soliton self-frequency shift. Opt Lett 31(9):1328
11. Sidorov-Biryukov DA, Serebryannikov EE, Zheltikov AM (2006) Time-resolved coherent anti-Stokes Raman scattering with a femtosecond soliton output of a photonic-crystal fiber. Opt Lett 31(15):2323
12. Deisseroth K (2011) Optogenetics. Nat Methods 8:26–29
13. Doronina LV, Fedotov IV, Voronin AA, Ivashkina OI, Zots MA, Anokhin KV, Rostova E, Fedotov AB, Zheltikov AM (2009) Tailoring the soliton output of a photonic crystal fiber for enhanced two-photon excited luminescence response from fluorescent protein biomarkers and neuron activity reporters. Opt Lett 34:3373–3375
14. Doronina-Amitonova LV, Fedotov IV, Ivashkina OI, Zots MA, Fedotov AB, Anokhin KV, Zheltikov AM (2011) Photonic-crystal-fiber platform for multicolor multilabel neurophotonic studies. Appl Phys Lett 98:253706
15. Liu H, Hu M, Liu B, Song Y, Chai L, Zheltikov AM, Wang C (2010) Compact high-power multiwavelength photonic-crystal-fiber-based laser source of femtosecond pulses in the infraredvisibleultraviolet range. J Opt Soc Am B 27:2284
16. Liu F, Song Y-J, Xing Q-R, Hu M-L, Li Y-F, Wang C-L, Chai L, Zhang W-L, Zheltikov A, Wang C-Y (2010) Broadband terahertz pulses generated by a compact femtosecond photonic crystal fiber amplifier. Photon Technol Lett IEEE 22:814–816
17. Fedotov AB, Voronin AA, Fedotov IV, Ivanov AA, Zheltikov AM (2009) Powerful wavelength-tunable ultrashort solitons in a solid-core photonic-crystal fiber. Opt Lett 34:851
18. Akhmanov S, Sukhorukov A, Chirkin A (1969) Nonstationary phenomena and space-time analogy in nonlinear optics. Sov Phys JETP 28(4):748–757

19. Comly J, Garmire E (1968) Second harmonic generation from short pulses. Appl Phys Lett 12
20. Kato K (1994) Temperature-tuned 90 deg; phase-matching properties of LiB_3O_5. Quantum Electron IEEE J 30:2950–2952
21. Marangoni M, Gambetta A, Manzoni C, Kumar V, Ramponi R, Cerullo G (2009) Fiber-format CARS spectroscopy by spectral compression of femtosecond pulses from a single laser oscillator. Opt Lett 34:3262–3264
22. Levenson M, Flytzanis C, Bloembergen N (1972) Interference of resonant and nonresonant three-wave mixing in diamond. Phys Rev B 6:3962–3965
23. Lanin AA, Fedotov AB, Zheltikov AM (2012) Ultrafast three-dimensional submicrometer-resolution readout of coherent optical-phonon oscillations with shaped unamplified laser pulses at 20 MHz. Opt Lett 37:1508–1510
24. Warren WS, Rabitz H, Dahleh M (1993) Coherent control of quantum dynamics: the dream is alive. Science (N. Y.) 259:1581–1589
25. Assion A, Baumert T, Bergt M, Brixner T, Kiefer B, Seyfried V, Strehle M, Gerber G (1998) Control of chemical reactions by feedback-optimized phase-shaped femtosecond laser pulses. Science 282(5390):919–922
26. Herek JL, Wohlleben W, Cogdell RJ, Zeidler D, Motzkus M (2002) Quantum control of energy flow in light harvesting. Nature 417:533–535
27. Dudovich N, Oron D, Silberberg Y (2002) Single-pulse coherently controlled nonlinear Raman spectroscopy and microscopy. Nature 418:512–514
28. Naumov AN, Materny A, Kiefer W, Motzkus M, Zheltikov AM (2001) Reversible computations and ultrafast logic gates by coherent multiwave mixing supplemented with quantum control. Laser Phys 11(12):1319–1323
29. De Greve K, McMahon PL, Press D, Ladd TD, Bisping D, Schneider C, Kamp M, Worschech L, Höfling S, Forchel A, Yamamoto Y (2011) Ultrafast coherent control and suppressed nuclear feedback of a single quantum dot hole qubit. Nat Phys 7:872–878
30. Weiner AM (2000) Femtosecond pulse shaping using spatial light modulators. Rev Sci Instrum 71:1929
31. Zheltikov AM (2007) Fiber-optic synthesizer of controlled sequences of ultrashort light pulses for single-beam coherent anti-Stokes Raman scattering microspectroscopy. JETP Lett 85:539–543
32. Ivanov A, Linik Y, Akimov D, Alfimov M, Siebert T, Kiefer W, Zheltikov A (2006) Coherent Raman spectroscopy with frequency-shifted and shaped pulses from a photonic-crystal fiber. Chem Phys Lett 418:19–23
33. Lanin AA, Fedotov AB, Zheltikov AM (2012) Broadly wavelength- and pulse width-tunable high-repetition rate light pulses from soliton self-frequency shifting photonic crystal fiber integrated with a frequency doubling crystal. Opt Lett 37:3618–3620
34. Knight JC (2003) Photonic crystal fibres. Nature 424:847–851
35. Lim H, Ilday FO, Wise FW (2003) Generation of 2-nJ pulses from a femtosecond ytterbium fiber laser. Opt Lett 28:660
36. Chong A, Renninger WH, Wise FW (2008) Properties of normal-dispersion femtosecond fiber lasers. J Opt Soc Am B 25(2):140
37. Lim H, Wise FW (2004) Control of dispersion in a femtosecond ytterbium laser by use of hollow-core photonic bandgap fiber. Opt Express 12(10):2231
38. Nielsen CK, Jespersen KG, Keiding SR (2006) A 158 fs 5.3 nJ fiber-laser system at 1 m using photonic bandgap fibers for dispersion control and pulse compression. Opt Express 14(13):6063
39. Liu X, Laegsgaard J, Turchinovich D (2010) Highly-stable monolithic femtosecond Yb-fiber laser system based on photonic crystal fibers. Opt Express 18:15475–15483
40. Tamura K, Jacobson J, Ippen EP, Haus HA, Fujimoto JG (1993) Unidirectional ring resonators for self-starting passively mode-locked lasers. Opt Lett 18:220

6
Prospects of Molecular Scale Logic Gates and Logic Circuits

Shammai Speiser

Abstract In recent years molecular electronics has emerged as a rapidly growing research field. The aim of this review is to introduce this subject as a whole with special emphasis on molecular scale potential devices and applications. As a particular example we will discuss all optical molecular scale logic gates and logic circuits based on molecular fluorescence and electronic excitation transfer processes. Charge and electronic energy transfer (ET and EET) are well-studied examples whereby different molecules can signal their state from one (the Donor, D) to the other (the Acceptor, A). We show how a half-adder logic circuit can be implemented on one molecule that can communicate its logic output as input to another half-adder molecule. This is achieved as an electronic energy transfer from a donor to an acceptor, thus implementing a molecular full adder. We discuss a specific pair, the rhodamine-azulene, for which there is considerable spectroscopic data, but the scheme is general enough to allow a wide choice of D and A pairs. We present results based on this pair, in which, for the first time, an all optical half-adder and full-adder logic circuits are implemented.

6.1 Introduction

The need for increasing miniaturization of logical circuits will soon reach the physical limit of MOSFET (Metal Oxide Semiconductor Field Effect Transistor) devices. The fabrication of small semiconductor based devices becomes more difficult and expensive as the nanometric scale is approached. Furthermore, bulk properties of semiconductors vanish at the nanometric level. Further miniaturization of electronic devices is expected to open the way to new technologies.

S. Speiser (✉)
Schulich Faculty of Chemistry, Technion – Israel Institute of Technology, Haifa 32000, Israel
e-mail: speiser@technion.ac.il

B. Di Bartolo et al. (eds.), *Nano-Structures for Optics and Photonics*, NATO Science for Peace and Security Series B: Physics and Biophysics,
DOI 10.1007/978-94-017-9133-5_6,

Since our research focuses on molecular-level devices e.g. molecular logic gates, there is the need to define such a device. When extending the macroscopic concept of a device to the molecular level one can define a molecular scale device as an assembly of a discrete number of molecular components designed to perform a specific function. Each molecular component performs a single act, while the entire supramolecular assembly performs a more complex function, which results from the cooperation of the various components. Such a molecular-level device operates via electronic and/or nuclear rearrangements and, like a macroscopic device, needs energy to operate and signals to communicate with the operator [1]. Single molecule based switches [2, 3], rectifiers [4, 5] and wires [6] have been reported already. The extension therefore to more complicated schemes, such as discussed in this paper is only natural.

Organic molecules are promising candidates for the realization of electronic and photonic devices. They have nanometric dimensions and their properties can be tuned through chemical modifications. Organic molecules can be easily produced and are relatively cheap to make, even in large quantities. It is important to note that the operating mechanisms of a device based on organic molecules have little in common with a conventional electronic one. Switching systems based on acid/base reactions, on redox processes, on photochromic behavior, on optical control of chirality, on conformational changes, on electrochemically and photochemically induced changes in liquid crystals, thin films and membranes and in supramolecular structures, and on photoinduced electron and energy transfer have been synthesized and studied [7, 8].

Logic gates of microprocessor systems are assembled by interconnecting transistors, and their input and output are electrical. However, the concepts of binary logic can be extended to chemical, optical or other types of signals. Thus, it may be necessary to design devices that would respond to signals in the same way transistors respond to the electrical stimulations. Data processing and communication require the encoding of information in electrical and optical signals in the form of binary digits. A threshold value and a logic convention are established for each signal. In a positive logic convention, a **0** is used to represent a signal that is below the threshold, and a **1** is employed to indicate a signal that is above it [9]. Each logic gate performs according to a set of rules arranged in so-called truth table, which is a list of gate outputs resulting from certain inputs. The **AND** gate executes a binary product, $\otimes$, while the **XOR** gate performs a binary sum, $\oplus$ [9].

The half adder is the basic component of computational arithmetic in digital computers. Its function is to add two one-digit binary numbers together. Thus, the **AND** and **XOR** are the two logic operations that need to be implemented on the two binary inputs in order to build a half adder, working in parallel. By concatenation of two such half adders one can build a full adder. A binary full adder accepts the two digits that are to be added, and also the third input, namely the "carry in" digit from a previous addition, where the summation is completed. A full adder needs to produce two outputs, the so-called "sum out", which is the **XOR** sum of the two inputs, and an output called the "carry out", which is the **AND** product of these inputs that becomes the carry in for the next addition cycle.

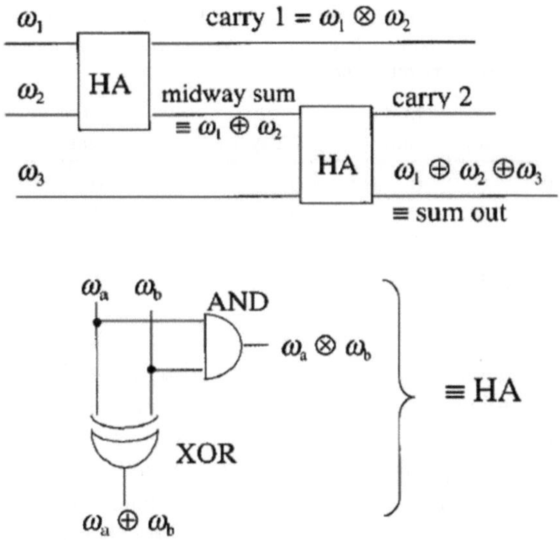

Fig. 6.1 Representation of a full adder constructed from two communicating half adders (**HA**) and the symbolic representation of a half adder

Table 6.1 Truth table for a full adder

ω_1	ω_2	Carry in $\equiv \omega_3$	Sum out	Carry out
0	0	0	0	0
1	0	0	1	0
0	1	0	1	0
0	0	1	1	0
1	1	0	0	1
0	1	1	0	1
1	0	1	0	1
1	1	1	1	1

The symbolic representations of the half and full-adder logic circuits are shown in Fig. 6.1. In Fig. 6.1 the full adder is made by concatenation of two half adders. Each half adder (**HA**) is constructed, using **AND** and **XOR** logic gates, where ω_1, ω_2 and ω_3 are Boolean input variables. In our study these will be laser inputs at specific frequencies exciting specific molecular electronic states. The **XOR** gate evaluates the Boolean function $\omega_1 \oplus \omega_2$, while the **AND** gate evaluates the Boolean product $\omega_1 \otimes \omega_2$, which is unity, if both lasers are on, and $\omega_1 \oplus \omega_2 \oplus \omega_3$ is defined as sum out. The full adder has three outputs: carry 1, carry 2 and the sum out. Either carry 1 or carry 2 will provide the value of output carry bit. The midway sum in our scheme will be communicated by energy transfer to the second **HA**. The corresponding truth tables for the above operations are given in Tables 6.1 and 6.2.

As mentioned above certain organic molecules change their structural and electronic properties due to stimulation by chemical (in the form of protons, metal ions, specific molecules, etc.), electrical (electrons or holes) or optical (photon) inputs. A molecule switches from one state to another when stimulated by an

Table 6.2 Truth table for the half-adder components of a full adder. (*Left*) First half adder that has the two subcommands as input. (*Right*) Second half adder that has the two the midway sum and the input carry bit as input

ω_1	ω_2	Midway sum	Carry 1
0	0	0	0
1	0	1	0
0	1	1	0
1	1	0	1

Midway sum	Carry in $\equiv \omega_3$	Sum out	Carry 2
0	0	0	0
1	0	1	0
0	1	1	0
1	1	0	1

input signal. In photochemical stimulation the most common switching processes are related to phoisomerization or photoinduced redox reactions. If the input is electrochemical, the induced processes are redox reactions [10–12]. Compared with chemical stimulation, photochemical and electrochemical stimulations have the advantage that they can be switched on and off easily and rapidly. Most of the transformations are reversible: the chemical system returns to its original state when the input signal is turned off. In some cases the molecular switch produces a chemical, electrical and/or optical output that varies in magnitude with the switching process.

Since both conventional logic gates and molecular switches convert input stimulations into output signals with intrinsic protocols, the principles of binary logic can be applied to the signal transduction operated by molecular switches [1, 7]. The analysis of their logical behavior requires, first of all, the assignment of threshold values and logic conventions to their input and output signals. Making an analogy to CMOS logic circuits, the signal transduction protocol of a molecular switch can be programmed to execute specific logic functions simply by selecting the initial logic assumptions. Various molecular based logic schemes have been discussed. However, no **all optical** logic gates and schemes have been demonstrated. The most trivial gate would be a **YES logic gate** [13]. The **NOT logic gate**, which inverts any signal received is easily achieved in chemical systems because of the ease with which luminescence (output) can be quenched by chemical input. An example of **NOT** chemical gate based on fluorescence quenching by photoinduced electron transfer (PET) [14] was reported.

The first example of molecular scale logic gate especially designed to implement **AND gate** was reported by de Silva et al. [15]. As in the case of **NOT** molecular logic gate, the operation of the AND gate is based on PET process, where the inputs are chemical and the output are fluorescence signals. Similarly OR logic gate based on requirement of non-selective luminescence enhancement, e.g. a set of non selective receptors giving a positive response upon binding has been demonstrated [16, 17]. An example of **XOR logic gate** which functions with two chemical inputs and an optical transmission output has also been reported [18]. Recently Margulies et al. [19] reported a fully reversible molecular half adder, based on controlling the absorption spectra of flourescein dye in solution by changing the pH conditions.

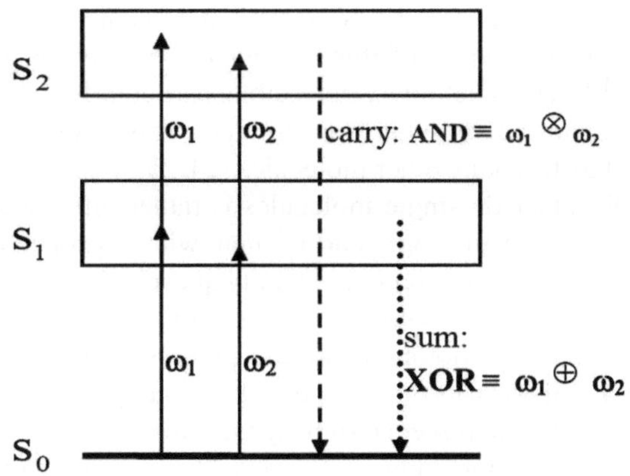

Fig. 6.2 Photophysical scheme of an **all optical** molecular half adder

Molecular full-adder logic circuit is one of the great challenges in design of molecular logic. A simulation of a molecular full adder based on 2-phenylethyl-N,N-dimethylamine (PENNA) molecule, where the inputs were provided optically, was recently reported by Remacle et al. [20]. Threshold values for output reading were obtained by a numerical solution of the equations of motion for the population as a function of laser pulse duration. In their scheme three distinct inputs are necessary to implement a full adder on PENNA molecule: two UV photons (x and y inputs) and carry in input (excitation by two green photons). Fluorescence from the S_1 state of the phenyl ring codes for carry out = 0, sum out = 1; ion fragmentation at the amine end codes for carry out = 1, sum out = 0; and ion fragmentation at the chromophore end for carry out = 1, sum out = 1. Most of the outputs are detected due to photoionization of the molecule, which means that the molecule self-destructs at the end of the computation.

We have suggested that a molecular half adder can be implemented on molecules that have detectable one photon absorption and a detectable two photon absorption [21]. Figure 6.2 depicts such a molecular half-adder scheme. Here the carry output is probed as $S_2 \rightarrow S_0$ fluorescence only when both ω_1 and ω_2 photons are present, whereas the sum output is probed as $S_1 \rightarrow S_0$ fluorescence for either ω_1 or ω_2 photons excitation. The degenerate case of such a half adder is also possible when $\omega_1 = \omega_1$. The implementation of the corresponding truth table (Table 6.2) for a single molecule is thus achieved. Molecular fluorescence from S_2 excited state seems to be against Kasha's rule. According to Kasha's rule [22] photon emission is expected only from lowest excited electronic state, S_1, of a molecule. This is because of extremely efficient nonradiative decay the S_2 state compared to the less efficient nonradiative relaxation S_1. Azulene (**Az**) and its derivates provide an exception to this general rule. For example, the rate of internal conversion ($S_2 \rightarrow S_1$) in **Az** is about $7 \times 10^8\ \mathrm{s}^{-1}$, which is exceptionally slow [23]. Actually, for this class of molecules the emission from the second electronically excited state is often stronger than the fluorescence from S_1 state. However, emission from S_2 state is not generally forbidden in most molecules, but due to the competing nonradiative processes it often has a low quantum yield. In spite of this low yield the emission from the second electronically excited state is definitely detectable, particularly so because it is blue shifted in comparison to the emission from S_1 state. This has been verified for rhodamine (**Rh**) as well as for many other molecules [24, 25].

Implementation of a half adder on single molecules is rather difficult so when employing a given number of molecules only part of them will be excited to either state. In a single molecule $S_1 \rightarrow S_0$ fluorescence is totally quenched upon absorbing both ω_1 and ω_2 photons. However, in solution only a portion of the molecules will be photoquenched by this two photon absorption process. Still by using the photoquenching model [24, 26] we can define a threshold pump intensity below which the laser induced $S_1 \rightarrow S_0$ fluorescence sum signal is low and can be defined as a binary 0 and that the corresponding induced $S_2 \rightarrow S_0$ fluorescence carry signal is detectable so that it can be defined as a binary **1**.

Experimentally the photoquenching effect is manifested by decrease of the fluorescence quantum yield with increasing exciting light intensity. As this intensity increases, the transient population in S_1 excited level decreases due to excitation to higher S_n states. Thus, this absorbed photon does not contribute to the fluorescence yield for S_1, causing the decrease in the S_1 relative fluorescence quantum yield. Kinetic analysis, assuming ultra-fast radiationless decay of the $S_n(n)$ state back to S_1, yields a "Stern-Volmer" type photoquenching relation (for $n = 2$) given by [26]

$$\phi_0/\phi = 1 + \tau_{10}\sigma_{12}I_p \quad (6.1)$$

where ϕ is the intensity dependent realtive fluorescence quantum yield, ϕ_0, is the realtive quantum yeild in the absence of laser pumping, τ_{10} is the fluorescence lifetime in the absence of laser pumping, σ_{12} is the absorption cross section for $S_1 \rightarrow S_1$ transition (cm^2/mol), and I_p is the laser pump intensity (photon/cm^2 s). In this scheme only determination of the **relative** values of the quantum yields is required. Equation (6.1) allows for **arbitrarily** setting the LIF intensity below which it can defined as **0** and define the value of I_p corresponding to it as input signal **1**. In a full adder the two half adders communicate with each other through the intermediate "midway sum" output. We have previously proposed a model for a molecular full adder [21] based on the utilization of two molecular half adders, where the midway sum is communicated to the second one via an intra- or inter-molecular electronic energy transfer (EET) [27] between donor and acceptor half adders. Such a general scheme is depicted in Fig. 6.3, where k_T is the transfer rate constant. This can be associated with an excitation transfer process such as electron transfer, EET, [27] or proton transfer process. The detected outputs of such a molecular full adder are: **carry 1** – $S_2 \rightarrow S_0$ fluorescence from the donor half adder, **carry 2** – $S_2 \rightarrow S_0$ fluorescence and **sum out** – $S_1 \rightarrow S_0$ fluorescence signals from the acceptor half adder. Carry 1 and carry 2 outputs are probed for each input combinations of laser frequencies.

We have suggested that the Rhodamine (**Rh**) – Azulene (**Az**) molecular system can be used as such a molecular full adder, where the communication between the **Rh** and **Az** half adder is by $S_1 \rightarrow S_1$ intermoecular EET [21]. Steer and coworkers

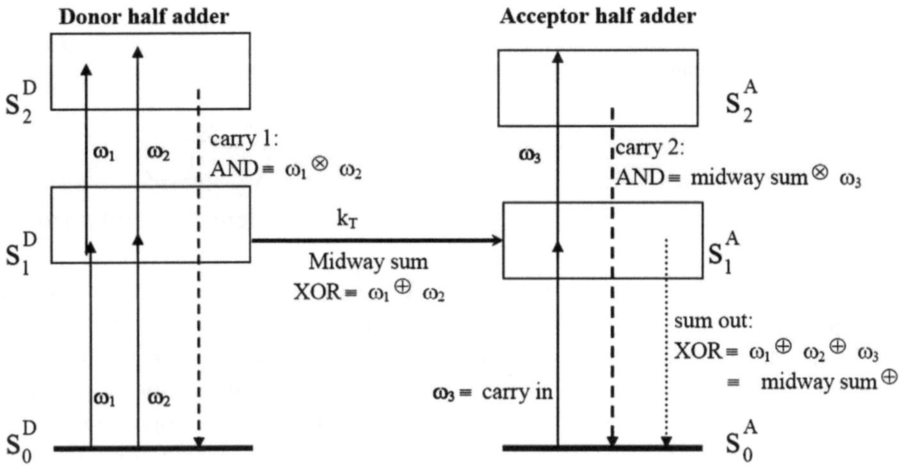

Fig. 6.3 Photophysical scheme of an **all optical** molecular full adder

attempted to realize the proposed molecular half adder and full-adder schemes utilizing a different molecular system [28–30]. They have chosen the azulene (donor)-zinc porphyrin (acceptor) couple and thoroughly studied the $S_2 \rightarrow S_2$ intermolecular EET process is solution and in micelles30 as well as the sequential forward $S_2 \rightarrow S_2$ and the back $S_1 \rightarrow S_1$ cyclic intramolecular EET in the **Az-zinc porphyrin** diad [29]. The feasibility of using **Az** as a half adder was discussed. Subsequently they have analyzed a molecular logic circuit based on this scheme where a computational dynamic simulation have indicated the possibility of implementing a molecular logic circuit in this system [30]. However, no experimental threshold values for the half adder input and output functions were established.

We have already discussed the full adder implemented on a novel **Rh-Az** bichromophoric molecule [31, 32]. It is the purpose of the present paper to provide evidence for the implementation of half-adder and full-adder functions in the **Rh-Az** intermolecular system, providing threshold values for the input and output signals, thus creating an experimental truth table that conforms to Table 6.2.

6.2 Experimental

Figure 6.4 shows the newly synthesized **4-azulene-1-ly-butylamine (Az-amine)** molecule used in this study. The synthesis of this molecule, and the bichromophric molecule **Rh-Az**, is described elsewhere [27]. Commercially available (Aldrich) spectral grade **RhodamineB isothiocyanate, Rh6G** (Sigma) and spectral grade methanol (Riedel-de-Haen) were used without further purification. Standard absorption measurements were made on Shimadzu UV-1601 UV-VIS spectrophotometer and fluorescence and excitation spectra and were taken on a Perkin-Elmer LS 50 spectrofluorimeter, using standard 1 cm length cuvettes.

Fig. 6.4 The newly synthesized 4-azulene-1-ly-butylamine (Az-amine)

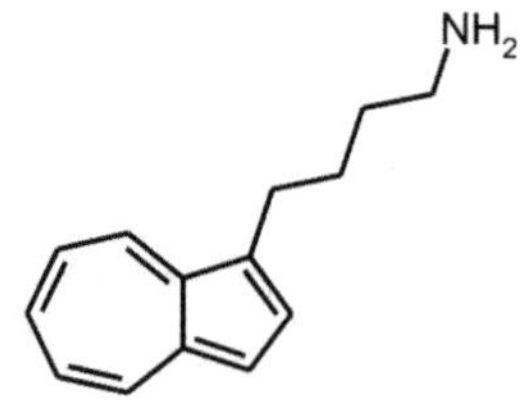

4-azulene-1-ly-butylamine

For most measurements of **Rh** and of **Az-ammine** emission spectroscopy, a methanol solution, at a concentration corresponding approximately to O. D. = 0.1 of the **Rh** moiety, at the 545 nm peak of the $S_0 \rightarrow S_1$ absorption, was kept. The same was applied for the other molecules such as **4-azulene-1-ly-butylamine (Az-amine)** molecule for which 6.5×10^{-6} M concentration, corresponding approximately to O. D. = 0.2 at 344 nm, of the $S_0 \rightarrow S_2$ absorption peak, was kept along most of the measurements. Highly concentrated solutions of 4×10^{-4}M were used for the characterization of the very weak absorption to the S_1 excited state of the **Az-amine** molecules. No aggregation effects were observed even in these high concentrations.

Laser induced fluorescence (LIF) experiments were conducted using either excitation by two different laser beams at 532 nm and at 1,064 nm, or in the degenerate excitation scheme, by just 532 nm excitation. The laser system (Continuum) operating at 10 Hz repletion rate used for the LIF experiments was based on Nd:YAG laser (Powerlite), which produces a Gaussian laser beam shape, with a 0.1 cm^2 cross section, of the fundamental frequency at 1,064.1 nm. The fundamental frequency was doubled to generate 532 nm second harmonic or doubled and mixed with the fundamental frequency to produce the 355 nm third harmonic. The different harmonics were separated using a set of dichroic mirrors. The pulse width of the fundamental was 6 ns. LIF spectra were recorded using a Spectra Pro 275, 0.275 m triple grating monochromator equipped with a stepper motor and a Hammamatsu R-298 photomultipler. The normalized fluorescence signal was processed using a Tektronix TDS 220 two channel digital oscilloscope, utilizing National Instruments LabView 5.0 software. The photoquenching experiments involved intensity dependent LIF of either utilizing a single frequency single beam or two intersecting beams of different laser frequencies. Laser intensities were determined by using an OPHIR model NOVA II power meter.

6.3 Results and Discussion

6.3.1 Spectroscopic Characterization of the Rohdamine-Azulene Molecular System

The S_2 state in **Az** is its dominant fluorescent state, thus EET studies of **Az** as a donor, only involving S_2, were carried out [28, 29, 33]. Since the S_1 state in **Az** has a low fluorescence yield its EET process to **Rh6G** was not studied before.

However, without such an efficient EET process to **Rh6G** S_0 state a molecular full adder can not be realized in the **Rh-Az** system. Therefore an investigation of this EET process was carried out in the course of this work, the details of which are published elsewhere [32]. In addition we have studied the relevant photophysical parameters involved in realizing a full adder based on this molecular system. Table 6.3 summarizes our previously-measured [32] spectroscopic parameters of rhodamine and of **Az-amine** half-adder systems and their intermolecular EET data, together with those measured in the present study (see next sections) that fit the schemes of Fig. 6.2, with Fig. 6.3. The following photophysical parameters were measured. relative emission intensities from S_1 and S_2 excited levels of **Rh6G** and **RB**, $S_1 \rightarrow S_0$ and $S_2 \rightarrow S_0$ relative fluorescence intensities of **Az-amine** [31, 32]. In addition we have measured Förster inter-EET critical radius, $\mathbf{R}_0$,27 for the **Rh-Az** EET pair. These parameters are shown in Table 6.3.

The photophysical characteristics of the **Rh6G** and **Az-amine** two half adders and of the **Rh-Az** intermolecular EET based full adder were elucidated by conducting laser photoquenching experiments in which the dependences of the relative fluorescence quantum yield as well as S_2 relative fluorescence intensity on the laser intensity were measured. The laser excitation frequencies were chosen to match the broad absorption bands corresponding to the $S_0 \rightarrow S_1$, $S_0 \rightarrow S_2$ and $S_1 \rightarrow S_2$ transitions in the rhodamine and azulene moieties, comprising the half adder and the full adder. Table 6.3 shows also the $S_1 \rightarrow S_2$ absorption cross sections, determined from the photoquenching experiments of the present study, discussed in detail in the next sections.

6.3.2 *Characterization of the Rh6G Half Adder*

As stated above a molecular half adder should have a fluorescent S_1 state and a two photon excited fluorescent S_2 state (Fig. 6.2). Implementation is demonstrated upon observation of simultaneous quenching of S_1 fluorescence (**XOR** gate operation) with the appearance of S_2 fluorescence (**AND** gate operation) and by setting the pumping threshold values for laser input **0** and **1** signals and fluorescence intensity output **0** and **1** signals. Figure 6.5 shows the $S_1 \rightarrow S_0$ LIF intensity dependence of **Rh6G** on the $S_1 \rightarrow S_2$ photoquenching laser intensity, I_p, (at 1,064 nm), at a constant intensity of the second harmonic (at 532 nm, $S_0 \rightarrow S_1$ excitation). The LIF $S_1 \rightarrow S_0$ spectrum, peaking at 550 nm, is basically the same as that for CW excitation [31, 32]. It is evident that an increase in laser intensity at 1,064 nm results in decreasing of the emission intensity from the **Rh6G** S_1 state, indicating photoquenching.

From the results represented in Fig. 6.5 we obtained a "Stern-Volmer" type plot for the photoquenching of the S_1 fluorescence relative quantum yield by absorption of 1,064 nm photons, shown in Fig. 6.6. Using Eq. (6.1) the absorption cross section for the $S_1 \rightarrow S_2$ transition at 1,064 nm is obtained from the slope of the photoquenching plot as $\sigma_{12} = (4.4 \pm 0.6) \times 10^{-19}$ cm^2/mol. The fit to Eq. (6.1) is reasonable with a correlation factor of 0.95. Any deviation from a perfect fit to the

Table 6.3 Previously measured [31, 32] EET parameters for **Rh-Az** molecular full adder and present study measured $S_1 \rightarrow S_2$ absorption cross sections for **Rh-Az** molecular full adder and for **Rh6G** and **Az-amine** molecular half adders

			Rh-Az	
Parameter	**Rh6G**	**Az-amine**	**Rh** moiety	**Az** moiety
$\tau_{10}(s)$	4.1×10^{-9c}	1.42×10^{-12c}	1.2×10^{-10e}	1.42×10^{-12c}
σ_{01} cm^2/mol At 18,797 cm^{-1}	4.0×10^{-16}	1.2×10^{-18}	4.0×10^{-16}	1.2×10^{-18}
σ_{12} cm^2/mol At 18,797 cm^{-1}	$(2.6 \pm 0.6) \times 10^{-16d}$	$(4.5 \pm 0.6) \times 10^{-15d}$	$(2.2 \pm 0.3) \times 10^{-16d}$	$(4.5 \pm 0.6) \times 10^{-15d}$
σ_{12} cm^2/mol At 9,398 cm^{-1}	$(4.4 \pm 0.9) \times 10^{-18d}$	–	$(2.2 \pm 0.6) \times 10^{-18d}$	–
$\omega_{S_1 \rightarrow S_0}$ cm^{-1b}	18,180	13,160	17,540	13,070
$\omega_{S_2 \rightarrow S_0}$ cm^{-1b}	27,930	25,970	27,930	27,860
$\mathbf{R}_0$(Å)	96.3		–	
$\mathbf{k}_{ET}$ (s^{-1})	–		7.8×10^9	

[a] Laser frequency choice, ω_2, depends upon input combination
[b] Band peak value
[c] Data taken from Refs. [23, 33, 34]
[d] Present study. The value for **Rh6G** at 18,797 cm^{-1} is given in Ref. [25]
[e] This value is derived from τ_{10} for **Rh6G** and the measured $\mathbf{k}_{ET}$ [32]

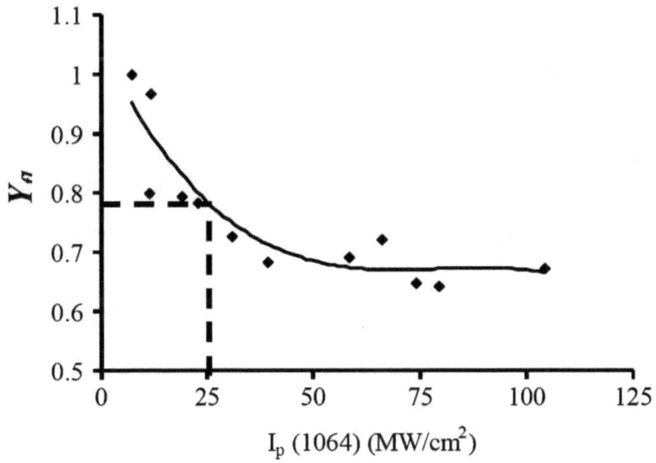

Fig. 6.5 The $S_1 \rightarrow S_0$ LIF intensity dependence of **Rh6G** on the $S_1 \rightarrow S_2$ photoquenching laser intensity, I_p, (at 1,064 nm), at a constant intensity of the second harmonic (at 532 nm, $S_0 \rightarrow S_1$ excitation)

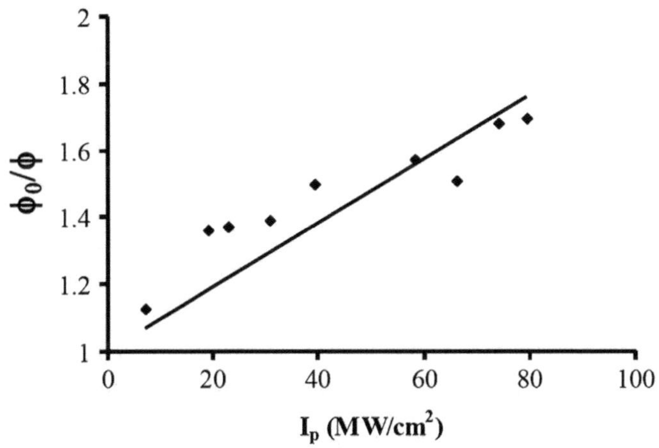

Fig. 6.6 "Stern–Volmer" photoquenching plot (Eq. (6.1)) for $S_2 \rightarrow S_0$ transition for **Rh6G** (2.86×10^{-6} M in methanol) vs. photoquenching laser intensity at 1,064 nm wavelength

linear "Stern Volmer" plot can be attributed to some nonlinear terms, ignored at the range of laser intensities employed here [35]. This mechanism is further supported by the observation of the simultaneous appearance of the $S_2 \rightarrow S_0$ fluorescence. For the same laser excitation intensity range the LIF of the $S_1 \rightarrow S_2$ transition of **Rh6G** was measured. Figure 6.7 shows the LIF $S_2 \rightarrow S_0$ spectrum of **Rh6G** molecule. The emission peak position for the intensity measurement was at 358 nm, in good agreement with the literature value [34]. The linear dependence of the S_2 level

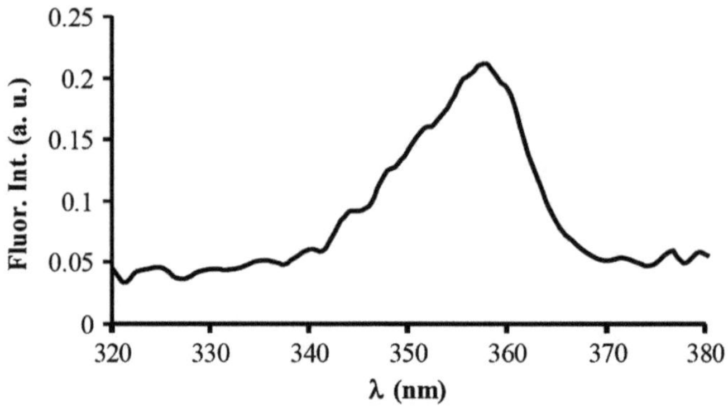

Fig. 6.7 LIF $S_2 \rightarrow S_0$ fluorescence spectrum of Rh6G (2.86×10^{-6} M in methanol)

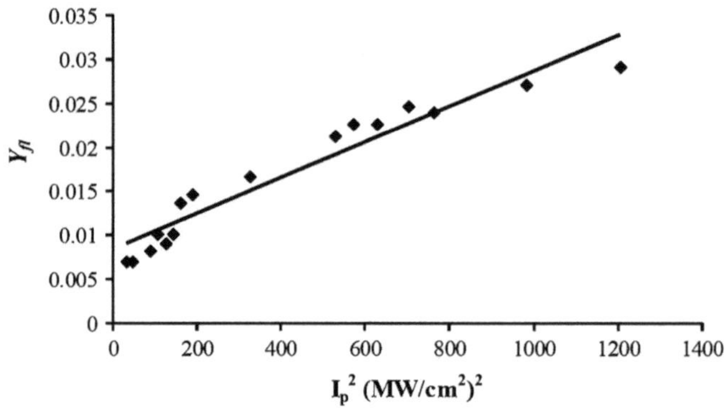

Fig. 6.8 LIF $S_2 \rightarrow S_0$ intensity of **Rh6G** (2.86×10^{-6} M in methanol), as function of the squared excitation laser intensity at 532 nm

emission, on the 1,064 nm excitation laser intensity, at a constant pump intensity for the $S_0 \rightarrow S_1$ excitation at 532 nm, is presented in Fig. 6.8, indicative of consecutive two-photon excitation of **Rh6G**.

We now arbitrarily set the threshold intensity for the input value of the $S_1 \rightarrow S_2$ excitation of the half adder (see Fig. 6.5) as 25 MW/cm^2, below this value the half-adder input intensity, for both ω_1 and ω_2, is defined as **1**. The measured relative intensity for the $S_1 \rightarrow S_0$ emission output is 0.79 at the threshold pump intensity, thus any output intensity lower than 0.79, is defined as sum output **0** for the half adder (**XOR** gate operation). At the defined the threshold pump intensity of 25 MW/cm^2 the measured relative intensity for the $S_2 \rightarrow S_0$ emission output is 0.31, any output intensity is lower than 0.31, it will be defined as **0** carry and any output above 0.31 a.u. is defined as carry **1** for the **Rh6G** half adder (**AND** gate operation). It should be note that these threshold values do not depend on the absolute values of the measured absorption cross sections derived from the fit to Eq. (6.1).

Similar results were obtained for the degenerate case of excitations with only 532 nm laser photons for both $S_0 \rightarrow S_1$ and $S_1 \rightarrow S_2$ transitions in R6G.

Table 6.4 The **Rh6G** half adder experimental truth table

ω_1 18,797 cm^{-1}	ω_2 18,797 cm^{-1} or 9,398 cm^{-1}	Sum $S_1 \rightarrow S_0$	Carry $S_1 \rightarrow S_0$
0	0	0	0
1 18,797 cm^{-1}	0	1	0
0	1 18,797 cm^{-1}	1	0
1 $I_p > 25$ MW/cm^2 18,797 cm^{-1}	1 $I_p > 25$ MW/cm^2 9,398 cm^{-1}	0 $(I_{fl} < 0.79)$	1 $(I_{fl} < 0.31)$

The same LIF $S_2 \rightarrow S_0$ fluorescence spectrum (Fig. 6.7) is observed and its intensity dependence is shown in Fig. 6.8. The good fit (0.97 correlation factor) of the $S_2 \rightarrow S_0$ fluorescence intensity to a quadratic dependence on the pumping laser intensity at 532 nm is indicative of two-photon absorption in **Rh6G** that also leads to photoquenching. The previously-measured cross section was $\sigma_{12} = (2.6 \pm 0.2) \times 10^{-16}$ cm^2/mol [26]. The intensity threshold data discussed above together with the spectroscopic data enable us to construct the **experimental** truth table for **Rh6G** half adder (Table 6.4). Similar results were obtained for the **RB** half adder.

In Table 6.4 we show how an **Rh6G** molecular half adder is implemented with 18,797 cm^{-1} (532 nm) and 9,398 cm^{-1} (1,064 nm) laser frequencies as inputs and $S_1 \rightarrow S_0$, and $S_2 \rightarrow S_0$ emissions as outputs. In the absence of inputs (laser off) no LIF is observed and both outputs are identically zero. When only 18,797 cm^{-1} frequency input is present, $S_1 \rightarrow S_0$ fluorescence output (sum) is observed. However, when both inputs are present (laser intensity above 25 MW/cm^2), $S_1 \rightarrow S_0$ fluorescence is suppressed due to the photoquenching (**XOR** gate implementation) with the simultaneous appearance of the $S_2 \rightarrow S_0$ LIF. At this laser input level $S_2 \rightarrow S_0$ LIF output intensity of $I_{fl} > 0.31$ is obtained, corresponding to logic **1** carry output, implementing an **AND** gate, while $S_1 \rightarrow S_0$ LIF output is below the defined threshold value ($I_{fl} < 0.79$), defining a **0** logic sum output , implementing the **XOR** gate of the half adder.

6.3.3 *Characterization of Az-Amine Half Adder*

In order to implement an all optical **Az-amine** molecular half adder, we have used the same methodology applied to the **Rh** half adder. Figure 6.9 shows fluorescence intensity dependence of **Az-amine** $S_1 \rightarrow S_0$ fluorescence, monitored at 760 nm, on the pumping laser intensity at 532 nm. The observed sub-linear dependence of the LIF on the pumping laser power (Fig. 6.10) is indicative of photoquenching [26]. Using the data of Fig. 6.11, the photoquenching plot, Fig. 6.11 was obtained. By employing Eq. (6.1) and using the reported 1.42 ps lifetime value for azulene [34],

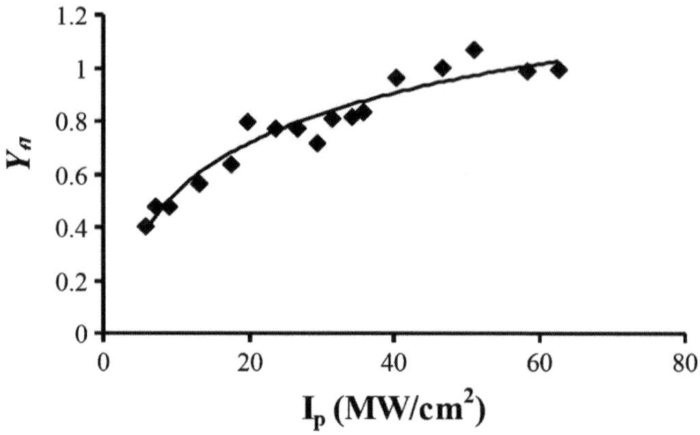

Fig. 6.9 $S_1 \rightarrow S_0$ LIF intensity dependence, for **Az-amine** (7×10^{-4} M in methanol), on the 532 nm pumping laser intensity

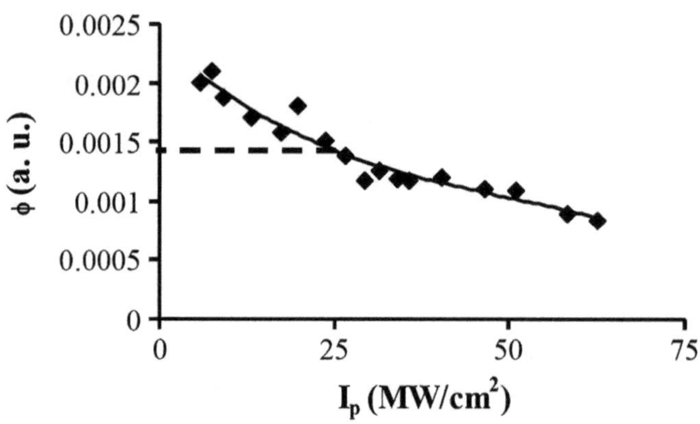

Fig. 6.10 Relative quantum yield intensity $S_1 \rightarrow S_0$ fluorescence, derived from Fig. 6.9. The *dashed lines* indicate the threshold intensities for laser input and fluorescence quantum yield output signals

we obtain from Fig. 6.12 the absorption cross section for the **Az-amine** $S_1 \rightarrow S_2$ transition. This value of $\sigma_{12} = (4.5 \pm 0.6) \times 10^{-15}$ cm^2/mol should be considered only as an estimate since it depends on the life time value of azulene which is probably different from that of **Az-amine** [23]. In addition, it is not clear that $S_1 \rightarrow S_2$ excitation fully describes the mechanism of 532 nm absorption by the azulene chromophore. The radiationless decay rates of the S_1 states of azulene (in the bichromophore and in the substituted monochromophore) are very rapid and the involvement of triplet states in maintaining a population of S_1 via $T_1 - S_1$ back-intersystem crossing, as described by Nickel and Kemp [36, 37] cannot be ruled out. While longer-lived triplet states will affect the absolute value of σ_{12} [34], the logic operations and the experimental threshold input values are not affected by it.

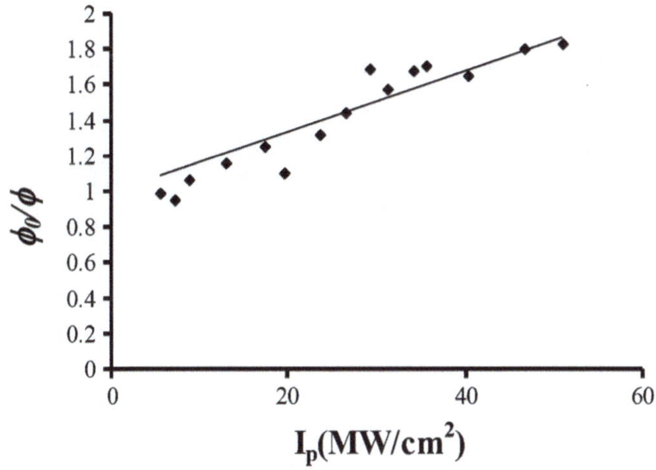

Fig. 6.11 "Stern-Volmer", photoquenching plot (Eq. (6.1)), for $S_1 \rightarrow S_0$ emission (**Az-amine** half adder) vs. 532 nm pump laser intensity

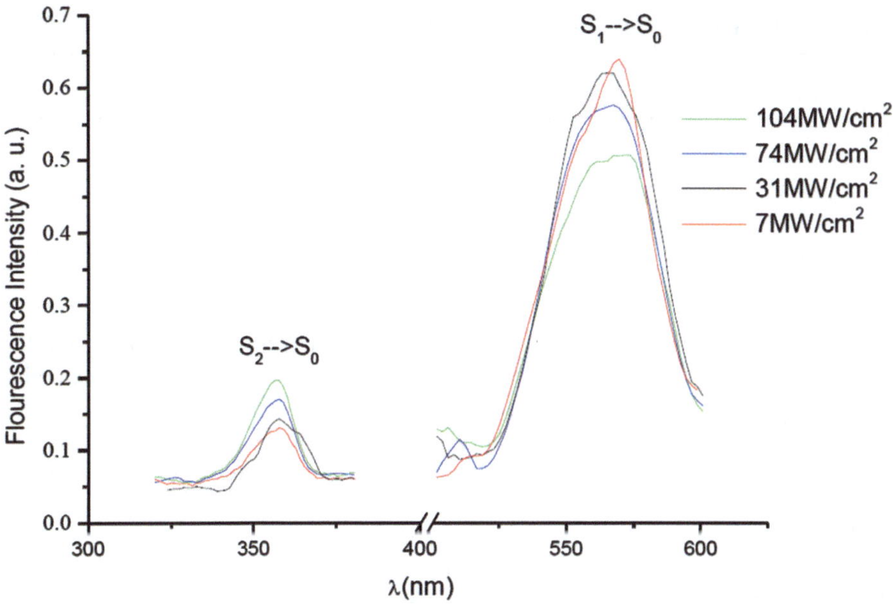

Fig. 6.12 The $S_2 \rightarrow S_0$ and $S_1 \rightarrow S_0$ transitions intensity dependence on 1,064 nm laser pumping intensity

The input and output threshold intensities for the $S_1 \rightarrow S_0$ emission are shown on the photoquenching plot. The threshold intensity for the input value is 25 MW/cm^2, below this value the input intensity is defied as **1**. The corresponding threshold value for the $S_1 \rightarrow S_0$ relative quantum yield is 0.0014, any $\phi < 0.0014$ value will be defined as **0** sumout logic value.

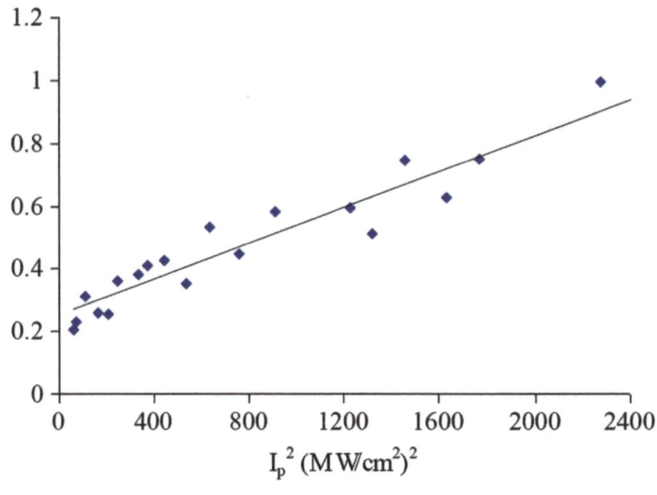

Fig. 6.13 $S_2 \rightarrow S_0$ **Az-amine** LIF intensity vs. the squared 532 nm pump laser intensity

Table 6.5 **Az-amine** half adder experimental truth table

ω_1 18,797 cm^{-1}	ω_2 18,797 cm^{-1}	Sum $S_1 \rightarrow S_0$	Carry $S_2 \rightarrow S_0$
0	0	0	0
1	0	1	0
0	1	1	0
1	1	0	1
$I_p > 25$ MW/cm^2	$I_p > 25$ MW/cm^2	$(\phi_{S_1 \rightarrow S_0} < 0.0014)$	$(Y_{S_2 \rightarrow S_0} < 0.43)$

Figure 6.12 shows the LIF $S_2 \rightarrow S_0$ spectrum of **Az-amine**, excited simultaneously at the same pumping laser intensity range used for the photoquenching experiments. The corresponding fluorescence intensity dependence on the squared laser intensity at 532 nm is shown in Fig. 6.13. The fit to a quadratic dependence of the fluorescence on laser intensity indicates a two-photon absorption in **Az-amine** molecule of 532 nm laser photons.

The input threshold intensity is set at 25 MW/cm^2, above this value the input intensity is defined as **1**. The corresponding output threshold intensity, for the $S_2 \rightarrow S_0$ relative emission intensity **Y** is 0.43, any **Y** > 0.43 is defined as **1** carry out value. These data enables us to construct the experimental truth table for **Az-amine** half adder (Table 6.5).

Az-amine molecular half adder is obtained with two identical 532 nm laser frequencies as inputs and $S_1 \rightarrow S_0$, $S_2 \rightarrow S_0$ emissions as outputs. In the absence of inputs (laser off) no LIF occurs. Thus both outputs are identically zero. When one of the inputs is present (laser intensity below 25 MW/cm^2), $S_1 \rightarrow S_0$ fluorescence output (sum) is above its threshold value, where $S_2 \rightarrow S_0$ fluorescence output (carry) falls below the threshold. However, when both inputs are present (laser intensity above 25 MW/cm^2), $S_2 \rightarrow S_0$ LIF becomes dominant due to the photoquenching effect. High $S_2 \rightarrow S_0$ LIF output is obtained resulting **AND** gate, and $S_1 \rightarrow S_0$ LIF output is below the defined threshold value.

6.3.4 The all Optical Intermolecular EET Based Rh-Az Full Adder

The characterization of the intermolecular EET based **Rh-Az** full adder follows that of the two moieties half-adder components discussed in the previous sections. These two half adders communicate via an efficient inter-EET process (Fig. 6.3). **Rh** $S_1 \rightarrow S_0$ fluorescence is highly quenched by inter-EET to azulene, resulting in $S_1 \rightarrow S_0$ for the mixture of **Rh6G** (1×10^{-7} M) and **Az** (1×10^{-3} M) [32].

In order to provide evidence for implementation of a molecular full adder, conforming with our model [21] (Fig. 6.3), we need to show that in addition to the half-adder functions of the two chromophores, the intremolecualr EET process in the **Rh-Az** pair provides the midway sum needed for the concatenation of the two half-adder moieties. Figure 6.13 shows the intensity dependence of **Az** $S_1 \rightarrow S_0$ fluorescence intensity monitored at 750 nm on the excitation laser intensity at 532 nm absorbed by the **Rh** donor molecule, indicative of this EET process, together with the fit to the theoretical model discussed in Ref. [38].

The performance of the **Rh-Az** intermolecular EET based full adder is thus achieved. Whenever S_1 of the **Rh** is excited, the efficient inter-EET to **Az** chromophore results in excitation of the **Az** moiety S_1 state, thus implementing the midway sum of the **XOR** part of the half adder (Fig. 6.4). Alternatively, the **Az** moiety can be excited to its S_1 state by ω_3, 18,797 cm^{-1} photons (carry in, Fig. 6.4). Absorption of a second ω_3 photon then excites it to the S_2 state. Thus, carry 2 output (fluorescence from the S_2 excited state of **Az** moiety) is detected. Threshold values for the inputs were determined by the photoqueching experiments. The threshold input values for ω_1 and ω_2 are set at 25 MW/cm^2. Threshold values for carry 1 outputs were set by the detected S_2 Rhodamine moiety fluorescence intensity, Y. The detected S_2 **Az** fluorescence provides the sum out output signal observed for any input intensity without a threshold. The observed performance of the **Rh–Az** pair full adder is summarized in terms of an experimental truth table, given in Table 6.6. The following notes are associated with Table 6.6:

(a) The **Az** sum out is always 1, without threshold input intensity, whenever S_1 of **Rh** is excited due to an efficient energy transfer to the **Az** chromophore (ϕ_{EET} 0.96) [32].
(b) If only ω_3 is present as an input, fluorescence from the S_1 excited state of **Az** may be observed even under steady state excitation conditions, without a threshold input intensity, where for example for input at $\omega_3 = 15{,}500$ cm^{-1} (645 nm) no fluorescence is observed from Rh that can mask the weak **Az** fluorescence.

Table 6.6 The **Rh-Az** inter-EET based full adder experimental truth table

Input combination	Input ω_1 18,797 cm^{-1}	ω_2 18,797 cm^{-1}	ω_3 18,797 cm^{-1}	Output Sum out (**Az** acceptor)	Carry Out Carry 1, (**Rh** donor) or Carry 2, (**Az** acceptor)
1	0	0	0	0	0
2	1	0	0	1[a]	0[a]
3	0	1	0	1[a]	0[a]
4	0	0	1	1[b]	0[b]
5	1 $I_p > 25$ MW/cm^2	1 $I_p > 25$ MW/cm^2	0	0	1, **Rh** moiety $(Y_{S_2 \rightarrow S_0} > 0.54)$
6	0	1	1	0	1[c], **Az** moiety
7	1	0	1	0	1[c], **Az** moiety
8	1 $I_p > 25$ MW/cm^2	1 $I_p > 25$ MW/cm^2	1	1[d]	1, **Rh** moiety $(Y_{S_2 \rightarrow S_0} > 0.54)$

(c) **Carry 2** output is always 1 whenever S_1 of the **Rh** is excited by 18,797 cm^{-1} photons due to efficient inter-EET to the **Az** acceptor molecule [32], thus exciting the **Az** S_1 state. Absorption of a second, ω_3 photon at 18,797 cm^{-1}, then excites it to the S_2 state. Thus, **carry 2** output (fluorescence from S_2 excited state of **Az**) is detected, without threshold output intensity.

(d) When all three input frequencies are present (input combination 8), the **sum out** signal (**Az** $S_1 \rightarrow S_0$ emission) value is always 1, without threshold input intensity. A consecutive two-photon absorption process: $S_0 \xrightarrow{532\,\text{nm}} S_1 \xrightarrow{1{,}064\,\text{nm}} S_1$ occurs at the **Rh** donor and LIF is observed its S_2 excited state. Implementation is also achieved for $\omega_1 = \omega_2 = \omega_3 = 18{,}797\,\text{cm}^{-1}$, as well as for other frequency combinations, due the broad bandwidth of the relevant transitions, by using a tuneable dye laser.

6.4 Summary

We have shown how an all optical half-adder logic circuit can be implemented on rhodamine and on the newly synthesized azulene-amine molecules, by utilizing combinations of one and two-photon absorptions for the input signals and LIF fluorescence from their S_1 and S_2 states as sum out and carry logic outputs. The rhodamine half adder can communicate its logic output as input to the azulene half-adder molecule, thus implementing a full adder. This is achieved as an intermolecular electronic energy transfer from a rhodamine-donor to azulene-acceptor. We present results on this pair, in which, **for the first time, an all optical** full-adder logic circuit is implemented. A general kinetic scheme applicable to both intramolecular and to intermolecular EET based full adders is discussed, with good agreement between theory and experiment

References

1. Balzani V, Credi A, Venturi M (2003) Molecular devices and machines: concepts and perspectives for the nanoworld. Wiley-VCH, Weinheim
2. Collier CP (1999) Electronically configurable molecular-based logic gates. Science 285:391–394
3. Collier CP (2000) A [2]Catenane-based solid state electronically reconfigurable switch. Science 289:1172–1175
4. Metzger RM (2000) All about (N-hexadecylquinolin-4-ium-1-yl)methylidenetricyanoquinodimethanide, a unimolecular rectifier of electrical current. J Mater Chem 10:55–62
5. Metzger RM (1999) Electrical rectification by a molecule: the advent of unimolecular electronic devices. Acc Chem Res 32:950–957
6. Tour JM (2000) Molecular wires for electronic applications. A Rev Polym News 25:329–336
7. Feringa BL (ed) (2001) Molecular switches. Wiley-VCH, Weinheim/Chichester
8. Ellenbogen J, Love J (2000) Architectures for molecular electronic computers. I. Logic structures and an adder designed from molecular electronic diodes. Proc IEEE 88:386–426
9. Tanenbaum AS (1999) Structured computer organization. Prentice-Hall International, Upper Saddle River
10. Irie M (2000) Diarylethenes for memories and switches. Chem Rev 100:1685–1716
11. Yokoyama Y (2000) Fulgides for memories and switches. Chem Rev 100:1717–1740
12. Berkovic G, Krongauz V, Weiss V (2000) Spiropyrans and spirooxazines for memories and switches. Chem Rev 100:1741–1754
13. Brown GJ, De Silva AP, Pagliari S (2002) Focus article: molecules that add up. Chem Commun 2461–2464
14. de Silva AP, de Silva SA, Dissanayake AS, Sandanayake KRAS (1989) Compartmental fluorescent ph indicators with nearly complete predictability of indicator parameters; molecular engineering of ph sensors. J Chem Soc Chem Commun 1054–1056
15. de Silva PA, Gunaratne NHQ, McCoy CP (1993) A molecular photoionic AND gate based on fluorescent signalling. Nature 364:42–44
16. Ghosh P, Bharadwaj PK, Mandal S, Ghosh S (1996) Ni(II), Cu(II), and Zn(II) cryptate-enhanced fluorescence of a trianthrylcryptand: a potential molecular photonic or operator. J Am Chem Soc 118:1553–1554
17. Ghosh P, Bharadwaj PK, Roy J, Ghosh S (1997) Transition metal (II)/(III), Eu(III), and Tb(III) ions induced molecular photonic or gates using trianthryl cryptands of varying cavity dimension. J Am Chem Soc 119:11903–11909
18. Prasanna de Silva A, McClenaghan ND (2000) Proof-of-principle of molecular-scale arithmetic. J Am Chem Soc 122:3965–3966
19. Margulies D, Melman G, Shanzer A (2005) Fluorescein as a model molecular calculator with reset capability. Nat Mater 4:768–771
20. Remacle F, Weinkauf R, Levine RD (2006) Molecule-based photonically switched half and full adder. J Phys Chem A 110:177–184
21. Remacle F, Speiser S, Levine RD (2001) Intermolecular and intramolecular logic gates. J Phys Chem B 105:5589–5591

22. Viswanath G, Kasha M (1956) Confirmation of the anomalous fluorescence of azulene. J Chem Phys 24:574
23. Wagner BD, Szymanski M, Steer RP (1993) Subpicosecond pump–probe measurements of the electronic relaxation rates of the S1 states of azulene and related compounds in polar and nonpolar solvents. J Chem Phys 98:301
24. Speiser S (1983) Photoquenching effects and S1 absorption in cryptocyanine. Opt Commun 45:84–86
25. Speiser S (1989) Observation of laser-induced off-resonance intermolecular electronic energy transfer. Appl Phys B Photophys Laser Chem 49:109–112
26. Speiser S, Shakkour N (1985) Photoquenching parameters for commonly used laser dyes. Appl Phys B Photophys Laser Chem 38:191–197
27. Speiser S (1996) Photophysics and mechanisms of intramolecular electronic energy transfer in bichromophoric molecular systems: solution and supersonic jet studies. Chem Rev 96(6):1953–1976
28. Yeow EKL, Steer RP (2003) Dynamics of electronic energy transfer from the S2 state of azulene to the S2 state of zinc porphyrin. Phys Chem Chem Phys 5:97–105
29. Yeow EKL, Ziolek M, Karolczak J, Shevyakov SV, Asato AE, Maciejewski A, Steer RP (2004) Sequential forward S 2-S 2 and back S 1-S 1 (cyclic) energy transfer in a novel azulene-zinc porphyrin dyad. J Phys Chem A 108:10980–10988
30. Yeow EK, Steer RP (2003) Energy transfer involving higher electronic states: a new direction for molecular logic gates. Chem Phys Lett 377:391–398
31. Salman H, Eichen Y, Speiser S (2006) A molecular scale full adder based on controlled intramolecular electron and energy transfer. Mater Sci Eng C 26:881–885
32. Kuznetz O, Davis D, Salman H, Eichen Y, Speiser S (2007) Intramolecular electronic energy transfer in rhodamine–azulene bichromophoric molecule. J Photochem Photobiol A Chem 191:176–181
33. Kaplan I, Jortner J (1978) Electronic energy transfer from the S2 state of rhodamine 6G. Chem Phys 32:381–396
34. Orner GC, Topp MR (1975) Biphotonic laser excitation of upper singlet state fluorescence in organic dyes. Chem Phys Lett 36:295–300
35. Speiser S, Van der Werf R, Kommandeur J (1973) Photoquenching: the dependence of the primary quantum yield of a monophotonic laser-induced photochemical process on the intensity and duration of the exciting pulse. Chem Phys 1:297–305
36. Nickel B, Klemp D (1993) The lowest triplet state of azulene-h8 and azulene-d8 in liquid solution. Chem Phys 174:319–330
37. Nickel B, Klemp D (1993) The lowest triplet state of azulene-h8 and azulene-d8 in liquid solution. I. Survey, kinetic considerations, experimental technique, and temperature dependence of triplet decay. Chem Phys 174:297–318
38. Kuznetz O, Salman H, Shakkour N, Eichen Y, Speiser S (2008) A novel all optical molecular scale full adder. Chem Phys Lett 451(1–3):63–67

7
Tunability of Plasmonic Devices

Dimitrios C. Zografopoulos and Romeo Beccherelli

Abstract An overview on tunable plasmonic components for guided-wave applications in integrated photonic circuitry is provided. Emphasis is given on a series of electro-optically controlled plasmonic elements, variable attenuators, phase modulators, and directional coupler switches, based on the use of nematic liquid crystalline materials. These extend to various plasmonic platforms, which target different light confinement scales in view of optical intra- and inter-chip interconnection applications.

Keywords Plasmonics • Liquid crystals • Integrated photonics • Directional couplers • Switching components

7.1 Introduction

Plasmonics is the scientific discipline dealing with the study of the properties of surface plasmons (SP), i.e., collective free electron oscillations that take place at the interfaces between metals and dielectrics as a response to impinging lightwave electromagnetic fields [1]. Research in plasmonics has experienced rapid expansion in the last decade, stimulated by significant advances in fabrication and characterizaion techniques, which allow the design and realization of metallic structures to be scaled down to the nanoscale. This provides an unprecedented control of SP properties and opens new pathways to tailor and control light-matter interactions, for instance, subwavelength light confinement that breaks the diffraction limit present in conventional dielectrics-based photonics. These possibilities have already found a key part in myriad applications spanning different technologies: nano-scaled

D.C. Zografopoulos (✉) • R. Beccherelli
Consiglio Nazionale delle Ricerche, Istituto per la Microelettronica e Microsistemi (CNR-IMM), Roma 00133, Italy
e-mail: dimitrios.zografopoulos@artov.imm.cnr.it; romeo.beccherelli@artov.imm.cnr.it

B. Di Bartolo et al. (eds.), *Nano-Structures for Optics and Photonics*, NATO Science for Peace and Security Series B: Physics and Biophysics,
DOI 10.1007/978-94-017-9133-5_7,

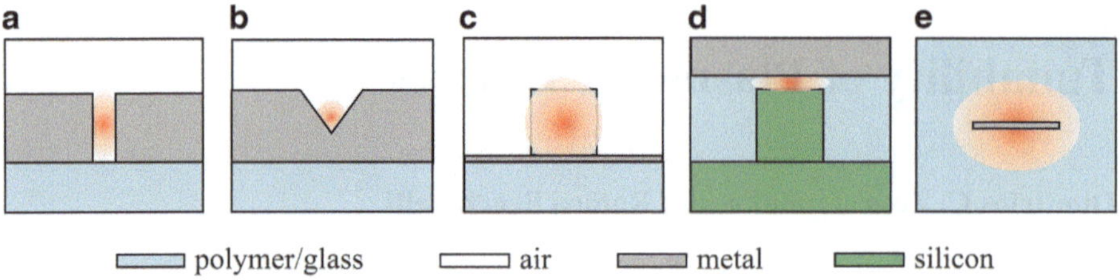

Fig. 7.1 Schematic layout of widely studied plasmonic waveguides: (**a**) slot, (**b**) channel, (**c**) dielectric-loaded, (**d**) metal-dielectric-semiconductor, and (**e**) long-range stripe waveguides

integrated photonics platforms [2–5], nano-antennas, superlenses and ultrafast nanoplasmonics [6] advanced chemical detection techniques like surface-enhanced Raman spectroscopy [7], biomedical sensor and treatment techniques [8], and solar energy conversion [9], to name but a few.

When metallic nanoparticles or nanoelements patterned on dielectric substrates are involved, SP manifest as localized fields, termed as localized surface plasmons (LSP). At certain frequencies, the conduction electrons in the metal respond collectively in resonance with the excitation field and lead to enhanced absorption and scattering of electromagnetic waves, as well as strongly enhanced near fields. Such "hot spots" play a fundamental role in plasmonic systems, e.g. for biodetection, light harvesting or nonlinear processes [10].

On the other hand, when metallic paths are formed, such as in nanoparticle arrays and metallic perforated or continuous films, SP manifest as surface propagating electromagnetic waves at the metal/dielectric interface, termed as surface plasmon polaritons (SPP). In addition, the metal electron oscillations increase the momentum of light at the interface and allow for field localization beyond the diffraction limit. Thus, light confinement in plasmonic circuitry can be deeply subwavelength. This enables the miniaturization of photonic components towards the size of their electronic counterparts, potentially providing the missing link between conventional microphotonics and nanoelectronics in future chip-scale technologies for optical interconnects. Indeed, light guidance at both visible and infrared wavelengths has been thus far demonstrated in a variety of plasmonic waveguides of different confinement scales. Figure 7.1 shows a set of the mostly investigated plasmonic waveguides as the core element for integrated photonics architectures.

For such a technology to become practical, tunable components, e.g. variable attenuators, phase modulators and switches, are needed for advanced light flow control functionalities, such as switches, switching matrices, optical routers and signal processing modules. These components demand for an external control mechanism, which is capable of dynamically tuning their waveguiding properties. Typical candidates in this respect are the nonlinear, thermo-optic and electro-optic effects, as extensively demonstrated in the case of purely dielectric integrated photonics. Although light guidance in SPP waveguides suffers from intrinsic dumping losses owing to metal conductivity, non-linear effects in plasmonic waveguides have recently started to be explored, as they offer very fast response times that stem from the instantaneous $\chi^{(3)}$ response. For instance, efficient non-linear switching has been predicted in hybrid metal-gap-semiconductor structures as the one shown in Fig. 7.1d, which offer extremely tight light confinement and high non-linear coefficients [11].

Thermo-optic components rely on the modulation of the dielectric (usually polymer) refractive index upon heating, which is induced by current injection. For typical materials one can achieve an index modulation in the range of 0.01, which can be sufficient for practical applications. The response speed ranges from sub-ms to a few μs and the required power consumption ranges in the order from a few to tens of mW. The metallic SPP waveguide can be used as the current injection electrode, although care must be taken to avoid detrimental effects, such as thermal crosstalk or electromigration. To date, the thermo-optic effect has been used as the standard mechanism for tunable plasmonic circuits, which are based on the dielectric-loaded SPP (DLSPP), schematically shown in Fig. 7.1c, [12–15], and the long-range SPP (LRSPP) platform, as in Fig. 7.1e, [16–19].

A third route towards tunable plasmonic devices involves the use of electro-optic materials, whose refractive index is dynamically tuned via an external control voltage. For instance, ferroelectric films, lithium niobate, and electro-optic polymers have been considered as candidate materials for electro-optic plasmonic components [20–23], although their efficient integration is hindered by technological challenges. In this work, we present a series of tunable plasmonic components, which are based on the use of nematic liquid crystals (LC), a class of organic materials that share physical properties with both solids and liquids. Their inherent large optical anisotropy, enhanced response to the application of external electric or magnetic fields, and very low driving power requirements have already favored their use in numerous conventional dielectrics-based photonic devices [24–28]. Recently, their use has been expanded also in the field nanoplasmonics [29] as a tuning mechanism of LSP in metallic nanostructures, such as nanodots [30], nanoholes [31, 32], nanodisk arrays [33], and nanoparticle composites [34–36]. Here, we demonstrate how liquid-crystalline materials can also be used for ultra-low power control of SPP propagation in various light confinement scales, aiming at integrated plasmonic components for optical interconnects.

7.2 Liquid-Crystal Tunable Plasmonic Devices

The use of nematic liquid crystals in tunable photonic and plasmonic structures involves a series of technological processes that facilitate their electro-optic control and induce the desired functionalities to the end device [24]. Liquid crystals are typically infiltrated in confined planar cells or cavities via capillary action, although the infiltration speed can be accelerated by using vacuum pumps. As a standard procedure, the cavity's walls are treated before infiltration in order to promote certain alignment conditions at the interfaces between LC and the surrounding materials. This is traditionally achieved by rubbed polyimide alignment layers, which promote LC molecular orientation along the rubbing direction. A promising alternative technique is photoalignment, where the LC preferred orientation is controlled by flashing a monolayer of optically responsive molecules, such as azobenzenes, with linearly polarized light [37, 38]. In addition to determining the LC properties in the rest case, i.e. when no external voltage is applied, LC confinement in microcavities also reduces significantly scattering losses with respect to bulk LC volumes [28, 39–41].

The control signal, which induces the reorientation of the LC molecules and thus the tuning of their dielectric and optical properties, is applied via electrodes made of optically transparent conductive materials, with indium tin oxide (ITO) being the standard choice. The position of the ITO electrodes is a degree of freedom in the design of LC-based components, as it impacts directly on the distribution of the electrostatic field and thus the switching of the nematic molecules. Care is taken to optically isolate the conductive ITO electrodes from the light guiding regions, in order to avoid excess dumping losses. In a complementary approach, ITO losses can be exploited in some switching devices to suppress optical power. In any case, plasmonic structures offer the possibility to use the same metallic film where optical propagation takes place as one of the electrodes for the application of the control voltage. This not only reduces the complexity and need for ITO definition, but it also applies the control signal directly at the plasmonic interface where maximum optical power is concentrated.

The switching speed of nematic materials depends on their material properties, namely elastic constants, dielectric anisotropy and viscosity, and the dimensions of the LC-cavity [42]. For standard materials in LC-photonic components, this leads to millisecond response, which is sufficient for light routing and switching applications. The main comparative advantage of the use of LC electro-optic devices are the extremely low switching power requirements, as these rely on a capacitive switching mechanism that avoids current flow and dissipation. The power consumption of a LC-based device can be estimated as $P \simeq C f V_{\mathrm{LC}}^2$, where C is the device capacitance and $f = 1 \div 10\,\mathrm{KHz}$ the LC-switching frequency, and V_{LC} the control voltage. For the length scales and materials involved in LC-tunable photonic and, in particular, the plasmonic components presented in this work, this translates into a power budget in the nW regime. This value is several orders of magnitude lower than the switching power needed in thermo-optic or nonlinear elements, which lies at least in the mW range.

7.2.1 Theoretical Investigation and Design of LC-Plasmonic Structures

The design of LC-plasmonic components relies on numerical techniques capable of rigorously describing the involved physics and simulating the optical properties of the target device. This overall analysis is divided into two types of studies: (a) the

LC orientation problem, that is the calculation of the spatial profile of the nematic molecular direction, and (b) the optical studies, namely the extraction of optical modes and the simulation of lightwave propagation. These two problems are not coupled, unless nonlinear phenomena are involved, where the LC orientation is influenced by the intensity of the optical field [27].

The first problem can be tackled by minimizing the total energy of the system, based on the Oseen-Frank continuum elastic theory [43–45]. The energy density in the LC bulk is expressed as a function of the local LC orientation, which is described by a unit vector, the nematic director, via tilt and twist angles. The electrostatic energy owing to the low-frequency control field is summed to the elastic energy contributions by the three main LC deformations, i.e. splay, twist, and bend. In order to calculate the spatial LC orientation profile that minimizes the total energy, two Euler-Lagrange equations are derived and in combination with Gauss' electric law, the final set of partial equations is formed. This can be solved, for instance, on the nodes of a finite-element mesh that discretizes the geometry under study, yielding the spatial distribution of the tilt and twist angles, and the electrostatic potential.

The information on the LC orientation profile is indispensable to the optical studies, as it determines the 3×3 optical dielectric permittivity tensor $\tilde{\varepsilon}$. Various electromagnetic computational techniques can be employed for the optical analysis, provided they allow for a fully-anisotropic description of $\tilde{\varepsilon}$. Full wave numerical tools simulate the optical response of the structure by discretizing its volume and solving for Maxwell's equations. Typical examples include the frequency-domain finite-element method (FEM) [46] and the finite-difference time-domain method [47]. In frequency-domain methods the structure is simulated assuming a monochromatic excitation. This permits the accurate description of material permittivity, which is provided as an input for each wavelength. Time-domain techniques simulate the propagation of a modulated pulse and thus provide the broadband response of the structure in a single run. Nevertheless, metals are highly dispersive in the visible and infrared spectrum and so are LCs, particularly in the visible. Thus, the rigorous time-domain investigation of LC-plasmonic structures demands for the implementation of advanced anisotropic and dispersive algorithms [48–50].

When waveguiding structures, such as directional couplers, are involved, the total longitudinal dimension of the component may well exceed hundreds of wavelengths, rendering the discretization in finite difference or finite element meshes of the whole three dimensional computational domain practically impossible in terms both of memory requirements and computational effort. In these cases, one resorts to specially adopted numerical methods, which take advantage of the existence of a primary waveguiding direction. Among these, the beam propagation method is a traditional tool for the study of waveguiding photonic and plasmonic structures [51] and it can be also adopted to the LC tensor anisotropy [44, 52, 53]. Another equivalent approach is to calculate the optical modes at the waveguide's cross-section [54] and then expand the excitation field in that eigenmode basis, calculate analytically light propagation to the coupled modes and reconstruct the optical field profile at the exit of the component [55–57].

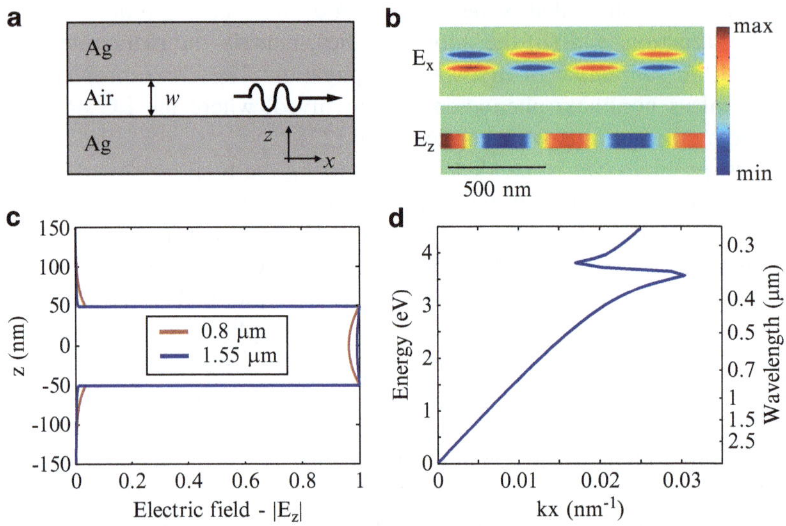

Fig. 7.2 (**a**) Schematic layout of the generic metal-insulator-metal plasmonic waveguide; a dielectric layer of thickness w is sandwiched between two semi-infinite metal layers. (**b**) Electric field snapshots calculated at $\lambda = 800\,\text{nm}$ for a 100-nm Ag-air-Ag waveguide. (**c**) Normalized profile of the z–component of the electric field at $\lambda = 800\,\text{nm}$ and $\lambda = 1{,}550\,\text{nm}$. (**d**) Dispersion curve of the fundamental mode of the Ag-air-Ag waveguide. Silver was modeled as a dispersive Drude-critical point medium [58]

Based on such numerical tools, in the following of this section the design and analysis of a series of LC-based plasmonic waveguiding components are presented, highlighting the key aspects regarding their operation and the functionalities they provide.

7.2.2 LC-Plasmonic Components for Sub-wavelength Confinement

As mentioned, surface plasmon polaritons are light waves sustained by collective charge oscillations at the boundary interface between a metal and a dielectric medium. This light guiding mechanism is fundamentally different from light confinement in high refractive index areas in dielectric waveguides and thus it does not suffer from the diffraction limit. In principle, the modal area of a propagating SPP can be deeply sub-wavelength, eventually limited by the decay length in the metal. One plasmonic waveguide is the metal-insulator-metal (MIM) structure shown in Fig. 7.2a. This is an approximation of slot-type waveguides such as the one of Fig. 7.1a. A dielectric layer is sandwiched between two semi-infinite metal planes

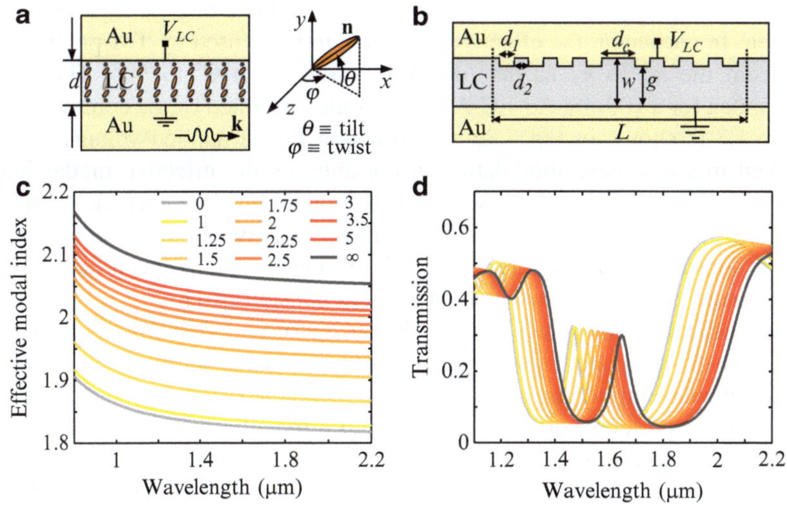

Fig. 7.3 (**a**) Layout of a generic metal-LC-metal plasmonic waveguide and definition of the tilt and twist LC angles. (**b**) Layout of a MLCM Bragg grating filter and definition of its geometrical parameters. (**c**) Effective modal index of the fundamental mode of the MLCM waveguide as a function of the applied voltage V_{LC} in Volt. (**d**) Electro-optic tuning of the MLCM Bragg filter response and resonance shift for applied voltage values as in (**c**)

placed at a subwavelength distance w. The MIM waveguide supports a fundamental TM-polarized mode, formed by the coupling of the individual SPPs at the two metal-dielectric interfaces [59]. For values of w around 100 nm and for common dielectric materials, such as air, silica, polymers, the MIM waveguide is single mode in the visible and infrared spectrum.

In the spectral window relevant to telecom applications, light propagates mostly in the dielectric, with a small part of optical power penetrating into the metal. Figure 7.2b shows the snapshot of the electric field propagating in a 100-nm Ag-air-Ag waveguide at a free space wavelength $\lambda_0 = 800$ nm and Fig. 7.2c the profile of its z-component in the guide's cross-section. In this spectral regime, the dispersion curve of the plasmonic mode resembles that of optical modes in dielectric waveguides, as evidenced in Fig. 7.2d [49].

The optical properties of MIM waveguides, such as modal effective index and propagation losses, depend strongly on the refractive index of the insulating layer, which is typically an isotropic dielectric. By replacing it with a nematic liquid crystal, one can take advantage of the large LC electro-optic effect and tune the properties of the waveguide by directly applying a control voltage V_{LC} between the metal layers. The configuration of such a metal-liquid-crystal-metal (MLCM) waveguide is shown in Fig. 7.3a. The molecules are aligned along the z-axis in the rest case and are progressively switched for higher V_{LC}. Given that the plasmonic optical mode is TM-polarized, the modal effective index depends on the tilt angle, which directly influences the LC effective index along the light polarization direction. In particular, the effective refractive index sensed by TM-polarized light rises from the n_o to n_e as the tilt angle varies from 0° to 90°. The effective modal index for a 100-nm Ag-LC-Ag waveguide infiltrated by the common nematic mixture E7 is shown in Fig. 7.3c as a function of the applied voltage [60]. It is remarked that extensive modulation is possible, as the effective modal index is tuned in an interval comparable with the LC optical birefringence $\Delta n = n_e - n_o$, which is equal to ~0.2 in the infrared. Liquid-crystal switching in such thin planar layers is not trivial from the fabrication point of view, yet it has been experimentally demonstrated in LC-infiltrated slots of these dimensions [61].

This modulation can be exploited in the design of both free-space and integrated MLCM based components, such as, respectively, phase-shifters in slit arrays for electro-optical beam steering and focusing [62] and the Bragg resonator filter shown in Fig. 7.3b. A periodic surface corrugation of a metallic interface creates a Bragg filter, which is interrupted by a defect cavity. The electro-optic tuning of the LC shifts the resonant frequency in an interval higher than 100 nm for an applied voltage change of only 1 V, as demonstrated in Fig. 7.3d. Strong light confinement leads to high sensitivity of the modal properties, and consequently of the overall component characteristics, on the refractive index of the dielectric material. Thus, liquid crystals provide a straightforward solution for the efficient tuning of such sub-wavelength plasmonic structures.

Even tighter confinement of surface propagating light waves can be achieved by passing from the vertical slot waveguides to grooved etched metallic structures, as the one shown in Fig. 7.1b. Such grooved geometries support SPPs that are confined around the tip of these triangular configurations, also termed as channel plasmon polaritons (CPP) [63]. Liquid-crystal infiltration in such structures is not expected to yield any advantage in terms of dynamical control, as the metallic film is continuous and therefore an equipotential surface would surround the waveguiding region in case of an applied field. However, by introducing a spacing between the two inclined metal films, as depicted in Fig. 7.4a, the control voltage can be directly applied between the two films and the voltage drop takes place in the light guiding region in the lower edge of the trapezoidal groove [43].

The profiles of the tilt and twist angles of a Au-E7-Au grooved waveguide with a 150-nm separation for $V_{LC} = 1.6\,V$ is shown in Fig. 7.4b. In this structure, it is the twist angle that mainly affects the SPP modal properties, since light is polarized perpendicular to the metallic film surface. Figure 7.4c shows that extensive tuning of the modal effective index is achieved for a voltage of a few Volt, with results shown up to the estimated LC dielectric breakdown voltage [64]. A phase shift $\Delta\phi = \pi$ can be accumulated after a propagation of less than $L = 10\,\mu m$, where $\Delta\phi = (2\pi/\lambda)\Delta n_{eff} L$. This modulation does not affect significantly neither the propagation losses nor the effective modal area of the plasmonic mode [43]. The propagation length, i.e. the distance at which optical power is reduced to $1/e$ of its input value, is ~$10\,\mu m$ and the modal area is ~$0.2\,\mu m^2$, which lies in the sub-wavelength regime.

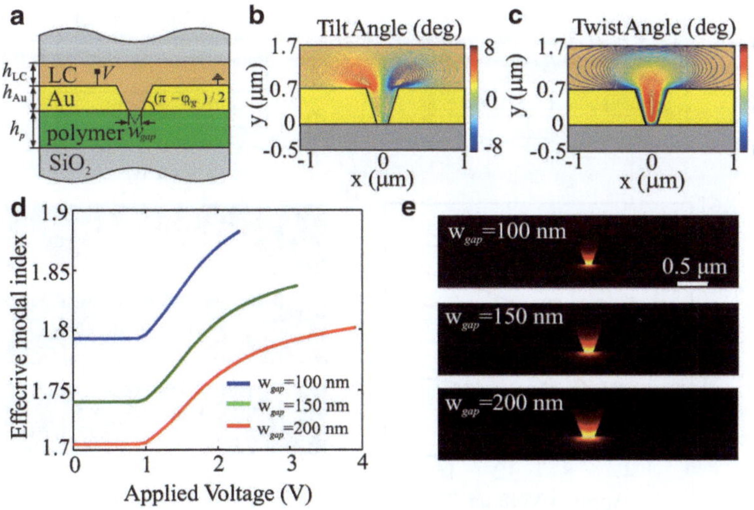

Fig. 7.4 (**a**) Schematic layout of a LC-CPP trapezoidal waveguide originating from the removal of a V-groove's tip. (**b**) Tilt and (**c**) twist angle profiles for $w_{gap} = 150$ nm and $V_{LC} = 1.6$ V. (**d**) Effective index of the fundamental LC-CPP mode as a function of the applied voltage and the width w. (**e**) Optical power profiles for $V_{LC} = 1.6$ V

7.2.3 Liquid-Crystal Loaded Surface Plasmon Polariton Components

A widely explored plasmonic waveguide in view of the integration in photonic circuitry architectures for optical intra-chip interconnection is the dielectric-loaded surface plasmon polariton (DLSPP) waveguide shown in Fig. 7.1c. A dielectric rib, typically a polymer such as PMMA, is patterned on-top of a metallic film with much larger width. This configuration, where the plasmonic SPP mode along the metal/cladding interface is confined by the dielectric loading, permits the definition of all essential elements in photonic circuitry, e.g. bends, splitters, directional couplers and ring resonators [13, 65–69].

Liquid-crystal infiltrated DLSPP waveguides can be designed based on the generic layout of Fig. 7.5a. A trench with polymer walls is defined above a metallic film and infiltrated with a nematic material. When the LC indices are higher than the polymer index, as in the combination of many common materials, such as PMMA and E7, the liquid-crystal cavity acts as a dielectric loading for the plasmonic guiding system. The control voltage is applied between the metal film and an ITO electrode placed above the trench with the introduction of a buffer polymer layer in order to isolate it from the light guiding region. The modal properties of propagating modes depend on the LC orientation and mostly on the tilt angle. Figure 7.5b shows a typical tilt angle profile where the LC is strongly anchored at the cavities' walls along the waveguide axis. In general, for sub-μm dimensions the LC-DLSPP guide

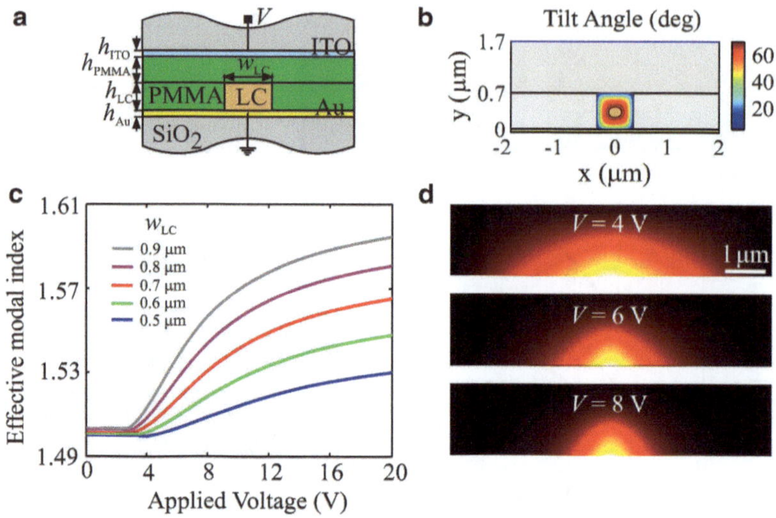

Fig. 7.5 (**a**) Schematic layout of a LC-tunable DLSPP waveguide, where the dielectric waveguide loading is realized by infiltrating a channel patterned in PMMA with the nematic material E7. (**b**) Tilt angle profile for $V_{LC} = 9\,V$ and $h_{LC} = w_{LC} = 700\,nm$. (**c**) Modal effective index of the LC-DLSPP mode as a function of applied voltage. The optical mode can be tuned at an extent capable of yielding significant phase changes at a propagation distance in the order of the waveguide's propagation length. (**d**) Electric field modal profile showing electro-optical light confinement control

supports only one mode, whose properties are controlled by the applied voltage. The modal index can be tuned in an interval $\Delta n_{eff} > 0.1$, as demonstrated in Fig. 7.5c. Higher voltages lead to a higher tilt angles and therefore effective LC index in the loading, which leads to stronger light confinement, as demonstrated in Fig. 7.5d [43].

Figure 7.6a shows a complementary structure for the design of LC-DLSPP waveguides. The nematic material forms a continuous layer above the metallic film, whereas the ITO control electrode is patterned. When the control voltage is applied, voltage drop and therefore tilting of the LC molecules takes place primarily in the region between the ITO and the metallic film, as shown in Fig. 7.6b. This effectively creates a high-index zone, which serves as the dielectric loading of the metallic film. Thus, a DLSPP waveguide is formed dynamically upon LC switching, whose properties are controlled by the value of the applied voltage. Extensive tuning of the effective modal index and area is achieved, as evidenced in Figs. 7.6c, d. This particular design facilitates fabrication, as it demands for the infiltration of a planar LC cell, instead of laterally confined geometries. Furthermore, by patterning the ITO electrode film a bus of densely packed DLSPP waveguides can be readily achieved, where each waveguide is induced only when the corresponding control voltage is applied.

By placing such LC-DLSPP waveguides in close proximity, more complex components that offer advanced functionalities can be designed, such as the direc-

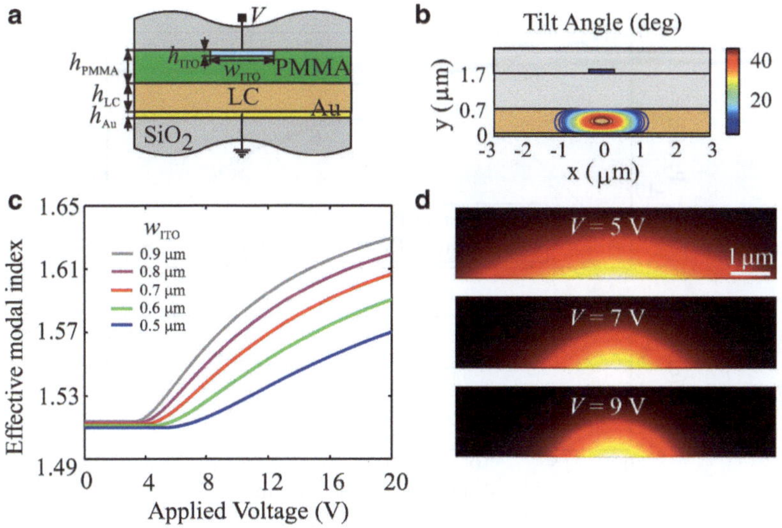

Fig. 7.6 (**a**) Schematic layout of the a LC-DLSPP waveguide, where the dielectric loading is effectuated by locally switching the LC molecules in a LC-slab via a patterned electrode on top. (**b**) Tilt angle profile for $V_{LC} = 9\,V$ and $h_{LC} = w_{LC} = 700\,nm$. (**c**) Modal effective index of the LC-DLSPP mode as a function of applied voltage. (**d**) Electric field modal profile showing electro-optical light confinement control

Fig. 7.7 (**a**) Schematic layout of a plasmonic directional coupler switch based on LC-DLSSP waveguides and definition of the geometrical and material parameters. (**b**) Effective modal index of both individual modes (*blue lines*) and supermodes (*green lines*) for $L_{gap} = 600\,nm$. The first electrode is kept fixed at $V = 4\,V$ and the voltage of the second electrode V_2 varies from 2 to 12 V

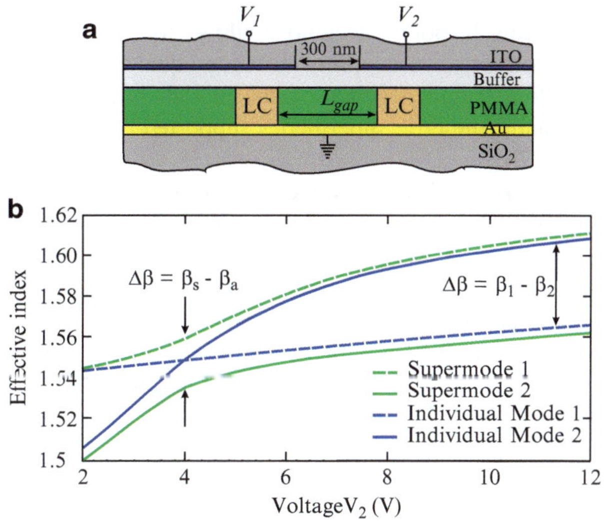

tional coupler switch of Fig. 7.7a [70]. Two LC-DLSPP are placed at a separation of 600 nm and are individually controlled by corresponding ITO electrodes on top. When $V_1 = V_2$ the structure is symmetric and the effective model index corresponding to each waveguide is the same. In this state, the coupler system

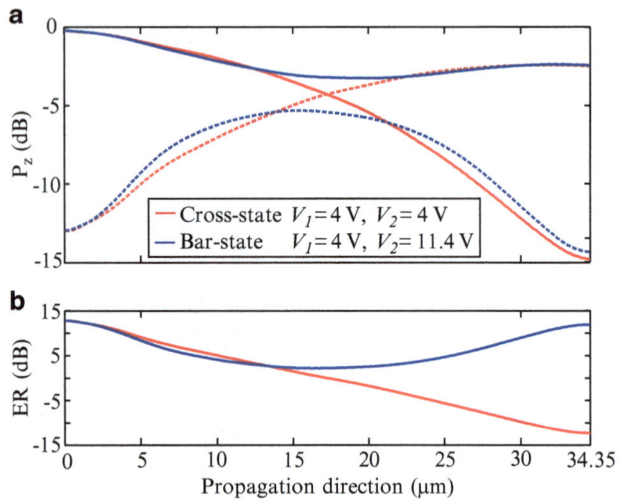

Fig. 7.8 (**a**) Power exchange between the two LC-DLSPP waveguides for the cross and bar states. *Red lines* correspond to the case of the synchronous LC-channels ($V_1 = V_2 = 4\,\text{V}$) cross-state and *blue lines* correspond to the case of the asynchronous LC-channels ($V_1 = 4\,\text{V}$ and $V_2 = 11.4\,\text{V}$) bar-state, while solid lines correspond to the power flow in the initially excited LC-channel and dashed lines correspond to the power flow in the cross port. (**b**) Extinction ratio calculated along the propagation direction

consisting in both waveguides allows for coupling of optical power between the two ports after a coupling distance equal to $L_c = 34.35\,\mu\text{m}$. The value of the coupling length depends on the modal index difference between the symmetric and antisymmetric supermodes supported by the directional coupler, as shown in Fig. 7.7b.

When one of the two waveguides is driven with a higher voltage the two ports are no longer synchronous and the light coupling mechanism depends on the detuning between them. When the propagation constants of the two supermodes differ by $\Delta\beta_{cr} = \sqrt{3}\pi/L_c$, light coupling after a propagation equal to L_c is minimal. Thus, by operating between the two states, cross and bar, switching with an extinction ratio higher than 10 dB is achieved, as shown in Fig. 7.8, where optical power coupling is monitored along lightwave propagation in the LC-DLSPP directional coupler switch.

7.2.4 *Long-Range Surface Plasmon Polariton Components*

One of the most promising plasmonic platforms for optical inter-chip interconnection is based on thin metallic stripe waveguides, termed as long-range surface plasmon polariton (LRSPP) waveguides, which are embedded in a dielectric

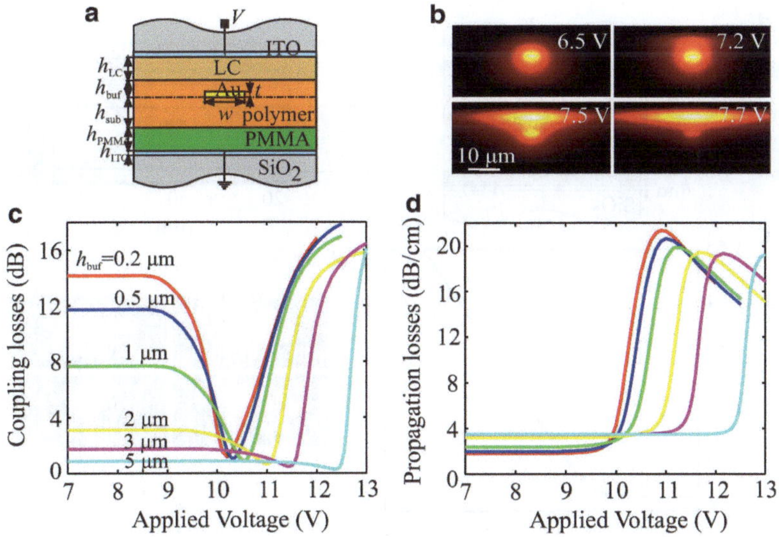

Fig. 7.9 (**a**) Schematic layout of the cross-section of a LC-LRSPP waveguide design for variable optical attenuators. (**b**) Electric field modal profile at four voltage values, calculated for $h_{\rm buf} = 3\,\mu$m and $n_p = 1.505$, describing the transition from a LRSPP to a lossy LC-dielectric slab mode. (**c**) Total coupling losses of the LC-LRSPP variable attenuator to a reference input and output isotropic symmetric waveguide as a function of the buffer layer thickness h_b and (**d**) corresponding propagation losses

medium, as shown in Fig. 7.1e. Their dimensions, typically a few microns wide and a few to tens of nanometers thick, allow for a fundamental plasmonic mode that shows reduced propagation losses, two to three orders of magnitudes lower than SPPs of the single individual metal/dielectric interfaces [71, 72]. Propagation lengths in LRSPP waveguides can reach several centimeters, and the spot size of the LRSPP mode matches well that of conventional polarization-maintaining single mode fibers, thus permitting low butt-coupling losses. Their fabrication involves only planar processing steps, i.e. polymer spin-coating, metal deposition, lithographic definition and liftoff [73, 74]. Finally, the waveguiding stripes can also serve as control electrodes by current injection [16–18] or the application of an external voltage. These characteristics favor the LRSPP-based platform for broadband optical signal routing applications between different photonic chips [75–77].

The properties of LRSPP waveguides depend highly on their environment, since optical power penetrates significantly into the dielectric surrounding medium. Symmetry breaking by the addition of other materials with different refractive indices can provide a means of controlling light confinement in LRSPP asymmetric structures [78]. Figure 7.9a shows such a structure, where a LC layer is introduced in the LRSPP waveguide's cladding. The LC is controlled by two ITO electrodes placed on-top of the layer and in the substrate. In the rest case, the planar alignment

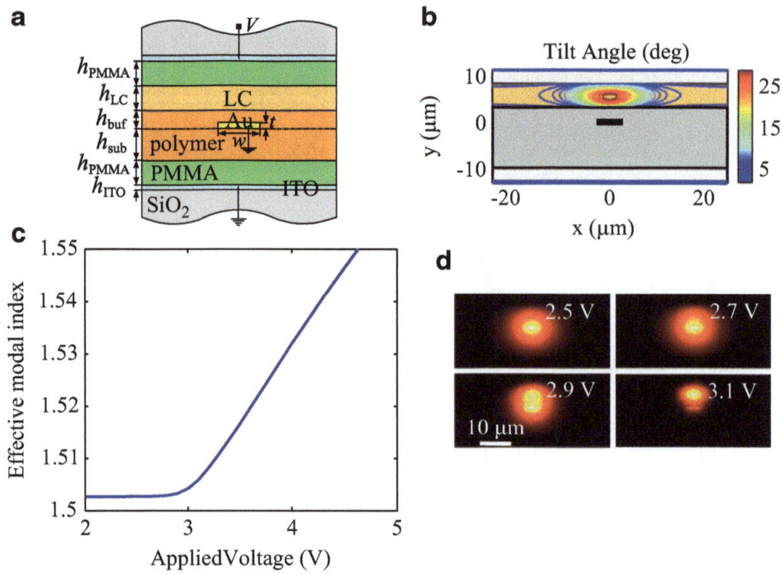

Fig. 7.10 (**a**) Schematic layout of the a LC-LRSPP waveguide design for in-line phase modulators. (**b**) Tilt angle profile at $V = 5\,\mathrm{V}$ for $h_{\mathrm{buf}} = 3\,\mu\mathrm{m}$. A high-tilt zone is formed above the grounded metal stripe. (**c**) Modal effective index of the fundamental LRSPP mode as a function of the applied voltage and (**d**) electric field profiles showing its progressive transition to a dielectric mode formed in the LC-switched area above the metal stripe

of the LC impones the orientation of the LC molecules along the waveguiding axis and thus the LRSPP senses the low ordinary index, which is selected lower than that of the polymer. In this configuration, light is propagating as an asymmetric LRSPP mode confined in the polymer region around the metal stripe. When the control voltage is applied, LC molecules are tilted in a uniform fashion in the LC layer, which becomes a high index cladding and induces light leakage. Such a transition from a well-guided spot to a leaky mode propagating inside the LC layer is shown in Fig. 7.9b [43]. By proper selection of material and geometrical parameters electro-optical variable attenuators are designed, where the on-off extinction ratio is controlled by two mechanisms: (a) coupling losses to the LC-LRSPP waveguide and (b) propagation losses, which become very large when light leaks into the LC layer and comes into contact with the lossy ITO layer. These in-line modulators show a few dB insertion losses and extinction ratios up to 50 dB, which are sufficient for light control applications [79]. LC-based plasmonic components of this type show advantages in terms of their fabrication and driving voltage and power requirements, compared to other proposed solutions relying on the electro-optic response of lithium niobate claddings [21].

Figure 7.10a shows a variation of this design, which targets the design of in-line phase modulators. The metal stripe is grounded and the ITO electrodes are isolated by buffer PMMA layers. In this configuration, the application of the control voltage leads to a voltage drop mainly between the grounded LRSPP waveguide and the

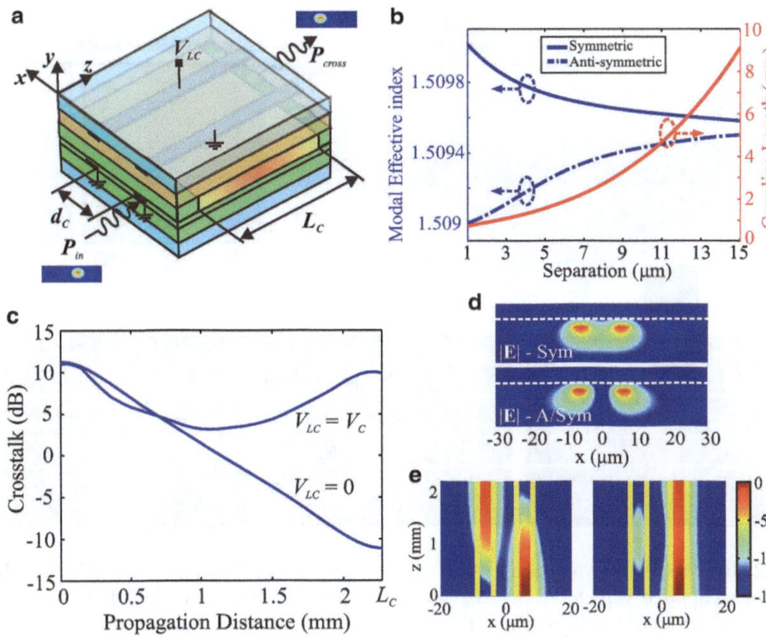

Fig. 7.11 (**a**) Three-dimensional view of a LC-LRSPP coplanar directional coupler switch. The coupling length is equal to L_c and the separation between the two metal stripes is d_C. (**b**) Modal effective indices for the two TM-polarized supermodes supported of the coupler structure in the rest state ($V_{LC} = 0$) as a function of the separation d_c, and corresponding coupling length L_c, defined as $L_c = 0.5\lambda_0/\Delta n$, where $\Delta n = n^{sym} - n^{asym}$, for $\lambda_0 = 1.55$ μm. (**c**) Crosstalk evolution for the two excitation scenarios at $V_{LC} = 0$ and $V_{LC} = V_c = 1.954$ V, which correspond to operation in the cross and bar state, respectively, for a propagation distance equal to the coupling length $L_c = 2.275$ mm. (**d**) Electric field modal profiles for the symmetric and anti-symmetric coupler supermodes. (**e**) Optical power propagation at 100 nm above the metal stripes for the cross and bar operation states

top electrode, thus providing a means to tune the refractive index of the guiding region. When the effective index sensed by the LRSPP voltage surpasses that of the polymer, light is coupled in the high-tilt dielectric waveguide formed inside the LC layer. The modal properties of the hybrid plasmon-dielectric mode can be tuned via the control voltage, for instance, the effective modal index as demonstrated in Fig. 7.10c. The modal effective area and spot location can also be electro-optically controlled, providing an extra degree of freedom in terms of in- and out-coupling as shown in Fig. 7.10d.

Directional couplers can be designed by bringing into proximity two adjacent LRSPP waveguides. Light is coupled periodically between the two ports at a beating length, which depends on the separation between these and the geometrical and material properties of the structure. By dynamically controlling the waveguiding properties of the two ports, switching functionalities are introduced and the optical power distribution between the two ports in the exit of the component can be tuned. Figure 7.11a shows the design of a LC-based coplanar directional coupler

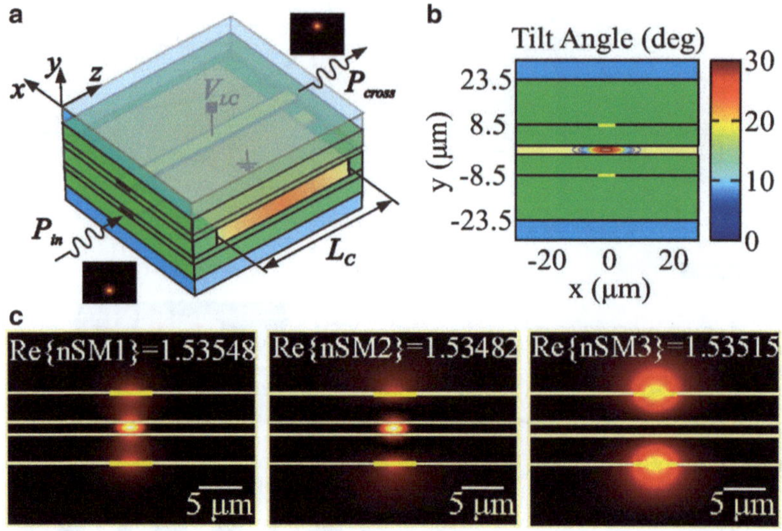

Fig. 7.12 (**a**) Perspective view of a vertical liquid-crystal LRSPP plasmonic switch. (**b**) Tilt angle profile for an applied voltage $V_{\mathrm{LC}} = 20.58$ V, which provides a coupling length $L_c = 2.348$ mm at 1.55 μm. (**c**) Electric field profiles $|\mathbf{E}|$ for the three supermodes supported by the coupler at $V_{\mathrm{LC}} = 20.58$ V, which satisfies at 1.55 μm the coupling phase-matching condition

switch. A LC layer is introduced in a LRSPP coupler configuration, extending a distance L_c along the propagation axis, typically a few millimeters [56, 80]. The metal stripes are grounded and two patterned ITO electrodes on top allow for the individual electro-optical control of each waveguide. In the rest case, the LC layer exhibits a low refractive index and light is guided mostly in the polymer region, coupling between the two waveguide ports. The coupling length L_c can be adjusted by controlling the separation distance d_c and thus the modal indices of the coupler supermodes, as shown in Fig. 7.11b. Typical electric field profiles for the symmetric and antisymmetric supermodes are shown in Fig. 7.11d, evidencing that a very small amount of optical power enters into the LC layer. The trade-off between switching crosstalk level, insertion losses, and coupling length can be selected by proper design, according to the required specifications.

When either of the two ports is driven with a control voltage applied between the respective ITO electrode and the grounded metal stripe, the two waveguides are desynchronized. At a critical decoupling value the coupler's operation is switched from the default cross to the bar state, thus permitting tunable light routing. Figure 7.12c shows the crosstalk between the two ports monitored along propagation through the LC-LRSPP switch for a coupling length of 2.275 mm. The optical power coupling at a plane 100 nm above the metal stripes is shown in Fig. 7.12e, also demonstrating the efficiency of the switching mechanism of the LC-LRSPP plasmonic switch.

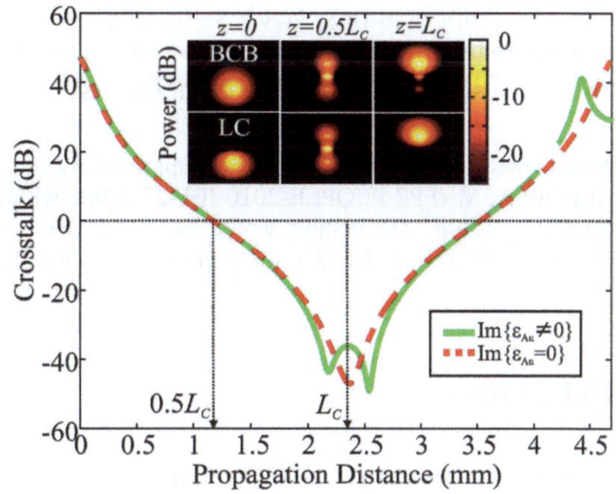

Fig. 7.13 Crosstalk evolution along propagation in the vertical LC-LRSPP directional coupler switch at 1.55 μm, where $L_c = 2.348$ mm. The ideal lossless case for the metal is also included for comparison. *Inset* shows modal power profiles at the input, output of the coupler and after a propagation distance $z = 0.5L_c$

Vertical coupling LC-LRSPP configurations can also be designed, in case integrated multi-planar photonic circuitry architectures are targeted [81]. Figure 7.12 shows such a structure, where the LC control layer is placed between two vertically spaced LRSPP waveguides [55]. The control voltage is directly applied between the two metal stripes, which define the ports of the directional coupler switch. In the rest case, the low index LC layer isolates the two waveguides, providing unconditional bar operation. When the LC is switched by the applied voltage a high-tilt zone is formed in the LC layer in-between the two plasmonic waveguides, as shown in the example of Fig. 7.12b. The high-tilt zone translates into high refractive index for the TM-polarized LRSPP mode, thus creating a dielectric waveguide.

The three-waveguide system, two plasmonic and one LC-dielectric, supports three optical supermodes, as in Fig. 7.12c. When a matching condition among their effective indices is met, light coupling between the two plasmonic ports is achieved with an intermediate transition through the LC-dielectric guide. Thus, cross switch operation is achieved, with millimeter long coupling lengths and very low crosstalk values, as demonstrated in Fig. 7.13. Such low crosstalk values are also predicted in more realistic excitation scenaria and by considering the LC scattering losses [55].

7.3 Conclusions

Nematic liquid crystalline materials have long been used in a wide range of applications for optics and photonics ranging from displays, spatial light modulators to tunable components for light manipulation in photonic circuitry. In this work, we presented an extension of their applicability to a class of plasmonic guided-wave components whose functionalities can be dynamically tuned by the electro-optic control of the nematic molecular orientation. In-line modulators and attenuators, phase shifters, filters, and directional coupler switches were demonstrated in various plasmonic platforms of radically different light confinement scales that target respective applications. Such switching LC-based elements are characterized by extremely low power demands, several orders of magnitude lower than conventional alternatives, e.g. thermo-optic components.

Acknowledgements This work was supported by the Marie-Curie Intra-European Fellowship ALLOPLASM (FP7-PEOPLE-2010-IEF-273528), within the 7th European Community Framework Programme. The authors would like to thank Dr. A. C. Tasolamprou, Dr. K. P. Prokopidis, and Assoc. Prof. E. E. Kriezis, for their contribution and helpful discussions.

References

1. Maier SA, Atwater HA (2005) Plasmonics: localization and guiding of electromagnetic energy in metal/dielectric structures. J Appl Phys 98:011101
2. Zia R, Schuller JA, Chandran A, Brongersma ML (2006) Plasmonics: the next chip-scale technology. Mater Today 9:20–27
3. Ebbesen TW, Genet C, Bozhevolnyi SI (2008) Surface-plasmon circuitry. Phys Today 61:44–50
4. Gramotnev DK, Bozhevolnyi SI (2010) Plasmonics beyond the diffraction limit. Nat Photonics 4:83–91
5. Sorger VJ, Oulton RF, Ma R-M, Zhang X (2012) Toward integrated plasmonic circuits. MRS Bull 37:728–738
6. Stockman MI (2011) Nanoplasmonics: past, present, and glimpse into future. Opt Express 19(22):22029
7. Le Ru E, Etchegoin P (2009) Principles of surface enhanced Raman spectroscopy and related plasmonic effects. Elsevier, Amsterdam
8. Atwater HA (2007) The promise of plasmonics. Sci Am 296:56–63
9. Atwater HA, Polman A (2010) Plasmonics for improved photovoltaic devices. Nat Mater 9:205–214
10. Kauranen M, Zayats AV (2012) Nonlinear plasmonics. Nat Photonics 6:737–748
11. Pitilakis AK, Kriezis EE (2013) Highly nonlinear hybrid silicon-plasmonic waveguides: analysis and optimization. J Opt Soc Am B 30:1954–1965
12. Giannoulis G, Kalavrouziotis D, Apostolopoulos D, Papaioannou S, Kumar A, Bozhevolnyi S, Markey L, Hassan K, Weeber J-C, Dereux A, Baus M, Karl M, Tekin T, Tsilipakos O, Pitilakis AK, Kriezis EE, Vyrsokinos K, Avramopoulos H, Pleros N (2012) Data transmission and thermo-optic tuning performance of dielectric-loaded plasmonic structures hetero-integrated on a silicon chip. IEEE Photonics Technol Lett 24:374–376
13. Pleros N, Kriezis EE, Vyrsokinos K (2011) Optical interconnects using plasmonics and Si-photonics. IEEE Photonics J 3:296–301
14. Hassan K, Weeber J-C, Markey L, Dereux A, Tsilipakos O, Pitilakis A, Kriezis EE (2011) Thermo-optic plasmo-photonic mode interference switches based on dielectric loaded waveguides. Appl Phys Lett 99(24):241110
15. Hassan K, Weeber J-C, Markey L, Dereux A (2011) Thermo-optical control of dielectric loaded plasmonic racetrack resonators. J Appl Phys 110:023106
16. Nikolajsen T, Leosson K, Bozhevolnyi SI (2004) Surface plasmon polariton based modulators and switches operating at telecom wavelengths. Appl Phys Lett 85:5833–5835
17. Nikolajsen T, Leosson K, Bozhevolnyi SI (2005) In-line extinction modulator based on long-range surface plasmon polaritons. Opt Commun 244:455–459
18. Gagnon G, Lahoud N, Mattiussi GA, Berini P (2006) Thermally activated variable attenuation of long-range surface plasmon-polariton waves. J Lightwave Technol 24:4391–4402

19. Park S, Kim M-S, Ju JJ, Kim JT, Park SK, Lee J-M, Lee W-J, Lee M-H (2010) Temperature dependence of symmetric and asymmetric structured Au stripe waveguides. Opt Commun 283:3267–3270
20. Liu SW, Xiao M (2006) Electro-optic switch in ferroelectric thin films mediated by surface plasmons. Appl Phys Lett 88:143512
21. Berini P, Mattiussi G, Lahoud N, Charbonneau R (2007) Wafer-bonded surface plasmon waveguides. Appl Phys Lett 90:061109
22. Sun X, Zhou L, Li X, Hong Z, Chen J (2011) Design and analysis of a phase modulator based on a metal–polymer–silicon hybrid plasmonic waveguide. Appl Opt 50:3428–3434
23. Randhawa S, Lachèze S, Renger J, Bouhelier A, de Lamaestre RE, Dereux A, Quidant R (2012) Performance of electro-optical plasmonic ring resonators at telecom wavelengths. Opt Express 20:2354–2362
24. Zografopoulos DC, Asquini R, Kriezis EE, d'Alessandro A, Beccherelli R (2012) Guided-wave liquid-crystal photonics. Lab Chip 12:3598–3610
25. Beeckman J, Neyts K, Vanbrabant PJM (2011) Liquid-crystal photonic applications. Opt Eng 50:081202
26. Gilardi G, Donisi D, Beccherelli R, Serpengüzel A (2009) Liquid crystal tunable filter based on sapphire microspheres. Opt Lett 34(21):3253–3255
27. d'Alessandro A, Asquini R, Trotta M, Gilardi G, Beccherelli R, Khoo IC (2010) All-optical intensity modulation of near infrared light in a liquid crystal channel waveguide. Appl Phys Lett 97:093302
28. Donisi D, Bellini B, Beccherelli R, Asquini R, Gilardi G, Trotta M, d'Alessandro A (2010) A switchable liquid-crystal optical channel waveguide on silicon. IEEE J Quantum Electron. 46:762–768
29. Abdulhalim I (2012) Liquid crystal active nanophotonics and plasmonics: from science to devices. J Nanophotonics 6:061001
30. Kossyrev PA, Yin A, Cloutier SG, Cardimon DA, Huang D, Alsing PM, Xu JM (2005) Electric field tuning of plasmonic response of nanodot array in liquid crystal matrix. Nano Lett 5:1978–1981
31. Liu YJ, Si GY, Leong ESP, Xiang N, Danner AJ, Teng JH (2012) Light-driven plasmonic color filters by overlaying photoresponsive liquid crystals on gold annular aperture arrays. Adv Mater 24:OP131–OP135
32. Hao Q, Zhao Y, Juluri BK, Kiraly B, Liou J, Khoo IC, Huang TJ (2011) Frequency-addressed tunable transmission in optically thin metallic nanohole arrays with dual-frequency liquid crystals. J Appl Phys 109:084340
33. Zhao Y, Hao Q, Ma Y, Lu M, Zhang B, Lapsley M, Khoo I-C, Huang TJ (2012) Light-driven tunable dual-band plasmonic absorber using liquid-crystal-coated asymmetric nanodisk array. Appl Phys Lett 100(5):053119
34. De Sio L, Caputo R, Cataldi U, Umeton C (2011) Broad band tuning of the plasmonic resonance of gold nanoparticles hosted in self-organized soft materials. J Mater Chem 21:18967–18970
35. DeSio L, Ferjani S, Strangi G, Umeton C, Bartolino R (2011) Universal soft matter template for photonic applications. Soft Mater 7:3739–3743
36. De Sio L, Cunningham A, Verrina V, Tone CM, Caputo R, Bürgi T, Umeton C (2012) Double active control of the plasmonic resonance of a gold nanoparticle array. Nanoscale 4:7619–7623
37. Chigrinov V, Kozenkov V, Kwok H-S (2008) Photoalignment of liquid crystalline materials. Wiley, Chichester/Hoboken
38. Beccherelli R, Manolis IG, d'Alessandro A (2005) Characterisation of photoalignment materials for photonic applications at visible and infrared wavelengths. Mol Cryst Liq Cryst 429(1):227–235
39. d'Alessandro A, Bellini B, Donisi D, Beccherelli R, Asquini R (2006) Nematic liquid crystal optical channel waveguides on silicon. IEEE J Quantum Electron 42:1084–1090
40. Bellini B, Larchanché J-F, Vilcot J-P, Decoster D, Beccherelli R, d'Alessandro A (2005) Photonic devices based on preferential etching. Appl Opt 44:7181–7186

41. Bellini B, d'Alessandro A, Beccherelli R (2007) A method for butt-coupling optical fibres to liquid crystal planar waveguides. Opt Mater 29:1019–1022
42. Steward IW (2004) The static and dynamic continuum theory of liquid crystals. Taylor & Francis, London
43. Zografopoulos DC, Beccherelli R, Tasolamprou AC, Kriezis EE (2013) Liquid-crystal tunable waveguides for integrated plasmonic components. Photonics Nanostruct Fundam Appl 11:73–84
44. Pitilakis AK, Zografopoulos DC, Kriezis EE (2011) In-line polarization controller based on liquid-crystal photonic crystal fibers. J Lightwave Technol 29:2560–2569
45. Bellini B, Beccherelli R (2009) Modelling, design and analysis of liquid crystal waveguides in preferentially@stringpr = Phys. Rep. etched silicon grooves. J Phys D Appl Phys 42:045111
46. Jin J (2002) The finite element method for electromagnetics. Wiley, New York
47. Taflove A, Hagness SC (2005) Computational electrodynamics: the finite-difference time-domain method, 3Rd edn. Artech House, Norwood
48. Prokopidis KP, Zografopoulos DC (2013) Efficient FDTD algorithms for dispersive Drude-critical points media based on the bilinear z-transform. Electron Lett 49:534–536
49. Prokopidis KP, Zografopoulos DC (2013) A unified FDTD/PML scheme based on critical points for accurate studies of plasmonic structures. J Lightwave Technol 31:2467–2476
50. Prokopidis KP, Zografopoulos DC, Kriezis EE (2013) Rigorous broadband investigation of liquid-crystal plasmonic structures using FDTD dispersive-anisotropic models. J Opt Soc Am B 30:2722–2730
51. Pitilakis A, Kriezis EE (2011) Longitudinal 2x2 switching configurations based on thermo-optically addressed dielectric-loaded plasmonic waveguides. J Lightwave Technol 29:2636–2646
52. Ziogos GD, Kriezis EE (2008) Modeling light propagation in liquid crystal devices with a 3-D full-vector finite-element beam propagation method. Opt Quant Electron 40:733–748
53. Vanbrabant PJ, Beeckman J, Neyts K, James R, Fernandez FA (2009) A finite element beam propagation method for simulation of liquid crystal devices. Opt Express 17:10895–10909
54. Beeckman J, James R, Fernăndez FA, De Cort W, Vanbrabant PJM, Neyts K (2009) Calculation of fully anisotropic liquid crystal waveguide modes. J Lightwave Technol 27:3812–3819
55. Zografopoulos DC, Beccherelli R (2013) Design of a vertically-coupled liquid-crystal long-range plasmonic optical switch. Appl Phys Lett 102:101103
56. Zografopoulos DC, Beccherelli R (2013) Long-range plasmonic directional coupler switches controlled by nematic liquid crystals. Opt Express 21:8240–8250
57. Zografopoulos DC, Pitilakis AK, Kriezis EE (2013) Dual-band electro-optic polarization switch based on dual-core liquid-crystal photonic crystal fibers. Appl Opt 52:6439–6444
58. Vial A, Laroche T, Dridi M, Le Cunff L (2011) A new model of dispersion for metals leading to a more accurate modeling of plasmonic structures using the FDTD method. Appl Phys A Mater Sci Process 103:849–853
59. Dionne JA, Sweatlock LA, Atwater HA, Polman A (2006) Plasmon slot waveguides: towards chip-scale propagation with subwavelength-scale localization. Phys Rev B 73:035407
60. Zografopoulos DC, Beccherelli R (2013) Liquid-crystal tunable metal-insulator-metal plasmonic waveguides and Bragg resonators. J Opt 15:055009
61. Pfeifle J, Alloatti L, Freude W, Leuthold J, Koos C (2012) Silicon-organic hybrid phase shifter based on a slot waveguide with a liquid-crystal cladding. Opt Express 20:15359–15376
62. Bahramipanah M, Mirtaheri SA, Abrishamian MS (2012) Electrical beam steering with metal-anisotropic-metal structure. Opt Lett 37:527–529
63. Volkov VS, Bozhevolnyi SI, Devaux E, Ebbesen T (2006) Compact gradual bends for channel plasmon polaritons. Opt Express 14:4494–4503
64. Dierking I (2001) Dielectric breakdown in liquid crystals. J Phys D Appl Phys 34(5):806–813
65. Holmgaard T, Bozhevolnyi SI (2007) Theoretical analysis of dielectric-loaded surface plasmon-polariton waveguides. Phys Rev B 75:245405

66. Chen Z, Holmgaard T, Bozhevolnyi SI, Krasavin AV, Zayats AV, Markey L, Dereux A (2009) Wavelength-selective directional coupling with dielectric-loaded plasmonic waveguides. Opt Lett 34:310–312
67. Holmgaard T, Chen Z, Bozhevolnyi SI, Markey L, Dereux A, Krasavin AV, Zayats AV (2009) Wavelength selection by dielectric-loaded plasmonic components. Appl Phys Lett 94:051111
68. Tsilipakos O, Pitilakis A, Yioultsis TV, Papaioannou S, Vyrsokinos K, Kalavrouziotis D, Giannoulis G, Apostolopoulos D, Avramopoulos H, Tekin T, Baus M, Karl M, Hassan K, Weeber J-C, Markey L, Dereux A, Kumar A, Bozhevolnyi SI, Pleros N, Kriezis EE (2012) Interfacing dielectric-loaded plasmonic and silicon photonic waveguides: theoretical analysis and experimental demonstration. IEEE J Quantum Electron 48(5):678–687
69. Papaioannou S, Vyrsokinos K, Tsilipakos O, Pitilakis A, Hassan K, Weeber J-C, Markey L, Dereux A, Bozhevolnyi SI, Miliou A, Kriezis EE, Pleros N (2011) A 320 Gb/s-throughput capable 2x2 silicon-plasmonic router architecture for optical interconnects. J Lightwave Technol 29(21):3185–3195
70. Tasolamprou AC, Zografopoulos DC, Kriezis EE (2011) Liquid crystal-based dielectric loaded surface plasmon polariton optical switches. J Appl Phys 110:093102
71. Berini P (2000) Plasmon-polariton waves guided by thin lossy metal films of finite width: bound modes of symmetric structures. Phys Rev B 61:10484–10503
72. Berini P (2009) Long-range surface plasmon polaritons. Adv Opt Photon 1:484–588
73. Charbonneau R, Scales C, Breukelaar I, Fafard S, Lahoud N, Mattiussi G, Berrini P (2006) Passive integrated optics elements based on long-range surface plasmon polaritons. J Lightwave Technol 24:477–494
74. Boltasseva A, Nikolajsen T, Leosson K, Kjaer K, Larsen MS, Bozhevolnyi SI (2005) Integrated optical components utilizing long-range surface plasmon polaritons. J Lightwave Technol 23:413–422
75. Ju JJ, Park S, Kim M-S, Kim JT, Park SK, Park YJ, Lee M-H (2007) 40 Gbits light signal transmission in long-range surface plasmon waveguides. Appl Phys Lett 91(17):171117
76. Ju JJ, Park S, Kim M-S, Kim JT, Park SK, Park YJ, Lee M-H (2008) Polymer-based long-range surface plasmon polariton waveguides for 10-Gbps optical signal transmission applications. J Lightwave Technol 26(11):1510–1518
77. Kim JT, Ju JJ, Park S, Kim M-S, Park SK, Lee M-H (2008) Chip-to-chip optical interconnect using gold long-range surface plasmon polariton waveguides. Opt Express 16(17):13133–13138
78. Berini P (2001) Plasmon-polariton waves guided by thin lossy metal films of finite width: bound modes of asymmetric structures. Phys Rev B 63:125417
79. Zografopoulos DC, Beccherelli R (2013) Plasmonic variable optical attenuator based on liquid-crystal tunable stripe waveguides. Plasmonics 8:599–604
80. Zografopoulos DC, Beccherelli R (2013) Liquid-crystal tunable long-range surface plasmon polariton directional coupler. Mol Cryst Liq Cryst 573:70–76
81. Won HS, Kim KC, Song SH, Oh C-H, Kim PS, Park S, Kim SI (2006) Vertical coupling of long-range surface plasmon polaritons. Appl Phys Lett 88:011110

8
Novel Plasmonic Probes and Smart Superhydrophobic Devices, New Tools for Forthcoming Spectroscopies at the Nanoscale

Andrea Giugni, Bruno Torre, Marco Allione, Francesco Gentile, Patrizio Candeloro, Maria Laura Coluccio, Gerardo Perozziello, Tania Limongi, Monica Marini, Raffaella Raimondo, Luca Tirinato, Marco Francardi, Gobind Das, Remo Proietti Zaccaria, Andrea Falqui, and Enzo Di Fabrizio

Abstract In this work we review novel strategies and new physical effects to achieve compositional and structural recognition at single molecule level. This chapter is divided in two main parts. The first one introduces the strategies currently adopted to investigate matter at few molecules level. Exploiting the capability of surface plasmon polaritons to deliver optical excitation at nanoscale, we introduce a technique relying on a new transport phenomenon with chemical sensitivity and nanometer spatial resolution. The second part describes how micro and nanostructured superhydrofobic textures can concentrate and localize a small number of molecules into a well-defined region, even when only an extremely diluted

A. Giugni • B. Torre • M. Allione • T. Limongi • M. Marini • L. Tirinato
M. Francardi • G. Das
PSE and BESE Divisions, King Abdullah University of Science and Technology, (KAUST), Thuwal, 23955 -6900 Kingdom of Saudi Arabia

R. Raimondo • E. Di Fabrizio (✉)
PSE and BESE Divisions, King Abdullah University of Science and Technology, (KAUST), Thuwal, 23955 -6900 Kingdom of Saudi Arabia

Bionem Lab, Department of Experimental Medicine, University Magna Graecia, Campus "Salvatore Venuta" Viale Europa 88100, Germaneto-Catanzaro, Italy
e-mail: Enzo.DiFabrizio@kaust.edu.sa

F. Gentile • P. Candeloro • M.L. Coluccio • G. Perozziello
Bionem Lab, Department of Experimental Medicine, University Magna Graecia, Campus "Salvatore Venuta" Viale Europa 88100, Germaneto-Catanzaro, Italy

R. Proietti Zaccaria
Italian Institute of Technology, Bolzaneto, Via Morego 30, 16163 Genova, Italy

A. Falqui
Dipartimento di Scienze Chimiche e Geologiche, Università di Cagliari, Strada Provinciale Monserrato-Sestu, km 0.700 - 09042 Monserrato (CA), Italy

B. Di Bartolo et al. (eds.), *Nano-Structures for Optics and Photonics*,
NATO Science for Peace and Security Series B: Physics and Biophysics,
DOI 10.1007/978-94-017-9133-5_8,

solution is available. Several applications of these devices as micro- and nano-systems for high-resolution imaging techniques, cell cultures and tissue engineering applications are also discussed.

Keywords Nano structured devices • Plasmonic • Hot electrons imaging • Superhydrophobicity • DNA imaging

8.1 Introduction

Raman is a key spectroscopy technique for molecular characterization and recognition. The Raman effect is a scattering phenomenon based on the interaction of incident light and molecular electronic clouds around the nuclei. The incoming electromagnetic wave induces a dipole moment modulated by the molecular thermal vibrations, being the polarizability a function of the quantum vibrational state. Classical Raman [1] is a linear optical process due to the linear dependence of the inelastically scattered intensity on the excitation intensity, though it depends on a wave mixing mechanism between a photon and a lattice vibration.

As a matter of fact, more than one optical phonon mode of the molecule is Raman active, originating the effective Raman scattering spectrum that appears as a fingerprint of the molecule involved in the interaction. These additional spectral features in the emission spectrum of the reradiated light appear as lower (Stokes) and higher energy (anti-Stokes) side bands. At room temperature most of the excited modes have energy $h\nu > k_B T$, thus due to their thermal population the probability of an anti-Stokes event is correspondingly small.

This vibrational spectroscopic technique has been widely and successfully used to identify and characterize with high accuracy both molecular structures and their interactions. Due to its structural sensitivity and label-free applicability, it has become of particular interest in biophysical and biochemical investigations despite the fact its signal is inherently weaker than other spectroscopic techniques such as elastic scattering or fluorescence emission. In fact, non-resonant Raman spectroscopy suffers from an extremely small cross section, between 10^{31} and 10^{29} cm^2/molecule, which is 12–14 orders of magnitude lower the typical fluorescence ones. When molecules are attached to specific nanometer-structured metallic substrates, Raman signal can be by far improved due to local field enhancement and molecules-substrate interaction, originating the so called Surface Enhanced Raman Scattering (SERS) [2–4]. Field enhancement is related to the physics of plasmons in metals and local chemical enhancement [5–7], which is an inherently broad band phenomenon. SERS cross section becomes proportional to the squared amplitude of the incident field and to the square of the scattering dipole moment (induced by the incident field itself). As a result, the signal is improved by 5–7 orders of magnitude, either exploiting a favorable superposition of surface enhancement and resonant Raman enhancement of the molecule. These effects allowed for the first observation of single-molecule Raman spectra in 1997 [8].

Similarly, the coupling of electromagnetic waves to surface oscillations of free electrons at the metal-dielectric interface, named Surface Plasmon Polariton (SPP), has been used to transport and focus the electromagnetic fields at the nanoscale. These achievements are nowadays widely recognized as keystone results and have stimulated quite innovative techniques and prospected novel research pathways at the nanoscale horizon.

Aside, advances in nanofabrication technology reached nowadays the ultimate limits of miniaturization, allowing the realization of multifunctional devices at single molecule level. These, exploiting the peculiar properties of matter at this scale, lead to important applications in mechanics, chemistry and optoelectronic. New photosynthetic devices, photonic cavity and metallic nanostructures are able to spatially manipulate at sub wavelength level optical beams, confining, guiding and enhancing the electric field well below the diffraction limit. Thenceforth, the possibility of control an enhanced local field with resolutions of few tens of nanometers has triggered improvements in different types of optical and charge transport spectroscopies, like in the case of SERS. Actual advancements in SERS, in fact, rely on engineered plasmonic structures [4] in which extremely intense electric fields are produced in very narrow regions, boosting the collected signal in an unprecedented way. Another important example is Tip Enhanced Raman Scattering (TERS) [9, 10] that allows Raman mapping with sub wavelength resolution by exploiting the local field enhancement amplified via a plasmon resonance at the apex of a nanometer-sized tip, used as a scanning "local hot spot". TERS overcomes most of the drawbacks of SERS due to the inherently local probing ability of Atomic Force Microscopes (AFM), while maintaining the field enhancement. Plasmon local confinement at the end of a very sharp tip, acting as a resonant optical antenna, in fact, circumvents the limitations due to heterogeneity of the probed substrates in SERS [11]. Due to the energy conservation within the optical diffraction limited focus volume, the tip-localized field enhancement is always associated with a correspondently decrease in field strength in the overall focused area, resulting in an improved ratio between the near-field signal and the far-field background.

Both enhanced Raman techniques allow the observation of individual properties not emerging for ensemble measurements due to averaging effects, and provide insights into the nature of the Raman scattering process itself.

The AFM was invented in 1986 [12] to extend the fields of applicability of Scanning Tunneling Microscopy (STM) to non-conductive substrates: sample topography is mapped by raster scanning an atomically sharp tip on its surface under a controlled force interaction, usually kept in a low-force non-destructive regime, to obtain a 3D, constant force, morphological map. The use of tip-sample force as feedback signal overcomes many of the limitations on sample requirements, specific of similar techniques, such as STM or Scanning Near-field Optical Microscope (SNOM). This, together with its high surface sensitivity, allows measurements on virtually any kind of sample, from soft biological molecules to hard inorganic materials, also in room and liquid environments. Beyond topographical capability, AFM sensitivity to interatomic forces down to pN range [13] provides an easily configurable and extremely powerful test bench for fundamental studies in bionanotechnology [14]. For example, modified probes such as magnetic [15–18], conductive [19–21] and bio-functionalized tips [22–25] are widely used for local functional detection of an endless number of properties at nanoscale [26, 27]. Most of the innovative applications of AFM rely on tip customization. In particular microfabricated tips have been designed to efficiently couple electromagnetic radiation to SPPs and to obtain SPP nanofocusing at the very apex of the tip [28]. A couple of relevant examples of this kind are provided by the implementation of a photon-to-SPP coupling in a tip placed in a photonic crystal-based cavity [29] and of a plasmonic microstructured SNOM recently demonstrated [30].

In an analogous way, conductive AFM technique [31,32] is now improved by hot-electron plasmonic assisted nanoscopy, recently demonstrated to be sensitive to the local electric properties at the nanoscale [33]. This technique can be strategically combined with the optical analysis of the scattered electromagnetic field for molecular sensing capability.

The use of all the above mentioned techniques gives access to few-molecule detection and spectroscopic investigation at the nanoscale, which certainly represents one of the greatest advancements in modern science and technology. However, their applicability in many fields such as medicine, biology, pharmacology, and chemistry, will certainly require also facing the challenge to localize and concentrate molecules and structures to be analyzed in a very confined region.

To achieve this further important step, nowadays researchers' efforts are addressed to the integration of local imaging techniques with special substrates which have specific ability such as superhydrophobicity. These systems allow circumventing the classic dilution limits [34] and to selectively extract, align, and manipulate specific samples at the microscale [35].

In this chapter we review recent advances in these fields. It is divided in two parts. In the first one the problem of achieving localized sub-wavelength resolution in optical and hot-electron spectroscopy is addressed. Section 8.2.1 introduces the main concept behind light-plasmon coupling and their propagation in a tapered nanostructure. In Sect. 8.2.2 we present a novel scanning technique for probing the topography and chemical properties of structured materials at the nanoscale. Based on existing atomic force microscopy, this new approach employs a tapered conical plasmonic tip to efficiently inject hot carriers into the sample and introduces key advances in terms of spatial resolution and chemical sensitivity toward single molecule detection and recognition. Section 8.2.3 underlines the advantages and perspective application of plasmonic nanofocusing when applied to ultrashort laser pulses.

In the second part, we focus on the matter of samples concentration technique via super-hydrophobic templates and its applications. Section 8.3.1 describes the recent implementation of micro and nano fabrication techniques applied to the realization of different kind of super-hydrophobic (SH) substrates for biocompatible scaffold design, spectroscopic and electron microscopy applications. In Sect. 8.3.2 the combination of super-hydrophobicity with plasmonic nanostructures is used to concentrate exosomes and perform studies to investigate tumor and healthy colon exosomes through Raman spectroscopy and X-ray scattering. Section 8.3.3 shows new application of textured SH surface to perform high resolution DNA imaging allowing fast sample localization at the nanometric scale by means of background-free transmission electron microscopy (TEM) imaging.

8.2 Advancements in Scanning Probe Structural Imaging

8.2.1 Metallic Cones and Plasmon Coupling

The particular nature of SPP as confined electromagnetic oscillation at a metal-dielectric boundary presents both an electromagnetic field evanescently decaying into the dielectric and a charge density fluctuation at the metal surface. This is the basis of their peculiar properties. The confinement of the electromagnetic wave quite below the classical diffraction limit and a favorably long propagation length are the most important for applications in engineered nanostructures [36,37]. SPPs can be exploited to efficiently transfer energy, outdoing other non-collective classical electronic transport phenomena, such as thermal or electric conduction. From the experimental point of view all these new techniques rely on a new physical phenomenology that has no analogue at the macro scale. This is made possible by the ability to control and manipulate materials structure, engineered at the nanoscale with unprecedented expertise, thanks to the level of accuracy and details of today fabrication technology.

A tapered metallic cone is ideally the most suited 3D structure to exploit the confinement and propagation properties of plasmons polaritons. It sustains the propagation of electromagnetic field as SPPs till the apex, either in lossy [38, 39] or in adiabatic mode [40, 41]. The propagation along conical structures has been widely studied in literature [42, 43]. Specifically, a conical plasmonic waveguide efficiently propagates radial modes [43–46], i.e. transverse magnetic modes (TM0) with electric radial symmetry with respect to the cone axis. The TM0 mode propagating in regime of adiabatic compression exhibits negligible dissipation and back reflection during the propagation. Both group and phase velocity tend to zero at the extreme apex of the cone allowing considerable increase of electric energy density, determining high radiation concentration confined in a nanometric spot [47].

Once generated, SPPs flow with a propagation length up to tens of micron, in the pre-established direction independently of the original optical wave, will depend only on the metal electronic properties. In real cases SPPs, while moving along the cone, could damp out in radiative mode due to surface roughness [36, 48], or eventually excite short living highly energetic charges with growing efficiency approaching the tip itself. There, these hot electrons, energized at level far above the thermal equilibrium, can be fruitfully used to probe electric transport properties of samples. Thanks to the relaxation of the momentum constraint that takes places at nano scale, when the cone cross section became comparable with the electron mean free path in the metal, such generation process become also appreciably efficient and can be used to obtain detailed conductive maps of samples surface. For the case of conical geometry a model to describe hot electrons generation by adiabatic SPPs decay at the nano-apex has been introduced. In particular, when the taper becomes thinner than the SPP skin depth and the constrains of the conservation of the linear momentum is relaxed by the "roughness of the contact", then the transfer efficiency η, i.e., the probability for an hot electron to be transferred across the Schottky junction from the metal into the semiconductor, increases significantly and can be expressed in terms of a nanoscale-adapted Fowler formula [49] simply as

$$\eta = \frac{\hbar\omega - e\phi_B}{2\hbar\omega}\Theta[\hbar\omega - e\phi_B] \tag{8.1}$$

where ϕ_B is the Schottky Barrier of the junction, e is the electron charge, ω the angular frequency of the photon propagating along the metal coating of the tip as a SPP and Θ is the Heaviside unit step-function.

Concerning coupling of electromagnetic radiation into SPP, due to the mismatch of the photon/plasmon momentum, light coupling to metal surfaces can be provided only supplying an external wave vector either via Kretschmann [50, 51] or Otto [52] configurations or structuring the metallic interface with an appropriate optical grating [53–57]. In this case, the grating behaves as unidirectional SPPs launcher, allowing the energy transfer between light and surface plasmons. Grating-coupled excitation of SPPs obeys the constraints of energy and momentum conservation laws. The first fixes the frequency, while the second, known the dispersion relation for SP in the metal [48, 58], determines the amount of momentum that must be provided to couple light and SP modes. Accordingly to the scattering geometry of Fig. 8.1, the grating period p can be obtained from

$$\frac{2\pi}{p} = K_z^{\mathrm{SPP}} - K_z^{\mathrm{ph}} = \frac{\omega}{c}\sqrt{\frac{\varepsilon_d\varepsilon_m}{\varepsilon_d + \varepsilon_m}} - \frac{2\pi}{\lambda}\sin\theta_{in}, \tag{8.2}$$

being ε_d and ε_m the frequency dependent permittivity of the metal and the dielectric material [59], λ the free space wavelength, θ_{in} the angle of incidence from the surface normal, and z the SPP propagation direction. The calculation, in the case of the gold surface, gives a grating period of ~2,150 nm, for an incidence of 36° and for the wavelength of 980 nm. Efficient coupling requires also the accomplishment of optimal generation strength that depends on the width and depth of the grating grooves. An estimation of such parameters can be obtained following a literature analytical model [60] that allows calculating the scattering coefficients between an incident plane wave and the SPP modes launched at the sub-wavelength metallic slit aperture. Alternatively, as in our case, an efficient coupling is obtained applying a 2D numerical optimization on the three significant parameters, see Fig. 8.1a, leading to an optimal width-to-wavelength ratio of about 0.3, in agreement with the literature analytical model.

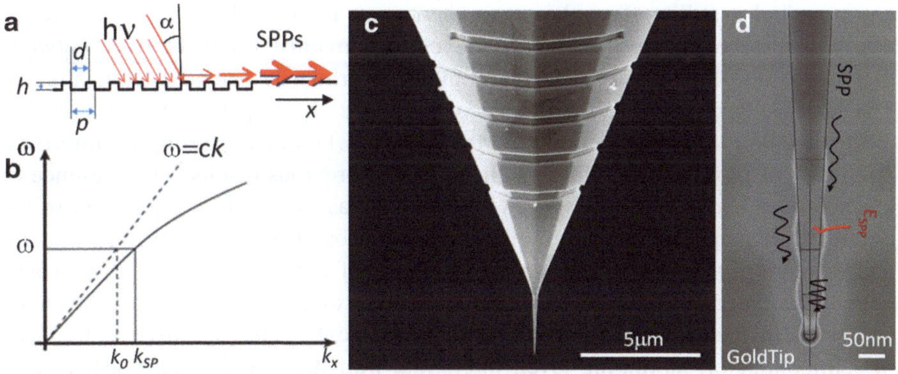

Fig. 8.1 Surface Plasmon generation, propagation and concentration along the device (**a**) Sketch of the grating coupling geometry where are indicated the four parameters involved in the coupling: incidence angle, groove width, height, and grating period. (**b**) The dispersion curves of the photons and for a SPP mode. It shows the momentum mismatch $k_{SP} - k_0$ that must be supplied by the grating to couple light to SPP modes. (**c**) Full device structure fabricated considering optimal value of the simulated parameters for a wavelength of 980 nm. (**d**) Adiabatic compression propagation represents the SPPs minimal loss propagation solution, and allows the propagation of SPPs until the apex of the cone, progressively slowing and pushing up the electromagnetic field amplitude as consequence of the focusing phenomenon. It shown the terminus part of an e.m. simulation of a TM_0 mode (radial mode) excited along a metallic conical structure (base = 300 nm and height = 2.5 μm, apex radius of 20 nm quite similar to the realized one)

8.2.2 *Hot Electrons Imaging*

Up to now hot electrons have been successfully applied in infrared photo detection [47, 61, 62], artificial photosynthesis [63], and catalysis applications [64], opening the way to new generation of optoelectronic devices. Here we describe a recent developed instrument in which we integrated a modified commercial AFM with an engineered plasmonic probe. This system allows the scanning of the surface of a sample obtaining at the same time topography and hot electron spectroscopy information with nanometric lateral resolution. The system was used to map semiconductor/insulator patterned surfaces. The hot electrons current source was obtained by the conversion of SPPs through an adiabatic nanotip in Schottky diode configuration.

The phenomenon driving this process, named internal photoemission, was described by Spicer in 1958 [65] to explain the photo-conductance observed at the metal-semiconductor interface. It states that energized electrons generated by the absorption of incident photons in the metal can transfer to the conduction band of the semiconductor, producing in this way a net current. In general this process is considered poorly efficient due to the short lifetime of the excited electrons and the relatively longer penetration depth of light in metals. In fact in few femtoseconds (that corresponds to a typical mean free path of tens of nanometers for hot electrons) electrons thermalize in the metal itself, too close to their generation place, losing any practical usability. The plasmonic device we realized, however, circumvents this limit promoting the hot electrons excitation mainly at the interface between the needle and the sample, with a momentum primarily oriented along the tip. The specific expression for the decay of adiabatically propagating SPPs into hot electrons along conical taper is reported in Eq. (8.1), demonstrating an improved efficiency of more than 10 times with respect to previous results. The presence of the Schottky barrier [66], see Fig. 8.2b, operates as the rectifying junction in the circuit. It is generally sensibly smaller than the band gap of the semiconductor and depends tightly on the electronic property of the materials at the interface, being inherently sensitive to the local difference of Fermi energies of the metal and semiconductor. Therefore, hot electrons current results sensitive to the type of semiconductor, its doping level, interstitial trap states and surface structure or impurities. To isolate the contribution of the sole hot electrons, we have performed photocurrent measurement on GaAs at three different wavelengths, namely 670, 980, and 1,060 nm. The first is above the semiconductor band gap, while the others are below, thus unable to directly generate electron-hole pairs.

Notably, also charge confinement at the tip requires a new physical description in terms of the metal-semiconductor junction potential geometry. The depletion region, in fact, resembles more the one of a point charge surrounded by a semiconductor than a classical plane junction. This "point contact geometry" problem was experimentally investigated firstly in 1964 [67] to explain unconventional results from the IV characteristic curves measured for Schottky junction at the micron scale. In that work, a more than linear current to voltage behavior due to space charge effect was found. More recently, with the scaling down of integrated electronic devices, renewed interest came out on this theme, and a lot of experimental [68] and theoretical [69, 70] works have been developed. According to these results, while in the classical 1D limit the depletion region in the semiconductor extents homogeneously for several hundred of nanometers to balance the potential across the junction, in a nanosized point contact geometry the electric field radiates from the tip and the isopotential surfaces assume spheroidal shape, see sketch in Fig. 8.2c. Also an electronic partial shielding effect takes place, spatially confining the depletion volume just around the tip, resulting in a barrier characterized by a small number of dopant ions. This fact improves the sensibility to different chemical species and the spatial resolution of the proposed scanning probe technique, as explained in Ref. [33]. To demonstrate this, we performed experiments on silicon and gallium arsenide, either oxidized or doped. The Au metal of the tip in contact with the samples realized the Au/Si Au/GaAs nanosized Schottky junction [71, 72]. We report here the specific case of GaAs surfaces locally patterned with ion implantation or locally oxidized with AFM lithography. The realized structures were subsequently measured with the SPPs nanoscopy setup at a laser excitation of 980 nm, to simultaneously obtain morphology and hot electrons current maps. GaAs surface proved to be an ideal candidate to our scope, either offering a quite chemically-stable interface with respect to the normal ambient oxidative condition, either displaying a peculiar photo absorption profile, that sharply falls down by orders of magnitude [73] just after the band gap threshold.

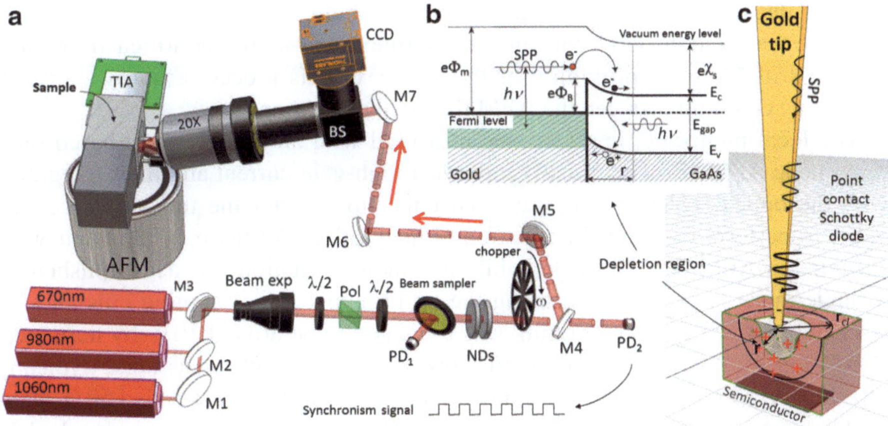

Fig. 8.2 Plasmonic current Atomic Force Microscope. (**a**) Schematics of the optical setup: three laser sources, 760, 980, and 1060 nm, share a common optical path to the sample. M1 M4 M5 M6 silver mirrors, M2 flip mirror, M3 beam splitter (80:20 R:T). Laser intensity and polarization state were controlled by means of zero order wave plates ($\lambda/2$) and a broadband polarizing beam-splitter cube (Pol). A calibrated photodiode was used to monitor the intensity of a sampled beam, about 8 %. Incoming laser was further attenuated with neutral density filters (NDs) and modulated by a mechanical chopper. A Mitutoyo 20x objective (BD Plan Apo SL20X 43 % at 980 nm) supply the laser excitation to the grating, while a confocal non polarizing beam splitter (BS) allows the collection of the scattered light to a CCD detector. (**b**) Band diagram at the metal/semiconductor interface showing the energy of the hot electrons with respect to the Fermi level of the metal and the energy of the semiconductor gap, E_{gap}, in the case of $h\nu \geq E_{\text{gap}}$. Φ_m and χ_s are the metal work function and semiconductor susceptivity, respectively, r is the depletion radius, and E_c and E_v are the energy of the conduction band edge and valence band edge, respectively. (**c**) Sketch of the depletion region profile in a nano-sized point contact geometry. The E field radiates from the tip while isopotential curves describe spheroidal surfaces. r indicates the actual confined radius while, circular *black line* indicates the expected extension of the depleted region for the corresponding classic 1D case

The n-type GaAs wafer (Si-doped, (100)-orientation, mobility μ = 3590–3850 cm^2/ V/s, ρ = 0.02 Ohm· cm) was wet etched for 3 min in a solution of HCl 37 % in water, in order to remove the native oxide layer, and subsequently rinsed in deionised water. Specimens of approximate area 0.25 cm^2 were cleaved from the wafer and a back (ohmic) contact was realized by using silver bonding-paste on the metallic AFM holder.

In Fig. 8.1c we report a SEM image of one of the devices realized on the body of an AFM commercial tip (μ-masch csc38, n-type silicon tip, height ~20 µm, full tip cone angle ~40°, lever force constant of 0.03 N/m, resonant frequency of 10 kHz). The tip has a pyramidal shape with octagonal base, while the base and height of the cone are 300 nm and 2.5 micron respectively. The grating was prepared by focused ion beam technique milling sub-wavelength grooves on the front side of a pyramidal AFM tip and successively coated by sputtering a 150 nm gold layer on it. We were able to control all grooves fabrication parameters within a precision of ±5 nm. After that a sharp micrometric Pt-C cone was realized using electron-beam-induced deposition, with an apex radius <10 nm. Finally the whole structure was gold coated by means of controlled Ar^+ sputtering to reach the final thickness of 25 nm in proximity of the tip apex. This process ensures a smooth plasmonic coupling between the AFM tip and the nanocone.

The local photoelectric effect was measured with an ad hoc setup based on a multimode AFM (Bruker, Veeco) and a fast high-gain current amplifier integrated with the optical system. This was used either to monitor the tip-sample contact either to perform the laser illumination of the grating. All the measurements were executed in AFM contact mode, while the photo excitation was accomplished by a single-mode laser, polarized along the tip principal axis. Coupling efficiency of laser light into SPPs at the grating was evaluated to be around 10 % by numerical simulation. To get rid of undesired slow drifts and mechanical noise typically affecting AFM measurements, we modulated the laser amplitude with a 50 % duty-cycle square wave at the frequency of 317 Hz. The arrangement is sketched in Fig. 8.2a. The typical current signal was in the range 10 pA–10 μA and it was amplified with a trans-impedance amplifier (FEMTO, LCA-M-50k-100M). A low-pass filter (24 dB/oct at ~50 kHz) was applied via a filtering unit (Stanford Research Systems, SIM965) to reduce all spurious noise features in the signal at frequencies higher than the accessible bandwidth of the low noise amplifier.

The measure of the current to voltage characteristic, or IV curve, has been demonstrated to be an extremely useful tools also to investigate the physics of a point contact junction revealing much of the local properties, such as chemisorbed layer and surface trapping states for example [68, 74]. Here an arbitrary function generator (Agilent, 81160A) was used to bias the sample in the ±10 V range and all the relevant signals (I, V_{Bias}, trigger signal, laser temporal pulse profile and z-AFM position) were recorded using a 16 bit ±10 V parallel acquisition board (National Instruments, USB-6366) driven at a sampling rate of 100 kHz. Figure 8.3a shows a short time interval, around zero applied bias, of the current waveform. Laser trigger-timestamps allowed to directly recognize each optical cycle so that it was easy to identify data intervals not affected by transient phenomena and to recover the current to voltage characteristics with and without illumination.

Direct scanning probe writing by high field discharge technique [75, 76] was used to write down local oxidized patterns at nanometric scale. During the writing processes, performed in contact mode at a defined set-point force below 10 nN, the tip scans the surface at constant speed ($1 \div 4$ μm/s) according to a pre-designed pattern. A positive tip-sample bias induces the GaAs local surface oxidation confined in a small water meniscus due to environmental humidity below the tip [77–80], as shown in Fig. 8.4a. Features with lateral resolution of about 50 nm and well-controlled height on the nanometric scale ($0.5 \div 4$ nm, in the $-2 \div -14$ V tip voltage range) are routinely obtained. The technique allows also for a multi-pass writing process up to a thickness of about $8 \div 10$ nm, even if exceptional thicknesses exceeding 100 nm has been recently demonstrated [20]. The photo induced current map shown in Fig. 8.3c was obtained via a real time demodulation of the current signal with the laser pulses temporal profile, using an AC-coupled lock-in amplifier (Stanford Research Systems, SR830). Synchronous signal was then time averaged

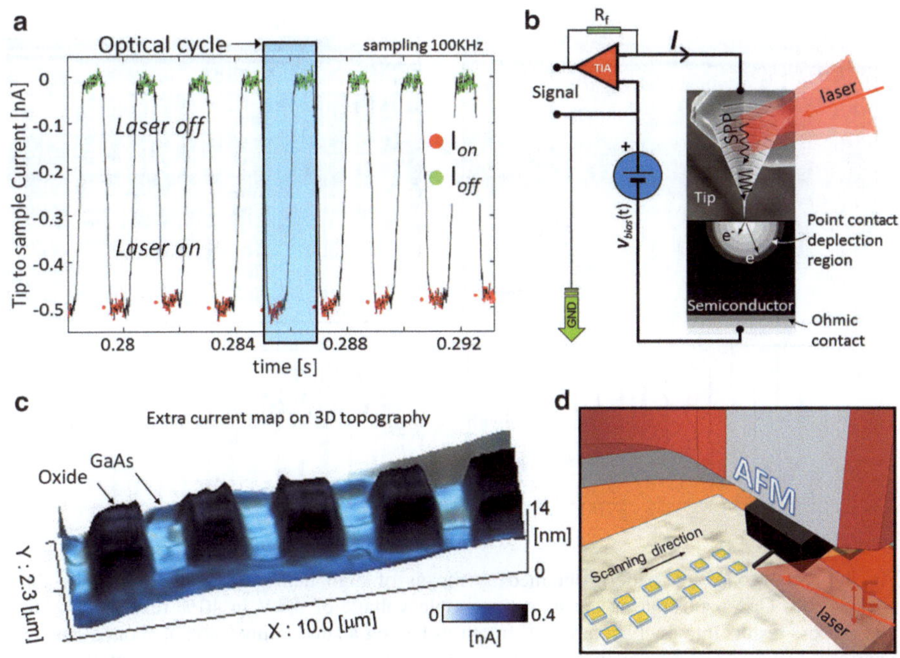

Fig. 8.3 (**a**) Laser induced hot electrons current at 0 V bias. Are indicated on and off current stated synchronized with the modulation of the laser. (**b**) Basic current circuitry scheme and experimental scattering geometry. The photo-generated current is acquired using a high gain bandwidth (10^8, 50 kHz respectively) trans-impedance amplifier, TIA, and subsequently demodulated with a lock in amplifier synchronized with AFM acquisition topography (**c**) Rendered 3D AFM topographic image of oxidized GaAs structure on semiconductive GaAs substrate. The superposed color map indicates the integrated intensity of the hot electron current acquired in parallel with topography. (**d**) 3D rendering of the optical setup

with an integration time of 20 ms and acquired in parallel to sample topography, thus resulting in a hot electrons map of the surface. The best lateral resolution achieved with this technique was of the order of 50 nm, corresponding to the extension of the depletion region in our Schottky barrier that defines the active region in our measurement. Reducing the depletion region might result in an improvement of the resolution down to the effective size of the contact, in the range of 10 nm.

Using focused ion beam gallium-ions implantation we prepared almost flat patterns with controlled local dopant content, i.e. inducing a local shift of the work function. At low doses, both surface amorphyzation and ion implantation occur, giving competitive opposite effects on sample conductance. Material properties can be therefore tuned with the applied ion dose, before entering the erosive regime. AFM topography of Fig. 8.5a shows a slightly visible chemically modified pattern on the GaAs substrate via implantation of 30 keV Ga ions at constant ion current along predefined stripes. In correspondence of stripes crossing, the implanted dose

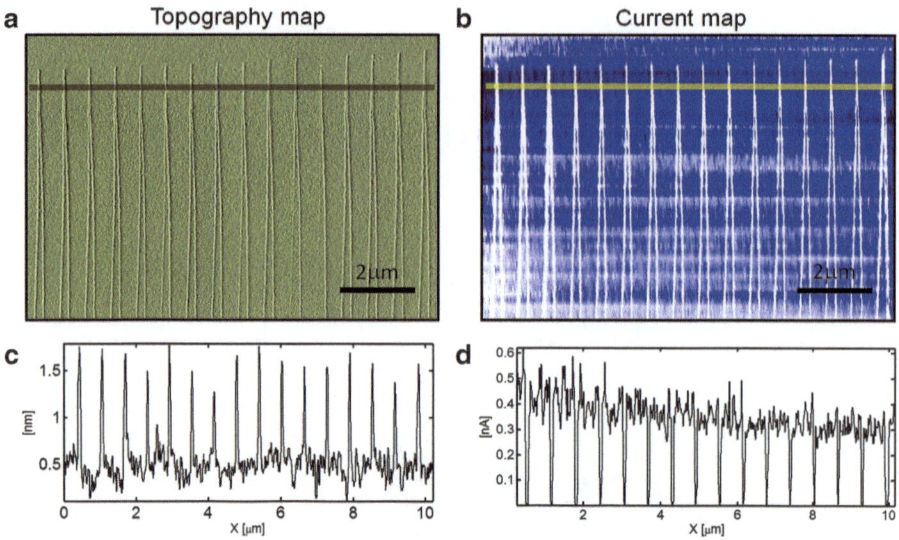

Fig. 8.4 Nano-oxidation pattern obtained by means of high field AFM nanolithography on a n-type (100) GaAs substrate at constant tip-sample voltage of −4 V in 40 % relative humidity. The process results into finely controlled structures having a lateral dimension of about 50 nm and a uniform height of about 1 nm over a length of several micrometers, as can be seen from the profile reported in the bottom panel. The two images show topography (**a**) and hot-electron current signal (**b**) acquired in parallel. (**c**) and (**d**) panels show the cross section signal taken in correspondence of the horizontal lines traced in the images (**a**) and (**b**) respectively

doubles, resulting in a higher degree of amorphyzation and thus in a very confined region of lower sample conductivity. This structure is highly contrasted by hot electron imaging of Fig. 8.5b, resulting in a low-current spot at the center of the line crossings, with dimension comparable with the final lateral resolution of the measurement, i.e. 50 nm. In conclusion, hot electrons imaging is regarded as an extremely powerful tool towards nanomaterial characterization due to its very high chemical sensitivity, unprecedented special resolution, and its capability to probe conductance at virtually zero bias, meaning with a much lower probability of sample oxidation. These results rely on the adiabatic compression of SPP whose decay rate in energetic electrons is inversely proportional to the apex radius of the plasmonic nanocone. Thus, a tapered waveguide not only adiabatically transforms a propagating SPP wave to quasi static intense local field but also efficiently transforms the SPPs into hot electrons, with a 30 % of conversion efficiency which outperforms previous results. This is the underlying idea of the high conversion efficiency of the proposed nanoscopy. Reducing the depletion region might result in an improvement of the resolution down to the effective size of the contact, in the range of 10 nm.

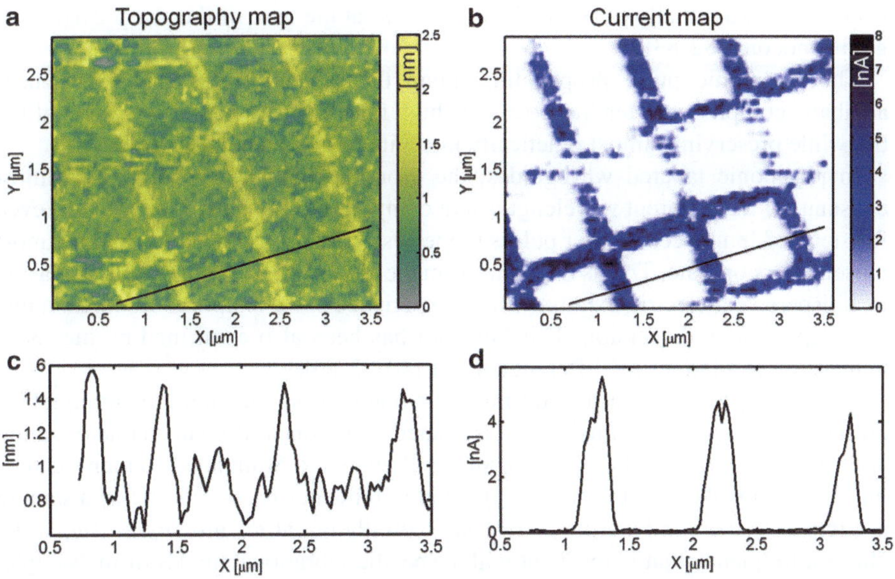

Fig. 8.5 Topography and hot-electron maps of custom locally patterned GaAs sample. (**a**) AFM topography. The sample was a GaAs wafer patterned by Ga ions implantation via focused ion beam process as single grid lines (40 pA, 100 ns point21, 30 keV, single pass). (**b**) Hot-electron AFM current signal through the n-type GaAs–Au tip junction (**c**) Cross section topography profile taken in correspondence of the *black line* on image (**a**). (**d**) Single line hot-electron intensity profiles taken in correspondence of the *black line* on image (**b**). Of the four peaks in topography profile in (**c**), just the three corresponding to implanted Ga ion features provide a clear contrast in the corresponding hot-electron signal, shown in (**d**): this is a clear proof of the high chemical sensitivity at the nanoscale of this new technique Topography map and profile are not deconvolved for the tip profile [81]

8.2.3 *Excitation with Ultra-short Laser Pulses*

An expected future development of the tapered plasmonic waveguides is related to the spectroscopic applications of ultra-short pulses to produce SPPs at the pico- and femto-second time-scale. The peculiarity of pulsed regime with respect to the continuous wave one is the broad frequency spectrum: each pulsed source has an associated frequency spectrum whose minimum width gets larger the shorter the time duration of the pulse.

SPP coupling in nanocones is a broadband phenomenon since it relies on propagation of SPPs, which is a non-resonant effect. In this case a direct coupling strategy such as the one obtained by focusing light on the back of a tapered nanocone is preferred to avoid the use of plasmonic gratings which are inherently wavelength selective. Theoretical calculations have demonstrated that with direct coupling mechanism an overall efficiency as high as 70 % can be obtained. This result refer to a radially polarized beam focused at the base of the sharp 2 μm long gold nanocone [82, 83].

The plasmonic pulse propagating along the tapered waveguide experiences adiabatic compression, leading to a very high field enhancement at the apex of the tip, while preserving almost practically intact its original temporal profile.

In plasmonic tapered waveguides, the group velocity turns out to be almost constant for the different wavelengths that compose the frequency spectrum of even few-tens-of-femtosecond laser pulses in the Vis-NIR spectral range, where the most pulsed lasers operate. This is different from the case of a laser pulses that propagate in a dispersive dielectric medium which experience pulse temporal broadening due to group velocity dispersion. This behavior has been also confirmed by means of numerical calculations by different authors [40, 82, 83].

This opens the way to a wide range of potential applications involving short pulses. The first most promising one is related to nonlinear Raman characterization at nanoscale such as Coherent Anti-Stokes Raman and Stimulated Raman Scattering. The first is the generation of a signal at a frequency $\omega_p + \Delta\omega$, when a sample is excited by the combination of two laser oscillating at ω_p and $\omega_p - \Delta\omega$, when $\Delta\omega$ is a frequency that is resonant with a specific vibrational modes of the sample. The second, obtained under the same experimental conditions, represents a small transfer of energy from the laser at $\omega_p - \Delta\omega$ to the laser at the pump frequency ω_p.

With respect to linear Raman and TERS, these techniques can provide much higher sensitivity. This is due to different concurrent factors. First of all these effects depend in a nonlinear way on the excitation intensity, hence the combination of the high peak intensity provided by pulsed lasers and the enhancement of the field at the tip apex can result in more favorable conditions. Moreover these techniques introduce further advantages like the suppression of fluorescence background that have already been well described in literature [84, 85].

Linear Raman with continuous wave laser excitation at few-nanometers lateral resolution form a tip of a tapered plasmonic waveguide was already demonstrated [29], but the coupling of pulsed lasers to the tapered nanocone to perform nonlinear Raman characterizations has still to be implemented.

Moreover, the high spatial resolution achievable at the apex of a plasmonic cone, thanks to the dispersion-free propagation of femtosecond broadband light pulses could be applied to nonlinear Raman excitation schemes of particular interest introduced recently. These new developments gives the advantage to obtain the full nonlinear Raman spectral characterization in a single measurement, and they rely either on spectral measurements obtained with broadband femtosecond laser pulses [86, 87] or on the nonlinear interaction of the laser pulses generated by two frequency detuned femtosecond oscillators, whose resulting interferogram is analyzed via Fourier transform-based techniques [88].

Recent articles [82, 83] have moreover proposed the use of this nano-focusing system to perform at the nanoscale more complex and powerful characterization, like multi-dimensional optical spectroscopy [89] to study molecular interaction at nanoscale and at single molecule level.

8.3 Super-hydrophobic Surfaces and Their Application

8.3.1 Basic Principle and Realization of the Structure

Super-hydrophobicity is a widely studied and a well-known phenomenon consisting in the ability of a surface to repel water drops, which assume a quasi-spherical shape with a contact angle greater than 150°. Here we introduce a new type of device that exploits the SH of a nanostructured surfaces to concentrate and localize molecules in ultra-diluted solutions. Due to the evaporation of the solvent, the solution progressively concentrates up to the critical concentrations when the dissolved molecules precipitate in an extremely localized pre-established region, roughly corresponding to interface area. This makes it much easier to find the originally dispersed solute, now aggregated at a high concentration, in a specific area of the surface, without long time-consuming wide-area scans of the sample to retrieve the molecules to be analyzed and/or imaged.

Two different types of Nanopatterned Superhydrophobic (NSH) surfaces were produced by means of microfabrication techniques: NSH surfaces without holes, to be used for optical spectroscopy and Scanning Electron Microscope (SEM) characterization, and NSH surfaces with holes, to be used for TEM.

Both types of sample were characterized by a texture of a periodic hexagonal lattice of cylindrical pillars. The diameter d and the distance l between the pillars were chosen following a criterion of design for optimal result described in Ref. [34] that gave $d = 10\,\mu\text{m}$ and $l = 20\,\mu\text{m}$; while the height of the pillars (h) was $10\,\mu\text{m}$. In this way, a small fraction of solid $f = \pi/4d^2/(l + d)^2 \sim 0.09$ [90] was obtained. This value provide an elevated hydrophobicity, with contact angles approaching 170°, and it is still sufficiently large to prevent the collapse of the drop at the early stage of evaporation. p-doped, (100) silicon wafers with resistivity of 5–10 Ohm cm were used as substrates; they were cleaned with acetone and isopropanol to remove possible contaminants and then etched with a 4 % hydrofluoric acid (HF) solution. The wafers were then rinsed with deionized water and dried with N_2. The masks for optical lithography were fabricated using direct laser writing (DWL66fs, HEIDELBERG Instruments). Standard optical lithography techniques (Mask Aligner, MA6/BA6, SUSS MICROTEC) were employed to generate regular patterns of disks within a layer of negative tone resist (AZ5214) that was spin-coated onto clean silicon wafers. The disks served as a mask in a Deep Reactive Ion Etching process (DRIE) (ICP-RIE, SI 500, SENTECH Instruments GmbH), whereby the final structures were obtained in the form of cylindrical pillars with an aspect ratio greater than 2. The DRIE process utilized was a pulsed, time-multiplexed etching that repeatedly alternated between three stages, namely (i) a deposition of a chemically inert passivation layer of C_4F_8; (ii) an isotropic plasma etch of SF_6; and (iii) a sample/chamber cleaning stage. Based on this alternated process, the pillars were fabricated with a nanothreads at the sidewalls, which were characterized by indentations with a pitch depending on the duration of the isotropic plasma etch and of the passivation phase. Being t_1, t_2, and t_3 the passivation, etching, and cleaning times respectively, the ratio $r = t_1/t_2$ was adjusted with high precision during the process optimization to ensure vertical side-walls. Any deviation from optimal r would unbalance the process: for a higher value of r, passivation dominates over etching and the resulting pillars would be tapered and resembling a frustum (truncated cone); for a smaller value, etching dominates over passivation and the pillars would be shaped into an upside-down frustum. The total time $t = t_1 + t_2 + t_3$ represents the total duration of a cycle. Provided that t_1/t_2 is held constant, the longer the time t, the wider is the pitch of the nanothread at the sidewalls. For the configuration used in this work, values of $t_1 = 8$ s, $t_2 = 7$ s, and $t_3 = 4$ s were used, thus giving $t = 19$ s and resulting in a 500 nm thread pitch. Upon removal of the residual resist with piranha solution (H_2SO_4:$H_2O_2 = 3:1$ v/v), a 5 nm layer of titanium and a 100 nm layer of gold were sequentially sputtered upon the substrates. The gold layer was finally covered by a thin (5 nm) film of a Teflon-like (C_4F_8) polymer to assure superhydrophobicity [35].

For the samples presenting holes, the fabrication process was different. In this case, double polished, (100), 50 μm, p-type thin silicon wafers were purchased from Si-Mat (Silicon Materials, Kaufering, Germany). They were cleaned with acetone and isopropanol to remove possible contaminant and then etched with a 4 % wet HF solution. The wafers were then rinsed with deionized water and dried with N_2. A 100 nm layer of chrome was deposited upon the back side of the substrates using a sputter coater (Q150TES Quorum Technologies, Dixon, CA). Standard optical lithography techniques (Suss Microtec MA6/BA6, Sunnyvale, CA) were employed to realize a regular pattern of disks within a layer of positive resist (SPR220), that was spin-coated onto the chrome layer. Therefore, the chrome was removed from the disks by exposing them for 50 s to a standard chrome etching solution (ETCH 18 from OSC-Organo Spezial Chemie, Bitterfeld, Germany). Upon removal of the residual resist with acetone and oxygen plasma, a Deep Reactive Ion Etching (DRIE) process (SI 500 Sentech Instruments GmbH, Berlin, Germany) was used to etch holes passing through the substrate, the patterned chrome layer serving as a mask. The samples were then immersed in a bath of chrome etching solution to dissolve the remaining chrome. The fabrication process for the pillars was pursued on the other side of the samples. A second lithography exposure step was used to realize a regular hexagonal pattern of disks within a layer of negative resist (AZ5214 from Microchemicals GmbH, Ulm, Germany). The disks were positioned exactly on the middle point of each subset of six holes. A DRIE process was therefore used, similar to the one described above for samples without holes, whereby the final pillars were obtained with a height h of about 12 μm, a diameter d of 10 μm and a distance l of 20 μm [91].

Patterning a surface at the micro and nanoscale can result in local features whose dimensions are compatible with those usually found in many biological systems, like lamellipodia, pseudopodia and cytoskeletal elements, the structures that sustain cellular membranes and are responsible for cellular division, motion and development. In our particular case, this property can be combined with the specific hydrophobic properties of these substrates, leading to new promising application in biology and cell scaffolding.

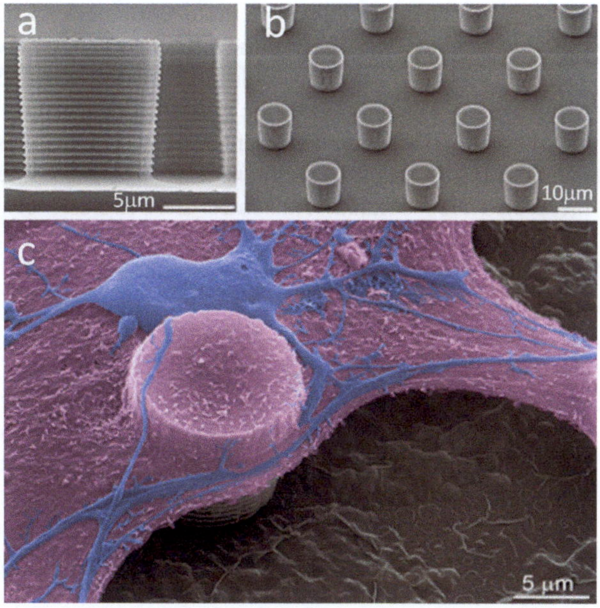

Fig. 8.6 SH devices for cell culture application: (**a**) pillar sidewalls detail and (**b**) low magnification of the surface. (**c**) False-color SEM micrograph showing a flat glial cell monolayer (*violet*) suspended between adjacent nanostructured pillars. A large neuron (*blue*) spreads upon the glial carpet with its multiple neuritic processes [92] (©2012 Wiley-VCH Verlag GmbH)

As an example, using this particular SH surfaces and implementing a new cell seeding procedure, we realized a 3D neuronal cell culture system, see Fig. 8.6. It resulted in an efficient tool for growing cells in an environment closely mimicking the in vivo one. This microfabricated substrate for cell culture resulted in an open scaffold which supports neuronal growth and differentiation and maximizes the physiological processes of cell–cell adhesion and synapse formation [35].

In conclusion we have demonstrated that ordered patterned surfaces, mainly obtained by means of micro and nano lithographic techniques, deposited with a layer of a Teflon-like polymer lead to highly reproducible SH surfaces. These devices can efficiently collect single molecules, sub cellular components and whole cells promoting also their self-organization resulting promising tools for innovative biotechnological applications.

8.3.2 *Exosomes Concentrations on Nanopatterned Super-hydrophobic Surfaces*

The combination of the superhydrophobic concept and plasmonic nanostructures can be an ideal platform to concentrate membranous extracellular organelles including exosomes, shedding microvesicles and apoptotic blebs. Rose Johnstone in the 1980s [93] described for the first time nanoscale membrane vesicles postulating that they were involved in cell-cell communication. After more than 30 years, their exact biological role and mechanism has yet to be fully elucidated. These nanovesiscles, released by cells upon fusion of multi-vesicular endosomes with cell surface, were called exosomes. They have dimension comprised between 30 and 100 nm and are generated through different biological mechanisms activated by different type of pathways involved in oncogenic transformation, micro environmental stimulation, cellular activation, immune-modulatory response, stress or death [94]. Because of their capacity to fuse into recipient cells and transfer them the repertoire of bioactive molecules they contain, exosomes are progressively recognized as mediator of intercellular communication [95, 96].

Recently, it has also been shown that tumor cells emit large quantities of exosomes containing pro-coagulant, growth regulatory and oncogenic cargo, which can be transferred throughout the cancer cell population to non-transformed stromal cells and also to the inflammatory infiltrates [95, 96]. These events seem to play a role in tumor invasion, angiogenesis, metastasis, drug resistance and cancer stem cell hierarchy. Many efforts are needed to identify and monitor the presence and the characteristics of such microvesicles. For all these reasons, it is of paramount importance the possibility to detect and characterize tumor exosomes at the beginning of the tumor initiation and, to do that, it's crucial to have the possibility to extract and concentrate vesicles also working with very low plasma concentrations.

Recently, two different kinds of NSH surfaces were used to investigate tumor and healthy colon exosomes through two different techniques: Raman spectroscopy and X-ray scattering [97, 98].

In the first one, conventional Silicon (Si) micropillars were covered with a randomly distributed silver nanograins pattern (Fig. 8.7). Solutions of about 0.2 ng/ml of exosomes were extracted from healthy and tumor colon cell lines. Several small drops (about 3 μl) were extracted from these solutions and placed upon NSH devices. During evaporation of the solvent, the solute was suitably conveyed into the active regions of the substrate thanks to the reduced friction coefficient, where the randomly distributed silver nanograins assemblies would yield SERS effects. Both the samples were thus analyzed and a minimum of 50 different Raman spectra were acquired for each sample. The Raman spectra were obtained upon excitation with the 514 nm line of an Ar^+ laser in backscattering geometry. Each spectrum was baseline corrected with a third order polynomial function after acquisition in order to take into account fluorescence effects.

Standard-normal variate normalization has been performed on the spectra, as earlier reported in literature [99]. From Raman spectra study it was possible to distinguish between healthy and tumor exosomes on the basis of a different expression of some subcellular component. More in details, exosomes derived from healthy colon cells showed a high presence of lipid signals whereas exosomes from tumor cells exhibited a high presence of RNA.

In the second approach [98], a Poly-methyl methacrylate (PMMA) surface (adapted for X-ray studies because it is optically transparent and shows a high X-ray transmission as it contains only light elements), with SH properties, was used to concentrate exosomes and, later on, to perform Small-angle X-ray scattering (SAXS) and Wade-angle X-ray Scattering (WAXS) measurements. Fabrication steps for PMMA-NHS devices are reported elsewhere [100].

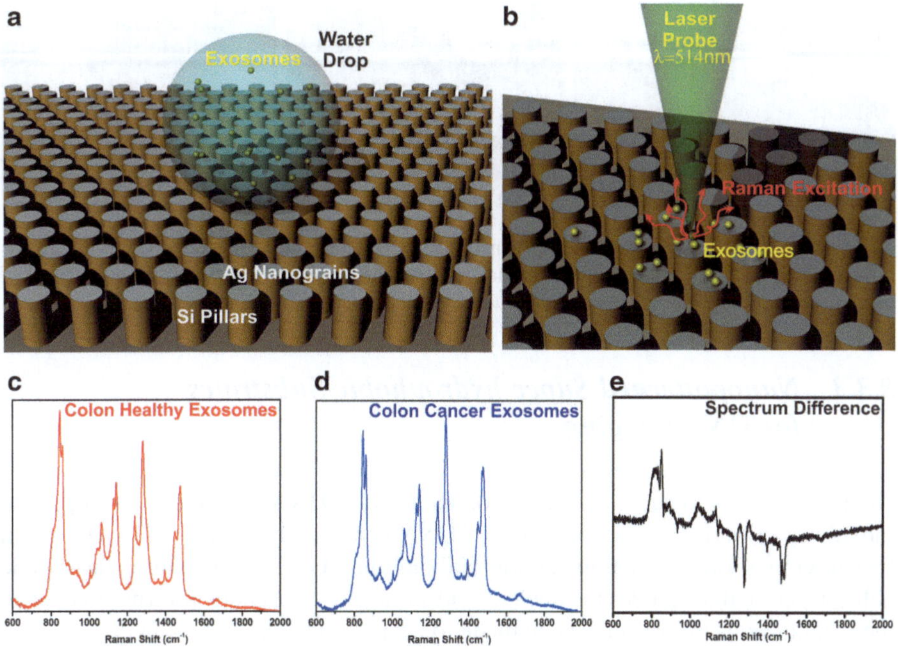

Fig. 8.7 Exosomes on NSH surfaces: (**a**) 3D rendering of a NHS surfaces with on top silver nanograins and above all a drop of water containing exosomes; (**b**) Schematics of Raman measurements on exosomes deposited on few pillars after water evaporation; (**c**) Averaged Raman spectra of exosomes derived from healthy colon cells; (**d**) Analogous spectra derived from cancerous colon cells. (**e**) Difference between the spectra in (**c**) and (**d**), see in line text considerations

Drops of about 5 μl containing exosomes were deposited on a nanostructured SHS PMMA surfaces, described by the dimensions and separation of the micro-cylinders fabricated (diameter d of 10 μm, distance l of 10 μm, and height h of 15 μm). After drying in air, the solid residue was attached by a small amount of fast glue to a glass capillary tip for X-ray scattering experiments. Both samples (healthy and tumor residues) appear to have the same microstructure, as supported by almost identical WAXS pattern obtained by synchrotron radiation X-ray technique [98]. Indeed, SAXS analysis, that reveals the lamellar morphology, showed differences in the number of orders, their periodicities and peak broadening. The reason could be the higher degree of disorder in the healthy exosome residue compared to the more regular organization for the tumor one. The observed variations could depend on the different compositions of the exosome membrane due to the variability of its inner lipid content, which influences the organization/distance of hydrophilic and hydrophobic regions of the membrane.

Both these works demonstrate how NHS surfaces can be useful for the investigation of a lot of complex biologically unsolved problems and represents a perfect example of fruitful interaction of nanotechnology and biology.

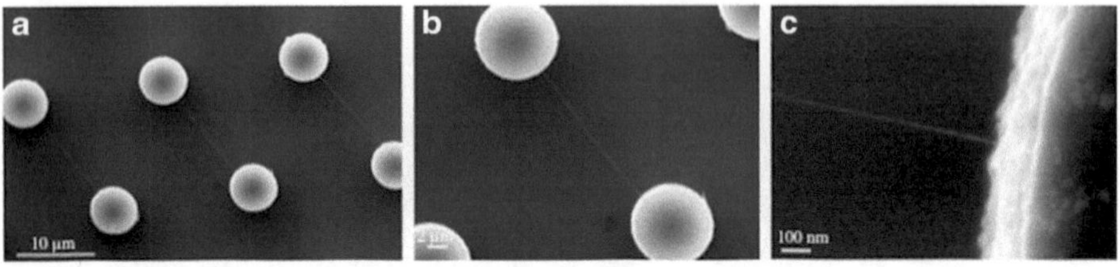

Fig. 8.8 SEM images of non-coated stretched nucleic acids well tense between micro-pillars. (**a**) top view, and (**b**) and (**c**) are zoomed-in images. In all the images the micro pillars diameter is of 6 μm and the pillar-pillar distance is of 12 μm

8.3.3 Nanopatterned Super-hydrophobic Substrates for DNA Imaging

Nucleic acids (NA) such as deoxyribonucleic acid (DNA), are the most important molecules in biology. The relevance of DNA became clearer in 1953 [101] when by studying X-ray diffraction patterns and building models, was figured the double helix structure that enables the nucleic acid storage of the genetic information. Since then, continuous efforts have been made to characterize DNA more in depth and to allow its detection with increasing sensibility, down to the single nucleotide. The most commonly used and reliable technique to understand the DNA response to environmental agents (temperature, pH) and the information about the structure of the double stranded molecule (in example the A- or Z-DNA conformation) is X-rays diffraction [102]. This technique needs a high molecule concentration and a complex and time expensive preparation steps to obtain crystalline samples, but the efficiency of the process is low and the quality of the obtained crystals insufficient. Other procedures were involved in the molecules DNA analysis and stretching: in recent years optical tweezers [103], forcing DNA into nanochannel [104] micro- [105] and nanofluidic [106] chips were widely used. Despite the fact that the evolving of DNA handling techniques allowed to obtain DNA wires [107] of various length, the development of alternative tools for the direct imaging of the nucleic acids molecules is an essential task. In this context the coupling of TEM and SH devices revealed to be an efficient tool to overcome the obstacles of traditional imaging techniques.

Using saline buffer solution containing DNA, the evaporation and concentration of the droplet results then in the control of the DNA molecules deposition and their consequent linkage mainly to the edges on the top of μ-pillars. Taking advantage of this approach, down to a femtomolar concentration of DNA solution was evaluated and imaged [34]. An example of this is shown in Fig. 8.8.

The same methodology was efficiently used with NSH surfaces with holes [91]. The pillars used in the work reported in this paragraph had a height h of about 10 μm, a diameter d of 6 μm and a distance l of 12 μm. With the devices patterned in this way, the suspended DNA corresponding to the hole can be orthogonally

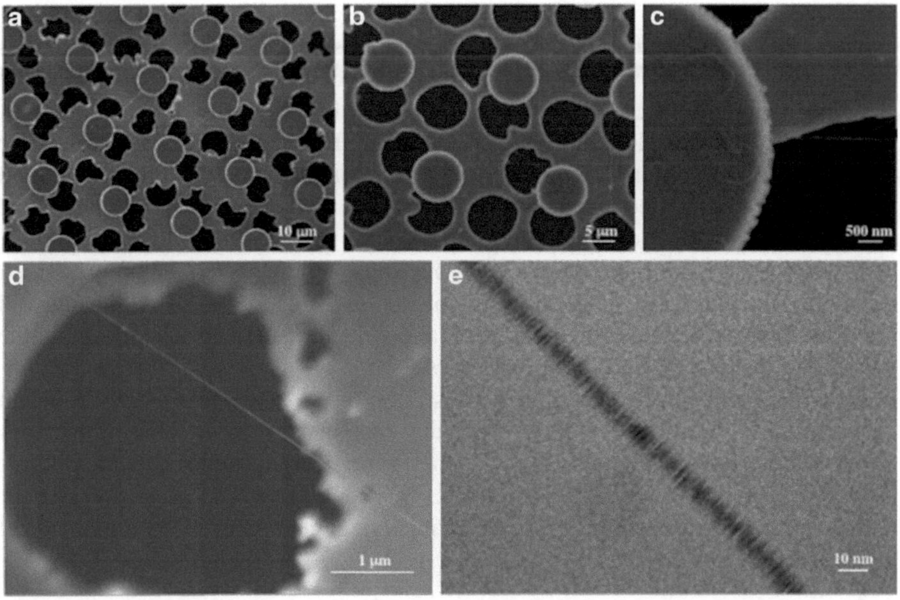

Fig. 8.9 Low magnification overview (**a**) and zoomed-in (**b**, **c**) SEM images of DNA bundles well tense over the holes on the μ-pillars lattice; is clearly visible that a useful DNA fibers percentage resulted suspended exactly in correspondence of the holes. In (**d**, **e**) high resolution TEM image of a nucleic acids bundle orthogonal to the electron beam. In all the images the micro pillars diameter is of 6 μm and the pillar-pillar distance is of 12 μm

exposed to the electron beam which passes freely through the apertures during TEM imaging. The bundles obtained after evaporation at our saline buffer conditions, are clearly visible in Fig. 8.9a–c. They are suspended and well stretched between adjacent pillars. In panel d is clearly visible one of the smallest fibers (5 nm) recently imaged by a 100 keV electron beam. The periodicity of the bundle is clearly observable just increasing the magnification (panel e) and the period is estimated to be 2.4 nm, confirming the previously reported data [91] inherent the A-conformation of DNA. With a very simple and fast preparation step that does not require additional functionalization or staining, the use of μ-pillars coupled with TEM imaging allowed the DNA-bundles direct imaging without background interference, allowing simpler and faster interpretation of the images acquired than previously reported methods.

In the previously reported experiments made by our group, the nucleic acid used was Lambda DNA, chosen as a reference due to the unambiguous imaging results and to its well-known base sequence. Recently we have worked on eukaryotic single stranded (ss) DNAs and the characterization of their interactions with proteins of biological importance such as repairing proteins. For example Rad51-repair protein, which binds to single and double stranded DNA, exhibits DNA-dependent ATPase activity and participates in DNA damage response pathway and double strand break repair [91], underwinds duplex DNA and originates helical nucleoprotein filaments. The understanding of the complex relationship occurring between peptides structurally associated with nucleic acids (the so called "nucleoproteins") can be achieved coupling different techniques [108]. On one hand the direct imaging of nucleoproteins suspended between pillars using TEM can be of great interest for the evaluation of conformational changes after proteins linkage, on the other hand Raman spectroscopy technique can be consider a fundamental tool to understand the spectral changes occurring between suspended and well tense fibers of ssDNA and DNA/protein. In this context, the coupling of these techniques represents a very efficient tool for genetic analysis, for the investigation of DNA damage and its related protein-mediated repair, also allowing the study of specific genomic alterations that can produce severe diseases.

Resuming, the properties of SH substrates on which DNAs were stretched determined a significative advance for the characterization of DNA molecules ultrastructure and its interactions with other biological molecules of wide interest such as transcription factor and proteins. This highlighted the great potential of this tool, ranging from single molecule characterization to more clinical biomedical applications.

References

1. Long DA (2001) The Raman effect: a unified treatment of the theory of Raman scattering by molecules, 1st edn. Wiley, Chichester
2. Kneipp K, Kneipp H, Itzkan I, Dasari RR, Feld MS (2002) Surface-enhanced Raman scattering and biophysics. J Phys Condens Matter 14:R597–R624
3. Ko H, Singamaneni S, Tsukruk VV (2008) Nanostructured surfaces and assemblies as SERS media. Small 4:1576–1599
4. Gopalakrishnan A, Malerba M, Tuccio S, Panaro S, Miele E, Chirumamilla M, Santoriello S, Dorigoni C, Giugni A, Proietti Zaccaria R, Liberale C, De Angelis F, Razzari L, Krahne R, Toma A, Das G, Di Fabrizio E (2012) Nanoplasmonic structures for biophotonic applications: SERS overview. Annalen der Physik 524:620–636
5. Campion A, Ivanecky JE, Child CM, Foster M (1995) On the mechanism of chemical enhancement in surface-enhanced Raman scattering. J Am Chem Soc 117:11807–11808
6. Doering WE, Nie S (2002) Single-molecule and single-nanoparticle SERS: examining the roles of surface active sites and chemical enhancement. J Phys Chem B 106:311–317
7. Maitani MM, Ohlberg DAA, Li Z, Allara DL, Stewart DR, Williams RS (2009) Study of SERS chemical enhancement factors using buffer layer assisted growth of metal nanoparticles on self-assembled monolayers. J Am Chem Soc 131:6310–6311
8. Kneipp K, Wang Y, Kneipp H, Perelman LT, Itzkan I, Dasari RR, Feld MS (1997) Single molecule detection using surface-enhanced Raman scattering (sers). Phys Rev Lett 78:1667–1670
9. Blum C, Schmid T, Opilik L, Weidmann S, Fagerer SR, Zenobi R (2012) Understanding tip-enhanced Raman spectra of biological molecules: a combined Raman, SERS and TERS study. J Raman Spectrosc 43:1895–1904
10. Stadler J, Schmid T, Zenobi R (2012) Developments in and practical guidelines for tip-enhanced Raman spectroscopy. Nanoscale 4:1856–1870

11. Pettinger B, Schambach P, Villagómez CJ, Scott N (2012) Tip-enhanced Raman spectroscopy: near-fields acting on a few molecules. Annu Rev Phys Chem 63:379–399
12. Binnig G, Quate CF, Gerber C (1986) Atomic force microscope. Phys Rev Lett 56:930–933
13. Israelachvili JN (2011) Intermolecular and surface forces, revised 3rd edn. Academic, Burlington
14. Kalinin SV, Gruverman A (eds) (2010) Scanning probe microscopy of functional materials: nanoscale imaging and spectroscopy. Springer, New York
15. Dietz C, Herruzo ET, Lozano JR, Garcia R (2011) Nanomechanical coupling enables detection and imaging of 5 nm superparamagnetic particles in liquid. Nanotechnology 22:125708
16. Torre B, Bertoni G, Fragouli D, Falqui A, Salerno M, Diaspro A, Cingolani R, Athanassiou A (2011) Magnetic force microscopy and energy loss imaging of superparamagnetic iron oxide nanoparticles. Sci Rep 1:202
17. Neves CS, Quaresma P, Baptista PV, Carvalho PA, Araújo JaP, Pereira E, Eaton P (2010) New insights into the use of magnetic force microscopy to discriminate between magnetic and nonmagnetic nanoparticles. Nanotechnology 21:305706
18. Schreiber S, Savla M, Pelekhov DV, Iscru DF, Selcu C, Hammel PC, Agarwal G (2008) Magnetic force microscopy of superparamagnetic nanoparticles. Small 4:270–278
19. Zhang L, Sakai T, Sakuma N, Ono T, Nakayama K (1999) Nanostructural conductivity and surface-potential study of low-field-emission carbon films with conductive scanning probe microscopy. Appl Phys Lett 75(22):3527
20. Lorenzoni M, Giugni A, Torre B (2013) Oxidative and carbonaceous patterning of Si surface in an organic media by scanning probe lithography. Nanoscale Res Lett 8(1):1–9
21. Tello M, Garcia F, Garcia R (2006) Fabrication of nanometer-scale structures by local oxidation nanolithography. In Bhushan B, Fuchs H (eds) Applied scanning probe methods IV: industrial applications, 2006 edn. Springer, Berlin/New York, pp 137–158
22. Hinterdorfer P, Baumgartner W, Gruber HJ, Schilcher K, Schindler H (1996) Detection and localization of individual antibody-antigen recognition events by atomic force microscopy. Proc Natl Acad Sci 93:3477–3481
23. Wildling L, Unterauer B, Zhu R, Rupprecht A, Haselgrübler T, Rankl C, Ebner A, Vater D, Pollheimer P, Pohl EE, Hinterdorfer P, Gruber HJ (2011) Linking of sensor molecules with amino groups to amino-functionalized AFM tips. Bioconjug Chem 22:1239–1248
24. Carvalho FA, Carneiro FA, Martins IC, Assunção Miranda I, Faustino AF, Pereira RM, Bozza PT, Castanho MARB, Mohana-Borges R, Da Poian AT, Santos NC (2012) Dengue virus capsid protein binding to hepatic lipid droplets (LD) is potassium ion dependent and is mediated by LD surface proteins. J Virol 86:2096–2108
25. Canale C, Petrelli A, Salerno M, Diaspro A, Dante S (2013) A new quantitative experimental approach to investigate single cell adhesion on multifunctional substrates. Biosens Bioelectron 48:172–179
26. Munday JN, Capasso F, Parsegian VA (2009) Measured long-range repulsive Casimir-Lifshitz forces. Nature 457:170–173
27. Kisiel M, Gnecco E, Gysin U, Marot L, Rast S, Meyer E (2011) Suppression of electronic friction on Nb films in the superconducting state. Nat Mater 10:119–122
28. De Angelis F, Proietti Zaccaria R, Francardi M, Liberale C, Di Fabrizio E (2011) Multi-scheme approach for efficient surface plasmon polariton generation in metallic conical tips on AFM-based cantilevers. Opt Express 19:22268–22279
29. De Angelis F, Das G, Candeloro P, Patrini M, Galli M, Bek A, Lazzarino M, Maksymov I, Liberale C, Andreani LC, Di Fabrizio E (2010) Nanoscale chemical mapping using three-dimensional adiabatic compression of surface plasmon polaritons. Nat Nanotechnol 5:67–72
30. Bao W, Melli M, Caselli N, Riboli F, Wiersma DS, Staffaroni M, Choo H, Ogletree DF, Aloni S, Bokor J, Cabrini S, Intonti F, Salmeron MB, Yablonovitch E, Schuck PJ, Weber-Bargioni A (2012) Mapping local charge recombination heterogeneity by multidimensional nanospectroscopic imaging. Science 338(6112):1317–1321
31. Xu D, Watt GD, Harb JN, Davis RC (2005) Electrical conductivity of ferritin proteins by conductive AFM. Nano Lett 5:571–577

32. Salomon A, Cahen D, Lindsay S, Tomfohr J, Engelkes V, Frisbie C (2003) Comparison of electronic transport measurements on organic molecules. Adv Mater 15:1881–1890
33. Giugni A, Torre B, Toma A, Francardi M, Malerba M, Alabastri A, Proietti Zaccaria R, Stockman MI, Di Fabrizio E (2013) Hot-electron nanoscopy using adiabatic compression of surface plasmons. Nat Nanotechnol 8:845–852
34. De Angelis F, Gentile F, Mecarini F, Das G, Moretti M, Candeloro P, Coluccio ML, Cojoc G, Accardo A, Liberale C, Proietti Zaccaria R, Perozziello G, Tirinato L, Toma A, Cuda G, Cingolani R, Di Fabrizio E (2011) Breaking the diffusion limit with super-hydrophobic delivery of molecules to plasmonic nanofocusing SERS structures. Nat Photonics 5:682–687
35. Limongi T, Cesca F, Gentile F, Marotta R, Ruffilli R, Barberis A, Dal Maschio M, Petrini EM, Santoriello S, Benfenati F, Di Fabrizio E (2013) Nanostructured superhydrophobic substrates trigger the development of 3D neuronal networks. Small 9:402–412
36. Kolomenski A, Kolomenskii A, Noel J, Peng S, Schuessler H (2009) Propagation length of surface plasmons in a metal film with roughness. Appl Opt 48:5683–5691
37. Nagaraj, Krokhin AA (2010) Long-range surface plasmons in dielectric-metal-dielectric structure with highly anisotropic substrates. Phys Rev B 81:085426
38. Sönnichsen C, Franzl T, Wilk T, von Plessen G, Feldmann J (2002) Drastic reduction of plasmon damping in gold nanorods. Phys Rev Lett 88:077402
39. Kats MA, Yu N, Genevet P, Gaburro Z, Capasso F (2011) Effect of radiation damping on the spectral response of plasmonic components. Opt Express 19:21748–21753
40. Stockman MI (2004) Nanofocusing of optical energy in tapered plasmonic waveguides. Phys Rev Lett 93:137404
41. Stockman MI (2011) Nanoplasmonics: past, present, and glimpse into future. Opt Express 19:22029
42. Issa NA, Guckenberger R (2007) Fluorescence near metal tips: the roles of energy transfer and surface plasmon polaritons. Opt Express 15(19):12131
43. Proietti Zaccaria R, De Angelis F, Toma A, Razzari L, Alabastri A, Das G, Liberale C, Di Fabrizio E (2012) Surface plasmon polariton compression through radially and linearly polarized source. Opt Lett 37:545
44. Proietti Zaccaria R, Alabastri A, De Angelis F, Das G, Liberale C, Toma A, Giugni A, Razzari L, Malerba M, Sun HB, Di Fabrizio E (2012) Fully analytical description of adiabatic compression in dissipative polaritonic structures. Phys Rev B 86:035410
45. Gramotnev DK, Vogel MW, Stockman MI (2008) Optimized nonadiabatic nanofocusing of plasmons by tapered metal rods. J Appl Phys 104(3):034311
46. Chen X-W, Sandoghdar V, Agio M (2010) Nanofocusing radially-polarized beams for high-throughput funneling of optical energy to the near field. Opt Express 18:10878–10887
47. Knight MW, Sobhani H, Nordlander P, Halas NJ (2011) Photodetection with active optical antennas. Science 332(6030):702–704
48. Raether H (1988) Surface plasmons on smooth surfaces, vol. 111 of Springer tracts in modern physics. Springer, Berlin/Heidelberg
49. Fowler RH (1931) The analysis of photoelectric sensitivity curves for clean metals at various temperatures. Phys Rev 38:45–56
50. Kretschmann E (1971) Die bestimmung optischer konstanten von metallen durch anregung von oberflächenplasmaschwingungen. Zeitschrift für Physik 241(4):313–324
51. Genchev Z, Nedelchev N, Mateev E, Stoyanov H (2008) Analytical approach to the prism coupling problem in the kretschmann configuration. Plasmonics 3(1):21–26
52. Otto A (1968) Excitation of nonradiative surface plasma waves in silver by the method of frustrated total reflection. Zeitschrift für Physik 216(4):398–410
53. Brueck SRJ, Diadiuk V, Jones T, Lenth W (1985) Enhanced quantum efficiency internal photoemission detectors by grating coupling to surface plasma waves. Appl Phys Lett 46(10):915
54. Baron A, Devaux E, Rodier J-C, Hugonin J-P, Rousseau E, Genet C, Ebbesen TW, Lalanne P (2011) Compact antenna for efficient and unidirectional launching and decoupling of surface plasmons. Nano Lett 11(10):4207–4212

55. Ropers C, Neacsu CC, Elsaesser T, Albrecht M, Raschke MB, Lienau C (2007) Grating-coupling of surface plasmons onto metallic tips: a nanoconfined light source. Nano Lett 7(9):2784–2788. PMID: 17685661
56. Neacsu CC, Berweger S, Olmon RL, Saraf LV, Ropers C, Raschke MB (2010) Near-field localization in plasmonic superfocusing: a nanoemitter on a tip. Nano Lett 10(2):592–596. PMID: 20067296.
57. Berweger S, Atkin JM, Xu XG, Olmon RL, Raschke MB (2011) Femtosecond nanofocusing with full optical waveform control. Nano Lett 11(10):4309–4313
58. Johnson PB, Christy RW (1972) Optical constants of the noble metals. Phys Rev B 6:4370–4379
59. Rakic AD, Djurišic AB, Elazar JM, Majewski ML (1998) Optical properties of metallic films for vertical-cavity optoelectronic devices. Appl Opt 37:5271
60. Lalanne P, Hugonin J, Rodier J (2005) Theory of surface plasmon generation at nanoslit apertures. Phys Rev Lett 95:263902
61. Goykhman I, Desiatov B, Khurgin J, Shappir J, Levy U (2012) Waveguide based compact silicon Schottky photodetector with enhanced responsivity in the telecom spectral band. Opt Express 20:28594–28602
62. Green MA, Pillai S (2012) Harnessing plasmonics for solar cells. Nat Photon 6:130–132
63. Mubeen S, Lee J, Singh N, Kramer S, Stucky GD, Moskovits M (2013) An autonomous photosynthetic device in which all charge carriers derive from surface plasmons. Nat Nanotechnol 8:247–251
64. Mukherjee S, Libisch F, Large N, Neumann O, Brown LV, Cheng J, Lassiter JB, Carter EA, Nordlander P, Halas NJ (2013) Hot electrons do the impossible: plasmon-induced dissociation of H_2 on Au. Nano Lett 13(1):240–247
65. Spicer W (1958) Photoemissive, photoconductive, and optical absorption studies of alkali-antimony compounds. Phys Rev 112:114–122
66. Böer K (2010) The schottky barrier. In: Introduction to space charge effects in semiconductors. Springer series in solid-state sciences, vol 160. Springer, Berlin/Heidelberg, pp 41–91
67. Lampert M, Many A, Mark P (1964) Space-charge-limited currents injected from a point contact. Phys Rev 135:A1444–A1453
68. Smit GDJ, Rogge S, Klapwijk TM (2002) Scaling of nano-Schottky-diodes. Appl Phys Lett 81:3852
69. Donolato C (1995) Electrostatic problem of a point charge in the presence of a semi-infinite semiconductor. J Appl Phys 78(2):684
70. Donolato C (2004) Approximate analytical solution to the space charge problem in nanosized Schottky diodes. J Appl Phys 95(4):2184
71. Hudait M, Krupanidhi S (2001) Doping dependence of the barrier height and ideality factor of Au/n-GaAs Schottky diodes at low temperatures. Phys B Condens Matter 307:125–137
72. Hardikar S, Hudait M, Modak P, Krupanidhi S, Padha N (1999) Anomalous current transport in Au/low-doped n-GaAs Schottky barrier diodes at low temperatures. Appl Phys A Mater Sci Process 68:49–55
73. Casey HC, Sell DD, Wecht KW (1975) Concentration dependence of the absorption coefficient for n- and p-type GaAs between 1.3 and 1.6 eV. J Appl Phys 46(1):250.
74. Hugelmann M, Schindler W (2004) Schottky diode characteristics of electrodeposited Au/n-Si(111) nanocontacts. Appl Phys Lett 85(16):3608
75. Vasko SE, Jiang W, Lai H, Sadilek M, Dunham S, Rolandi M (2013) High-field chemistry of organometallic precursors for direct-write of germanium and silicon nanostructures. J Mater Chem C 1:282
76. Lorenzoni M, Torre B (2013) Scanning probe oxidation of SiC, fabrication possibilities and kinetics considerations. Appl Phys Lett 103(16):163109
77. Sugimura H, Nakagiri N (1995) Chemical approach to nanofabrication: modifications of silicon surfaces patterned by scanning probe anodization. Jpn J Appl Phys 34: 3406–3411

78. Morimoto K, Araki K, Yamashita K, Morita K, Niwa M (1997) Si nanofabrication using AFM field enhanced oxidation and anisotropic wet chemical etching. Appl Surf Sci 117–118:652–659
79. Tello M, García R (2001) Nano-oxidation of silicon surfaces: comparison of noncontact and contact atomic-force microscopy methods. Appl Phys Lett 79(3):424
80. García R, Calleja M, Rohrer H (1999) Patterning of silicon surfaces with noncontact atomic force microscopy: Field-induced formation of nanometer-size water bridges. J Appl Phys 86(4):1898
81. Canale C, Torre B, Ricci D, Braga P (2011) Recognizing and avoiding artifacts in atomic force microscopy imaging. In: Braga PC, Ricci D (eds) Atomic force microscopy in biomedical research. Methods in molecular biology vol 736. Humana Press, New York, pp 31–43
82. Chen X-W, Mohammadi A, Ghasemi AHB, Agio M (2013) Ultrafast coherent nanoscopy. Mol Phys 111:3003–3012
83. Mohammadi A, Agio M (2012) Light scattering under nanofocusing: towards coherent nanoscopies. Opt Commun 285:3383–3389
84. Evans CL, Xie XS (2008) Coherent anti-stokes Raman scattering microscopy: chemical imaging for biology and medicine. Annu Rev Anal Chem 1:883–909
85. Min W, Freudiger CW, Lu S, Xie XS (2011) Coherent nonlinear optical imaging: beyond fluorescence microscopy. Annu Rev Phys Chem 62(1):507–530. PMID: 21453061
86. Fu D, Holtom G, Freudiger C, Zhang X, Xie XS (2013) Hyperspectral imaging with stimulated raman scattering by chirped femtosecond lasers. J Phys Chem B 117(16):4634–4640
87. Freudiger CW, Min W, Holtom GR, Xu B, Dantus M, XieX S (2011) Highly specific label-free molecular imaging with spectrally tailored excitation-stimulated Raman scattering (STE-SRS) microscopy. Nat Photonics 5:103–109
88. Ideguchi T, Holzner S, Bernhardt B, Guelachvili G, Picque N, Hansch TW (2013) Coherent Raman spectro-imaging with laser frequency combs. Nature 502:355–358
89. Tian P, Keusters D, Suzaki Y, Warren WS (2003) Femtosecond phase-coherent two-dimensional spectroscopy. Science 300:1553–1555
90. Cassie ABD, Baxter S (1944) Wettability of porous surfaces. Trans Faraday Soc 40:546
91. Gentile F, Moretti M, Limongi T, Falqui A, Bertoni G, Scarpellini A, Santoriello S, Maragliano L, Proietti Zaccaria R, Di Fabrizio E (2012) Direct imaging of DNA fibers: the visage of double helix. Nano Lett 12:6453–6458
92. Limongi T, Cesca F, Gentile F, Marotta R, Ruffilli R, Barberis A, Dal Maschio M, Petrini EM, Santoriello S, Benfenati F, Di Fabrizio E (2013) 3D cell cultures: nanostructured superhydrophobic substrates trigger the development of 3D neuronal networks (small 3/2013). Small 9:334–334
93. Johnstone RM, Adam M, Hammond JR, Orr L, Turbide C (1987) Vesicle formation during reticulocyte maturation. Association of plasma membrane activities with released vesicles (exosomes). J Biol Chem 262:9412–9420
94. Keller S, Sanderson MP, Stoeck A, Altevogt P (2006) Exosomes: from biogenesis and secretion to biological function. Immunol Lett 107(2):102–108
95. Muralidharan-Chari V, Clancy JW, Sedgwick A, D'Souza-Schorey C (2010) Microvesicles: mediators of extracellular communication during cancer progression. J Cell Sci 123: 1603–1611
96. Skog J, Wurdinger T, van Rijn S, Meijer DH, Gainche L, Curry WT, Carter BS, Krichevsky AM, Breakefield XO (2008) Glioblastoma microvesicles transport RNA and proteins that promote tumour growth and provide diagnostic biomarkers. Nat Cell Biol 10:1470–1476
97. Tirinato L, Gentile F, Mascolo DD, Coluccio M, Das G, Liberale C, Pullano S, Perozziello G, Francardi M, Accardo A, Angelis FD, Candeloro P, Di Fabrizio E, {SERS} analysis on exosomes using super-hydrophobic surfaces. Microelectron Eng 97(0):337–340 (2012) Micro- and Nano-Engineering (MNE) 2011, selected contributions: Part I

98. Accardo A, Tirinato L, Altamura D, Sibillano T, Giannini C, Riekel C, Di Fabrizio E (2013) Superhydrophobic surfaces allow probing of exosome self organization using X-ray scattering. Nanoscale 5:2295–2299
99. Notingher I, Green C, Dyer C, Perkins E, Hopkins N, Lindsay C, Hench LL (2004) Discrimination between ricin and sulphur mustard toxicity in vitro using Raman spectroscopy. J R Soc Interface 1:79–90
100. Accardo A, Gentile F, Mecarini F, De Angelis F, Burghammer M, Di Fabrizio E, Riekel C (2010) In situ x-ray scattering studies of protein solution droplets drying on micro- and nanopatterned superhydrophobic pmma surfaces. Langmuir 26(18):15057–15064
101. Watson JD, Crick FHC (1953) The structure of DNA. Cold Spring Harb Symp Quant Biol 18:123–131
102. Richmond TJ, Davey CA (2003) The structure of DNA in the nucleosome core. Nature 423:145–150
103. Smith SB, Cui Y, Bustamante C (1996) Overstretching B-DNA: the elastic response of individual double-stranded and single-stranded DNA molecules. Science 271:795–799
104. Riehn R, Lu M, Wang Y-M, Lim SF, Cox EC, Austin RH (2005) Restriction mapping in nanofluidic devices. Proc Natl Acad Sci U S A 102:10012–10016
105. Lim SW, Abate AR (2013) Ultrahigh-throughput sorting of microfluidic drops with flow cytometry. Lab Chip 13:4563–4572
106. Marie R, Pedersen JN, Bauer DLV, Rasmussen KH, Yusuf M, Volpi E, Flyvbjerg H, Kristensen A, Mir KU (2013) Integrated view of genome structure and sequence of a single DNA molecule in a nanofluidic device. Proc Natl Acad Sci U S A 110:4893–4898
107. Wohlgamuth CH, McWilliams MA, Slinker JD (2013) DNA as a molecular wire: distance and sequence dependence. Anal Chem 85:8634–8640
108. Poma A, Spanò L, Pittaluga E, Tucci A, Palladino L, Limongi T (2005) Interactions between saporin, a ribosome-inactivating protein, and DNA: a study by atomic force microscopy. J Microsc 217:69–74

9
Narrow-Linewidth Lasers on a Silicon Chip

Edward H. Bernhardi and Markus Pollnau

Abstract Diode-pumped distributed-feedback (DFB) channel waveguide lasers were demonstrated in Er^{3+}-doped and Yb^{3+}-doped Al_2O_3 on standard thermally oxidized silicon substrates. Uniform surface-relief Bragg gratings were patterned by laser-interference lithography and etched into the SiO_2 top cladding. The maximum grating reflectivity exceeded 99 %. Monolithic DFB cavities with Q-factors of up to 1.35×10^6 were realized. The Er^{3+}-doped DFB laser delivered 3 mW of output power with a slope efficiency of 41 % versus absorbed pump power. Single-longitudinal-mode operation at a wavelength of 1545.2 nm was achieved with an emission line width of 1.70 ± 0.58 kHz, corresponding to a laser Q-factor of 1.14×10^{11}. Yb3+-doped DFB lasers were demonstrated at wavelengths near 1,020 nm with output powers of 55 mW and a slope efficiency of 67 % versus launched pump power. An Yb^{3+}-doped dual-wavelength laser was achieved based on the optical resonances induced by two local phase shifts in the DFB structure. A stable microwave signal at ~15 GHz with a –3-dB width of 9 kHz and a long-term frequency stability of ±2.5 MHz was created via the heterodyne photo-detection of the two laser wavelengths. Interaction of the intra-cavity evanescent laser field with micro-particles in contact with the grating surface induces changes in the microwave beat signal, whose detection enabled real-time detection and accurate size measurement

E.H. Bernhardi (✉) • M. Pollnau
Integrated Optical Microsystems Group, MESA+ Institute for Nanotechnology, University of Twente, P.O. Box 217, 7500 AE Enschede, The Netherlands
e-mail: M.Pollnau@utwente.nl

B. Di Bartolo et al. (eds.), *Nano-Structures for Optics and Photonics*, NATO Science for Peace and Security Series B: Physics and Biophysics, DOI 10.1007/978-94-017-9133-5_9, © Springer Science+Business Media Dordrecht 2015

of single micro-particles with diameters ranging between 1 and 20 μm, which represents the typical size of many fungal and bacterial pathogens. A limit of detection of ~500 nm was deduced.

9.1 Introduction

Monolithic lasers find numerous applications in spectroscopy, optical communications, and optical sensing. These applications often place stringent requirements with respect to efficiency, output power, spectral characteristics, compactness, robustness, and thermal stability on the laser. Waveguide lasers are of particular importance for the realization of compact, rigid, and robust optical devices, since the entire laser cavity along with the optical feedback elements can be fabricated on the same substrate. Besides, when an application requires a laser array, e.g. for simultaneous operation over a specific wavelength range, cost issues strongly favor an on-chip integrated approach. In this work, surface-corrugated Bragg gratings are integrated with Al_2O_3 channel waveguides to realize a variety of monolithic distributed-feedback (DFB) and distributed-Bragg-reflector (DBR) cavities in rare-earth-ion-doped Al_2O_3 waveguides.

9.2 Waveguide Fabrication, Spectroscopy, and Gain

The Al_2O_3 ridge channel waveguides were fabricated in 1-μm-thick Al_2O_3 layers which were deposited onto standard thermally oxidized silicon wafers [1]. For the Er^{3+}-based laser device a doping concentration of $3 \times 10^{20}\,cm^{-3}$ was used, while the Yb^{3+}-based laser devices had a doping concentration of $5.8 \times 10^{20}\,cm^{-3}$. All the waveguides supported single-transverse-mode operation at the respective pump and laser wavelengths and were 10 mm long, 2.5–3.0 μm wide, and etched to a depth of ~0.1 μm via a chlorine-based reactive ion etching process [2]. For Er^{3+} concentrations in the range of $1–2 \times 10^{20}\,cm^{-3}$, internal net gain was obtained over a wavelength range of 80 nm (1,500–1,580 nm) and a peak gain of 2.0 dB/cm was measured at 1,533 nm [3]. The influence of energy migration and energy-transfer upconversion among neighboring Er^{3+} ions on luminescence decay and steady-state population densities in $Al_2O_3 : Er^{3+}$ thin films was investigated by means of photoluminescence decay measurements under quasi-CW excitation. A fast quenching process induced by, e.g., active ion pairs and clusters, undesired impurities, or host material defects such as voids, that was not revealed by any particular signature in the luminescence decay curves because of negligible emission by the quenched ions under quasi-CW excitation, was verified by pump-absorption experiments [4]. This quenching process strongly affects device performance of $Al_2O_3 : Er^{3+}$ as an amplifier, to a lesser extent also of $Al_2O_3 : Yb^{3+}$, but it affects a lasing device only by increasing its threshold [5]. The first laser in an Al_2O_3 layer on a silicon chip was demonstrated utilizing a micro-ring cavity [6].

9.3 High-Q Passive Cavities

For the fabrication of Bragg gratings [7], a plasma-enhanced chemical vapor deposition (PECVD) SiO_2 cladding layer was deposited on top of the ridge waveguides. The uniform surface-relief Bragg gratings were fabricated on the top surface of the PECVD SiO_2 cladding by means of laser interference lithography (LIL). A grating pattern was defined in a 120-nm-thick negative resist layer on top of the PECVD cladding and was etched into the SiO_2 layer using a CHF_3 : O_2 reactive ion plasma. The grating period varied between 316 nm for the Yb^{3+}-doped devices, with a Bragg wavelength of ~1,020 nm, to 507 nm for the passive cavities with a Bragg wavelength of ~1,590 nm. Transverse and axial cross-sectional views of the waveguide laser structure are displayed in Fig. 9.1.

Cavities with exceptionally high Q-factors are required for the realization of narrow-linewidth lasers and highly sensitive integrated optical sensors. Fifteen different DBR cavities were fabricated in order to investigate their grating reflectivity, finesse, and Q-factor. The length of the Bragg gratings on either side of the DBR cavities was varied from 1.25 to 4.75 mm in steps of 0.25 mm, with the distance between the Bragg reflectors chosen such that the total physical cavity length was 10 mm in all cases, with a grating coupling coefficient of $\kappa = 6.5\,\text{cm}^{-1}$. The reflectivity of the Bragg reflectors could be inferred from the measured finesse of each respective cavity [7]. The Bragg gratings deliver reflectivities higher than 99 % for TE polarization, as shown in Fig. 9.2, due partly to the waveguide propagation losses (including grating-induced scattering losses) being as low as 0.14 dB/cm. Measured Q-factors exceed 1.02×10^6 (TE polarization) for these DBR cavities. In addition to the DBR cavities, numerous DFB cavities were also fabricated and characterized.

In order to induce quarter-wavelength phase-shifts in the 10-mm-long uniform Bragg gratings, as is required for the DFB cavities, a 1-mm-long localized adiabatic sinusoidal tapering of the waveguide width was fabricated in the center region of

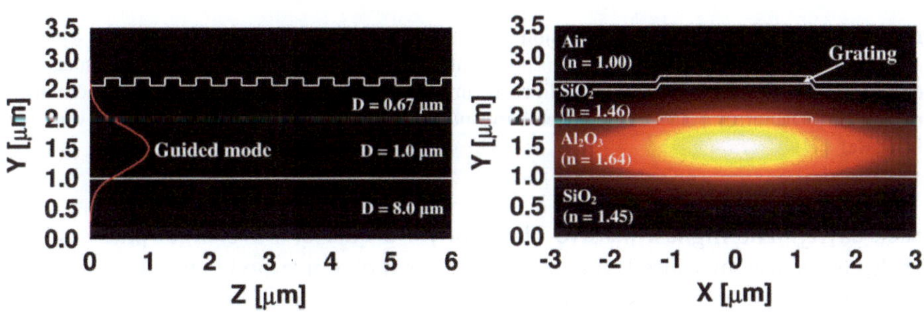

Fig. 9.1 (*Left*) A transverse cross-sectional view of the waveguide layer structure showing the calculated TE mode profile. (*Right*) An axial cross-sectional view of the waveguide structure showing the thickness D of each layer (Figure taken from Ref. [7])

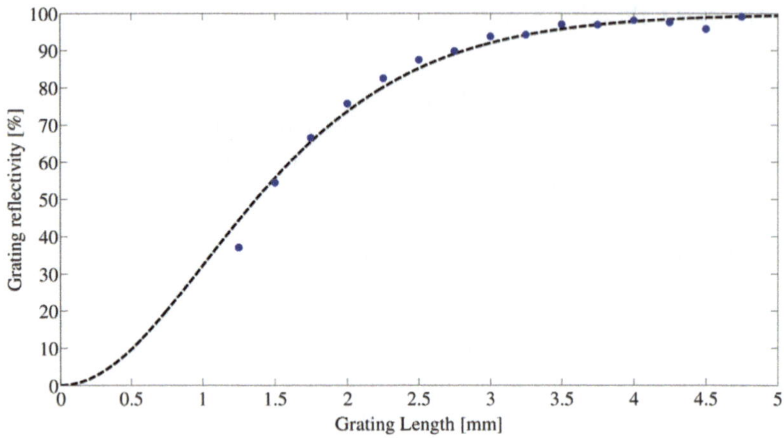

Fig. 9.2 Grating reflectivity at the Bragg wavelength for TE polarization as a function of grating length. The *blue dots* represent the reflectivity as determined from the measured finesse, while the *dashed line* is the predicted reflectivity according to coupled mode theory (Figure taken from Ref. [7])

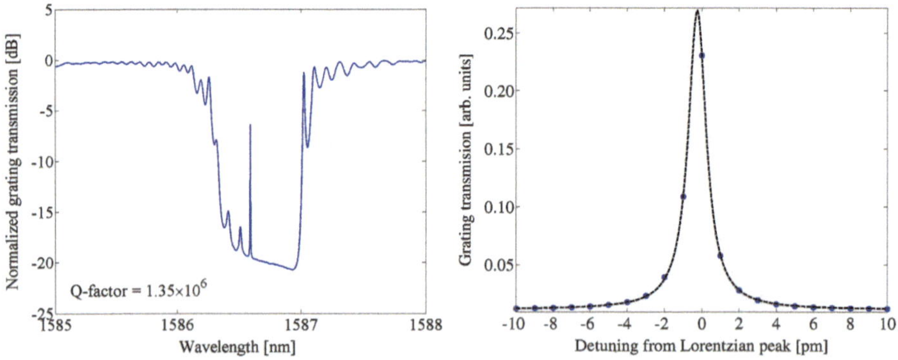

Fig. 9.3 (*Left*) Measured TE transmission spectrum for the passive DFB cavity. (*Right*) Enlargement of the single Fabry-Pérot transmission peak in (a), where the *blue dots* are the measured data points, while the *dashed line* is a Lorentzian fit to the data. This peak represents the highest measured Q-factor of 1.35×10^6, which corresponds to a linewidth of 1.17 pm (Figure modified from Ref. [7])

each cavity. The highest passive Q-factor for a quarter-wavelength phase-shifted DFB cavity was measured to be 1.35×10^6, which corresponds to a line-width of 1.17 pm. Figure 9.3 shows the measured normalized transmission spectrum of this particular cavity where the high-quality resonance is clearly visible.

9.4 Narrow-Linewidth Waveguide Lasers

Making use of the Bragg grating-based cavity configurations and distributed quarter-wavelength phase shifts which were described in the previous section, DFB cavities were realized in Er^{3+}-doped as well as Yb^{3+}-doped Al_2O_3 waveguides. The 1,480-nm-diode-pumped $Al_2O_3 : Er^{3+}$ DFB laser exhibits a threshold of 2.2 mW absorbed pump power and a maximum laser output power of more than 3 mW at the maximum available pump power of 67 mW, resulting in a slope efficiency of 41 % versus absorbed pump power [8], see Fig. 9.4. The laser operates TE-polarized at a wavelength of 1,545.2 nm with an emission linewidth of 1.7 kHz, as inferred from a self-heterodyne coherence measurement whose result is shown in Fig. 9.5. This value corresponds to a laser Q-factor [9] of 1.14×10^{11}.

Recently, these results have been picked up by research groups at the Massachusetts Institute of Technology and the University of California in Santa Barbara. Both groups exploited a low-loss SiN/SiO_2 waveguide structure [10], where the SiN core included a side-relief grating, and deposited an $Al_2O_3 : Er^{3+}$ layer on top, thereby demonstrating a laser [11] as well as a laser array [12]. An $Al_2O_3 : Yb^{3+}$ DFB laser [13], which was diode-pumped at 976 nm, exhibited a threshold of

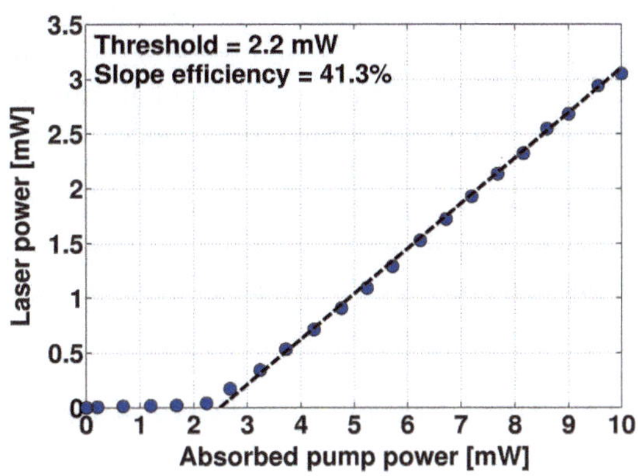

Fig. 9.4 Laser output power of the DFB laser as a function of absorbed pump power (*circles*). The *dashed line* represents a linear fit with a slope efficiency of 41.3 % (Figure taken from Ref. [8])

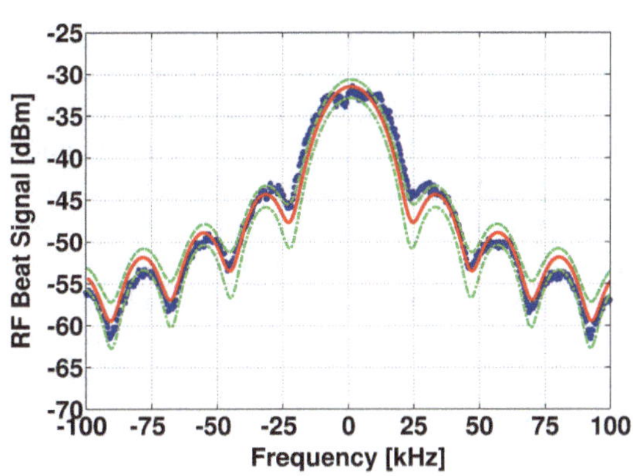

Fig. 9.5 Measured RF beat signal (*circles*) of the DFB laser, along with the best fitted theoretical RF power spectrum of a 1.70 kHz Lorentzian linewidth (*solid line*). The calculated RF power spectrum curves for Lorentzian linewidths of 1.70 + 0.58 kHz (*dashed line*) and 1.70 − 0.58 kHz (*dash-dotted line*) enclose 68 % of the measured data (Figure taken from Ref. [8])

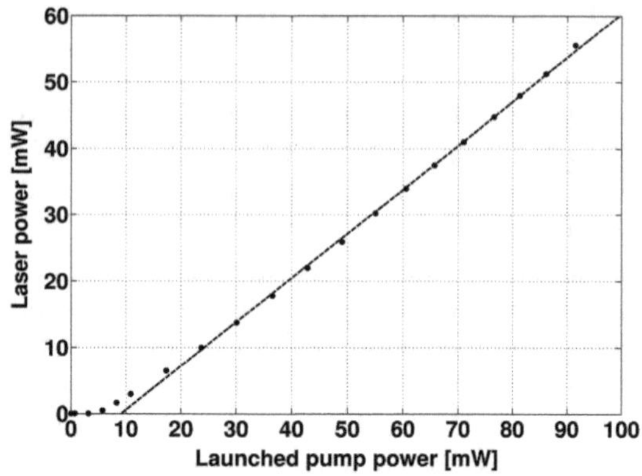

Fig. 9.6 Measured input-output power characteristics of the $Al_2O_3 : Yb^{3+}$ DFB channel waveguide laser (Figure modified from Ref. [13])

5 mW, see Fig. 9.6. The laser operated at a wavelength of 1,022.2 nm where the maximum measured laser output power exceeded 55 mW at a launched pump power of 91.5 mW, resulting in a slope efficiency of 67 %. The differences in the performance of the two lasers are explained largely by the higher absorption cross-section and the absence of energy-transfer upconversion in the Yb^{3+} compared to the Er^{3+} system. A monolithic distributed-Bragg-reflector channel waveguide laser in Yb^{3+}-doped aluminum oxide was also demonstrated [14]. Single-longitudinal-mode and single-polarization operation was achieved at a wavelength of 1,021.2 nm. Continuous-wave output powers of up to 47 mW and a launched pump power threshold of 10 mW resulted in a slope efficiency of 67 %. The high output power and high efficiency of the $Al_2O_3 : Yb^{3+}$ waveguide lasers along with the absorption coefficient of water, which is about 50 times smaller at a wavelength of ~1,020 nm as compared to Er^{3+} emission wavelengths around 1,550 nm [15], make such monolithic $Al_2O_3 : Yb^{3+}$ waveguide lasers ideal for bio-sensing applications.

9.5 Dual-Wavelength Waveguide Lasers

Aside from these monolithic single-longitudinal-mode Al_2O_3 waveguide lasers, monolithic dual-wavelength, or dual-longitudinal-mode, Al_2O_3 waveguide lasers were also realized. The main purpose for using dual-wavelength lasers, in this work, is to create a stable electrical beat signal at the output of a photodetector, with the frequency of the beat signal corresponding to the wavelength spacing of the two optical waves. The operation of the dual-wavelength cavity is based on two localized quarter-wavelength phase shifts in a DFB cavity [16], see Fig. 9.7. When two phase shifts are induced in a uniform waveguide Bragg grating, two resonance peaks appear in the transmission stop-band of the device. The two resonances share a common cavity which consists of both phase-shift regions, and the wavelength spacing between these resonances and longitudinal field distributions depend on the spatial separation and values of the respective phase shifts.

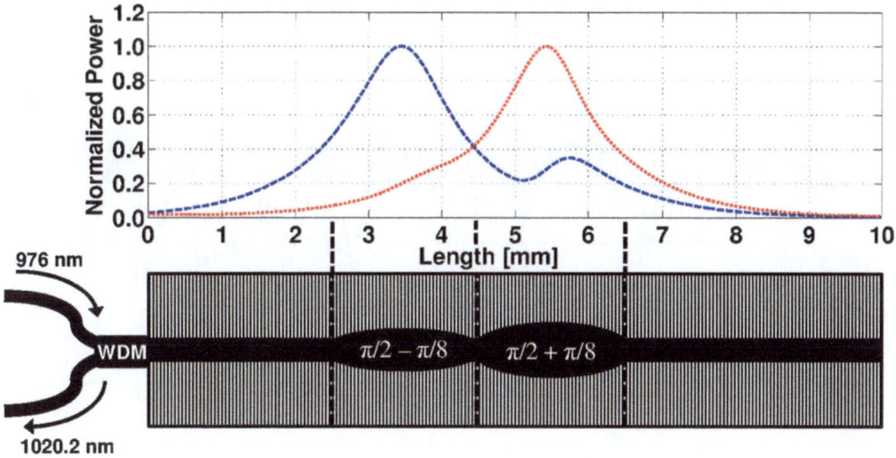

Fig. 9.7 Schematic of the dual-wavelength DFB cavity, along with the calculated longitudinal field distribution of the two respective laser wavelengths (Figure modified from Ref. [17])

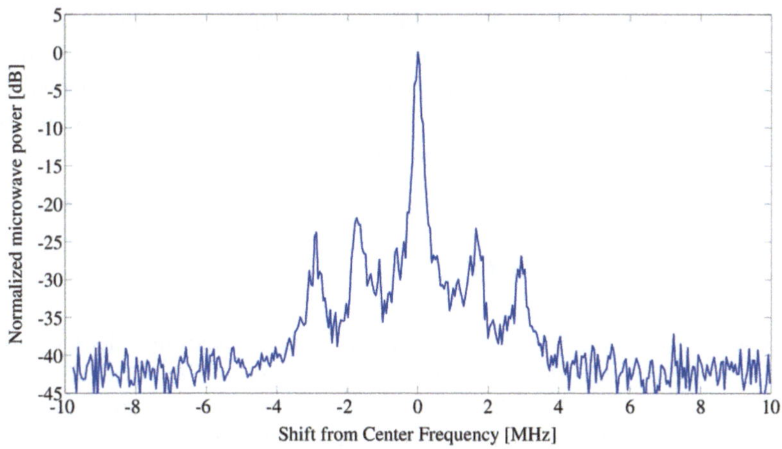

Fig. 9.8 Electrical spectrum of the microwave beat signal centered at 15.0426 GHz (Figure modified from Ref. [17])

A 10-mm-long Yb^{3+}-doped DFB cavity with two 2-mm-long distributed phase shifts were fabricated, with the phase shifts centered at 3.5 and 5.5 mm (as measured from the end-facet from where the light is launched) [17]. To confirm that the laser was indeed operating on two longitudinal modes, the laser output was measured with a photodetector and an electrical spectrum analyzer, which confirmed a microwave beat signal at 15.0426 GHz, which is shown in Fig. 9.8. The beat signal frequency implies a wavelength separation of 52 pm between the two individual longitudinal

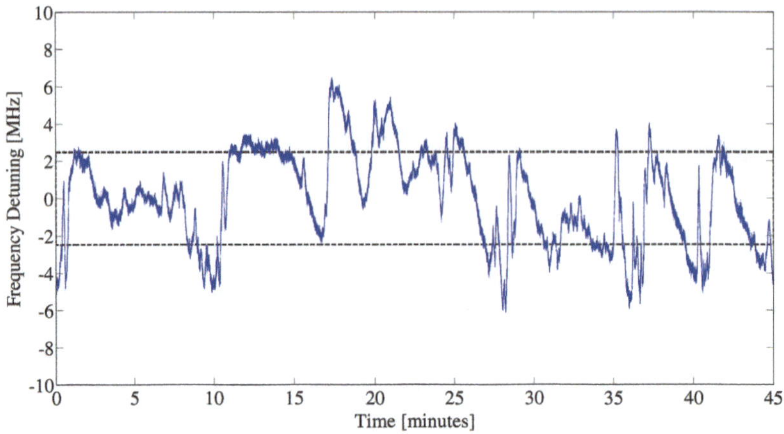

Fig. 9.9 Measured frequency stability of the microwave signal during a period of 45 min. The standard deviation of the center frequency during this period was ±2.5 MHz, indicated by the *black horizontal dashed lines* (Figure modified from Ref. [17])

laser modes. The two sidebands on either side of the main microwave peak are produced by relaxation oscillations from the two respective longitudinal modes. The long-term frequency stability of the microwave beat signal was measured over a period of 45 min with a 100 ms interval. The standard deviation of the microwave frequency during this period was found to be ±2.5 MHz; see Fig. 9.9.

9.6 Micro-particle Sensor

An integrated intra-laser-cavity micro-particle sensor based on a dual-wavelength distributed-feedback channel waveguide laser in $Al_2O_3 : Yb^{3+}$ was demonstrated [18]. We achieved real-time detection and accurate size measurement of single micro-particles with diameters ranging between 1 and 20 μm, which represent the typical sizes of many fungal and bacterial pathogens as well as a large variety of human cells. The sensing principle relies on measuring changes in the frequency difference between the two longitudinal laser modes as the evanescent field of the dual-wavelength laser interacts with micro-sized particles on the surface of the waveguide.

We demonstrated optical sensing by systematically probing the intra-cavity evanescent laser fields with various borosilicate glass microspheres of diameters ranging between 1 and 20 μm. Each microsphere was attached to an atomic force microscope (AFM) cantilever of low stiffness (spring constant 0.01–0.02 N/m). The cantilever was mounted on a 3-dimensional computer-controlled translation stage, as shown in Fig. 9.10. This experimental setup allowed us to scan a microsphere in

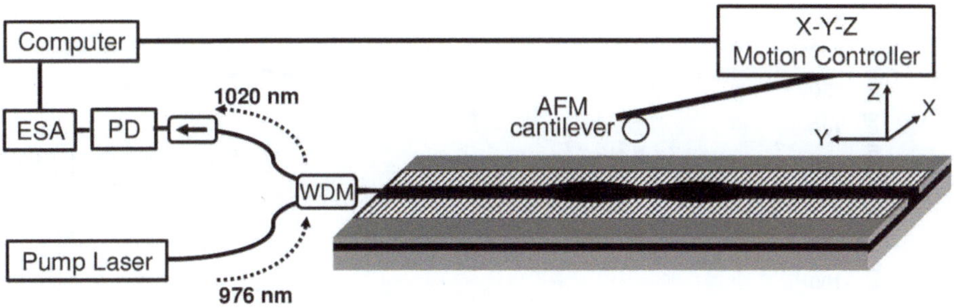

Fig. 9.10 Experimental setup used to characterize the intra-laser-cavity micro-particle dual-wavelength laser sensor. *ESA*: Electronic spectrum analyzer; *PD*: Photodetector; *WDM*: Wavelength division multiplexing fiber; *AFM*: Atomic force microscope (Figure taken from Ref. [18])

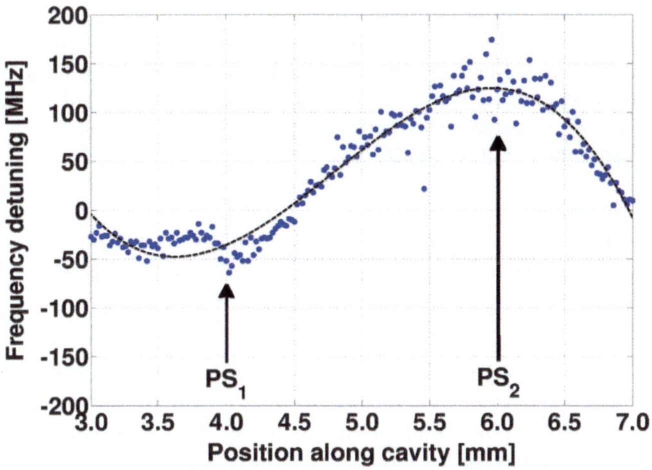

Fig. 9.11 Measured beat frequency detuning (*blue dots*), as a 10-μm-diameter borosilicate glass microsphere is scanned on the waveguide top surface along the length of the laser cavity. The *dashed line* represents a third-order polynomial fit to the data and serves as a guide for the eye. PS_1 and PS_2: center of the distributed phase shift on the pumped and unpumped side of the cavity, respectively (Figure taken from Ref. [18])

contact mode across the top surface of the waveguide laser with a lateral resolution of 25 nm, while recording the center frequency of the microwave beat signal for each position of the microsphere.

To illustrate the impact of the particle-induced scattering loss on the microwave beat frequency, we scanned a 10-μm-diameter borosilicate glass microsphere on the waveguide top surface along a 4-mm-long section of the laser cavity, covering both distributed phase-shift regions, see Fig. 9.11. Inside the phase-shift region located on the pumped side of the cavity, a negative detuning of the beat signal was observed, while probing the other phase-shift region resulted in a positive detuning [18].

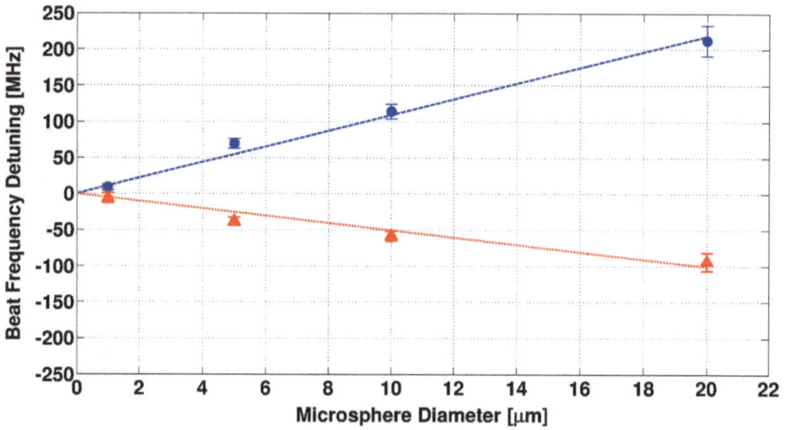

Fig. 9.12 Laser microwave-beat-frequency detuning as a function of microsphere diameter. The *red triangles* were measured in the center of the phase shift on the pumped side, while the *blue circles* were measured in the center of the phase shift on the unpumped side of the laser cavity. The *red* and *blue lines* represents linear fits through the origin with slopes of −5 and 11 MHz/μm, respectively (Figure taken from Ref. [18])

The interaction between the intra-laser-cavity light fields and the microspheres was further analyzed by performing a 2-dimensional scan on the waveguide surface inside the center of each of the two respective phase-shift regions, using a 5-μm-diameter borosilicate glass microsphere. By averaging along the y-direction (along the length of the waveguide), a single curve could be constructed for the two 2-dimensional surface scan, representing the beat frequency detuning as the microsphere is scanned perpendicular across the waveguide [18].

In order to investigate the size-measurement capability of our dual-wavelength laser sensor, the effect of various sizes of microspheres on the microwave beat frequency was measured. Each microsphere, attached to an AFM cantilever, was scanned across the waveguide laser in the center region of each of the two respective phase shifts, while the frequency detuning of the microwave beat signal was observed, see Fig. 9.12. Inside the first phase-shift region, the largest investigated microsphere with a diameter of 20 μm induced a 94 MHz decrease in the microwave beat signal, while the smallest microsphere, with a diameter of 1 μm, induced a frequency decrease of 6 MHz. In the second phase-shift region, the largest and smallest microspheres induced an increase in the microwave beat signal of 212 and 9 MHz, respectively. Assuming that the frequency detuning of the laser-generated beat signal has a linear dependence on microsphere size, the resolution of the sensor is currently limited to particles of ~500 nm diameter. This size limitation is due to the free-running frequency stability of the laser of ~5 MHz, which is most likely due to fluctuations in the pump power as well as some optical back-reflections into the laser cavity [17].

9.7 Summary

Over the past few years we have demonstrated DFB and DBR channel waveguide lasers in rare-earth-ion-doped Al_2O_3 on silicon substrates. In an Er^{3+}-doped DFB laser, single-longitudinal-mode operation was achieved with an emission line width of 1.7 kHz. Yb^{3+}-doped DFB and DBR lasers were demonstrated with output powers of 55 mW and a slope efficiency of 67 %. An Yb^{3+}-doped dual-phase-shift DFB laser with dual-wavelength emission was operated and a stable microwave beat signal at ~15 GHz was created via heterodyne photo-detection. By measuring changes in the microwave signal, we could detect single micro-particles with diameters between 1 and 20 μm on the waveguide surface.

References

1. Worhoff K, Bradley JDB, Ay F, Geskus D, Blauwendraat TP, Pollnau M (2009) Reliable low-cost fabrication of low-loss $Al_2O_3 : Er^{3+}$ waveguides with 5.4-dB optical gain. IEEE J Quantum Electron 45:454–461
2. Bradley J, Ay F, Wörhoff K, Pollnau M (2007) Fabrication of low-loss channel waveguides in Al_2O_3 and Y_2O_3 layers by inductively coupled plasma reactive ion etching. Appl Phys B 89:311–318
3. Bradley JD, Agazzi L, Geskus D, Ay F, Wörhoff K, Pollnau M (2010) Gain bandwidth of 80 nm and 2 dB/cm peak gain in $Al_2O_3 : Er^{3+}$ optical amplifiers on silicon. J Opt Soc Am B 27:187
4. Agazzi L, Wörhoff K, Pollnau M (2013) Energy-transfer-upconversion models, their applicability and breakdown in the presence of spectroscopically distinct ion classes: a case study in amorphous $Al_2O_3 : Er^{3+}$. J Phys Chem C 117:6759–6776
5. Agazzi L, Bernhardi EH, Wörhoff K, Pollnau M (2012) Impact of luminescence quenching on relaxation-oscillation frequency in solid-state lasers. Appl Phys Lett 100:011109
6. Bradley JD, Stoffer R, Agazzi L, Ay F, Wörhoff K, Pollnau M (2010) Integrated $Al_2O_3 : Er^{3+}$ ring lasers on silicon with wide wavelength selectivity. Opt Lett 35:73–75
7. Bernhardi E, Lu Q, van Wolferen H, Wörhoff K, de Ridder R, Pollnau M (2011) Monolithic distributed Bragg reflector cavities in Al_2O_3 with quality factors exceeding 106. Photonics Nanostruct Fundam Appl 9:225–234
8. Bernhardi EH, van Wolferen HAGM, Agazzi L, Khan MRH, Roeloffzen CGH, Wörhoff K, Pollnau M, de Ridder RM (2010) Ultra-narrow-linewidth, single-frequency distributed feedback waveguide laser in $Al_2O_3 : Er^{3+}$ on silicon. Opt Lett 35:2394–2396
9. Eichhorn M, Pollnau M (submitted 2014) The theory of continuous-wave lasers in the spot light of the vacuum photon
10. Bauters JF, Heck MJR, John DD, Barton JS, Bruinink CM, Leinse A, Heideman RG, Blumenthal DJ, Bowers JE (2011) Planar waveguides with less than 0.1 dB/m propagation loss fabricated with wafer bonding. Opt Express 19:24090–24101
11. Purnawirman, Sun J, Adam TN, Leake G, Coolbaugh D, Bradley JDB, Hosseini ES, Watts MR (2013) C- and l-band erbium-doped waveguide lasers with wafer-scale silicon nitride cavities. Opt Lett 38:1760–1762
12. Belt M, Huffman T, Davenport ML, Li W, Barton JS, Blumenthal DJ (2013) Arrayed narrow linewidth erbium-doped waveguide-distributed feedback lasers on an ultra-low-loss silicon-nitride platform. Opt Lett 38:4825–4828

13. Bernhardi E, van Wolferen H, Wörhoff K, de Ridder R, Pollnau M (2011) Distributed feedback channel waveguide lasers in Erbium- and Ytterbium-Doped Al_2O_3 on silicon – OSA technical digest (CD). In CLEO/Europe and EQEC 2011 conference digest, p. CJ10_4, Optical Society of America, Munich
14. Bernhardi EH, van Wolferen HAGM, Wörhoff K, de Ridder RM, Pollnau M (2011) Highly efficient, low-threshold monolithic distributed-Bragg-reflector channel waveguide laser in $Al_2O_3 : Yb^{3+}$. Opt Lett 36:603–605
15. Hale GM, Querry MR (1973) Optical constants of water in the 200-nm to 200-microm wavelength region. Appl Opt 12:555–563
16. Villanueva GE, Perez-Millan P, Palaci J, Cruz JL, Andres MV, Marti J (2010) Dual-wavelength DFB Erbium-Doped fiber laser with tunable wavelength spacing. IEEE Photonics Technol Lett 22:254–256
17. Bernhardi EH, Khan MRH, Roeloffzen CGH, van Wolferen HAGM, Wörhoff K, de Ridder RM, Pollnau M (2012) Photonic generation of stable microwave signals from a dual-wavelength $Al_2O_3 : Yb^{3+}$ distributed-feedback waveguide laser. Opt Lett 37:181–183
18. Bernhardi EH, van der Werf KO, Hollink AJF, Wörhoff K, de Ridder RM, Subramaniam V, Pollnau M (2013) Intra-laser-cavity microparticle sensing with a dual-wavelength distributed-feedback laser. Laser Photonics Rev 7:589–598

10
Integrated Optomechanics: Opportunities for Tunable Nanophotonic Devices

Wolfram H.P. Pernice

Abstract Nanomechanical devices are of immense interest for measuring minuscule quantities with ultimate detection responsivity. Due to their very small mass such devices are extremely sensitive to small fluctuations in the environment and thus allow for making superior sensors. However, because of their small dimensions efficient transduction mechanisms are challenging to develop, both to readout their mechanical motion and to drive them into oscillation. An intriguing property of nanoscale resonators is the fact that they are size-matched to nanophotonic components. Therefore they can be seamlessly integrated into on-chip photonic circuits which enables using optical methods for readout and actuation. In this case the link between mechanical and optical domains is the gradient optical force which can be harnessed for integrated optomechanical interactions. Because gradient forces are a generic optical concepts they can be generated in a wide range of material systems and thus also provide a way to give active functionality to optical circuits where alternative tuning mechanisms do not exist.

Keywords Nanophotonics • Nanomechanics • Optomechanics • Integrated optics

10.1 Introduction

The field of integrated optics or integrated photonics was born out of the need to realize reliable and fast telecommunication devices. Initially the transmission of information was carried out in free space using radio or microwaves. The technology to transmit signals changed rapidly with the invention of the laser in 1960, allowing

W.H.P. Pernice (✉)
Institute of Nanotechnology (INT), Karlsruhe Institute of Technology (KIT), 76344 Karlsruhe, Eggenstein-Leopoldshafen, Germany
e-mail: wolfram.pernice@kit.edu

B. Di Bartolo et al. (eds.), *Nano-Structures for Optics and Photonics*,
NATO Science for Peace and Security Series B: Physics and Biophysics,
DOI 10.1007/978-94-017-9133-5_10,

for the use of much higher carrier frequencies and bandwidth. However, propagation of optical frequencies in free space is plagued by a variety of noise sources, which severely limit such optical communication devices. Effects of the environment on the communication link, such as humidity, turbulence in the transmission medium and diffraction/dispersion effects can be readily avoided by moving from free space to guided waves. The fundamental building block to achieve propagation of lightwaves in a deterministic fashion on chip is the waveguide. Waveguides both determine the path light will take and also protect the propagating wave from external influences. The most widely known waveguide structure is the optical fiber, which allows transmitting light over thousands of kilometers with very low optical loss and high data rates. While the invention of the optical fiber enabled the modern information age, the need to manipulate the transmitted optical data signals remains. Even though this can be achieved with traditional optical components such devices usually require a large footprint. As optical communication systems become more and more complex, filtering and processing equipment becomes unscalable and thus the only way to move beyond bulky equipment consists in miniaturization. Employing the principles of nanoscience and nanotechnology allows for scaling devices to dimensions on the order of the wavelength of light. Originally driven by the needs of the computer industry, the tools of nanotechnology were developed to produce smaller and smaller electronic components. The same procedures to realize miniature electronic circuits can be used to design and fabricated miniature photonic circuits that allow for dense packing of optical functionality on a chip and thus the realization of integrated photonic circuits. A photonic integrated circuit or integrated optical circuit is a device that integrates multiple photonic functions in analogy to an electronic integrated circuit. The major difference between the two is that a photonic integrated circuit provides functionality for information signals imposed on optical wavelengths, nowadays typically in the near infrared wavelength regime. Besides the benefit of high integration density and reduced device footprint, using nanophotonic elements also allows for the exploitation of new physical effects that only occur on a nano- or even quantum scale. Nanophotonic elements can be used to tightly focus light into tiny structures and cavities, which leads to tremendous field intensities and thus strong non-linear optical effects. Novel photonic structures such as photonic crystals and high quality resonators are being developed to modify the propagation of light in specific wavelength regimes, allowing us to tailor optical materials to particular needs. Difficulties arise when implementing tunable photonic circuits whose functionality needs to be reconfigured or tailored during operation. Traditionally tunability is obtained by employing intrinsic material effects that couple to the optical properties of nanophotonic waveguides. Frequently used is the thermo-optical effect which modifies the refractive index upon a change of the operation temperature. However, the thermo-optical effect is a slow effect and thus not suitable for manipulation of fast optical signals. In materials that show optical non-linearity fast switching of the refractive index is possible under strong driving optical fields. Nevertheless, the majority of materials does not provide the desired second order Pockel's effect and thus only allows for Kerr non-linearity, limiting tunability to a small refractive index change. Other options such as the electro-optical effect may also be employed, but only in materials which are not centro-symmetric. A recent way to overcome such limitations is the use of movable mechanical elements that are brought into close proximity of a waveguide and thus alter the properties of a propagating mode. Within the framework of nanophotonics and nanomechanics such a platform is a natural choice as both technologies are fabrication compatible and lead to devices that have a similar footprint. In order to link optical to mechanical degrees of freedom in a chipscale environment gradient optical forces can be employed which provide broadband actuation and readout independent of a particular material choice. In the following a brief introduction to gradient force actuation and the use in integrated photonic circuits will be given as a promising route towards all-optically tunable photonic devices.

10.2 Gradient Optical Forces

In general two related optical forces can be employed for allowing mechanical devices to interact with electromagnetic fields. The more well-known optical force is the radiation pressure force which results from momentum transfer of a photon to a mechanical element. When a photon is reflected from a mirror surface it transfers twice its momentum onto the surface as a result of changing its direction under normal incidence. Thus the momentum transfer also results in a phase change of the incoming wave and thus radiation pressure forces may be understood as a result of a gradient profile in the phase of a propagating wave. In a chipscale framework traditionally only two-dimensional structures are employed which prevent access to out-of-plane fields and thus free-space beam propagation. Therefore radiation pressure forces are not necessarily the method of choice for optical waveguide actuation. Alternatively, optical-tweezer like forces can be harnessed to drive free-standing mechanical structures into motion. In this case momentum transfer results from a gradient in the optical intensity in the near-field of propagating modes which manifests in the generation of optical forces that are strong enough to serve as an actuation mechanism for nanoscale devices. In order to understand the origin of the gradient optical force it is instructive to look at the example of a free-standing waveguide in air that is brought into the vicinity of a material with higher refractive index such as glass for example. Such a structure can be realized by fabricating an integrated nanophotonic waveguide and removing the underlying buffer layer in order to realize a mechanically movable resonator. Here we consider a silicon waveguide of 500 nm width which is realized on a silicon-on-insulator (SOI) wafer. The thickness of the waveguiding silicon layer is 110 nm and the silicon layer is separated from the bulk silicon wafer by a silicon dioxide layer of 3 μm thickness. In order to evaluate the optical force we consider a cross section of the waveguide as shown in Fig. 10.1a. In agreement with an experimental implementation the waveguide beam is mechanically clamped between two multi-mode interference couplers (MMI), which are the larger rectangular waveguide sections at the ends of the beam. They act as optical lenses that refocus the

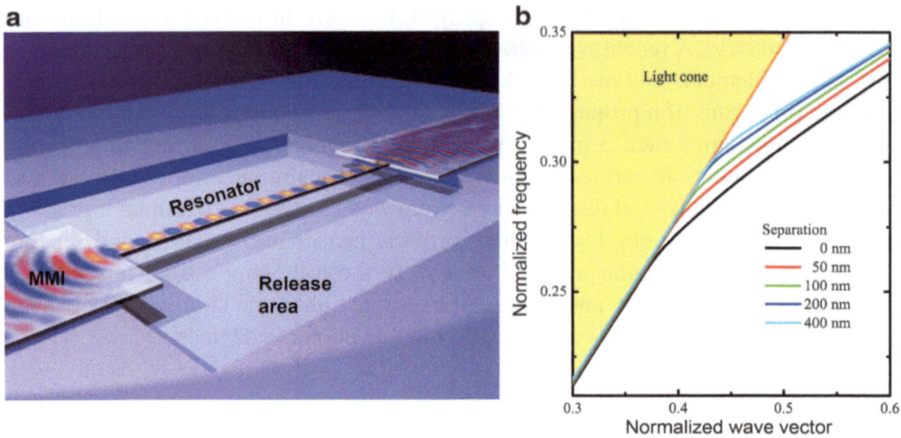

Fig. 10.1 (**a**) Illustration of a free-standing silicon waveguide which acts as both a light rail and a mechanical resonator. The beam is clamped between two multi-mode interference devices for mechanical stability. (**b**) The calculated dispersion diagram for the propagating optical mode on the beam in dependence of separation to the underlying oxide layer. Guided optical modes can be found below the light line

multi-mode beam expansion back into the free-standing waveguide structure. The fundamental mode of the waveguide is excited in TE polarization. Neglecting scattering losses, the fundamental mode will propagate along the waveguide beam without changing its profile. Therefore the investigation of a two-dimensional cross-section is a valid simplification. The waveguide is separated from the substrate by a gap of size g. When the waveguide is brought closer to the substrate, the waveguide mode will be coupled evanescently to the dielectric. With decreasing gap, the propagation constant of the fundamental mode decreases. This is demonstrated in the dispersion diagram obtained from finite-element simulations as shown in Fig. 10.1b.

Above the cut-off frequency, the dispersion curves corresponding to varying gap sizes show a local maximum before converging towards the propagation constant of the free waveguide. From the dispersion diagram we are able to derive the optical force, which acts on the waveguide. We calculate the optical forces by considering the frequency shift of the optical mode with decreasing gap distance as [1]

$$F = -\frac{1}{\omega}\frac{\partial \omega}{\partial g}U \tag{10.1}$$

In the above equation the optical frequency ω is related to the effective refractive index n_{eff} of the waveguide as $\omega = \omega_0/n_{eff}$, where ω_0 is the free space frequency. Because in integrated photonic circuits we are usually measuring the effective index of the fundamental mode of the waveguide rather than the frequency, the above equation can be more conveniently expressed in terms of effective mode indices as [2]

$$F = -\frac{1}{n_{eff}} \frac{\partial n_{eff}}{\partial g} U \tag{10.2}$$

The refractive index or equivalently the propagation constant of a waveguide can be numerically obtained when a particular waveguide geometry has been chosen. Assuming that the energy on the beam is given by $U = PLn_g/c$ where P is the optical power, L is the length of the beam, and n_g is the group index of the mode, we get the optical force normalized to length and optical power as

$$F_{norm} = F/PL = -\frac{n_g}{cn_{eff}} \frac{\partial n_{eff}}{\partial g} \tag{10.3}$$

The above equation can be conveniently evaluated using finite element simulations to obtain the effective refractive index in dependence of gap g. From finite element simulations the group index can also be quickly obtained by further calculating the effective refractive index as a function of wavelength and thus the gradient optical force can be easily derived. From the negative sign in Eq. (10.2) it is apparent that the force is attractive, thus pulling the beam towards the substrate. When the gap size is zero, the optical force reaches a maximum value on the order of 10 pN/μm/mW. It decreases monotonically towards zero when the gap is increased towards infinity. Based on the calculations, we find a typical strength of the force on the order of pN when assuming device dimensions in the micrometer range. Therefore given the high sensitivity and small mass of the free standing waveguide this force value is significant for the actuation the nanomechanical motion of the waveguide [3, 4].

When applying an optical input field to the waveguide the free-standing mechanical structure thus experiences a force which will lead to a deflection of the beam from its resting position towards the substrate. Thus the actual distance between waveguide and substrate can be controlled by changing the propagating optical power as evident from Eq. (10.2). Depending on the stiffness of the mechanical device the displacement can be used to alter the optical properties of probe optical modes propagating along the beam. At the same time, the displacement also provides a convenient way to detect mechanical motion optically. This is because the change in the resting position in turn also induces a change of the effective refractive index, which translates into a phase change of any optical field propagating along the beam waveguide. Thus phase-sensitive detection mechanisms as inherent to interferometers [5, 6] or on-chip optical resonators [7] can be used to monitor mechanical vibrations. Generally, the displacement is very small which necessitates the use of highly sensitive optical detection schemes. However, several techniques can be employed to further tailor the properties of the optical force drive.

10.3 Enhancing Gradient Optical Forces

While the above illustration already shows that strong driving optical fields can be realized by coupling a waveguide to a substrate such a geometry is limited by practical challenges. For strong gradient forces a small gap between the waveguide and the substrate is required which is difficult to achieve from a fabrication point of view.

Instead, coupling between several optical modes may be employed as a route to overcome fabrication induced challenges. In this case the propagating optical mode is placed in the near-field of a neighboring optical waveguide such that the evanescent fields of both waveguide modes overlap and mutually interact. When using dielectric waveguides such a geometry leads to a significant enhancement of the gradient in the electrical field and thus an increase of the driving optical force [8]. When both waveguides are brought together below a distance of say 100 nm, the electrical field component is forced into the region with a lower dielectric constant between the waveguides, leading to a so-called slot waveguide, originally proposed in Cornell by M. Lipson's group [9]. Slot waveguides are ideal waveguide geometries for gradient force actuation because they allow gradient field enhancement and also the realization of waveguide beams with smaller cross-section and thus smaller spring constant. Because of the resulting strong optical forces this waveguide geometry is also suitable for the actuation of stiff mechanical resonators for high frequency operation [7].

Besides gradient enhancement resonant feedback can be used to further increase optical forces on chip. Rather than increasing the gradient term in (10.3), this concept aims at increasing the optical power on the mechanical beam. Within optical cavities light fields are trapped at the cavity resonance frequency, leading to significant build-up of optical intensity when the cavity quality factor is high. In integrated optical circuits several cavity geometries can be readily employed for feed-back enhancement such as photonic crystal defect cavities [10], optical ring resonators [11] or disk resonators [12]. Ring resonators are also directly compatible with the slot waveguide concept and thus provide both force and sensitivity enhancement.

10.4 Mechanically Tunable Nanophotonic Circuits

By exploiting a circuit approach for fabricating joint mechanical-optical devices complex systems can be realized in a chip-scale framework. Integrated photonic circuits have been developed for several decades and thus a rich toolbox of optical components is available to design specialized photonic devices for a desired application. Among those are optical filters, interferometers and power splitters, to name just a few. In principle, all such devices can be made tunable by co-integrating mechanical resonators with the rest of the circuit.

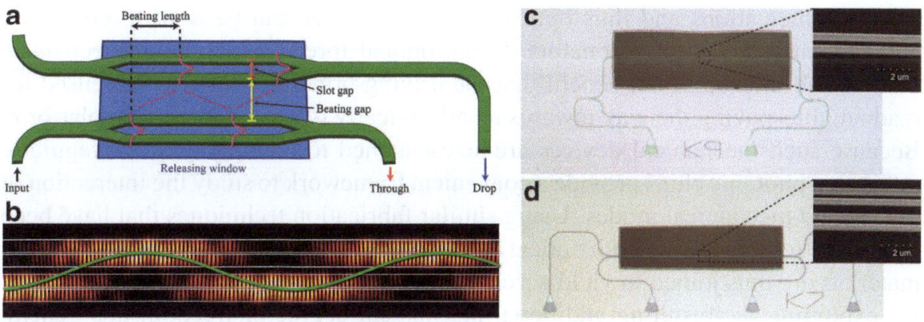

Fig. 10.2 (**a**) Illustration of a mechanically tunable directional coupler. Two slot waveguides are coupled together, allowing light to be transferred from the input port to either the drop or the through port. (**b**) Numerical simulation of the mode profile along the beam waveguides. A characteristic beating pattern shows the exchange of optical intensity between the top and the bottom waveguides. (**c**) Physical realization of a tunable Mach-Zehnder interferometer with a single slot waveguide (*inset*). (**d**) Physical implementation of a tunable directional coupler as in (**a**) with two slot waveguides (*inset*) [13]

In nanophotonic devices this can be easily accomplished by creating free-standing waveguide structures through wet chemical release. By removing the underlying substrate the resulting waveguide acts as both an optical wire as well as a movable mechanical component that can be actuated with gradient optical forces.

A potential application is shown in Fig. 10.2a for building an all-optically tunable coupler or switch [13]. In this case slot waveguides are employed to realize tunable photonic components. The two waveguides are coupled together in the optical near-field, leading to a characteristic beating pattern along the direction of propagation, as shown in Fig. 10.2b. The beating length depends on the propagation constant of the two slot waveguides. By applying an optical tuning light, the slot gap can be decreased using attractive gradient optical forces. This in turn changes the beating length of the device and thus transfers light either into the through or the drop port.

Physical implementations of the concept are shown in Fig. 10.2c, d for a tunable Mach-Zehnder interferometer (MZI) or a tunable directional coupler as sketched in Fig. 10.2a. In this case the circuits are realized from tensile stressed silicon nitride which allows for the fabrication of long free-standing waveguide beams. In the inset of the two optical micrographs the slot waveguides are shown with a nanoscale gap between the two dielectric nanostrings. This gap can be controlled with optical forces and thus forms the basis of optical tunability.

10.5 Conclusions

Even though the momentum transfer of photons to matter is usually too weak to be observed in a macroscopic context, the picture changes when moving towards nanoscale mechanical resonators. Due to their small mass they are susceptible to small fluctuations and thus optically induced forces can be used to efficiently actuate nanomechanical resonators. Using optical forces as actuation mechanism comes with the additional benefit that the driving principle can also be reused for readout, thus paving the way towards an all-optically mediated integrated platform. Because such mechanical devices are size-matched to nanophotonic waveguides, integrated photonic chips provide a convenient framework to study the interaction of optical and mechanical modes. Using similar fabrication techniques that have been developed for fabricating both optical and mechanical components as well as similar materials are thus joined to yield a rich toolbox for next-generation optical circuits. By exploiting the resulting additional mechanical degree of freedom this general concept can be ported to a wide range of materials and thus also gives tunability to otherwise passive material systems.

References

1. Povinelli ML, Loncar M, Ibanescu M, Smythe EJ, Johnson SG, Capasso F, Joannopoulos JD (2005) Evanescent-wave bonding between optical waveguides. Opt Lett 30:3042–3044
2. Pernice WHP, Li M, Tang HX (2009) Theoretical investigation of the transverse optical force between a silicon nanowire waveguide and a substrate. Opt Express 17(3):1806–1816
3. Rugar D, Budakian R, Mamin H, Chui B (2004) Single spin detection by magnetic resonance force microscopy. Nature 430:329–332
4. Ekinci KL, Roukes ML (2005) Nanoelectromechanical systems. Rev Sci Instrum 76:061, 101
5. Li M, Pernice WHP, Xiong C, Baehr-Jones T, Hochberg M, Tang HX (2008) Harnessing optical forces in integrated photonic circuits. Nature 456:480–484
6. Li M, Pernice WHP, Tang HX (2009) Broadband all-photonic transduction of nanocantilevers. Nat Nanotechnol 4(6):377–382
7. Li M, Pernice WHP, Tang H (2010) Ultrahigh-frequency nano-optomechanical resonators in slot waveguide ring cavities. Appl Phys Lett 97(18):183110
8. Li M, Pernice WHP, Tang HX (2009) Tunable bipolar optical interactions between guided lightwaves. Nat Photon 3(8):464–468
9. Almeida VR, Xu Q, Barrios CA, Lipson M (2004) Guiding and confining light in void nanostructure. Opt Lett 29(11):1209–1211
10. Eichenfield M, Chan J, Camacho RM, Vahala KJ, Painter O (2009) Optomechanical crystals. Nature 462(7269):78–82
11. Bagheri M, Poot M, Li M, Pernice WPH, Tang HX (2011) Dynamic manipulation of nanomechanical resonators in the high-amplitude regime and non-volatile mechanical memory operation. Nat Nanotechnol 6(11):726–732
12. Li M, Pernice WHP, Tang HX (2009) Reactive cavity optical force on microdisk-coupled nanomechanical beam waveguides. Phys Rev Lett 103(22):223901
13. Fong KY, Pernice WHP, Li M, Tang HX (2011) Tunable optical coupler controlled by optical gradient forces. Opt Express 19(16):15098–15108

11
Fluorescence-Based Studies of Nanoparticles

Gokhan Bilir, Gönül Özen, John Collins, Maura Cesaria, and Baldassare Di Bartolo

Abstract Fluorescence spectroscopy is particularly adept at uncovering the effects of spatial confinement on the spectral properties of transition metals and rare earth ions that are used as dopant in nano-particles. We have investigated a number of nanocrystal systems, such as YAG ($Y_2Al_5O_{12}$) and Yttria (Y_2O_3) of different sizes doped with Chromium, Praseodymium and Neodymium ions. We have chosen to present the results we have obtained on a particular system, Y_2O_3 doped with Neodymium ion at different concentrations and different nano-sizes that may be considered representative of all the systems we have investigated.

Keywords Dynamical characteristics • Fluorescence • Nanoparticles • Nd-doped • Thermal line broadening and line shifting • Y_2O_3

G. Bilir
Department of Physics, Istanbul Technical University, Istanbul, Turkey

Department of Physics, Boston College, Chestnut Hill, MA, USA

G. Özen
Department of Physics, Istanbul Technical University, Istanbul, Turkey

J. Collins
Department of Physics and Astronomy, Wheaton College, Norton, MA 02766, USA
e-mail: jcollins@wheatonma.edu

M. Cesaria
Department of Physics, Universita del Salento, Lecce, Italy

B. Di Bartolo (✉)
Department of Physics, Boston College, Chestnut Hill, MA, USA
e-mail: dibartob@bc.edu

B. Di Bartolo et al. (eds.), *Nano-Structures for Optics and Photonics*,
NATO Science for Peace and Security Series B: Physics and Biophysics,
DOI 10.1007/978-94-017-9133-5_11, © Springer Science+Business Media Dordrecht 2015

11.1 Motivation

Collective excitations in solid materials are altered significantly by the dimensional confinement of the materials' structures. These effects are similar to those of an electron in a quantum well.

Consider the wavefunction of an electron in a deep quantum well with width L_z in the z-direction. The eigenstates of this system are plane wave states in the direction parallel to the interfaces and bound states in the infinitely deep quantum well in the z-direction:

$$\Psi_n(z) = \sqrt{\frac{2}{AL_z}} e^{i\mathbf{k}_{\|}\cdot\mathbf{r}_{\|}} \sin(k_z z)$$

where $\mathbf{r}_{\|}$ is the position vector in a plane parallel to the interface, $\mathbf{k}_{\|}$ the wave vector component in a plane parallel to the interface, A the area of the interface, and

$$k_z = \frac{n\pi}{L_z} \qquad n = (1, 2, 3 \ldots)$$

The energy eigenstates are labeled by n

$$E_n = \frac{\hbar^2 k_{\|}^2}{2m} + \frac{\hbar^2 \pi^2 n^2}{2mL_z^2}$$

The major effect of the spatial confinement in the z-direction is that the z component of the bulk vacuum wavevector is restricted integral multiple of π/L_z.

In the case of a three-dimensional confinement the values of the wave vector are restricted in the three directions. In addition, if the confinement is that produced by nano-dimensions, the phonon spectrum can no longer be treated as a continium. This is particularly true at the low frequencies.

What happens to phonons exemplifies the dramatic effects of spatial confinement on extended wavefunctions. In semiconductor materials, the reduction of the dimensions leads to an increase of the band gap. Localized centers in nano-particles present an interesting situation. The localization of the wavefunction prevents the spatial confinement to produce any relevant effect on the energy levels. But the confinement can affect the centers in other ways. Before describing them we recall the occurrence of the following situation in quantum systems. A perturbation that has negligible effect on the energy levels of a system may have a relevant effect on the probability of transition between the apparently unperturbed levels.

This is best exemplified by the energy levels of rare earth ions in solids. Due to the shielding of the optically active electron that reside in the $4f$ shell by the 8 electron that reside in the $5s$ and $5p$ shells the rare earth ions are not strongly affected by the crystalline field of the coordinating ions or by the time variation of this field. They present spectra which are seemingly independent of the host lattice and are sharp

($\Delta\omega \ll \omega_D$) at room temperature. However, such characteristics as the transition probability generally depend on the host lattice.

Localized centers in nano-particles may differ from centers in bulk in what we could call the "dynamical" characteristics:

(a) Probability of radiative transitions,
(b) Probability of non-radiative transitions,
(c) Lifetime of excited states
(d) Thermal line broadening and thermal line shifting.

These occurrences are sensitive to the small changes of the wavefunctions of the states involved in a transition that affect the coupling of the emitting ion to the optical oscillators of the vacuum and to the lattice vibrations. We may also note that while the spectrum of the vacuum oscillators is in general the same in all cases, the spectrum of the lattice vibrations of the host material depends on the material itself.

With this background information let us look now at what we can expect from the spectroscopic investigation of the emission spectra of localized centers in nano-particles.

11.2 Choice of a Representative System

We have investigated a large number of nano-crystal systems, such as YAG and Y_2O_3 of different sizes doped with Cr, Pr and Nd at different concentrations. Synthesization of these samples was achieved by the Pechini method (YAG) and by the thermal decomposition method (Y_2O_3). The particle sizes were determined by using XRD patterns and confirmed by SEM and TEM measurements. A representative system among the ones we have examined is Y_2O_3 doped with Neodymium ions at different concentrations, and with different nano-sizes.

The Nd ion enters the nano-crystals in its trivalent state; pumping it with light at ~800 nm (12,500 cm^{-1}) results in the excitation of the levels $^2H_{9/2} -^4 S_{5/2}$ and subsequently of the metastable doublet $^4F_{3/2}$. The energy gap between the $^4F_{3/2}$ doublet and the lower level $^4I_{15/2}$ is ~5,000 cm^{-1}, wide enough to hinder any radiationless decay. Indeed the lifetime of the metastable $^4F_{3/2}$ doublet is generally found to be practically independent of temperature. On the other hand the lifetime is host-dependent: it is 480 μs in YLF [1], 230 μs in YAG [2, 3], 33 μs in YVO_4 [1,2], and 260 μs in Y_2O_3 [3]. The host-dependence of the lifetime indicates a clear dependence of the wavefunction representing the metastable state on the environment in which the Nd^{3+} ion resides.

Before moving to the presentation of the experimental data we want to make the point that nano-powders doped with optically active ions seem to be able to luminescence even when the dopant concentration exceeds the maximum concentration at which solid of the same nature can emit luminescence radiation. This occurrence has been verified in the present case of Y_2O_3:Nd^{3+} nano-powders.

11.3 Experimental Setup

11.3.1 Preparation of the Nd^{3+} Doped Y_2O_3 Powders

Nanosized Y_2O_3 samples doped with 1.0 % mol Nd^{3+} ions concentrations were prepared by thermal decomposition of yttrium-neodymium alginate. Yttrium nitrate hexahydrate $Y(NO_3)_3 \cdot 6\,H_2O$, neodymium nitrate hydrate $Y(NO_3)_3 \cdot H_2O$ and low-viscosity (250 cps of 2 % solution) alginic acid sodium salt of an analytic grade were purchased from the Sigma Aldrich Company.

The yttrium-neodymium alginate beads were prepared by the thermal decomposition method according to the prescription given in the literature [4]. 0.2 M 100 ml yttrium-neodymium nitrate solution were prepared by dissolving the appropriate amount of the yttrium nitrate and neodymium nitrate salts in ultra pure water. Also 50 ml alginate solutions with 1 % w/w concentration were prepared by dissolving an appropriate amount of the sodium alginate salt in the ultra sure water under magnetic stirring.

Alginate is a biopolymer which is extracted from three species of brown alagae. It is a linear heteropolysaccaride composed of D-mannuronic acid and L-guluronic acid. Gelation of alginate is possible but the interaction of the carboxylate group with divalent ions and the formation of beads can be achieved by dropwise addition of sodium alginate into divalent solution by using a syringe with a needle or pipette [4,5]. The formation of the yttrium-neodymium alginate beads was achieved by the dropwise addition of the sodium alginate solution into yttrium-neodymium salts solution by using a syringe.

The prepared beads were kept in their gelling medium for 30 min under gentle stirring; then they were removed from their gelling medium, placed in a porcelain crucible and heated to different temperatures ranging from 350 to 500 °C for 24 h with a heating rate of 10 °C/min in an electric furnace in air medium. some of the products obtained were annealed at 600, 800, 1,000, and 1,400 °C, to investigate the particles size dependence on the annealing temperature.

11.3.2 Structural Characterization

X-ray diffraction investigations were carried out with a Bruker AXS D* Model (Cu-Kα radiation) diffractometer at 40 kV and 30 mA setting in the 2θ range from 20° to 70° with a scanning step of 0.02°. A JEOL 6335F model scanning electron microscope was used to get SEM images of the samples. Elemental analysis was made by using an energy dispersive spectroscopy (EDS) attached to SEM. TEM images of the samples were obtained by using a JEOL 2010F model transmission electron microscope operated at 200 kV with a field emission gun. TEM specimens were prepared by depositing a few drops of a sample dispersed in ethanol by using an ultrasonic bath, on a carbon coated copper grid.

11.3.3 Spectroscopic Measurements

The continuous emission spectra of the samples were obtained by pumping the samples with the output of a Laser Drive Inc. Model LDI-820 laser diode, that resulted in the excitation of the levels $^2H_{9/2} -^4 S_{5/2}$. The luminescence signal was directed toward the entrance slit of a 1 m McPherson Model 2051 monochromator and chopped at a frequency of 250 Hz before entering the slit. The monochromator provided a resolution 0.8 Å with the slits set at 50 μm and a wavelength reproducibility of 0.1 Å. The optical signal was detected by a Hamamatsu 7102 photomultiplier tube with an S1 response, sent to a EG&G Model 5210 lock-in amplifier and recorded in a computer.

The same arrangement was used for the line-width and line-shift measurements with the entrance and output slits of the monochromator set at 80–150 μm, which correspond to 1–2 Å spectral width and a 5 Å/min scanning rate.

The experimental data on the responses of the samples to pulsed excitation were obtained by using a Schwarz Electro-Optics Inc. Model Titan-P Model Titanium-Sapphire laser and a Tektronix Model TDS3052B oscilloscope.

For the experiments in the temperature range 34–300 K, the samples were mounted on the cold finger of a closed cycle Helium refrigerator. This system uses a Janis Research Model RD dewar connected with a Leybold Model RW2 compressor. The temperature was controlled by using the Lake Shore Cryotronics 805 Model controller. For the experiments at higher temperatures the samples were placed on a hot plate which provided up to 700 K temperatures and the surface temperatures of the samples were precisely measured by using a J-type thermocouple.

The XRD patterns of the 1 % Nd^{3+} doped Y_2O_3 samples are given in Fig. 11.1. The crystal structure and phase of all samples were analyzed by using these patterns. The cubic phase of the yttria without any other phases was identified by comparing the peak positions and the intensities with those in the Joint Committee on Powder Diffraction Standards (JCPDS) data files. It was clearly seen that the peak positions correspond well to the standard card with number 41-1105.

11.4 Experimental Results

11.4.1 Structural Characterization

Figure 11.2 shows the SEM images of the as-prepared (500 °C) and annealed at 1,000 °C Y_2O_3 samples. The nanopowders with spherical shapes are relatively uniform and have an average diameter of ~30 and ~50 nm in as-prepared (500 °C) and annealed (1,000 °C) samples, respectively. It is clearly seen from the SEM images that the particle sizes of the samples are strongly dependent on the synthesis and annealing temperature and show an increasing tendency with the increasing annealing temperature. Also SEM images confirmed the particle sizes estimated

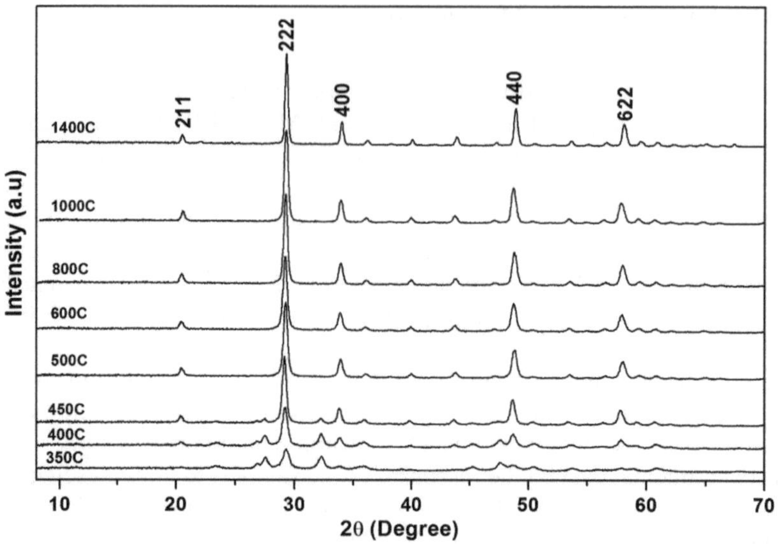

Fig. 11.1 XRD patterns of the 1 % Nd^{3+} doped Y_2O_3 nanopowders synthesized and annealed at different temperatures

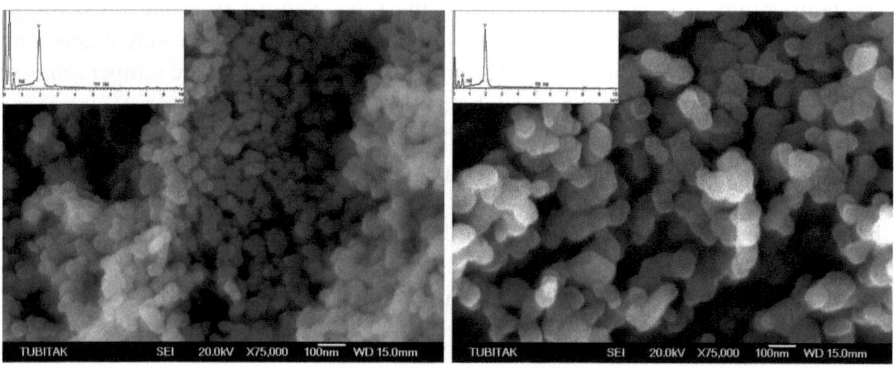

Fig. 11.2 (*Left*) SEM image of as-prepared 1 % Nd^{3+} doped Y_2O_3 nanopowders (*inset*: corresponding EDS spectrum); (*Right*) sample synthesized at 350 °C b) SEM image of 1 % Nd^{3+} doped Y_2O_3 nanopowders annealed at 1,000°C (*inset*: corresponding EDS spectrum)

from XRD measurements by using the Scherrer Equation. The inset of Fig. 11.2 shows the EDS (Energy Dispersive X-Ray Spectroscopy) spectra of the samples.

We also conducted a series of TEM measurements to confirm the particle sizes and the morphologies of the synthesized samples. Representative TEM images are given in Fig. 11.3. It is clearly seen from the micrographs that the particle sizes are in good agreement with the results obtained from both of XRD and SEM measurements.

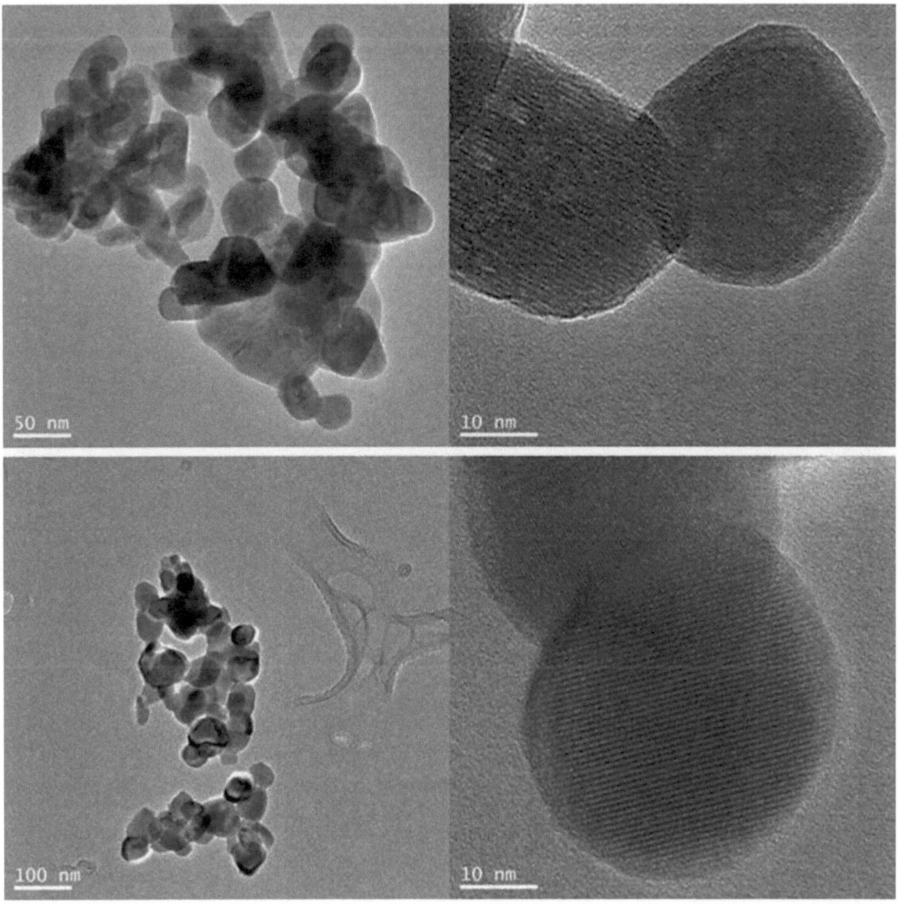

Fig. 11.3 Representative TEM micrographs of 1 % Nd^{3+} doped Y_2O_3. (*Top*) sample synthesized at 350 °C. (*Bottom*) sample annealed at 1,000 °C

11.4.2 Spectroscopic Characterization

The measurements of continuous luminescence spectra were conducted from 35 to 300 K and in the 850–1,150 nm wavelength range. The luminescence spectra of all the samples at 35 and 300 K and the corresponding energy level scheme are given in Figs. 11.4 and 11.5, respectively. The emission spectra of the samples consist of two group of emission peaks due to the ${}^4F_{3/2} \rightarrow {}^4I_{9/2}$ and ${}^4F_{3/2} \rightarrow {}^4I_{11/2}$ transitions. We note that the general aspect of the spectra is the same for all the samples. The intensity of the emission lines increases for all samples with the decreasing temperature; it also increases with the increasing particle sizes. The spectra with greatest intensity are those of the crystal. Examining the spectra in detail we see that

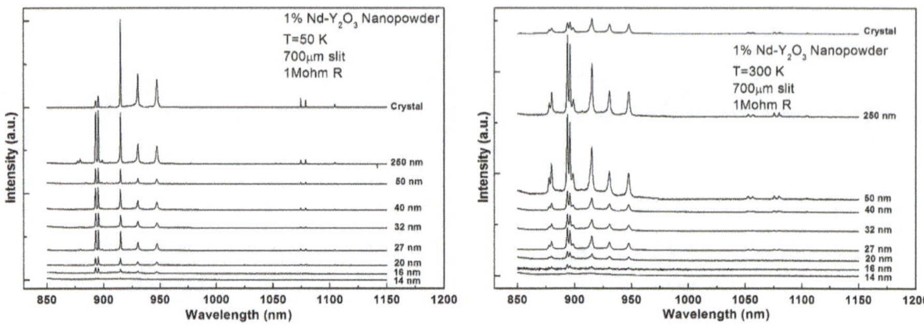

Fig. 11.4 (*Left*) Continuous emission spectra of samples at 35 K. (*Right*) Continuous emission spectra of samples at 300 K

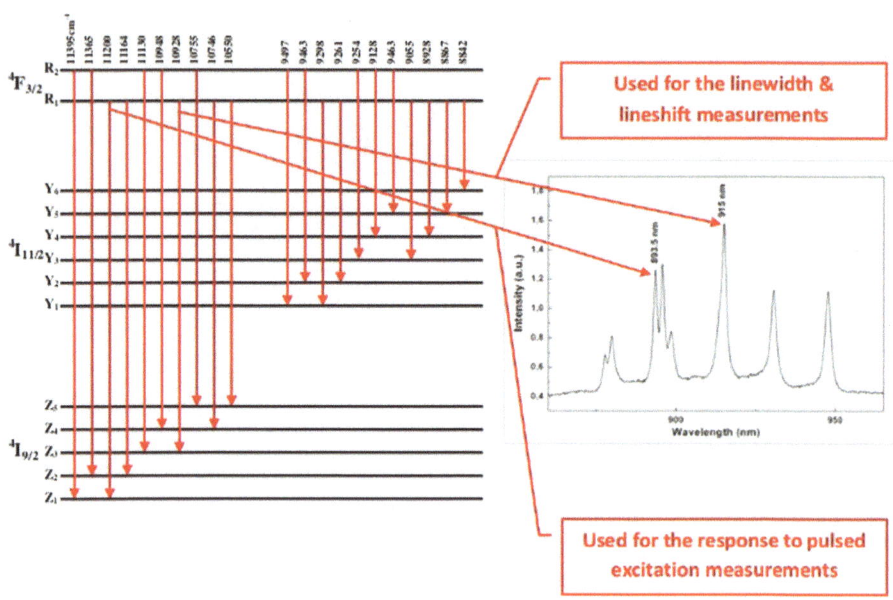

Fig. 11.5 Energy level scheme of Nd^{3+} in Y_2O_3

the smaller is the size of the particles, the wider are the spectral lines and the greater is the shift of each line toward longer wavelengths with respect to the corresponding line in the crystal.

Figure 11.6 presents the shift of a well isolated line at 914.5 nm with respect to the corresponding line in the crystal. This is the line we have chosen in order to study the effects of temperature on the width and position of the spectral lines of the samples. The results of these measurements are reported in Figs. 11.7 and 11.8.

The measurements of the response of the samples to pulsed excitation were carried out in the 35–700 K temperature range. The results of these measurements

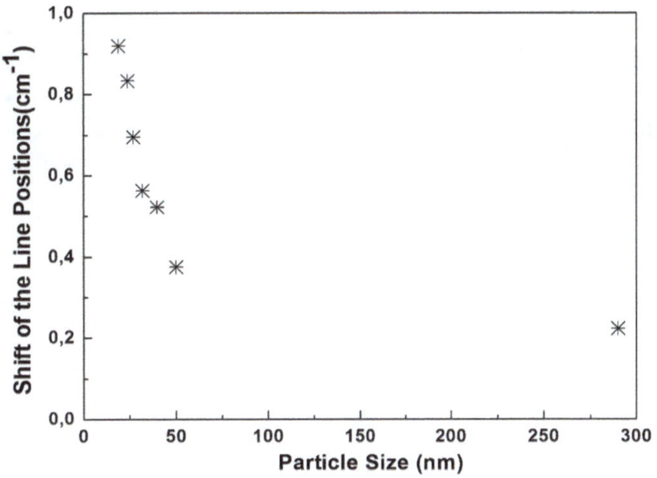

Fig. 11.6 Red-shift of a line position of different samples with respect to the crystal line

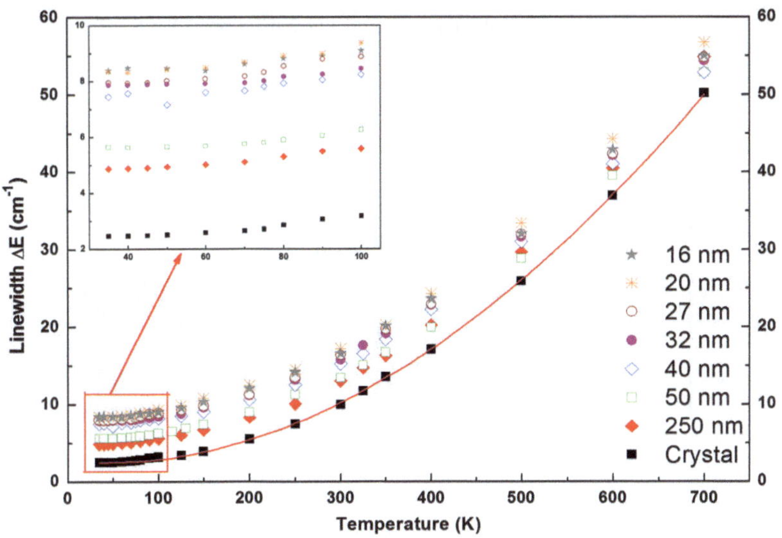

Fig. 11.7 Variation of linewidth as a function of temperature in 1 % Nd^{3+} doped Y_2O_3. The *dots* are experimental results and the *solid curve* is the theoretical fittings with respects to the crystal line

at 33 K are reported in Fig. 11.9. Since the luminescence output of the Nd-doped samples consists of spectral lines related to transitions that initiate in the doublet $^4F_{3/2}$, the response of these systems to pulsed excitation is the same, regardless of the spectral line at which it is measured. We have chosen to measure this response at $\lambda = 892.5$ nm (see Fig. 11.5). We have found that the lifetime of the $^4F_{3/2}$ doublet in the Y_2O_3 crystal the value ~250 μs, almost independent of temperature.

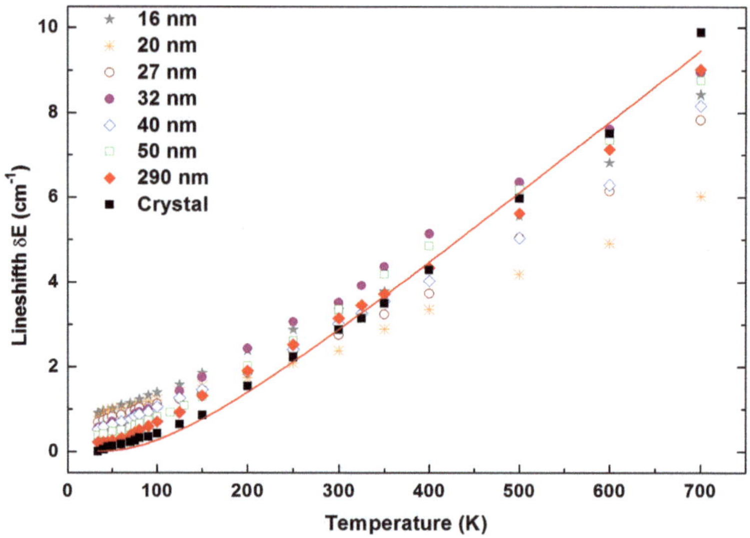

Fig. 11.8 Line-shifts as a function of temperature in 1 % Nd^{3+} doped Y_2O_3. The *solid red line* is the theoretical fitting to the crystal' experimental data

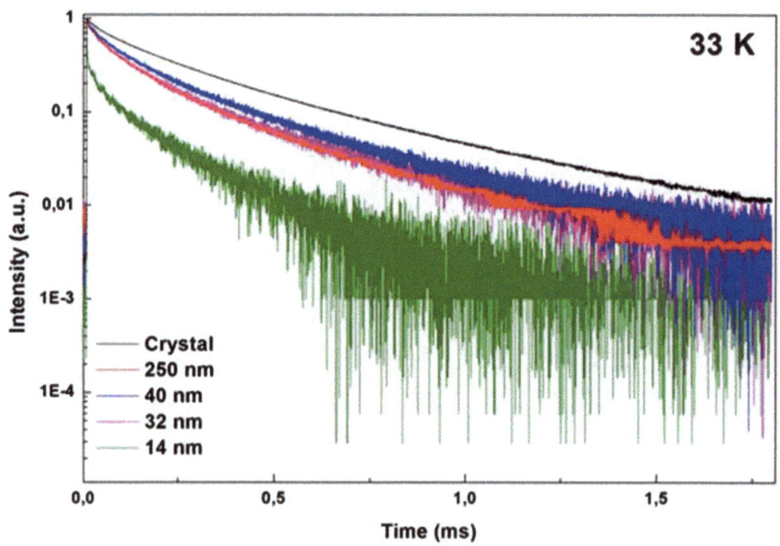

Fig. 11.9 Response to pulsed excitation results of all powders at 33 K

The decay pattern of the crystal is close to an exponential, more so at high temperatures. The response of the powder samples deviates from exponential at early times and then becomes exponential with a decay constant greater than that of the crystal, so that at a time after the initial response the two patterns

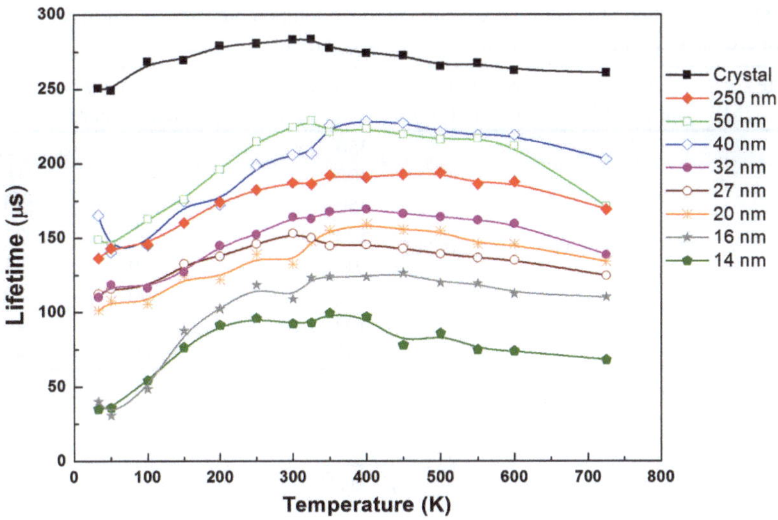

Fig. 11.10 Variation of lifetimes with the temperature

(the pattern of the crystal and that of the sample) cross out. The smaller is the particle and the lower is the temperature the greater is the deviation of the initial pattern from exponentiality and the longer is the time at which the two patterns cross. Figure 11.10 presents the lifetimes of all the samples as function of temperature. When a decay pattern was exponential, the lifetime was calculated from an exponential fit. When a pattern was not exponential, we took as measure of the lifetime the area under the decaying signal, having set the intensity of the initial signal equal to one.

11.5 Discussion of the Results

11.5.1 Structural Properties

The cubic phase of the yttria was determined by using XRD patterns. It was observed that the lattice parameter increases with the addition of dopant ions and found to be as large as 10.6051 Å as a consequence of the incorporation of the Nd^{3+} ions into Y^{3+} sites, due to the larger ionic radius of Nd^{3+} (112.3 pm) with respect to that of Y^{3+} (104 pm) [6, 7]. It was observed that the widths of the diffraction lines broadened with the decreasing synthesis and/or annealing temperature because of the decreasing crystalline sizes. The most narrow lines were observed for the sample annealed at 1,400 °C. The particle sizes of the powders were estimated by using the Scherrer equation [8] reported below

Table 11.1 Formation of yttrium-neodymium alginate solution prepared, synthesis and annealing temperatures and mean particle sizes from Scherrer equation

Alginate (% w/w) (M)	Yttrium nitrate (M)	Neodymium nitrate (°C)	Synthesis temperature (°C)	Annealing temperature	Mean size (nm)
			350	–	13.8
			400	–	16.0
			450	–	20.4
1	99 % of 0.2 M	1 % of 0.2 M	500	–	27.1
			500	600	32.3
			500	800	40.0
			500	1,000	49.7
			500	1,400	248.2

$$L = \frac{K\lambda}{\beta \cos\theta}$$

where L is the crystalline length, K is a constant that varies with the method of taking the breadth ($0.89 < K < 1$), λ is the wavelength of the incident X-ray beam, β is the width of the peak at half maximum intensity of a specific phase (hkl) in radians and θ is the center angle of the peak. The (222) peaks were used to estimate the particles' size. As seen in Fig. 11.1 and our calculations given in Table 11.1 the particle size of the samples increases with increasing annealing temperature. Also the breadth of the XRD peaks decreases with increasing annealing temperature.

The SEM and TEM images confirmed the particle size calculate by using the Scherrer equation. A nebulous structure which we attributed tot he the existence of organic compound (alginate) was seen in some samples' SEM images. This structure disappears with increasing annealing temperature. On the other hand the EDS results confirm the presence of Nd^{3+} content. All of the structural characterization measurement showed that the particle sizes can be effectively controlled by the synthesis and annealing temperature, and increased with increasing treatment temperature.

11.5.2 Spectroscopic Response

The general aspect of the spectral output of the powder samples strongly resembles the spectra of the crystal confirming the fact that the confinement provided by the restrictions of the particles' sizes does not affect by much the energy levels of the Nd^{3+} ion.

The shifts reported in Fig. 11.6 indicate that their magnitude becomes greater the smaller is the dimension of the particle. It seems then legitimate to attribute this behavior to the increasingly important role played by the surface in smaller particles. The Nd ions at the surface are in a different environment with respect to the ions in the interior of the material. In a crystal the ions in the surface are a negligible

Table 11.2 Best fit parameters ($\alpha, \bar{\alpha}, T_D, \Delta E_0$) obtained from the line-shift and line-width measurements

	Line shifts		Line widths		
Particle size	T_D (K)	α (cm^{-1})	ΔE_0 (cm^{-1})	T_D (K)	$\bar{\alpha}$ (cm^{-1})
Crystal	429	22	2.46	440	95
248.2 nm	285	12	4.88	397	80
49.7 nm	263	11	5.65	360	62
40 nm	258	9.1	7.46	305.7	44
32.3 nm	207	8.3	7.88	305.5	46
27.1 nm	352	12	7.96	210	21
20.4 nm	236	5.7	8.36	109	5.9
16 nm	272	9.5	8.40	194	18

minority with respect to the ions in the bulk of the material, but in a particle it is not so. In very small particles going from the surface toward the center of the particle there is gradation in the Nd-O bond lengths, approaching the condition of Nd in a single bulk crystal. It is possible that, even in the interior of the particle, the conditions of the environment in the bulk will never be reached. This may account for the fact that the shift of he lines increases with decreasing of the particles' sizes. The results concerning the thermal line broadening were handled as follows.

The experimental data were compared with the following equation [9]

$$\Delta E(\text{cm}^{-1}) = \Delta E_0 + \Delta E^R \tag{11.1}$$

where

$$\Delta E^R = \bar{\alpha}\left(\frac{T}{T_D}\right)^7 \int_0^{T_D/T} \frac{x^6 e^x}{e^x - 1} \mathrm{d}x$$

In (11.1) ΔE_0 is the temperature independent-residual width due to random crystal strains and spontaneous phonon emission processes and ΔE^R is the temperature-dependent contribution to the width due to the Raman scattering of phonons: $\bar{\alpha}, T_D$ and ΔE_0 are treated as adjustable parameters to get the best fit (a least square fit) to the experimental line-width data shown in Fig. 11.7.

The values of the parameters that were used to produce the fitting of the experimental data with the theoretical formulas are listed in Table 11.2. It is of interest to note that the residual line-width at low temperature increases with decreasing particle size.

The effect of temperature on the position of the chosen sharp line is presented in Fig. 11.8. We determined that the line position is red-shifted with increasing temperature. The experimental data were compared with Eq. (11.2) that expresses the shift as due to the process of emission and absorption of virtual phonons [9]:

$$\delta E(\text{cm}^{-1}) = \delta E^R = \alpha\left(\frac{T}{T_D}\right)^4 \int_0^{T_D/T} \frac{x^3}{e^x - 1} \mathrm{d}x \tag{11.2}$$

where α and T_D are treated as adjustable parameters to get a best fit. The line position at T = 0 K was estimated by extrapolating the experimental data to zero temperature. The corresponding fitting values of α and T_D for the chosen line are listed in Table 11.2.

The fitting values of the coupling coefficients $\bar{\alpha}$ and α obtained for the line-widths and line-shifts are typical for rare-earth ion doped laser crystals. They are usually less than $100\,\text{cm}^{-1}$. In case of transition metal ion the coupling coefficients $\bar{\alpha}$ and α are usually about $500\,\text{cm}^{-1}$ [10, 11].

The decay patterns of the luminescent Nd^{3+} ions embedded in the nanoparticles are expected to deviate from an exponential decay of the same ions in the crystal for two reasons:

1. The role played by the surfaces will be increasingly important as the radius of the particle is reduced, producing a deviation from exponentiality. Molecular groups such as O-H residing on the surface may enhance the decay process due to their stiff vibrations that rob the ion of its excitation energy. We reasonably expect a deviation from exponentiality at the beginning of the decay pattern.
2. The ensemble of particles in the sample differs also from the bulk material for the presence of the interstices between the particles. In what follows we shall describe how we can handle the problem presented by this difference.

The rate of spontaneous relaxation of an atomic system is given by [12]

$$A(\Psi J, \Psi' J') = \frac{64\pi^2(\Delta E)^3}{3h(2J+1)}\eta F^2$$

where F is the matrix element of the electric dipole operator, ΔE the energy gap between the states $|\Psi J\rangle$ and $|\Psi' J'\rangle$, and $\eta = n(n^2+2)^2/9$ is the Lorentz correction for the local field with n, the index of refraction of the host material.

Applying these considerations to the ions embedded in the nanoparticles, we can say that the rate of their spontaneous emission (inverse of the radiative lifetime) depends on an effective index of refraction n_{eff} which consists of a combination of the index of refraction of the nanoparticle n_{NP} and of the index of refraction of the surrounding medium n_{med}:

$$n_{\text{eff}} = \chi n_{NP} + (1-\chi) n_{\text{med}}$$

where χ is the filling factor indicating the fraction of space that is occupied by the nanoparticles. The validity of the usage of n_{eff} in calculating the probability of decay rests on the fact that the size of the particles is much smaller than the wavelength of light [13].

Since the ratio χ is independent of the size of the nanoparticles the tails of the response patterns of the particles with different sizes should be parallel after they cross the pattern corresponding to the crystal response to pulsed excitation.

The responses of the samples to pulsed excitation present the following interesting aspects:

1. At each temperature the decay patterns depend on the size of the particles with increasingly greater deviation from exponentiality for smaller particles (see Fig. 11.9)
2. The patterns that deviate from exponentiality eventually cross the exponential pattern of the crystal. The greater is the deviation the more distant in time is the crossing. A possible explanation of these experimental finding is the existence of large amount of O-H groups at the surface of the each particle which, considering the stiff O-H vibration, could provide a quenching of the luminescence and the shortening of the lifetime.
3. All the decay patterns of the particles of different sizes become parallel after they cross the decay pattern of the ions in the crystal. This behavior indicates that the effective index of refraction is lower for the nanoparticles than for the single crystal, hence decreasing the radiative decay rate, and the filling factor χ is independent of the particles' size.

11.6 Conclusions

This study was focused on the synthesis and characterization of Y_2O_3 nanophosphors doped with 1 % Neodymium (Nd^{3+}) ions. Nanoparticles were synthesized by using the thermal decomposition technique and annealed at different temperatures to control particle sizes. The synthesized particles were structurally characterized by using X-ray diffraction (XRD), scanning electron microscope (SEM) and transmission electron microscope (TEM). The cubic phase of the Y_2O_3 was determined and the size of the particles was found to range from ~15 to 300 nm.

A series of spectroscopic measurements on a number of these nano-particle samples were performed. For comparison we also characterized spectroscopically an Y_2O_3 crystal sample.

For each sample the emission spectrum was accurately measured in a wide range of temperatures ranging from 30 to 300 K. The spectra consisted of the Nd^{3+} ion's narrow spectral lines corresponding to transitions from the $^4F_{3/2}$ doublet level to the $^4I_{11/2}$ and $^4I_{9/2}$ many-folds.

We also measured the width and the position of a well isolated line in the temperature range 30–700 K. The results indicate that the residual width at low temperatures is larger for powders with smaller particle sizes.

Finally measurements were performed on the decay patterns of Nd^{3+} ions following a pulsed excitation. These measurements were conducted in the temperature range of 30–700 K. Such patterns vary considerably with the size of the nanoparticle, so that they may be considered signatures of these sizes. When comparing the decay patterns of the powders with the exponential decays of the $Y_2O_3 : Nd^{3+}$ crystal we found larger deviations from exponentiality for the powders of smaller particles' sizes, and smaller deviations from exponentiality at high temperatures.

Acknowledgements One of the authors (Gokhan Bilir) would like to thank the Scientific and Technological Research Council of Turkey (TUBITAK) for its support. All the authors would like to thank Dr. Dezhi Wang of Boston College for his support with the XRD and TEM measurements.

References

1. Koechner W (2006) Solid state laser engineering, 6th edn. Springer, Berlin, p 57
2. Moncorgé R (2005). In: Liu G, Jacquier B (eds) Spectroscopic properties of rare earths in optical materials. Springer, Berlin/Heidelberg/New York, p 333
3. Fouassier C (2005). In: Bruce King R (ed) Encyclopedia of inorganic chemistry, 2nd edn. Wiley, Chichester
4. Baskoutas S, GiaBouranis P, Yannopoulos SN, Dracopoulos V, Toth L, Chrissanthopoulos A, Bouropoulos N (2007) Thin Solid Films 515:8461
5. Gomboltz WR, Wee SF (1998) Adv Drug Deliv Rev 31:267
6. Jia YQ (1991) J Solid State Chem 95:184–187
7. Srinivasan R, Yogamalar NR, Elanchezhiyan J, Joseyphus RJ, Bose AC (2010) J Alloys Compd 496:472
8. Scherrer P (1918) Gött Nachr Gesell 2:98
9. Di Bartolo B (2010) Optical interactions in solids, 2nd edn. World Scientific, Singapore, p 385
10. Karpick JT, Di Bartolo B (1972) Il Nuovo Cim 7B:62
11. Chen X, Di Bartolo B, Barnes N, Walsh B (2004) Phys Status Solidi (b) 241:1957
12. Reid MF (2000). In: Newman DJ, Ng B (eds) Crystal field handbook. Cambridge University Press, Cambridge, p 190
13. Liu G, Chen X (2007). In: Gschneidner KA Jr, Bünzli J-CG, Pecharsky VK (eds) Handbook on the physics and chemistry of rare earths, vol 37. Elsevier/North Holland, Amsterdam, p 90. Chapter 233

12
Incandescent Lamp-Like White-Light Emission from Doped and Undoped Oxide Nanopowders

Maura Cesaria, Gokhan Bilir, Gönül Özen, Marco Bettinelli, John Collins, Fabio Piccinelli, and Baldassare Di Bartolo

Abstract We report the production of a broad band emission (ranging from 400 to 900 nm) following the monochromatic infrared light (803.5 nm) continuous wave excitation of either nominally un-doped or Nd-doped up to 20 % yttrium oxide (Y_2O_3) nanopowders, $Y_3Al_5O_{12}$ (YAG) and Cr^{3+} doped $Gd_3Ga_5O_{12}$ (GGG) nanocrystallites. Our experimental results indicate that such emission feature is (i) a nano-scale phenomenon, (ii) demands a threshold pumping power, (iii) cannot be ascribed to an overlap of sharp emission bands in the un-doped case and, (iv) even if assisted by the dopant presence, is a host matrix-related process. In the case of the Y_2O_3-based samples, we demonstrate the possibility to obtain "warm" white light with high efficiency and color rendering index approaching the theoretical limit with an alternative approach. Our experimental results make our white light emission very interesting at both fundamental and applicative levels and may open the way to an alternative route with respect to incandescent lamps.

M. Cesaria
Department of Mathematics and Physics "E. De Giorgi", Universitá del Salento, Lecce, Italy
e-mail: maura.cesaria@le.infn.it

G. Bilir
Department of Physics, Istanbul Technical University, Istanbul, Turkey

Department of Physics, Boston College, Chestnut Hill, MA, USA

G. Özen
Department of Physics, Istanbul Technical University, Istanbul, Turkey

M. Bettinelli • F. Piccinelli
Luminescent Materials Laboratory, Universitdi Verona, Verona, Italy

J. Collins
Department of Physics, Wheaton College, Norton, MA, USA
e-mail: jcollins@wheatonma.edu

B. Di Bartolo (✉)
Department of Physics, Boston College, Chestnut Hill, MA, USA
e-mail: dibartob@bc.edu

B. Di Bartolo et al. (eds.), *Nano-Structures for Optics and Photonics*,
NATO Science for Peace and Security Series B: Physics and Biophysics,
DOI 10.1007/978-94-017-9133-5_12, © Springer Science+Business Media Dordrecht 2015

12.1 Introduction

Since the revolutionary introduction of the incandescent bulb by Thomas A. Edison in 1880 [1], the most important challenges to be met by science and technology in the field of artificial lighting for indoors and outdoors are the development of sources yielding high luminous efficiency, low power consumption, reliability, durability, fast switching, thermal stability, environmental friendliness and, very importantly, matching the light quality of incandescent lamps. Even if up to now many strategies and artificial lighting sources (neon lights, discharge lamps and, very recently, semiconductor-based devices) have been developed, an "ideal" lighting source has not been found. Notably, while incandescent lamps suffer from low luminous efficiency (roughly 5 % of the delivered electric power is converted into light), they are still unmatched with respect to their color quality, meaning their inherent resembling the sun light which is the most comfortable for human perception. To design functional lighting devices a number of problems must be overcome accounting for the industry's request indicating the indoor lighting quality based on a few basic parameters defined by assuming Planckian radiators as reference light source [2–4]: the CIE (International Commission on Illumination) coordinates, the luminous efficiency (determined by the conversion efficiency from electrical-to-optical power and the luminous efficacy (conversion factor from optical power to luminous flux)), the correlated color temperature (CCT) (the chromaticity of a spectrum off of the Planckian locus in the CIE chromaticity diagram) and the color rendering index (CRI). One important CIE Standard Illuminant is the so-called "warm white point", having CIE coordinates (0.448, 0.408) and CCT of 2,856 K, that marks the chromaticity of tungsten incandescent lamps, namely the artificial light sources most comfortable for human eye.

Power efficiency at least of 90 $\mathrm{lmW^{-1}}$, CCT of less than 4.000 K (lamps with CCT rating below 3,200 K are usually considered "warm" sources, while those with a CCT above 4,000 K are usually considered "cool" in appearance) and a minimal value of 80 for CRI are requested for indoor lighting to replicate natural light. As a matter of interest, the nomenclature "warm" and "cool" source refers to the overall color tone of a white light source based on the way color groups are perceived by human eyes. While colors from the violet/blue end of the spectrum are seen as slightly bluish, meaning as "cool", colors of the red/orange/yellow side of the electromagnetic spectrum have "warm" appearance.

Accounting for performance with respect to the internationally agreed-upon metric, energy efficiency has been the primary driving force to replace wasteful incandescent lamps by solid state lighting sources such as light emitting diodes (LEDs). Such devices dominate many technological fields (optoelectronics, automobile light, traffic light, street lighting, landscape decoration, backlight of liquid crystal display (LCD) for TV sets, computers and mobile telephones, et al.) due to their being cheap, durable, stable, highly efficient, usable in integrated electronics, long-lived, mercury-free and energy-saving illumination sources.

Apart inorganic materials, the LED working principle has been recently applied to organic films made of small molecular-weight molecules and conducting polymers, leading to devices briefly termed OLEDs (Organic LEDs) and PLEDs (Polymer LEDs), respectively [5, 6]. Presently, whereas OLEDs and PLEDs are widely investigated due to low-cost fabrication and hold great promise [7–16], remaining technological challenges are the knowledge of the parameters controlling and limiting organic electroluminescence, large area fabrication, and the device lifetimes for deep-blue devices [7]. While OLED-active layers can be fabricated based on high-vacuum vapor deposition techniques, PLEDs can be assembled completely by solution-based processes. In any case, balanced charge carrier injection and charge carrier transport leading to efficient emission demand multilayer device structures. In this respect, there is a serious problem mainly for PLEDs. In fact, the solution processing of different layers needs carefully selection of solvents and materials to prevent a polymer layer to be re-dissolved when covered by an other polymer layer. Therefore, presently conventional-LED technology is more mature and competing for developing efficient WL sources.

Being LED an inherently single-color emitter several strategies have been developed to obtain a white LED (WLED) [17, 18]. Based on the basic recipe that white light (WL) emission can result by mixing two complementary colors or three primary colors, mainly the LED technology has been combined with phosphorescent emitters. Such materials, also termed phosphors (PSPs), are mostly rare-earth (RE)-doped inorganic crystalline materials and exhibit up-conversion (UP) behavior (UC materials convert lower energy near-IR radiations into higher energy visible radiations). RE elements are attractive because of their having many energy levels and emitting narrow lines. Moreover, the intensity of the UC spontaneous emission may be controlled by the RE-dopant concentration.

Most of the methods for producing WL by UC-processes exploit a combination of several $4f^n - 4f^n$ $(0 < n < 14)$ transitions of RE ions. In this approach a suitable near-infrared (near-IR) source, delivering high power (above a certain threshold), is used to get visible emission by exciting nanocrystalline powders of a transparent host material doped with lanthanide ions [19–25]. The quantum efficiency depends upon the phonon frequency of the host lattice (low cut-off phonon frequency is favorite because it makes inefficient the nonradiative relaxation of excited states), [26–28].

Such nano-PSPs-based lighting sources suffer the drawback that the luminescence spectra do not cover the whole visible-light range [19,20,22–25], thus leading to luminescence spectra not optimal for indoor lighting or displaying [29]. The outlined scenario shows that finding the optimal converting PSP-system (meaning the one able to produce warm WL with high CRI and high efficiency) still represents a challenge.

Therefore, even if due to several advantages (long operational lifetime, compactness, lower energy-wasting, good material stability, high luminous efficiency and environmental protection) WLEDs are widely studied and are believed to be the “green lighting source of the 21st century” with wide applications and prospective market [30–37], the laboratory prototypes do not match the “ideal” WL for indoor lighting. Moreover, the recent rising cost of the RE elements and the lossy process for expanding a narrow wavelength source into a wider spectrum by PSPs strongly demand an alternative to be found beyond REs. In this respect, being able to obtain suitable WL emission by an undoped matrix material and/or a host material doped with transition metal(s) rather than RE-PSPs would bypass the critical issues of color balance, thermal stability, inherent loss mechanisms of photon Stokes-shifting between optical excitation and emission wavelengths involved in RE-PSPs. We have investigated such an idea and found that a bright broadband (from ~400 to ~900 nm) WL emission can be produced upon monochromatic IR-excitation of nanopowders of the following materials: yttrium oxide (Y_2O_3) singly doped with Nd ions, nominally un-doped Y_2O_3, nominally un-doped $Y_3Al_5O_{12}$ (YAG) and $Gd_3Ga_5O_{12}$ (GGG) singly doped with Cr ions.

In the literature, it is reported: (i) anti-Stokes wide band emission by exciting nano-crystals of compounds in which Nd was one of the dopants [38] or a full stoichiometric part of the system [24, 39, 40]; (ii) bright up-converted WL emission from $LiNdP_4O_{12}$ [38,39], $NdAlO_3$ [24] and $LiYbP_4O_{12}$ [23] nano-crystals under low pressure and infrared (IR) excitation; (iii) production of WL by using just lanthanide oxides (Yb_2O_3, Sm_2O_3, CeO_2) [19] and $Yb_3Al_5O_{12}$, $(Yb,Y)_2O_3$ crystals [20], with the broadband emission not related to $4f^n - 4f^n$ transitions, and (iv) broadband emission from SiC nano-particles and mechanically milled Si [41,42].

Notably, the originality of our results resides in our being able to observe broadband visible emission: (i) from several nominally undoped oxide materials, while previously reported visible broadband production was yielded by materials at least doubly doped with different RE-ions or including them as stoichiometric components, and (ii) for the first time, when the dopant is a transition metal ion (Cr in our case).

This report summarizes the experimental results of our studies with a focus on the experimental findings leading to questions to be addressed at either fundamental or experimental level. Even if the inherent origin of the observed phenomenon is still a challenge to our knowledge, our experimental findings introduce several basically important points of discussion.

12.2 Experimental

12.2.1 Choice of Materials

Y_2O_3 is an excellent host matrix for RE-dopants in optical applications [43–47] due to its high melting point (2,380 °C), optical isotropy, chemical and thermal stability, wide transparency range (from 0.2 to 8–9 μm) [48–50] wide band gap (5.6 eV), high refractive index (1.8), its Y^{3+} ion having similar ionic radius with respect to RE ions [43,46,51], relatively low cut-off phonon energy (380 cm^{-1}) [52], and low luminescence quenching rate. Particularly, the Nd ion, besides being very efficient for solid-state laser applications, is a very good candidate to up-convert IR-light to visible one [53, 54]. Absorption of a self-trapped exciton (5.88 eV at 300 K) is reported in literature for Y_2O_3 and the Stokes shift of such optical features is discussed in term of an increase of the lattice relaxation energy [55]. A broad band luminescence spectrum with a maximum near 2.3 eV and consisting of a number of overlapping bands with their maxima near 3.40, 2.90, 2.60, 2.35, 2.10, 1.90, and 1.70 eV is reported for Y_2O_3 thick films either in vacuum or oxygen atmosphere but on excitation in the forbidden band (5.6 eV) [56]. In literature, no UC emission by Y_2O_3 is reported in vacuum, being such finding ascribed to absence of metastable initial or intermediate levels in the IR or visible spectral regions to act as UC steps for this compound.

Notably, even if Y_2O_3 is of interest either at experimental and theoretical level, it is mainly investigated upon doping/co-doping with RE ions for applicative reasons. This caused the direction of research focused mainly on nature of luminescence and mechanisms of the excitation energy transfer from lattice to luminescence centers (dopants).

The garnets YAG and GGG are well known in nanocrystalline form [57] and are among the most widely used laser host materials [58–65] because of their hardness, general stability against chemical and mechanical changes, optical isotropy, good thermal conductivity and low-thermal expansion, high-optical transparency, high threshold for optical damage, high receptiveness to RE- ions. While YAG has slightly better thermal and optical properties than GGG, however GGG is available more easily and in better crystalline quality than YAG.

In regard to the luminescence of undoped YAG crystals, most reports focus on the UV emission and identification of the main luminescence in the YAG structure and the main defect center competing with Ce^{3+}ions in Ce doped YAG scintillators. However, upon excitation at 270 nm, multiple strong luminescence peaks have been observed in undoped YAG single crystals, with peaks at 700 and 800 nm ascribed to low level of iron impurities [66].

12.2.2 Sample Synthesis and Characterization Methods

Dealing with the Nd-doped Y_2O_3 (Y_2O_3:Nd) and nominally un-doped Y_2O_3 samples, nano-sized Y_2O_3 particles doped with Nd^{3+}-concentration of 2, 5, 10 and 20 % were prepared by thermal decomposition of yttrium-neodymium alginate. The synthesis temperature was set to 500 °C and annealing treatments of the as-synthesized samples at 600, 800, 1,000 and 1,400 °C made the NP-size to increase from 20 to 248 nm.

Turning to the garnet-based nanopowders, the co-precipitation approach [67, 68] and the Pechini method [69, 70] were exploited to produce the samples termed $(YAG)_{copr}$ and $(GGG{:}Cr)_{Pech}$ according to the nomenclature $(M)_{synth}$ where M is the material (YAG and Cr-doped GGG) and the subscript indicates the synthesis approach.

The continuous emission spectra were collected by pumping the samples with the output of a Laser Drive Inc. Model LDI-820 laser diode operating at 803.5 nm. The luminescence signal was directed toward the entrance slit of a 1 m McPherson Model 2051 monochromator and chopped at a frequency of 250 Hz before entering the slit. The monochromator provided a resolution 0.8 Åwith the slits set at 50 μm and a wavelength reproducibility of 0.1 Å. The optical signal was detected by Hamamatsu R1387, sent to a EG&G Model 5210 lock-in amplifier and recorded in a computer. The maximum pumping power delivered by the above exciting source was 3 W.

An Allied Scientific Pro ASP-MK350 Model Illuminance Meter was used to measure the CIE coordinates, the power efficiency and the CRI of our WL-emitting samples.

12.3 Experimental Results

In regard to the Y_2O_3-based samples, a broadband emission (from 400 to 900 nm) was observed for content of Nd-dopant ranging from 0 (nominally un-doped Y_2O_3) to 20 % (sample Y_2O_3: Nd (20 %)) with CIE coordinates (0.45, 0.40) and (0.47, 0.39), respectively. Since the CIE coordinates of our samples are close to the ones of the "warm white point", in the following the observed broadband emission will be termed "WL emission". Moreover, high power efficiency (864 versus 327 lx/W for the un-doped and doped sample, respectively) and CRI very close to the maximum value (100) were estimated. The detected WL emission was inspected depending on nature and pressure of the background gas as well as temperature. In addition, it was found that while the WL emission can be easily obtained under low-pressure conditions, under atmospheric pressure it can be detected only by increasing the pumping power above a certain threshold and depends on the nanopowder sizes. Even if Nd-dopant makes easier to induce WL emission as compared to the un-doped case, however the Nd presence is not decisive in causing WL emission. A deeper understanding of the emission mechanism needs to be gained that accounts for interplay/coupling between the manifold of Nd-levels and the host matrix dispersion.

Spectra of the doped and undoped case differ slightly in the spectral shape and due to the presence of three dips ascribed to strong absorption of Nd ions. Importantly, our experiments strongly indicate that the observed WL emission cannot be attributed to Nd but is inherently related to the host oxide Y_2O_3. Definitively, even if the mechanisms leading to the WL emission by our undoped and Nd-doped Y_2O_3 samples are still unknown and need understanding, we have obtained performance advantages (emission spectra strongly similar to incandescent lamps) with respect to the presently available WL sources trying to mimic incandescent lamps.

Turning to the $(YAG)_{copr}$ and $(GGG{:}Cr)_{Pech}$ samples, they were found to be both single-phase with average particle size of 55 and 78 nm, respectively. The production of a broadband visible emission was found to be a threshold process as a function of the pumping power (0.20 W for $(GGG{:}Cr)_{Pech}$ and 1.05 W for $(YAG)_{copr}$) and the onset of a broad emission band (extending from about 400 to 900 nm) was observed in the visible and NIR regions, peaking in the anti-Stokes side of the exciting laser line and with analogous features for both materials.

The measured CIE coordinates ((0.55, 0.42) for $(GGG{:}Cr)_{Pech}$ and (0.53, 0.44) for $(YAG)_{copr}$) lie in the greenish yellow region of the CIE diagram.

Dealing with the origin of the observed broadband emission, we rule out the contribution of low level of iron impurities. In fact, even if Cr^{3+}, Cu^{2+}, Fe^{3+} and other transition metal ions are common metallic impurities in nominally pure compounds containing Al^{3+} [71,72], the similar behavior of the broadband emission spectra of our $(YAG)_{copr}$ and $(GGG{:}Cr)_{Pech}$ samples, the non-decisive Cr-role and the absence of Fe impurities in precursors exploited for synthesizing $(GGG{:}Cr)_{Pech}$, suggest a different mechanism. On the other hand, we have found no report showing visible broadband emission by an oxide doped with a transition metal ion (Cr in our case). Therefore, our experimental findings suggest that the detected visible broadband emission cannot be uniquely ascribed to the Cr presence and Fe impurities, that is the general emission mechanism may be also host-dependent and likely the same in both materials (YAG and GGG).

Definitively, while visible broadband emission is reported in the literature by materials at least doubly doped with different RE-ions or including them as stoichiometric components, our study demonstrates it can be obtained by nominally undoped oxide materials and, for the first time, by transition-metal-doped insulating oxide materials.

Moreover, in our experiments, we observed partial melting of our samples after their exposure to the laser beam and such effect was the more evident the smaller the nanopowder size was.

12.4 Concluding Remarks: Unsolved Questions and Future Perspectives

Several problems demand a solution and require further investigations in order to be able to make the observed phenomenon reproducible, fully explainable and useful, as a long term goal, for device applications. Understanding the inherent physical process leading to the broadband emission by our samples can enable deeper understanding of the interplay among exciting source, excitation regime (linear or non-linear) and material dispersion (both electronic and phonon density of states). Such conscious knowledge is the key to suggest strategies to find new materials beyond the specific systems we have investigated. Moreover, at the experimental level, being able to produce warm WL in a highly controllable way, with high CRI (approaching to the maximum value 100!) and power efficiency, can also impact on industry and lightening market.

In regard to the laser-induced observed melting-effects, such evidence can be accounted for warming up due to the exciting source. Being the melting temperature of the used host materials quite high (for example, 2,698 °C for Y_2O_3), so high sample temperature can be ruled out and experimental findings can be explained by invoking the modified thermal transport [73, 74] and lowered melting temperature [75–77] induced by the nanoscale regime.

All of this also confirms that the NP-sample temperature differ from the color temperature estimated by the black-body spectrum method applied to the sample emissivity. The occurrence of quantum size effects makes the Plank's black-body model unable to describe any WL-emission spectrum [78, 79], implying that the CT calculated by the Stefan-Boltzmann law (i.e. by using Planckian radiators as typical reference light) is not the sample temperature [80]. The sample temperature estimated by the theoretical blackbody emissive power spectrum may differ from the one estimated by the emissive power spectrum of nanomaterials because the two spectra may have different spectra profile even if the total integrated power is the same [81].

Definitively, the basic issues suggested by our experimental findings to gain insight in the observed phenomena are listed in the following:

(i) the processes by which the exciting light is absorbed by the nanopowder in the absence of a dopant ion and the processes leading to the detected WL-emission, meaning its inherent nature;
(ii) the WL-emission dependence on centers/defects and their coupling/interplay with the host material energy levels, meaning the importance of using high purity rather than nominally un-doped precursor materials and fabrication techniques getting high control of crystallinity and purity;
(iii) the role played by exciting wavelength and pumping power;
(iv) the dopant-role as well as the different roles of transition metal ions and RE ions in assisting the broadband-emission by the host crystal;
(v) the dependence of the broadband emission on nanopowder size, size dispersion and packing degree;
(vi) the temporal evolution of the broadband-emission intensity (emission rise-time and decay-time);
(vii) the influence of particle size and pumping power on the temporal evolution of the WL emission
(viii) the general class of materials or other materials yielding "ideal" WL-emission for future device applications.

12.5 Summary

Our recent experiments have shown that a bright broadband (from ~400 to ~900 nm) warm white light (WL) can be produced upon monochromatic IR-excitation of nanopowders of the following materials: Y_2O_3 singly doped with Nd ions, nominally un-doped Y_2O_3, nominally un-doped YAG and GGG singly doped with Cr ions. The originality of these results resides in detecting WL emission by several nominally undoped oxide materials, while previously reported WL production was yielded by materials at least doubly doped with different RE ions or including them as stoichiometric components. Fundamental processes leading to WL emission and their relationship with the choice of materials and experimental parameters (exciting wavelength, pumping power, nanopowder size and purity, background pressure, temperature) are the challenging task of our investigation.

In this article we described and discussed the existing scenario about WL lighting materials, the strategies developed to design ideal lighting sources and corresponding unsolved problems, our choice of materials, the present state of progress of our experiments, the unsolved basic questions of our research activity and the key research issues to be investigated.

References

1. Edison T (1880) Electric lamp. US Patent 223,898
2. The IES Nomenclature Committee and American National Standards Institute (2010) Nomenclature and definitions for illuminating engineering. Illuminating Engineering Society of North America
3. Hunt RW (1991) Measuring colour. Ellis Horwood, Hemel Hempsted, pp 38–109
4. Central Bureau Of the CIE, Vienna, Austria (1995) Method of measuring and specifying colour rendering properties of light sources, CIE Publication 13.3-1995. Color Res Appl 20:212–212
5. Tang CW, VanSlyke SA (1987) Organic electroluminescent diodes. Appl Phys Lett 51(12):913
6. Burroughes JH, Bradley DDC, Brown AR, Marks RN, Mackay K, Friend RH, Burns PL, Holmes AB (1990) Light-emitting diodes based on conjugated polymers. Nature 347:539–541
7. Zhu M, Yang C (2013) Blue fluorescent emitters: design tactics and applications in organic light-emitting diodes. Chem Soc Rev 42:4963–4976
8. Kido J, Kimura M, Nagai K (1995) Multilayer white light-emitting organic electroluminescent device. Science 267(5202):1332–1334
9. Reineke S, Lindner F, Schwartz G, Seidler N, Walzer K, Lussem B, Leo K (2009) White organic light-emitting diodes with fluorescent tube efficiency. Nature 459:234–238
10. Reineke S, Thomschke M, Lüssem B, Leo K (2013) White organic light-emitting diodes: status and perspective. Rev Mod Phys 85:1245–1293
11. Zhou G, Wong W-Y, Suo S (2010) Recent progress and current challenges in phosphorescent white organic light-emitting diodes (WOLEDs). J Photochem Photobiol C Photochem Rev 11(4):133–156
12. Kawamura Y, Goushi K, Brooks J, Brown JJ, Sasabe H, Adachi C (2005) 100% phosphorescence quantum efficiency of Ir(III) complexes in organic semiconductor films. Appl Phys Lett 86(7):071104
13. Baldo MA, O'Brien DF, You Y, Shoustikov A, Sibley S, Thompson ME, Forrest SR (1998) Highly efficient phosphorescent emission from organic electroluminescent devices. Nature 395:151–154
14. Holder E, Langeveld BMW, Schubert US (2005) New trends in the use of transition metal-ligand complexes for applications in electroluminescent devices. Adv Mater 17:1109–1121
15. Wong W-Y, Ho C-L (2009) Functional metallophosphors for effective charge carrier injection/transport: new robust oled materials with emerging applications. J Mater Chem 19: 4457–4482

16. Jeong HS, Kim SH, Lee KS, Jeong JM, Yoo TW, Kwon MS, Yoo KH, Kim TW (2013) Optical properties of white organic light-emitting devices fabricated utilizing a mixed $CaAl_{12}O_{19}$:Mn^{4+} and $Y_3Al_5O_{12}$:Ce^{3+} color conversion layer. J Nanosci Nanotechnol 13(6):4394–4397. 2013-06-01T00:00:00
17. Mottier P (2010) LED for lighting applications. Wiley
18. Lin CC, Liu R-S (2011) Advances in phosphors for light-emitting diodes. J Phys Chem Lett 2(11):1268–1277
19. Wang J, Tanner PA (2010) Upconversion for white light generation by a single compound. J Am Chem Soc 132(3):947–949. PMID: 2002(5211)
20. Wang J, Hao JH, Tanner PA (2010) Luminous and tunable white-light upconversion for yag (yb3al5o12) and (yb,y)2o3 nanopowders. Opt Lett 35:3922–3924
21. Mahalingam V, Mangiarini F, Vetrone F, Venkatramu V, Bettinelli M, Speghini A, Capobianco JA (2008) Bright white upconversion emission from Tm^{3+}/Yb^{3+}/Er^{3+}-doped $Lu_3Ga_5O_{12}$ nanocrystals. J Phys Chem C 112(46):17745–17749
22. Redmond S (2004) Multiple scattering and nonlinear thermal emission of Yb^{3+}, Er^{3+}:Y_2O_3 nanopowders. J Appl Phys 95(8):4069
23. Strek W, Marciniak L, Bednarkiewicz A, Lukowiak A, Wiglusz R, Hreniak D (2011) White emission of lithium ytterbium tetraphosphate nanocrystals. Opt Express 19:14083–92
24. Strek W, Marciniak L, Hreniak D, Lukowiak A (2012) Anti-Stokes bright yellowish emission of NdAlO3 nanocrystals. J Appl Phys 111(2):024305
25. Verma R, Rai S (2013) Continuum emission in Nd^{3+}/Yb^{3+} co-doped $Ca_{12}Al_{14}O_{33}$ phosphor: charge transfer state luminescence versus induced optical heating. Chem Phys Lett 559(0):71–75
26. Cao C, Qin W, Zhang J, Wang Y, Wang G, Wei G, Zhu P, Wang L, Jin L (2008) Up-conversion white light of Tm^{3+}/Er^{3+}/Yb^{3+} tri-doped CaF_2 phosphors. Opt Commun 281(6):1716–1719
27. Yang LW, Han HL, Zhang YY, Zhong JX (2009) White emission by frequency up-conversion in Yb^{3+}-Ho^{3+}-Tm^{3+} triply doped hexagonal nayf4 nanorods. J Phys Chem C 113(44):18995–18999
28. Kaiser W, Garrett CGB (1961) Two-photon excitation in CaF_2:Eu^{2+}. Phys Rev Lett 7:229–231
29. Pimputkar S, Speck JS, DenBaars SP, Nakamura S (2009) Prospects for LED lighting. Nat Photonics 3:180–182
30. Hashimoto T, Wu F, Speck JS, Nakamura S (2007) A GaN bulk crystal with improved structural quality grown by the ammonothermal method. Nat Mater 6:568–571
31. Tu C-C, Zhang Q, Lin LY, Cao G (2011) Brightly photoluminescent phosphor materials based on silicon quantum dots with oxide shell passivation. Opt Express 20:A69
32. Peng M, Wondraczek L (2010) Photoluminescence of Sr(2)P(2)O(7):Bi(2+) as a red phosphor for additive light generation. Opt Lett 35:2544–6
33. Kim J-P, Jang M-S, Kim W-H, Joo J-Y, Cho J-H, Kim D-W, Song S-B (2012) Improvement in the color uniformity of {LED} by microspheres generated from phase separation. Opt Mater 34(9):1614–1617
34. Bredol M, Kynast U, Ronda C (1991) Designing luminescent materials. Adv Mater 3:361–367
35. Zhang X, Zhang J, Huang J, Tang X, Gong M (2010) Synthesis and luminescence of eu2+-doped alkaline-earth apatites for application in white {LED}. J Lumin 130(4):554–559
36. Lei Z, Xia G, Ting L, Xiaoling G, Ming LQ, Guangdi S (2007) Color rendering and luminous efficacy of trichromatic and tetrachromatic led-based white {LEDs}. Microelectron J 38(1):1–6
37. Graves H, Ticleanu C (2011) LED lighting: a review of the current market and future developments. Building Research Establishment
38. Marciniak L, Strek W, Bednarkiewicz A, Lukowiak A, Hreniak D (2011) Bright upconversion emission of Nd^{3+} in $LiLa_{1-x}Nd_xP_4O_{12}$ nanocrystalline powders. Opt Mater 33:1492–1494
39. Strek W, Marciniak L, Bednarkiewicz A, Lukowiak A, Hreniak D, Wiglusz R (2011) The effect of pumping power on fluorescence behavior of $LiNdP_4O_{12}$ nanocrystals. Opt Mater 33:1097–1101
40. Atabaev T, Hwang Y-H, Kim H-K (2012) Color-tunable properties of Eu^{3+}- and Dy^{3+}-codoped Y_2O_3 phosphor particles. Nanoscale Res Lett 7(1):1–7

41. Roura P, Costa J (2002) Radiative thermal emission from silicon nanoparticles: a reversed story from quantum to classical theory. Eur J Phys 23(2):191
42. Roura P, Costa J, López-de Miguel M, Garrido B, Fort J, Morante J, Bertran E (1998) Black-body emission from nanostructured materials. J Lumin 80:519–522
43. Atabaev T, Hwang Y-H, Kim H-K (2012) Color-tunable properties of Eu^{3+}- and Dy^{3+}-codoped Y_2O_3 phosphor particles. Nanoscale Res Lett 7(1):1–7 (2012)
44. Qin X, Yokomori T, Ju Y (2007) Flame synthesis and characterization of rare-earth (Er^{3+}, Ho^{3+}, and Tm^{3+}) doped upconversion nanophosphors. Appl Phys Lett 90(7):073104
45. Tu D, Liang Y, Liu R, Li D (2011) Eu/Tb ions co-doped white light luminescence Y2O3 phosphors. J Lumin 131:2569–2573
46. Atabaev TS, Vu HHT, Piao Z, Hwang Y-H, Kim H-K (2012) Tailoring the luminescent properties of Gd_2O_3:Tb^{3+} phosphor particles by codoping with Al^{3+} ions. J Alloy Compd 541(0):263–268
47. Flores-Gonzalez M, Ledoux G, Roux S, Lebbou K, Perriat P, Tillement O (2005) Preparing nanometer scaled Tb-doped Y_2O_3 luminescent powders by the polyol method. J Solid State Chem 178:989–997
48. Wickersheim KA, Lefever RA (1961) Infrared transmittance of crystalline yttrium oxide and related compounds. J Opt Soc Am 51:1147
49. Nigara Y (1968) Measurement of the optical constants of yttrium oxide. Jpn J Appl Phys 7(4):404–408
50. Tomiki T, Tamashiro J, Tanahara Y, Yamada A, Fukutani H, Miyahara T, Kato H, Shin S, Ishigame M (1986) Optical spectra of Y_2O_3 single crystals in vuv. J Phys Soc Jpn 55(12): 4543–4549
51. Atabaev TS, Lee JH, Han D-W, Hwang Y-H, Kim H-K (2012) Cytotoxicity and cell imaging potentials of submicron color-tunable yttria particles. J Biomed Mater Res Part A 100A(9):2287–2294
52. Das GK, Tan TTY (2008) Rare-earth-doped and codoped Y_2O_3 nanomaterials as potential bioimaging probes. J Phys Chem C 112(30):11211–11217
53. Balda R, Sanz M, Mendioroz A, Fernández J, Griscom LS, Adam J-L (2001) Infrared-to-visible upconversion in Nd^{3+}-doped chalcohalide glasses. Phys Rev B 64:144101
54. Russell D, Holliday K (2001) Upconversion and energy transfer dynamics in Nd^{3+}:$KLiYF_5$. Opt Commun 191:277–294
55. Konrad A, Herr U, Tidecks R, Kummer F, Samwer K (2001) Luminescence of bulk and nanocrystalline cubic yttria. J Appl Phys 90(7):3516
56. Bordun O (2002) Influence of oxygen vacancies on the luminescence spectra of Y_2O_3 thin films. J Appl Spectrosc 69(3):430–433
57. Speghini A, Piccinelli F, Bettinelli M (2011) Synthesis, characterization and luminescence spectroscopy of oxide nanopowders activated with trivalent lanthanide ions: the garnet family. Opt Mater 33:247–257
58. Geusic JE, Marcos HM, Van Uitert LG (1964) Laser oscillations in nd-doped yttrium aluminum, yttrium gallium and gadolinium garnets. Appl Phys Lett 4(10):182–184
59. Zych E, Brecher C, Glodo J (2000) Kinetics of cerium emission in a YAG:Ce single crystal: the role of traps. J Phys Condens Matter 12(8):1947
60. Rotman SR, Warde C (1985) Defect luminescence in cerium-doped yttrium aluminum garnet. J Appl Phys 58(1):522
61. Dubinskii MA, Schepler KL, Semashko VV, Abdulsabirov RY, Korableva SL, Naumov AK (1998) Spectroscopic analogy approach in selective search for new Ce^{3+} -activated all-solid-state tunable ultraviolet laser materials. J Modern Opt 45:221–226
62. Selim FA, Solodovnikov D, Weber MH, Lynn KG (2007) Identification of defects in $Y_3Al_5O_{12}$ crystals by positron annihilation spectroscopy. Appl Phys Lett 91(10):104105
63. Lupei A, Lupei V, Gheorghe C, Ikesue A, Enculescu M (2011) Spectroscopic characteristics of Dy^{3+} doped $Y_3Al_5O_{12}$ transparent ceramics. J Appl Phys 110(8):083120

64. Stevenson AJ, Bittel BC, Leh CG, Li X, Dickey EC, Lenahan PM, and Messing GL (2011) Color center formation in vacuum sintered $Nd_3Y_{3-3x}Al_5O_{12}$ transparent ceramics. Appl Phys Lett 98(5):051906
65. Chao W-H, Wu R-J, Wu T-B (2010) Structural and luminescent properties of YAG:Ce thin film phosphor. J Alloy Compd 506:98–102
66. Varney CR, Reda SM, Mackay DT, Rowe MC, Selim FA (2011) Strong visible and near infrared luminescence in undoped YAG single crystals. AIP Adv 1(4):042170 (reference therein)
67. Zhao G, Li T, He X, Xu J (2002) Preparation of gadolinium gallium garnet polycrystalline material by coprecipitation method. Mater Lett 56:1098–1102
68. Ozen G, Collins J, Bettinelli M, Bartolo BD (2013) Luminescence of $\{Y_3AL_5O_{12}\}$ nanoparticles doped with praseodymium ions. Opt Mater 35(7):1360–1365
69. Pechini P (1967) Method of preparing lead and alkaline earth titanates and niobates and coating method using the same to form a capacitor. US Patent 3,330,697
70. Mazur P, Hreniak D, Niittykoski J, Strek W, Hölsä J (2005) Formation of nanostructured Tb^{3+}-doped yttrium aluminium garnets by the glycol route. Mater Sci Pol 23(1):253–260
71. Hirata E, Tamagawa K, Ohki Y (2010) Cr^{3+} impurities and photoluminescence in $LaAlO_3$. Jpn J Appl Phys 49(9R):091102
72. Olenovich N, Gromadskaya G, Anbinder I (1975) Spectral determination of impurities in some pure aluminum compounds. J Appl Spectrosc 23(3):1161–1163
73. Eapen J, Rusconi R, Piazza R, Yip S (2010) The Classical Nature of Thermal Conduction in Nanofluids. J Heat Transf 132:102402
74. Desai TG (2011) Thermal transport in nanoclusters. Appl Phys Lett 98(19):193107
75. Pawlow P (1909) Über die Abhängigkeit des Schmelzpunktes von der Oberflächenenergie eines festen Körpers. Z Phys Chem 65:1–35
76. Wang J, Duan H, Huang Z, Karihaloo B (2006) A scaling law for properties of nano-structured materials. Proc R Soc A Math Phys Eng Sci 462(2069):1355–1363 (reference therein)
77. Takagi M (1954) Electron-diffraction study of liquid-solid transition of thin metal films. J Phys Soc Jpn 9(3):359–363
78. Martynenko Y, Ognev L (2005) Thermal radiation from nanoparticles. Tech Phys 50(11): 1522–1524
79. Wuttke C, Rauschenbeutel A (2013) Thermalization via heat radiation of an individual object thinner than the thermal wavelength. Phys Rev Lett 111:024301
80. Planck M (1901) Ueber das Gesetz der Energieverteilung im Normalspectrum. Annalen der Physik 309(3):553–563
81. Ferguson L, Dogan F (2001) Spectrally selective, matched emitters for thermophotovoltaic energy conversion processed by tape casting. J Mater Sci 36(1):137–146

13
An Approach in the Structural and Spectroscopic Analysis of Yb^{3+}-Doped YAG Nano-ceramics by Conjugation of TEM-EDX and Optical Techniques

George Boulon, Yannick Guyot, Malgorzata Guzik, Thierry Epicier, Pawel Gluchowski, Dariusz Hreniak, and Wieslaw Strek

Abstract We show our approach in the structural and spectroscopic analysis of Yb^{3+}-doped YAG nano-ceramics prepared by the low temperature-high pressure sintering technique (LTHP) by conjugation of both TEM-EDX and optical techniques. Pressure sintering dependences of absorption, emission and decays are analyzed and interpreted. The sample pressurized at 8 GPa for sintering is characterized by the highest transparency and confirms the $Y_3Al_5O_{12}$ garnet structure of the grains of ~21 nm average size. Yb^{3+} ion distribution has been analyzed by both TEM-EDX evaluation in grains and grain boundaries and spectroscopy of Yb^{3+} pairs of small population from the cooperative luminescence phenomenon. EDX analysis at the TEM scale provides unambiguous results on a clear tendency of almost uniform Yb^{3+} distribution. An important new observation has been made at 4 K and room temperature with the $^2F_{7/2} \rightarrow ^2F_{5/2}$ 0-phonon absorption line at 975.7 nm in addition of the 0-phonon line of the YAG structure of grains at 968 nm similar to that of bulky YAG single crystals. We have discussed the origin of this new 0-phonon line relaxing only by non-radiative transitions and conclude that this line might be assigned to Yb^{3+} distorted sites on the grain surfaces.

G. Boulon (✉) • Y. Guyot
Institute Light Matter (ILM), UMR 5306 UCB Lyon 1-CNRS, University of Lyon, Bât. Kastler, 69622 Villeurbanne, France
e-mail: georges.boulon@univ-lyon1.fr

M. Guzik
Faculty of Chemistry, Chemistry Department, University of Wroclaw, 14 F. Joliot-Curie, PL-50-383 Wroclaw, Poland

T. Epicier
MATEIS, UMR 5510 CNRS, INSA of Lyon, University of Lyon, Bât. B. Pascal, 69621, Villeurbanne, France

P. Gluchowski • D. Hreniak • W. Strek
Institute of Low Temperature and Structure Research, Polish Academy of Sciences, ul. Okolna 2, PL-50-422 Wroclaw, Poland

B. Di Bartolo et al. (eds.), *Nano-Structures for Optics and Photonics*,
NATO Science for Peace and Security Series B: Physics and Biophysics,
DOI 10.1007/978-94-017-9133-5_13,

Keywords Nano-ceramics; YAG; Yb^{3+} dopant distribution; Spectroscopic properties; TEM-EDX

13.1 Introduction

Yttrium aluminum garnet $Y_3Al_5O_{12}$ (YAG) is very well known compound since 1964, widely applied as a matrix for lanthanide based phosphors [1–3]. Great interest of this compound is a result of its special properties such as high chemical stability (melting point at 1,940 °C), high thermal conductivity ($14\,W\,m^{-1}\,K^{-1}$), isotropic cubic structure and large energy gap (6.5 eV), so that this material is a good electric insulator. YAG is an active laser crystal in the near infra-red with such rare earth dopants as Nd^{3+}, Ho^{3+}, Er^{3+}, Tm^{3+}, Yb_3^+ and it is even used for a window material substituting of sapphire. However, there are some limitations in the growing of rare earth-doped YAG crystals, especially by the Czochralski technique merged with the increasing of the concentrations of dopants, with the fabrication of potential shapes and finally with the cost. The need of prospect other types of materials lead to transparent polycrystalline ceramics of cubic structure as a new generation of optical materials. Huge efforts were necessary to succeed to fabricate high quality rare earth-doped $Y_3Al_5O_{12}$ laser ceramics. Before 1995, it was considered that ceramic materials cannot be used for lasers, because the optical quality of conventional ceramics was quite low. Laser ceramics require indeed no optical scattering and perfect homogeneity and finally Nd^{3+}-doped YAG laser ceramic was emerged from pioneer works of A. Ikesue (World Lab. Co., Ltd., Nagoya) [4, 5] and advanced researches [6, 7]. Another Japanese group (H. Yagi's group) from Konoshima Co. has succeeded to get nice advances with laser ceramics [8–11]. Nowadays laser ceramics are commercially available. The already described YAG ceramics have a grain size in the range of a few micrometers.

There is also interest in synthesis and optical properties of rare-earth-doped YAG nanocrystalline ceramics [12, 13]. YAG nanopowders may be synthesized by several methods including sol–gel [14], Pechini [15], solvothermal [16], combustion [17] and co-precipitation [18]. For preparation of high optical quality ceramics both structurally pure and non-aggregated nanopowders are required. However, less work has been done to sinter nanocrystalline ceramics [19–31]. Dense ceramics with nano-grain size may display new structural, magnetic, electric or electronic properties with great potential for applications. However, sintering usually causes grain growth to micrometer range. There are three main methods to prevent grain growth during sintering [19]: application of hot-pressing or sintering forging, addition of dopant to modify diffusion process, and sintering powders in metastable crystallographic phases that during sintering act as seeds for the final phases. The third one is used for such materials like TiO_2 and Al_2O_3 [20, 21]. For example nanoscale glass-ceramic alumina doped by rare-earth oxides (Nd_2O_3, Eu_2O_3,

Er_2O_3) with transparency in the mid-infrared range can be achieved by heating for a short time bulk glasses [22]. Also non-conventional sintering methods like microwave sintering [23,24], shock or dynamic consolidation [25,26], field assisted sintering [27, 28], high pressure low temperature sintering [29–31] are applied. These materials are potentially attractive for many applications since the reduction of the grain size to the nanometer scale can improve their physical and mechanical properties.

Naturally, the first nanoceramics developed were Nd:YAG nanoceramics [12,13]. They were obtained by using the high pressure low temperature (HPLT) technique. This method allows the synthesis of transparent YAG nano-ceramics composed of the nano-metric grains comparable with the sizes of starting nano-powders. These materials are potentially attractive for many applications since reduction of the grain size to the nanometer scale can improve their physical and mechanical properties. In particular the effect of applied pressure on preparation of transparent YAG nanoceramics was studied. The optimal sintering conditions for obtaining optically transparent nanoceramics are temperature of 450 °C and pressure of 8 GPa. As result, a nanoporous and transparent sample with 37 nm grains was obtained with the extinction coefficient equal to 13.6 cm^{-1} at 1,064 nm. Uniform distribution of nano-sized pores in the ceramics has been observed, which contributes to scattering of transmitted light.

We have continued this research by selecting Yb^{3+} ion used either as sensitizer for other rare earth ions or for direct laser ion or still as a structural probe ion. The state-of-art of Yb^{3+}-doped inorganic materials in both basic and applied research was developed in the following articles [32,33]. Especially YAG host we made a detailed structural and spectroscopic analysis of Yb^{3+}-doped YAG of bulky shape [34], nanocrystallite shape [35], transparent ceramic shape [36, 37] and also transparent nanoceramic shape [38] allowing comparison of these type of materials.

Recently, white anti-Stokes emission was observed in Yb:YAG nano-crystalline ceramics and investigated under focused beam of 976 nm laser diode (LD) excitation [39]. The emission intensity increased significantly with concentration of active ions. This effect was more pronounced in vacuum compared to atmospheric pressure. Only for full concentrated YbAG nano-ceramic white emission could be observed in normal atmosphere pressure. These new phenomena should be understood and we think we should first investigate both structural and spectroscopic analysis of these nano-ceramic samples fabricated from the nano-crystalline powders, by combining X-Ray, TEM-EDX, absorption, emission and decay measurements. In particular, new important results on the effect of applied pressure during sintering process on the absorption spectra at low temperature (4 K) and emission spectra will be investigated. In addition, Yb^{3+} ion distribution in nano-ceramics will be also analyzed by both spectroscopy of Yb^{3+} pairs from the cooperative luminescence spectra and TEM-EDX evaluation of Yb^{3+} concentrations in grains and grain boundaries.

13.2 Experimental Setup

13.2.1 Synthesis of Transparent Nano-ceramics at the Polish Academy of Sciences in Wroclaw

The initial YAG nano-powders doped with a nominal ytterbium ions molar concentration between 0.5 and 100 % vs. yttrium ions were prepared using modified Pechini method described elsewhere [15]. Nano-ceramics composed of nano-sized grains were fabricated by the low temperature-high pressure sintering technique (LTHP) [12]. Before sintering, pellets of diameter 5 mm and height about 2.5 mm were formed from the nano-powder by cold pressing under pressure of 50 MPa. The pellet was placed in the container in graphite tube used as a resistance heater. Green body was separated from graphite by boron nitride (BN). The force exerted by the anvils creates a quasi-isostatic pressure between 2 and 8 GPa. The samples were sintered at temperature 450 °C during 1 min. The scheme of the equipment used for fabrication of the samples is presented in Fig. 13.1a, b. Figure 13.2 shows the photos of the ceramics obtained under different pressures i.e. green body (without any pressure) and 2, 4, 6, 8 GPa, respectively. As one can see the transparency is changing with the pressure and the best is observed for the samples 6 and 8 GPa.

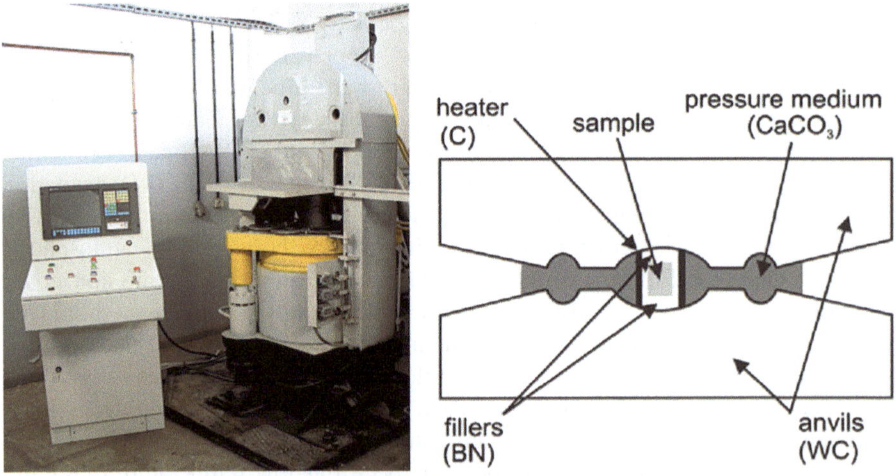

Fig. 13.1 The equipment used for fabrication of the ceramics (**a**) Press specification: Pressure (1–8 GPa), Temperature (up to 1,800 °C), Time (1–240 min). (**b**) Scheme of the cell used in this work: Pressure 8 GPa, Temperature 450 °C, Time ∼2 min, YAG nanopowder obtained at 900 °C [12]

Fig. 13.2 Photos of green body (GB) and nano-ceramics samples under 2, 4, 6, 8 GPa pressures, respectively. The best transparencies are seen for 6 and 8 GPa

13.2.2 X-Ray Powder Diffraction (XRD)

Data have been obtained at the "Diffractometry Center Henri Longchambon" of the University Lyon1. The powdered crystal diffractograms were recorded by using a Bruker D8 ADVANCE powder diffractometer with nickel-filtered Cu Kα radiation ($\lambda = 1.5418$ Å) and a Vantec detector. The measurements were performed within 2θ range of 10–70° with the scan rate 0.008° per step and the counting time 4 s per step.

13.2.3 Transmission Electron Spectroscopy

TEM experiments were conducted on thin foils classically prepared by ion beam thinning (PIPS Gatan), using a JEOL2010F field emission gun transmission electron microscope operating at 200 kV. The microscope was fitted with an EDX (energy dispersive X-ray) analyzer for elemental chemical analysis (Fig. 13.3). Detector of the latest generation was used, a SDD (Silicon Drift Detector) X-MAX 80 from Oxford Instruments; this technology provides a much better counting rate and sensibility limit than conventional detectors based on a Si–Li diode. The set-up is the same as that used in our recent paper on the spatial distribution of the Yb^{3+} rare earth ions in $Y_3Al_5O_{12}$ and Y_2O_3 optical ceramics as analyzed by TEM [40, 41]. This offers a great advantage in the present context of measuring small quantities (and their possible small variations) of rare earth elements in optical ceramics. Since one of major goals of this work is to study the homogeneity of the dopant distribution near and at grain-boundaries, most EDX measurements reported in the following were performed with a nano-probe, the size of which will be given through the value of its full-width at half maximum "FWHM" (in nm). The quality of High

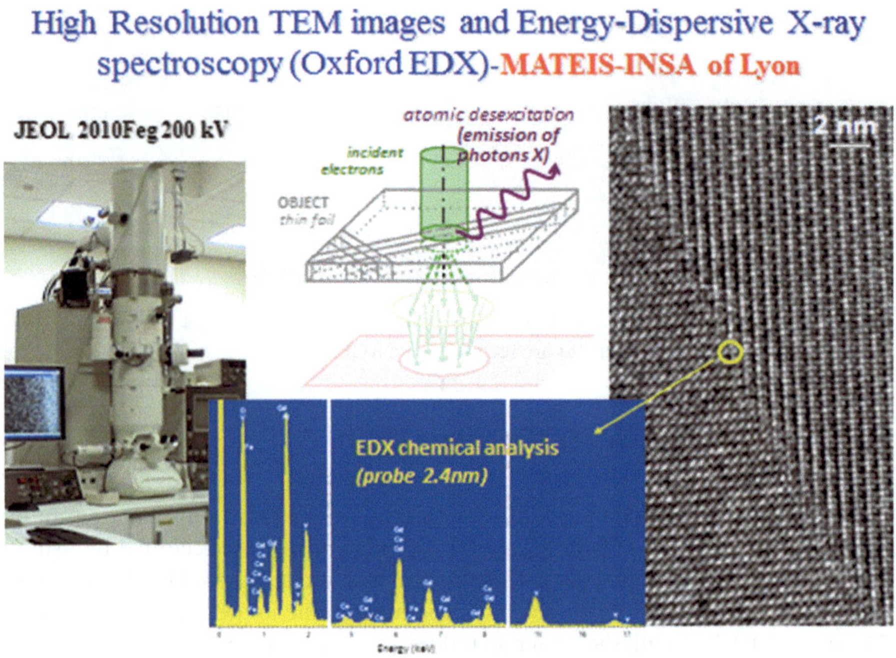

Fig. 13.3 Schematic view of the high resolution TEM images and energy-dispersive X-ray spectroscopy (Oxford EDX) at MATEIS-INSA of Lyon

Resolution TEM images recorded on most analyzed grain-boundary areas indicates that the thickness is not nominally greater than 20 nm; this gives a beam broadening of the order of 2.4 nm in YAG. These numeric values indicate that with an incident nano-metric probe, the spatial resolution of EDX analysis remains reasonably good. As usual in EDX-TEM, all results will be given in atomic % (at.%), with the total composition included all atomic species (O, Y, Al and the doping rare earth element) normalized to 100 %.

13.2.4 Spectroscopic Measurements

Absorption Spectra

Absorption spectra in the 800–1,150 nm spectral range were recorded at 4 and 293 K with a Cary-Varian 5000 Scan spectrometer equipped with an Oxford CF 1204 helium flow cryostat at the Chemistry Department of the University of Wroclaw.

Emission Spectra

Emission spectra and decays of fluorescence were recorded at the Institute Light Matter of the University f Lyon1. Nd:YAG impulse laser equipped with

Ti:sapphire oscillator were used as excitation sources for the emission spectra and decays measurements. The emission spectra were measured with a Jobin–Yvon THR 1000 monochromator and Synapse Horiba Jobin Yvon CCD camera. Emission decays were recorded with a LeCroy WaveSurfer 400 oscilloscope. All spectra were corrected for the sensitivity and wavelength of the experimental setup.

13.3 Recent Results and Discussion

13.3.1 Structural Characterization from XRD Pattern

Previous characterizations of semi-transparent nano-crystalline YAG ceramics doped with 1 mol% Nd^{3+} and 5 mol% Yb^{3+} have been done by the Group of the Polish Academy of Sciences from Wroclaw [12]. The grain size of the sintered pellets was in the range of 25 nm. It was observed that during high pressure sintering YAG decomposes into YAP and alumina. The decomposition of the YAG particles and transformation to the much larger YAP crystals is confirmed by the coexistence of the broad peaks originating from YAG (JCPDS No. 33-40) and the sharp reflexes from YAP structure (JCPDS No. 16-219), which is observed for samples pressed at 550, 600 °C and also at higher temperatures.

As can be seen in Fig. 13.4 the decomposition process can start even at 550 °C (visible trace of YAP reflex at 24°). On the basis of these results, to avoid decomposition, all samples analyzed in this paper were sintered at 450 °C. Figure 13.5 presents the conditions of decomposition of YAG to YAP according to the reaction:

$$Y_3Al_5O_{12}(YAG) \rightarrow 3YAlO_3(YAP) + Al_2O_3$$

as deducted from on line X-ray diffraction as a function of temperature and pressure.

Figure 13.6 confirms the garnet structure of the 5 mol% Yb^{3+}-doped $Y_3Al_5O_{12}$ nano-ceramics and the insert shows the evolution of the structural data of the cell parameter (in Å), and the average grain sizes (in nm), calculated from the usual Sherrer equation. If pressure is increasing from green body to 8 GPa, the cell parameter is increasing whereas the grain size is decreasing but seems to increase again between 6 and 8 GPa (see insert Table in Fig. 13.6). Comparison of the refined values of the lattice parameters for pure $Y_3Al_5O_{12}$ ($a = 12.009$ Å) shows the increase of the lattice parameters for Yb^{3+}-doped YAG, which proves a partial substitution of ytterbium ions (ionic radii of 86.8 pm by Shannon and Prewitt) for yttrium (ionic radii of 90 pm) sites in the garnet lattice like for Nd^{3+}-doped YAG [42]. Confrontation of the resulting diffraction patterns with the reference standard of the YAG confirms the presence of only one $Y_3Al_5O_{12}$ garnet cubic crystalline phase.

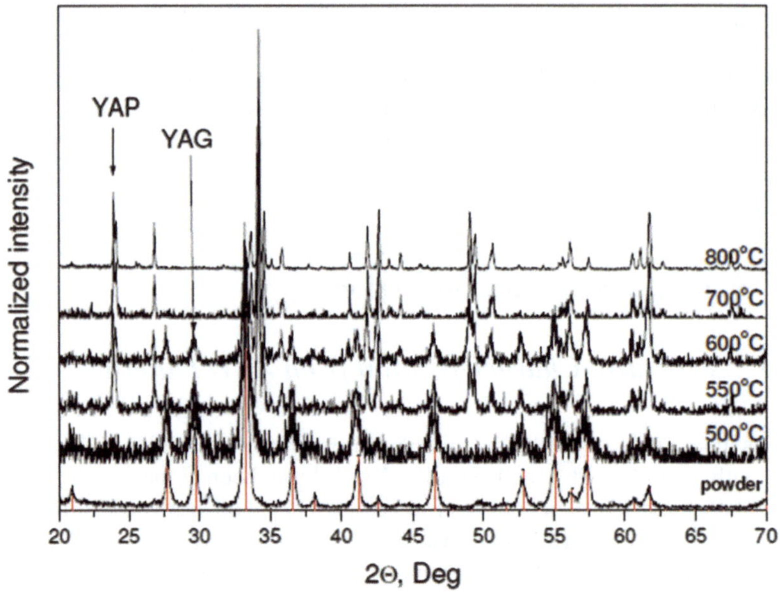

Fig. 13.4 XRD patterns of YAG powders co-doped with 5 % Yb^{3+} and 1 % Nd^{3+} sintered under 7.7 GPa at different temperatures; (Average size of nano-grains ~25 nm almost like initial nano-powders) [12]

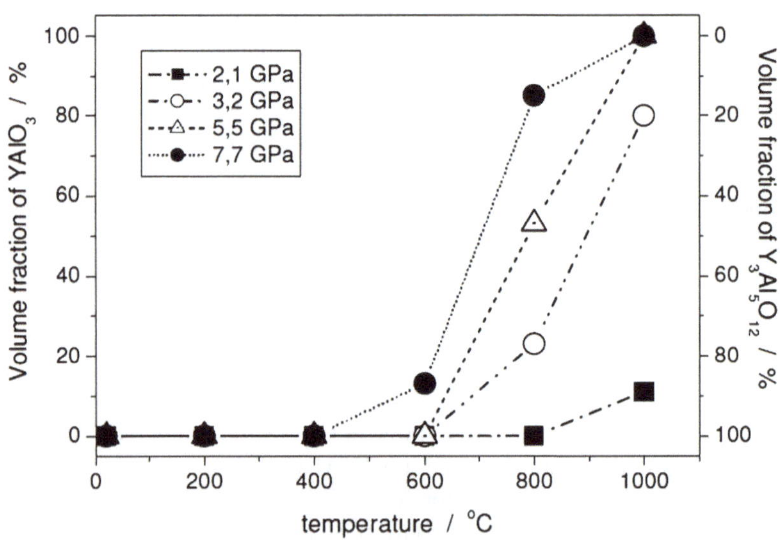

Fig. 13.5 Decomposition diagram for samples of nano-crystalline YAG to micro-crystalline YAP during hot pressing as a function of temperature at different pressures [12]

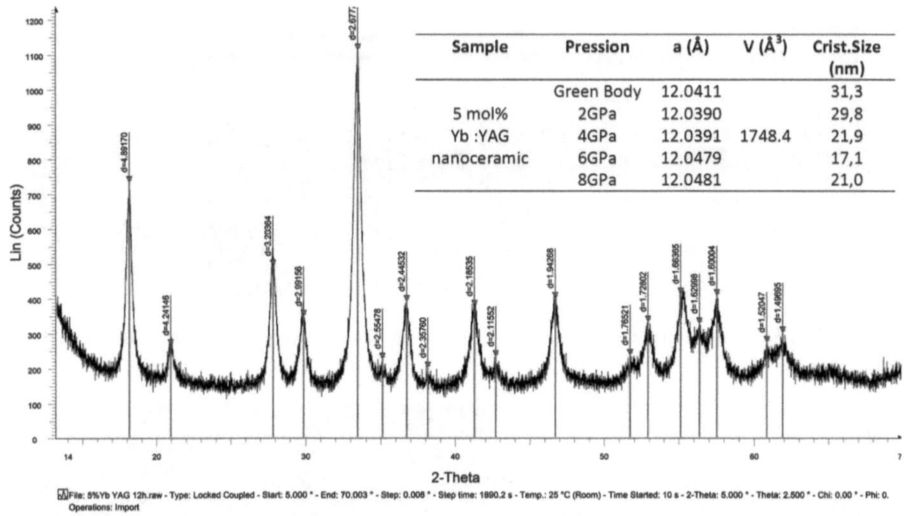

Sample	Pression	a (Å)	V (Å^3)	Crist.Size (nm)
	Green Body	12.0411		31,3
5 mol%	2GPa	12.0390		29,8
Yb :YAG	4GPa	12.0391	1748.4	21,9
nanoceramic	6GPa	12.0479		17,1
	8GPa	12.0481		21,0

Fig. 13.6 XRD pattern of 5 mol% Yb^{3+}-doped YAG nano-ceramic sintered at 450 °C. *Insert*: Table Pressure dependence of the cell parameter and the average grain size

Table 13.1 The contents (Atomic %) of O, Al, Fe, Y and Yb for six different positions in the grain and along the grain boundary

No.	Spectra (Atomic %)	O	Al	Fe	Y	Yb
1.	Grain	60.02	19.95	0.11	15.65	3.89
2.	Grain boundary	60.03	21.75	0.22	14.43	3.38
3.	Grain	60.52	20.43	0.18	15.80	3.08
4.	Grain center	60.07	25.25	0.08	11.04	3.36
5.	Grain near the GB	59.99	22.54	0.32	13.30	3.65
6.	Grain boundary	60.04	20.74	0.13	15.13	3.75

13.3.2 Characterization of the Yb^{3+} Ions Spatial Distribution by TEM-EDX Technique

We have selected the sample pressurized at 8 GPa which is characterized by the highest transparency (see Fig. 13.2). First we should mention a lower Yb^{3+} content than expected. In the example of 20 mol% Yb^{3+}, shown here to illustrate the TEM properties, an average value of 3.5 ± 0.5 at.% Yb^{3+} (17.4 ± 2.5 weight %) was recorded for over more than 20 analysis points inside grains and at grain boundaries. The punctual values in Table 13.1 vary indeed between 3.08 and 3.89 at.% Yb^{3+}.

The sample was first polished to a thickness of about 100 μm and then mechanically grinded to produce smaller debris. These thinner parts were then sandwiched between initial larger pieces and crushed into a quartz mortar (such sandwich procedure prevents from any undesirable contamination from the mortar). The resulting agglomerates were then deposited on a classical holey carbon grid.

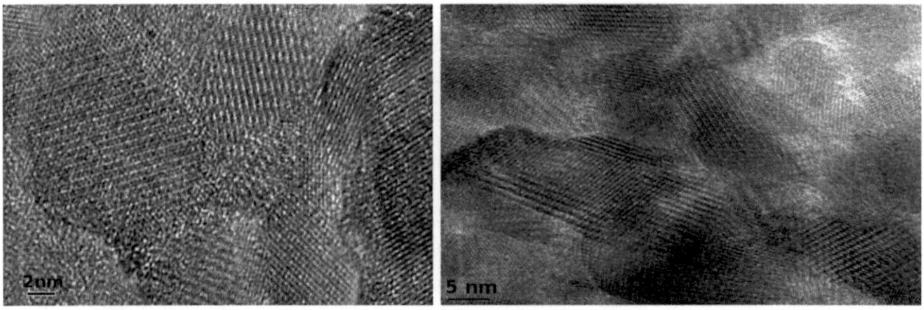

Fig. 13.7 TEM images of the 20 mol% Yb^{3+}-doped YAG nano-ceramic fabricated at 450 °C under 8 GPa

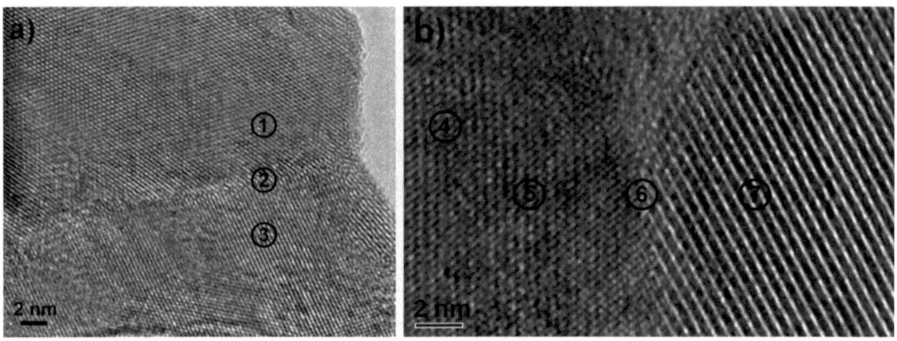

Fig. 13.8 TEM images of different 20 mol% Yb^{3+}: YAG nano-ceramic areas: grain, grain-boundary and close to grain boundary

The same procedure in Yb^{3+}-doped nano-ceramic samples like for the Yb^{3+}-doped YAG bulky ceramics [36, 37] and the Ce^{3+}-doped YAG bulky ceramics [40, 41] was applied. Visualization of grains and grain boundaries can be seen in the TEM micrographs in Figs. 13.7–13.9, respectively.

The nanostructure consists of small crystallites in the range 5–20 nm (rough estimate). At some grain boundaries, the "amorphous-type" contrast is frequently revealed in High Resolution micrographs but its concentration was not high enough to determine the composition.

The most striking feature of the EDX measurements results from the quasi uniform distribution of Yb^{3+} cations between grains and grain boundaries like in bulky YAG garnet and Y_2O_3 sesquioxides [36, 37] as can be seen in Table 13.1 containing data from the study of the grain boundaries shown in Figs. 13.7 and 13.8, respectively. Yb^{3+} cations were not detected as being concentrated on the grain boundaries in contrary to Ce^{3+} cations [40, 41] or Nd^{3+} cations [43].

EDX analysis with a probe of 1 nm is reported in Table 13.1. Yb^{3+} ions concentrations correspond to the positions of the probe on the grain boundaries and at the upper, down, left and right sides of the grain boundaries. Some content of Fe arises from the specimen holder and environment in the microscope.

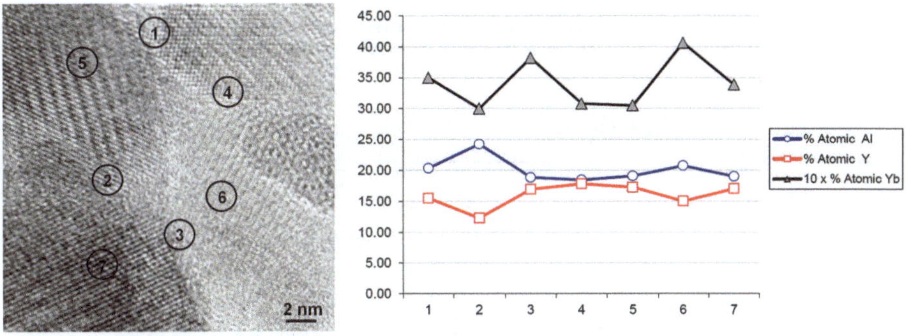

Fig. 13.9 Analysis of several grain-boundaries (GB) and grains (G) within the same region and positions of the EDX analysis with a probe of 2 nm and graphical dependences of Al (*blue line*), Y (*red line*) and Yb^{3+} (*black line*). To improve the visibility of the minor Yb^{3+} contents, Yb^{3+} concentrations have been multiplied by a factor 10

These results clearly show that there is very little variation in the Yb^{3+} content at the grain-boundary and within the adjacent grains. For comparison, differences of a factor of 4 were detected for the Ce^{3+} concentration in Ce^{3+}-doped YAG ceramics on a similar nano-metric size variation of 20 nm. Due to the near infrared emission of Yb^{3+} rare earth ions, it was not possible to observe such segregation by using Confocal Microscopy technique as it was used with Ce^{3+} ions emitting an intense luminescence band in blue, green and yellow ranges easy to detect with photomultipliers [40, 41]. However, EDX analysis at the TEM scale provides unambiguous results and we can conclude here a clear tendency of almost uniform Yb^{3+} distribution between grains and grain boundaries in nano-ceramics in the similar way of bulky ceramics.

Although fluctuations occur, no evidence is clearly found for any Yb^{3+} segregation at interfaces of both "grain" and "grain boundary" where Yb^{3+} contents exhibit a similar scattering inside the 2 nm value of the probe beam. Then the main conclusion from TEM-EDX measurements is related to the absence of Yb^{3+} aggregates either in grains or in grain boundaries which excludes the concept of the non-radiative transition of the Yb^{3+} $^2F_{5/2}$ (5) excited level located at 975 nm as it will be shown later on, by resonant migration of $^2F_{7/2}$ (1) $\longleftrightarrow$ $^2F_{5/2}$ (5).

All the previous experimental results correlate well with low segregation coefficients of Ce^{3+} and Nd^{3+} large rare earth cations, but higher segregation coefficients for small rare earth cations like Yb^{3+} in the garnet structure observed in the melt crystal growth and thin film liquid phase epitaxy from flux [44–48]. The data on segregation in melt/flux crystal growth are widely available and can be used for preliminary estimation of the dopant distribution in solid-state ceramic processing. The lower is the segregation coefficient of dopant in the melt growth, the greater is amount of dopant collected on the grain boundaries of the ceramic polycrystalline materials. This could help to predict the highest uniformity of Yb^{3+} dopant distribution. The large difference of values of the segregation coefficients (0.08 for

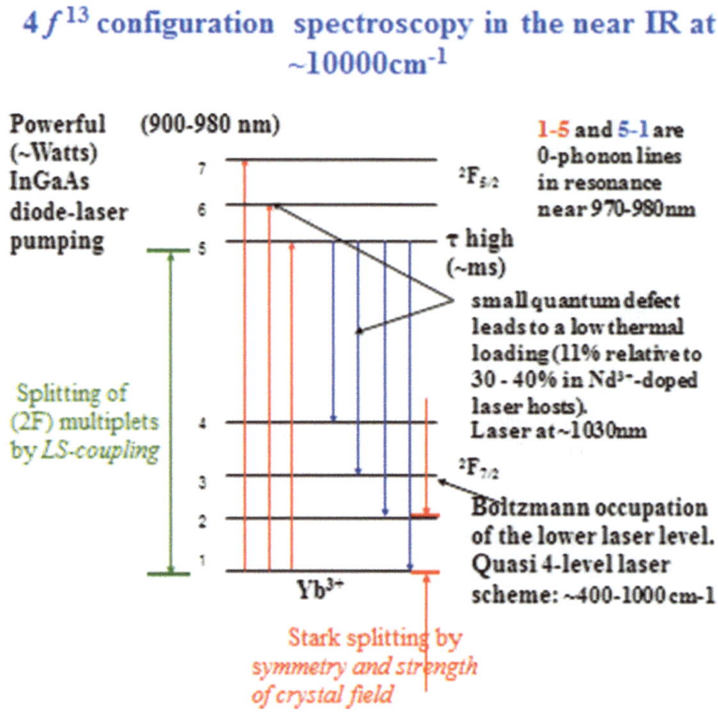

Fig. 13.10 The Yb^{3+} energy level diagram for the 4f13 configuration [32]

Ce^{3+}, 0.18 for Nd^{3+} and 1.09 for Yb^{3+} respectively) in YAG ceramics [49] explains also the respective behavior of the distribution of these luminescent cations inside the volume of nano-ceramics with especially a high segregation for Ce^{3+} and Nd^{3+} and much more regular distribution for Yb^{3+} ions.

13.3.3 Optical Spectroscopic Properties of Yb^{3+} Nano-ceramics

13.3.3.1 Energy Level Diagram

First, let's remind the energy level diagram for $^4f_{13}$ configuration of Yb^{3+} ions as shown in [32]. This is the simplest diagram of the rare earth ions. All transitions are located in the near infrared spectral region: 1 → 5, 6, 7 absorption lines between 900 and 1,000 nm and 5, 6, 7 → 1, 2, 3, 4 emission lines between 940 and 1,100 nm (see Fig. 13.10). The resonance lines 1 → 5 and 5 → 1 so-called 0-phonon lines in both absorption and emission spectra are the usual structural probes to assign all transitions in materials, with intensities depending of the radiative energy transfer inside the volume of the materials.

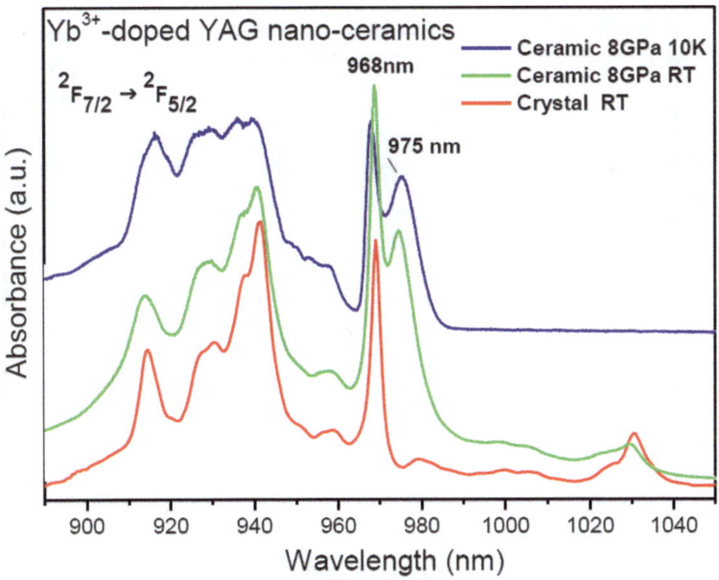

Fig. 13.11 Comparison of the absorption spectra in 5 mol% Yb^{3+}-doped YAG nano-ceramic at RT and 10 K and Yb^{3+}-doped YAG crystal at RT

13.3.3.2 Absorption Spectra and the Observation of a Specific Line of Nano-ceramics

The main feature to be noticed is the first observation in all nano-ceramics of the double 0-phonon line transition $^2F_{7/2}$ (1) → $^2F_{5/2}$(5) at 968.3 and 975.9 nm respectively at RT, 968 and 975 nm respectively at 10 K, as an example in 5 mol% Yb^{3+}-doped YAG nano-ceramics (Fig. 13.11). The highest energy line is characteristic of Yb^{3+} ions in the YAG garnet structure. The lowest energy line is only specific of nano-ceramic samples, not observed in single crystals as clearly seen in this Figure. We have especially analyzed the sintering pressure dependence of the absorption spectra at RT (Fig. 13.12) and that is more relevant at 4 K (Fig. 13.13) for which the new line is more intense and more resolved. Two important observations have to be mentioned: the band-width is much higher than that of the garnet structure and the intensity of this new absorption line at RT is increasing with the sintering pressure between 4 and 8 GPa. It means this new 0-phonon line might be connected especially with strains associated with the application of the pressure.

If the 0-phonon line at 968 nm can be undoubtedly assigned to regular Yb^{3+} ions in the grain garnet structure, at the same energy location that this of bulky YAG single crystals, the question arises about the origin of the new 0-phonon line at the lowest energy (975.7 nm at 4 K)?

The first answer can be related with Yb^{3+} distorted sites located in the surface of the nano-ceramics, as it is typical for nano-materials. The pressure dependence

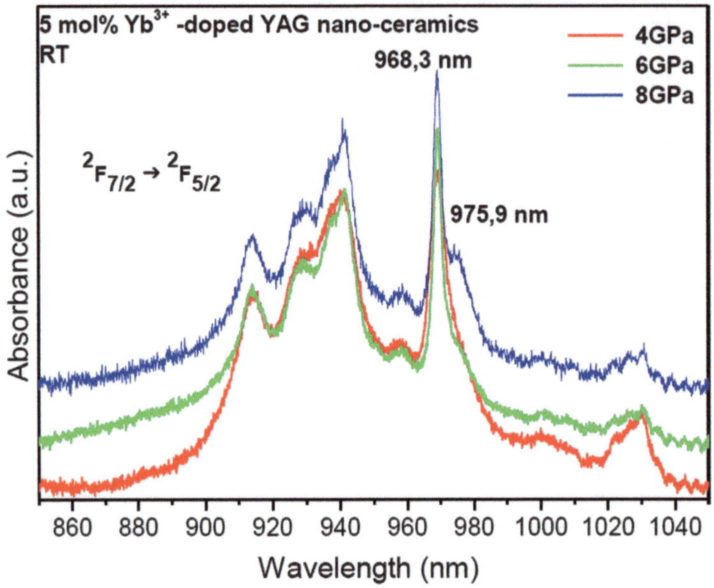

Fig. 13.12 Absorption spectra in 5 mol%Yb^{3+}-doped YAG nano-ceramics under 4, 6, 8 GPa sintering pressures at RT

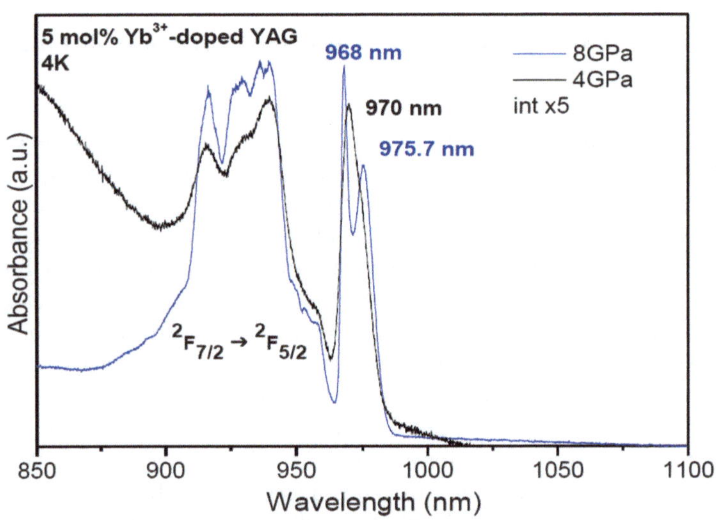

Fig. 13.13 Absorption spectra in 5 mol%Yb^{3+}-doped YAG nano-ceramic at 4 K

recorded for 5 mol% Yb^{3+}-doped nano-ceramic in Fig. 13.12 is in favor of the increasing of the population of the sites responsible of this absorption when pressure increases from 2 to 8 GPa. However, another hypothesis should be considered. This line might be also partially connected to additional traces of crystalline phases, more

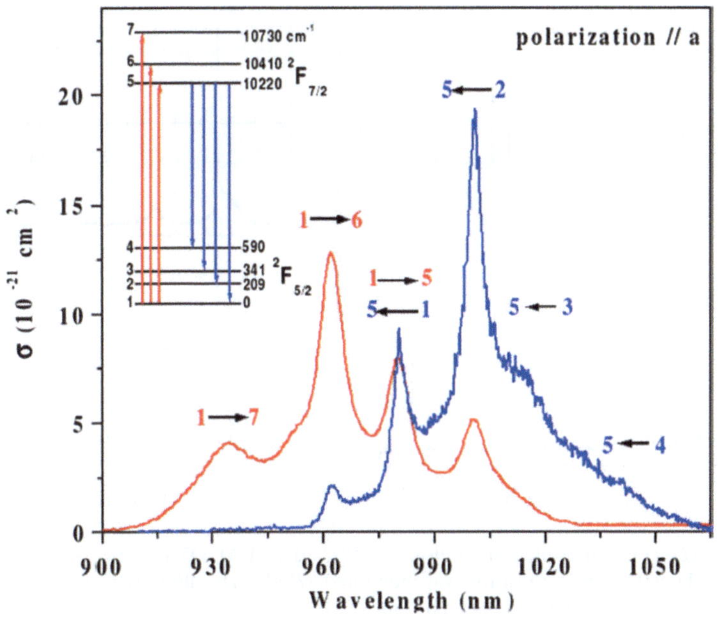

Fig. 13.14 Absorption and emission spectra of $Y_{1\text{-}x}Yb_xAlO_3$ (with $x = 0.005$) at room temperature. The 0-phonon line $1 \rightarrow 5$ and $5 \rightarrow 1$ is located at 978.5 nm [50]

precisely both Al_2O_3 and $YAlO_3$ phases as a consequence of the decomposition reaction:

$$Y_3Al_5O_{12}(YAG) \rightarrow 3YAlO_3(YAP) + Al_2O_3$$

This is why we have analysed absorption spectra of Yb^{3+}-doped $YAlO_3$ [50] and Yb^{3+}-doped Al_2O_3 [51, 52] shown in Figs. 13.14 and 13.15. The 0-phonon line of $Y_{1\text{-}x}Yb_xAlO_3$ is positioned at 978.5 nm at much lower energy than the 975 nm of the Yb^{3+}-doped nano-ceramic and so cannot be retained with detectable amount in these samples as a coherent conclusion of the temperature sintering at 450 °C explained earlier to avoid formation at least of YAP phase.

As for Yb^{3+}-doped Al_2O_3 thin films [51] or nano-powders $Jusza_{2011}$ large differences can be seen in spectra of the 0-phonon lines either at 982 nm or at 970 nm. This discrepancy might be due to the presence of the two phases α and θ depending on the temperature [53].

The best fit is indeed observed with Yb^{3+}-doped α-Al_2O_3 single crystals grown by LHPG technique [54]. It has been noted that only a small amount of Yb^{3+} ions enter the lattice by substitution of Y^{3+} cations and that higher dopant concentration leads to the presence of a variety of Yb^{3+} containing phases like garnet ones. Three Stark components of the $^2F_{5/2}$ manifold are clearly identified in Fig. 13.15 with the highest intensity of the 0-phonon line at 10,252 cm^{-1} (975.4 nm). The center

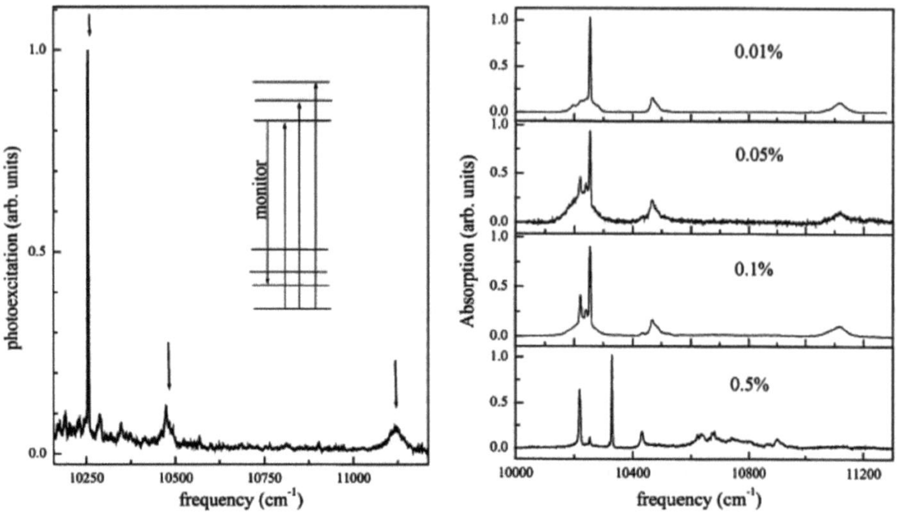

Fig. 13.15 (*Left*) Photo-excitation spectrum of Yb^{3+}-doped Al2O3 single crystal at 78 K grown by LHPG. The 9,920 cm^{-1} transition was monitored. (*Right*) The absorption spectrum of ceramic corundum with four concentrations of Yb^{3+} ions [53]

frequencies of the lines relative to the ground state are located at 337, 550, and 1,020 cm^{-1}. Additional peaks appear for $0.01 < c < 0.05$ at%. $Yb_3Al_5O_{12}$ garnet clusters (10,330 cm^{-1} peak at 968.1 nm) form for $0.1 < c < 0.5$ at%. We can conclude about the possibility to assign this 0-phonon line of Yb^{3+}-doped α-Al_2O_3 single crystals with the new absorption line observed in nano-ceramics and also with the possibility to form clusters of $Yb_3Al_5O_{12}$.

We insist again on the weak concentration of Yb^{3+}-doped α-Al_2O_3 single crystals which is almost impossible to be detected by XRD and was not detected by TEM so that XRD spectra characterize only YAG phase. If this hypothesis is retained it means there is a coincidence between 0-phonon lines of Yb^{3+} distorted sites of YAG located in the surface of the nano-ceramics and that of Yb^{3+}-doped α-Al_2O_3 single crystals but with a much higher concentration of Yb^{3+} distorted sites in YAG phase.

Another hypothesis can be formulated concerning the presence of traces of SiO_2 silica glass in the samples not from the raw materials but incorporated during polishing operations. In Yb^{3+}-doped silica glasses we observed this line located at 976 nm in CAS and LSCAS glasses [54] almost in coincidence with this one in nano-ceramics. Nevertheless, this is difficult to think that Yb^{3+} dopants can be introduced in the silicate structure by polishing operations and giving high absorption peak equivalent to the absorption peak of the regular Yb^{3+} ions in the garnet lattice of grains. Consequently, we didn't support this hypothesis.

In conclusion, the new absorption line peaking at 975.7 nm at 4 K seem to point out Yb^{3+} distorted sites on the grain surfaces of nano-ceramics with probably additional small amount of Yb^{3+}-doped α-Al_2O_3 single crystals almost impossible

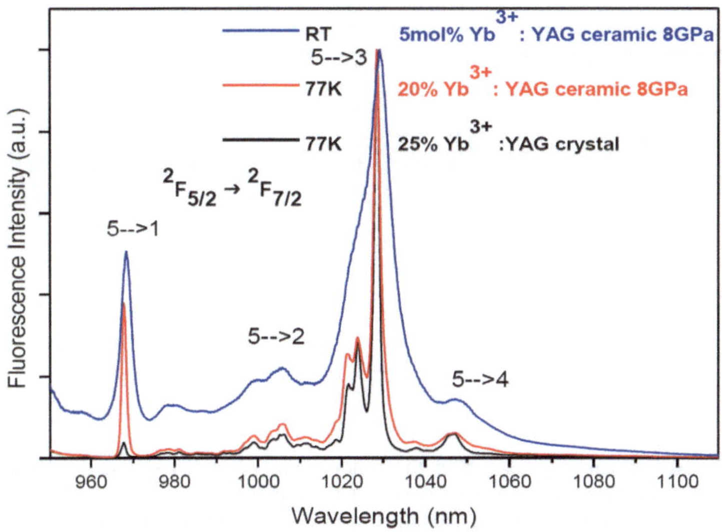

Fig. 13.16 Emission spectra of 5 mol% Yb^{3+}-doped YAG nano-ceramic (8 GPa) at RT and 77 K under excitation by 917 nm of a Ti-sapphire laser. The comparison can be done with 25 % Yb^{3+}-doped YAG single crystal at 77 K. The 0-phonon lines in emission are 968.4 nm at RT and 967.2 nm at 77 K for the nano-ceramic sample and 967.6 nm for the single crystal

to detect by XRD. In order to get complementary information on this assignment, we will look at the emission spectra and decay times of Yb^{3+} ions in these YAG nano-ceramics.

13.3.3.3 Near IR Emission Spectra

The emission spectra of 5 mol% Yb^{3+}-doped YAG nano-ceramic were measured at RT and 77 K and compared with bulky single crystal under pumping with a tuneable Ti-sapphire laser (Fig. 13.16). The re-absorption phenomenon seen by the intensity of the resonant 0-phonon, line at 967.6 nm of ${}^2F_{5/2}$ (5) $\rightarrow$ ${}^2F_{7/2}$(1) is very strong in the large volume of the single crystal whereas does not exist with the 0-phonon line at 967.2 nm in the small volume of the nano-ceramic characterized by an average size of 21 nm (see insert in Fig. 13.6). This re-absorption is associated with the radiative energy transfer, or self-trapping process, by resonance between Yb^{3+} ions, and so can be affected by the non-radiative energy transfer, or self-quenching process, when the effective diffusion average length is high inside the lattice. This lack of re-absorption is confirmed in Fig. 13.17a, b by both pressure and concentration dependences in which we can see always clearly the ${}^2F_{5/2}$ (5) $\rightarrow$ ${}^2F_{7/2}$(1) 0-phonon lines.

We tried to identify more precisely the two types of centers by pumping selectively the two 0-phonon lines with a tuneable Ti-Sapphire laser of $\sim$1 cm^{-1}

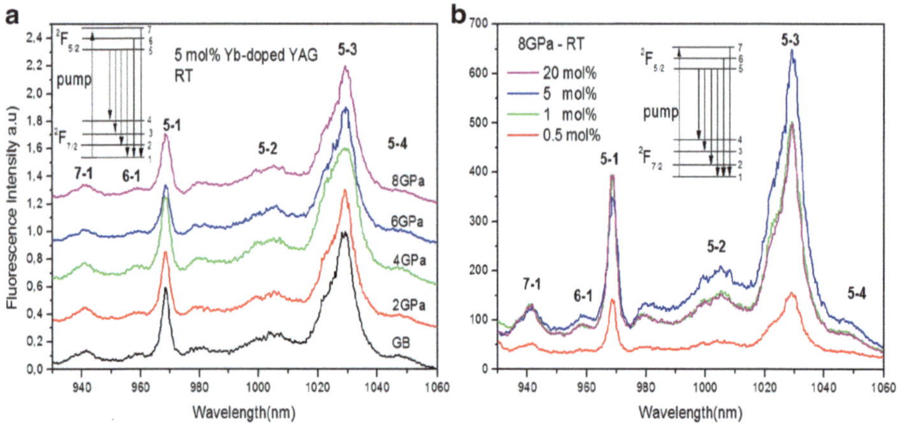

Fig. 13.17 (**a**) Pressure dependence of the 5 mol% Yb^{3+}-doped YAG nano-ceramic emission spectra at RT. (**b**) Concentration dependence of the *x* mol% Yb^{3+}-doped YAG nano-ceramics emission spectra at RT under 8 GPa. Excitation by 917 nm of a Ti-sapphire laser

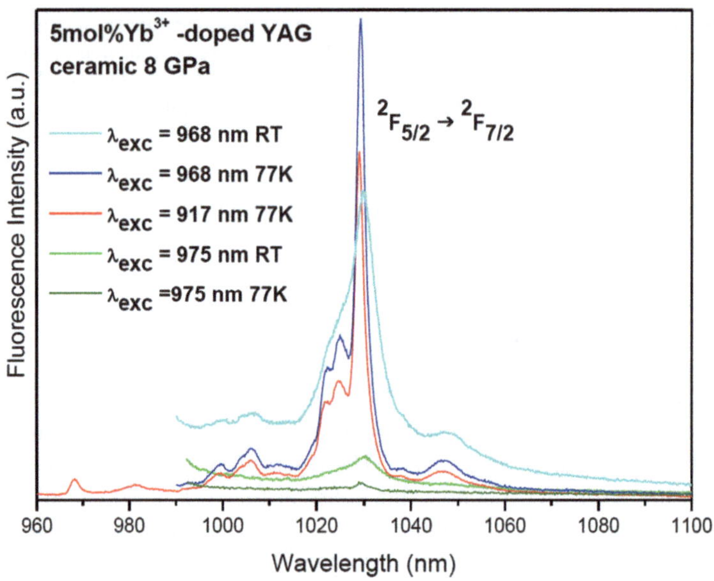

Fig. 13.18 Selective laser excitation of the two 0-phonon lines of 5 mol% Yb^{3+}-doped YAG under 8 Pa at RT and 77 K

narrow line-width used as pumping source. The results are shown in Fig. 13.18 under the excitation of both:

- 917 nm in the $^2F_{5/2}$ (7) highest energy level to display all lines of the emission spectrum,
- 968 nm in the $^2F_{5/2}$ (5) level of the Yb^{3+} ions in the YAG garnet grain,

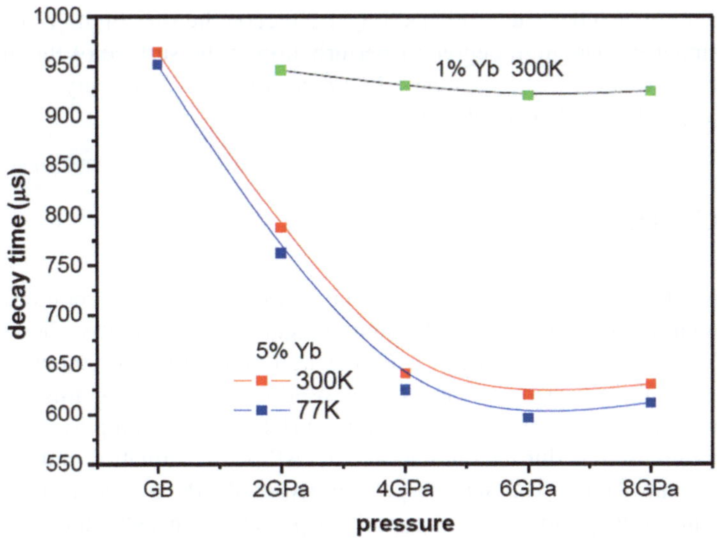

Fig. 13.19 Pressure dependence of the $^2F_{5/2}$ (5) decay times in 1 and 5 mol% Yb^{3+}-doped nano-ceramics

- 975 nm in the $^2F_{5/2}$ (5) in resonance with this new absorption line assigned previously to Yb^{3+} distorted sites on the grain surfaces of nano-ceramics.

The 917 and 968 nm excitations give rise to the same emission spectra related with $^2F_{5/2}$ (5)$\rightarrow$ $^2F_{7/2}$ (1, 2, 3, 4) transitions, whereas under 975 nm the radiative probability of the $^2F_{5/2}$ (5) level is almost zero. The very weak intensity of the usual highest emission transition $^2F_{5/2}$ (5) $\rightarrow$ $^2F_{7/2}$ (3) of Yb^{3+} ions at around 1,030 nm is probably due to the direct excitation of the foot of the 968 nm absorption line of the garnet structure due to the overlapping of the two 0-phonon absorption lines.

The absence of Yb^{3+} aggregates from TEM-EDX measurements either in grains or in grain boundaries exclude the explanation of the quenching of Yb^{3+} $^2F_{5/2}$ (5) excited level by resonant migration of $^2F_{7/2}$ (1)$\longleftrightarrow$ $^2F_{5/2}$ (5). Consequently this is another evidence of the role played by the distorted sites on the surface leading directly to non-radiative dissipation of the energy in the surroundings. These non-radiative transitions are favored by the observation in TEM of a few amorphous phases associated with grain boundaries.

13.3.3.4 Decays of the Yb^{3+} $^2F_{5/2}$ (5) Excited Level

The decay profiles of the $^2F_{5/2}$ (5) excited level have exponential profile for 1 and 5 mol% Yb^{3+}-doped nano-ceramics independently of the sintering pressure. Figure 13.19 displays the notable decrease of the decay times when the sintering pressure is increasing. This quenching phenomenon means there is an increase of the non-radiative links due to the increasing of the pressure that is to say the increasing of perturbation on Yb^{3+} sites with a stronger effect on the sites in the vicinity of the grain boundaries. The high degree of perturbation at the surface of the grains can explain the presence of the 0-phonon line observed at 975 nm only in absorption without any radiative de-excitation.

13.4 Conclusion

Our approach in the structural analysis of Yb^{3+}-doped YAG nano-ceramics samples by conjugation of both TEM-EDX and optical techniques was shown. Yb^{3+}-doped YAG nano-powders were prepared using modified Pechini method with the objective to fabricate nano-ceramics composed of nano-sized grains by the low temperature-high pressure sintering technique (LTHP). The samples were sintered at temperature 450 °C during 1 min to avoid YAP phase formation and pressurized between 2 and 8 GPa. The sample pressurized at 8 GPa is characterized by the highest transparency and XRD confirms the $Y_3Al_5O_{12}$ garnet structure of the grains of ~21 nm average size.

Yb^{3+} ion distribution has been analyzed by both spectroscopy of Yb^{3+} pairs from the cooperative luminescence phenomenon and TEM-EDX evaluation in grains and grain boundaries. EDX analysis at the TEM scale provides unambiguous results on a clear tendency of almost uniform Yb^{3+} distribution between grains and grain boundaries in nano-ceramics like in bulky ceramics and in contrary with Ce^{3+} and Nd^{3+} activator results. Yb^{3+} ion pairs, with small population, observed by cooperative luminescence belong only to the Yb^{3+}-doped YAG nano-grains of garnet structure containing radiative Yb^{3+} isolated centers and not to the surface of the grains which play the role of the grain boundaries since we have shown that Yb^{3+} distorted sites of the surface do not give rise to any radiative transition.

An important new observation has been made with the absorption line $^2F_{7/2}$ (1)→ $^2F_{5/2}$ (5) at 975.7 nm at 4 K and room temperature, in addition of the 0-phonon line of the YAG structure of grains at 968 nm similar to that of bulky YAG single crystals. We have discussed the question concerning the origin of this new 0-phonon line relaxing only by non-radiative transitions and conclude that this line might be assigned rather to Yb^{3+} distorted sites on the grain surfaces of YAG nano-ceramics in contact with an amorphous phase, with possibly additional traces of Yb^{3+}-doped α-Al_2O_3 single crystals of too weak population not enough to be detected either by XRD or by TEM. The evidence of the distorted sites at the surface leading directly to non-radiative dissipation of energy in the surroundings is in agreement with the TEM observation of a few amorphous phases in grain boundaries.

The absence of Yb^{3+} aggregates with high concentration from TEM-EDX measurements in both grains and grain boundaries excludes the explanation of the pressure quenching of Yb^{3+} $^2F_{5/2}$ (5) excited level by resonant migration of $^2F_{7/2}(1)\longleftrightarrow{}^2F_{5/2}$ (5). This quenching should be rather associated to an increase of the non-radiative links due to increasing of the pressure during sintering on Yb^{3+} sites with a stronger effect in the vicinity of the grain boundaries.

Acknowledgements The research exchange program between the Polish Academy of Sciences (PAS) and CNRS (France) is acknowledged. Thanks are due to the Department of Chemistry of the University of Wroclaw for the access of low temperature absorption measurements and to the CLYM (Centre Lyonnais de Microscopie- http://clym.insa-lyon.fr) for the access to the JEOL 2010F 200 kV microscope

References

1. Geusic JE, Marcos HM, Van Uitert LG (1964) Laser oscillations in Nd-doped yttrium aluminum, yttrium gallium and gadolinium garnets. Appl Phys Lett 4(10):182
2. Ikesue A (2002) Polycrystalline Nd:YAG ceramics lasers. Opt Mat 19(1):183–187
3. Simmonds MC, Huang YZ, Rodenburg JM (2003) Synthesis of nanosize powders and thin films of yb-doped yag by sol–gel methods. Chem Mat 15(18):3474–3480
4. Ikesue A, Kinoshita T, Kamata K, Yoshida K (1995) Fabrication and optical properties of high-performance polycrystalline Nd:YAG ceramics for solid-state lasers. J Am Ceram Soc 78(4):1033–1040
5. Ikesue A, Furusato I, Kamata K (1995) Fabrication of polycrystal line, transparent YAG ceramics by a solid-state reaction method. J Am Ceram Soc 78(1):225–228
6. Ikesue A, Aung YL, Yoda T, Nakayama S, Kamimura T (2007) Fabrication and laser performance of polycrystal and single crystal nd:yag by advanced ceramic processing. Opt Mater 29(10):1289–1294. In: Proceedings of the 1st international laser ceramic symposium E-MRS fall meeting 2005 1st international laser ceramic symposium
7. Shoji I, Taira T, Ikesue A (2007) Thermally-induced-birefringence effects of highly Nd3+-doped Y3Al5O12 ceramic lasers. Opt Mat 29(10):1271–1276
8. Lu J, Prabhu M, Xu J, Ueda K-i, Yagi H, Yanagitani T, Kaminskii AA (2000) Highly efficient 2% Nd:yttrium aluminum garnet ceramic laser. Appl Phys Lett 77(23):3707
9. Lu J, Song J, Prabhu M, Xu J, Ueda K-i, Yagi H, Yanagitani T, Kudryashov A (2000) High-power Nd:Y3Al5O12 ceramic laser. Jpn J Appl Phys 39(Part 2, No.10B):L1048–L1050
10. Lu J, Prabhu M, Song J, Li C, Xu J, Ueda K-i, Yagi H, Yanagitani T, Kaminskii AA (2001) Highly efficient Nd:Y3Al5O12 ceramic laser. Jpn J Appl Phys 40(Part 2, No.6A):L552–L554
11. Yagi H, Yanagitani T, Yoshida H, Nakatsuka M, Ueda K (2007) The optical properties and laser characteristics of Cr3+ and Nd3+ co-doped Y3Al5O12 ceramics. Opt Laser Technol 39(6):1295–1300
12. Hreniak D, Gierlotka S, Łojkowski W, Stręk W, Mazur P, Fedyk R (2005) High-pressure induced structural decomposition of RE-doped YAG nanoceramics. Solid State Phenom 106:17–22
13. Fedyk R, Hreniak D, Łojkowski W, Stręk W, Matysiak H, Grzanka E, Gierlotka S, Mazur P (2007) Method of preparation and structural properties of transparent YAG nanoceramics. Opt Mat 29(10):1252–1257
14. Goldburt E, Kulkarni B, Bhargava R, Taylor J, Libera M (1997) Size dependent efficiency in Tb doped Y2O3 nanocrystalline phosphor. J Lumin 72–74:190–192
15. Pechini P (1967) Method of preparing lead and alkaline earth titanates and niobates and coating method using the same to form a capacitor. US Patent 3,330,697
16. Liu Y, Zhan J, Ren M, Tang K, Yu W, Qian Y (2001) Hydrothermal synthesis of square thin flake CdS by using surfactants and thiocarbohydrate. Mater Res Bull 36(7-8):1231–1236
17. Sun L, Yao J, Liu C, Liao C, Yan C (2000) Rare earth activated nanosized oxide phosphors: synthesis and optical properties. J Lumin 87–89:447–450
18. Bhargava R, Gallagher D, Racz J (1996) Method of manufacturing encapsulated doped particles. US Patent 5,525,377
19. Bowen P, Carry C (2002) From powders to sintered pieces: forming, transformations and sintering of nanostructured ceramic oxides. Powder Technol 128(2–3):248–255

20. Kear B, Colaizzi J, Mayo W, Liao S-C (2001) On the processing of nanocrystalline and nanocomposite ceramics. Scripta Materialia 44(8-9):2065–2068
21. Liao S, Mayo W, Pae K (1997) Theory of high pressure/low temperature sintering of bulk nanocrystalline TiO2. Acta Materialia 45(10):4027–4040
22. Rosenflanz A, Frey M, Endres B, Anderson T, Richards E, Schardt C (2004) Bulk glasses and ultrahard nanoceramics based on alumina and rare-earth oxides. Nature 430(7001):761–4
23. Xie Z, Yang J, Huang X, Huang Y (1999) Microwave processing and properties of ceramics with different dielectric loss. J Eur Ceram Soc 19(3):381–387
24. Panneerselvam M, Subanna GN, Rao KJ (2001) Translucent yttrium aluminum garnet: microwave-assisted route to synthesis and processing. J Mater Res 16:2773–2776
25. Hokamoto K, Tanaka S, Fujita M, Itoh S, Meyers M, Chen H-C (1997) High temperature shock consolidation of hard ceramic powders. Phys B Condens Matter 239(1-2):1–5
26. Kaszuw Ara W, Leonowicz M, Januszewski D, Mendoza G, Davies H, Paszula J (1998) Consolidation of magnetic powders by shock compression. J Mater Sci Mater Electron 9(1):17–23
27. Groza JR (1999) Nanosintering. Nanostructured Mater 12(5-8):987–992
28. Groza J, Zavaliangos A (2003) Nanostructured bulk solids by field activated sintering. Rev Adv Mater Sci 5(1):24–33
29. Liao S-C, Chen Y-J, Kear B, Mayo W (1998) High pressure/low temperature sintering of nanocrystalline alumina. Nanostructured Mater 10(6):1063–1079
30. Liao S-C, Pae KD, Mayo WE (1995) High pressure and low temperature sintering of bulk nanocrystalline TiO2. Mater Sci Eng A 204(1-2):152–159
31. Choudhury S, Gandhi AS, Jayaram V (2003) Bulk, dense, nanocrystalline yttrium aluminum garnet by consolidation of amorphous powders at low temperatures and high pressures. J Am Ceram Soc 86(2):247–251
32. Boulon G (2008) Why so deep research on Yb3+-doped optical inorganic materials? J Alloy Compd 451(1-2):1–11
33. Boulon G, Guyot Y, Yoshikawa A (2009) Optimization of the gain in Yb3+-doped cubic laser crystals of 99.99% purity. J Rare Earths 27(4):616–618
34. Yoshikawa A, Boulon G, Laversenne L, Canibano H, Lebbou K, Collombet A, Guyot Y, Fukuda T (2003) Growth and spectroscopic analysis of Yb[sup 3+]-doped Y[sub 3]Al[sub 5]O[sub 12] fiber single crystals. J Appl Phys 94(9):5479
35. Amami J, Hreniak D, Guyot Y, Zhao W, Boulon G (2010) Size-effect on concentration quenching in Yb3+-doped Y3Al5O12 nano-crystals. J Lumin 130(4):603–610
36. Esposito L, Epicier T, Serantoni M, Piancastelli A, Alderighi D, Pirri A, Toci G, Vannini M, Anghel S, Boulon G (2012) Integrated analysis of non-linear loss mechanisms in Yb:YAG ceramics for laser applications. J Eur Ceram Soc 32(10):2273–2281
37. Epicier T, Boulon G, Zhao W, Guzik M, Jiang B, Ikesue A, Esposito L (2012) Spatial distribution of the Yb3+ rare earth ions in Y3Al5O12 and Y2O3 optical ceramics as analyzed by TEM. J Mater Chem 22(35):18221
38. Zhao W, Hreniak D, Boulon G, Strek W, Brenier A, Yin M, Gluchowski P, Lukowiak A, Wiglusz R, Epicier T (2010) Spectroscopic properties of Yb3+-doped Y3Al5O12 nano-ceramics obtained under different sintering pressures. Radiat Meas 45(3-6):304–306
39. Strek W, Marciniak L, Gluchowski P, Hreniak D (2013) Infrared laser stimulated broadband white emission of Yb3+:YAG nanoceramics. Opt Mat 35(11):2013–2017
40. Zhao W, Mancini C, Amans D, Boulon G, Epicier T, Min Y, Yagi H, Yanagitani T, Yanagida T, Yoshikawa A (2010) Evidence of the inhomogeneous Ce 3+ distribution across grain boundaries in transparent polycrystalline Ce 3+ -doped (Gd,Y) 3 Al 5 O 12 garnet optical ceramics. Jpn J Appl Phys 49(2):022602
41. Zhao W, Anghel S, Mancini C, Amans D, Boulon G, Epicier T, Shi Y, Feng X, Pan Y, Chani V, Yoshikawa A (2011) Ce3+ dopant segregation in Y3Al5O12 optical ceramics. Opt Mat 33(5):684–687

42. Zhydachevskii Y, Syvorotka I, Vasylechko L, Sugak D, Borshchyshyn I, Luchechko A, Vakhula Y, Ubizskii S, Vakiv M, Suchocki A (2012) Crystal structure and luminescent properties of nanocrystalline YAG and YAG:Nd synthesized by sol–gel method. Opt Mat 34(12):1984–1989
43. Ramirez MO, Wisdom J, Li H, Aung YL, Stitt J, Messing GL, Dierolf V, Liu Z, Ikesue A, Byer RL, Gopalan V (2008) Three-dimensional grain boundary spectroscopy in transparent high power ceramic laser materials. Opt Express 16(9):5965
44. Xu X, Zhao Z, Song P, Zhou G, Xu J, Deng P (2004) Structural, thermal, and luminescent properties of Yb-doped Y3Al5O12 crystals. J Opt Soc Am B 21(3):543
45. Chani VI, Yoshikawa A, Kuwano Y, Hasegawa K, Fukuda T (1999) Growth of Y3Al5O12:Nd fiber crystals by micro-pulling-down technique. J Cryst Growth 204(1-2):155–162
46. Chani VI (1990) Iron garnet films for optical isolators in wavelength range 800–1300 nm. SPIE 1125:7. doi:10.1117/12.961362
47. Chani V (2004) Crystal-chemistry and fiber crystal growth of optical oxide materials. In: Fukuda T, Rudolph P, Uda S (eds) Fiber crystal growth from the melt. Advances in materials research 6. Springer, Berlin/Heidelberg, pp 129–184
48. Simura R, Yoshikawa A, Uda S (2009) The radial distribution of dopant (Cr, Nd, Yb, or Ce) in yttrium aluminum garnet (Y3Al5O12) single crystals grown by the micro-pulling-down method. J Cryst Growth 311(23-24):4763–4769
49. Jiang B, Gong Z, Chen M, Li J, Liu W, Pan Y (2012) Comparative spectroscopic investigation of Yb3x Y3(1 − x)Al5O12 (x = 3, 5, 10 and 15%) transparent ceramics. Bull Russ Acad Sci Phys 76(6):643–647
50. Boulon G, Guyot Y, Canibano H, Hraiech S, Yoshikawa A (2008) Characterization and comparison of Yb^3+-doped YAlO_3 perovskite crystals (Yb:YAP) with Yb^3+-doped Y_3Al_5O_12 garnet crystals (Yb:YAG) for laser application. J Opt Soc Am B 25(5):884
51. Song Q, Li C-R, Li J-Y, Ding W-Y, Li S-F, Xu J, Deng X-L, Song C-L (2006) Photoluminescence properties of the Yb:Er co-doped Al2O3 thin film fabricated by microwave ECR plasma source enhanced RF magnetron sputtering. Opt Mat 28(12):1344–1349
52. Jusza A, Anders K, Jastrzębska A, Polis P, Olszyna A, Kuś M, Kunicki A, Piramidowicz R (2011) Luminescent and structural properties of Yb3+-doped Al2O3 nanopowders. Opt Mat 33(10):1487–1491
53. Krebs J, Happek U (2001) Yb3+ energy levels in α-Al2O3. J Lumin 94–95:65–68
54. Guyot Y, Steimacher A, Belançon MP, Medina AN, Baesso ML, Lima SM, Andrade LHC, Brenier A, Jurdyc A-M, Boulon G (2011) Spectroscopic properties, concentration quenching, and laser investigations of Yb^3+-doped calcium aluminosilicate glasses. J Opt Soc Am B 28(10):2510

14 Taking Microcavity Label-Free Single Molecule Detection Deep into the Protein Realm: Cancer Marker Detection at the Ultimate Sensitivity

Stephen Arnold, Stephen Holler, and Xudong Fan

Abstract Finding a method for label-free sensing of individual bio-nanoparticles is considered the "Holy Grail" in the bio-sensing field. An ideal technology that could do this would be able to follow the sensing of biological antigen-antibody interactions in their native form and in real-time without interfering tags. It has been over 10 years since the possibility of label-free microcavity detection of single virus or single protein binding by a reactive frequency shift was theorized, and over 5 years since the non-specific detection and sizing of Influenza A (InfA) was demonstrated using this mechanism. The signal to noise ratio in that experiment was only 3. Detecting the smallest virus MS2 with a mass only one hundredth of InfA, therefore seemed hopeless. The prospect of detecting an intermediate size protein such BSA was anticipated to be even further out of reach, since its mass is 5,000× smaller than InfA. However, within the last 2 years both were detected with an extraordinary microcavity that marries micro-cavity photonics with nano-plasmonic receptors. The following article chronicles this advance.

Keywords Single molecule detection • Whispering gallery mode • Plasmonic enhancement • Microcavity • Gold nanoshell • Biosensing • Reactive sensing principle • Smallest virus • Cancer marker • Microresonator • Ring resonator • WGM • LSP • Plasmonic epitope • WGM-nanoplasmonic-hybrid resonator • WGM-nph • Label-free single molecule detection • Thyroglobulin • Bovine serum albumin • BSA

S. Arnold (✉)
MicroParticle PhotoPhysics Laboratory (MP^{3}L), New York University Polytechnic School of Engineering, Brooklyn, NY 11201, USA
e-mail: sarnold935@aol.com

S. Holler
Department of Physics, Fordham University, Bronx, NY, 10458 USA

X. Fan
Department of Biomedical Engineering, University of Michigan, Ann Arbor, MI 48109, USA

B. Di Bartolo et al. (eds.), *Nano-Structures for Optics and Photonics*,
NATO Science for Peace and Security Series B: Physics and Biophysics,
DOI 10.1007/978-94-017-9133-5_14,

14.1 Introduction

Finding a method for label-free sensing of individual biomolecules is considered the "Holy Grail" in the bio-sensing field. An ideal technology that could do this would be able to follow the sensing of biological antigen-antibody interactions in their native form and in real-time without interfering tags [1]. At the NATO School of Spectroscopy at Erice in 2011 an approach for label-free detection of single protein molecules was proposed [2]. It was a "long shot", driven by the hypothetical belief that the measurement of the frequency perturbation of a microcavity resonance upon protein binding could succeed as it had in the measurement of InfluenzaA virus (InfA) [3]. However, this would be no ordinary resonator like the bare Whispering Gallery Mode (WGM) device used in Ref. [3], since the signal to noise ratio (S/N) in that experiment was $\sim$3, and the frequency shift signal is expected to be reduced in proportion to the analyte mass, which for protein is $\sim$5,000$\times$ less than InfA. Researchers at Caltech attempted to reduce the noise by using a bulky reference interferometer, but were only able to push the S/N to $\sim$38 for InfA, thereby falling short of single protein detection by more than a factor of 100 [4]. At the NYU-Polytechnic we took the opposite approach, instead of reducing the noise we attempted to boost the signal. One of the researchers, S.I. Shopova, discovered that a gold nano-shell receptor bound to the sensing region of a microspherical resonator increased the frequency shift signal for an InfA sized polystyrene particle by 4x, [5] and although this enhancement was small, the mechanism by which it occurred opened a new window for single protein detection [6]. The device was now dubbed a WGM-nanoplasmonic-hybrid resonator (WGM-nph). As compared with free space methods for ultra-sensitive detection,[7, 8] the WGM-nph is particularly appealing since it operates with no free-space components, thereby allowing for simplicity and miniaturization. With refinement, the nanoplasmonic enhanced microcavity has enabled the limit of detection (LOD) in the InfA experiment of 192 attograms (192 ag) to be reduced to 10 zeptograms (10 zg, 5 kDa), thereby making the detection of individual protein binding events visible [9]. This chapter describes the label-free single protein quest which consumed our last 2 years.

14.2 Competition and the Way Forward

We were not alone in our pursuit. A number of papers appeared a few months after Ref. [6] with similar ideas, but without demonstrating plasmonic-enhanced detection of a single dielectric nano-particle [10, 11]. By the time a new postdoctoral fellow V.R. Dantham arrived at the MP^3L on September 1, 2011, our direction was clear. First we would attempt single virus detection for the smallest single RNA virus MS2, with a mass only $\sim$1 % of InfA ($\sim$6 ag), and if successful, use the knowledge gained to detect single molecules of protein. Before describing the virus detection experiment we will briefly review the sensing principle responsible for our frequency shift signals.

14.3 Sensing Principle

A great deal of the physics behind detection and light transport by a WGM involves the Maxwell stress tensor, however the sensing of a dielectric particle from the frequency shift of the cavity requires looking for a conservative force, for which its work around a closed loop is zero. That force arises from the gradient of the intensity in the reactive field (evanescent field), and is known as the gradient force [12]. In a dipole approximation the gradient force takes on a familiar form [12]:

$$\mathbf{F} = \frac{\mathrm{Re}[\alpha_{ex}]}{4} \nabla E_0^2(\mathbf{r}). \tag{14.1}$$

where $E_0^2(\mathbf{r})/2$ is the time averaged intensity, and $\mathrm{Re}[\alpha_{ex}]$ is the real part of the excess polarizability of the analyte. The frequency shift may be arrived at from a simple thought experiment.

Imagine a gedanken experiment in which a single photon resides within the cavity and is not allowed out (adiabatic invariance). It pulls on an external dielectric object at position $\mathbf{r}$ (Fig. 14.1) with its reactive field (e.g. the evanescent field). The work done to move the object from $\mathbf{r}$ to the surface at $\mathbf{r}_s$ costs the photon energy, and thereby lowers its frequency by

$$\hbar \Delta\omega = -\frac{\mathrm{Re}[\alpha_{ex}]}{4} \int_r^{r_s} \nabla E_0^2(\mathbf{r}) \cdot d\mathbf{r} = -\frac{\mathrm{Re}[\alpha_{ex}]}{4} \left[E_0^2(\mathbf{r}_s) - E_0^2(\mathbf{r}_s) \right]. \tag{14.2}$$

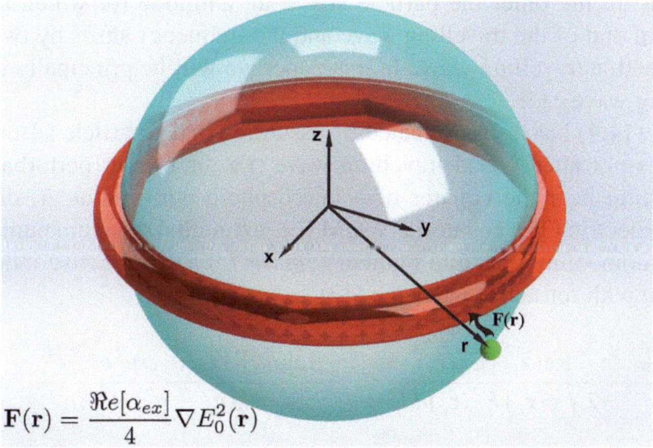

Fig. 14.1 Illustration of a particle being drawn toward the equator of a WGM by the gradient force. The ring around the equator represents the time-averaged intensity of a traveling wave. Vector **r** is the position vector of the particle

With **r** taken to be just beyond the evanescent field

$$\hbar \Delta\omega = -\frac{\mathrm{Re}[\alpha_{ex}]}{4} E_0^2(\mathbf{r}_s). \tag{14.3}$$

So the photon energy shift in absolute terms is the time averaged energy required to polarize the dielectric particle. In a classical sense Planck's constant is a place saver for the energy in the cavity divided by the frequency. By expressing the cavity energy as a volume integral over the sum of the electric and magnetic energy densities, and recognizing that for a very high Q cavity these energy densities are equal we get

$$\frac{\Delta\omega}{\omega} = -\frac{\frac{\mathrm{Re}[\alpha_{ex}]}{4} E_0^2(\mathbf{r}_s)}{\frac{1}{2}\int \epsilon(\mathbf{r}_c) E_0^2(\mathbf{r}_c) dV} = -\frac{\mathrm{Re}[\alpha_{ex}] E_0^2(\mathbf{r}_s)}{2\int \epsilon(\mathbf{r}_c) E_0^2(\mathbf{r}_c) dV} \tag{14.4}$$

The result on the right is now independent of the number of photons, since the number of photons would have multiplied the intensities in both the numerator and denominator. This result, known as the Reactive Sensing Principle (RSP), [13] applies to all cavities, so long as the fields are known.

For a travelling wave such as one that circulates around a microsphere in one direction, the time averaged intensity is independent of the particle's azimuthal position [13]. However, if the light-matter interaction is strong enough, the degeneracy corresponding to clockwise and counter-clockwise modes can be lifted by the nanoparticle's presence, and two states corresponding to symmetric and antisymmetric standing waves (relative to the particle's position) will form [14, 15]. In one, the particle sits at a node, and in accordance with Eq. (14.4) no frequency shift is seen. In the other the particle sits at an antinode for which the intensity doubles from that of the travelling wave and the frequency shifts by twice the shift associated with a travelling wave. In this paper we will be principally interested in the travelling wave case.

Equation (14.4) has been evaluated by assuming that a particle adsorbing on the equator does not alter WGM travelling wave (i.e. first order perturbation). For a protein binding near the equator of a microsphere with radius R, driven into a Transverse Electric (TE) equatorial WGM (i.e. azimuthal quantum number m equal to angular momentum quantum number $l; m = l$) and refractive index n_s, in an environment with refractive index n_e, [13]

$$\frac{\Delta\omega}{\omega} - \frac{\mathrm{Re}[\alpha_{ex}] E_0^2(\mathbf{r}_s)}{2\int \epsilon(\mathbf{r}_c) E_0^2(\mathbf{r}_c) dV} \approx -\frac{\mathrm{Re}[\alpha_{ex}] |Y_{l,l}(\theta,\phi)|^2 e^{-\frac{r_s - R}{L}}}{\epsilon_0 (n_s^2 - n_e^2) R^3} \tag{14.5}$$

where $(r_s - R)$ is the distance from the center of the nanoparticle from to the surface, and L is the characteristic length of the evanescent intensity. For a spherical nanoparticle bound to the surface $(r_s - R)$ is just the radius a of the particle. The exponential factor can be safely set to 1 for a protein, since the protein size (~3 nm) is much less than the length of the evanescent field (~150 nm)[13]. The expression on the right in Eq. (14.5) shows a distinct dependence on radius R. With careful consideration the signal strength is found to go as $R^{5/2}$. It was this equation that directed researchers to reduce the resonator size for the first detection of individual nanoparticles [3]. However reduction of the microcavity size leads to diffractive losses, which eventually spoil Q. For a silica cavity, the smallest reasonable size, for excitation by a free space wavelength of $\lambda_0 = 780\,\text{nm}$, is 35 µm [16].

For an extended object such as a bacterium or large virus the statement below Eq. (14.3) "the photon energy shift in absolute terms is the time averaged energy required to polarize the dielectric particle" still holds true, but treating the target as a point dipole is inappropriate since there can be a substantial variation in evanescent intensity within the structure. The problem is overcome by replacing the polarization energy in the numerator of Eq. (14.4) by the integral over the permittivity difference in the volume of the target analyte,[17]

$$\mathrm{Re}[\alpha_{ex}]E_0^2(\mathbf{r}_s) \rightarrow \mathrm{Re}\int \Delta\epsilon(r)E_b(r)\cdot E_a^*(r)dV, \tag{14.6}$$

where $\Delta\epsilon(r) = \epsilon_d - \epsilon_m$, ϵ_d is the permittivity of the analyte, ϵ_m is the permittivity of the medium, $E_b(r)\cdot E_a^*(r)$ is the product of the fields before and after analyte insertion, and dV is a differential volume element of the analyte.

14.4 Detection of the Smallest Virus

Viruses attack cells, and hijack their biochemical machinery in order to produce clone viruses. Early detection of virus is necessary in order to start early treatment, and thereby limit damage. The WGM biosensor is sensitive to polarizability (Eq. (14.4)), which is proportional to mass. To detect virus at the ultimate sensitivity requires single virus sensitivity, and to ensure that no virus arriving at the detector is missed, we set our sights on one of the smallest viruses; if the smallest can be detected, then all larger viruses will be easier to detect. One of the smallest viruses that attack bacteria, MS2 is smaller than any of the viruses that attack human cells.

The MS2 detection should have been a very challenging since the LOD obtained from the InfA experiments was 192 ag, so attempting to detect a bio-particle such as MS2 with a mass of ~6 ag was well below what we had been able to achieve using a bare microspherical cavity. However, we were optimistic that this might be possible from calculations made at the end of Ref. [6]. These calculations, based on the Reactive Sensing Principle (RSP), suggested that a nanoplasmonic-receptor attached at the equator of the microsphere, and driven into a dipole plasmon resonance, should generate a hybrid mode that would be orders of magnitude more sensitive to receptor binding than binding on a bare microcavity.

We chose as our nanoplasmonic-receptor a gold nanoshell surrounding a silica core, and decided to attempt to attach one to a defect site on a silica microsphere surface, using an approach called Light Force Functionalization (LFF) [18]. To bring

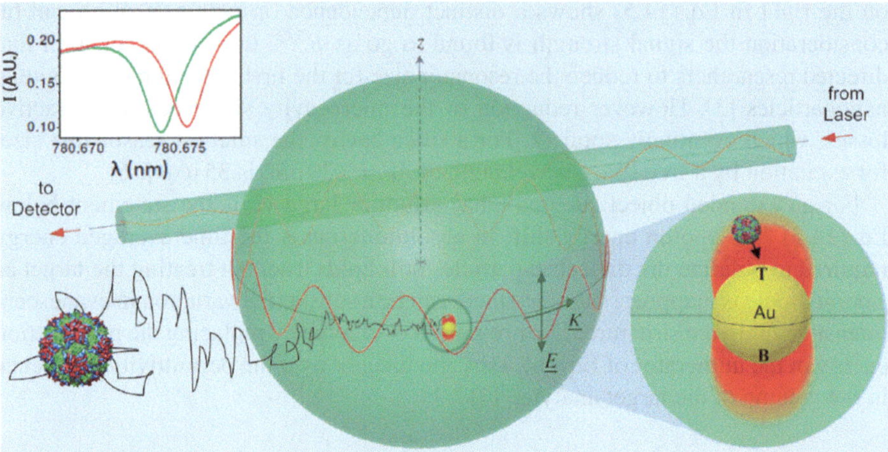

Fig. 14.2 Illustration of an MS2 virus diffusing toward a WGM-nph resonator. A Carousel Trap is formed within the evanescent intensity of the WGM that propels the MS2 virus along the equator where it meets a Plasmonic Epitope driven into a dipole mode

the nanoshell to the equator we used the gradient force [12] provided by the evanescent field of the WGM.[19] In our case the gradient is radially inward, so the Re[α_{ex}] must be positive in order that the nanoshell be drawn to the equator. Since the nanoshell is essentially an electro-mechanical oscillator, its dipole moment reverses direction relative to the driving field as it is driven beyond its resonance frequency. Consequently to pull a nanoshell toward the equator, the WGM must be on the low frequency side of plasmon resonance (i.e. longer wavelength). Choosing a colloid of shells with resonant frequencies slightly above the WGM easily does this. The optimal inward force occurs half a linewidth below the plasmon resonance. Since the attached shell protrudes from the surface and will act as a receptor for virus and protein, it will be called a "plasmonic epitope" [20].

Guiding the virus to this nanoshell requires other forces. Figure 14.2 shows the assembled detector supporting a WGM around the microcavity equator with a virus particle diffusing within $\sim$200 nm of the surface. Unlike the nanoshell, the virus is dielectric and therefore has a polarizability that is positive, independent of frequency. Consequently the virus is drawn toward the equator of the resonator by a radially directed gradient force. At neutral pH a net negative charge develops on the protein coat of the virus, causing it to be repelled by the silica microsphere due to surface bound negative charges associated with the ionized silanol groups. The combination of this repulsive Debye interaction and the attractive gradient force leads to a radial potential well. At the same time the circulating momentum flux propels the particle toward the plasmonic epitope principally by scattering WGM photons. Overall, the virus is said to be under the influence of Carousel-trap forces [19]. The motion that ensues carries the virus toward the plasmonic epitope [21]. The gradient force near the plasmonic hot spots is orders or magnitude greater than near the microcavity surface, thereby increasing the probability for capture by the epitope.

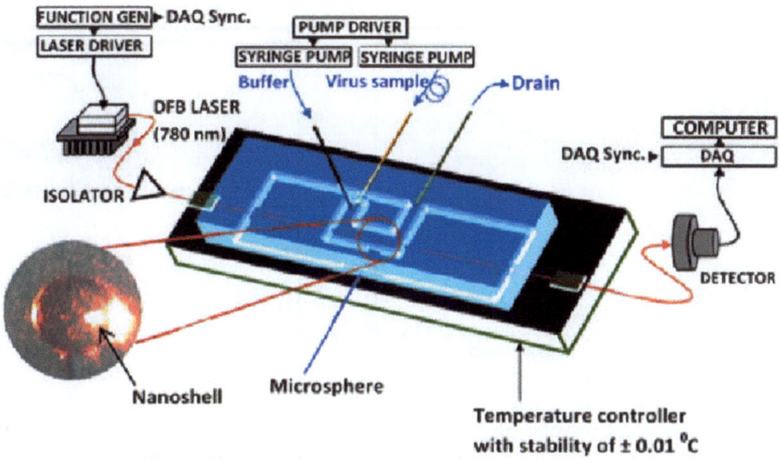

Fig. 14.3 Microfluidic WGM biosensor with the image of an assembled WGM nanoplasmonic hybrid (WGM-nph) resonator

The WGM-nph assembly and subsequent MS2 detection experiments were carried out in a home-made microfluidic cell shown in Fig. 14.3 [16]. The cell included a tapered fiber that guided light from a tunable distributed feedback (DFB) laser with a free space wavelength near 780 nm. A WGM within a microsphere (R $\sim$45 μm) was stimulated by contact with this fiber, and detected from a dip in the transmission through the fiber. The MS2 binding experiments are described in what follows.

Before assembling the hybrid system we repeated binding curves (i.e. wavelength shift vs. time following MS2 injection) on a bare cavity as we had done earlier in 2007 [16]. The results were essentially the same as those in our earlier work. The binding curves were smooth with no steps. This is consistent with our calculations based on the RSP that showed that binding an MS2 virus particle on the equator of a 45 μm radius SiO_2 microcavity should give an optimal wavelength shift of $\sim$0.25 fm which was far less than our LOD of 2 fm. Next we washed the cell and attached a Gold nanoshell to the equator (as shown in Fig. 14.3). The nanoshell colloid was picked to contain shells that resonate a few nanometers below 780 nm.

Upon injecting the MS2 virus in the presence of the WGM-nph the binding curves were remarkably different from our earlier work, [16] as shown in Fig. 14.4a. In addition to smooth portions there were now steps with heights as much as 68x what could be expected on the bare silica equator. These steps had a signal to noise ratio S/N $\sim$8. Over the entire experiment 28 steps were recorded that were associated with binding to the plasmonic nanoshell, and the resonance that we were tracking had a modest Q of 4×10^5 [inset in Fig. 14.4a]. Lowering the salt concentration in solution can eliminate binding to the silica surface, by increasing the Debye repulsion between the negatively charged silica surface and like charged virus, but we were impressed that the nanoshell binding results could easily be detected

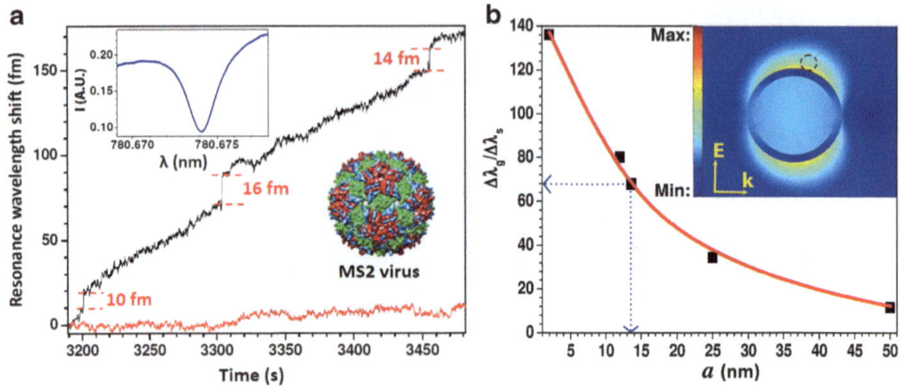

Fig. 14.4 (**a**) Resonance wavelength shift curve shows steps associated with binding to the gold nano-shell built up on a gradual increase due to nonspecific binding to the silica surface of the microsphere. *Inset* shows the resonance spectrum; (**b**) Simulation of the ratio of wavelength shifts for attachment of a virus particle of radius a to the gold hotspot of the nanoshell to that for attachment to the silica surface of the microsphere. *Inset* shows Comsol simulation of the near-field intensity with a particle of 10 nm radius

even in the presence of so much non-specific binding to the rest of the resonator. Light assisted binding to the nanoshell is clearly associated with the steps we have detected

A key question remained: Can the size of the virus be extracted from the measured signal? The answer will be affirmative for the largest wavelength shift. However there are some limitations to obtaining a simple analytic expression imposed by using commercial nanoshells. Unfortunately a shell having an outer circumference to wavelength ratio of $2\pi r_2/\lambda = 2\pi(71.5\,\text{nm})/780\,\text{nm} = 0.58$ can be treated as a Rayleigh particle only in a gross approximation. This led us to simulate the intensity using the Finite Element Method (FEM, Comsol). Figure 14.4b shows a FEM simulation result at 780 nm for a shell with a core radius of 60 nm and a shell thickness of 11.5 nm having a virus (refractive index = 1.5, typical of virus) of radius $a = 12.5$ nm on its point of highest intensity. The dipole pattern is apparent along with some distortion caused by the electromagnetic wave changing phase across the structure. This causes the highest intensity point to be slightly in the forward direction. The enhancement we are interested in is the wavelength shift for the virus seated on the gold hot spot to that on the silica equator $\xi = \Delta\lambda_g/\Delta\lambda_s$ which according to the RSP is the polarization energy of the virus in contact with the nanoshell in comparison with that on the bare silica equator. Because of the non-uniform field the substitution in Eq. (14.6) is needed, which requires integration over the body of the virus [17];

$$\xi = \frac{\Delta\lambda_g}{\Delta\lambda_s} = \frac{\int \Delta\epsilon \left(E_b \cdot E_a^*\right)_g dV_{analyte}}{\int \Delta\epsilon \left(E_b \cdot E_a^*\right)_{sl} dV_{analyte}} \tag{14.7}$$

We modeled this enhancement for the gold nanoshell having a smooth outer surface ξ_s as a function of the virus radius a using FEM and found an analytical expression using the shell parameters in Fig. 14.4b;

$$\xi = \Delta\lambda_g/\Delta\lambda_s = A[r_2/(r_2 + fa)]^6 \tag{14.8}$$

with $A = 155.56$ and $f = 0.76$. The wavelength shift ratio reaches 155.56 for an infinitesimally small adsorbate and falls rapidly as the adsorbate increases in size (Fig. 14.4b) due to the attenuation of the near field with range. Equation (14.8) is the key to finding the size of the virus from the experimental wavelength shift $\Delta\lambda_g$ but requires working out $\Delta\lambda_s$. The latter is proportional to the polarizability from Eq. (14.5), which is in turn proportional the volume of the virus; $\Delta\lambda_s \propto a^3$. With the appropriate asymptotic expression for the rest of Eq. (14.5) [3] incorporated into Eq. (14.7), we arrived at an estimate for the virus size

$$a \approx \frac{G}{1 - \frac{2Gf}{r_2}}, \quad \text{where } G = \frac{R^{5/6}}{D^{1/3}A^{1/3}\lambda_r^{1/6}}(\Delta\lambda_g)^{1/3}, \tag{14.9}$$

the dimensionless parameter D which includes the dielectric properties of the virus, the microsphere and water has a value of 1.50 [3]. For the largest measured shift of 17 fm we obtain a viral radius of 13.3 nm that is in excellent agreement with a neutron diffraction analysis of MS2 for which the reported outer radius is 13.6 ± 1.0 nm [22].

It should be cautioned that this "excellent agreement" may be fortuitous, since only the average physical properties of the gold shell colloid are known. For a better understanding, spectroscopy of the nanoshell epitope must also be done. In addition, as we will see in the next section, the nanoshells do not have a smooth coating of gold. Bumps on the gold surface would tend to increase the intensity on the nanoshell above what is calculated for a smooth shell.

With a S/N of 8, the possibility of detecting even smaller bio-particles is realistic. This actually goes below a factor of 1/8 in the detectable mass, since there is a distinct increase in the projected enhancement $\Delta\lambda_g/\Delta\lambda_s$ as the bioparticles becomes smaller as revealed in Fig. 14.4b; the field is more intense as we come closer to the nanoshell's surface. Based on the smooth shell hypothesis the detection of a bio-particle having a mass down to 0.4 ag becomes possible. This mass is in the single protein realm. For example, the Thyroid Cancer marker, Thyroglobulin (Tg), has a mass of 1 ag. We made a decision to attempt to detect single molecules of Tg.

14.5 Single Molecule Detection of Thyroid Cancer Protein and Beyond [22]

At this juncture it is best to translate mass into molecular weight. With each ag = 6×10^5 Da, the mass of Tg is 600 kDa. It is an important protein. An operation to remove a cancerous thyroid should remove all traces of new Tg production.

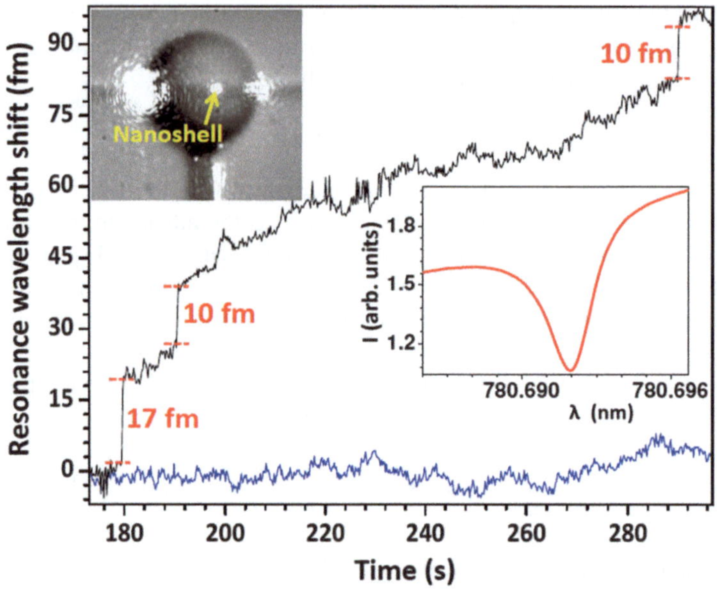

Fig. 14.5 Resonance wavelength shift curve shows steps associated with Tg protein adsorbing to the gold nanoshell attached to the equator of a WGM-nph resonator with R ∼35 μm

Consequently it is carefully monitored, and although single molecule detection of Tg is currently not available on a point-of-care basis, it would represent the ultimate sensitivity.

Based on our projections from the MS2 measurements we expected to possibly be able to detect Tg with a S/N ∼2, but we were in for a surprise.

First we built a new bare resonator as we had done in our experiments on MS2. Next Tg was injected at 300 fM concentration and a smooth binding curve was observed. This seemed reasonable for the particular resonator (R ∼35 μm) for which the arrival of a single Tg protein was expected to produce a signal no larger than 0.06 fm, which was well below our LOD of ∼1 fm. At this point we put together a WGM-nph resonator and the surprises began. We easily observed steps due to Tg, and they were not small (Fig. 14.5). To be more precise they were as large as 22 fm for a total enhancement beyond a bare cavity of ∼360×. This represented 240 % beyond our projection for a smooth shell epitope; for the smooth shell $\xi = 150$. For a month or so we thought we might have clusters of molecules, and drove the concentration down to see if the signal got smaller, but it did not. So we hypothesized that the problem was in the physical makeup of our system. Perhaps the shells were not smooth.

At our TEM facility we were able to see the actual structure of nanoshells in our colloid (Fig. 14.6). The shells were covered with bumps up to 10 nm in radius (Fig. 14.5) [22]. It was hypothesized that this might lead to a further field enhancement.

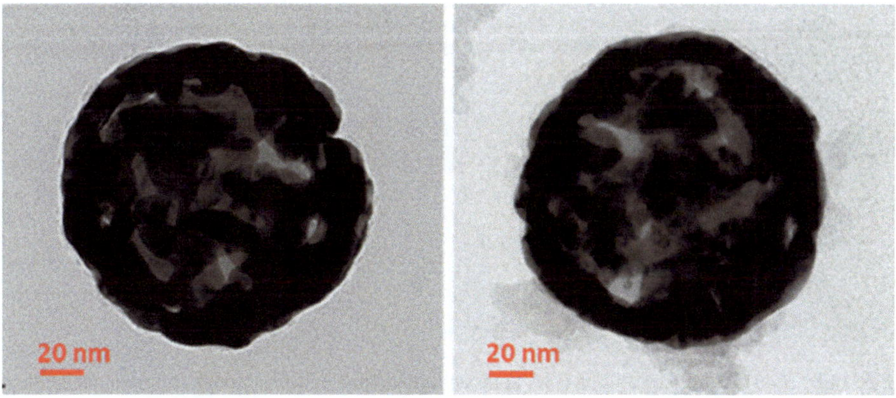

Fig. 14.6 TEM images of silica core-gold nanoshells

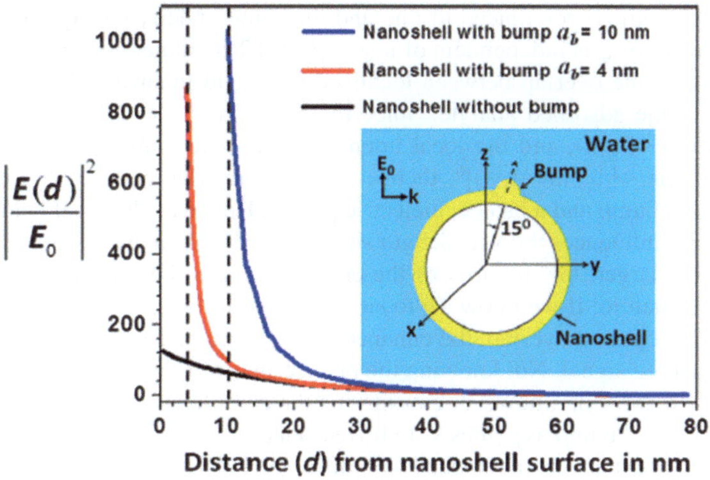

Fig. 14.7 Enhancement of intensity just above hemispherical bumps of different radii (a_b) on a gold nanoshell (inner radius $r_1 = 60$ nm and outer radius $r_2 = 71.5$ nm) at the dipole resonance of the nanoshell at 780 nm (*blue* and *red*) and enhancement above the smooth nanoshell (*black*). All separations d are measured from the nanoshell surface along the *dashed line* shown in the *inset*

To understand the extent to which these bumps might lead to a further enhancement we performed Finite Element Method (FEM, Comsol) calculations on a shell with a hemi-spherical bump in the region where the largest intensity had been seen on the smooth shell (Fig. 14.4b). The results are shown in Fig. 14.7. Just above a 10 nm radius bump the intensity was calculated to increase by 840 % over the intensity above the bare shell. Of course not all of this 840 % can be realized since the field above the bump is not uniform. Now the major question is: Do the signal levels for Tg make sense? This can be answered based on the reactive sensing principle (RSP), by evaluating Eq. (14.7) including the bump.

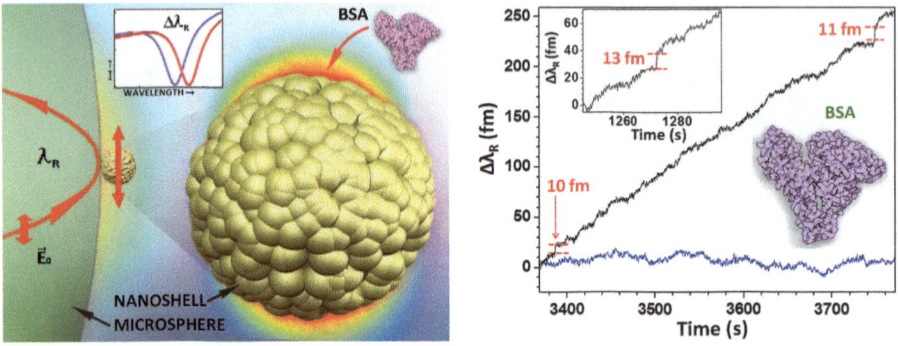

Fig. 14.8 (*Left*) An illustration of BSA approaching a nanoshell plasmonic epitope attached to a microsphere with R ~35 nm. (*Right*) The resulting signal

Since this shift is referenced to one and the same underlying high quality (Q) bare microcavity, ξ is independent of losses [6, 21] The numerator of the Eq. (14.7) depends upon the overlap between localized near field intensity of the bump and the form of the adsorbed analyte. Since the bump radius a_b is on the order of a protein size (~5 nm), and the local intensity a distance s above the bump drops off roughly as $\sim[a_b/(a_b + s)]^6$, there is effective overlap between the near field intensity of a bump and a protein dielectric profile. However, the local field near the bump hardly influences the wavelength shift signal due to an MS2 virus, since it is considerably larger; the majority of the enhancement in this case comes from the nanoshell structure. If we allow Tg to sit on the 10 nm bump in an orientation that produces the optimal overlap, the enhancement $\xi = 245$. This explains two thirds of the measured signal. Not knowing the precise shape of the gold surface hampers our ability to say more, however considering the complexity that we are deal with, we consider the "bump hypothesis" to be reasonable.

We conclude that the nano-shell has different levels within its near field. At the 20–30 nm level we are dealing principally with the smooth shell enhancement, whereas at the 5–10 nm level the smooth shell fields polarize the bump causing an additional enhancement. Figure 14.7 reveals that the closer the analyte is to the bump the larger the expected enhancement. With a smaller molecule such as Bovine Serum Albumin (BSA) with a mass of only 0.11 ag, the characteristic thickness is ~3.4 nm and the overlap with the bump field is considerable greater than for Tg, for which the thickness is ~11 nm. Our path to the next challenge was clear, and the results clearer (Fig. 14.8) [23].

The goals we set for ourselves are fulfilled, although with initially unexpected twists and turns. The signal to noise ratios for our best signals for Tg and BSA are 22:1 and 13:1, respectively. From the S/N in the BSA experiment we conservatively estimate the protein limit of detection (LOD) in our WGM-h to be the mass of BSA divided by 13, or ~0.008 ag (5 kDa). Consequently it should be easy to detect single protein markers of various dangerous cancers such as breast cancer (CA15-3, 290 kDa), lung cancer (ALK, 176 kDa), liver cancer (AFP,

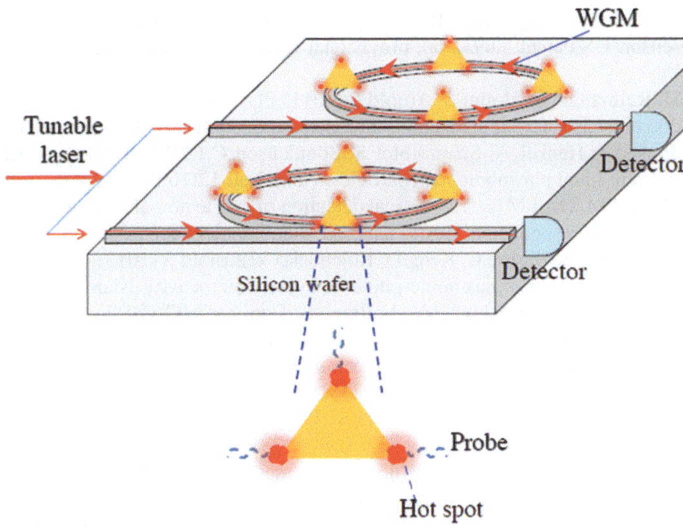

Fig. 14.9 A multiplexed ring resonator sensor on a chip

70 kDa) and melanoma (BRAF, 84 kDa). The final step is to integrate this plasmonic enhancement mechanism into a sensor that can efficiently and specifically capture target molecules.

Our answer to integration is to add plasmonic epitopes to silicon photonic WGM ring resonators as illustrated in Fig. 14.9 [20]

Acknowledgements Steve Arnold thanks Rino DiBartolo for his kind invitation to lecture at this NATO advanced study institute. He would also like to think all of the other instructors for teaching him about their research and especially for introducing him to the interesting world of plasmonics. In this respect he would also like to especially thank Mark Stockman for many conversations over pasta and wine. The research described herein was supported by the National Science Foundation grants CBET 0933531 and EECS 1303499.

References

1. Fan X, White IM, Shopova SI, Zhu H, Suter JD, Sun Y (2008) Sensitive optical biosensors for unlabeled targets: a review. Analytica Chimica Acta 620:8–26
2. Arnold S, Dantham VR, Rivilis N, Holler S (2013) Taking whispering gallery mode biosensing to the single protein limit: the birth of nanoplasmonic-microcavity hybrid sensors. In: DiBartolo B, Collins J (eds) Nano-optics for enhancing light-matter interactions on a molecular scale. NATO science for peace and security series B: physics and biophysics. Springer, Dordrecht, pp 271–285
3. Vollmer F, Arnold S, Keng D (2008) Single virus detection from the reactive shift of a whispering-gallery mode. Proc Natl Acad Sci U S A 105:20701–20704
4. Lu T et al (2011) High sensitivity nanoparticle detection using optical microcavities. Proc Natl Acad Sci U S A 108:5976–5979

5. Shopova SI, Arnold S, Rajmangal R (2013) Plasmonic enhancement of a whispering gallery mode biosensor. US Patent 8,493,560, provisional application no. 61/372,010, filed on 9 Aug 2010
6. Shopova SI, Rajmangal R, Holler S, Arnold S (2011) Plasmonic enhancement of a whispering gallery mode biosensor for single nanoparticle detection. Appl Phys Lett 98:243104
7. Ament I, Prasad J, Henkel A, Schmachtel S, Sönnichsen C (2012) Single unlabeled protein detection on individual plasmonic nanoparticles. Nano Lett 12:1092–1095
8. Zijlstra P, Paulo PM, Orrit M (2012) Optical detection of single non-absorbing molecules using the surface plasmon resonance of a gold nanorod. Nat Nanotechnol 7:379–382
9. Dantham VR, Holler S, Barbre C, Keng D, Kolchenko V, Arnold S (2013) Label-free detection of single protein using a nanoplasmonic-photonic hybrid microcavity. Nano Lett 13:3347–3351
10. Santiago-Cordoba MA, Boriskina SV, Vollmer F, Demirel MC (2011) Nanoparticle-based protein detection by optical shift of a resonant microcavity. Appl Phys Lett 99:073701
11. Swaim JD, Knittel J, Bowen, WP (2011) Detection limits in whispering gallery biosensors with plasmonic enhancement. Appl Phys Lett 99:243109
12. Ashkin A, Dziedzic JM, Bjorkholm JE, Chu S (1986) Observation of a single-beam gradient force optical trap for dielectric particles. Opt Lett 11:288–290
13. Arnold S, Khoshsima M, Teraoka I, Holler S, Vollmer F (2003) Shift of whispering-gallery modes in microspheres by protein adsorption. Opt Lett 28:272–274
14. Teraoka I, Arnold S (2009) Resonance shifts of counterpropagating whispering-gallery modes: degenerate perturbation theory and application to resonator sensors with axial symmetry. JOSA B 26:1321–1329
15. Zhu J, Ozdemir SK, Xiao YF, Li L, He L, Chen DR, Yang L (2009) On-chip single nanoparticle detection and sizing by mode splitting in an ultrahigh-Q microresonator. Nat Photonics 4: 46–49
16. Arnold S, Ramjit R, Keng D, Kolchenko V, Teraoka I (2008) Microparticle photophysics illuminates viral bio-sensing. Faraday Discuss 137:65–83
17. Teraoka I, Arnold S (2009) Theory of resonance shifts in TE and TM whispering gallery modes by nonradial perturbations for sensing applications. JOSA B 23:1381–1389
18. Arnold S, Holler S, Keng D, Shopova SI (2010) Functionalizing a sensing ribbon on a whispering gallery mode microresonator using light force to fabricate a whispering gallery mode sensor. US Patent application 12/783,367, filed 19 May 2010
19. Arnold S, Keng D, Shopova SI, Holler S, Zurawsky W, Vollmer F (2009) Whispering gallery mode carousel. Opt Express 17:6230–6238
20. Arnold S, Dantham VR, Barbre C, Garetz BA, Fan X (2012) Periodic plasmonic enhancing epitopes on a whispering gallery mode biosensor. Opt Express 20:26147–26159
21. Dantham VR, Holler S, Kolchenko V, Wan Z, Arnold S (2012) Taking whispering gallery-mode single virus detection and sizing to the limit. Appl Phys Lett 101:043704–043704
22. Kuzmanovic DA, Elashvili I, Wick C, O'Connell C, Krueger S (2003) Bacteriophage MS2: molecular weight and spatial distribution of the protein and RNA components by small-angle neutron scattering and virus counting. Structure 11:1339–1348
23. Dantham VR, Holler S, Barbre C, Keng D, Kolchenko V, Arnold S (2013) Label-free detection of single protein using a nanoplasmonic-photonic hybrid microcavity. Nano Lett 13:3347–3351

15
Fluorescence and Raman Scattering in Plasmonic Nano-structures: From Basic Science to Applications

Sergei Gaponenko

Abstract Enhanced light–matter interaction in presence of metal nanobodies opens an avenue towards practical applications of plasmonically enhanced fluorescence and Raman scattering in medicine, biology, forensic research, cultural heritage examination. An overview of experimental results and an outlook is provided in terms of practical implementation of plasmonic effects in medicine and in other fields.

Keywords Plasmonics • Nanosensors • Fluorescence • Raman scattering • Biomedicine

15.1 Introduction

Metal nanostructures capable to support plasmonic excitation by incident light essentially modify light–matter interaction resulting in local concentration of incident light, modification of radiative and non-radiative transition rates, and the probability of photon scattering. Thus all types of secondary emission (spontaneous emission and scattering of photons) may experience considerable enhancement under certain conditions [1–5]. A rather general consideration of incident field concentration and photon density of states concentration provides a rationale for huge enhancement factors for Raman scattering [1, 5–7] and noticeable enhancement factors for luminescence [5, 8]. Experimental performance of enhanced secondary emission for atomic, molecular systems, semiconductor quantum dots and inorganic microcrystals using randomized, multilayer and spatially organized metal-dielectric nanostructures are extensively studied by many groups ([1–5] and Refs. therein).

S. Gaponenko (✉)
National Academy of Sciences, Nezavisimosti Ave 68, Minsk 220072, Belarus
e-mail: s.gaponenko@ifanbel.bas-net.by

B. Di Bartolo et al. (eds.), *Nano-Structures for Optics and Photonics*,
NATO Science for Peace and Security Series B: Physics and Biophysics,
DOI 10.1007/978-94-017-9133-5_15,

In this contribution modeling and experimental performance of plasmonically enhanced luminescence and Raman scattering for simple but instructive systems is overviewed aimed at optimizing the design of metal nanostructures to get high net enhancement for desirable practical tasks. Special attention is given to luminescence and Raman scattering for semiconductor nanocrystals (quantum dots) provided these are considered as promising fluorescent and Raman markers in biomedical application. An overview of potential applications is provided towards extensive implementation of plasmonic structures in analytical instrumentation including material analysis and biomedical studies with fluorescent and Raman testers.

15.2 Fluorescence Enhancement

15.2.1 Modeling

For a system in question (a molecule or a nanocrystal near a metal nanobody(ies)), the intensity of fluorescence is proportional to the absorbed light intensity and the quantum yield Q [2, 5, 8]. Supposing the absorption coefficient of light by a molecule (nanocrystal) in a plasmonic nanostructure remains the same as in vacuum and intensity independent, and supposing the vacuum fluorescence features fully radiative type of downward transition (i.e. quantum yield equals 1), the overall factor of luminescence intensity modification F at a point with r vector reads

$$F(\omega, \omega', \mathbf{r}) = G(\omega, \mathbf{r})Q(\omega', \mathbf{r}) = \frac{|\mathbf{E}(\omega, \mathbf{r})|^2}{|\mathbf{E}_0(\omega)|^2} \frac{\gamma_{\mathrm{rad}}(\omega', \mathbf{r})}{\gamma_{\mathrm{rad}}(\omega', \mathbf{r}) + \gamma_{\mathrm{nonrad}}(\omega', \mathbf{r})} \tag{15.1}$$

where ω, (ω') is excitation (emission) radiation frequency, $\mathbf{E}(\mathbf{E}_0)$ is the electric field of light wave with (without) a metal nanobody.

$$G = \frac{|\mathbf{E}(\omega, \mathbf{r})|^2}{|\mathbf{E}_0(\omega)|^2} \tag{15.2}$$

is the intensity enhancement facto for excitation light, and

$$Q = \frac{\gamma_{\mathrm{rad}}(\omega', \mathbf{r})}{\gamma_{\mathrm{rad}}(\omega', \mathbf{r}) + \gamma_{\mathrm{nonrad}}(\omega', \mathbf{r})} \tag{15.3}$$

is the luminescene quantum yield in presence of a nanobody. Here γ_{rad} and γ_{nonrad} are radiative and nonradiative transition probabilities, respectively.

Equation (15.1) gives no chance to offer an a priori estimation of fluorescence enhancement. Though noticeable overlap of metal nanostructure extinction spectrum with fluorescence excitation spectrum is the primary precondition of the very plasmonic effect to come into play, there are three principal factors, G, γ_{rad} and γ_{nonrad} to be examined explicitly to get the net enhancement (or quenching) factor. Notably, each of the three above factors features spectral and distance dependence. Moreover, incident field enhancement factor is calculated based on classical electrodynamics and here the mutual orientation if incident electric field and wave vector are important whereas modification of quantum yield is essentially the quantum electrodynamics problem with the probe emitting dipole orientation and its displacement versus nanoparticle to be the major parameters. Radiative downward transitions in metal nanostructures experience modification because of the modified photon local density of states (DOS) since the transition rate is proportional to local DOS. Enhanced local DOS results in enhanced radiative decay rate. Simultaneously, non-radiative decay is also promoted resulting in heating of a metal nanobody [3, 5].

To summarize, the calculation of the overall enhancement (or quenching) breaks into the two separate problems, each depending on the specific list of parameters. The first problem is classical analysis of light intensity near a metal nanobody. For a few simple systems this problem can be solved analytically. However, the comparison of analytical and numerical modeling showed that for a metal body bigger than 20 nm scattering becomes important which cannot be calculated analytically [8]. The second problem is quantum electrodynamical consideration of downward quantum transitions, both radiative and non-radiative for an excited molecule (nanocrystal) located in the vicinity of metal nanobody(ies). To develop certain intuition in the field of metal enhanced fluorescence numerical modeling is helpful with spectral and topological mapping of the desirable enhancement factors over the geometrical space (molecule coordinates and dipole moment orientation) and the spectral space including emission wavelength, excitation wavelength and polarization of light. One can see the F factor readily becomes the function of three geometrical coordinates, light polarization, dipole moment orientation, incident (excitation) wavelength and emission wavelength.

To make a reasonable presentation of the modeling results, a number of parameters should be fixed. In what follows we shall fix the emission wavelength at 530 nm which is e.g. the case of fluorescein, a common commercial fluorescent molecule. Then we shall consider the simplest nanobody type, a single sphere, choose silver as the most promising and widely used plasmonic material, fix light polarization along geometrical alignment of a body and an emitter (see insert in Fig. 15.1). The variable parameters will be the metal sphere size, the excitation wavelength, the spacing between an emitter and a body. To get closer to the real experimental situation, averaging over dipole moment orientation of an emitter will be performed.

It was found that factor F strongly depends on nanoparticle size, excitation wavelength and spacing between a molecule and the metal nanoparticle. The highest enhancement factor was found to be equal to 50 and occurs for a 50 nm particle at excitation wavelength 368 and 6 nm spacing. Representative data are given in Figs. 15.1–15.3. Figure 15.1 shows dependence of enhancement factor versus spacing for one selected excitation wavelength 410 nm. Notably, this figure shows that enhancement is actually an interplay of growing intensity and falling quantum yield.

Figure 15.2 presents the maximal values of enhancement calculated for different size of silver nanosphere. The optimal distance between molecule and metal particle in every case was about 6–8 nm. However the excitation wavelength where the

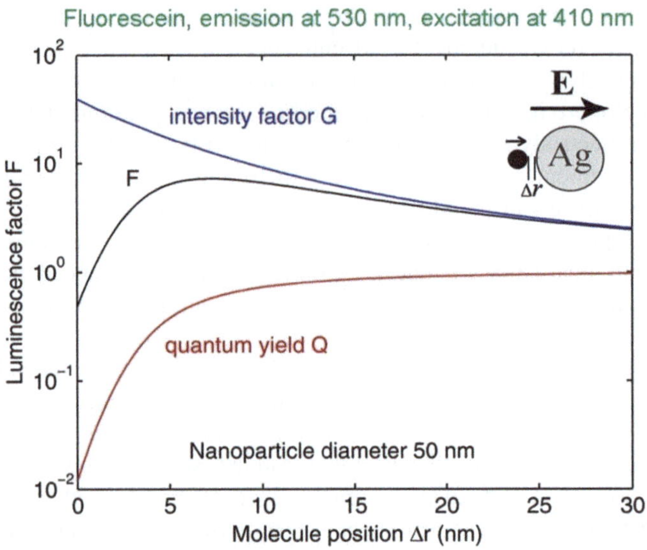

Fig. 15.1 Dependencies of the incident intensity modification factor G, quantum yield Q, and the total luminescence intensity factor F on a molecule distance Δr to a silver nanoparticle surface with diameter 50 nm. The excitation wavelength is 410 nm, the emission wavelength is 530 nm. The molecule dipole moment is normal to a silver particle surface and is parallel to the electric field E of the excitation light (Adapted from [8])

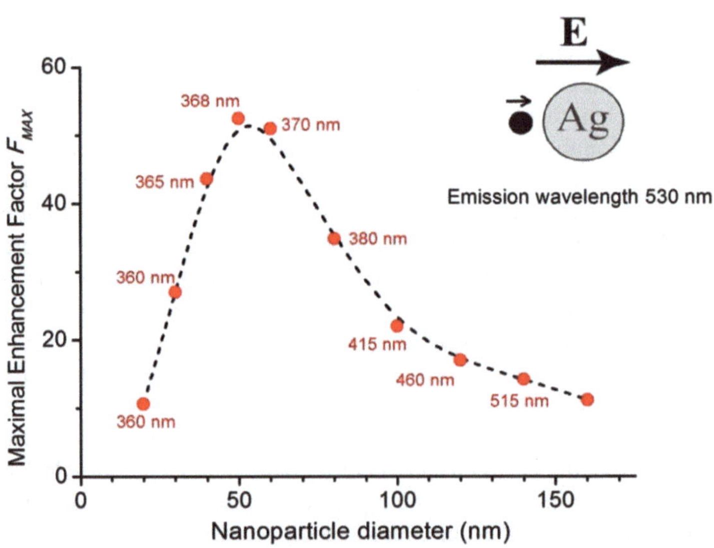

Fig. 15.2 Calculated maximal values of luminescence enhancement factor FMAX versus silver nanoparticle diameter for optimized combination of excitation wavelength and molecule–nanoparticle distance. Emission wavelength is 530 nm everywhere. Excitation wavelength for which the enhancement factor obeys its maximal value is indicated for every size value (Adapted from [8])

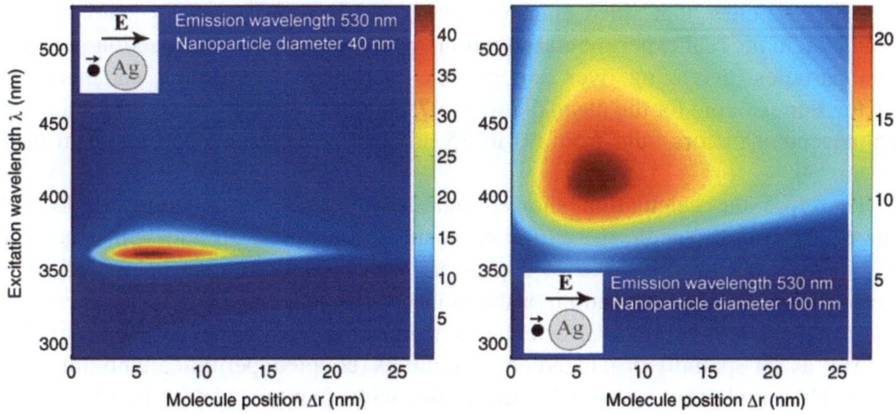

Fig. 15.3 Mapping of the fluorescence enhancement factor for the emission wavelength 530 nm (fluorescein) and two different values silver nanoparticle diameter, 40 nm (*left*) and 100 nm (*right*) versus excitation wavelength and molecule nanoparticle spacing (Adapted from [8])

extreme enhancement occurs strongly differs for different size of nanoparticles (numbers in the graph in Fig. 15.2). Notably, the excitation wavelength for extreme enhancement monotonically shifts with size in correlation with size dependent exctinction spectrum. Larger particles feature higher extinction in the long-wave side because of growing contribution from scattering. Note, for practical application one should bear in mind that a molecule is to be efficiently excited, i.e. the molecule absorption spectrum should fit the spectral band where enhancement factor is high.

Figure 15.3 shows representative mapping of fluorescence enhancement in terms of enhancement factor versus excitation wavelength and molecule position. First, one can see that the optimal spacing of 6–8 nm remains the same for rather different nanoparticle size (100 versus 40 nm). Second, one can see that though 40 nm particle offers higher enhancement (about 50), the spectral range of enhancement with respect to excitation wavelength is pretty narrow and measures just a few nm. In many cases it may not fit the absorption spectrum. Bigger particles offer much wider spectral range for excitation by an expense of lower enhancement. For example, 100 nm particles promise more than 10-fold enhancement over excitation range from 370 to 520 nm, and more than 20 in the excitation range 400–430 nm. This property makes bigger particles more reasonable when applied for molecules emitting in the green.

It is clear that since enhancement factor depends on orientation (alignment) of incident light polarization and dipole moment of emitter with respect to the line connecting an emitter and the nanoparticle center, possible randomization of physically adsorbed molecules on the metal surface will result in lower enhancement. For the representative example of randomly oriented fluorescein molecules near silver nanoparticles we found 3–4-fold drop in enhancement [8].

It is also clear that since enhancement factor is sensitive to the nanoparticle diameter, in the real experimental situations with inevitable size dispersion of metal nanoparticles, averaging over nanoparticle diameters reduces the enhancement factor. For example for the mean diameter 47 nm and dispersion 30 nm, the maximal enhancement factors falls from 50 to 15 as the size dispersion grows from 0 to 30 nm [9].

Therefore the calculation showed that for a single silver nanoparticle and the fluorescein molecule the extreme enhancement equals 50 and will occur for 50 nm diameter at a distance of approx. 5–7 nm for optimal alignment and light polarization. Otherwise, random orientation and size dispersion make enhancement to drop down below 10. Higher enhancement is expected for ellipsoidal particles as well as for spatially organized nanostructures (coupled, periodic, nano-antennas etc.). This prediction is valid for molecules with quantum yield equal to 1. For molecules with low quantum yield possessing intrinsic nonradiative path, the additional nonradiative decay promoted by metal proximity is not valuable and huge factors like 10^3 can be obtained [10].

15.2.2 Experiments

In medicine, cancer development in a human body can be traced through analysis of presenting antigens which are specific for cancer diseases and are termed tumor markers. Tumor markers, in turn, can be identified when being coupled to antibodies which are specific for every marker. Antibodies can be identified by means of fluorescence detection when being chemically labeled with highly luminescence molecule, such as fluorescein. This approach is used both for diagnostic purposes and for tracing cancer patients after treatment. Enhanced detection of labeled antibodies in medicine is the possible most valuable application of surface enhanced fluorescence with respect to highly luminescing labels. Many groups have reported on successful implementation of fluorescence of molecules deposited on a silver or gold substrate with controllable spacing [11–23]. Since quantum dots are considered as potentially more stable fluorescent labels to replace organic labels, the activity on plasmonically enhanced quantum dot luminescence is also performed [24–30]. Recently several groups purposefully applied plasmonic enhancement to labeled antibodies [31–38]. Typical enhancement factors are about 10 times with only a few reports on higher enhancement values. Notably, higher enhancements are typically reported for labels featuring low quantum yield. In this case, though enhancement factors get higher because induced quenching is masked by the intrinsic one, high enhancement should not be overemphasized since the overall light harvested by a detector is defined by the product of the enhancement factor and original non-enhanced intensity which appears to be very low for molecules with intrincicaly low quantum yield Q_0. One can understand that e.g. 10-fold enhancement of fluorescence for molecules with $Q_0 = 1$ gives the same intensity as 1,000-fold

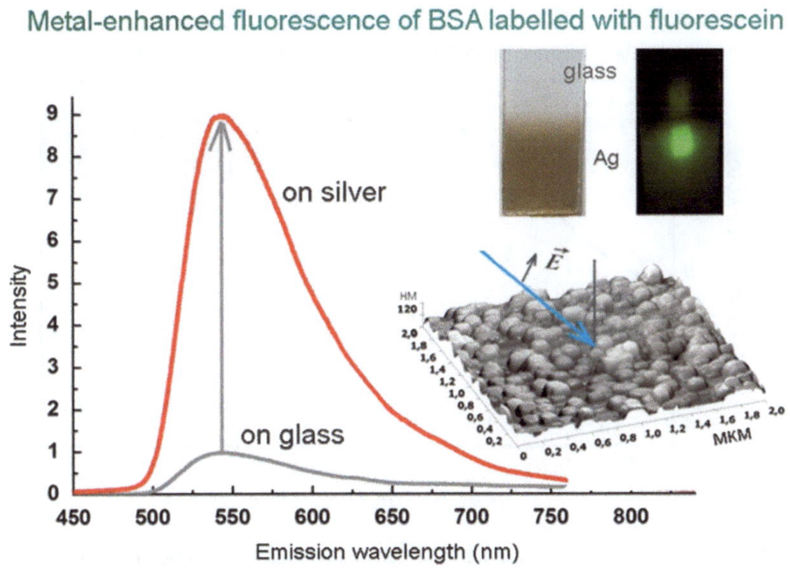

Fig. 15.4 Fluorescence enhancement for labeled bowine serum albumin with a silver substrate

enhancement for molecules with $Q_0 = 0.01$. Nevertheless, there are a number of claims for huge enhancement where authors avoid reporting on the known low Q_0 values.

Figure 15.4 presents a typical example of plasmonically enhanced fluorescence from a model biomolecule – bowine serum albumin (BSA) labeled with a fluorescein-based label (FITC). The luminescence intensity is found to depend on the incident light polarization being higher for p-polarized light versus s-polarized or unpolarized version (Fig. 15.5). The optimal spacing between labeled molecules and metal surface is controlled by a few alternating polyelectrolytes layers to get a few nm spacing [39]. Similar regularities are also obtained for human antibodies (Fig. 15.6) [40].

15.2.2.1 The Notion of Nanoantenna: De Fide QED Processes are Calculated De Facto with Classical Techniques

Fluorescence enhancement and Raman scattering enhancement in plasmonic nanostructure occurs owing to the two basic events: (i) purely classical electrodynamics (CED) phenomenon of light intensity rise up near a metal nanobody, and (ii) entirely quantum electrodynamics (QED) effect on modification of quantum transition rate due to modified photon local density of states (DOS) (see, e.g. [5–7] and Refs. therein). During the last decade a vast collection of publications has been related to the introduction of the nanoantenna notion in the consideration of these phenomena as well as in nanophotonic engineering [21,41–45]. The antennae notion

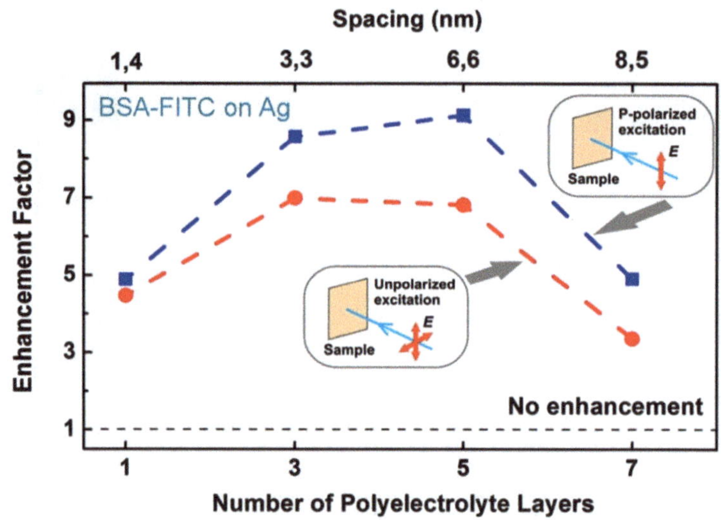

Fig. 15.5 Fluorescence enhancement for FITC labeled bowine serum albumin for various spacing and light polarization

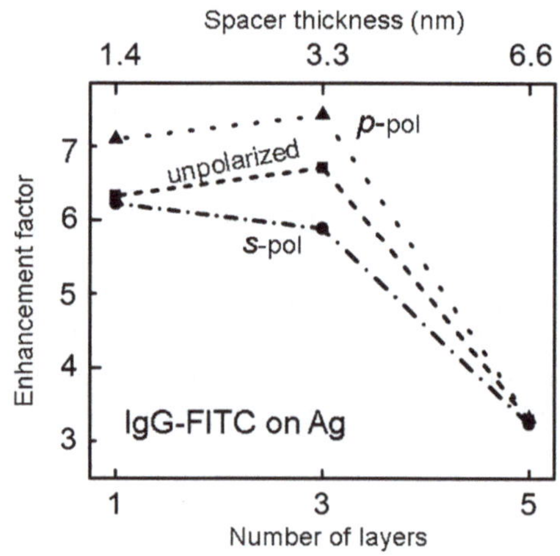

Fig. 15.6 Fluorescence enhancement for FITC labeled human immunoglobulin for various spacing and light polarization

is the entirely CED object designed to control the radiation patterns of a classical oscillator emitting electromagnetic waves. It was developed in radiophysics and is extensively used in radioengineering. Therefore it seems reasonable to discuss possible "bridging" of CED with QED via nano-antenna consideration.

In terms of the *quantum electrodynamics*, the probability of a quantum downward transition with spontaneous photon emission is proportional to the electromagnetic mode density (photon density of states, DOS). In a spatial structure with complicated distribution of dielectric permittivity (including plasmonic structures), modification of the spontaneous transition probability is treated as a consequence of the enhancement/depletion of the photon local DOS (see, e.g. [2, 5]). However, the very definition of the photon local DOS faces certain methodological issue. Here the reasonable and meaningful *operational* approach by D'Aguanno et al. [46] appears to be helpful. Namely, modification of the photon local DOS is considered to be equal to the modification of electromagnetic emission rate by a *classical* oscillator placed at the point of interest with respect to the rate in vacuum.

Therefore, taking *de fide* the conception of photon emission in the course of *quantum jumps* we use *de facto classical* equations to calculate modification of transition rate (probability) in a system with complicated topology. In other words, while the classical electrodynamics cannot explain the very event of a photon emission in the course of atomic or molecular transition, it is capable to offer the *computational technique* for its rate calculation.

Therefore the nanoantenna based approach should be treated as a useful *computational procedure* provided that we understand that photon creation in the course of a quantum system downward relaxation by no means can be derived or explained by classical radiophysics techniques. Finally, we note such a computational quantum-classical "duality" has been examined in detail in the context of light emission by a quantum versus a classical emitter in front of a mirror [47].

15.3 Hyperbolic Materials

In optics, distinction between materials with either fully positive (dielectrics) or fully negative (metals) dielectric permittivity ϵ is important. Dielectrics feature positive ϵ, though its components ϵ_x, ϵ_y, ϵ_z may not be equal each other, i.e. dielectric can feature anisotropy. Positive ϵ offers electromagnetic wave to propagate. Metals have $\epsilon < 0$ and only evanescent electromagnetic oscillations occur while the incident electromagnetic wave experience strong reflection. In very thin (subwavelength) metal films light tunneling occurs offering a way for electromagnetic wave to get through a metal plate. Hyperbolic materials are the new type of composite materials which feature different sign of dielectric permittivity in orthogonal directions, say, $\epsilon_x, \epsilon_y > 0$, $\epsilon_z < 0$. In such materials, the constant frequency surface in **k**-space (**k** is the wave vector) obeys hyperbolical shape unlike dielectric materials which feature either a sphere (iso[1-5]tropic case, $\epsilon_x = \epsilon_y = \epsilon_z > 0$) or an ellipsoid (anisotropic case, $\epsilon_x, \epsilon_y, \epsilon_z > 0$ but $\epsilon_x = \epsilon_y = \epsilon_z$ does not hold). This property gives rise to the notion of hyperbolic materials. The open character of the constant frequency surface offers enhancement on photon density of states thus providing an option to enhance radiative transition rates.

Notably, the very idea of a medium with sign-different permittivity in orthogonal directions dates back to a few decades and is related to the properties of plasma. The extensive consideration of such an option in optics has been performed lately [48–52].

To acquire the desirable properties, material should comprise a mixture of metal and dielectric inclusions with definite anisotropy (alignment) of subwavelength

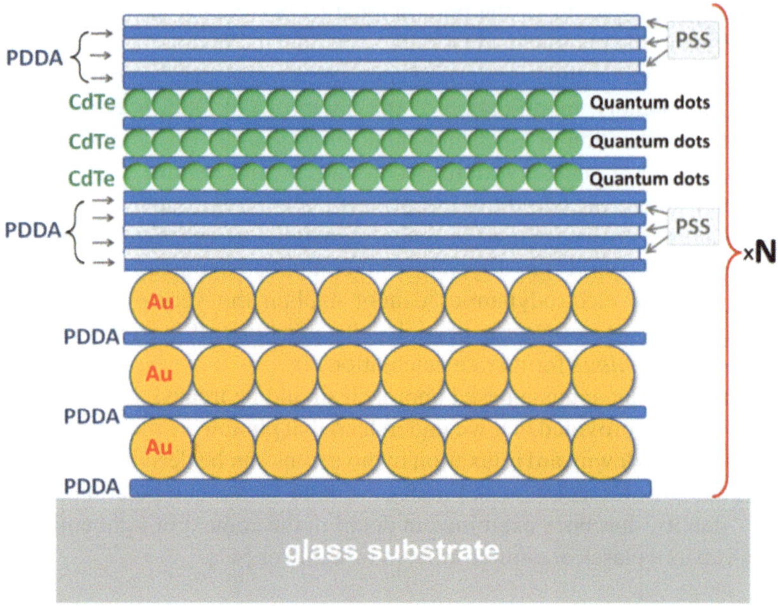

Fig. 15.7 A multilayer metal-semiconductor structure featuring the properties of a hyperbolic metamaterial. PDDA and PSS denote the polyeletrolyte used to prevent fluorescence quenching of quantum dots to make possible their use both as material-making component and as fluorescent probes whose properties are modified by the metamaterial

scale components. An example of such a system is presented in Fig. 15.7. Here a multilayer stack of metal nanoparticles and semiconductor quantum dots spaced by a number of polyelectrolyte layer forms a periodic structure where quantum dot serve both as material component with positive permittivity and fluorescent probes whereas polyelectrolyte layers measures just a few nanometers and desired to prevent fluorescence quenching. This structure examined in Ref. [53] has shown multiple enhancement of quantum dots fluorescence along with increase in the decay rate. These features offers a hint towards plausible treatment of the structure in terms of hyperbolic metamaterial which has been lately proved in modeling [54].

15.4 Plasmonically Enhanced Raman Scattering

15.4.1 Application to Inorganic Crystallites

Plasmonic enhancement of Raman scattering (referred to as Surface Enhanced Raman Scattering, SERS) has been extensively studied for molecular systems

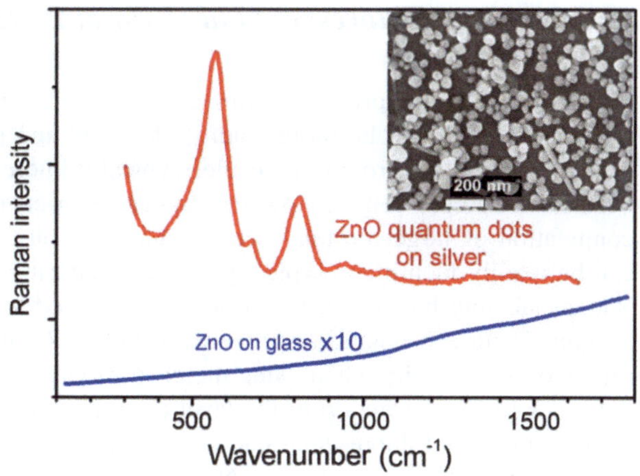

Fig. 15.8 Plasmonically enhanced Raman scattering observed for ZnO quantum dots on a silver substrates. The *inset* shows scanning electron microscopy image of the substrate. The estimated enhancement factor exceeds 10^4 times

during several decades ([1] and Refs. therein) and is considered as important driving phenomenon in nanoscience towards ultrasensitive analysis in medicine, forensic, environmental, and cultural heritage research. These studies are concentrated dominantly on molecular spectroscopy. In this context, the question arises how far the SERS experience gained in molecular spectroscopy can be extended towards inorganic species including microcrystallites and nanocrystals (quantum dots). Our recent studies have shown that inorganic crystallites are responsive to SERS effect and even in spite of its near field character 2–3 orders of the magnitude are affordable for enhancement of scattering from micrometer size inorganic crystalline pigments [55–57]. Semiconductor quantum dots have been tried for this phenomenon as well. A few groups reported on pronounced enhancement of Raman scattering though absolute numbers of enhancement factors are much lower than for molecules being in the range of 10^2–10^3, with possibly highest value of the order of 10^4 observed for ZnO dots [58–61]. Lower SERS enhancement factor for dots versus molecules are at first glance surprising since dots measure the size of 2–5 nm and in many instances reproduce properties close to molecules. However, a close insight provide a hint that the typically applied core-shell structure of the dots or capping by organic groups as stabilizers give rise to certain unavoidable spacing thus reducing the Raman enhancement factors.

The pronounced enhancement of non-resonant Raman scattering for ZnO nanocrystals adsorbed on surfaces formed by silver colloidal nanoparticles is shown in Fig. 15.8 [61]. The effect was found for the ZnO longitudinal optical phonon line (560–570 cm^{-1}) with a red laser (632 nm), the photon energy of a laser being well below the ZnO absorption onset (350 nm). The enhancement for silver substrates measures up to 10^4 times as estimated for an overall harvested light from 1.5 μm spots. The enhancement for gold substrates was about 10^2 times. The estimated ZnO detection limit is about 10^{18} mole.

15.4.2 From Fluorescent Label Towards Raman Labels

Many medically important proteins can be labeled by conjugation with smaller molecules which can be more readily identified and thus serve as reporters with respect to the body protein molecule. Typically fluorescent labels are used as has been discussed previously. However Raman reporters can also be implemented by conjugation of target proteins with a smaller molecule whose Raman spectrum can be readily identified. Typically Raman scattering cross-section is so low that Raman labeling has not been seriously considered to date. In our opinion, great progress in surface enhanced spectroscopy brings Raman reporters into the realm of immunoassay activity when using metal substrates to enhance Raman signal. The latter is known to grow up to the level close to typical fluorescent markers (note, many commercial fluorescent markers in immunoassays feature quantum yield of the order of 10 % or even lower). Noteworthy, sharpness and specificity of Raman spectra versus fluorescence emission spectra promise better recognition and enable multiplexing. Furthermore, photostability including luminescence quenching and absorption degradation upon illumination known for molecular fluorophore will be readily excluded when Raman labels are used. Finally, non-resonant character of Raman effect enables using of a single laser to excite multiple markers, not necessary overlapping with the absorption spectrum. Noteworthy that semiconductor quantum dots are promising Raman markers since they possess a small number of well defined lines. Therefore Raman scattering enhancement for quantum dots goes well far beyond basic research since Raman labels can been proposed to substitute fluorescent labels in high-sensitive biomolecule detection in medicine, e.g. identification of tumor markers in bioliquids [61].

15.4.3 From Raman Spectrometry Towards Raman Testers

Plasmonic enhancement of Raman scattering can be extended towards development of simple Raman testers to perform routine tests based on Raman signatures instead of complicated and expensive Raman spectroscopy [62]. A multiband filter, combined with a very simple photodetector, can be used to replace complex and expensive Raman spectrometers with gratings and cooled CCD detectors. This will reduce the cost of a diagnostic device, as well as the time needed to perform the test. As a result, the proposed diagnostic device [61] will be much cheaper and easier both in production and in service. Such tester can as as small as a mobile phone or an I-pad with a SERS substrate to be inserted therein similar as a flash memory card. It would facilitate more cost-effective express cancer tests, as well as increase the laboratory throughput. In turn, this would make early-stage cancer detection feasible for routine patient screening, including remote locations. An example of designed and fabricated targeted mutiband filter for oral cancer to be checked by means of saliva examination has been given in our recent paper [61]. In case of using Raman labels conjugated with target molecules (antibodies), filters for detection of marker's lines can be readily developed.

15.5 Conclusions

The progress in understanding of physical processes in light-matter interaction in presence of metal nanobodies opens an avenue towards practical applications of plasmonically enhanced fluorescence and Raman scattering in medicine, biology, forensic research, cultural heritage examination. Typical improvement in sensitivity for fluorescence techniques is feasible of the order of 10 for highly luminescent species to 10^3 for species with originally low quantum yield. For Raman based techniques more than 10^3 improvement in sensitivity is readily possible even for inorganic crystalline species and much higher effect is feasible for molecular probes. Plasmonic enhancement of Raman scattering offers a substitution of fluorescent labels in immunoassay by Raman labels and makes development of simple and cheap Raman testers probable to substitute expensive spectrometers.

Acknowledgements Cooperation with D.V. Guzatov, S.V.Vaschenko, V.V.Stankevich, A.A.Ramanenka, S.G.Kozhukh are greatly acknowledged. The work has been supported by the Program "Convergency" and Belarussian Republican Foundation for Basic Research. Specially acknowledged is promotion by the NATO for Peace Program.

References

1. Kneipp K, Moskovits M, Kneipp H (eds) (2006) Surface-enhanced Raman scattering, vol 103 of Topics in applied physics, Springer, Berlin/Heidelberg
2. Novotny L, Hecht B (2006) Principles of nano-optics. Cambridge University Press, Cambridge
3. Klimov V (2010) Nanoplasmonics, FIZMATLIT
4. Geddes CD (ed) (2010) Metal-enhanced fluorescence. Wiley-Blackwell, Hoboken
5. Gaponenko SV (2010) Introduction to nanophotonics. Cambridge University Press, Cambridge
6. Gaponenko S, Guzatov D (2009) Possible rationale for ultimate enhancement factor in single molecule Raman spectroscopy. Chem Phys Lett 477:411–414
7. Gaponenko S (2002) Effects of photon density of states on Raman scattering in mesoscopic structures. Phys Rev B 65:140303
8. Guzatov DV, Vaschenko SV, Stankevich VV, Lunevich AY, Glukhov YF, Gaponenko SV (2012) Plasmonic enhancement of molecular fluorescence near silver nanoparticles: theory, modeling, and experiment. J Phys Chem C 116:10723–10733
9. Vaschenko SV, Ramanenka AA, Guzatov DV, Stankevich VV, Lunevich AY, Glukhov YF, Sveklo IF, Gaponenko SV (2012) Plasmon-enhanced fluorescence of labeled biomolecules on top of a silver sol-gel film. J Nanophotonics 6:061710
10. Kinkhabwala A, Yu Z, Fan S, Avlasevich Y, Müllen K, Moerner WE (2009) Large single-molecule fluorescence enhancements produced by a bowtie nanoantenna. Nat Photonics 3: 654–657

11. Weitz DA (1983) The enhancement of Raman scattering, resonance Raman scattering, and fluorescence from molecules adsorbed on a rough silver surface. J Chem Phys 78:5324
12. Kümmerlen J, Leitner A, Brunner H, Aussenegg F, Wokaun A (1993) Enhanced dye fluorescence over silver island films: analysis of the distance dependence. Mol Phys 80:1031–1046
13. Sokolov K, Chumanov G, Cotton TM (1998) Enhancement of molecular fluorescence near the surface of colloidal metal films. Anal Chem 70:3898–3905
14. Strekal N, Maskevich A, Maskevich S, Jardillier JC, Nabiev I (2000) Selective enhancement of Raman or fluorescence spectra of biomolecules using specifically annealed thick gold films. Biopolymers 57:325–328
15. Ditlbacher H, Felidj N, Krenn J, Lamprecht B, Leitner A, Aussenegg F (2001) Electromagnetic interaction of fluorophores with designed two-dimensional silver nanoparticle arrays. Appl Phys B Lasers Opt 73:373–377
16. Aslan K, Lakowicz JR, Szmacinski H, Geddes CD (2004) Metal-enhanced fluorescence solution-based sensing platform. J Fluoresc 14:677–679
17. Kulakovich O, Strekal N, Artemyev M, Stupak A, Maskevich S, Gaponenko S (2006) Improved method for fluorophore deposition atop a polyelectrolyte spacer for quantitative study of distance-dependent plasmon-assisted luminescence. Nanotechnology 17:5201–5206
18. Cheng D, Xu Q-H (2007) Separation distance dependent fluorescence enhancement of fluorescein isothiocyanate by silver nanoparticles. Chem Commun (Cambridge, England):248–250
19. Bharadwaj P, Novotny L (2007) Spectral dependence of single molecule fluorescence enhancement. Opt Express 15(21):14266
20. Zhang J, Fu Y, Chowdhury MH, Lakowicz JR (2007) Metal-enhanced single-molecule fluorescence on silver particle monomer and dimer: coupling effect between metal particles. Nano Lett 7:2101–2107
21. Kühn S, Hå kanson U, Rogobete L, Sandoghdar V (2006) Enhancement of single-molecule fluorescence using a gold nanoparticle as an optical nanoantenna. Phys Rev Lett 97:017402
22. Bek A, Jansen R, Ringler M, Mayilo S, Klar TA, Feldmann J (2008) Fluorescence enhancement in hot spots of AFM-designed gold nanoparticle sandwiches. Nano Lett 8:485–490
23. Tam F, Goodrich GP, Johnson BR, Halas NJ (2007) Plasmonic enhancement of molecular fluorescence. Nano Lett 7:496–501
24. Shimizu K, Woo W, Fisher B, Eisler H, Bawendi M (2002) Surface-enhanced emission from single semiconductor nanocrystals. Phys Rev Lett 89:117401
25. Kulakovich O, Strekal N, Yaroshevich A, Maskevich S, Gaponenko S, Nabiev I, Woggon U, Artemyev M (2002) Enhanced luminescence of CdSe quantum dots on gold colloids. Nano Lett 2:1449–1452
26. Song J-H, Atay T, Shi S, Urabe H, Nurmikko AV (2005) Large enhancement of fluorescence efficiency from CdSe/ZnS quantum dots induced by resonant coupling to spatially controlled surface plasmons. Nano Lett 5:1557–1561
27. Ito Y, Matsuda K, Kanemitsu Y (2007) Mechanism of photoluminescence enhancement in single semiconductor nanocrystals on metal surfaces. Phys Rev B 75:033309
28. Chen Y, Munechika K, Jen-La Plante I, Munro AM, Skrabalak SE, Xia Y, Ginger DS (2008) Excitation enhancement of CdSe quantum dots by single metal nanoparticles. Appl Phys Lett 93:053106
29. Okamoto K, Vyawahare S, Scherer A (2006) Surface-plasmon enhanced bright emission from CdSe quantum-dot nanocrystals. J Opt Soc Am B 23(8):1674
30. Komarala VK, Rakovich YP, Bradley AL, Byrne SJ, Gun'ko YK, Gaponik N, Eychmüller A (2006) Off-resonance surface plasmon enhanced spontaneous emission from CdTe quantum dots. Appl Phys Lett 89:253118
31. Matveeva E, Gryczynski Z, Malicka J, Gryczynski I, Lakowicz JR (2004) Metal-enhanced fluorescence immunoassays using total internal reflection and silver island-coated surfaces. Anal Biochem 334:303–311
32. Matveeva EG, Gryczynski I, Barnett A, Leonenko Z, Lakowicz JR, Gryczynski Z (2007) Metal particle-enhanced fluorescent immunoassays on metal mirrors. Anal Biochem 363:239–245

33. Matveeva EG, Gryczynski Z, Lakowicz JR (2005) Myoglobin immunoassay based on metal particle-enhanced fluorescence. J Immunol Methods 302:26–35
34. Yamaguchi T, Kaya T, Takei H (2007) Characterization of cap-shaped silver particles for surface-enhanced fluorescence effects. Anal Biochem 364:171–179
35. McDonagh C, Stranik O, Nooney R, Maccraith BD (2009) Nanoparticle strategies for enhancing the sensitivity of fluorescence-based biochips. Nanomedicine (London, England) 4:645–656
36. Nooney R, Clifford A, Leguevel X, Stranik O, McDonagh C, Maccraith BD (2010) Enhancing the analytical performance of immunoassays that employ metal-enhanced fluorescence. Anal Bioanal Chem 396:1127–1134
37. Tabakman SM, Lau L, Robinson JT, Price J, Sherlock SP, Wang H, Zhang B, Chen Z, Tangsombatvisit S, Jarrell JA, Utz PJ, Dai H (2011) Plasmonic substrates for multiplexed protein microarrays with femtomolar sensitivity and broad dynamic range. Nat Commun 2:466
38. Zhou L, Ding F, Chen H, Ding W, Zhang W, Chou SY (2012) Enhancement of immunoassay's fluorescence and detection sensitivity using three-dimensional plasmonic nano-antenna-dots array. Anal Chem 84:4489–4495
39. Zhang R, Wang Z, Song C, Yang J, Sadaf A, Cui Y (2013) Immunoassays based on surface-enhanced fluorescence using gap-plasmon-tunable Ag bilayer nanoparticle films. J Fluoresc 23:71–77
40. Ramanenka AA, Vaschenko SV, Stankevich VV, Lunevich AY, Glukhov YF, Gaponenko SV (2014) Plasmonic enhancement of luminescence of fluorescein isothyocyanate and human immunoglobulin conjugates. J Appl Spectrosc 81:228
41. Mühlschlegel P, Eisler H-J, Martin OJF, Hecht B, Pohl DW (2005) Resonant optical antennas. Science (New York) 308:1607–1609
42. Rogobete L, Kaminski F, Agio M, Sandoghdar V (2007) Design of plasmonic nanoantennae for enhancing spontaneous emission. Opt Lett 32(12):1623
43. Fischer H, Martin OJF (2008) Engineering the optical response ofplasmonic nanoantennas. Opt Express 16:9144–9154
44. Bakker RM, Yuan H-K, Liu Z, Drachev VP, Kildishev AV, Shalaev VM, Pedersen RH, Gresillon S, Boltasseva A (2008) Enhanced localized fluorescence in plasmonic nanoantennae. Appl Phys Lett 92:043101
45. Guzatov DV, Klimov VV (2011) Optical properties of a plasmonic nano-antenna: an analytical approach. New J Phys 13:053034
46. D'Aguanno G, Mattiucci N, Centini M, Scalora M, Bloemer MJ (2004) Electromagnetic density of modes for a finite-size three-dimensional structure. Phys Rev E 69:057601
47. Hinds E (1993) Pertubative cavity qed. In: Berman PR, Bates DR, Bederson B (eds) Cavity quantum electrodynamics. Advances in atomic, molecular and optical physics. Academic, Boston, pp. 1–54.
48. Smith DR, Schurig D, Mock JJ, Kolinko P, Rye P (2004) Partial focusing of radiation by a slab of indefinite media. Appl Phys Lett 84:2244
49. Noginov MA, Li H, Barnakov YA, Dryden D, Nataraj G, Zhu G, Bonner CE, Mayy M, Jacob Z, Narimanov EE (2010) Controlling spontaneous emission with metamaterials. Opt Lett 35:1863–1865
50. Jacob Z, Kim J-Y, Naik GV, Boltasseva A, Narimanov EE, Shalaev VM (2010) Engineering photonic density of states using metamaterials. Appl Phys B 100:215–218
51. Drachev VP, Podolskiy VA, Kildishev AV (2013) Hyperbolic metamaterials: new physics behind a classical problem. Opt Express 21:15048–15064
52. Kidwai O, Zhukovsky SV, Sipe JE (2011) Dipole radiation near hyperbolic metamaterials: applicability of effective-medium approximation. Opt Lett 36:2530–2532
53. Ozel T, Nizamoglu S, Sefunc MA, Samarskaya O, Ozel IO, Mutlugun E, Lesnyak V, Gaponik N, Eychmuller A, Gaponenko SV, Demir HV (2011) Anisotropic emission from multilayered plasmon resonator nanocomposites of isotropic semiconductor quantum dots. ACS Nano 5:1328–1334

54. Zhukovsky SV, Ozel T, Mutlugun E, Gaponik N, Eychmuller A, Lavrinenko AV, Demir HV, Gaponenko SV (2014) Hyperbolic metamaterials based on quantum-dot plasmon-resonator nanocomposites. Opt Expr 22 (in press, accepted)
55. Klyachkovskaya E, Strekal N, Motevich I, Vaschenko S, Harbachova A, Belkov M, Gaponenko S, Dais C, Sigg H, Stoica T, Grützmacher D (2011) Enhanced Raman scattering of ultramarine on Au-coated Ge/Si-nanostructures. Plasmonics 6:413–418
56. Klyachkovskaya EV, Strekal ND, Motevich IG, Vashchenko SV, Valakh MY, Gorbacheva AN, Belkov MV, Gaponenko SV (2011) Enhancement of Raman scattering by ultramarine using silver films on surface of germanium quantum dots on silicon. Opt Spectrosc 110:48–54
57. Klyachkovskaya EV, Guzatov DV, Strekal ND, Vaschenko SV, Harbachova AN, Belkov MV, Gaponenko SV (2012) Enhancement of Raman scattering of light by ultramarine microcrystals in presence of silver nanoparticles. J Raman Spectrosc 43:741–744
58. Hugall JT, Baumberg JJ, Mahajan S (2009) Surface-enhanced Raman spectroscopy of CdSe quantum dots on nanostructured plasmonic surfaces. Appl Phys Lett 95:141111
59. Chursanova MV, Dzhagan VM, Yukhymchuk VO, Lytvyn OS, Valakh MY, Khodasevich IA, Lehmann D, Zahn DRT, Waurisch C, Hickey SG (2009) Nanostructured silver substrates with stable and universal SERS properties: application to organic molecules and semiconductor nanoparticles. Nanoscale Res Lett 5:403–409
60. Milekhin AG, Sveshnikova LL, Duda TA, Surovtsev NV, Adichtchev SV, Azhniuk YM, Himcinschi C, Kehr M, Zahn DRT (2010) Resonance effects in Raman scattering of quantum dots formed by the Langmuir-Blodgett method. J Phys Conf Ser 245:012045
61. Rumyantseva A, Kostcheev S, Adam P-M, Gaponenko SV, Vaschenko SV, Kulakovich OS, Ramanenka AA, Guzatov DV, Korbutyak D, Dzhagan V, Stroyuk A, Shvalagin V (2013) Nonresonant surface-enhanced Raman scattering of ZnO quantum dots with Au and Ag nanoparticles. ACS Nano 7:3420–3426
62. Kozhukh S, Zhukovsky S, Yurkevich I, Stepkova S, Gaponenko S, Lunevich A, Glukhov Y (2012) Design of high-transmission multiband multilayer filters for Raman spectroscopy. J Nanophotonics 6:061704

16
Principles and Applications of Rare Earth Ion-Doped Nanoparticles

John Collins

Abstract This article presents some fundamental properties of nanoparticles of insulating materials doped with rare earth ions as optically active centers. The optical properties nanoparticle semiconductors are well established; the particle diameter dictates the band gap, and hence the particle's absorption and emission properties. In the case of insulators, we show that such an effect is unlikely. The optical properties of insulating nanoparticles doped with optically active ions are affected by the particle size mainly through (1) the reduced phonon density of states and (2) the role of the surface sites occupied by the dopant ions. In this article we examine the fundamental physics of these systems, and call attention to experiments in which the effects of particle size have been observed. We also discuss the application of rare earth ion doped nanoparticles to the field of nanothermometry.

16.1 Introduction to Rare Earth Ions in Solids

The optically active dopant ions of interest in this article are the lanthanide series of the rare earth ions. In this section we summarize some of their fundamental properties, particularly those relevant to their optical properties. We pay special attention to those properties that may (or may not) be affected by such a system when placed in a nanoparticle.

The rare earth elements range from lanthanum (atomic number 57) to lutetium (atomic number 71). When placed in a solid, the rare earth ions usually substitute in the trivalent state. The ground state configuration of the trivalent rare earth ions is $[\mathrm{Xe}]f^n$, where we call attention to the fact that the $4f$ shell is partially filled $(1 < n < 13)$ and is optically active. It is also important to note that

J. Collins (✉)
Department of Physics and Astronomy, Wheaton College, Norton, MA 02766, USA
e-mail: jcollins@wheatonma.edu

B. Di Bartolo et al. (eds.), *Nano-Structures for Optics and Photonics*, NATO Science for Peace and Security Series B: Physics and Biophysics,
DOI 10.1007/978-94-017-9133-5_16,

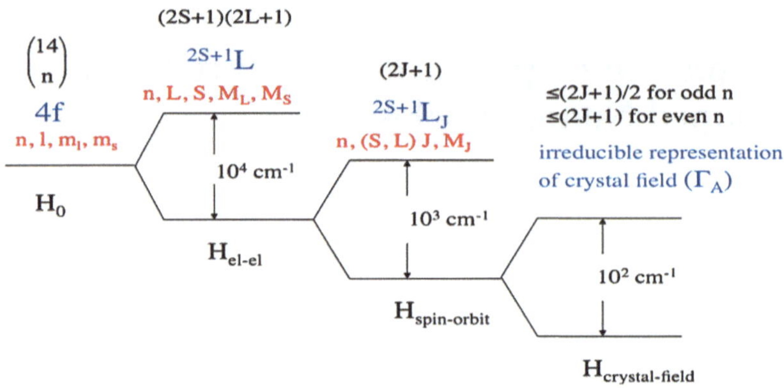

Fig. 16.1 The interactions experienced by the $4f$ electrons (of decreasing importance left-to-right), the labels of the states, the "good" quantum numbers, and the degeneracies. The diagram also shows the order of magnitude of the energy splittings (in cm^{-1}) that result from each interaction

Xe (atomic number 54) has filled shells, including the $5s^2$ and $5p^6$. Because the wavefunctions of the $4f$ electrons lie closer to the nucleus than those of the $5s$ and $5p$ electrons, the $4f$ electrons are shielded from the crystalline field by the $5s$ and $5p$ electrons. This limits the interaction between the ion and the crystalline field to a small perturbation.

16.1.1 States of Rare Earth Ions in Solids

The optical activity of the rare earth ions in solids occurs mainly between electronic states within the $4f$ configuration. In this section we give a brief introduction to the nature of the states associated with the $4f$ electrons, their labeling, and how they vary from one host to another.

The Hamiltonian of an atomic system can be written as follows:

$$H = H_0 + H_{\text{el-el}} + H_{SO} + H_{CF}. \tag{16.1}$$

In Eq. (16.1), H_0 is the Hamiltonina in the central field approximation, $H_{\text{el-el}}$ is the electron-electron interaction, H_{SO} is the spin orbit interaction, and H_{CF} is the crystal field interaction. We briefly discuss each of these separately. Their affects are summarized in Fig. 16.1.

In a perturbative approach, the most important term in the Hamiltonian, H_0, is considered first. It contains the kinetic energies of the electrons, the electrostatic interaction of the electrons with the nucleus, as well as those parts of the interactions between electrons that contribute to a central field. The eigenstates of H_0 lead to all the $4f$ electrons being degenerate, because each $4f$ electron experiences the same central field. The total orbital angular momentum (l) of each $4f$ electron and its z-component (m_l) are conserved separately, as is the z-component of the spin of each electron (m_s). Thus, in addition to n, the principal quantum number, l, m_l, and m_s are all good quantum numbers. The degeneracy is given by the number of ways that the spins of the n electrons can be arranged in the $4f$ shell, constrained, of course, by the Pauli Exclusion Principle. Such a set of states is known as a configuration.

$H_{\text{el–el}}$ is the electrostatic interaction between the $4f$ electrons. (This is not important for Ce^{3+} or Yb^{3+}, which have one electron and one hole in the $4f$ shell, respectively.) This interaction reduces the symmetry so that the orbital and spin angular momenta of the individual electrons are not conserved. Instead, the total and z-components of orbital and spin angular momentum of the system of $4f$ electrons are good quantum numbers. The configuration is accordingly split into different spectral terms, each term having a specific total angular momentum, and a total spin S, and designated by ^{2S+1}L. Such a label is called a spectral term. The z-components of the spin and orbital angular momenta run from $M_S = -S, -S+1\ S$ and $M_L = -L,\ -L+1,\ L$, respectively, and so each spectral term is $(2L+1)(2S+1)$ degenerate.

The spin-orbit interaction, H_{SO}, couples the spin and orbital angular momenta together, so the total angular momentum of the system of $4f$ electrons is conserved. Assuming that Russell-Saunders coupling is valid, the allowed total angular momentum, designated by J, runs from $J = L + S$ to $J = L - S$. Thus, good quantum numbers are J and its z-component, M_J, which can take on the values $M_J = -J,\ -J+1,\ J$. The states of the free ion are labeled as $^{2S+1}L_J$, and are $2J+1$ degenerate. Note that the labeling scheme retains the S and L designations, as if they were still good quantum numbers. This is because the states retain much of their S and L character, even after the spin-orbit coupling is included. Finally, we considered the electrostatic interaction between the ion and the static crystalline field due to the ion's nearest neighbors. At this point, spherical symmetry is lost, so that angular momentum as we normally envision it is no longer a good quantum number. Instead the quantum numbers are associated with the symmetry of the crystalline field. Generally speaking, the higher the local symmetry, the higher will be the degeneracies of the states. The degeneracy of these crystal field levels is less than or equal to $2J+1$ for an ion with an even number of electrons. For systems with an odd number of electrons, Kramer's degeneracy demands that the degeneracy of the levels must be less than or equal to $(2J+1)/2$. (In this case, a full splitting of the $2J+1$ levels is done with a magnetic field.) These crystal field levels should be labeled, strictly speaking, according to the irreducible representations of the crystalline field. However, the crystal field splitting of the J-manifolds is usually much smaller than the spin-orbit splitting, and so the $^{2S+1}L_J$ notation is often retained.

Because of the small crystal field splitting, the $4f$ energy levels of the rare earth ions relative to the ground state are not very sensitive to the host crystal. Thus, it is possible to develop an energy level diagram that is mainly valid for the trivalent rare earth ions in any host material. Such a diagram is often referred to as the Dieke diagram, and is shown in Fig. 16.2.

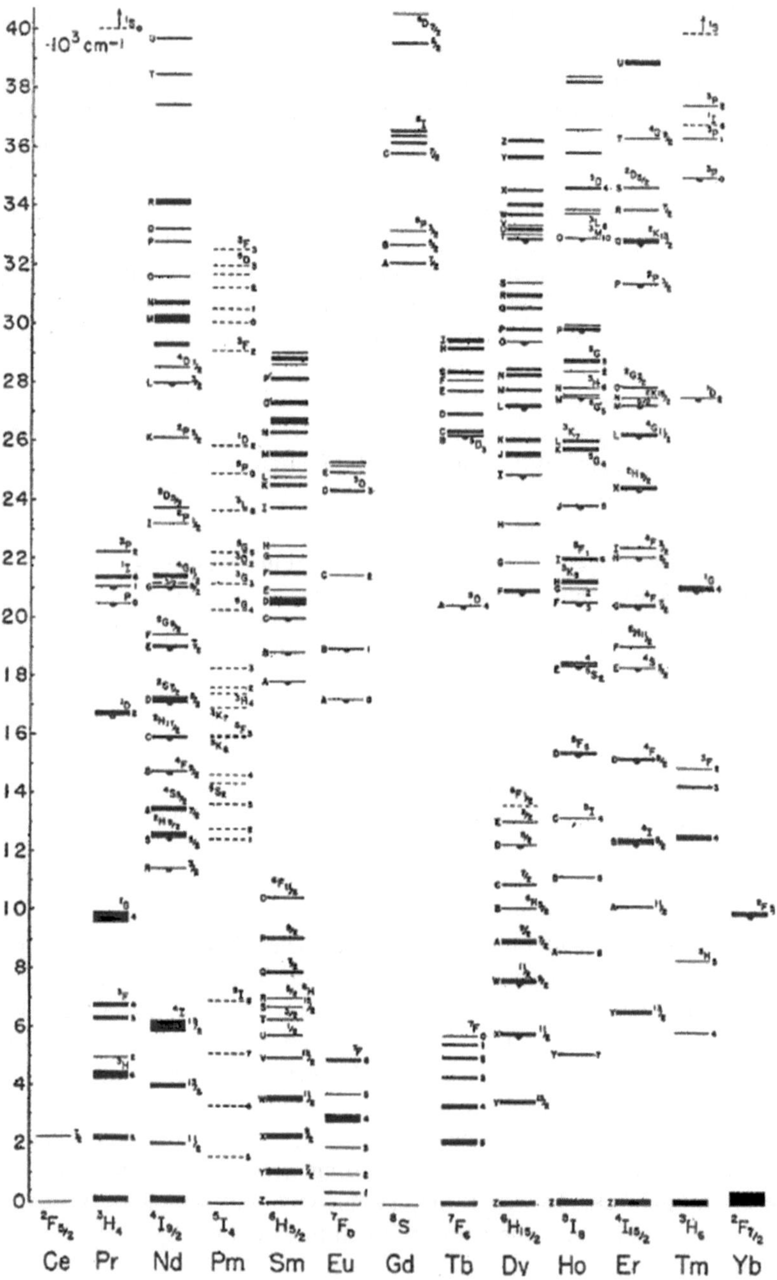

Fig. 16.2 The energy levels of the trivalent rare earth (lanthanide) ions in solids [1]

16.1.2 Radiative Transitions Among 4f Levels

The purely radiative transitions among these $4f$ states are given by Fermi's golden rule, with the matrix elements responsible for the transitions given by $\langle\psi_f^*|er|\psi_i\rangle$. ψ_f and ψ_i are the final and initial $4f$ electronic wavefunctions, respectively, and er is the electric dipole operator. The wavefunctions of the $4f$ levels of the free ion have a definite parity given by:

$$parity = (-1)^{\sum_d l_d} = -1^{3n}$$

where l is the angular momentum of each electron and the sum is over all the $4f$ electrons. (Note $l = 3$ for an electron in a f-shell). The definite parity of the states and the odd nature of the electric dipole operator implies that an electric dipole transition is forbidden between two states of the $4f$ configuration in a free ion.

When placed in a crystal, the mixing of the $4f$ states with those from other configurations of opposite parity (e.g. the $5d$ configuration) relaxes this selection rule somewhat. Consequently, the transitions between $4f$ states are weak, with f-numbers on the order of 10^{-6}, and radiative lifetimes in the microsecond and millisecond regimes. The absorption and emission lines are sharp, and the Stokes shifts are small. These qualities of rare earth ions in solids (sharp lines, weak transitions, small Stokes shifts) are all related to the fact that the electronic charge distribution of the $4f$ states are shielded from the nearby ligands, and so the ion retains much of the character of the free ion. Due to the weak effects of the neighboring ions on the $4f$ states, non-radiative decay from many excited states is also generally weak. Thus, in spite of the weak radiative rates, radiative emission may still be the dominant decay process, so that these systems can be efficient emitters. Applications of rare earth ions to optical systems (e.g. lasers and lamp phosphors) are numerous, and usually capitalize on this high radiative efficiency.

16.2 Confinement in Nanoparticles

It is well known that when the size of the particle is reduced into the nano regime, effects of confinement may be observed. This is most commonly observed in semiconductors, where tuning the bandgap of the material, and hence the emission wavelength, can be altered by changing the particle size. The goal of this section is to discuss the prominent features of confinement in rare earth ion-doped insulators. We shall discover that the confinement effect observed in semiconductors will be less important in insulators, and that the main effect of particle on the luminescence properties of the systems of interest here will be related to the surface states and to the phonon density of states.

16.2.1 Effects of Confinement on Electronics States

In semiconductor materials (e.g. CdS and CdSe), reducing the size of the particles to tens of nanometers or less results in a noticeable widening of the bandgap. To understand whether or not this effect occurs in insulating systems, we review the basic physics of the bandgap dependence on the particle size.

For simplicity, we will consider a crystal as a three-dimensional cube with sides L with infinite potential at the walls. A free charge carrier occupies a state of the entire crystal, and obeys the following relationships:

$$E = \frac{p^2}{2m^*} = \frac{\hbar^2}{2m^*}k^2 \quad \text{dispersion relation}$$

$$E_e = \left(\frac{h^2}{8m^*L^2}\right)n^2 \quad \text{allowed energies} \tag{16.2}$$

where $n^2 = n_x^2 + n_y^2 + n_z^2$ and L is the characteristic size of the crystal, m^* is the effective mass of the charge carrier, and is related to the curvature of the dispersion curve:

$$m^* = \hbar\frac{d^2k}{d\omega^2}$$

Note that the effective mass of a hole is negative, due to the fact that the dispersion curve for the valence band is concave down. For charge carriers in solids, it is customary to use the top of the valence band as the zero energy. For an electron in the conduction band, the energy in Eq. (16.2) must be shifted up by the bandgap energy, E_{gap}. With these adjustments, the energies of the particles are:

$$E_e = E_{\text{gap}} + \left(\frac{\hbar^2}{8m_e^*L^2}\right)n^2 \quad \text{electron in conduction band} \tag{16.3}$$

$$E_h = \left(\frac{\pi^2\hbar^2}{2m_h^*L^2}\right)n^2 \quad \text{hole in valence band} \tag{16.4}$$

Note that Eq. (16.4) is inherently negative. The bandgap, $E_e - E_h$, is simply determined by the kinetic energies of the electron and of the hole and on the energy gap in bulk materials. For example, when L is large, the minimum allowed kinetic energy ($n = 1$ state) of the electron (or hole) is basically zero. (The interaction between an electron and a hole is neglected here).

As L decreases, the absolute value of the energy of these lowest kinetic energy states increases. It should be noted that, as L decreases, the number of kinetic energy states is not changing, the states are merely shifted up in energy. Thus, as the size of

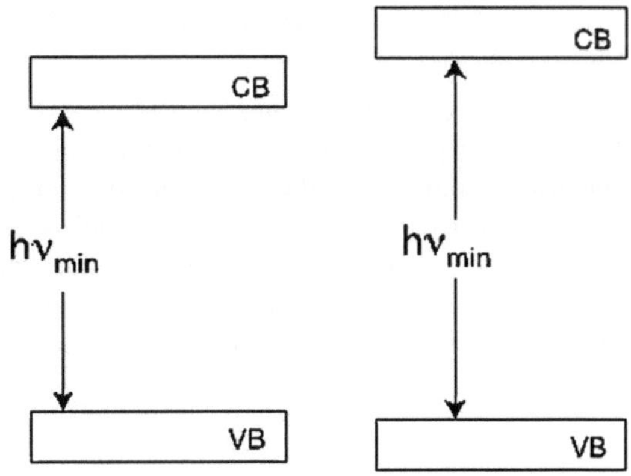

Fig. 16.3 The energies of electrons in the conduction band and holes in the valence band for large crystals (*left diagram*) and nanoparticles (*right diagram*). Note that in this representation, the bottom of the conduction band and the top of the valence band are located at the lowest available values of the kinetic energy

the particle is reduced, and so the energy required to bridge the gap increases. This is shown in Fig. 16.3. This analysis of the increase in the bandgap as the particle size decreases is predicated on the fact that the electrons are free, and so can be described by highly non-localized states of the entire crystal. We make the following observations:

1. In insulators and large gap semiconductors the effective mass of the electrons and holes is much larger than in semiconductors. (The parabolas on the E vs. k diagrams, are flatter). Since the kinetic energy is inversely proportional to the effective mass (Eqs. (16.3) and (16.4)), the kinetic energies of the charge carriers are much smaller in insulators. Thus, the widening of the gap with decreasing particle size is not observed in these systems.
2. The states of the rare earth ions in solids that lie within the $4f^n$ configuration are highly localized to the ion, and so will not exhibit confinement effects.
3. For REI doped semiconductors, the REI states in the vicinity of the conduction band could be affected by particle confinement if the shift in the bandgap caused a substantial change in the position of the REI level relative to the band edge. Consider, for example, a system in which a rare earth ion energy level is located just below the bandgap in a bulk crystal. Such a level could luminesce, but thermal excitation into the conduction band often tends to decrease the luminescence at higher temperatures. As the particle size decreases and the conduction band is raised, the REI level remains fixed. Consequently, the rare earth ion level can ultimately be located well below the conduction band, so that decreasing the thermal excitation from the excited REI level into the conduction band and the increasing the radiative efficiency of the particles.

In summary, we expect to see little or no electronic confinement effects on the optical properties of rare earth ion-doped insulators as the size of the particles goes from the macroscopic to the nano regime.

16.2.2 *Effects of Confinement on the Density of Phonon States*

Though the electronic states of the rare earth ion-doped insulators are highly localized, the phonon states are extended states of the entire solid. Thus, the phonon spectrum should show the effect of confinement. In this section we consider the behavior of phonon states when the particle size is reduced.

Consider a simple cubic solid with side length d and atomic spacing a. The wavelengths of the phonons vary from roughly twice the atomic spacing to twice the diameter of the particle. The energy of a phonon is

$$E_{ph} = \frac{hc_s}{\lambda} = \frac{hc_s}{2L}n,$$

where $a < L < d$ and $n^2 = n_x^2 + n_y^2 + n_z^2$. The speed of sound in the solid, c_s, generally is on the order of a few thousand meters per second. For $d \sim 10^{-2}$ m and a speed of sound of 3,000 m/s, the minimum energy (longest wavelength) phonons have energies on the order of

$$E_{min} = h\frac{c_s}{\lambda} = h\frac{c_s}{2d} \approx 6 \times 10^{-10}\,\text{eV}$$

and the maximum energy phonon is approximately

$$E_{max} = h\frac{c_s}{\lambda} = h\frac{c_s}{2a} \approx 0.03\,\text{eV} \tag{16.5}$$

where we set $a = 0.2$ nm. A simplified phonon density of states is given be the Debye approximation:

$$\rho_{ph}(\nu) = \frac{3\pi}{2}\left(\frac{2d}{c_s}\right)^3 \nu^2.$$

Note that the maximum phonon energy, given by Eq. (16.5), is determined by the interatomic spacing, and so is independent of the particle size. The low frequency phonons, however, are determined by the size of the particle: as the particle size decreases, the energy of the lowest allowed phonon mode increases. Thus, the low frequency phonons that exist in the bulk are no longer supported in a nanoparticle. Given the prominent role of phonons in the relaxation of an atomic system from an excited electronic state, this effect is of potential importance to the luminescence of rare earth ions in solids.

There is another important way in which the nanoparticle's phonon spectrum differs from that of the bulk crystal. For a bulk crystal, the density of phonon states can be reasonable treated as a continuous function of frequency, as is done in the Debye approximation. For a nanoparticle, however, the density of phonon states must be treated as a discrete function. To see why this happens, consider first that the total number of phonon modes is given by $3N$, where N is the total number of

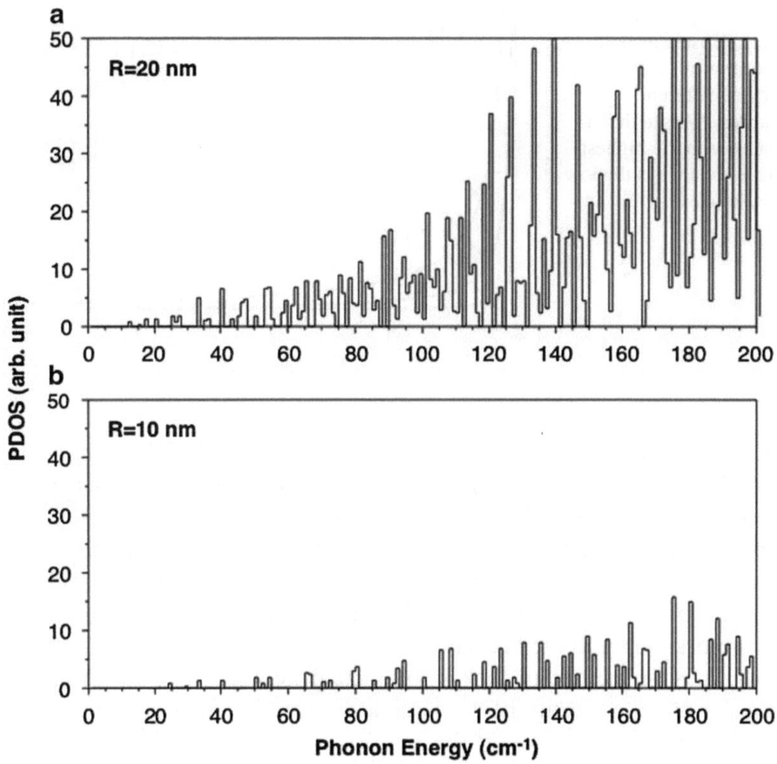

Fig. 16.4 The phonon density of states of the spherical and torsional modes of spherical nanoparticles of radius (**a**) 20 nm and (**b**) 10 nm. The velocity of sound was assumed to be 3,500 m/s [2]

particles in the crystal. As one moves from the bulk to the nano, N decreases as the cube of the diameter, and the number of phonon modes can be estimated as:

$$3N \sim \left(\frac{d}{a}\right)^3 \tag{16.6}$$

For a bulk crystal with $d = 0.25$ cm and $a = 0.25$ nm, $3N \sim 3 \times 10^{21}$, whereas when $d = 2.5$ nm, $3N \sim 3 \times 10^3$. Thus, going from a particle size of 0.25 cm to 2.5 nm the total number of allowed phonon modes decreases by 18 orders of magnitude! The result is that the phonon spectrum can no longer be treated as a continuous function of frequency. The discrete nature of the density of states is especially evident at the lower phonon frequencies.

Figure 16.4 shows the results of calculations by Liu et al. who modeled the nanoparticle as a sphere and calculated all the spherical and torsional modes [2, 3]. The speed of sound was assumed to be 3,500 m/s, and the calculations were carried

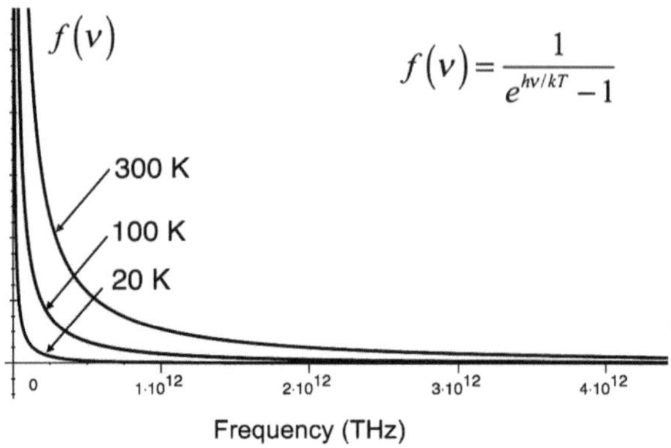

Fig. 16.5 The phonon occupancy (Bose-Einstein distribution) as a function of frequency at different temperatures. We see that as those states with the highest occupancy are always the low frequency modes

out for nanoparticles of diameters 20 and 10 nm. The spectra show the two effects of decreasing the particle size; (1) the disappearance of the low frequency phonons, and (2) the movement from the continuous to the discrete. We note that these effects become very pronounced for nanoparticles less than 10 nm. At particle sizes 50 nm or above, a nearly continuous function re-emerges, so that for particles with diameter greater than about 50 nm confinement effects on the phonon modes are likely not to play an important role in the luminescence from the rare earth ions in solids.

Finally, it is instructive to look at the actual number of phonons in a crystal. Treating the phonon density of states as a discrete function of frequency, we let the function $\phi_{nano}(\nu_i)$ represent the number of phonon modes at frequency ν_i. The number of phonons at this frequency is given by the number of modes at frequency ν_i weighted by the occupancy of that mode as given by the Bose-Einstein distribution.

$$n(\nu_i) = \phi_{nano}(\nu_i)\frac{1}{e^{h\nu/kT} - 1} \tag{16.7}$$

For a nanoparticle, the normal modes of vibration $\phi_{nano}(\nu_i)$ must be calculated (e.g. see Ref. [2]). The number of phonons in each mode depends on energy of the mode and on the temperature, and is shown in Fig. 16.5. Figure 16.5 clearly shows that the highest occupancy numbers are largest for those phonon modes with low phonon frequencies. Given that these highly occupied, low frequency modes do not exist in the nanoparticle, then the total number of phonons per unit volume in a nanoparticle is much less than that in the bulk. Note that although the number of allowed phonon modes per unit volume ($3N/V$) is independent of particle size, the actual number of phonons per unit volume decreases as the particle size decreases.

Thus, we see that decreasing the particle size leads to (1) an elimination of the lowest energy phonon modes, (2) a discrete phonon spectrum with many phonon modes no longer supported, and (3) a decrease in the total number of phonons/volume. Any of these effects can be expected to affect the luminescence properties of rare earth ion-doped solids.

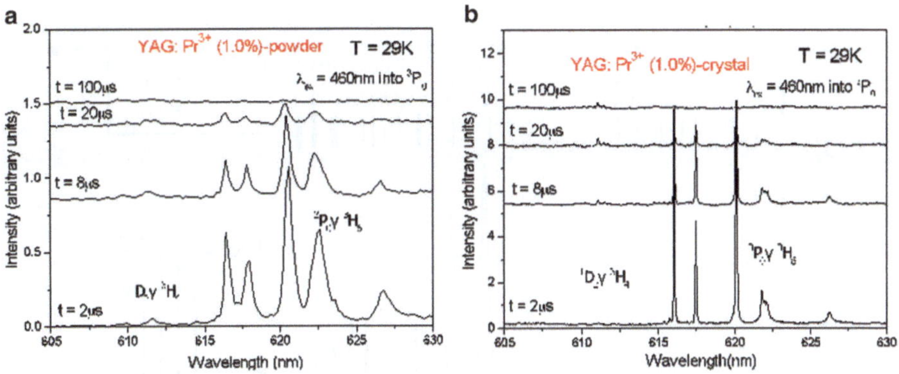

Fig. 16.6 Time resolved luminescence spectra of the nanocrystalline YAG:Pr^{3+} (*left*) and bulk YAG:Pr^{3+} (*right*) with the time delays 2, 8, 20, and 100 μs in the 605 and 630 nm wavelength range at 29 K (The excitation was into the $^3P_0$ level with the dye laser tuned at 460 nm) [5]

16.2.3 Placement of Rare Earth Ions and Surface States

In bulk crystals, the incorporation of the rare earth ion during crystal growth is often assumed to be random and that each ion site, to a first approximation is identical. Though this never the case, it is often true that in a bulk crystal the variations in "equivalent" sites in the crystal is small. Variation in the sites occupied by the rare earth ions lead to emission at slightly different energies, which contribute to the linewidth of an observed spectral line [4]. Insights into the site-to-site variations can be obtained by looking at the linewidth of the emission (or absorption) at low temperatures, when thermal broadening is not contributing to the linewidth.

In nanocrystals, the linewidths of the $4f^n$ transitions at low temperature are much broader than in bulk crystals, indicating a larger variety of sites (i.e. local crystalline fields) at which the rare earth ion is located [4, 5]. This effect is seen in Fig. 16.6, which shows emission spectra Pr-doped YAG in a bulk single crystal and in nanoparticles. The large variety of sites in a nanoparticle is often explained as being due to the surface-to-volume ratio as the particle size decreases or, differently stated, to the presence of surface states. The idea is that the rare earth ions located at the surface will experience a significantly different crystalline field as those near the center of the nanoparticle, causing the energy levels to be shifted, resulting in a broader spectral line. (More examples of the broadening of spectral lines can be found in an article by Di Bartolo et al. in this volume.)

Such reasoning is, however, only qualitative, and does not explain any details of the observed linewidth. Several questions may be asked, such as: Exactly how do the local fields change as one moves from the center of the nanoparticle to the surface?; Where is the rare earth ion most likely to be situated, closer to the center of the nanoparticle or closer to its surface? Despite the vast amount of work that has been done on rare earth ion-doped nanoparticles, these questions remain open.

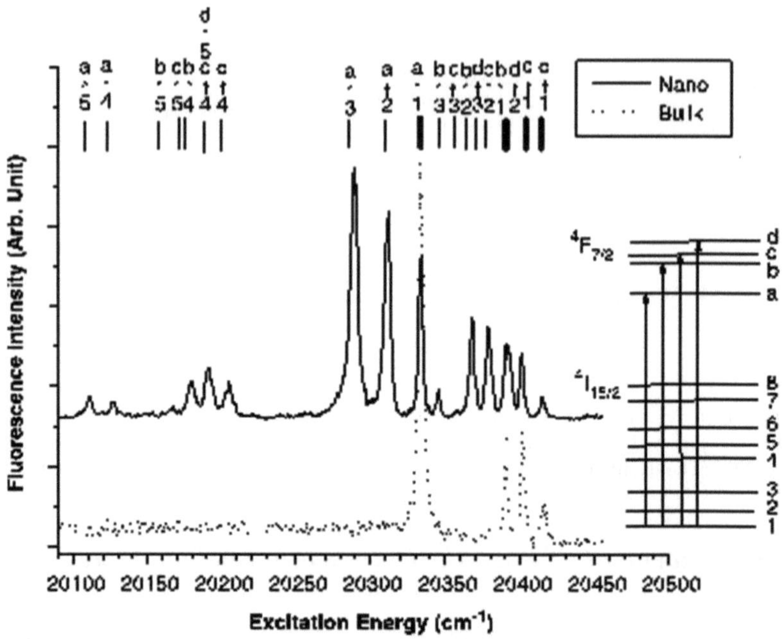

Fig. 16.7 Excitation spectrum of the $^7F_2$ to $^4I_{15/2}$ emission in bulk (*dotted line*) and in 10–40 nm diameter nanocrystals (*solid line*) of Er:Y_2O_2S at 2.6 K [2]

16.2.4 Affect of Phonon Density of States on the Luminescence of Er-Doped Y_2O_2S Nanoparticles

In this section we give experimental evidence of the expected reduced density of phonon states (see Sect. 16.2.2) as one moves from the bulk to the nano regime.

The work presented here was conducted by G. Liu at Argonne National Laboratories, and, to this author's knowledge, is the only data of its kind, and represents perhaps the clearest evidence indicating that the phonon density of states is indeed reduced due to confinement. The excitation and emission spectra were obtained using a tunable pulsed laser as a source. This laser was operated at 10 Hz and had a pulsewidth of 5 ns. Acquisition of the data was done with a boxcar integrator set a delay of 5 μs after the pulse and a gate width of 1 μs. Figure 16.7 shows the excitation spectrum of Er-doped Y_2O_2S at low temperature (2.6 K) in particles of diameter of ∼400 nm and of diameter ∼25 nm ± 15 nm [2]. We first assume that before the experiment starts the energy levels of the system are in thermal equilibrium, so that at 2.6 K all the ions are in their ground state. That is, all ions are in the crystal field level labeled "1" in Fig. 16.7. In the larger particles (dotted line) the excitation spectrum shows four absorption transitions (1 →a, b, c, d) that lead to emission. This behavior is completely unsurprising, and is consistent with the assumption of thermal equilibrium among the energy levels of the system.

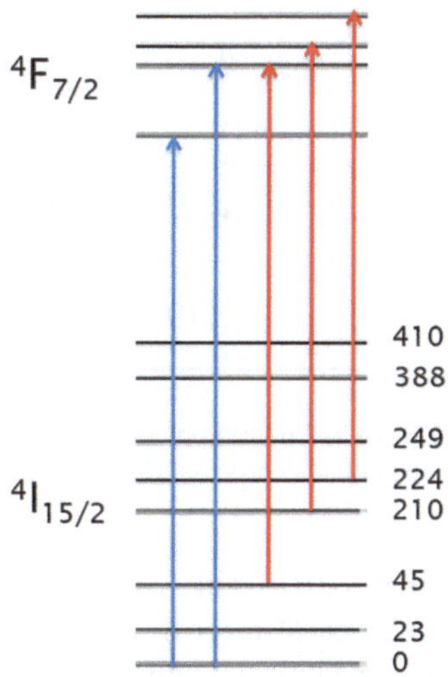

Fig. 16.8 Energy level diagram of the $^4I_{15/2}$ and $^4F_{7/2}$ manifolds of Er in Y_2O_2S. The energies of each level are given in cm^{-1}

In the nanocrystals (solid line), however, the excitation spectrum shows numerous lines in addition to the four lines observed in the bulk spectrum. An analysis of the energies of these lines shows that they represent transitions from crystal field levels 1–5 of the ground $^4I_{15/2}$ manifold to the crystal field levels a – d of the excited $^4F_{7/2}$ manifold. This is rather surprising, since thermal equilibrium demands that only the lowest energy level be occupied at $T = 2.6$ K. The data clearly indicate, however, that the upper crystal field levels of the ground state manifold have are occupied. Evidently, thermal equilibrium among the various energy levels is not readily achieved in these nanoparticles because, as we saw in Sect. 16.2.2, the nanoparticles cannot support the long wavelength, low energy phonons. These spectral lines that result from the occupation of levels not in thermal equilibrium with the phonon bath are sometimes called "hot bands".

The results shown in Fig. 16.7 can be explained as follows. In the excitation experiment, absorption of the exciting laser light into the $^4F_{7/2}$ level is followed by a fast relaxation to the lower $^4S_{3/2}$ level. This relaxation, which is accompanied by the emission of phonons, can be fast enough to occur before the laser pulse has ended. These phonons can be absorbed by a nearby Er ion in its ground state, causing that Er ion to become excited to one of the excited levels in the ground state manifold. The absorption or emission of a single phonon that brings an ion from one energy level to another is known as a direct process.

To understand how this happens, it is useful to refer to Fig. 16.8, which gives the energy levels of the ground state $^4I_{15/2}$ manifold, and also to Fig. 16.9, which shows three phonon process that can cause some of the upper levels of the $^4I_{15/2}$ manifold to become occupied. These processes include the absorption of a single phonon, known as a direct process and two two-phonon processes that involve the emission and absorption of phonons. These are the Raman process and the Orbach process [6], the difference between them being that the Orbach process proceeds through a real

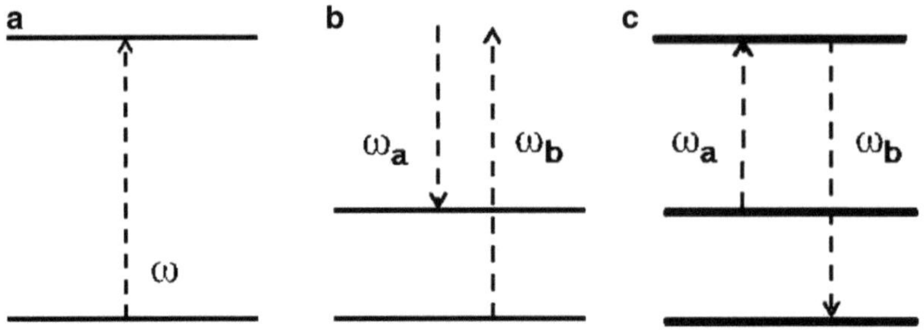

Fig. 16.9 Phonon processes that can cause the ion to transition between energy levels: (**a**) Direct (one phonon) process, (**b**) Raman (two-phonon) process, and (**c**) Orbach (two-phonon) process

intermediate state, whereas the Raman process involves a virtual intermediate state. We must also recall that in nanoparticles the phonon density of states is different than the bulk, showing a low phonon cutoff energy, a discrete spectrum, and a drastic reduction in number of low energy phonons.

For the sake of this argument, we assume that the low phonon energy cutoff frequency is 50 cm^{-1}. In this scenario, levels at 23 and 45 nm cannot become occupied by absorbing a single phonon because the required phonon energies are below the cutoff energy, and hence do not exist in the nanoparticle. We make the following points.

1. Only those energy levels of the $^4I_{11/2}$ manifold above 50 cm^{-1} can be occupied by a direct process.
2. The energy levels at 23 and 45 cm^{-1} will require a two phonon process to become excited, since phonons of those frequencies are not supported in small particles. Both Orbach and Raman processes are possible, providing the appropriate phonon energies are available.
3. Following excitation with a photon, relaxation of the ion to a lower state will produce phonons, possibly of several different frequencies. Additionally, a phonon at a particular frequency can produce phonons of different frequencies during scattering if anharmonic terms are present in the vibrational potential energy curves. Thus, we conclude that phonons of many frequencies are present within a ns following excitation, making Orbach and/or Raman processes possible.
4. The radiative decay from the excited electronic states (e.g. $^4S_{3/2}$) to the upper levels of the $^4I_{15/2}$ manifold can populate direction these levels without involving any phonon participation.

The points outlined above show that the upper states of the ground $^4I_{15/2}$ manifold can become populated following excitation into a higher level. In bulk, at 2.6 K, the ions in these "hot" levels will relax to the ground state, reaching thermal equilibrium on a ps time scale. In nanoparticles, these levels remain occupied for a much longer period of time, a few ns at least, giving them time to absorb a photon before the laser

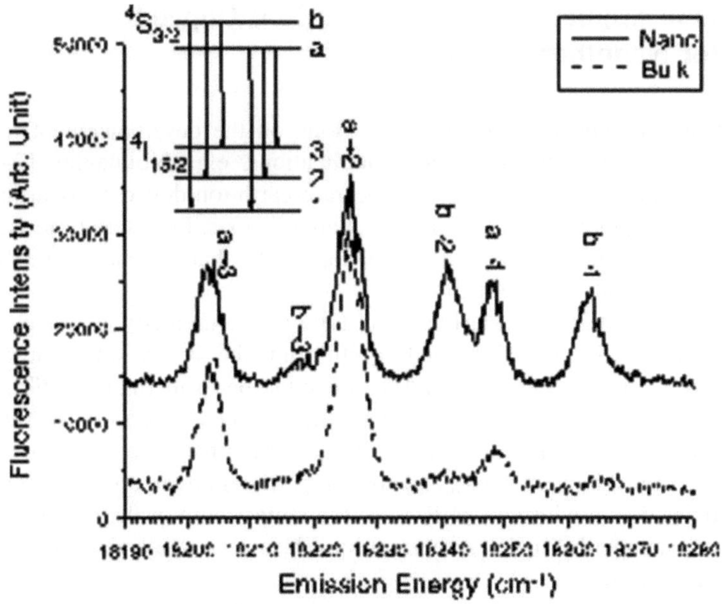

Fig. 16.10 The luminescence of the $^4S_{3/2}$ to $^4I_{15/2}$ transition of bulk (*dotted line*) and of 10–40 nm diameter nanocrystals (*solid line*) of Er:Y_2O_2S at 2.6 K [2]

pulse (5 ns FWHM) is over. This explains how the hot absorption bands appear in the excitation spectrum in Fig. 16.7.

Exactly which process or processes are responsible for the population of the upper levels of the $^4I_{15/2}$ manifold in unknown, but the observation of excitation lines from these levels at low temperature clearly indicates the effect of confinement on the density of phonon states in nanoparticles. We note that once the temperature is raised above 7 K in this system, the hot bands disappear. Apparently, even at such low temperatures the two-phonon processes are fast enough so the thermal equilibrium can be established on a sub-ns timescale. Further evidence of this non-equilibrium condition in nanoparticles is shown in the emission lines representing the $^4S_{3/2}$ to $^4I_{15/2}$ transition in Y_2O_2S:Er^{3+} (Fig. 16.10). At 2.6 K in the bulk, the emission originates from only the lowest energy level of the $^4S_{3/2}$ manifold. In the nanoparticles, however, hot emission bands are observed. Pumping into the upper levels, relaxation to the $^4S_{3/2}$ results in the population of the upper crystal field level of that manifold in both bulk and in nanoparticles. However, in the bulk there is fast relaxation from level (b) of the $^4S_{3/2}$ manifold to its lower level (a), whereas in nanoparticles there is no available mode to accept a phonon of that low frequency ($\sim$25 cm^{-1}). Thus, one phonon processes are forbidden, and consequently level (b) remains populated long enough to emit a photon.

16.3 One Application: Rare Earth Ions Doped Systems for Nanothermometry

A significant amount of work has been done on the development of rare earth ion doped nanoparticles, for various applications, e.g., bioimaging [7–10] and scintillators [11, 12]. Despite much work, rare-earth ion-doped nanoparticles have not yet made significant inroads into these applications. In this section, we discuss one application of rare earth ion-doped nanoparticles that have had some success in the marketplace: nanothermometry.

The physical principle underlying any type of thermometry, and even the idea of a temperature, is that of thermal equilibrium: the system under investigation must be in thermal equilibrium, and also it must be in thermal equilibrium with the "thermometer". The process of nanothermometry refers to measuring the temperature of a very small region, say in the certain part of a cell. Obviously the thermometer must be very small, and so nanoparticles are an obvious solution. For nanoparticles doped with rare earth ions, the temperature of the nanoparticle, and hence the system with which it is thermal equilibrium, is usually determined through optical measurements.

To discuss how optical measurements can be used for this purpose, we consider a simplified energy level diagram of an ion in a solid, as shown in Fig. 16.11. In this model, we assume the system has two energy levels, A and B, that can emit to the lower levels, that are nearby in energy to one another, and that are far from any other levels. (By the phrase "far from any other levels" we mean that these two levels are separated from other levels by energies $\gg kT$). According to the Boltzmann distribution, the ratio of the population of the two levels, A and B, thermal equilibrium is given by:

$$\frac{N_A}{N_B} = \frac{g_A}{g_B} \exp^{\left\{\frac{E_B - E_A}{kT}\right\}} \tag{16.8}$$

where g_A and g_B represent the degeneracies of levels A and B, respectively, and E_A and E_B are their respective energies. Under the condition whereby neither the energies nor the lifetimes of the two levels show any temperature dependence, then the ratio of the intensities of the emissions from levels A and B will show the same

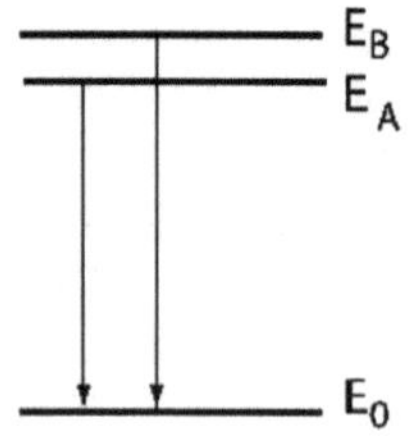

Fig. 16.11 An energy level diagram depicting a simplified version of an energy levels scheme used in nanothermometry: two levels, E_A and E_B, in thermal equilibrium, where each emit to a lower level

temperature dependence as in Eq. (16.8). Thus of the ratio of any two emission lines, one level A and one from level B, will behave according to:

$$\frac{I_A}{I_B} = C \exp\left\{\frac{E_B - E_A}{kT}\right\} \tag{16.9}$$

It should be noted that the aforementioned condition, that neither the energies nor the lifetimes of the two levels have any temperature dependence, is rarely true. Energy levels are known to shift with temperature, which could change E_A and E_B. Also, radiative transitions of energy levels of ions in solids often compete with non-radiative transitions, and the nonradiative transitions often show a temperature dependence. Finally, since an excited ion is coupled to the phonon field, each spectral line is accompanied by sidebands due to so-called vibronic transitions. If the coupling of the ion to the lattice is dependent on the excited state, then the radiative lifetimes of the two levels can change with temperature.

Because these effects are always present in any system, Eqs. (16.8) and (16.9) will not strictly hold. These equations, however, approximately hold if the temperature range over which the experiment is carried out is small enough so that these effects can be neglected. For biological applications, the temperature range is usually limited to temperatures within a few tens of degrees of body temperature, and the model described above works quite well.

In examining Eq. (16.9) it is obvious that when $\Delta E \ll kT$ and $\Delta E \gg kT$, the temperature dependence of the intensities is very small. The system will be most temperature-sensitive in a temperature range close to where the first derivative of Eq. (16.9) with respect to T is a maximum, that is when $T \sim \Delta E/k$. Thus, if one wishes to measure the temperature in a range near T_0, then it is optimum to select a system in which $\Delta E \sim kT_0$. Working in vivo for the human body, for example, $T_0 \sim 312$ K, which corresponds to an energy difference $\Delta E = (0.695\,\text{cm}^{-1}/\text{K})(312\text{K}) = 217\text{cm}^{-1}$. As an example, we present some recent work by Vetrone et al. on rare earth ion-doped nanoparticles designed as nanothermometers for biological applications [13]. The system is $NaYF_4$ doped with two rare earth ions, Yb and Er. An energy level diagram of these systems are shown in Fig. 16.12. In this application, the Yb is excited with a near IR laser operating at ~1 μm. Following excitation with the laser, one excited Yb ion transfers its energy to an Er ion, bringing Er from its ground state to its excited $^4I_{11/2}$ level. This is followed by a second energy transfer from an excited Yb ion to the Er, bringing it to the excited $^4F_{7/2}$ level. From the $^4F_{7/2}$ level, the Er ion decays non-radiatively to the $^2H_{11/2}$ and the $^4S_{3/2}$, which become thermalized according to the Boltzmann factor, and which emit in the visible region. This process of converting the initial IR laser light into visible photons is called upconversion. In this system, the Er was not pumped directly because the biological tissue is relatively transparent near 1 μm as opposed to the visible region.

Because this system is meant for biological applications at body temperature, the previous discussion suggests that the value of this system would be most sensitive if the energy difference between the $^4S_{3/2}$ and $^2H_{11/2}$ levels were on the order of a few

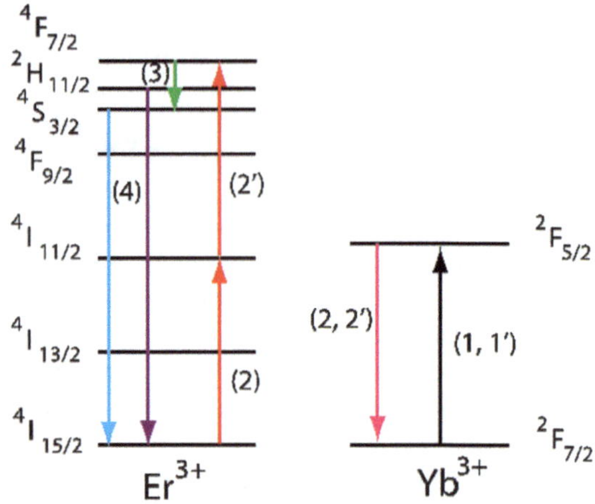

Fig. 16.12 Diagram depicting the two-step upconversion energy transfer process from Yb to Er, and the subsequent visible emission from Er. (*1*) First Yb absorption, (*2*) energy transfer to Er, (*1'*) second Yb absorption, (*2'*) energy transfer to Er, elevating it to the $^4F_{7/2}$ level, (*3*) non-radiative relaxation to/thermalzation of $^4S_{3/2}$ and $^2H_{11/2}$ levels, and (*4*) emission from those levels

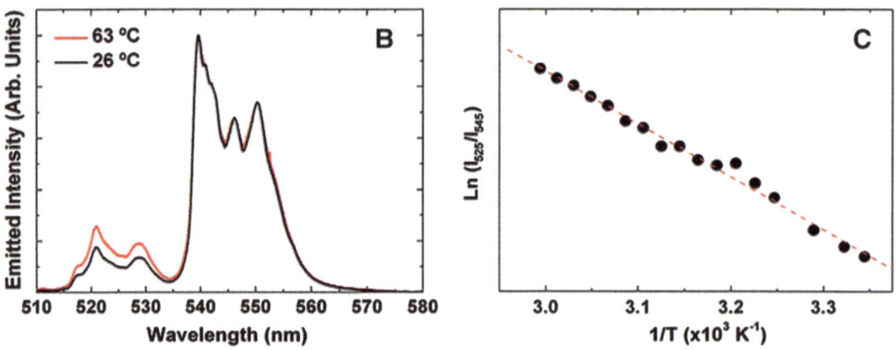

Fig. 16.13 (*Left*) Fluorescence spectra of $NaYF_4$:Er,Yb at different temperatures, and (*right*) a calibration curve for the ratio of the intensities as a function of temperature [13]

hundred cm^{-1}. In fact, the emission from the $^4S_{3/2}$ and $^2H_{11/2}$ levels are centered at 545 and 525 nm, respectively, we which corresponds to an energy separation of about 700 cm^{-1}. Though this is slightly high and will not have temperature maximum sensitivity, it was shown by Ventrone et al. that between 25 and 45 °C this system is accurate to within 1 °C (See Fig. 16.13). We note that there are other examples of nanothermometry using rare earth ions. For example, nanocrystals of this same system ($NaYF_4$: Er, Yb) are being used as nanothermometers in laser writing applications [14]. It is observed that when the laser overexposes the photoresist the temperature increases rapidly, and irreparable damage done to the photoresist. Thus, it is of great interest to monitor the temperature of the photoresist local to where it is being written on. In preparing the photoresist, the nanoparticles can be mixed in and hence dispersed throughout the photoresist. Then, using one laser to write on the photoresist and another laser to excite the nanoparticles, the temperature of the nanoparticles can be continuously monitored, and the writing can be accomplished without damage to the photoresist.

References

1. Dieke GH (1968) Spectra and energy levels of rare earth ions in crystals. Interscience, New York
2. Liu G, Chen X, Zhuang H, Li S, Niedbala R (2003) Confinement of electronphonon interaction on luminescence dynamics in nanophosphors of Er3+:Y2O2S. J Solid State Chem 171: 123–132
3. Liu GK, Zhuang HZ, Chen XY (2002) Restricted phonon relaxation and anomalous thermalization of rare earth ions in nanocrystals. Nano Lett 2:535–539
4. Meltzer RS, Hong KS (2000) Electron-phonon interactions in insulating nanoparticles: Eu_2O_3. Phys Rev B 61:3396–3403
5. Ozen G, Collins J, Bettinelli M, Di Bartolo B (2013) Luminescence of Y3AL5O12 nano-particles doped with praseodymium ions. Opt Mater 35:1360–1365
6. Baldassare DB (1968) Optical interactions in solids. Wiley, New York
7. Naccache R, Rodríguez EM, Bogdan N, Sanz-Rodríguez F, de la Cruz MDCI, de la Fuente AJ, Vetrone F, Jaque D, Solé JG, Capobianco JA (2012) High resolution fluorescence imaging of cancers using lanthanide ion-doped upconverting nanocrystals. Cancers 4:1067–1105
8. Vetrone F, Boyer J-C, Capobianco JA, Speghini A, Bettinelli M (2003) Concentration-dependent near-infrared to visible upconversion in nanocrystalline and bulk Y 2 O 3 :Er 3+. Chem Mater 15:2737–2743
9. Bogdan N, Vetrone F, Ozin GA, Capobianco JA (2011) Synthesis of ligand-free colloidally stable water dispersible brightly luminescent lanthanide-doped upconverting nanoparticles. Nano Lett 11:835–840
10. Zhou J, Yu M, Sun Y, Zhang X, Zhu X, Wu Z, Wu D, Li F (2011) Fluorine-18-labeled Gd3+/Yb3+/Er3+ co-doped NaYF4 nanophosphors for multimodality PET/MR/UCL imaging. Biomaterials 32:1148–1156
11. Jacobsohn LG, Sprinkle KB, Roberts SA, Kucera CJ, James TL, Yukihara EG, DeVol TA, Ballato J (2011) Fluoride nanoscintillators. J Nanomater 2011:1–6
12. Yukihara E, Jacobsohn LG, Blair MW, Bennett BL, Tornga SC, Muenchausen RE (2010) Luminescence properties of Ce-doped oxyorthosilicate nanophosphors and single crystals. J Lumin 130:2309–2316
13. Vetrone F, Naccache R, Zamarr6n A, Juarranz de la Fuente A, Sanz-Rodríguez F, Martinez Maestro L, Martín Rodriguez E, Jaque D, García Solé J, Capobianco JA (2010) Temperature sensing using fluorescent nanothermometers. ACS Nano 4:3254–3258
14. Mueller JB, Fischer J, Mange YJ, Nann T, Wegener M (2013) In-situ local temperature measurement during three-dimensional direct laser writing. Appl Phys Lett 103:123107

References

17
Nanostructures in the Terahertz Range

John W. Bowen

Abstract With advances in technology, terahertz imaging and spectroscopy are beginning to move out of the laboratory and find applications in areas as diverse as security screening, medicine, art conservation and field archaeology. Nevertheless, there is still a need to improve upon the performance of existing terahertz systems to achieve greater compactness and robustness, enhanced spatial resolution, more rapid data acquisition times and operation at greater standoff distances. This chapter will review recent technological developments in this direction that make use of nanostructures in the generation, detection and manipulation of terahertz radiation. The chapter will also explain how terahertz spectroscopy can be used as a tool to characterize the ultrafast carrier dynamics of nanomaterials.

Keywords Terahertz • Nanostructures • Nanoscale sensing • Nanomaterial characterisation

17.1 Introduction

Compared to other regions of the electromagnetic spectrum, the terahertz range has only relatively recently become accessible for imaging and spectroscopy. This is due to the considerable technological challenges that have had to be overcome in order to generate, manipulate and detect terahertz radiation. A few years ago, this state of affairs gave rise to the notion of the "terahertz gap" in the spectrum between the infrared and microwave regions, generally considered to cover frequencies from 100 GHz to 10 THz, corresponding to wavelengths from 3 mm down to 30 μm. However, while the field of terahertz science and technology is still in its infancy, the

J.W. Bowen (✉)
School of Systems Engineering, University of Reading, Whiteknights, Reading, RG6 6AY, UK
e-mail: j.bowen@reading.ac.uk

B. Di Bartolo et al. (eds.), *Nano-Structures for Optics and Photonics*,
NATO Science for Peace and Security Series B: Physics and Biophysics,
DOI 10.1007/978-94-017-9133-5_17,

technology has advanced to such an extent that terahertz imaging and spectroscopy are now no longer confined to the laboratory, and are beginning to find real-world applications in a diverse range of areas. Many of these applications are a consequence of terahertz radiation being able to penetrate through materials which are otherwise opaque in other regions of the electromagnetic spectrum, combined with the ability to achieve three-dimensional and depth-resolved imaging, ultra-fast time resolution, identification of materials via molecular resonances, and improved safety over X-ray imaging due to the non-ionising nature of the radiation. As an example, the development of relatively portable terahertz imaging systems has enabled the first use of terahertz imaging in situ at an archaeological site [1]. This work was carried out in the remote, hot and dusty environment of the Turkish desert, to image Neolithic wall paintings obscured by covering layers of plaster. Other examples of in situ terahertz imaging of historic wall paintings and of applications in cultural heritage in general are given in [2] and [3]. There is also much interest in the use of terahertz imaging and spectroscopy for the detection of weapons, explosives or contraband concealed about the body or hidden in luggage [4], the detection of trace elements of substances [5], in medical imaging and diagnosis [6], and in a variety of industrial non-destructive testing applications [7]. However, despite this pioneering work, the portable systems available to date are still rather large and cumbersome, are relatively fragile and easily upset by extremes of temperature and moisture, and, depending on the spatial and spectral resolution required, can have rather slow data acquisition rates. Therefore, with these applications in mind, there is a need for terahertz systems with greater compactness and robustness, enhanced spatial resolution, more rapid data acquisition times and the ability to operate at greater standoff distances. This in turn requires the development of higher power terahertz sources, more responsive detectors and imaging arrays, all of which can be operated without the need for cryogenic cooling. This chapter will explore recent developments in this direction made possible by the use of nanostructures in the generation, manipulation and detection of terahertz radiation. Topics covered include the nanostructure enhancement of sources and detectors, terahertz near-field microscopy, terahertz sensing of molecules and nanoparticles, and terahertz molecular imaging. In addition, it will be explained how terahertz radiation can be used as an important tool to characterize the ultrafast carrier dynamics of nanomaterials. As a precursor to the discussion of these new developments, the next section of this chapter gives an overview of the principles of operation of some of the terahertz imaging and spectroscopy systems in current use.

17.2 Terahertz Systems

There are several techniques available to carry out terahertz imaging and spectroscopy. As this chapter will discuss the application of nanostructures to extend and improve the performance of some of these techniques, as well as discussing the use of terahertz spectroscopy to characterize nanomaterials, this section will give a brief overview of the principles behind their operation.

17.2.1 Terahertz Time Domain Spectroscopy and Imaging

Terahertz time domain spectroscopy (TDS) is currently the most widely used laboratory based terahertz spectroscopy technique, and is also employed in the portable imaging system used for the work on wall paintings reported in [1] and mentioned above. TDS is based upon generating short, typically several hundred femtosecond long, pulses of broadband terahertz radiation and recording the resulting reflected or transmitted pulses from a sample or object of interest. The short pulse length makes it possible to resolve reflections due to interfaces between different layers within a stratified object, as these will appear as individual pulse reflection signatures occurring after different time delays within the recorded time domain signal, corresponding to the time of flight to each layer. This depth resolution can be further improved by using a deconvolution technique to achieve sub-pulse length and sub-wavelength depth resolution [8]. Moreover, the incident terahertz pulses are so short that they contain a wide range of frequencies, spanning at least 100 GHz to 3 THz (wavelengths of 3 mm down to 100 μm) in typical systems. Therefore, broadband spectroscopy is possible by carrying out a fast Fourier transform of the time domain signals. As the technique is coherent, the resulting spectra contain both amplitude and phase information, enabling direct recovery of both the refractive index and absorption coefficient of a sample, without recourse to the Kramers-Kronig relations.

The TDS technique can be extended to enable imaging by focusing the terahertz beam to a small spot through which the sample can be raster scanned, or, if more convenient, in fibre coupled systems the beam can be raster scanned across the sample. The scanning is usually accomplished by mounting either the sample or the generator and detector heads on translation stages.

As there is a complete time domain signal behind each pixel in the resulting image, there are multiple parameters that can be used to construct an image, each of which can provide complementary information about the object being imaged. Typical imaging parameters include the peak amplitude in the time domain, the time delay, the integrated amplitude over a given range of frequencies, and the impulse response function [8].

The terahertz pulses are generated using a near-infrared femtosecond pulsed laser. Although the output pulses from the laser are short enough to have a bandwidth of several terahertz, their centre frequency is in the infrared range, and so a terahertz emitter has to be used to convert the output into the terahertz range. There are two main approaches to achieving this conversion. In photoconductive generation, the beam from the femtosecond laser generates photocarriers in a piece of semiconductor spanning the feed gap of a terahertz planar antenna. As a bias voltage is applied to the two halves of the antenna, a photocurrent is generated through the antenna for the duration of the incident infrared pulse and persists for the carrier lifetime, resulting in a short pulse of terahertz radiation being emitted from the antenna. In the second approach, known as optoelectronic generation, the terahertz pulses are generated by difference frequency mixing of the frequency components in the infrared pulse in a non-linear crystal with a non-zero second order susceptibility.

Detection of the terahertz pulses is achieved using a gated detection system. Again there are two main approaches. In photoconductive detection, the terahertz pulse is incident on an antenna structure similar to that used for photoconductive generation. This time, however, the antenna is not connected to a bias supply but is connected to equipment to record the current induced by the incident terahertz pulse. As current only flows when photocarriers are produced in the semiconductor due to an infrared pulse from the femtosecond laser being incident during the time of arrival of the terahertz pulse, the detector is gated by the femtosecond laser. In order to achieve this, a beam splitter is used to separate the output from the laser into pump and probe pulses to excite the generator and gate the detector respectively. Because the carrier lifetime results in the terahertz pulse being significantly longer than the infrared pulse, the time delay between the terahertz pulse and the infrared pulse can be scanned using an optical delay line so that the time domain waveform of the terahertz pulse can be mapped out over a series of successive output pulses from the femtosecond laser.

In the second approach, electro-optic detection, the terahertz pulse and the infrared gating probe pulse are incident on a crystal that displays the linear Pockel's effect. If no terahertz field were present, the incident linearly polarized infrared beam would have its polarization converted by the crystal, so that it would become circularly polarized. However, due to the Pockel's effect, the incident terahertz field results in an elliptically polarized output infrared beam. A Wollaston prism and pair of photodetectors are used to analyse the polarization of the infrared beam into linearly polarized components, the difference in intensity of which is directly proportional to the terahertz field amplitude. Again, both the terahertz pulse and infrared pulse need to be incident simultaneously for an output signal to be recorded, and so the terahertz signal can be recorded in the time domain by scanning the delay of the infrared probe pulse.

In portable systems, both the pump and probe pulses can be delivered to the emitter and detector using fibre optical umbilicals, allowing a high degree of flexibility for scanning different objects in situ.

17.2.2 Photomixer Systems

An alternative technique that is beginning to be employed in commercially available terahertz spectroscopy systems is based around mixing together the output from two visible or infrared lasers to generate terahertz radiation at the difference frequency. The mixing is achieved within a semiconductor, usually GaAs, in which the intensity modulation of the laser beams due to their interference results in carrier generation with a terahertz periodicity. Often these laser beams are focused to a small spot in the semiconductor, and the resulting carrier currents feed a terahertz planar antenna integrated on its surface. Typically, photomixers are pumped by distributed feedback diode lasers operating at a wavelength around 1,550 nm, which can be amplified by erbium doped fibre amplifiers. This approach has resulted in continuous wave terahertz outputs of between 20 mW at 100 GHz and 25 μW at 0.914 THz. The output frequency can be swept over a wide range by sweeping the wavelength of the diode lasers, so that spectroscopy can be accomplished and depth information can be retrieved using frequency modulated continuous wave (FMCW) techniques. Again, imaging is possible by raster scanning. The resulting systems are very compact and portable. An in-depth review of terahertz photomixing is given in [9].

17.2.3 Quasi-optics

In order to keep terahertz systems as compact as possible, and bearing in mind that the wavelengths involved are a few millimeters to fractions of a millimetre, the terahertz beams produced by these systems are typically only a few wavelengths or tens of wavelengths across. As a consequence, just as with nano-optics at visible wavelengths, diffraction becomes a significant aspect of the propagation and similar techniques to those used to analyse nano-optics can be applied to terahertz optics, albeit scaled up in wavelength. Moreover, again for reasons of compactness, any lenses or mirrors used to control the spread of terahertz radiation tend to be used in the near to far-field transition region of the beam and have to be designed to take this into account. As a more general treatment than geometrical optics is needed to describe the behavior of such optical systems, they are generally referred to as being quasi-optical.

One convenient way to describe an arbitrary diffractively spreading paraxial beam is as a superposition of a set of beam-modes, each of which maintains a characteristic form as it propagates, albeit with a scaling width and changing phase-front curvature. These are usually Hermite-Gaussian or Laguerre-Gaussian beam-modes, each having a Gaussian transverse amplitude distribution modulated by either Hermite or Laguerre polynomials of different order. These beam-modes provide an analytical basis for modelling the propagation of a beam away from its narrowest point, termed the beam-waist, and enable the design of lens and mirror profiles to refocus the beam to a new beam-waist at a set distance. A full discussion of quasi-optics is given in [10].

The diffractive spreading imposes a number of size constraints on terahertz optical systems. Firstly there is a maximum distance from a lens at which a beam-waist may be formed, which depends purely on the width of the beam as it emerges from the lens and is independent of the lens focal length. Secondly, any apertures in a quasi-optical system have to have a diameter which is at least three times the $1/e$ amplitude half-width for a fundamental Gaussian beam-mode in order to avoid significant truncation. Truncation is undesirable, both because it would result in a loss of power and in the generation of higher order beam-modes, giving rise to dispersion. Finally, diffraction limits the smallest spot size to which a beam may be focused to around one half of the wavelength, at which point the beam would be extremely non-paraxial. Although such extremely converging beams are not used in practice, the spot size in most imaging systems is typically a little larger than the wavelength and decreases in size with the wavelength across the system bandwidth. However, as will be discussed in Sect. 17.4, there are ways in which the radiation can be concentrated to much smaller dimensions, allowing nanoscale imaging and spectroscopy.

17.3 Nanostructure Enhancement of Sources and Detectors

Recent work has shown that the incorporation of nanostructures in terahertz sources and detectors can offer considerable benefits in improving their efficiency. These advances will be discussed in this section.

17.3.1 Plasmonic Enhancement of Photoconductive Emitters and Detectors

One of the problems with both conventional photoconductive emitters and detectors and with photomixers is that it is difficult to obtain ultrafast and high quantum efficiency operation simultaneously. In a conventional emitter, the gap between the antenna electrodes cannot be smaller than the diffraction limited spot size of the pump laser, typically of the order of 2 μm. Given the carrier drift velocity, only a small number of the photocarriers can reach the antenna electrodes on a sub-picosecond timescale and so contribute to the terahertz radiation. The remaining carriers just result in a dc current, which lowers the bias field, in turn reducing the carrier acceleration between the electrodes. In order to suppress this unwanted dc current, most conventional emitters use short carrier lifetime semiconductors to ensure that carriers which are slow to reach the electrodes recombine before they get there. However, this means that only a small proportion of the photoexcited carriers contribute to the terahertz emission, limiting the efficiency.

Berry and Jarrahi [11] have shown how it is possible to overcome this problem by incorporating nanoscale plasmonic gratings into the emitter electrodes, eliminating the need for a short carrier lifetime semiconductor. Even though the gaps within the plasmonic gratings are very much smaller than the diffraction limited laser spot, the intensely localized surface plasmon fields generate carriers in the close vicinity of the gratings, enabling the carriers to quickly contribute to the current within the metal grating/antenna electrode structure. A very thin photoabsorbing semiconductor layer is used to ensure that carriers are not generated deeper within the semiconductor, as they would not reach the grating electrodes on a sufficiently short timescale. While electrons generated immediately under the positively biased grating will be drawn towards that grating and holes immediately under the negatively biased grating are drawn towards that grating to contribute to the terahertz emission, the carriers of opposite sign need to be collected by a grounded central contact between the two electrodes. Using this scheme, carriers that would take a relatively long time to reach the opposite electrode, and so contribute to an unwanted dc current, are removed, while around half of the electrons and holes that are generated contribute to the terahertz emission, which is a significant improvement on conventional emitters.

In addition, as the pump laser radiation can be absorbed over the entire plasmonic grating area rather than just the gap between the electrodes, the active area can be increased without significantly increasing the capacitance of the antenna. This, combined with the fact that normal semiconductors have a higher thermal conductivity than those with a short carrier lifetime, means that higher pump power levels can be used.

As they have similar structures and are based on similar operating principles, the incorporation of plasmonic gratings within photoconductive detectors and antenna coupled photomixers can lead to similar improvements in quantum efficiency.

17.3.2 Terahertz Generation from Plasmonic Nanoparticle Arrays

An alternative to conventional terahertz generators may be provided by the illumination of nanostructured metal surfaces by femtosecond lasers. Different research teams have presented evidence for different mechanisms behind this phenomenon.

Kajikawa et al. [12] have demonstrated terahertz emission from surface-immobilized gold nanospheres. The emission was seen to have a quadratic amplitude dependence on the excitation field, suggesting it is a second-order nonlinear optical process giving rise to optical rectification.

On the other hand, Polyushkin et al. [13] have demonstrated a non-quadratic dependence in terahertz emission from metal nanostructures, including semicontinuous metal films and ordered arrays of nanoparticles. In this case, based on a model by Welsh et al. [14], they suggest that the mechanism is plasmon mediated multiphoton ionization and ponderomotive acceleration of electrons in the uneven surface plasmon evanescent field, giving rise to terahertz emission.

Greater understanding of these processes may eventually lead to the development of more efficient terahertz emitters and may even lead to the interesting possibility of using a single metal nanoparticle as a local terahertz emitter [13].

17.4 Terahertz Near-Field Microscopy

As discussed in Sect. 17.2.3, diffraction limits the smallest spot size to which a terahertz beam can be focused to around half a wavelength. However, a variety of techniques based around sub-wavelength apertures, probes, plasmonic parallel plate waveguides, aperture-less near-field scattering microscopy and nanojunctions can dramatically increase the spatial resolution possible in terahertz imaging and spectroscopy systems, to the extent that nanometre-scale resolution is possible. This section gives an overview of these techniques.

17.4.1 Imaging with Sub-wavelength Apertures and Probes

Imaging with sub-wavelength apertures and probes can be used to detect the evanescent terahertz fields in the vicinity of an object. The apertures or probes can be integrated with the terahertz detector to yield resolutions of a few micrometres [15, 16] and there is even a commercially available terahertz integrated microprobe range with a 3 μm spatial resolution [17], which can easily be added to existing terahertz TDS systems. Images can be constructed by scanning the probe or aperture across the surface of the object.

Rather than using a physical aperture, it is also possible to tightly focus the probe pulse onto an electro-optic crystal to form a so-called dynamic aperture, as terahertz detection will only take place in the crystal within the confines of the probe pulse spot [18, 19].

17.4.2 Plasmonic Focusing

An alternative terahertz microscopy approach is to concentrate terahertz radiation to sub-wavelength dimensions in tapered plasmonic parallel plate waveguides. Focusing down to 10 μm by 18 μm has been demonstrated at 0.155 THz [20], although, theoretically, losses limit the smallest spot size to only 100–250 nm [21].

17.4.3 Aperture-Less Near-Field Scattering Techniques

Rather than integrating a detector with a scanning probe, it is possible to insert a sharply pointed probe tip into a terahertz beam and measure the scattered evanescent field as the tip is scanned across an object. With this technique, it is necessary to have some means to distinguish the scattered field from the incident field. In one approach, the tip is scanned near to the surface of an electro-optic detection crystal with the correct crystallographic orientation to detect just the scattered fields from the tip which are polarized normal to the crystal surface, while being insensitive to the incident terahertz radiation which is polarized parallel to the crystal surface [22,23]. Alternatively, a vibrating probe tip can be used with lock-in detection [24].

Huber et al. [25] have achieved very impressive results by using metalised atomic force microscopy tips as the scattering probe. Because of non-linearity in the field near the tip, they have been able to achieve background suppression of the incident radiation by detecting the higher harmonics of the scattered field using a Michelson interferometer. Using this technique with radiation from a 2.54 THz laser they have managed to produce terahertz images of single nano-transistors, differentiating regions of different carrier density, with a spatial resolution of 40 nm, which is 1/3,000 of the wavelength.

17.4.4 Oxide Nanojunctions

Ma et al. [26] have demonstrated terahertz generation and detection at 10 nm oxide nanojunctions, opening up the possibility of terahertz spectroscopy and control of individual nanoparticles and molecules. The nanojunctions are formed by writing nanowires at the interface between layers of $LaAlO_3$ and $SrTiO_3$ using an atomic force microscope tip. This writing can be accomplished reversibly, so the junctions can be formed or reformed at will, potentially enabling the object to be firstly located with the atomic force microscope and then a nanojunction formed around it for terahertz nano-spectroscopy . The terahertz fields are generated and detected by $\chi^{(3)}$ processes under illumination from a femtosecond laser, and the same junction can potentially be used for both generation and detection at the nanoscale, four orders of magnitude smaller than the diffraction limit.

17.5 Terahertz Sensing of Molecules and Nanoparticles

Because molecules and nanoparticles are considerably smaller than the wavelength, their absorption cross-sections are quite small at terahertz frequencies, making it difficult to detect them unless they are present in large quantities. However, it has recently been demonstrated that their absorption cross sections can be enhanced by over three orders of magnitude by placing them across or inside nanoslot resonators. This yields a dramatically increased absorption coefficient, allowing much smaller quantities to be detected.

Using a slot with a length of 90 μm and width of 50 nm, formed in a 50 nm gold layer on a quartz substrate, Park et al. [27] have used TDS to demonstrate detection of as little as 40 ng of the explosive RDX drop-cast over an area of 10 mm^2. This corresponds to only 22 fg of RDX within the slot. Without the nanoslot resonator, it is necessary to use of the order of 1 mg to see a noticeable difference in transmission through the RDX on the bare quartz substrate.

As the radiation is funneled through the nanoslot, the electric field becomes enhanced within the slot by orders of magnitude compared to the magnetic field. Because the absorption cross-section of the molecules is directly proportional to the wave impedance, which is the ratio of the electric to the magnetic field, the absorption cross-section increases accordingly. In the case of the RDX experiment mentioned above, the molecular absorption cross-section was enhanced by a factor of 2,800 times over that with the bare substrate.

The absorption enhancement can be tuned by changing the length of the slot so that it resonates at different frequencies, which can correspond to specified absorption bands in the target molecules. Park et al. [27] have demonstrated this by carrying out a second experiment with a 150 μm long slot, resonant at 0.5 THz, which shows a similar absorption cross-section enhancement, but for lactose molecules rather than RDX.

The same group have also shown that similar nanoslot resonators are sensitive to even single nanorods placed across them, the nanorod changing the resonant conditions of the slot, depending its size and placement [28, 29].

17.6 Terahertz Molecular Imaging

A number of studies have indicated that conventional terahertz imaging may be a useful medical diagnostic tool, particularly for skin cancer [30] or tumours near the surface of organs [31]. However, terahertz molecular imaging (TMI) [32], which uses nanoparticle probes as contrast agents, offers enhanced performance by increasing contrast and enabling micrometre spatial resolution. Additionally, it offers the scope to track nanoparticle drug delivery.

The mechanism by which TMI operates is as follows. After injection into the biological tissue to be imaged, the nanoparticle probes target cancer cells by antibody phase conjugation, whereupon they become absorbed inside the cell by a process known as endocytosis. The tissue is illuminated with a chopped near-infrared laser beam which gives rise to surface plasmon polaritons on the nanoparticles, inducing a water temperature increase in the cells. As the refractive index of water in the terahertz range is temperature dependent, this results in a terahertz reflection change which can be detected using terahertz time domain imaging. Modulating the near-infrared laser enables differential detection, which is necessary to achieve the required sensitivity.

The TMI technique has been compared with magnetic resonance imaging (MRI) by transfecting superparamegnetic iron oxide (SPIO) nanoparticles into ovarian cancer cells in a mouse. Good correlation was seen between the TMI and MRI images, implying that TMI may be a useful technique in the operating theatre to complement pre- and post-operative MRI [32].

17.7 Terahertz Characterisation of Nanomaterials

Terahertz spectroscopy is an important probe of the ultrafast carrier dynamics of nanomaterials because it provides information about charge transport on sub-picosecond to nanosecond timescales and nanometre length scales which cannot be provided by other techniques. Terahertz frequencies closely match typical carrier scattering rates of 10^{12}–10^{14} s, making terahertz spectroscopy ideal for studying conductivity.

To give ultrafast time resolution, a modification of the TDS technique, known as time-resolved terahertz spectroscopy (TRTS) or optical pump/terahertz probe (OPTP) spectroscopy is used [33]. In this implementation, as well as providing the pulses to drive the terahertz generator and detector, the femtosecond laser provides another near-infrared pump beam to induce transient photoconductivity in the sample. A further optical delay line is included so that the terahertz behavior of the sample can be monitored at different times after photoexcitation of the sample.

The relaxation dynamics of the sample's transient photoconductivity can be probed by carrying out a "pump scan", wherein the delay in the pump beam is scanned while holding the terahertz detector gating delay at the peak of the terahertz time-domain waveform. The photoconductive lifetime can be determined from this type of measurement. Example measurements of the photoconductivity decay of GaAs, InAs and InP nanowires are given in Joyce et al. [34].

Conductivity can be determined by carrying out a "pump scan", wherein the pump beam delay is held constant while the terahertz gating delay is scanned to record the full terahertz waveform at a fixed time after photoexcitation. After fast Fourier transformation to the frequency domain and normalizing to a reference spectrum without the sample, the frequency dependent absorption coefficient and refractive index n of the sample can be found from the transmission amplitude and phase. In order to determine these quantities for nanostructures rather than bulk media, it may be necessary to make use of an effective medium approximation [35]. The frequency dependent complex conductivity can then be found from α and n, although in doing so, it is usually necessary to include a component of the permittivity representing the contribution from bound modes, i.e. lattice vibrations and core electrons [36]. Once the conductivity has been determined experimentally, it can be fitted with an appropriate conductivity model (of which there are a number) to determine parameters such as the mobility, carrier scattering time and plasma frequency. A detailed overview of the whole procedure is given in [36]. Examples of terahertz characterization of a variety of nanostructures can be found in [35, 37–39].

17.8 Conclusions

This chapter has shown how, despite the relatively large size of the wavelengths involved, nanostructures are beginning to make a significant impact in improving the performance of terahertz systems and devices applicable to a wide range of situations. In the case of terahertz sources and detectors excited by femtosecond lasers operating in the near-infrared, the terahertz performance improvements are the result of nanostructure enhancements of the interaction of the near-infrared laser radiation with the device. Recent developments in terahertz near-field microscopy mean that it is now possible to achieve nanoscale spatial resolution at terahertz frequencies. The use of nanoslot resonators can dramatically increase the terahertz absorption cross-section of molecules to allow the detection of small quantities of target substances, as well as providing a means to sense individual nanoparticles. Terahertz molecular imaging makes use of nanoparticle contrast agents to enhance the contrast and improve the spatial resolution of terahertz medical imaging for cancer diagnosis and nanoparticle drug delivery. Furthermore, terahertz spectroscopy provides unique information to characterize the ultrafast behavior of nanomaterials.

In addition to the new developments that have been covered in this chapter, work is underway on terahertz detection and generation via plasma waves in nanometre field effect transistors [40], on nano-hot electron bolometers [41], plasmonic and graphene based terahertz modulators [42, 43], terahertz polarisers based on carbon nanotudes and nanowire arrays [44, 45], terahertz vacuum nanoelectronics [46], graphene lasers [47], terahertz-driven molecular surface rotors for molecular machines [48], and wireless nanosensor networks for nanorobots, nanodiagnostic techniques and co-operative drug delivery systems [49]. It is clear that nanostructures are likely to play an increasing role in the future development of terahertz systems for many real-world applications.

References

1. Walker GC, Bowen JW, Matthews W, Roychowdhury S, Labaune J, Mourou G, Menu M, Hodder I, Jackson JB (2013) Sub-surface terahertz imaging through uneven surfaces: visualizing Neolithic wall paintings in Çatalhöyük. Opt Express 21:8126–8134
2. Jackson JB, Bowen J, Walker G, Labaune J, Mourou G, Menu M, Fukunaga K (2011) A Survey of terahertz applications in cultural heritage conservation science. IEEE Trans Terahertz Sci Technol 1:220–231
3. Walker GC, Jackson JB, Giovannacci D, Bowen JW, Delandes B, Labaune J, Mourou G, Menu M, Detalle V (2013) Terahertz analysis of stratified wall plaster at buildings of cultural importance across Europe. In: Pezzati L, Targowski P (eds) *Optics for Arts, Architecture and Archaeology IV,* Proc. SPIE vol. 8790. Munich, Germany, p 87900H
4. Kemp MC (2011) Explosives detection by terahertz spectroscopy-a bridge too far? IEEE Trans Terahertz Sci Technol 1:282–292
5. Brown E, Khromova T, Globus T, Woolard D, Jensen J, Majewski A (2006) Terahertz-regime attenuation signatures in bacillus subtilis and a model based on surface polariton effects. IEEE Sens J 6:1076–1083
6. Pickwell E, Wallace VP (2006) Biomedical applications of terahertz technology. J Phys D: Appl Phys 39:R301–R310
7. Morita Y (2005) Terahertz technique for detection of microleaks in the seal of flexible plastic packages. Opt Eng 44:019001
8. Walker GC, Bowen JW, Labaune J, Jackson J-B, Hadjiloucas S, Roberts J, Mourou G, Menu M (2012) Terahertz deconvolution. Opt Express 20:27230–27241

9. Preu S, Döhler GH, Malzer S, Wang LJ, Gossard AC (2011) Tunable, continuous-wave Terahertz photomixer sources and applications. J Appl Phys 109:061301
10. Martin D, Bowen J (1993) Long-wave optics. IEEE Trans Microw Theory Tech 41(10):1676–1690
11. Berry CW, Jarrahi M (2012) Terahertz generation using plasmonic photoconductive gratings. New J Phys 14:105029
12. Kajikawa K, Nagai Y, Uchiho Y, Ramakrishnan G, Kumar N, Ramanandan GKP, Planken PCM (2012) Terahertz emission from surface-immobilized gold nanospheres. Optics Lett 37:4053–4055
13. Polyushkin DK, Hendry E, Stone EK, Barnes WL (2011) THz generation from plasmonic nanoparticle arrays. Nano Lett 11:4718–4724
14. Welsh G, Hunt N, Wynne K (2007) Terahertz-pulse emission through laser excitation of surface plasmons in a metal grating. Phys Rev Lett 98:026803
15. Mitrofanov O, Lee M, Hsu J, Brener I, Harel R, Federici J, Wynn J, Pfeiffer L, West K (2001) Collection-mode near-field imaging with 0.5-THz pulses. IEEE J Sel Top Quantum Electron 7(4):600–607
16. Kawano Y (2011) Highly sensitive detector for on-chip near-field thz imaging. IEEE J Sel Top Quantum Electron 17:67–78
17. TeraSpike, AMO GmbH. Available at http://www.amo.de/thz_tip.0.html
18. Seo MA, Adam AJL, Kang JH, Lee JW, Ahn KJ, Park QH, Planken PCM, Kim DS (2008) Near field imaging of terahertz focusing onto rectangular apertures. Opt Express 16:20484
19. Adam AJL, Brok JM, Seo MA, Ahn KJ, Kim DS, Kang JH, Park QH, Nagel M, Planken PC (2008) Advanced terahertz electric near-field measurements at sub-wavelength diameter metallic apertures. Opt Express 16:7407
20. Zhan H, Mendis R (2010) Mittleman DM, Superfocusing terahertz waves below lambda/250 using plasmonic parallel-plate waveguides. Opt Express 18:9643–9650
21. Rusina A, Durach M, Nelson KA, Stockman MI (2008) Nanoconcentration of terahertz radiation in plasmonic waveguides. Opt Express 16:18576
22. van der Valk NCJ, Planken PCM (2004) Towards terahertz near-field microscopy. Philos Trans Ser A Math Phys Eng Sci 362:315–319; discussion 319–321
23. Planken PCM, van Rijmenam CEWM, Schouten RN (2005) Opto-electronic pulsed THz systems. Semicond Sci Technol 20:S121–S127
24. Astley V, Zhan H, Mittleman D, Hao F, Nordlander P (2007) Plasmon-enhanced Terahertz near-field microscopy – OSA technical digest series (CD). In: Conference on lasers and electro-optics/quantum electronics and laser science conference and photonic applications systems technologies, Baltimore. Optical Society of America, p CTuJJ5
25. Huber AJ, Keilmann F, Wittborn J, Aizpurua J, Hillenbrand R (2008) Terahertz near-field nanoscopy of mobile carriers in single semiconductor nanodevices. Nano Lett 8:3766–3770
26. Ma Y, Huang M, Ryu S, Bark CW, Eom C-B, Irvin P, Levy J (2013) Broadband terahertz generation and detection at 10 nm scale. Nano Lett 13:2884–2888
27. Park H-R, Ahn KJ, Han S, Bahk Y-M, Park N, Kim D-S (2013) Colossal absorption of molecules inside single terahertz nanoantennas. Nano Lett 13:1782–1786
28. Park H-R, Bahk Y-M, Choe JH, Han S, Choi SS, Ahn KJ, Park N, Park Q-H, Kim D-S (2011) Terahertz pinch harmonics enabled by single nano rods. Opt Express 19:24775–24781
29. Park H-R, Bahk Y-M, Ahn KJ, Park Q-H, Kim D-S, Martín-Moreno L, García-Vidal FJ, Bravo-Abad J (2011) Controlling terahertz radiation with nanoscale metal barriers embedded in nano slot antennas. ACS Nano 5:8340–8345
30. Woodward RM, Wallace VP, Pye RJ, Cole BE, Arnone DD, Linfield EH, Pepper M (2003) Terahertz pulse imaging of ex vivo basal cell carcinoma. J Investig Dermatol 120:72–78
31. Ashworth P, O'Kelly P, Purushotham A, Pinder S, Kontos M, Pepper M, Wallace V (2008) An intra-operative THz probe for use during the surgical removal of breast tumors. In: 2008 33rd international conference on infrared, millimeter and terahertz waves, Pasadena
32. Son J-H (2013) Principle and applications of terahertz molecular imaging. Nanotechnology 24:214001

33. Beard MC, Turner GM, Schmuttenmaer CA (2002) Terahertz spectroscopy. J Phys Chem B 106:7146–7159
34. Joyce HJ, Docherty CJ, Gao Q, Tan HH, Jagadish C, Lloyd-Hughes J, Herz LM, Johnston MB (2013) Electronic properties of GaAs, InAs and InP nanowires studied by terahertz spectroscopy. Nanotechnology 24:214006
35. Baxter JB, Schmuttenmaer CA (2006) Conductivity of ZnO nanowires, nanoparticles, and thin films using time-resolved terahertz spectroscopy. J Phys Chem B 110:25229–25239
36. Lloyd-Hughes J, Jeon T-I (2012) A review of the terahertz conductivity of bulk and nano-materials. J Infrared Millim Terahertz Waves 33:871–925
37. Jung GB, Myung Y, Cho YJ, Sohn YJ, Jang DM, Kim HS, Lee C-W, Park J, Maeng I, Son J-H, Kang C (2010) Terahertz spectroscopy of nanocrystal-carbon nanotube and -graphene oxide hybrid nanostructures. J Phys Chem C 114:11258–11265
38. Parkinson P, Joyce HJ, Gao Q, Tan HH, Zhang X, Zou J, Jagadish C, Herz LM, Johnston MB (2009) Carrier lifetime and mobility enhancement in nearly defect-free core-shell nanowires measured using time-resolved terahertz spectroscopy. Nano Lett 9:3349–3353
39. Strait JH, George PA, Levendorf M, Blood-Forsythe M, Rana F, Park J (2009) Measurements of the carrier dynamics and terahertz response of oriented germanium nanowires using optical-pump terahertz-probe spectroscopy. Nano Lett 9:2967–2972
40. Shur M (2010) Plasma wave terahertz electronics. Electron Lett 46:s18–s21
41. Karasik BS, Sergeev AV, Prober DE (2011) Nanobolometers for THz photon detection. IEEE Trans Terahertz Sci Technol 1:97–111
42. Vasić B, Gajić R (2012) Broadband and subwavelength terahertz modulators using tunable plasmonic crystals with semiconductor rods. J Phys D Appl Phys 45:095101
43. Weis P, Garcia-Pomar JL, Höh M, Reinhard B, Brodyanski A, Rahm M (2012) Spectrally wide-band terahertz wave modulator based on optically tuned graphene. ACS Nano 6:9118–9124
44. Ren L, Pint CL, Arikawa T, Takeya K, Kawayama I, Tonouchi M, Hauge RH, Kono J (2012) Broadband terahertz polarizers with ideal performance based on aligned carbon nanotube stacks. Nano Lett 12:787–790
45. Cartwright AN (2011) Wavelength-independent optical polarizer based on metallic nanowire arrays. IEEE Photon J 3:1083–1092
46. Jang K-H, Park SH, Lee K, Park G-S, Jeong YU (2012) High-order photonic bandgap reflex klystron using carbon nanotube multi-beam cathode. Electron Lett 48:707
47. Otsuji T, Boubanga Tombet SA, Satou A, Fukidome H, Suemitsu M, Sano E, Popov V, Ryzhii M, Ryzhii V (2012) Graphene-based devices in terahertz science and technology. J Phys D Appl Phys 45:303001
48. Neumann J, Gottschalk KE, Astumian RD (2012) Driving and controlling molecular surface rotors with a terahertz electric field. ACS Nano 6:5242–8
49. Jornet JM, Akyildiz IF (2011) Channel modeling and capacity analysis for electromagnetic wireless nanonetworks in the terahertz band. IEEE Trans Wirel Commun 10:3211–3221

18 Nanomaterials for Lighting and Solar Energy Conversion

Filippo Monti, Eleonora Pavoni, and Nicola Armaroli

Abstract Luminescence and photoinduced energy and electron transfer are fundamental processes that are acquiring an increasing importance for a number of applications in several fields. Our group is particularly concerned with those related to lighting and solar energy conversion, which entail the design, preparation and characterization of suitably designed molecules, metal complexes, supramolecular arrays and nanomaterials. Some of our recent results in this area will be briefly illustrated, particularly on (i) transition metal complexes, (ii) luminescent hybrid materials made of inorganic luminophores and carbon nanotubes, and (iii) photoinduced processes in multichromophoric systems containing fullerene as electron acceptor combined with suitable electron donors. In an effort to keep a tutorial approach and a broader view, we will introduce the data presented in light of their potential applications and, more generally, in the context of the energy transition to be accomplished in the present century.

18.1 Introduction

The massive exploitation of the energy of fossil fuels is the key feature that distinguishes modern society from earlier epochs. This trend started out during the nineteenth century and is still ongoing (Fig. 18.1). Currently, over 80 % of the energy used by our civilization comes from fossil fuels, not a big difference compared to one century ago, when the fossil share was around 90 % (mainly coal). The massive exploitation of fossil fuel resources has been the main factor causing the stunning

F. Monti • E. Pavoni • N. Armaroli (✉)
Istituto per la Sintesi Organica e la Fotoreattività, Via Piero Gobetti 101, 40129 Bologna, Italy
e-mail: nicola.armaroli@isof.cnr.it

B. Di Bartolo et al. (eds.), *Nano-Structures for Optics and Photonics*,
NATO Science for Peace and Security Series B: Physics and Biophysics,
DOI 10.1007/978-94-017-9133-5_18,

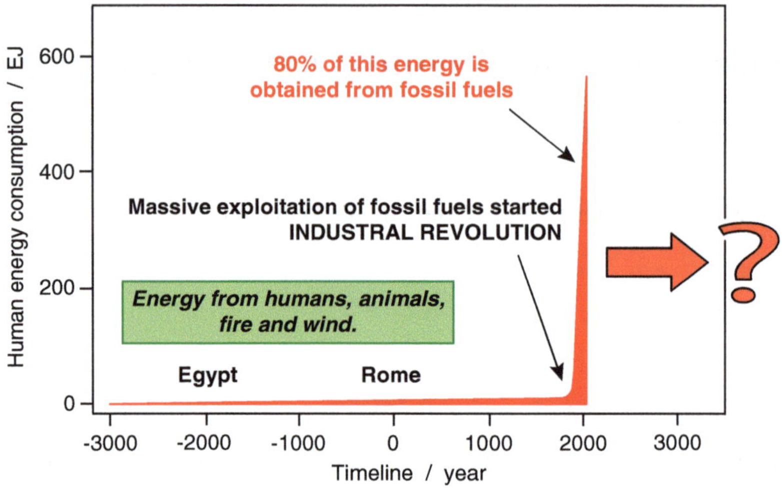

Fig. 18.1 Trend of primary energy consumption in the history of human civilization

improvement of the quality of life of millions of people on Earth during the last 100 years, a time interval that we may term as the "fossil fuel era". It must be pointed out, however, that these lucky individuals and communities still constitute a minority of the world global population [1–4].

Presently, about 10 Gt of fossil carbon (as oil, gas and coal) is taken from the lithosphere and primarily injected into the atmosphere as CO_2. This molecule is colorless, odorless and non toxic, and its accumulation in the atmosphere (only about 50 % of anthropogenic CO_2 is processed through the global carbon cycle) is perhaps the largest and most unnoticed environmental human footprint on nature [5]. CO_2 is a very stable molecule that can remain intact in the atmosphere for several decades, whilst the Earth's atmosphere is completely mixed in about 1 year. Thus CO_2 concentration is substantially identical all over the planet and recently reached the landmark value of 400 ppm, the highest level of the last 800,000 years [5, 6]. The international scientific community is now virtually unanimous in considering the increase of greenhouse gases in the atmosphere the main reason for the ongoing warming of the planet [7].

The burning of fossil fuels entails not only the release of CO_2 but also of many other waste products into the biosphere, with various effects on the environment and human health [8]. Such effects are typically not integrated into the pricing of energy and are termed external costs or *externalities*. They include for instance immediate and short term impacts related to the discovery, extraction, transportation, distribution or burning of energy resources, for example accidents in coal mines and oil spills during sea transportation or extraction. Such direct effects are complemented by others that are spread over time to future generations (e.g., the long-term storage of wastes) or over space to the entire planet, such as transboundary air pollution [9] or alteration of the carbon cycle that triggers climate change [10]. The latter consequences are the most ethically problematic, because affect also those who cannot enjoy the benefit of energy abundance, and even future generations [8].

After World War II, the most important source of fossil energy has been conventional oil. However, in the last 30 years, the gap between crude oil production and discovery has been widening. In other words, it is increasingly difficult to cover the rising demand of oil with new discoveries. Some geologists believe that the so-called *peak oil* has come [11], and this would be the very reason for the sudden increase of the oil price occurred in the early twenty-first century. In this uncertain context, the exploitation of unconventional oil and gas reserves (e.g., shale oil and gas and tar sands) is undergoing a spectacular development, particularly in the US [12]. It has to be emphasized, however, that these new approaches to fossil fuel exploitation are much more challenging from the environmental and economic point of view, when compared to conventional hydrocarbons [13]. For instance, the activities related to shale gas and oil extraction have been linked to groundwater contamination [14, 15], induced seismicity [16], and methane release in the atmosphere [17]. On the other hand, the economic viability of these resources, particularly in terms of energy return on energy investment (EROI) is still questioned [18]. Another route that is being attempted to expand hydrocarbon production is the exploration of remote, hostile and environmentally sensible areas such as the Arctic region [19].

The picture briefly outlined above shows that times are mature to progressively phase out fossil fuels for at least two reasons: (i) they are not regenerated in the timescale of human civilization and they are going to be exhausted; (ii) their use causes severe problems to human health and enormous damages to the environment and the climate stability. The remaining stock of fossil fuel energy should be progressively used for creating the conditions for a transition toward the development of new alternative energy sources and technologies, particularly those based on solar energy [2].

Solar energy occurs in several direct and indirect forms (e.g., electromagnetic radiation, air and water flows, biomass, etc.) and it is by far the most abundant, reliable, sustainable, and safe energy source that can qualitatively and quantitatively replace fossil fuels [2]. The transition to a solar-powered civilization will be a long and difficult process in which at least three key strategies must be implemented: (1) increase of efficiency in energy production and reduction of energy consumption in affluent countries; (2) development of renewable (or perennial) energy sources; (3) reduction of CO_2 dumping in the atmosphere and its utilization as a carbon source [2, 3].

In the following paragraphs we will briefly overview some results of our group in the *fundamental* study and testing of molecular and supramolecular materials that can be of interest for the above listed objectives in the field of energy materials.

18.2 Luminescent Materials for New Lighting Technologies

18.2.1 Modern Lighting

Electric lighting is one of the most important technological and social achievements of the twentieth century. However, still today, the rhythm of everyday life is beaten by the daily light-dark cycle in many poor regions of the world, where at sunset most human activities are stopped. Artificial lighting was generated by open fire for millennia; eventually, the fire was tamed through the use of lamps, as testified by archaeological findings all around the world. Until the mid nineteenth century, the majority of dwellings utilized fuels of vegetal or animal origin such as beeswax or whale oil. In 1879, Thomas Edison patented the incandescent carbon filament lamp, the milestone of modern lighting; its diffusion benefited from the almost concomitant spreading of electricity, particularly in urban areas. Edison's lamp converted only 0.2 % of electricity into light, nonetheless it was 20 times more efficient than a candle in converting chemical energy into photons [20].

Throughout the twentieth century, several electric lighting devices were introduced such as the tungsten lamp (1906), that dominated residential lighting for 100 years, the sodium vapor lamp (1930s), now utilized for street illumination in its high-pressure variant, the fluorescent tubes (1940s), widely used in large internal settings such as offices and factories [21]. The evolutions of these fundamental designs were introduced more recently: the halogen lamp, an advanced filament system, entered the market in the 1960s and the compact fluorescence lamp, a hybrid between bulbs and fluorescent tubes, appeared in the 1980s. In Fig. 18.2 is depicted the trend of the efficiency of some lighting devices in converting electricity power

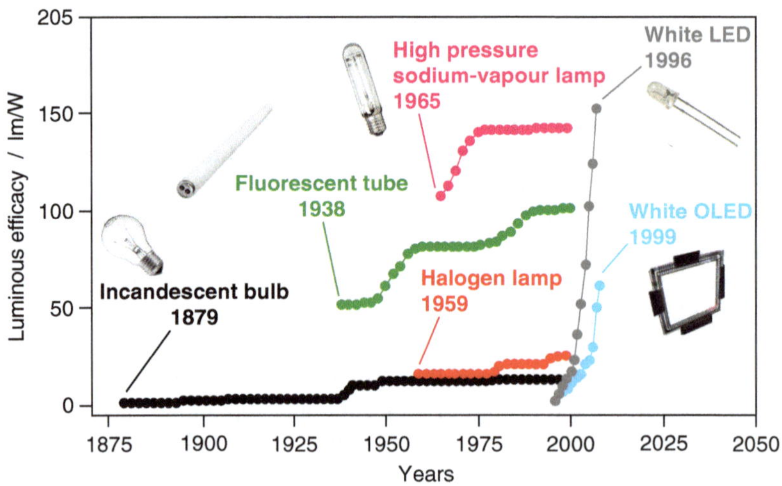

Fig. 18.2 Historical trend of luminous efficacy for some light sources

into visible (Vis) electromagnetic radiation that can be sensed by the human eye. This parameter is termed luminous efficacy and is measured in lm/W (lumen/watt).

Currently, the world lighting market is over 70 billion euros per year; the related industry provides over 150,000 jobs in Europe only. However, lighting still brings about a heavy environmental burden, which must be drastically reduced to afford a truly sustainable service that may enhance the quality of life of citizens worldwide. In particular, it has to be emphasized that (i) the overall efficiency of conversion of primary energy entering power plants into a lighting service for end users is still below 10 %, an unacceptably low level; (ii) the electricity consumption for lighting is about 20 % of the world total, with an estimated production of over 2 Gt/year of CO_2 at power plants; (iii) 1.5 billion people around the world still lack access to electricity and get illumination by burning fuels in poor dwellings—including one million barrels of oil a day—with unacceptable safety and health threats [22,23].

The picture illustrated above shows that there is still a long road towards the achievement of environmentally sustainable lighting for a nine billion people planet by 2050.

All of the technologies that have dominated the lighting market so far are incandescent- or discharge-based, where light generation is an *indirect effect* that occurs upon heating or discharging. Incandescent sources emit up to 95 % of the energy as heat (i.e., infrared photons) hence their efficiency is intrinsically low. On the other hand, discharge lamps generate ultraviolet (UV) radiation, which is then converted into visible light by solid or gaseous compounds; they can be substantially more efficient than incandescent lamps but their color quality tends to be poorer. There is a variety of both type of lamps but these two established lighting approaches have been exploited near to their limit. Therefore, radically new lighting concepts have emerged in the last 20 years, which are based on solid luminescent materials that generate electronic excited states in an electric field (electroluminescence) and *directly* deactivate by emitting photons in the Vis spectral region. This approach is termed solid-state lighting (SSL), and requires suitably engineered devices where the transport of charge occurs in one specific direction (diodes). At present, there are two main families of SSL devices, namely light-emitting diodes (LEDs, point sources) [24] and organic light-emitting diodes (OLEDs, flat and diffuse sources [25,26]. LEDs are based on inorganic phosphors and are now taking the market lead for general illumination. OLEDs are based on organometallic or organic compounds and are used only for display applications, but not yet for lighting. The main bottlenecks on the way to OLED lighting are the high economic and energetic costs of deposition technologies and the lack of efficient, stable and cheap emitters over the whole colour gamut.

18.2.2 OLEDs and LECs

OLEDs are flat electroluminescent devices where the photon output is generated within a multilayer stack a few hundred nanometers thick (Fig. 18.3a) [25–27].

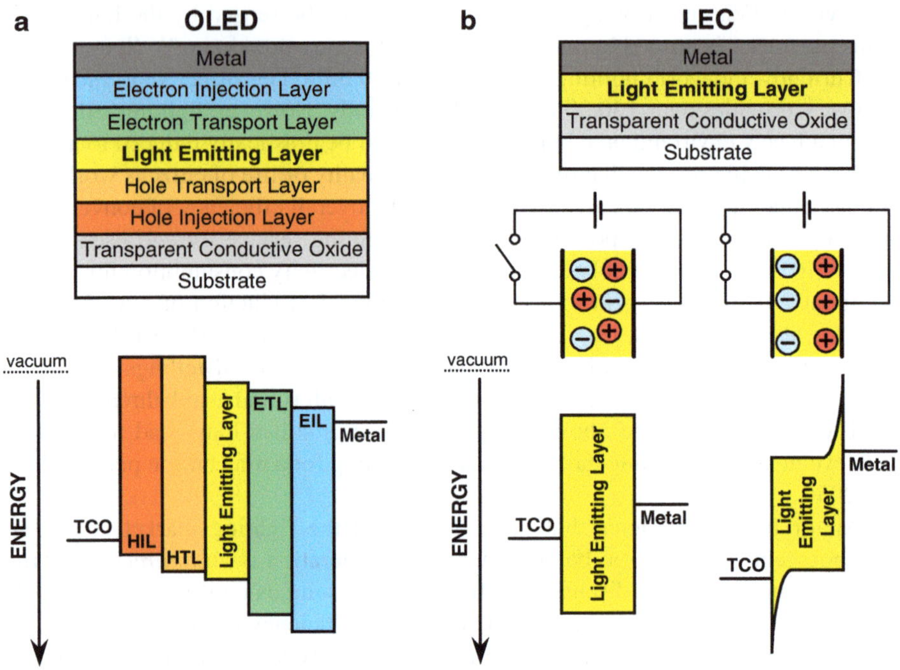

Fig. 18.3 Schematic representations of standard OLED (**a**) and LEC (**b**) device architectures. An OLED is made of multiple layers that are in most cases deposited stepwise by thermal vacuum evaporation. The injection of electrons is achieved by the use of (i) a low work function metal or (ii) a chemically n-doped electron injection layer, both of which are unstable in air and require demanding encapsulation. By contrast, a LEC consists of only one opto-electronically active layer in which ions are displaced when external bias is applied. This simple architecture enables an efficient hole and electron injection from air-stable metals

They are now extensively used as displays in portable applications and start to be available also in prototype TVs. The key layer of the device is that containing the luminescent material, which can be polymer, a fluorescent molecule or a phosphorescent transition-metal complex within a charge transporting matrix [25–27]. White OLEDs are based on multilayer stacks sometimes made of over 15 individual layers [27, 28] that are prepared by vacuum sublimation, which implies the use of thermally stable non-ionic materials. However, since luminescent transition-metal complexes are typically ionic species, the choice of suitable compounds is dramatically restricted [29, 30]. Moreover, the high economic and energetic cost of the multi-layer evaporation process, along with the need to ensure a sealed encapsulation of the devices have so far prevented a significant penetration of this technology in the lighting market, due to high price.

The above limitations of OLED technology have prompted the exploration of new concepts for flat electroluminescent devices [31, 32], among which LECs (light-emitting electrochemical cells) are emerging as the most popular [31, 33].

LECs have a much simpler architecture compared to OLEDs (Fig. 18.3b). They are processed from solution, do not need air-sensitive charge-injection layers or metals for electron injection and consequently require less severe packaging procedures. LECs are essentially made of an ionic luminescent material in an ionic environment, sandwiched between two electrodes. The light emitting material can be a conjugated light-emitting polymer or an ionic transition-metal complex (iTMC). The first type is termed polymer-LECs (PLECs) [34]. The second type is even simpler as it entails an ionic emitter that enables one-component devices; these LECs are referred to as iTMC-LECs [31, 33, 35]. Very recently, LECs based on organic ionic luminescent molecules have also been proposed [36].

In this paper we focus on recent advances in luminescent materials that can be utilized in iTMC-LECs, particularly Ir(III) and Cu(I) complexes, and also present new hybrid materials based on carbon nanotubes that may be suitable for future exploitation. We are not going to present details on the mechanism of LEC functioning, for which we suggest the reading of specialized recent articles [31–33].

18.2.3 Ionic Luminescent Ir(III) Complexes

The first metal complex tested in a device with LEC configuration was $[Ru(bpy)_3]^{2+}$ (bpy = 2,2'-bipyridine) [37]. However, LECs made of ruthenium-based complexes exhibit limited color tuning because the emission band of these compounds is confined in the orange-red part of the visible spectrum. This constraint strongly limits the applicative potential for lighting and display technologies, which requires wider color tunability in order to achieve white emission based on red, green and blue (RGB) luminescence.

The by far most versatile family of iTMCs is that of Ir(III) (Ir-iTMCs) [38]. These compounds exhibit a unique combination of physical and chemical properties and offer a vast variety of stable complexes that cover the entire visible spectrum, all the way from blue to red [39]. In particular, by passing from a second-row (e.g., Ru) to a third-row (i.e., Ir) transition element of the periodic table, improved stability of the complexes is achieved thanks to (i) increased metal–ligand bond strength and (ii) higher ligand-field splitting energy (LFSE) that makes dissociative metal-centered (MC) excited states less thermally accessible compared to Ru(II) analogues [38]. LFSE is further enhanced, compared to Ru(II) complexes, thanks to the higher electric charge of the iridium ion and by the presence of anionic cyclometalating ligands (C-N), typically used with Ir(III) centers. The combination of these features in Ir-iTMCs can lead to very high luminescence quantum yields of virtually any color, while providing at the same time good photochemical stability [38–41].

The typical structure of an iridium complex utilized for LEC devices is depicted in Fig. 18.4. It consists of two negatively charged cyclometalating ligands (C-N) and one neutral ligand (usually N-N, but not only, *vide infra*). This configuration, besides providing the electronic properties described above, grants the monocationic nature of the complex that is a necessary and sufficient condition for LEC functioning.

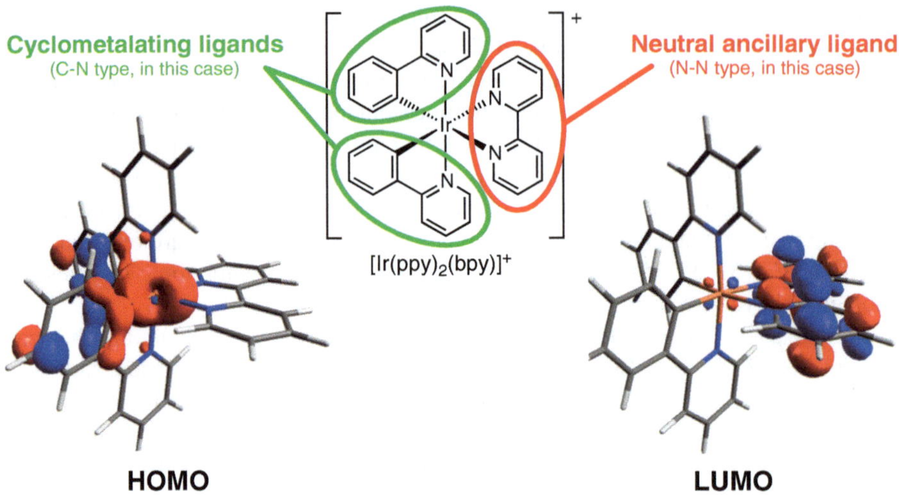

Fig. 18.4 Chemical structure of the archetypal complex $[Ir(ppy)_2(bpy)]^+$, together with its frontier molecular orbitals calculated at the B3LYP/6-31G(d,p)&LanL2DZ level of theory

The rationalization of the photophysical properties of Ir-iTMCs (and particularly the emission features) requires the determination of the energy and the nature of the highest-occupied and the lowest-unoccupied molecular orbitals (HOMO and LUMO, respectively) [31]. For typical iridium(III) complexes with (C-N) and (N-N) ligands, the HOMO is an admixture of Ir d_π orbitals (t_{2g}) and phenyl π orbitals of the cyclometalated (C-N) ligands, whereas the LUMO is usually located on the neutral (N-N) ancillary ligand. The HOMO-LUMO distribution of the archetypal complex $[Ir(ppy)_2(bpy)]^+$ (ppy = 2-phenylpyridine), as determined by DFT calculations at the B3LYP level of theory, is depicted in Fig. 18.4. From this representation, it is clear that the emitting triplet state of Ir-iTMCs exhibits a mixed ^{3}MLCT/^{3}LLCT character (MLCT = metal-to-ligand-charge-transfer; LLCT ligand-to-ligand-charge-transfer). Luminescence color tuning is obtained through modification of the HOMO-LUMO gap, therefore emission blue shift is obtained by stabilization of the HOMO and destabilization of the LUMO, while red shifting is obtained with the opposite approach [31].

In Fig. 18.5 are shown two examples of complexes exhibiting blue-shifted and red-shifted emission, compared to the archetypal compound $[Ir(ppy)_2(bpy)]^+$. Blue shifting is obtained by stabilizing the HOMO with electron withdrawing fluorine residues or by replacing the pyridine with pyrazole units; LUMO destabilization is accomplished by placing electron-donating *t*-butyl groups on the ancillary ligand. Red shifting is obtained by stabilizing the LUMO with electron withdrawing substituents, or through replacement of bpy with more extended conjugated aromatic ligands.

The choice of luminescent Ir-iTMCs in the green and even in the red is ample, whereas stable blue emitting complexes, particularly in solid state and under

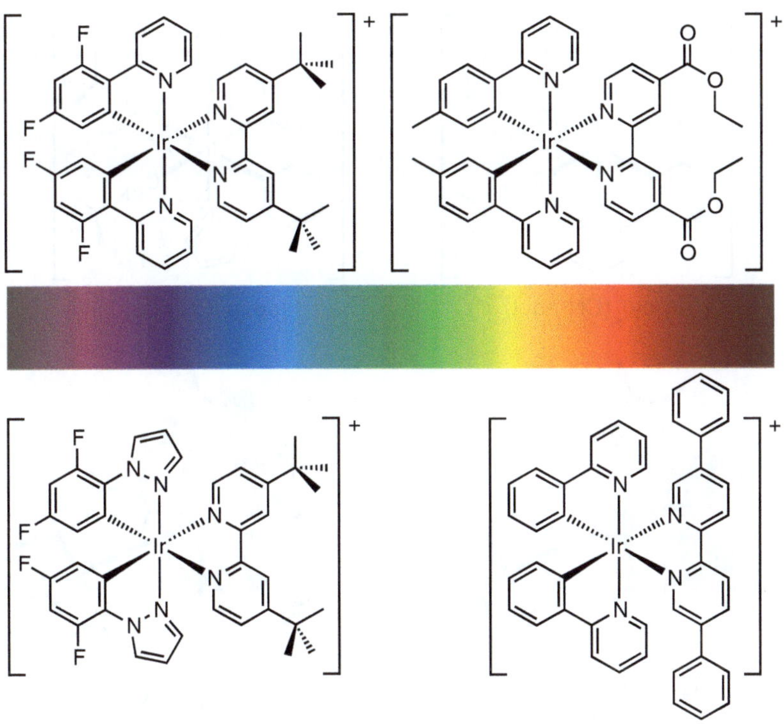

Fig. 18.5 Chemical structures of different cationic Ir(III) complexes emitting in different regions of the visible spectrum. The emission energy can be fine-tuned by playing (almost) independently on the cyclometalated (HOMO) or on the ancillary ligands (LUMO), or even on both of them

the electric field, is limited [31]. Blue-phosphorescent bis-cyclometalated cationic Ir(III) complexes with several neutral ligands such as diimines [42, 43], carbenes [44, 45] and carbon monoxide [46] have been prepared. They typically exhibit problems such as (i) low phosphorescence quantum yields that result from quenching of high-energy emissive excited states by metal-centered states of similar energy and (b) red-shift and quenching of phosphorescence in going from solutions to neat solids owing to aggregation and triplet triplet interaction. Recently alkyl isocyanides complexes have been proposed as a viable route to strongly luminescent blue Ir-iTMCs [47–49]. These ligands provide a strong field around the coordination center, enhance the energy of charge transfer electronic transition involving the metal center (i.e., MLCT) and make ligand centered levels accessible. This concept can be viewed is orbital terms on Fig. 18.6 where the HOMO and LUMO of $[Ir(ppz)_2(CN\text{-}t\text{-}but)_2]^+$ (CN-*t*-but = *tert*-butyl isocyanide) are depicted. Both orbitals are centered on the isocyanide ligand, therefore the electronic transition related to the lowest electronic excited state has a ligand centered (LC) character. This is evidenced in the spectral shape, which exhibits the typically vibronically structured profile of aromatic ligands.

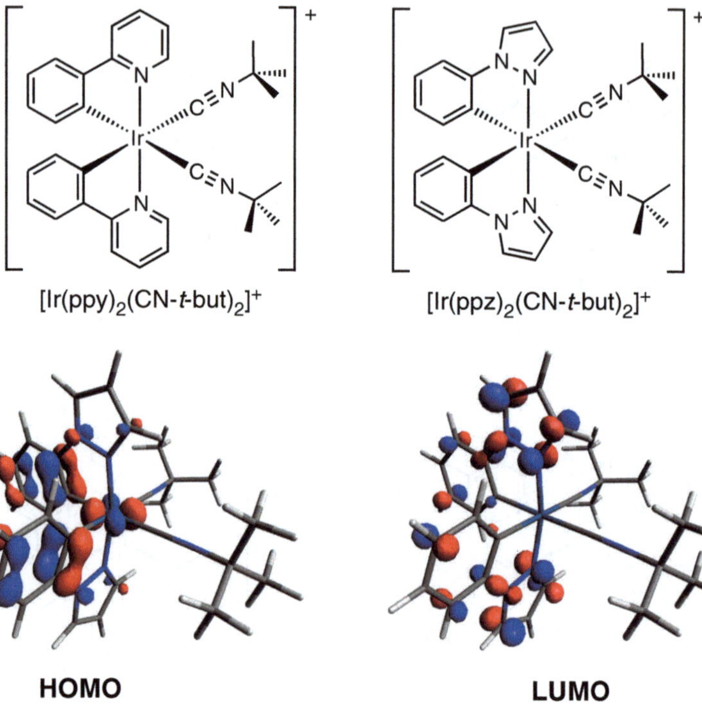

Fig. 18.6 Two prototypical examples of cationic Ir(III) complexes having neutral isocyanides as ancillary ligands: $[Ir(ppy)_2(CN\text{-}t\text{-}but)_2]^+$ (*top left*) and $[Ir(ppz)_2(CN\text{-}t\text{-}but)_2]^+$ (*top right*). The frontier molecular orbitals are reported for the latter, calculated at the B3LYP/6-31G(d,p)&LanL2DZ level of theory

Recently, we have investigated several cationic Ir-iTMCs complexes with neutral isocyanide ligands, opening an interesting route towards stable and brightly emitting blue complexes [47–49]. A selection of such compounds is reported in Fig. 18.7 along with emission spectra in solution and in the solid state. The isocyanide complexes having phenyl-pyridine cyclometalating ligands display high quantum yields (PLQY $\approx$ 50–80 %) and long emission lifetimes ($\tau \approx$ 20–60 µs) [47,48]. On the other hand, if cyclometalating phenyl-pyrazoles are present, the complexes are poorly emissive (PLQY < 5 %) due to the presence of lower-lying ^{3}MC states [49].

In an attempt to shorten excited-state lifetimes and reduce saturation problems in devices, cationic Ir-iTMC complexes with strong field carbene-based ancillary ligands (:C-N) were designed and investigated, Fig. 18.8 [50]. The idea proved correct because the emission spectral onset of these compounds is still in the blue region of the visible spectrum ($\lambda \approx$ 450 nm), but lifetimes are much shorter if compared to isocyanide analogues (4–5 µs vs. 20–60 µs). However, they turned out to be very poor emitters in solution at room temperature (PLQY < 1 %) due to deactivation to lower lying metal-centered levels, as determined by temperature dependent photophysical studies and DFT calculations [50]. This drawback is

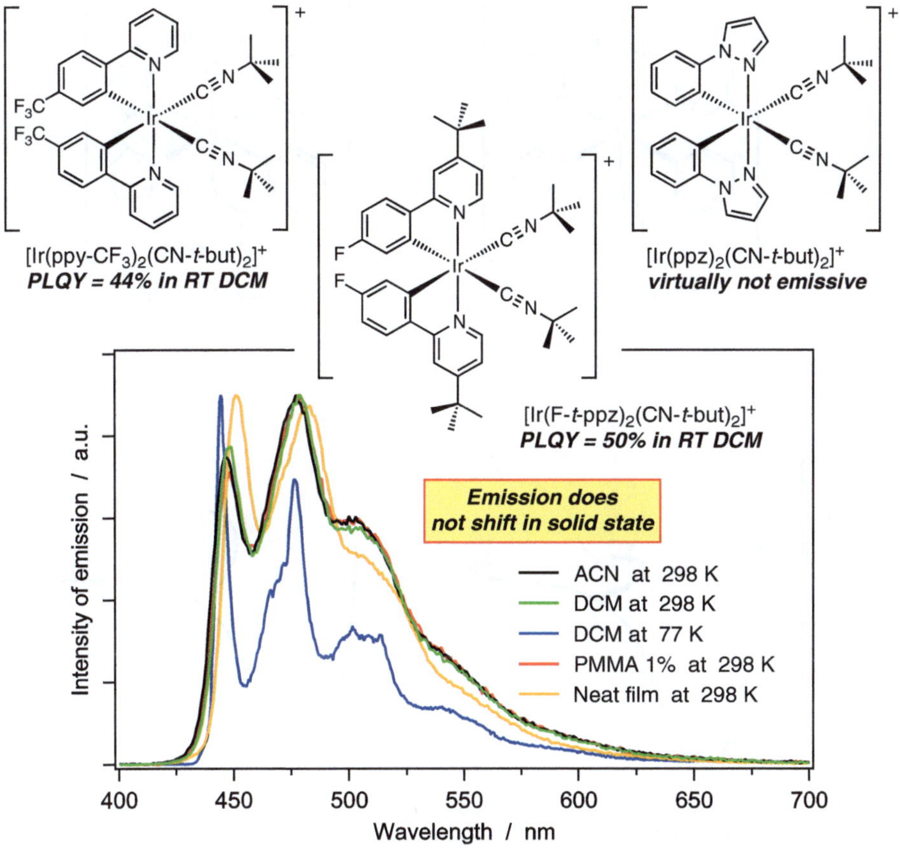

Fig. 18.7 Some selected examples of cationic Ir(III) complexes having isocyanides as ancillary ligands [47–49]. While complexes having phenyl-pyrazoles as cyclometalating ligands are virtually not emissive, other complexes based on different cyclometalating moieties lead to bright luminescence

prevented in carbene compounds of (:C-C:) type (Fig. 18.8), which are instead good emitters also in solution at 298 K (PLQY $\approx$ 30 %). On the other hand, both (:C-N) and (:C-C:) type complexes are bright emitters in the solid state (1 % PMMA matrix) and at 77 K, where the population of the ^{3}MC level is prevented [50].

The best luminophores among the above-described blue emitting Ir-iTMC complexes are now under testing in LEC to verify their viability as blue emitters in electroluminescent devices, an important target for LEC technology.

18.2.4 Ionic Luminescent Cu(I) Complexes

The very high chemical stability accompanied by the possibility to tune the emission color across the entire Vis spectral region make Ir(III) complexes a formidable

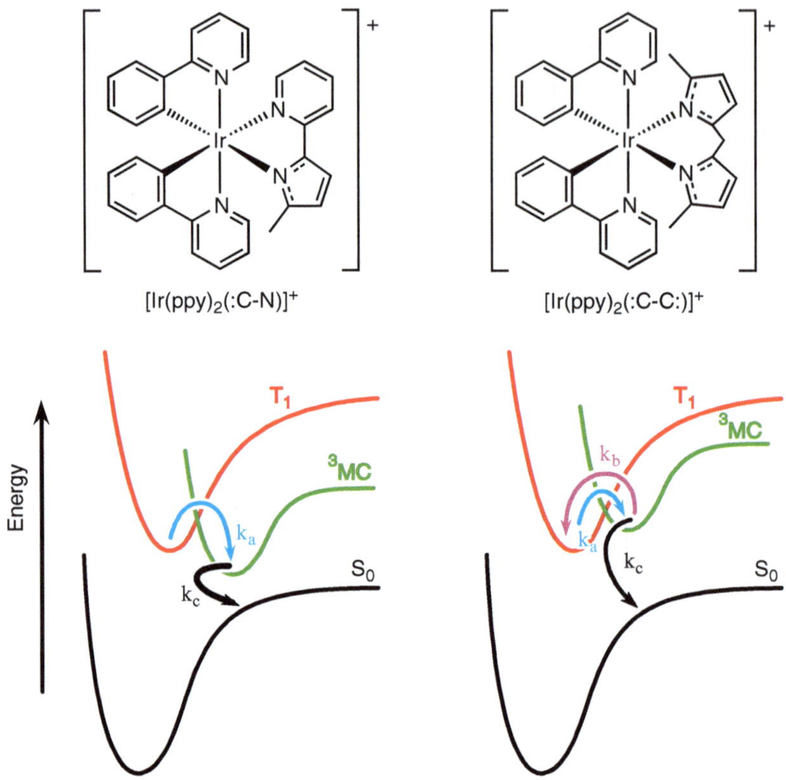

Fig. 18.8 Two representative examples of cationic Ir(III) complexes having chelating carbenes as ancillary ligands: the complex on the *left* is showing a :C-N type ligand, the one on the *right* a :C-C: bis-carbene. In the *bottom* part of the figure, the relative energy position of the excited states of these complexes is reported (scaled according to the experimental and the DFT data) in order to elucidate their photophysical behavior

benchmark for potential competitors among luminescent coordination compounds. Also iridium, however, has its own drawback: it is the rarest element on the Earth's crust with a yearly world production of about 4 t and a constantly high, yet strongly fluctuating price (Jan 2004–Jan 2014 average, 558 $/oz) [51]. Although the amount of iridium to be used in a single flat electroluminescent device with an active layer of a few tens of nm is tiny, illuminating the world with Ir(III) materials might prove unrealistic. For this reason, alternative luminophores based on more abundant and cheaper elements are intensively pursued all around the world.

Cu(I) complexes and clusters are at present the largest class of luminescent metal compounds based on a relatively abundant element [29]. Copper may exhibit two oxidation states: +1 and +2. Cu(II) complexes are characterized by a d^9 electronic configuration and show relatively intense metal-centered (MC) absorption bands in the Vis and near infrared (NIR) spectral region, which deactivate via ultrafast non-radiative pathways. Accordingly, Cu(II) complexes do not have any interest for luminescence-related applications and all the investigations in the area of copper luminescent materials are concentrated on Cu(I) [52].

18.2.4.1 $[Cu(N\text{-}N)_2]^+$ Complexes

Historically, the most extensively investigated family of Cu(I) complexes had the general formula $[Cu(N\text{-}N)_2]^+$, where N-N indicates a chelating bisimine ligand, typically a substituted 1,10-phenanthroline (phen) [52–56]. Cu(I)-bisphenanthroline complexes absorb in the UV region due to the $\pi - \pi^*$ transitions of the ligands. Most importantly, they absorb throughout the whole visible spectral region due to an envelope of metal-to-ligand-charge-transfer (MLCT) bands. The particular spectral shape of a given complex in the visible window is dictated by the substitution pattern of the ligands that affect the dihedral angle between the two phenanthroline planes and the extent of π-delocalization of the accepting orbital [52]. These peculiar effects dramatically influence the probabilities of MLCT electronic transitions, yielding complexes exhibiting a variety of colors (e.g., orange, deep red or magenta) [57]. In Fig. 18.9 are reported the absorption and emission spectra of three selected $[Cu(N\text{-}N)_2]^+$ complexes as exemplary cases.

Upon photoexcitation, tetrahedral Cu(I)-bisphenanthroline complexes undergo molecular rearrangements to attain a square planar geometry, eventually followed by the formation of a five-coordinated exciplex with the pick-up of external nucleophiles (i.e., solvent molecules or counteranions) [52, 53]. This process is related to the generation of relatively long-lived (ns–μs timescale) thermally equilibrated MLCT excited states in which the metal center is formally Cu(II) and thus prefers a flattened square planar geometry. Such mechanism was initially postulated by Mc Millin by means of classical experiments of photochemical quenching [58] and was confirmed several years later through the concerted effort of theoretical calculations (DFT) [59–62] and experimental techniques such as ultrafast transient absorption [59,63,64], time-correlated single photon counting [59,61], fluorescence upconversion [64, 65], light-initiated time-resolved X-ray absorption spectroscopy (LITR-XAS) in solution [59,63,66,67] and in the solid state (photocrystallography) [68,69].

The MLCT emission band of $[Cu(N\text{-}N)_2]^+$ complexes is very weak for the vast majority of compounds reported so far and stems from thermally equilibrated ^{1}MLCT and ^{3}MLCT levels [53]. In nucleophilic solvents, luminescence is barely detectable and measurements are almost exclusively made in CH_2Cl_2, which allows prompt solubilization without providing nucleophilic interactions. Luminescence bands of $[Cu(N\text{-}N)_2]^+$ complexes in CH_2Cl_2 at ambient temperature are broad and centered between 680 and 740 nm; related PLQY are typically below 0.1 % and lifetimes between 50 ns and a few hundred ns [53]. No benefit is obtained at 77 K, where excited state thermal equilibration is prevented and excitation is trapped on the lowest ^{3}MLCT level that is characterized by a very low radiative constant [52]. A couple of notable exceptions to this trend has been found with

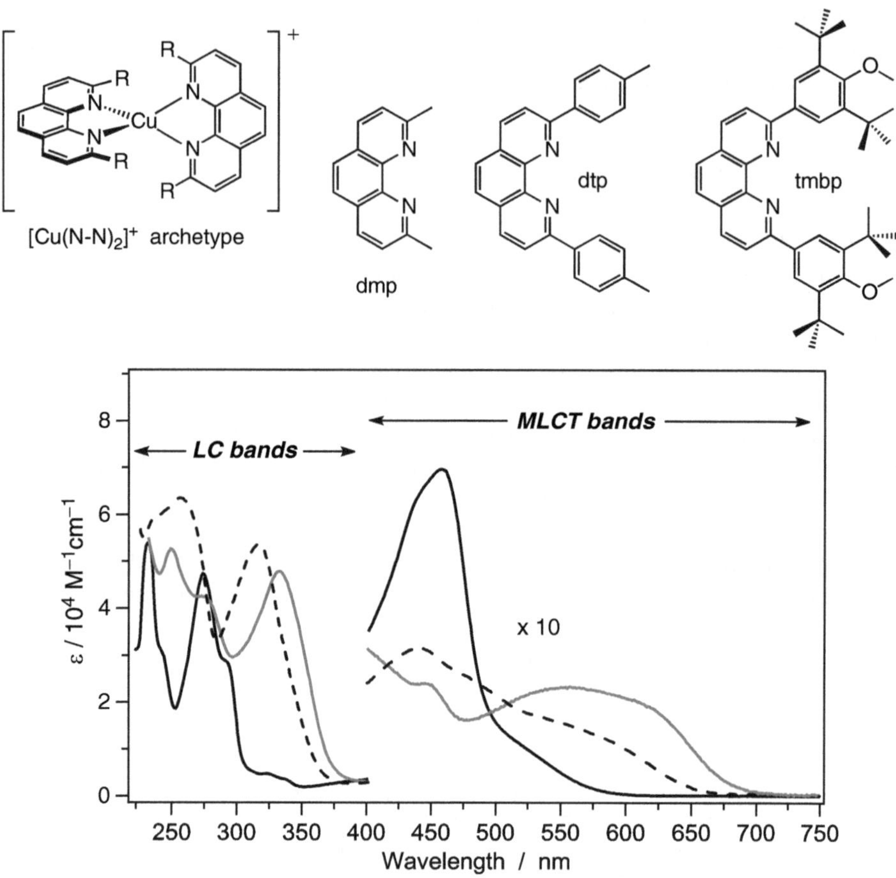

Fig. 18.9 Absorption spectra of $[Cu(dmp)_2]^+$ (*full black line*), $[Cu(dtp)_2]^+$ (*dashed line*) and $[Cu(tbmp)_2]^+$(*grey line*) in CH_2Cl_2 solution at room temperature. dmp = 2,9-dimethyl-1,10-phenanthroline; dtp = 2,9- 2,9-di-p-tolyl-1,10-phenanthroline; tmbp = 2,9-bis(3,5-di- *tert*-butyl-4-methoxyphenyl)-1,10-phenanthroline. The three complexes are orange, deep red or magenta

complexes having 2,9-substituted phenanthrolines with long alkyl chains or very bulky aromatic ligands, which exhibit a bright orange intense luminescence at 77 K, due to specific structural constraints in the low temperature matrix [57, 70].

In more recent times, the utilization of phenanthroline ligands with bulky *tert*-butyl (*t*-butyl) substituents has afforded remarkable PLQY in solution at room temperature also for $[Cu(N\text{-}N)_2]^+$ complexes. The three strongest emitters reported to date are depicted in Fig. 18.10 with related photophysical parameters: $[Cu(dmp)(dtbp)]^+$ [71], $[Cu(dtbp)_2]^+$ [72, 73], and $[Cu(dsbtmp)_2]^+$ [74] (where dtbp = 2,9-di-*tert*-butyl-1,10- phenanthroline; dsbtmp = 2,9-di(*sec*-butyl)-3,4,7,8-tetra-methyl-1,10-phenanthroline).

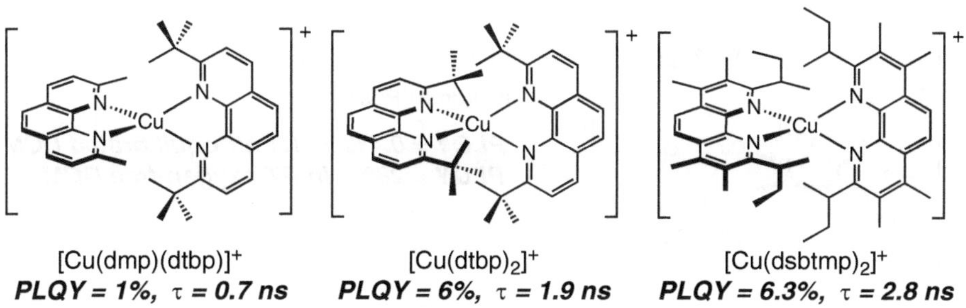

Fig. 18.10 Chemical structures of the three strongest emitters belonging to the class of $[Cu(N\text{-}N)_2]^+$ complexes, together with their PLQY and related lifetimes in aerated CH_2Cl_2

$[Cu(dsbtmp)_2]^+$ emits also in strongly coordinating solvents and exhibits excellent thermodynamic and photochemical stability that makes it an outstanding photosensitizing coordination compound made of an earth-abundant element [74]. At present, it constitutes the most performing $[Cu(N\text{-}N)_2]^+$ available.

18.2.4.2 $[Cu(N\text{-}N)(P\text{-}P)]^+$ Complexes

Heteroleptic $[Cu(N\text{-}N)(P\text{-}P)]^+$ complexes containing both N- and P-coordinating ligands, have been pursued starting from the late 1970s [75]. Since the replacement of one N-N ligand with a P-P unit may bring about an improvement of the emission properties, the quest for highly luminescent metal complexes that has grown in recent years sparked revived interest towards this class of compounds [76–78].

The absorption and luminescence spectra of $[Cu(dmpp)(POP)]^+$ (where dmpp = 2,9-dimethyl-4,7-diphenyl-1,10-phenanthroline and POP = bis[2-(diphenyl phosphino)phenyl]ether) is reported in Fig. 18.11, as a representative example [79]. Substantial blue-shifts of the lower-energy bands are observed compared to typical spectra of $[Cu(N\text{-}N)_2]^+$ compounds.

Spectral features above 350 nm and in the 350–450 nm window are due to LC and MLCT transitions. $[Cu(N\text{-}N)(P\text{-}P)]^+$ complexes undergo dramatic oxygen quenching, as deduced from the strong difference in excited state lifetimes passing from air-equilibrated to oxygen-free CH_2Cl_2 solution, 0.4 and 17.3 μs in the case of $[Cu(dmp)(POP)]^+$ [79]. The character of the emitting state in $[Cu(N\text{-}N)(P\text{-}P)]^+$ complexes is MLCT, as established experimentally and theoretically [76, 80, 81]. The electron-withdrawing effect of the P-P unit on the metal center tends to disfavor the Cu(I) → N-N electron donation, as also reflected by the higher oxidation potential of the Cu(I) center compared to $[Cu(N\text{-}N)_2]^+$ compounds [82], leading to a blue shift of MLCT transitions. Accordingly, primarily due to the energy gap law [83], PLQY enhancement occurs, with emission bands typically observed in the green spectral window, unlike $[Cu(N\text{-}N)_2]^+$ complexes which are typically red and near infrared emitters [79].

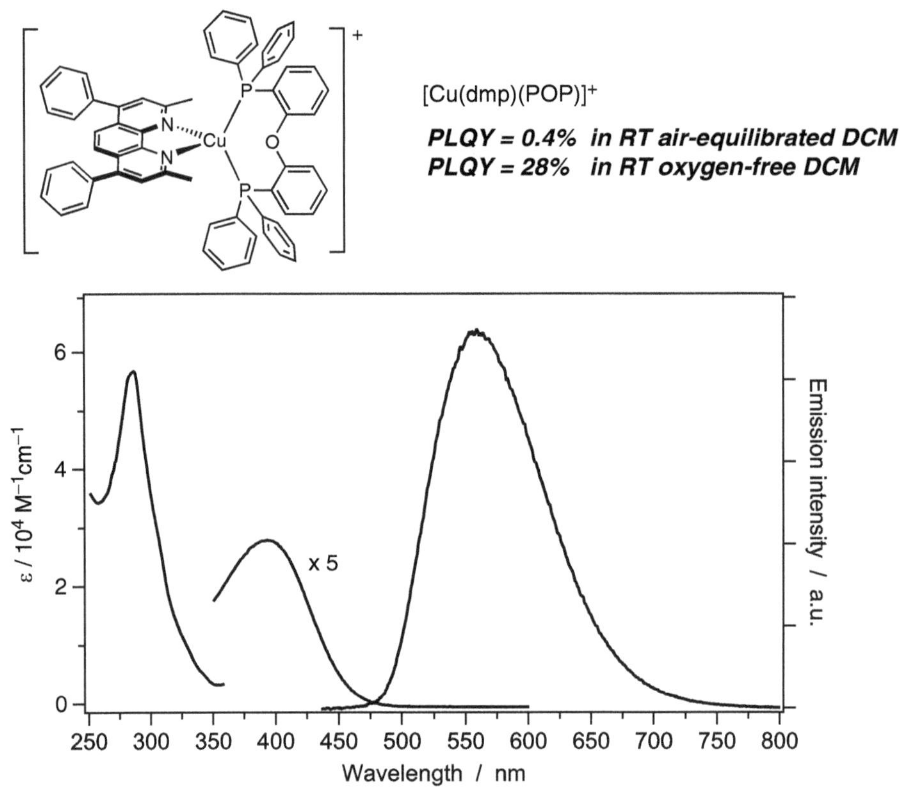

Fig. 18.11 Chemical structure of $[Cu(dmpp)(POP)]^+$ and the associated room-temperature absorption end emission spectra (in CH_2Cl_2)

The strong solvent and oxygen dependence of the PLQY of MLCT excited states of $[Cu(N\text{-}N)(P\text{-}P)]^+$ compounds is related to exciplex quenching [84, 85], likewise $[Cu(N\text{-}N)_2]^+$ analogues. Therefore, the geometry of $[Cu(N\text{-}N)(P\text{-}P)]^+$ complexes plays a central role in addressing the extent of luminescence output.

A variety of mono- and bidentate phosphine ligands has been prepared to coordinate Cu(I) in combination with phenanthroline-type units: triphenylphosphine (PPh_3), bis(diphenyl-phosphino)methane (dppm), bis(diphenylphosphino) ethane (dppe) and bis[2-(diphenylphosphino)phenyl]ether (POP), represent some recent examples [76, 78, 79, 86, 87]; their chemical structures are reported in Fig. 18.12 together with other representative bis-phosphines.

$[Cu(phen)(POP)]^+$ complexes proposed by Mc Millin, where phen denotes a variably substituted 1,10 phenanthroline, may show remarkable PLQY compared to $[Cu(N\text{-}N)_2]^+$ compounds [76]. For instance, on passing from pristine phenanthroline to dimethyl- or diphenyl-substituted analogues, and also thanks to the efficient steric and electron-withdrawing effects of the POP ligand, remarkable emission

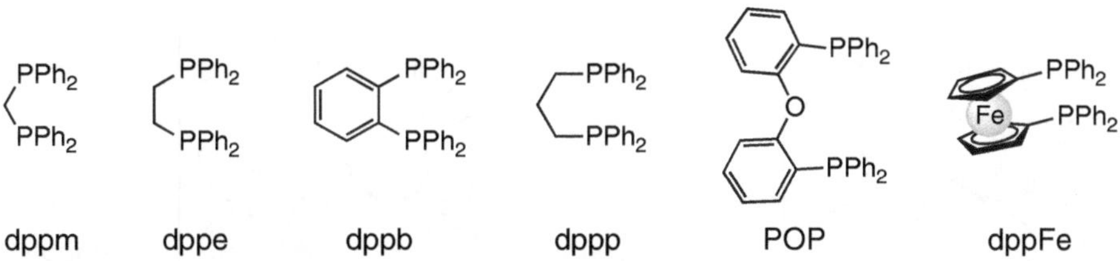

Fig. 18.12 Some of the most typical bidentate phosphine ligands (P-P) used for the synthesis of heteroleptic $[Cu(N\text{-}N)(P\text{-}P)]^+$ complexes

quantum yields (15 % in CH_2Cl_2 oxygen-free solution) and long lifetimes (≈15 μs) have been measured [76]. By contrast, the replacement of the POP ligand with two PPh_3 units, affords weaker emission due to the lower geometric rigidity that leads to weak and red-shifted luminescence comparable to those of bis-phenanthroline-type complexes.

One of the major issues in the implementation of heteroleptic complexes with both phenanthroline and phosphine ligands has been their unpredictable stability in solution, which may also prevent easy solution processing for LEC device fabrication. A very recent paper, however, has provided the definitive rationalization of the stability of $[Cu(N\text{-}N)(P\text{-}P)]^+$ complexes, showing the key role played by steric factor related to substituents on the phenanthroline ligand, particularly in the 2,9 positions [88]. In fact, $[Cu(N\text{-}N)(P\text{-}P)]^+$ complexes undergo a dynamic equilibrium in solution with ligand exchange reactions. Accordingly, the formation of the different possible complexes ($[Cu(N\text{-}N)_2]^+$, $[Cu(P\text{-}P)_2]^+$, $[Cu(N\text{-}N)(P\text{-}P)]^+$) is mainly influenced by the relative thermodynamic stability of each of them, in turn crucially dependent on the steric constraints of the various combinations, as well as the chelating bite angles of the phosphine ligand [88]. This remarkable finding is opening rational routes for the design of highly luminescent Cu(I) complexes that we anticipate will be reported soon.

18.2.4.3 $[Cu(P\text{-}P)_2]^+$ Complexes

Some combinations of phosphine ligands may afford relatively stable Cu(I) complexes with the coordination sphere made only of phosphorus atoms. The first example was reported some years ago with the heteroleptic complex $[Cu(POP)(dppb)]^+$ (dppb = 1,2-bis(diphenylphosphino)-benzene, Fig. 18.13) that exhibit a relatively intense emission band in solution (PLQY = 2 % in CH_2Cl_2, $\tau = 2.4$ μs), which was assigned to a MLCT transition. This compound, in combination with polyvinyl carbazole as matrix, has been employed to fabricate a white light electroluminescent device [89].

It is interesting to compare the absorption and luminescence spectra (Fig. 18.14) in CH_2Cl_2 solution of the three complexes ($[Cu(N\text{-}N)_2]^+$, $[Cu(N\text{-}N)(P\text{-}P)]^+$ and

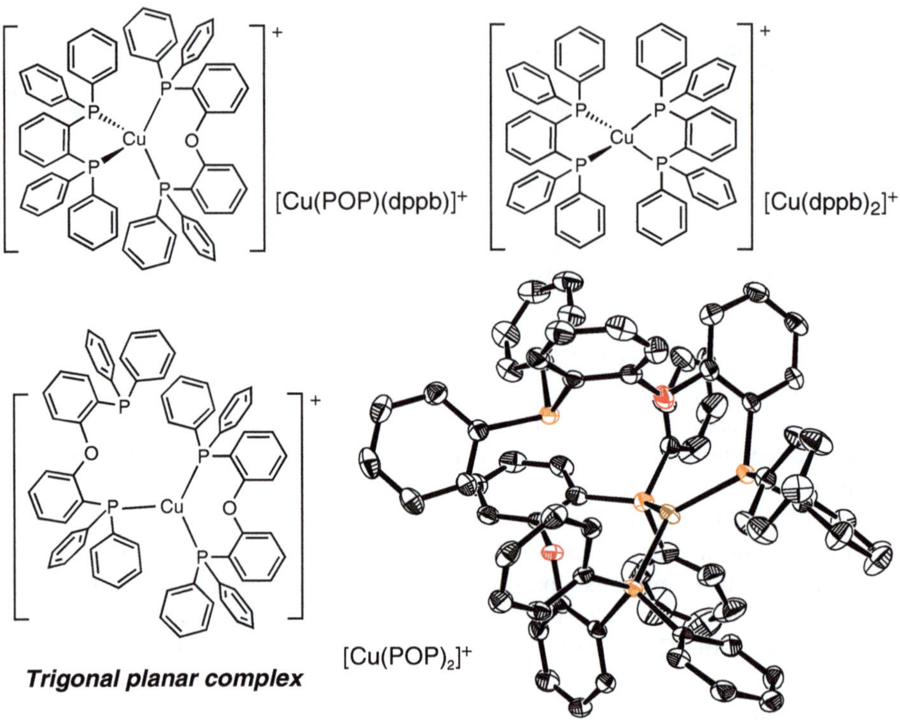

Fig. 18.13 The heteroleptic complex $[Cu(POP)(dppb)]^+$, together with the homoleptic $[Cu(dppb)_2]^+$ and $[Cu(POP)_2]^+$ analogues [90, 91]. The X-ray crystal structure of the trigonal $[Cu(POP)_2]^+$ complex is also reported for clarity

$[Cu(P\text{-}P)_2]^+$). A progressive shift to the high energy side is observed along the series, which is reflected in emission color tuning from red to green to blue-greenish. This trend shows that, also for Cu(I) complexes, it is possible to tune the emission color across the visible spectral region, although much further progress is needed to reach the stability and tunability of Ir(III) complexes.

More recently, also some homoleptic $[Cu(P\text{-}P)_2]^+$ complexes were proposed such as $[Cu(POP)_2]^+$ [90,91] and $[Cu(dppb)_2]^+$ [90], Fig. 18.13. The former, due to steric constraints, exhibits trigonal geometry with an uncoordinated phosphorus atom; in the solid state it exhibits a PLQY of 2 %. On the other hand, $[Cu(dppb)_2]^+$ turns out to be one of the strongest emitting Cu(I) complexes with a PLQY of 56 % in KBr matrix [90].

The design and implementation of stable $[Cu(P\text{-}P)_2]^+$ complexes is only at the beginning and it is conceivable that this field of investigation will considerably expand in the years to come, with interesting perspectives in the area of electroluminescent devices.

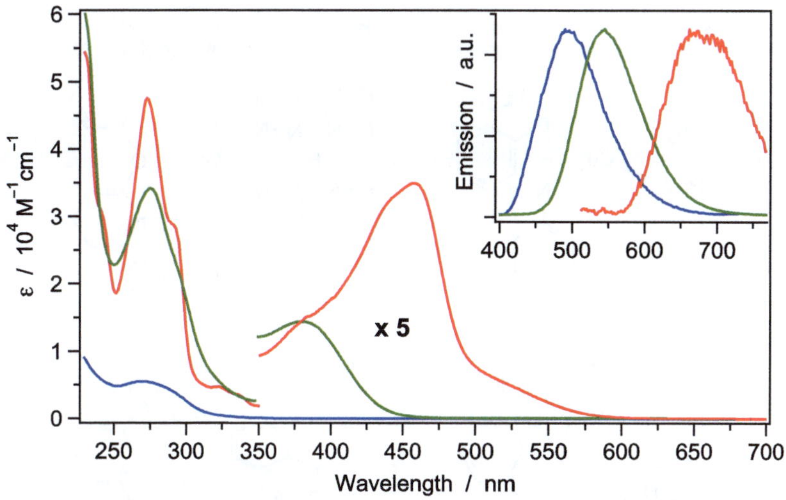

Fig. 18.14 Absorption and (*inset*) normalized luminescence spectra of $[Cu(dmp)_2]^+$ (*red*), $[Cu(dbp)(POP)]^+$ (*green*) and $[Cu(POP)(dppb)]^+$ (*blue*) in CH_2Cl_2 at 298 K. The former two complexes exhibit more intense UV absorption than $[Cu(POP)(dppb)]^+$ due to strong ligand centered absorption of the phenanthroline ligands. 2,9-di-*n*-butyl-1,10-phenanthroline

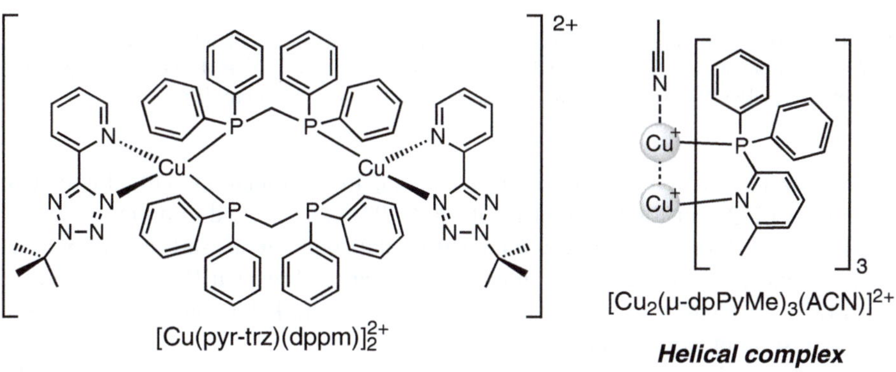

Fig. 18.15 Highly luminescent dinuclear charged Cu(I) complexes using ligands other than standard phenanthroline-based ones

18.2.4.4 Other Emitting Cu(I) Complexes

Other types of highly luminescent charged Cu(I) complexes have been obtained recently. A couple of examples from our laboratories are depicted in Fig. 18.15, which exhibit completely different structures. The first one is a $[Cu(N\text{-}N)(P\text{-}P)]^+$ dinuclear complex with alternative N-N ligands compared to what illustrated

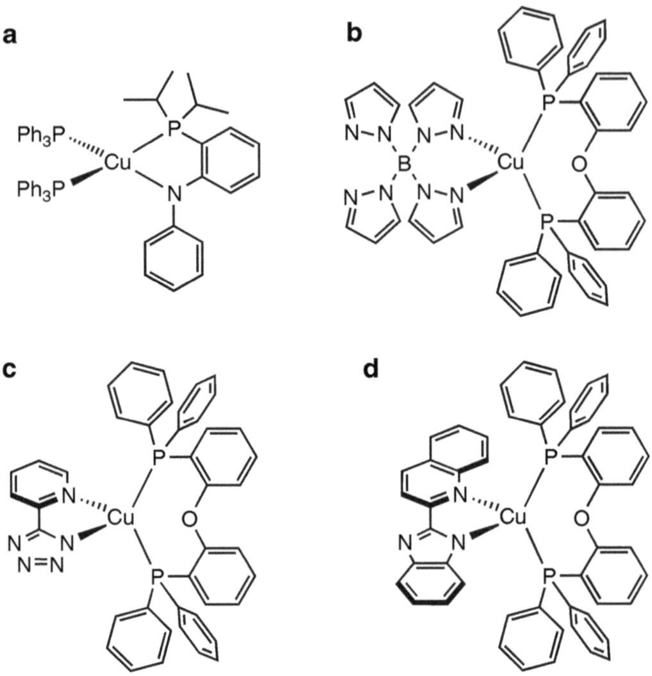

Fig. 18.16 Highly luminescent neutral Cu(I) complexes with (**a**) amidophosphine ligands and triphenylphosphine [95]; (**b**) POP and 5-(2-pyridyl)-tetrazolate [96]; (**c**) POP and tetrakis(pyrazol-1-yl)borate (pz4B) [97]; (**d**) POP and 2-(2'-quinolyl)benzimidazole (qbm) [98]

previously (i.e., a tetrazole-pyridine chelator), in combination with dppm, that exhibits a PLQY in PMMA matrix of 35 % [92]. The second complex is a very stable helical dinuclear structure obtained by the spontaneous self-assembly of 2-diphenylphosphino-6-methylpyridine (dpPyMe) around two Cu(I) ions; in this case the PLQY in PMMA diluted matrix reaches the remarkable value of 46 % [93]. Device testing in LEC structures is ongoing.

Notably, highly emitting *neutral* Cu(I) complexes started to appear in the literature some years ago, made of amidophosphine ligands and having the general formula [Cu(P-N)(P-P)] [94, 95], Fig. 18.16a. They exhibit PLQY in the range between 16 and 70 % in benzene.

Other neutral mononuclear [Cu(N-N)(P-P)] complexes have been recently reported, exhibiting PLQY as high as 90 % in the solid state, Fig. 18.16b–d [96–98]. Dinuclear cluster-like halide systems have also been prepared that were tested in OLED electroluminescent devices, showing relatively small luminance values (1,800 cd/m^2) and limited stability [99].

18.3 Hybrid Materials Made of Eu(III) Complexes and Carbon Nanotubes

In recent years, we have undertaken a research program on hybrid luminescent materials made of an organic scaffold and suitably chosen inorganic luminophores. Based on the experience accumulated over the years, we focussed our attention on carbon nanotubes as scaffolds and luminescent complexes of trivalent rare-earths (or lanthanide) ions as emitters, particularly those based on the Eu(III) ion, which emits in the red spectral region [100].

Presently, rare earth (RE) elements are vital for several technological applications (i.e., lighting, telecommunications, lasers, magnets, batteries, and imaging) [101–109] and are considered among the most strategically relevant materials for the entire world economy [110, 111]. Trivalent ions of RE elements such as Eu, Tb, Yb, Nd, and Er exhibit line-like emission bands from $f - f$ electronic transitions that yield an unparalleled color purity [112–114]. The coordination of properly selected organic chelating ligands around RE ions [115] can extend their use beyond the vast field of inorganic phosphors [116, 117]. This strategy offers several advantages such as (i) saturation of the coordinating positions of the metal centers and protection from quenching interactions, (ii) light-harvesting "antenna" effect for the sensitization of the ion itself upon excitation of the surrounding ligands, and (iii) possibility of binding a RE emitting ion to virtually any organic [118–120] or inorganic scaffold.

To attain a wider applicability of RE luminophores, a great deal of work has been done to harness and enhance their luminescence output, taking advantage of several scaffolding materials [112, 113, 121]. Among these, carbon nanotubes (CNTs) turned out to be particularly attractive platforms, thanks to their unique physical properties. The final goal is to combine the distinctive luminescence properties of RE complexes with the exceptional features of CNTs (e.g., high aspect ratio, thermal and electrical conductivity, physical robustness, chemical stability, cell permeability) [122], in order to prepare strongly emissive carbon-based nanomaterials with potential technological applications [100, 123]. Herein we summarize some of our work in this area.

18.3.1 External Physical Adsorption

The first preparative strategy we undertook to prepare hybrid luminescent carbon nanotubes was very simple. It entailed the in situ adsorption in MeOH solution of an Eu(III) complex onto oxidized SWCNTs, via hydrophobic interactions between the CNT walls and the skeleton of the aromatic ligands [124].

The hybrid was readily prepared using the highly luminescent and commercially available tris-dibenzoylmethane mono-1,10-phenanthroline europium(III) complex ($[Eu(DBM)_3(phen)]$) and mildly oxidized SWCNTs (Fig. 18.17). The simple mixing

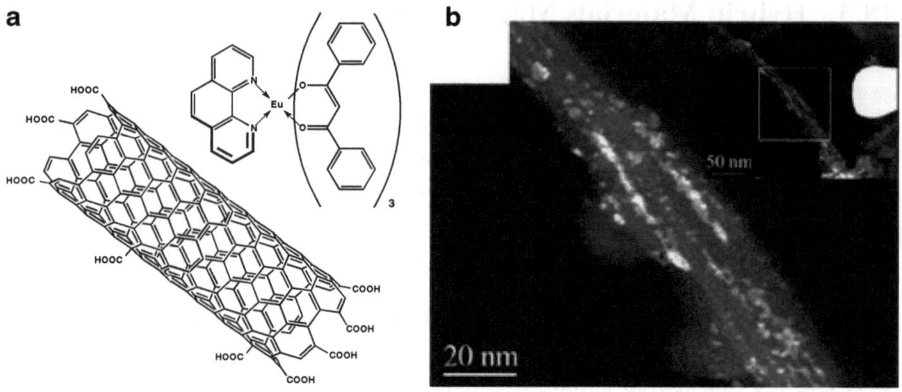

Fig. 18.17 (**a**) Schematic representation of the interaction between SWCNTs and [Eu(DBM)$_3$(phen)]. (**b**) Z-contrast STEM images of the hybrid material, taken at different magnifications showing the presence of nanoparticles of the luminescent complex onto the CNT walls

of both units in methanol (MeOH) was followed by sonication, stirring, and filtration. The filtrate was washed until the material showed no traces of Eu(III) luminescence and then dried under high vacuum. The structural characterization of the SWCNTs–Eu(III) composite was accomplished using thermogravimetric analysis (TGA), Raman spectroscopy, atomic force microscopy (AFM) and transmission electron microscopy (TEM). The combined results from these measurements unambiguously confirmed the presence of a Eu(III)-based material on the SWCNTs surface. In particular, the scanning transmission electron microscopy (STEM) images (Fig. 18.17) show high-contrast spherical particles located on the walls of CNTs, with an average width of 3 nm. Such particles contain the Eu element as shown by localized energy-dispersive X-ray spectroscopy (EDS) measurements.

Eventually, light excitation of the composite material suspended in MeOH and in a polystyrene matrix shows the characteristic sharp emission band of the Eu(III) ion centered at 612 nm. Importantly, the emission decay profiles of the hybrid nanomaterial dispersed in different media match with those of the [Eu(DBM)$_3$(phen)] complex alone under the same experimental conditions, due to the environment-insensitive luminescence properties of Ln(III) complexes (Fig. 18.18). This indicates negligible quenching interactions between SWCNTs and the adsorbed Eu(III) complex, despite the fact that the latter is attached directly on the carbon scaffold.

18.3.2 Binding by Electrostatic Interactions

Despite the promising emission properties of the above-mentioned CNT materials, limited and uncontrolled loading and possible desorption of the luminophores

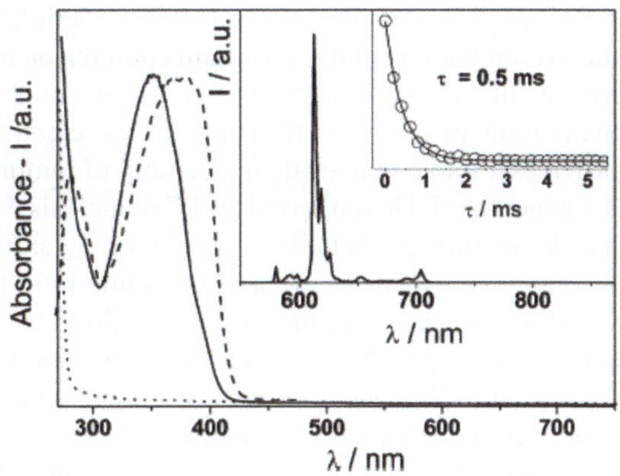

Fig. 18.18 Electronic absorption spectra of SWCNTs–Eu(III) (*full line*) and oxidized SWCNTs (*dotted line*) and excitation profile of SWCNTs–Eu(III) (*dashed line*, $\lambda_{em} = 614$ nm) in a polystyrene matrix. *Inset*: emission spectra ($\lambda_{ex} = 350$ nm) and decay profiles of Eu(III)–SWCNTs in a polystyrene matrix at 298 K

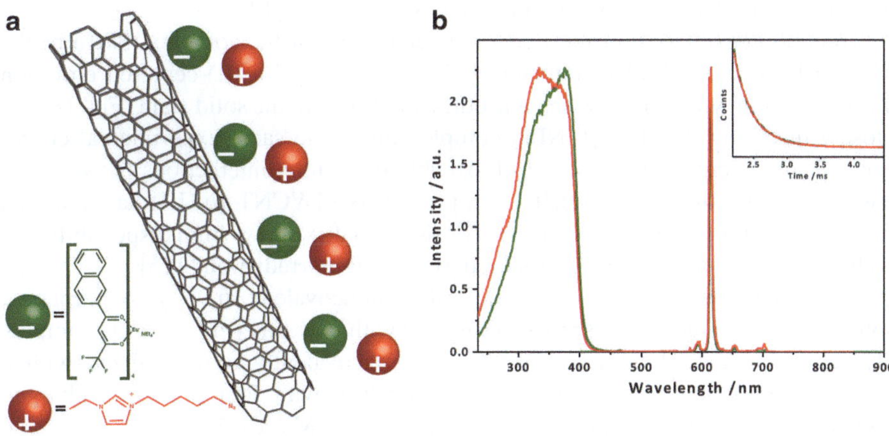

Fig. 18.19 (**a**) Schematic representation of the structure of the Eu(III)–MWNT nanomaterial assembled through electrostatic interactions and chemical structures of the ionic liquid (IL) and the $[EuL_4] \cdot NEt_4$ complex. (**b**) Luminescence spectra ($\lambda_{ex} = 370$ nm) and (*inset*) decay profiles of Eu(III)–MWNT hybrid (*green*) and $[EuL_4] \cdot NEt_4$ complex alone (*red*) in the solid state (KBr matrix)

from the CNT surface may limit the scale-up of preparative protocols and hence the possibility of implementing practical applications. Therefore, an alternative way to prepare a strongly emissive and more stable CNT–Eu(III) hybrid material has been proposed, in which a positively charged imidazolium-based ionic liquid (IL) was utilized as a passive intermediate layer between the surface of multiwall carbon nanotubes (MWCNTS) and Eu(III) luminophores to keep them intact [125]. The IL is equipped with a terminal azido group that spontaneously assembled on the surface of MWCNTs at high temperature, by means of strong hydrophobic interactions (Fig. 18.19). This affords a positively charged imidazolium layer on the MWCNT surface that attracts negatively charged tetrakis(2-napthoyltrifluoroacetonato)europate(III) units through electrostatic forces, upon simple addition (Fig. 18.19).

Notably, $\pi - \pi$ interactions between naphthoyl and imadazolium units increase the overall loading of the europium complex on the MWCNT surface. Therefore, the loading of the emitter is proportional to the number of IL units on the nanotube, and the combination of attractive $\pi - \pi$ and electrostatic IL/luminophore interactions provides a stable noncovalent assembly of luminophores on CNTs. The structure of the nanomaterial is supported by TGA analysis that shows 60 % weight loss from the hybrid, mainly attributable to the heavy Eu atoms. Moreover, X-ray photoelectron spectroscopy (XPS) shows a high atomic ratio for F atoms (11 %) in combination with Eu, confirming the presence of the fluorine-containing Eu(III) compound on the surface of the MWNT. By using a MWCNTs–IL analogue, in which fullerene C_{60} replaces the CNT and allows conventional photochemical experiments due to good solubility, the electrostatic interaction between the positively charged imidazole units and the anionic Eu(III) coordination sphere is definitively demonstrated, assuming a similar interaction in the MWCNT analogue [125].

Upon photoexcitation of the ligand-centered band of the complex (370 nm) the MWCNT–Eu(III) hybrid nanomaterial shows a strong Eu(III)-centered emission with a quantum yield of 44 % and a lifetime of 290 μs in the solid state (Fig. 18.19), closely matching the $[EuL_4] \cdot NEt_4$ complex under the same experimental conditions. This underpins the absence of sizeable quenching interactions between the nanotube scaffolds and the Eu(III) complex. This MWCNT–Eu(III) nanomaterial mixed with a polymethylmethacrylate (PMMA) matrix shows high dispersibility as well as a strong, uniform red emission upon UV light irradiation [125].

We further extended the above-described noncovalent strategy to prepare a solution-processable luminescent CNT hybrid with the same $[EuL_4] \cdot NEt_4$ complex [126]. In this case, however, MWCNTs were covalently functionalized with a polyamidoamine-based second-generation dendron with four positively charged alkylated ammonium groups as terminal units (d-MWCNTs), Fig. 18.20. The material was prepared with a similar procedure as described (Fig. 18.20) above and characterized by TGA, XPS, TEM, and SEM to support the functionalization of MWCNTs with the alkylated branching units. Thanks to the long dendronic arms attached to the carbon scaffold, d-MWCNTs–$[EuL_4]$ is highly soluble in several organic solvents of different polarity such as toluene, MeOH, and DMSO. Photoluminescence studies on d-MWCNTs–$[EuL_4]$ in toluene afford emission quantum yield and lifetime values (37 % and 0.47 ms, respectively) similar to those of $[EuL_4] \cdot NEt_4$ alone. Accordingly, this hybrid material affords optically transparent and strongly luminescent solutions of CNTs and corroborates the concept of a passive intermediate layer to effectively separate the CNT scaffold and the emitting center in order to preserve the luminescence performance [126].

The strong luminescence of CNTs exohedrally functionalized with Ln(III)-based complexes may enable the successful exploitation of CNT hybrids as luminescent materials in optoelectronic devices. The route is now open to explore further combinations of organic and inorganic emitters in order to tune the range of the absorption and emission properties.

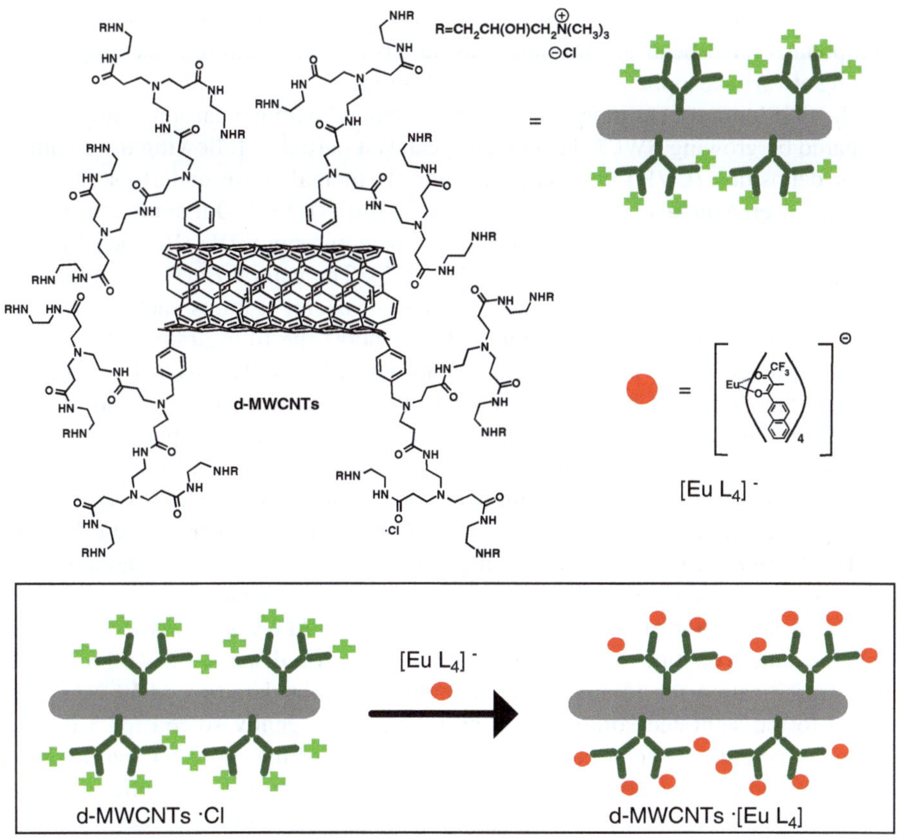

Fig. 18.20 Schematic structure of the dendrimer d-MWCNTs.Cl and sketching of the electrostatic coupling process leading to d-MWCNTs.$[EuL4]$

18.3.2.1 Endohedral Encapsulation

The possibility of driving a given organic or inorganic compound inside the hollow space of CNTs has been intensively explored as an alternative way for the non-destructive functionalization of nanotubes [127–130]. The encapsulation strategy offers several advantages over the chemical functionalization of the outer surface, whereas the confinement of molecular units into CNTs may afford brand new hybrid materials with unprecedented electronic and conductive properties. Moreover, the walls of CNTs may preserve guests for an unlimited period of time, which can be of great relevance for many applications.

Several organic molecules have been confined inside CNTs, but almost invariably they underwent photoinduced quenching due to energy- or electron transfer by the MWCNTs, which exhibit low-lying electronic levels (MWCNTs absorb down to the NIR region) and are good electron acceptors [100, 131]. On the other hand, several inorganic materials such as metal oxides, nanoparticles, and coordination compounds have been encapsulated inside CNTs using different techniques and aiming at various applications [127, 130, 132–134].

The first luminescent inorganic material confined within a carbon nanotube was prepared by growing SWCNTs on Eu_2O_3 catalyst particles, following the chemical vapor deposition (CVD) method [135]. Shortly thereafter we undertook the challenge to insert a metal complex, again based on Eu(III) [136]. Our approach entailed the use of neutral tris-hexafluoroacetylacetonate europium(III) ($Eu{-}acacF_6$) into tip-opened oxidized MWCNTs (Eu@MWCNT). The neutral Eu(III) complex was selected as the guest moiety due to a number of useful features such as (i) small size, which facilitates encapsulation and may boost the filling rate, (ii) electrical neutrality that prevents repulsive interactions with CNT walls, (iii) lack of aromatic units in the ligands that prevent stacking interactions with the CNT outer walls and favor removal of adsorbed luminophores, (iv) strong hydrophobic character of fluorine atoms granting compatibility for the internal cavity of CNTs, (v) high sensitivity of XRD and EDS techniques for fluorine atoms, which facilitate the tracing of the complex in the hybrid, and (vi) strong and characteristic Eu(III)-centered emission bands of $Eu{-}acacF_6$ with a photoluminescence quantum yield of 17 % and an excited-state lifetime of 0.70 ms in the solid state.

The encapsulation of the europium complex into ox-MWCNTs was accomplished with the nanoextraction method [137]. The complex was dispersed in 4 ml of $CHCl_3$ by ultrasonication for 15 min; eventually, 10 mg of powdered ox-MWCNTs were added to the saturated solution and the mixture was gently stirred for 3 days at room temperature. Filtered and meticulously washed Eu@MWCNT hybrids were characterized using analytical and optical techniques, always in parallel with two reference materials: i.e., $Eu{-}acacF_6$ adsorbed on tip-closed MWCNT both (i) washed (Eu–MWCNT-washed) and (ii) unwashed (Eu–MWCNT-unwashed).

In addition to the features of the reference Eu(III) complex and ox-MWCNTs, TGA analysis of Eu@MWCNT shows an additional band attributed to the encapsulated complexes, as typically found for C_{60}–SWCNTs peapods [138]. This demonstrates enhanced thermal stability of the guest due to the protective walls of the CNTs. This rationale is corroborated by the absence of such band in a control experiment with the Eu–MWCNT-washed sample. Additionally, XRD analysis confirms the presence of fluorine and europium in the hybrid, which were not found in Eu–MWCNT-washed. The HR-TEM images show the presence of material that closely follows the curvature of CNTs (Fig. 18.21), and the EDS spectrum within a small area shows the fingerprints of the F and Eu atoms (Fig. 18.21). Taken together, these results strongly indicate the successful encapsulation of $Eu{-}acacF_6$ into MWCNTs.

Photophysical studies provide further experimental evidence for encapsulation [136]. Light excitation of Eu@MWCNT at the absorption maximum of $Eu{-}acacF_6$ (300 nm) shows the characteristic sharp Eu(III)-centered emission feature, like the Eu–MWCNT-unwashed sample; by contrast, no luminescence is detected for the Eu–MWCNT-washed sample. The excitation spectrum recorded at the Eu(III) ion emission maximum for both Eu@MWCNT and Eu–MWCNT-unwashed matches

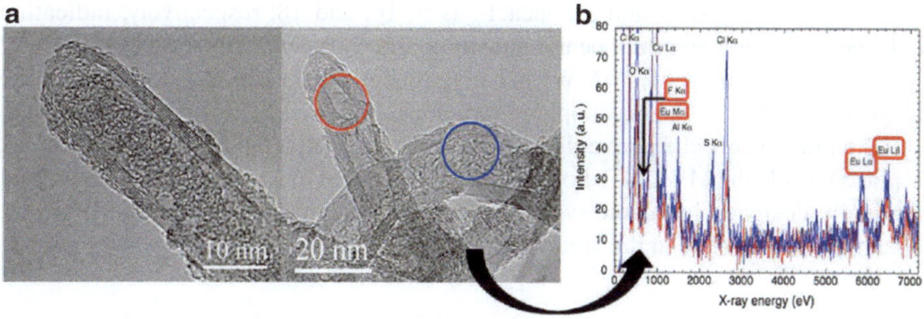

Fig. 18.21 (**a**) HR-TEM images of the Eu@MWCNT hybrid nanomaterial and (**b**) EDS analysis spectra corresponding to *blue* and *red circles* in panel (**a**), showing the presence of europium and fluorine atoms

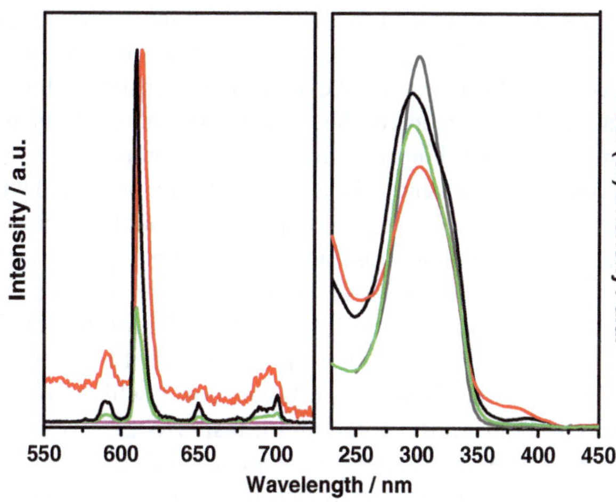

Fig. 18.22 Emission (*left*, $\lambda_{ex} = 300$ nm) and excitation (*right*, $\lambda_{ex} = 616$ nm) spectra of Eu@MWCNT (*red*), Eu−acacF_6 (*black*), Eu–MWCNT-washed (*magenta*), and Eu–MWCNT-unwashed (*green*) in the solid state. The *gray line* in the *left panel* corresponds to the absorption spectrum of the Eu−acacF_6 complex in CH_2Cl_2, reported for comparison purposes

that of Eu−acacF_6 (Fig. 18.22), showing that the complex has maintained its integrity upon encapsulation. Nonetheless, the PLQYs and excited state lifetimes (τ) of Eu@MWCNT and Eu–MWCNT-unwashed are significantly different (PLQY = 7 and 20 % and $\tau = 0.45$ and 0.82 ms, respectively) [136]. This is attributed to self absorption of the luminescence of the Eu(III) complex by the CNT scaffold and to weak electronic interactions of Eu−acacF_6 with the inner walls of MWCNTs.

A detailed analysis of the luminescence spectra, further corroborates the encapsulation of the complex within MWCNT [136]. The strongest Eu(III) emission band is located at 610 nm and arises from the "hypersensitive" electric dipole transition that varies its intensity and peak position due to the local environment, symmetry, and changes of the dipole moment in the coordination sphere. The emission maximum of Eu@MWCNT is red-shifted by 3 nm relative to the pristine reference complex and the Eu–MWCNT-unwashed sample. Further, the ratio between the area of the electric dipole and magnetic dipole emission peaks of Eu@MWCNT vs.

Eu–MWCNT-unwashed and Eu–acacF_6 is 9, 16, and 18, respectively, indicating reduced emission maximum peak intensity in the encapsulated hybrid [136]. These spectral peculiarities of Eu@MWCNT may reflect changes in symmetry and in the local environment of the luminophores when in close vicinity to the nanotube walls.

Taken in concert, the above-described results show that the Eu–acacF_6 complex is placed inside the MWCNTs hollow space. This proof of concept demonstrates that rationally designed inorganic systems can keep the emission output inside CNTs, opening a new avenue in the area of luminescent nanomaterials.

18.4 Supramolecular Systems Mimicking Fundamental Photosynthetic Processes

Technical evolution in the history of mankind has often made progress by observing and mimicking Nature, but implementing hugely simpler devices than those developed through natural evolution. For instance, human beings move by walking, mankind has invented the wheel; Nature exploits the brain for recording and computing, mankind has invented the computer. To exploit sunshine, Nature has developed photosynthesis, a very complex process developed over billions of years that produces high energy compounds through an intricate combination of chemical reactions. Mankind, instead, has invented photovoltaic devices, which convert sunlight into electricity (a highly versatile form of energy) with an efficiency at least an order of magnitude higher than that of the natural photosynthetic processes. Over geologic times, Nature has also transformed the products of photosynthesis into fossil solar fuels (i.e., oil, gas, coal), which are concentrated forms of energy that can be usefully stored and transported. However, as we have discussed in the beginning of this chapter, there are several urgent reasons to phase out fossil fuels, therefore scientists are now strongly engaged in finding ways to harness sunlight for producing artificial solar fuels, to be made at the industrial scale [139–141]. To this end, they have thoroughly investigated the complex mechanism of natural photosynthesis and are now trying to create artificial photosynthetic processes much simpler than the natural ones and capable of producing the simplest solar fuels (i.e., hydrogen and methanol) [142, 143].

18.4.1 Basic Components of an Artificial Photosynthetic System

The construction of artificial photosynthetic arrays for the production of solar fuels (i.e., hydrogen through water splitting, Fig. 18.23), can take as a blueprint the molecular and supramolecular organization of natural photosynthetic organisms. Accordingly, light harvesting must lead to charge separation, which has to be followed by charge transport that delivers oxidizing and reducing equivalents

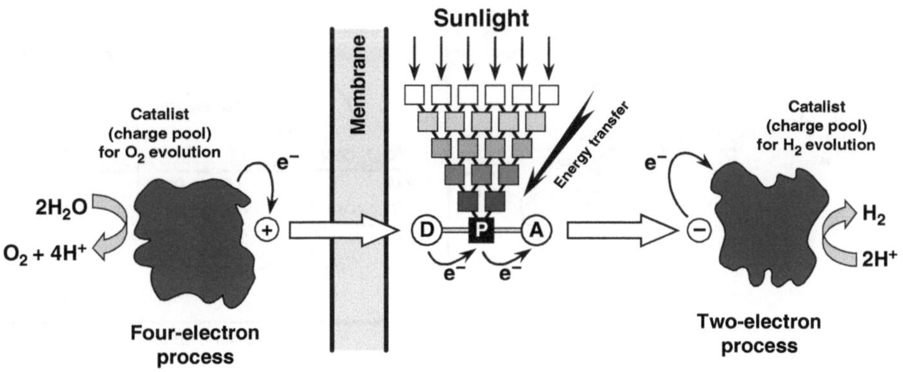

Fig. 18.23 Schematic representation of photochemical water splitting (artificial photosynthesis). Five fundamental components are needed: an antenna for light harvesting, a charge separation multicomponent system D–P–A (donor-photosensitizer-acceptor), a catalyst for hydrogen evolution, a catalyst for oxygen evolution, and a membrane separating the reductive and the oxidative processes

at catalytic sites, where oxygen and hydrogen evolution must occur separately [142]. Therefore, a conceivable artificial photosynthetic system should include the following fundamental components: (1) an antenna for light harvesting; (2) a reaction center undergoing charge separation; (3) reduction and (4) oxidation catalysts as interfaces between the charge separation center and the final products, and (5) a membrane to provide physical separation of the products.

After a period of expansion following the first oil shock in the 1970s [144], research in this area has plummeted for about 30 years, during which the low price of fossil fuels strongly discouraged the search for alternative energy sources and innovative technologies [2]. However, at the beginning of the twenty-first century when the oil price sudden increased and stabilized around 100 $/bbl, research in artificial photosynthesis has remarkably revived, becoming one of the most active and promising area of multidisciplinary investigation in the science arena [139–142]. In this context, and with reference to Fig. 18.23, our group has carried out research in the area of light harvesting and charge separation systems, which are the key interface between sunlight and the final desired fuels [145,146]. A few examples will be illustrated.

18.4.2 Fullerodendrimers

Fullerenes are among the most investigated families of molecules over the last three decades [147]. One of the most relevant properties of fullerenes is their strong electron-accepting character, which has been extensively exploited in photochemistry [148]. In particular, a vast number of multicomponent systems in which fullerene electron acceptors are coupled with suitable electron donors have been

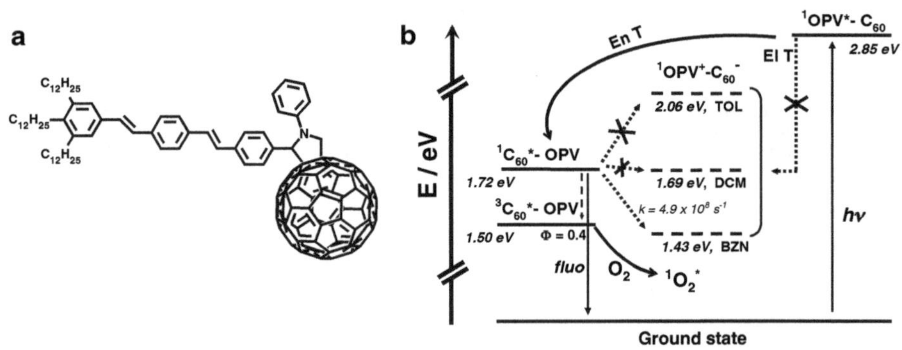

Fig. 18.24 C_{60}–OPV dyad (**a**) and energy level diagram (**b**) indicating the lowest singlet and/or triplet excited states of fullerene (*left*) and OPV (*right*) moieties, with related photoinduced processes (*En T* energy transfer, *El T* electron transfer)

synthesized and investigated. The final goal is the implementation of rationally designed multicomponent arrays featuring photoinduced energy and electron transfer, which are the fundamental processes that trigger the conversion of sunlight into chemical energy (Fig. 18.23) [145,147–151]. C_{60} derivatives and organic conjugated molecules such as oligophenylenevinylenes (OPV) and oligophenyleneethynylenes (OPE) can be covalently linked in a facile way and a number of "dyad" systems such as C_{60}–OPV in Fig. 18.24a have been prepared over the years [152,153]. C_{60}–OPV exemplifies the typical photophysical behavior of such systems [154] can be described through the energy level diagram depicted in Fig. 18.24b.

Upon selective excitation of the OPV moiety of C_{60}–OPV, quenching of its intense fluorescence is detected in solvents of any polarity. This is attributed to ultrafast singlet-singlet OPV $\rightarrow$ C_{60} energy transfer that prevails over electron transfer. The latter is highly exoergonic ($-\Delta G^o = 1.13$ eV) and results to be much slower than OPV $\rightarrow$ C_{60} energy transfer, since located in the so-called Marcus inverted region [154]. On the other hand, upon fullerene excitation of C_{60}–OPV in the Vis region, no quenching of C_{60} singlet and triplet states is detected in toluene (TOL) and in dichloromethane (DCM), as inferred by fluorescence and sensitized singlet oxygen luminescence spectra (the latter allows quantitative detection of fullerene triplets [155]). Both types of spectra exhibit the same intensity as an individual fulleropyrrolidine unit taken as reference. On the contrary, in benzonitrile (BZN) a 60 % loss of singlet and triplets is found for C_{60}–OPV, indicating that an OPV $\rightarrow$ C_{60} electron transfer quenching is active in this polar medium: no other deactivation pathways are thermodynamically allowed (Fig. 18.24b). Electron transfer is only possible in polar BZN because the charge separated level is made thermodinamically accessible due to the increased polarity of the medium, that stabilizes the charge separated state, unlike the case of less polar TOL and DCM (Fig. 18.24b). Two important experimental findings have to be emphasized: (i) the quenching trend of fullerene singlet and triplet excited states is the same by exciting either the C_{60} or the OPV moiety, indicating that the latter simply acts as an excitation energy transfer antenna for the carbon sphere and (ii) the fullerene triplet lifetime of C_{60}–OPV is not quenched in BZN, showing that the quenching process is originated only from the fullerene singlet level while the parallel loss of triplet is just a consequence of singlet depletion, followed by regular $^1C_{60}^* \rightarrow {}^3C_{60}^*$ intersystem crossing [154].

The above results show that tracing of fullerene excited states by Vis-NIR luminescence spectroscopy is a simple yet powerful tool to monitor photoinduced electron transfer in hybrids made of C_{60} fullerenes and organic conjugated oligomers. This prompted us to investigated more complex nanostructures such as dendrimers containing fullerene units at peripheral units 2–4 (Fig. 18.25) [156].

The photophysical properties and photoinduced processes of **1–3** are essentially the same as in the simpler dyad C_{60}–OPV. However, the fluorescence and singlet oxygen luminescence spectra of **1–3** in BZN show reduced electron transfer efficiency upon increasing the dendrimer structure. This can be attributed only to structural factors since the electron donor (OPV) and acceptor (fullerene) are exactly the same for the three compounds. Progressive isolation of the central OPV electron donor by the surrounding floppy dendronic wedges tends to disfavor solvent induced stabilization of the transient ionic species upon electron transfer, thereby reducing charge separation efficiency in larger structures. The dendritic architecture is therefore able to isolate the central core and influences substantially its light-induced properties [156].

This rationale is corroborated by another study on a family of rigid dendrimers based on OPV and C_{60}, where the core/periphery position is reversed: the center of the structure is a fullerene unit [157]. In Fig. 18.26 is depicted the largest representative of this family of fullerodendrimers (**4**). In these systems, electron transfer in benzonitrile is more efficient along with the dendrimer size, just the opposite as in **1–3**. This result is promoted by at least three factors (i) the impossibility of attaining a globular structure isolating the core from the solvent, as in **1–3**, due to rigidity; (ii) a progressively reduced effective donor-acceptor distance; (iii) a partial delocalization of the positive charge among the acceptor units when the dendrimer structure becomes bigger and bigger, which lowers external reorganization energy and provides kinetic advantage to electron transfer [157].

In summary, in complex electron donor-acceptor nanosystems containing fullerenes the extent of electron transfer can be finely tuned by molecular design, thanks to pure structural and physical factors.

18.4.3 A Supramolecular Photosynthetic Antenna-Charge Separation Module

Very recently, we have proposed a sophisticated multicomponent architecture with 12 light harvesting Zn-porphyrin units grafted onto an exafunctionalized fullerene

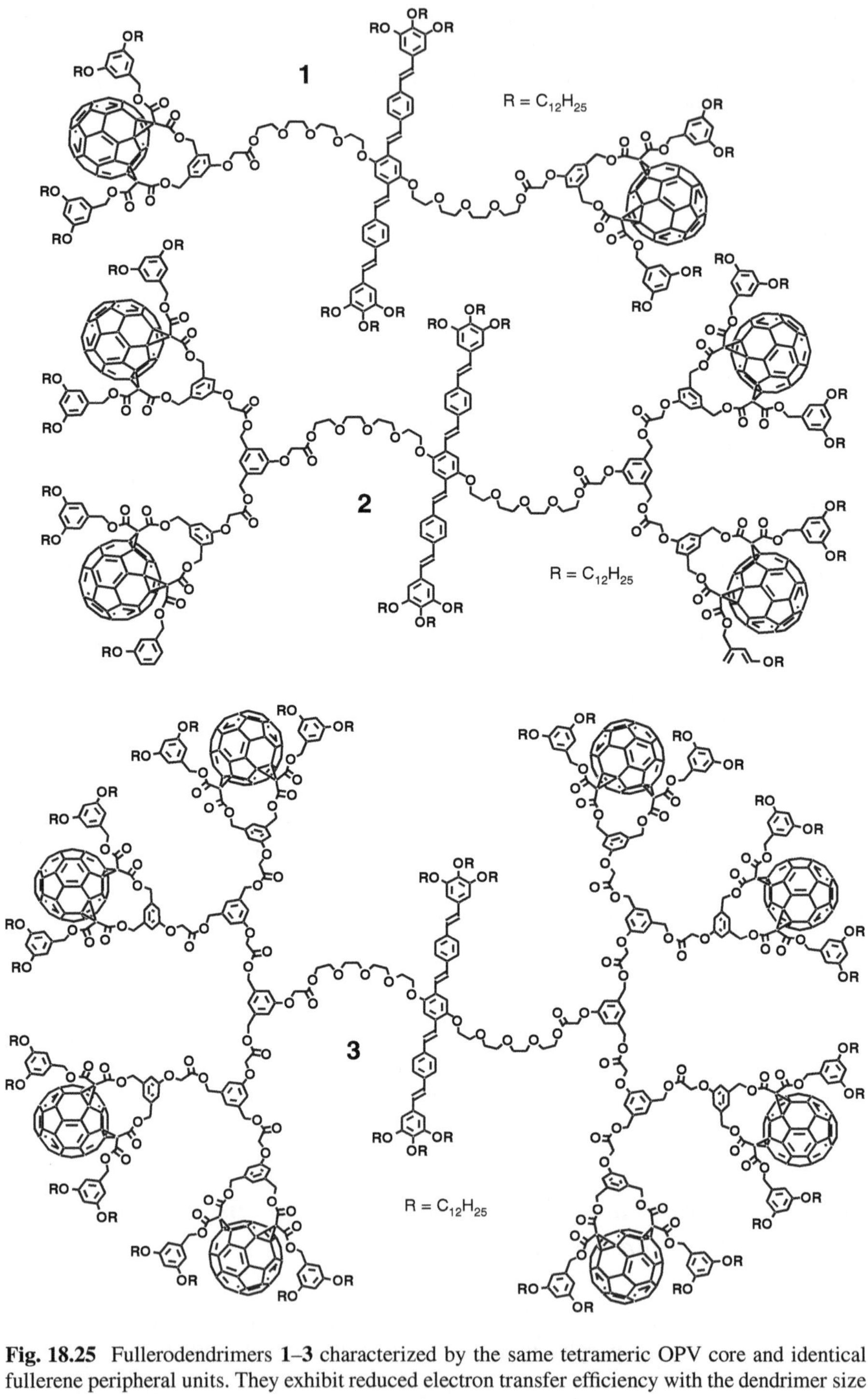

Fig. 18.25 Fullerodendrimers **1–3** characterized by the same tetrameric OPV core and identical fullerene peripheral units. They exhibit reduced electron transfer efficiency with the dendrimer size

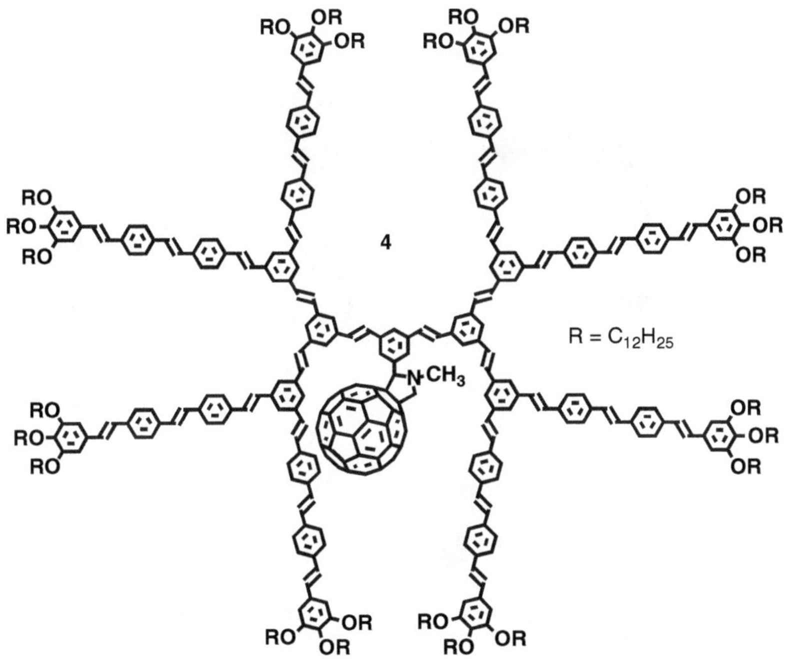

Fig. 18.26 Chemical structure of the fullerodendrimer **4** with a fullerene central core. It exhibits enhanced electron transfer efficiency compared to its smaller analogues

center ($C_{60}(ZnP)_{12}$). A monosubstituted fullerene (F) is then supramolecularly attached to the Zn(II) ion of the porphyrin units through a pyridine residue, creating the self-assembled supramolecular nanostructure ($C_{60}(ZnP)_{12}F_n$) schematically depicted in Fig. 18.27 [158].

In this system, only the mono-substituted C_{60} derivative plays an active role (i.e., electron acceptor) in the cascade of photoinduced events. The hexa-substituted C_{60} moiety serves as a nanoscaffold allowing to generate a cluster of porphyrins behaving at the same time as light harvesting antennas and electron donors. In benzonitrile, a charge separated state living 300 ns is observed. Such unexpectedly long lifetime is attributed to hole hopping over the porphyrin electron donor antennas, a process that occurs in natural photosynthetic arrays, of which $C_{60}(ZnP)_{12}F_n$ is an advanced lab-made model by design.

18.5 Conclusion

The design, preparation and testing of new materials with tailored physical and chemical properties is an area of extraordinary importance to successfully undertake the complex and challenging transition to a civilization powered by perennial energy

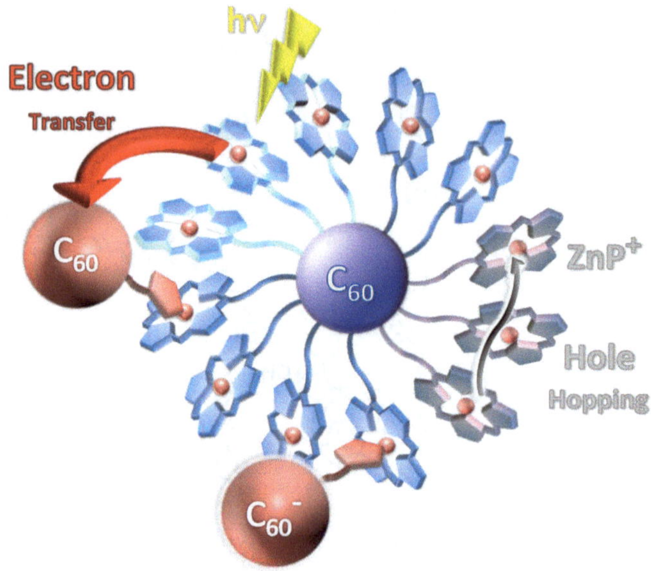

Fig. 18.27 Schematic representation of the events occurring in $C_{60}(ZnP)_{12}F_n$ in benzonitrile solution: light harvesting, charge separation and hole hopping

sources. New materials are necessary for enhancing energy efficiency of end user usage, allow electricity storage at any scale, afford cheaper and more effective insulating and conducting materials, produce more efficient and sustainable batteries, convert solar energy, just to mention some key applications.

Here we have focussed on photoactive materials exhibiting light harvesting, luminescence, and photoinduced energy/electron transfer processes, which are needed in the technological areas of lighting and solar energy conversion. We have illustrated recent results in the relentless search for new luminescent materials for flat electroluminescent devices, highlighting the importance of molecular design and the intrinsic difficulty to implement materials that are tuneable, cheap and easy to prepare at the same time. On the other hand, we have presented some nanomaterials that exhibit fundamental processes needed to assemble artificial photosynthetic arrays. A lot of work still needs to be done for practical market applications, but the amount of know-how accumulated in recent years through fundamental research shows that the final goal is getting closer.

Acknowledgements We thank CNR (PM. P04. 010, MACOL), EC (FP7-STRP 248043—CELLO and PITN-GA-2008-215399—FINELUMEN), MIUR-PRIN 2010 INFOCHEM (contract no. CX2TLM), MIUR-FIRB Futuro in Ricerca SUPRACARBON (contract no. RBFR10DAK6) for the financial support of our research. We are also grateful to the colleagues who contributed to the work presented here and particularly Henk Bolink and Enrique Ortí (Valencia, Spain), M.K. Nazeeruddin (EPFL Lausanne), Jean-François Nierengarten (Strasbourg, France) and Davide Bonifazi (Namur, Belgium), along with the members of their teams.

References

1. Armaroli N, Balzani V (2007) The future of energy supply: challenges and opportunities. Angew Chem Int Ed 46(1–2):52–66
2. Armaroli N, Balzani V (2011) Energy for a sustainable world – from the oil age to a sun powered future. Wiley-VCH, Weinheim
3. Armaroli N, Balzani V, Serpone N (2013) Powering planet earth – energy solutions for the future. Wiley-VCH, Weinheim
4. Smil V (2008) Energy in nature and society. General energetics of complex systems. MIT, Cambridge
5. Volk T (2008) CO_2 rising. The world's greatest environmental challange. MIT, Cambridge
6. Monastersky R (2013) Global carbon dioxide levels near worrisome milestone. Nature 497(7447):13–14
7. IPCC (2013) Fourth assessment report: climate change 2013, United Nations Intergovernmental Panel on Climate Change
8. Armaroli N, Balzani V (2011) The legacy of fossil fuels. Chem Asian J 6(3):768–784
9. Wohrnschimmel H, MacLeod M, Hungerbuhler K (2013) Emissions, fate and transport of persistent organic pollutants to the arctic in a changing global climate. Environ Sci Technol 47(5):2323–2330
10. Olah GA, Prakash GKS, Goeppert A (2011) Anthropogenic chemical carbon cycle for a sustainable future. J Am Chem Soc 133(33):12881–12898
11. Miller RG, Sorrell SR (2014) The future of oil supply. Philos Trans R Soc A 372(2006)
12. Kerr RA (2010) Natural gas from shale bursts onto the scene. Science 328(5986):1624–1626
13. Hughes JD (2013) A reality check on the shale revolution. Nature 494(7437):307–308
14. Vidic RD, Brantley SL, Vandenbossche JM, Yoxtheimer D, Abad JD (2013) Impact of shale gas development on regional water quality. Science 340(6134):6134
15. Osborn SG, Vengosh A, Warner NR, Jackson RB (2011) Methane contamination of drinking water accompanying gas-well drilling and hydraulic fracturing. Proc Natl Acad Sci USA 108(20):8172–8176
16. NAS (2013) Induced seismicity potential in energy technologies. The National Academies Press, Washington, DC
17. Tollefson J (2013) Methane leaks erode green credentials of natural gas. Nature 493(7430):12–12
18. Dale AT, Khanna V, Vidic RD, Bilec MM (2013) Process based life-cycle assessment of natural gas from the marcellus shale. Environ Sci Technol 47(10):5459–5466
19. Schiermeier Q (2012) The great arctic oil race begins. Nature 482(7383):13–14
20. Smil V (2005) Creating the twentieth century: technical innovations of 1867–1914 and their lasting impact. Oxford University Press, Oxford
21. Brox J (2010) Brilliant: the evolution of artificial light. Houghton Mifflin Harcourt, New York
22. Martin WJ, Glass RI, Balbus JM, Collins FS (2011) A major environmental cause of death. Science 334(6053):180–181
23. Mills E (2005) The specter of fuel-based lighting. Science 308(5726):1263–1264
24. Crawford MH (2009) Leds for solid-state lighting: performance challenges and recent advances. IEEE J Sel Top Quantum 15(4):1028–1040
25. Wang J, Zhang F, Zhang J, Tang W, Tang A, Peng H, Xu Z, Teng F, Wang Y (2013) Key issues and recent progress of high efficient organic light-emitting diodes. J Photochem Photobiol C 17(0):69–104
26. Müllen K, Scherf U (eds) (2006) Organic light emitting devices: synthesis, properties, and applications. Wiley-VCH, Weinheim
27. Gather MC, Köhnen A, Meerholz K (2011) White organic light-emitting diodes. Adv Mater 23(2):233–248
28. Reineke S, Lindner F, Schwartz G, Seidler N, Walzer K, Lussem B, Leo K (2009) White organic light-emitting diodes with fluorescent tube efficiency. Nature 459(7244):234–238

29. Barbieri A, Accorsi G, Armaroli N (2008) Luminescent complexes beyond the platinum group: the d^{10} avenue. Chem Commun 19(19):2185–2193
30. Evans RC, Douglas P, Winscom CJ (2006) Coordination complexes exhibiting room-temperature phosphorescence: evaluation of their suitability as triplet emitters in organic light emitting diodes. Coord Chem Rev 250(15–16):2093–2126
31. Costa RD, Ortí E, Bolink HJ, Monti F, Accorsi G, Armaroli N (2012) Luminescent ionic transition-metal complexes for light-emitting electrochemical cells. Angew Chem Int Ed 51(33):8178–8211
32. Sessolo M, Bolink HJ (2011) Hybrid organic-inorganic light-emitting diodes. Adv Mater 23(16):1829–1845
33. Hu T, He L, Duan L, Qiu Y (2012) Solid-state light-emitting electrochemical cells based on ionic iridium(iii) complexes. J Mater Chem 22:4206–4215
34. Pei Q, Yu G, Zhang C, Yang Y, Heeger AJ (1995) Polymer light-emitting electrochemical cells. Science 269(5227):1086–1088
35. Slinker J, Bernards D, Houston PL, Abruña HD, Bernhard S, Malliaras GG (2003) Solid-state electroluminescent devices based on transition metal complexes. Chem Commun 22(19):2392–2399
36. Pertegás A, Tordera D, Serrano-Pérez JJ, Ortí E, Bolink HJ (2013) Light-emitting electrochemical cells using cyanine dyes as the active components. J Am Chem Soc 135(48):18008–18011
37. Maness KM, Terrill RH, Meyer TJ, Murray RW, Wightman RM (1996) Solid-state diode-like chemiluminescence based on serial, immobilized concentration gradients in mixed-valent poly$[Ru(vbpy)_3](PF_6)_2$ films. J Am Chem Soc 118(43):10609–10616
38. Flamigni L, Barbieri A, Sabatini C, Ventura B, Barigelletti F (2007) Photochemistry and photophysics of coordination compounds: iridium. Top Curr Chem 281:143–203
39. Lowry MS, Bernhard S (2006) Synthetically tailored excited states: phosphorescent, cyclometalated iridium(iii) complexes and their applications. Chem Eur J 12(31):7970–7977
40. Slinker JD, Rivnay J, Moskowitz JS, Parker JB, Bernhard S, Abruña HD, Malliaras GG (2007) Electroluminescent devices from ionic transition metal complexes. J Mater Chem 17(29):2976–2988
41. Ulbricht C, Beyer B, Friebe C, Winter A, Schubert US (2009) Recent developments in the application of phosphorescent iridium(iii) complex systems. Adv Mater 21(44):4418–4441
42. Bolink HJ, Cappelli L, Cheylan S, Coronado E, Costa RD, Lardies N, Nazeeruddin MK, Ortí E (2007) Origin of the large spectral shift in electroluminescence in a blue light emitting cationic iridium(iii) complex. J Mater Chem 17(48):5032–5041
43. He L, Duan L, Qiao J, Wang R, Wei P, Wang L, Qiu Y (2008) Blue-emitting cationic iridium complexes with 2-(1h-pyrazol-1-yl)pyridine as the ancillary ligand for efficient light-emitting electrochemical cells. Adv Funct Mater 18(14):2123–2131
44. Yang C-H, Beltran J, Lemaur V, Cornil J, Hartmann D, Sarfert W, Fröhlich R, Bizzarri C, De Cola L (2010) Iridium metal complexes containing n-heterocyclic carbene ligands for blue-light-emitting electrochemical cells. Inorg Chem 49(21):9891–9901
45. Kessler F, Costa RD, Di Censo D, Scopelliti R, Ortí E, Bolink HJ, Meier S, Sarfert W, Gratzel M, Nazeeruddin MK, Baranoff E (2012) Near-uv to red-emitting charged bis-cyclometallated iridium(iii) complexes for light-emitting electrochemical cells. Dalton Trans 41(1):180–191
46. Shik Chin C, Eum M-S, yi Kim S, Kim C, Kwon Kang S (2007) Blue-light-emitting complexes: cationic (2-phenylpyridinato)iridium(iii) complexes with strong-field ancillary ligands. Eur J Inorg Chem 2007(3):372–375
47. Shavaleev NM, Monti F, Costa RD, Scopelliti R, Bolink HJ, Ortí E, Accorsi G, Armaroli N, Baranoff E, Grätzel M, Nazeeruddin MK (2012) Bright blue phosphorescence from cationic bis-cyclometalated iridium(iii) isocyanide complexes. Inorg Chem 51(4):2263–2271
48. Shavaleev NM, Monti F, Scopelliti R, Armaroli N, Grätzel M, Nazeeruddin MK (2012) Blue phosphorescence of trifluoromethyl- and trifluoromethoxy-substituted cationic iridium(iii) isocyanide complexes. Organometallics 31(17):6288–6296

49. Shavaleev NM, Monti F, Scopelliti R, Baschieri A, Sambri L, Armaroli N, Grätzel M, Nazeeruddin MK (2013) Extreme tuning of redox and optical properties of cationic cyclometalated iridium(iii) isocyanide complexes. Organometallics 32(2):460–467
50. Monti F, Kessler F, Delgado M, Frey J, Bazzanini F, Accorsi G, Armaroli N, Bolink HJ, Ortí E, Scopelliti R, Nazeeruddin MK, Baranoff E (2013) Charged bis-cyclometalated iridium(iii) complexes with carbene-based ancillary ligands. Inorg Chem 52(18):10292–10305
51. http://www.platinum.matthey.com/
52. Armaroli N, Accorsi G, Cardinali F, Listorti A (2007) Photochemistry and photophysics of coordination compounds; copper. Top Curr Chem 280:69–115
53. Armaroli N (2001) Photoactive mono- and polynuclear cu(i)-phenanthrolines. A viable alternative to ru(ii)-polypyridines? Chem Soc Rev 30(2):113–124
54. Lavie-Cambot A, Cantuel M, Leydet Y, Jonusauskas G, Bassani DM, McClenaghan ND (2008) Improving the photophysical properties of copper(i) bis(phenanthroline) complexes. Coord Chem Rev 252(23–24):2572–2584
55. McMillin DR, McNett KM (1998) Photoprocesses of copper complexes that bind to DNA. Chem Rev 98(3):1201–1219
56. Scaltrito DV, Thompson DW, O'Callaghan JA, Meyer GJ (2000) MLCT excited states of cuprous bis-phenanthroline coordination compounds. Coord Chem Rev 208:243–266
57. Kalsani V, Schmittel M, Listorti A, Accorsi G, Armaroli N (2006) Novel phenanthroline ligands and their kinetically locked copper(i) complexes with unexpected photophysical properties. Inorg Chem 45(5):2061–2067
58. Everly RM, McMillin DR (1989) Concentration-dependent lifetimes of $[Cu(NN)_2]^+$ systems – exciplex quenching from the ion-pair state. Photochem Photobiol 50(6):711–716
59. Chen LX, Shaw GB, Novozhilova I, Liu T, Jennings G, Attenkofer K, Meyer GJ, Coppens P (2003) MLCT state structure and dynamics of a copper(i) diimine complex characterized by pump-probe x-ray and laser spectroscopies and dft calculations. J Am Chem Soc 125(23):7022–7034
60. Robertazzi A, Magistrato A, de Hoog P, Carloni P, Reedijk J (2007) Density functional theory studies on copper phenanthrolines. Inorg Chem 46(15):5873–5881
61. Siddique ZA, Yamamoto Y, Ohno T, Nozaki K (2003) Structure-dependent photophysical properties of singlet and triplet metal-to-ligand charge transfer states in copper(i) bis(diimine) compounds. Inorg Chem 42(20):6366–6378
62. Zgierski MZ (2003) Cu(i)-2,9-dimethyl-1,10-phenanthroline: density functional study of the structure, vibrational force-field, and excited electronic states. J Chem Phys 118(9):4045–4051
63. Chen LX, Jennings G, Liu T, Gosztola DJ, Hessler JP, Scaltrito DV, Meyer GJ (2002) Rapid excited-state structural reorganization captured by pulsed x-rays. J Am Chem Soc 124(36):10861–10867
64. Shaw GB, Grant CD, Shirota H, Castner EW, Meyer GJ, Chen LX (2007) Ultrafast structural rearrangements in the MLCT excited state for copper(i) bis-phenanthrolines in solution. J Am Chem Soc 129(7):2147–2160
65. Iwamura M, Takeuchi S, Tahara T (2007) Real-time observation of the photoinduced structural change of bis(2,9-dimethyl-1,10-phenanthroline)copper(i) by femtosecond fluorescence spectroscopy: a realistic potential curve of the jahn-teller distortion. J Am Chem Soc 129(16):5248–5256
66. Chen LX (2004) Taking snapshots of photoexcited molecules in disordered media by using pulsed synchrotron x-rays. Angew Chem Int Ed 43(22):2886–2905
67. Chen LX (2005) Probing transient molecular structures in photochemical processes using laser-initiated time-resolved x-ray absorption spectroscopy. Annu Rev Phys Chem 56:221–254
68. Coppens P (2003) What can time-resolved diffraction tell us about transient species?: excited-state structure determination at atomic resolution. Chem Commun 12(12):1317–1320

69. Coppens P, Vorontsov I, Graber T, Kovalevsky AY, Chen YS, Wu G, Gembicky M, Novozhilova IV (2004) Geometry changes of a cu(i) phenanthroline complex on photoexcitation in a confining medium by time-resolved x-ray diffraction. J Am Chem Soc 126(19):5980–5981
70. Felder D, Nierengarten JF, Barigelletti F, Ventura B, Armaroli N (2001) Highly luminescent cu(i)-phenanthroline complexes in rigid matrix and temperature dependence of the photophysical properties. J Am Chem Soc 123(26):6291–6299
71. Miller MT, Gantzel PK, Karpishin TB (1999) A highly emissive heteroleptic copper(i) bis(phenanthroline) complex: $[Cu(dbp)(dmp)]^+$ (dbp=2,9-di-tert-butyl-1,10-phenanthroline; dmp=2,9-dimethyl-1,10-phenanthroline). J Am Chem Soc 121(17):4292–4293
72. Gothard NA, Mara MW, Huang J, Szarko JM, Rolczynski B, Lockard JV, Chen LX (2012) Strong steric hindrance effect on excited state structural dynamics of cu(i) diimine complexes. J Phys Chem A 116(9):1984–1992
73. Green O, Gandhi BA, Burstyn JN (2009) Photophysical characteristics and reactivity of bis(2,9-di-tert-butyl-1,10-phenanthroline)copper(i). Inorg Chem 48(13):5704–5714
74. McCusker CE, Castellano FN (2013) Design of a long-lifetime, earth-abundant, aqueous compatible cu(i) photosensitizer using cooperative steric effects. Inorg Chem 52(14):8114–8120
75. Buckner MT, McMillin DR (1978) Photo-luminescence from copper(i) complexes with low-lying metal-to-ligand charge-transfer excited-states. J Chem Soc Chem Commun 17(17):759–761
76. Cuttell DG, Kuang SM, Fanwick PE, McMillin DR, Walton RA (2002) Simple cu(i) complexes with unprecedented excited-state lifetimes. J Am Chem Soc 124(1):6–7
77. McCormick T, Jia WL, Wang SN (2006) Phosphorescent cu(i) complexes of 2-(2'-pyridylbenzimidazolyl)benzene: impact of phosphine ancillary ligands on electronic and photophysical properties of the cu(i) complexes. Inorg Chem 45(1):147–155
78. Tsukuda T, Nakamura A, Arai T, Tsubomura T (2006) Luminescence of copper(i) dinuclear complexes bridged by diphosphine ligands. Bull Chem Soc Jpn 79(2):288–290
79. Armaroli N, Accorsi G, Holler M, Moudam O, Nierengarten JF, Zhou Z, Wegh RT, Welter R (2006) Highly luminescent cu-i complexes for light-emitting electrochemical cells. Adv Mater 18(10):1313–1316
80. Howell SL, Gordon KC (2004) Spectroscopic and density functional theory studies of 1,10-phenanthroline, its radical anion, and $[Cu(phen)(PPh_3)_2]^+$. J Phys Chem A 108(13): 2536–2544
81. Yang L, Feng JK, Ren AM, Zhang M, Ma YG, Liu XD (2005) Structures, electronic states and electroluminescent properties of a series of cu(i) complexes. Eur J Inorg Chem 10(10): 1867–1879
82. Kuang SM, Cuttell DG, McMillin DR, Fanwick PE, Walton RA (2002) Synthesis and structural characterization of cu(i) and that ni(ii) complexes that contain the bis[2-(diphenylphosphino)phenyl]ether ligand. Novel emission properties for the cu(i) species. Inorg Chem 41(12):3313–3322
83. Englman R, Jortner J (1970) The energy gap law for radiationless transitions in large molecules. Mol Phys 18:145–164
84. Palmer CEA, Mc Millin DR, Kirmaier C, Holten D (1987) Flash-photolysis and quenching studies of copper(i) systems in the presence of lewis-bases – inorganic exciplexes. Inorg Chem 26(19):3167–3170
85. Rader RA, McMillin DR, Buckner MT, Matthews TG, Casadonte DJ, Lengel RK, Whittaker SB, Darmon LM, Lytle FE (1981) Photostudies of $[Cu(bpy)(PPh_3)_2]^+$, $[Cu(phen)(PPh_3)_2]^+$, and $[Cu(dmp)(PPh_3)_2]^+$ in solution and in rigid, low-temperature glasses – simultaneous multiple emissions from intraligand and charge-transfer states. J Am Chem Soc 103(19):5906–5912
86. Saito K, Arai T, Takahashi N, Tsukuda T, Tsubomura T (2006) A series of luminescent cu(i) mixed-ligand complexes containing 2,9-dimethyl-1,10-phenanthroline and simple diphosphine ligands. Dalton Trans 37(37):4444–4448

87. Tsubomura T, Takahashi N, Saito K, Tsukuda T (2004) Structure and luminescence of a dinuclear copper complex bridged by a diphosphine ligand. Chem Lett 33(6):678–679
88. Kaeser A, Mohankumar M, Mohanraj J, Monti F, Holler M, Cid J-J, Moudam O, Nierengarten I, Karmazin-Brelot L, Duhayon C, Delavaux-Nicot B, Armaroli N, Nierengarten J-F (2013) Heteroleptic copper(i) complexes prepared from phenanthroline and bis-phosphine ligands. Inorg Chem 52(20):12140–12151
89. Moudam O, Kaeser A, Delavaux-Nicot B, Duhayon V, Holler M, Accorsi G, Armaroli N, Séguy I, Navarro J, Destruel P, Nierengarten JF (2007) Electrophosphorescent homo- and heteroleptic copper(i) complexes prepared from various bis-phosphine ligands. Chem Commun 19(29):3077–3079
90. Kaeser A, Moudam O, Accorsi G, Séguy I, Navarro J, Belbakra A, Duhayon C, Armaroli N, Delavaux-Nicot B, Nierengarten JF (2014) Copper(i), silver(i) and gold(i) homoleptic bis-phosphine complexes. Eur J Inorg Chem. doi:10.1002/ejic.201301349
91. Yuasa J, Dan M, Kawai T (2013) Phosphorescent properties of metal-free diphosphine ligands and effects of copper binding. Dalton Trans 42(45):16096–16101
92. Femoni C, Muzzioli S, Palazzi A, Stagni S, Zacchini S, Monti F, Accorsi G, Bolognesi M, Armaroli N, Massi M, Valenti G, Marcaccio M (2013) New tetrazole-based cu(i) homo- and heteroleptic complexes with various p-p ligands: synthesis, characterization, redox and photophysical properties. Dalton Trans 42(4):997–1010
93. Cid J-J, Mohanraj J, Mohankumar M, Holler M, Accorsi G, Brelot L, Nierengarten I, Moudam O, Kaeser A, Delavaux-Nicot B, Armaroli N, Nierengarten J-F (2013) A stable and strongly luminescent dinuclear cu(i) helical complex prepared from 2-diphenylphosphino-6-methylpyridine. Chem Commun 49(9):859–861
94. Miller AJM, Dempsey JL, Peters JC (2007) Long-lived and efficient emission from mononuclear amidophosphine complexes of copper. Inorg Chem 46(18):7244–7246
95. Harkins SB, Peters JC (2005) A highly emissive Cu_2N_2 diamond core complex supported by a PNP^- ligand. J Am Chem Soc 127(7):2030–2031
96. Bergmann L, Friedrichs J, Mydlak M, Baumann T, Nieger M, Brase S (2013) Outstanding luminescence from neutral copper(i) complexes with pyridyl-tetrazolate and phosphine ligands. Chem Commun 49(58):6501–6503
97. Czerwieniec R, Yu J, Yersin H (2011) Blue-light emission of cu(i) complexes and singlet harvesting. Inorg Chem 50(17):8293–8301
98. Min J, Zhang Q, Sun W, Cheng Y, Wang L (2011) Neutral copper(i) phosphorescent complexes from their ionic counterparts with 2-(2'-quinolyl)benzimidazole and phosphine mixed ligands. Dalton Trans 40(3):686–693
99. Zink DM, Volz D, Baumann T, Mydlak M, Flügge H, Friedrichs J, Nieger M, Bäse S (2013) Heteroleptic, dinuclear copper(i) complexes for application in organic light-emitting diodes. Chem Mater 25(22):4471–4486
100. Mohanraj J, Armaroli N (2013) Luminophores and carbon nanotubes: an odd combination? J Phys Chem Lett 4(5):767–778
101. Carlos LD, Ferreira RAS, Bermudez VD, Julian-Lopez B, Escribano P (2011) Progress on lanthanide-based organic-inorganic hybrid phosphors. Chem Soc Rev 40(2):536–549
102. Crow JM (2011) Unsung elements. New Sci 210(2817):36–38
103. Eliseeva SV, Bunzli JCG (2011) Rare earths: jewels for functional materials of the future. New J Chem 35(6):1165–1176
104. Hanson DJ (2011) Concern grows over rare-earths supply. Chem Eng News 89(20):28–29
105. Kido J, Okamoto Y (2002) Organo lanthanide metal complexes for electroluminescent materials. Chem Rev 102(6):2357–2368
106. Kuriki K, Koike Y, Okamoto Y (2002) Plastic optical fiber lasers and amplifiers containing lanthanide complexes. Chem Rev 102(6):2347–2356
107. Montgomery CP, Murray BS, New EJ, Pal R, Parker D (2009) Cell-penetrating metal complex optical probes: targeted and responsive systems based on lanthanide luminescence. Acc Chem Res 42(7):925–937

108. van der Ende BM, Aarts L, Meijerink A (2009) Lanthanide ions as spectral converters for solar cells. Phys Chem Chem Phys 11(47):11081–11095
109. Bunzli J-CG, Eliseeva SV (2013) Intriguing aspects of lanthanide luminescence. Chem Sci 4(5):1939–1949
110. Jacoby M, Jiang J (2010) Securing the supply of rare earths. Chem Eng News 88(35):9–12
111. Nakamura E, Sato K (2011) Managing the scarcity of chemical elements. Nat Mater 10(3):158–161
112. Binnemans K (2009) Lanthanide-based luminescent hybrid materials. Chem Rev 109(9):4283–4374
113. Bunzli JCG, Eliseeva SV (2010) Lanthanide nir luminescence for telecommunications, bioanalyses and solar energy conversion. J Rare Earth 28(6):824–842
114. Quici S, Cavazzini M, Marzanni G, Accorsi G, Armaroli N, Ventura B, Barigelletti F (2005) Visible and near-infrared intense luminescence from water-soluble lanthanide [tb(iii), eu(iii), sm(iii), dy(iii), pr(iii), ho(iii), yb(iii), nd(iii), er(iii)] complexes. Inorg Chem 44(3):529–537
115. Leonard J, Nolan C, Stomeo F, Gunnlaugsson T (2007) Photochemistry and photophysics of coordination compounds: lanthanides. Top Curr Chem 281:1–43
116. Ronda C (ed) (2008) Luminescence. From theory to applications. Wiley-VCH, Weinheim
117. Höppe H (2009) Recent developments in the field of inorganic phosphors. Angew Chem Int Ed 48(20):3572–3582
118. Armelao L, Quici S, Barigelletti F, Accorsi G, Bottaro G, Cavazzini M, Tondello E (2010) Design of luminescent lanthanide complexes: from molecules to highly efficient photo-emitting materials. Coord Chem Rev 254(5–6):487–505
119. Parker D, Williams JAG (1996) Getting excited about lanthanide complexation chemistry. J Chem Soc Dalton Trans 18(18):3613–3628
120. Sabbatini N, Guardigli M, Lehn JM (1993) Luminescent lanthanide complexes as photochemical supramolecular devices. Coord Chem Rev 123(1–2):201–228
121. Eliseeva SV, Bunzli JCG (2010) Lanthanide luminescence for functional materials and bio-sciences. Chem Soc Rev 39(1):189–227
122. Rao CNR, Satishkumar BC, Govindaraj A, Nath M (2001) Nanotubes. ChemPhysChem 2(2):78–105
123. De Volder MFL, Tawfick SH, Baughman RH, Hart AJ (2013) Carbon nanotubes: present and future commercial applications. Science 339(6119):535–539
124. Accorsi G, Armaroli N, Parisini A, Meneghetti M, Marega R, Prato M, Bonifazi D (2007) Wet adsorption of a luminescent euiii complex on carbon nanotubes sidewalls. Adv Funct Mater 17(15):2975–2982
125. Maggini L, Traboulsi H, Yoosaf K, Mohanraj J, Wouters J, Pietraszkiewicz O, Pietraszkiewicz M, Armaroli N, Bonifazi D (2011) Electrostatically-driven assembly of MWCNTs with a europium complex. Chem Commun 47(5):1625–1627
126. Maggini L, Toma FM, Feruglio L, Malicka JM, Da Ros T, Armaroli N, Prato M, Bonifazi D (2012) Luminescent blooming of dendronic carbon nanotubes through ion-pairing interactions with an euiii complex. Chem Eur J 18(19):5889–5897
127. Eder D (2010) Carbon nanotube-inorganic hybrids. Chem Rev 110(3):1348–1385
128. Khlobystov AN, Britz DA, Briggs GAD (2005) Molecules in carbon nanotubes. Acc Chem Res 38(12):901–909
129. Wang Z, Shi Z, Gu Z (2010) Chemistry in the nanospace of carbon nanotubes. Chem Asian J 5(5):1030–1038
130. Pan X, Bao X (2011) The effects of confinement inside carbon nanotubes on catalysis. Acc Chem Res 44(8):553–562
131. Kalbac M, Kavan L, Gorantla S, Gemming T, Dunsch L (2010) Sexithiophene encapsulated in a single-walled carbon nanotube: an in situ Raman spectroelectrochemical study of a peapod structure. Chem Eur J 16(38):11753–11759
132. Chen W, Fan ZL, Zhang B, Ma GJ, Takanabe K, Zhang XX, Lai ZP (2011) Enhanced visible-light activity of titania via confinement inside carbon nanotubes. J Am Chem Soc 133(38):14896–14899

133. Fernandes GE, Tzolov MB, Kim JH, Liu ZJ, Xu J (2010) Infrared photoresponses from PBS filled multiwall carbon nanotubes. J Phys Chem C 114(51):22703–22709
134. Pradhan BK, Kyotani T, Tomita A (1999) Nickel nanowires of 4 nm diameter in the cavity of carbon nanotubes. Chem Commun (14):1317–1318
135. Sitharaman B, Rajamani S, Avti PK (2011) Time-resolved red luminescence from europium-catalyzed single walled carbon nanotubes. Chem Commun 47(5):1607–1609
136. Maggini L, Mohanraj J, Traboulsi H, Parisini A, Accorsi G, Armaroli N, Bonifazi D (2011) A luminescent host–guest hybrid between a euiii complex and MWCNTs. Chem Eur J 17(31):8533–8537
137. Yudasaka M, Ajima K, Suenaga K, Ichihashi T, Hashimoto A, Iijima S (2003) Nano-extraction and nano-condensation for C_{60} incorporation into single-wall carbon nanotubes in liquid phases. Chem Phys Lett 380(1–2):42–46
138. Zhang M, Yudasaka M, Bandow S, Iijima S (2003) Thermogravimetric analysis for the array of C_{60} molecules formed in single-wall carbon nanotube. Chem Phys Lett 369(5–6):680–683
139. Barber J (2009) Photosynthetic energy conversion: natural and artificial. Chem Soc Rev 38(1):185–196
140. Cogdell RJ, Gardiner AT, Cronin L (2012) Learning from photosynthesis: how to use solar energy to make fuels. Philos Trans R Soc A 370(1972):3819–3826
141. Faunce T, Styring S, Wasielewski MR, Brudvig GW, Rutherford AW, Messinger J, Lee AF, Hill CL, deGroot H, Fontecave M, MacFarlane DR, Hankamer B, Nocera DG, Tiede DM, Dau H, Hillier W, Wang L, Amal R (2013) Artificial photosynthesis as a frontier technology for energy sustainability. Energy Environ Sci 6(4):1074–1076
142. Balzani V, Credi A, Venturi M (2008) Photochemical conversion of solar energy. ChemSusChem 1(1–2):26–58
143. Olah GA, Goeppert A, Prakash GKS (2006) Beyond oil and gas: the methanol economy. Wiley-VCH, Weinheim
144. Balzani V, Moggi L, Manfrin MF, Bolletta F, Gleria M (1975) Solar-energy conversion by water photodissociation. Science 189(4206):852–856
145. Accorsi G, Armaroli N (2010) Taking advantage of the electronic excited states of [60]-fullerenes. J Phys Chem C 114(3):1385–1403
146. Tancini F, Monti F, Howes K, Belbakra A, Listorti A, Schweizer WB, Reutenauer P, Alonso-G6mez J-L, Chiorboli C, Urner LM, Gisselbrecht J-P, Boudon C, Armaroli N, Diederich F (2014) Cyanobuta-1,3-dienes as novel electron acceptors for photoactive multicomponent systems. Chem Eur J 20(1):202–216
147. Nierengarten JF, Langa F (eds) (2007) Fullerenes: principles and applications. Royal Society of Chemistry, Cambridge
148. Fukuzumi S, Ohkubo K (2013) Long-lived photoinduced charge separation for solar cell applications in supramolecular complexes of multi-metalloporphyrins and fullerenes. Dalton Trans 42(45):15846–15858
149. Gust D, Moore TA, Moore AL (2001) Mimicking photosynthetic solar energy transduction. Acc Chem Res 34(1):40–48
150. Guldi DM (2002) Fullerene-porphyrin architectures; photosynthetic antenna and reaction center models. Chem Soc Rev 31(1):22–36
151. Imahori H, Fukuzumi S (2004) Porphyrin- and fullerene-based molecular photovoltaic devices. Adv Funct Mater 14(6):525–536
152. Peeters E, van Hal PA, Knol J, Brabec CJ, Sariciftci NS, Hummelen JC, Janssen RAJ (2000) Synthesis, photophysical properties, and photovoltaic devices of oligo(p-phenylene vinylene)-fullerene dyads. J Phys Chem B 104(44):10174–10190
153. Sanchez L, Herranz MA, Martin N (2005) C_{60}-based dumbbells: connecting C_{60} cages through electroactive bridges. J Mater Chem 15(14):1409–1421
154. Armaroli N, Accorsi G, Gisselbrecht JP, Gross M, Krasnikov V, Tsamouras D, Hadziioannou G, Gomez-Escalonilla MJ, Langa F, Eckert JF, Nierengarten JF (2002) Photoinduced processes in fullerenopyrrolidine and fullerenopyrazoline derivatives substituted with an oligophenylenevinylene moiety. J Mater Chem 12(7):2077–2087

155. Armaroli N, Accorsi G, Felder D, Nierengarten JF (2002) Photophysical properties of the re(i) and ru(ii) complexes of a new C_{60}-substituted bipyridine ligand. Chem Eur J 8(10):2314–2323
156. Gutierrez-Nava M, Accorsi G, Masson P, Armaroli N, Nierengarten JF (2004) Polarity effects on the photophysics of dendrimers with an oligophenylenevinylene core and peripheral fullerene units. Chem Eur J 10(20):5076–5086
157. Armaroli N, Accorsi G, Clifford JN, Eckert JF, Nierengarten JF (2006) Structure-dependent photoinduced electron transfer in fullerodendrimers with light-harvesting oligophenylenevinylene terminals. Chem Asian J 1(4):564–574
158. Yoosaf K, Iehl J, Nierengarten I, Hmadeh M, Albrecht-Gary A-M, Nierengarten J-F, Armaroli N (2014) A supramolecular photosynthetic model made of a multiporphyrinic array constructed around a C_{60} core and a C_{60}–imidazole derivative. Chem Eur J 20(1):223–231

Part II
Interdisciplinary Lecture

19
The Discovery of the Higgs Particle

Giovanni Costa

Abstract About 50 years ago, the Higgs mechanism was introduced in the Standard Model of particle physics in order to reconcile the presence of massive particles with the requirement of gauge invariance.This mechanism predicts the existence of the Higgs boson and, while the Standard Model has been successfully tested by all experimental data, this element was missing and hunted for since long time. Only the construction of the most complex experimental facilities of the Large Hadron Collider at CERN led in 2012 to the discovery of the Higgs particle by two independent experimental teams. The main steps of these investigations are reviewed; the properties of the Higgs particle, its production and decay modes are briefly discussed; moreover, open questions about possible extensions of the Standard Model are mentioned together with their experimental implications.

Keywords Particle and nuclear physics • Theory of particle physics

19.1 Introduction

On 4 July 2012 CERN announced the discovery of a new particle with mass about 125 GeV/c^2, which was interpreted as the Higgs boson. The existence of this particle was predicted almost 50 year ago and this prediction led to long experimental searches at the existing large accelerators, and to the construction of the Large Hadron Collider (LHC) at CERN in Switzerland.

The test of the existence of the Higgs boson is a crucial point for the confirmation of the so-called Standard Model (SM), the theoretical model that describes the fundamental interactions (electromagnetic, strong and weak) of the elementary

G. Costa (✉)
Department of Physics, University of Padova, Via Marzolo 8, 35131 Padova, Italy
e-mail: costa@pd.infn.it

B. Di Bartolo et al. (eds.), *Nano-Structures for Optics and Photonics*,
NATO Science for Peace and Security Series B: Physics and Biophysics,
DOI 10.1007/978-94-017-9133-5_19,

particles. In fact, the Higgs particle is the missing piece of the SM that plays the essential role of generating the masses of all the other elementary particles. For this reason, if the Higgs boson were not detected, one should have to look for new ingredients and replace the SM with a new and possibly more complicated theory.

Before considering the new experimental results, we shall recall very briefly the properties of the fundamental particles and their interactions. Then we shall examine the structure of the SM, which is a Quantum Field Theory including strong, weak and electromagnetic interactions. We disregard the gravitational interactions which are not relevant for our considerations; their effects in particle physics become important only at extremely high energies, not reachable in laboratories. In this lecture, we shall concentrate on the analysis of the electromagnetic and weak interactions, which are unified in the SM model and called *electroweak*.

In particular, we shall discuss the idea of the spontaneous symmetry breaking (SSB), which was essential for the construction of the SM, and which led to the prediction of the Higgs boson.

After these theoretical considerations, we shall give some information about the experimental searches. Finally, we shall give an outlook on the future investigations.

19.2 Fundamental Particles and Their Interactions

It is useful to make a very short introduction and remind what are the *fundamental elementary particles* and their interactions. The present situation is represented in Table 19.1, where we have listed the quarks (u, d, c, s, t, b) and the leptons (e, μ, τ and three kind of neutrinos), and specified their electric charges Q (in units of the proton charge); all these particles are *fermions* with spin $1/2$. They are grouped into three families; the matter around us is made only by the member of the first family: the quarks *up* and *down* are the constituents of neutrons and protons (each made of three quarks) and, together with the electrons, form all the atoms. The other two families are replicas of the first one; they appear in high energy interactions in states which rapidly decay into those of the first family. (For a little more detail see e.g. Ref. [1]). We remind that, for each particle, there is a corresponding *antiparticle* with opposite electric charge and opposite internal quantum numbers.

The particles listed in Table 19.1 are characterized by their interactions, and Table 19.2 summarizes the fundamental interactions and their messengers: photons, gluons and weak vector bosons. All electrically charged particles are subject to electromagnetic (EM)interactions; all leptons to weak interactions; quarks , besides EM and weak interactions, participate in strong interactions. All interactions (with

Table 19.1 Quarks and leptons

Family	$Q = 2/3$	$Q = -1/3$	$Q = -1$	$Q = 0$
1st	u (up)	d (down)	e (electron)	ν_e (e-neutrino)
2nd	c (charm)	s (strange)	μ (muon)	ν_μ (μ-neutrino)
3rd	t (top)	b (bottom)	τ (tau)	ν_τ (τ-neutrino)

Table 19.2 Fundamental interactions

Interaction	Quantum	Spin	Mass
EM	γ (photon)	1	0
Weak	$W^{\pm}$ and Z^0	1	Big
Strong	Gluon	1	0
Gravitational	Graviton	2	0

the exception of the gravitational ones) are mediated by vector particles, i.e. *bosons with spin 1*. Specifically, the EM and the strong interactions are mediated by massless vector bosons, photons and gluons, respectively. The weak interactions present a peculiar feature: the quanta that mediate the interactions (the weak vector bosons) are massive and very heavy.

The gravitational interactions, mediated by massless spin 2 bosons, called gravitons, are very important at macroscopic scales; however they are extremely weak with respect to the other types of interactions, and they play practically no role in the domain of elementary particles, at least in the energy range reached in laboratories.

The special property of the bosons that mediate weak interactions is related to the fact that their range is extremely short. This can be argued from a very simple argument, based on the uncertainty Heisenberg principle:

$$r \approx c\Delta t \approx \frac{c\hbar}{\Delta E} \approx \frac{\hbar}{Mc} \tag{19.1}$$

where r is the range of the interaction and M is the mass of the weak boson. With the value $r \approx 10^{-17}$ cm, one gets $M \approx 80\,\mathrm{GeV/c^2}$ ($\hbar = 6.6 \times 10^{-22}$ MeV.s; 1 GeV = 10^3 MeV = 10^9 eV).

The appropriate theoretical scheme for the treatment of the fundamental particles and their interactions is a relativistic quantum field theory (QFT), in which all particles, not only the quanta mediating the interactions, but also quarks and leptons (fermion particles with spin $1/2$) are described by fields. What we call particles are the quanta of the fields.

Electromagnetic interactions are described by *Quantum Electrodynamics* (QED), which is a very successful theory that reproduces with great accuracy all electromagnetic processes.

Strong interactions are described by *Quantum Chromodynamics* (QCD), that provides a very good quantitative estimates at high energies; its applications at lower energy are somewhat limited by the complexity of the calculations, due to the fact that the interactions are much stronger than the electromagnetic ones and the approximation methods employed in QED are not reliable at low energies.

19.3 The Standard Model

The field theory which includes QED, QCD and the weak interactions is called *Standard Model* (SM), and it is based on the symmetry group $SU(3)\times SU(2)\times U(1)$.

In this lecture we shall limit ourselves to the analysis of the weak interactions. The construction of a consistent field theory was rather complicated, due to the fact that the massive nature of the mediating vector bosons forbids a simple extension of QED. New ideas were necessary for building the final theory, which is the sector of the Standard Model related to the symmetry $SU(2) \times U(1)$, based on the unification of weak and electromagnetic (*electroweak*) interactions. The construction of this theoretical model was based on the works of Glashow, Salam and Weinberg [2–4].

The first phenomenological theory of weak interactions was due to Fermi in 1934: it was based on a four-fermion contact (zero range) interaction. This theory was expressed in a general form by Feynman and Gell-Mann with the Hamiltonian [5]:

$$H_w = \frac{G_F}{\sqrt{2}} J^\mu J_\mu \tag{19.2}$$

where J^μ is a four-vector (electrically charged) current and G_F is the coupling constant introduced by Fermi. Later on the idea of *vector boson* messengers (denoted by $W^\pm$) coupled to the vector current was expressed in analogy with the formulation of QED, in terms of the famous Feynmam graphs. For example, the electron-proton scattering

$$e^- + P \to e^- + P \tag{19.3}$$

is described, to lowest order approximation, in terms of a photon exchange, which means that a photon, emitted by the electron, is absorbed by the proton, and vice versa. Similarly, one can describe the neutron beta decay

$$N \to P + e^- + \bar{\nu}_e \tag{19.4}$$

in terms of the exchange of a vector boson: the neutron, transforming into a proton, emits a W^- boson that decays into an electron and an antineutrino. The exchanged boson has to be charged since we are dealing with charged currents. In 1983, the existence of neutral weak currents was discovered , so that also a neutral vector boson, denoted by Z^0, was required. Both charged and neutral bosons were discovered in 1983 at CERN by the group leaded by Carlo Rubbia [6–8].

The fact that the messengers of the weak interactions are massive gives rise to a serious difficulty; while it is very simple to calculate the lower order contributions (corresponding to the so-called tree level Feynman graphs), the higher order corrections are not calculable. This difficulty arises because the theory is not renormalizable. The requirement of *renormalization*, which is fulfilled by QED and QCD, must be satisfied by any QFT, since this property is necessary in order to be able to go beyond the lowest approximation and evaluate higher order corrections.

Let us consider the case of QED, specifically the field theory of electrons and photons. The property of this theory of being renomalizable is based on the fact that it possesses a symmetry, the so-called *gauge symmetry*. Here we limit ourselves to some qualitative considerations, and define the gauge symmetry by a simple example. The Lagrangian of a system of electrons and photons is given by (here and in the following we consider, for the sake of simplicity, the Lagrangian density):

$$\mathscr{L} = \bar{\psi}(x)(i\gamma^{\mu}\partial_{\mu} - m)\psi(x) - \frac{1}{4}F^{\mu\nu}F_{\mu\nu} + ej^{\mu}(x)A_{\mu}(x) \tag{19.5}$$

where $\psi(x)$ is the electron field, $F_{\mu\nu} = \partial_{\mu}A_{\nu} - \partial_{\nu}A_{\mu}$ is the electromagnetic tensor (A_{μ} is the EM four-vector potential) and

$$j^{\mu}(x) = \bar{\psi}(x)\gamma^{\mu}\psi(x) \tag{19.6}$$

is the electron current coupled to $A_{\mu}(x)$.

The above Lagrangian contains only two free parameters: the mass m of the electron and its electric charge e; their values are not predicted by the theory and are fixed by experiments.

The Lagrangian of the free (i.e. non interacting) electrons is given by the first part of $\mathscr{L}$:

$$\mathscr{L}_0 = \bar{\psi}(x)(i\gamma^{\mu}\partial_{\mu} - m)\psi(x) \tag{19.7}$$

which is invariant under the phase transformation

$$\psi(x) \rightarrow exp(i\alpha)\psi(x) \tag{19.8}$$

where α stands for an arbitrary real constant. However, $\mathscr{L}_0$ is no longer invariant if the constant α is replaced by a function $\alpha(x)$ of the space-time variable x:

$$\psi(x) \rightarrow exp(i\alpha(x))\psi(x) \tag{19.9}$$

The invariance is restored with the introduction of a four-vector field, as in Eq. (19.5), which transforms as;

$$A_{\mu}(x) \rightarrow A_{\mu}(x) + \frac{1}{e}\partial_{\mu}\alpha(x) \tag{19.10}$$

The invariance under the transformations (19.9) and (19.10) is called *gauge invariance*.

All processes among electrons (together with their anti-particles, the positrons) and photons can be evaluated with the powerful method of the Feynman graphs. Higher order corrections imply the evaluation of Feynman graphs which contain one or more loops with *internal lines* that correspond to the propagation of *virtual* electrons, positrons and photons. In principle, the momenta of the virtual particles can range from zero to infinity, and then one should perform integrals extended to infinite ranges: most often it happens that these integrals are divergent. How can one cure this divergence? One has to understand its origin. It is reasonable to think that it is due to our approximation. Going to very high energies, other new unknown interactions may appear, the EM coupling may vary, the particles may not be point-like as they are considered in QFT, etc. We believe that the correct description of the situation can eliminate all the divergence, so that the appearance of divergent terms in the present theory is simply due to our ignorance. Then we can proceed as follows: we introduce a cut-off, i.e. we restrict the integration to a finite upper limit Λ. One can show that it is possible to separate in a unique way the terms that contain Λ by those that are independent. The former become divergent in the limit $\Lambda \rightarrow \infty$, while the latter remain finite. Due to gauge invariance one can show that there are only two Λ-dependent terms that can be isolated and included into two constants: they can be reabsorbed by the parameters m and e, that become *renormalized* and can be identified with the experimental values. The finite terms, independent of Λ, represent the corrections given by the higher order calculations.

Adopting this procedure, it was shown that all QED calculations are in perfect agreement with the experimental values; in fact, QED is the most satisfactory quantum field theory which is verified with extremely great accuracy. (See e.g. Ref. [9]).

Unfortunately, there are theories that are not renormalizable: in this kind of theories the number of Λ – dependent terms increases in going to higher and higher order, and the resulting arbitrary constants cannot be reabsorbed by the very limited number of free parameters (such as masses and coupling constants). As a consequence, these theories cannot give any predictable results.

As quoted above, the renormalization of QED is made possible because the theory is gauge invariant. The gauge transformations correspond to a local symmetry, i.e. independent transformations occur in different space-time points or, in other words, the transformation parameters are functions of space-time. We remark that the requirement of gauge invariance has very strong effects. We repeat that, in a field theory of free electrons, the gauge invariance requires the presence of a massless vector field, which is interpreted as the EM field that is, in particle language, the photon. This property is generalized to other QFT's; gauge invariance requires, in general, the introduction of a number of massless vector fields.

19.4 The Prediction of the Higgs Boson

The theoretical prediction of the existence of a scalar particle (spin zero and positive parity, denoted by $J^P = 0^+$), later called Higgs boson, is related to the inclusion of the weak interactions in the SM. The problem is how to implement gauge symmetry in the case of weak interactions, and reconcile gauge symmetry with massive vector fields. In fact, the mass term of a vector field would have the form

$$M^2 A^\mu(x) A_\mu(x) \tag{19.11}$$

which, in general, is not invariant under gauge transformations. With the exception of an Abelian symmetry, as in the case of the one-parameter group $U(1)$, massive terms of the type given above are not allowed by gauge invariance.

This problem can be solved making use of the symmetry breaking mechanism (SSB): it was considered in the frame of condensed matter physics, specifically in the theory of superconductivity, by Anderson [10], and in elementary particle physics by Nambu and Jona-Lasinio [11].

Consider a simple example of classical mechanics: suppose that one has a straight rod that stands vertically. It looks the same from any horizontal direction, so that the rotational symmetry around the rod is perfect. But if one squeezes it down, it will bend in some arbitrary direction and the symmetry is lost.

A second example is that of a ferromagnet. As a simple model, let us consider the case in which the Lagrangian that describes the spin-spin interactions is rotationally invariant (at high temperature): all spins are oriented at random, so that no magnetization is present. If we put the ferromagnet in a uniform magnetic field, the spins align along the direction of the field. However, even in the absence of a magnetic field, below a given critical temperature T_c, the spins get aligned along one direction. Then the ground state is no longer rotational invariant and, therefore it does not display the symmetry of the Lagrangian: there is a SSB but, since any direction of alignment is equally good, one says that the symmetry is not lost but it is hidden. At the critical temperature T_c, there is a phase transition from the symmetric state in which all spins are directed at random, to a state in which all spins are aligned. The breaking is not of the so-called Wigner-Weyl kind, that implies the introduction of a symmetry breaking term in the Lagrangian.

This mechanism is based on the fact that, starting with a physical system which possesses a symmetry (in other words, its Lagrangian is invariant under the transformations of a symmetry group $\mathscr{G}$), it may happen that the equations of motion derived from the Lagrangian have a lower symmetry (their symmetry group is a subgroup of $\mathscr{G}$). In general this symmetry breaking causes a transition between two different phases. This situation can be realized in a QFT, where all fermions and vector fields are massless beyond some energy scale; below the energy scale of the phase transition, some of them acquire mass, and this corresponds to a system with a lower symmetry.

We avoid the mathematical formalism and, in the case of the SM, we limit ourselves to the presentation of the results. In general, for any physical system, one can write a Lagrangian which contains a term T that can be interpreted as the kinetic energy, and a term V as the potential energy:

$$L = T - V \tag{19.12}$$

Also in the case of a field theory, one can individuate two terms with similar interpretations. In order to give a specific example, we consider a toy model based on a scalar complex field. Its Lagrangian can be written as

$$\mathscr{L} = (\partial_\mu \phi)^* (\partial^\mu \phi) - \mu^2 \phi^* \phi - \lambda (\phi^* \phi)^2 \tag{19.13}$$

This Lagrangian is invariant under the global transformations of the group $U(1)$

$$\phi(x) \to e^{i\alpha}\phi(x), \qquad \phi^*(x) \to e^{-i\alpha}\phi^*(x) \tag{19.14}$$

It is useful to define the last two terms as the *potential* of the system

$$V(\phi) = \mu^2|\phi|^2 + \lambda|\phi|^4 \tag{19.15}$$

In the case $\mu^2 > 0$, the potential has only one minimum, i.e. $|\phi|_{min} = 0$, that corresponds to the ground state, i.e. the vacuum state, where no particles are present. However, in the case $\mu^2 < 0$, a situation that can be realized in QFT, one gets a big change: the minimum of the previous case becomes a local maximum, and the effective minimum corresponds to

$$|\phi|^2_{min} = -\frac{\mu^2}{2\lambda} \tag{19.16}$$

In fact, the above relation represents a continuum set of solutions, since the phase of ϕ at the minimum is completely arbitrary

$$\phi_{min} = |\phi|_{min}\, e^{i\gamma} \tag{19.17}$$

The case $\mu^2 < 0$ is not unrealistic, since the renormalization procedure replace the constant μ^2 by a function of energy which can change sign at a certain energy scale.

In conclusion, the above solution represents an infinite set of degenerate vacuum states: the choice of a particular vacuum state, corresponding to a fixed value of the phase γ in Eq. (19.17), gives rise to SSB since the $U(1)$ symmetry is lost. In other words, there a continuous set of minima that represents possible ground states and the system goes into one of them. Let us consider, for the sake of simplicity, the case $\gamma = 0$. Then, by writing

$$\phi(x) = \frac{1}{\sqrt{2}}[\phi_1(x) + i\phi_2(x)] \tag{19.18}$$

the minimum of the potential is given by

$$(\phi_1)_{min} = \sqrt{\frac{-\mu^2}{\lambda}} \equiv v \tag{19.19}$$

The quantity v is called *vacuum expectation value* (v.e.v.) of the scalar field. It is convenient to redefine the two fields ϕ_1 and ϕ_2 as

$$\phi_1(x) = \eta(x) + v, \qquad \phi_2(x) = \xi \tag{19.20}$$

Then the Lagrangian $\mathcal{L}$ takes the form (neglecting the explicit expressions of the terms that represent the couplings among the different fields)

$$\mathcal{L}' = \frac{1}{2}\partial_\mu \eta \partial^\mu \eta + \frac{1}{2}\partial_\mu \xi \partial^\mu \xi - \frac{1}{2}m^2\eta^2 + coupling \;\; terms \tag{19.21}$$

where

$$m^2 = -2\mu^2 \tag{19.22}$$

Let us compare the two expressions of $\mathcal{L}$ and $\mathcal{L}'$. The first expression, considered in the case $\mu^2 > 0$, contains a complex scalar field or, equivalently, two real scalar fields ϕ_1 and ϕ_2 with squared mass μ^2. The second expression, corresponding to the case $\mu^2 < 0$, contains a real scalar field η with squared mass $m^2 = -2\mu^2$, and a *massless* real scalar field ξ.

If we consider the case in which the parameter μ^2 changes continuously with the energy scale from a positive to a negative value, we can describe the change of the behaviour of the system as a *phase transition*. The transition occurs from a state that is invariant under the transformation (19.14) of the group $U(1)$ to a new state in which the invariance is lost. This is a simple example of SSB; it is accompanied by the appearance of a massless scalar boson called *Goldstone boson*. This is a specific case of an important general result due to Goldstone and Nambu: the SSB implies always the appearance of massless scalar bosons, called Nambu-Goldstone bosons (NG) [12, 13].

In general, if the SSB implies a breaking that takes the system from a state invariant under the symmetry group $\mathcal{G}$ to one invariant under a subgroup $\mathcal{H}$ of $\mathcal{G}$, the number of NG bosons is equal to the difference $N - N'$ between the number N of generators of $\mathcal{G}$ and the number N' of generators of $\mathcal{H}$.

A second important results is that, if the SSB process occurs in the case of a gauge field theory, then a set of massless vector fields, equal to the number of the NG bosons, acquires masses different from zero.

We can go on with our toy model, promoting the global symmetry (19.14) to a local one

$$\phi(x) \to e^{i\alpha(x)}\phi(x), \qquad \phi^*(x) \to e^{-i\alpha(x)}\phi^*(x) \tag{19.23}$$

The invariance under these transformations requires the introduction of the *covariant derivative*

$$\partial_\mu \phi(x) \to D_\mu \phi(x) = [\partial_\mu + ieA_\mu(x)]\phi(x) \tag{19.24}$$

where $A_\mu(x)$ is the vector field with the transformation property defined by Eq. (19.10). Then the Lagrangian (19.13) is replaced by:

$$\mathcal{L}(\phi, A_\mu) = (D_\mu \phi)^* D^\mu \phi - V(\phi) - \frac{1}{4}F_{\mu\nu}F^{\mu\nu} \tag{19.25}$$

where, as usual, $F_{\mu\nu} = \partial_\mu A_\nu - \partial_\nu A_\mu$. The above equation can be rewritten in the form

$$\mathscr{L}(\phi, A_\mu) = \partial_\mu \phi^* \partial^\mu \phi - V(\phi) - \frac{1}{4} F_{\mu\nu} F^{\mu\nu} - e j^\mu A_\mu \tag{19.26}$$

in terms of the current

$$j^\mu = i(\phi^* \partial^\mu \phi - \phi \partial^\mu \phi^*) \tag{19.27}$$

Eq. (19.26) corresponds, for $\mu^2 > 0$, to the Lagrangian of the scalar electrodynamics, i.e. of a charged scalar field coupled with the electromagnetic field. Since the first two terms are the same as those in Eq. (19.13), the mechanism of the SSB can occur also in the present case. Making use of the same notation of Eq. (19.21), the Lagrangian can be expressed in the form (for details see e.g. Ref. [14])

$$\mathscr{L}' = \frac{1}{2} \partial_\mu \eta \partial^\mu \eta - \frac{1}{2} m^2 \eta^2 - \frac{1}{4} F_{\mu\nu} F^{\mu\nu} + \frac{1}{2} e^2 v^2 A_\mu A^\mu + coupling \;\; terms \tag{19.28}$$

We point out an important result: the vector field A_μ, which is massless before SSB, acquires a mass different from zero, specifically

$$M = ev \tag{19.29}$$

and moreover the massless field ξ has disappeared. One can check that there is the same number of degrees of freedom before and after the phase transition. In the first phase, the field A_μ is massless and transversal (two degrees of freedom), while in the second phase, it acquires a mass and a longitudinal component (three degrees of freedom). This additional component corresponds to the ξ field that becomes unphysical and it is "eaten" by the vector field: in other words, the Goldstone boson disappears and it is replaced by the longitudinal component of the vector field. On the other hand, the η field is physical and and with a mass different from zero: its quanta are *scalar* (i.e. spin zero and positive parity) bosons.

After this example, we consider the general case limiting ourselves to quote the main results without giving any proof. In the case of a local gauge theory in which the Lagrangian is invariant under the local gauge group $\mathscr{G}$ with N generators, there are N massless vector fields. In the process of SSB, in which the unbroken symmetry corresponds to the subgroup $\mathscr{H}$ of $\mathscr{G}$ with N' generators, the $N - N'$ Goldstone bosons are eaten by the $N - N'$ vector fields that get masses different from zero, while the remaining N' vector fields remain massless.

Now we apply these considerations to the case of the electroweak Lagrangian. In order to accommodate the phenomenological situation, we need to start from a Lagrangian which is invariant under the gauge transformations of the group

$$SU(2) \times U(1) \tag{19.30}$$

This group has four generators (since the groups $SU(2$ and $U(1)$ have dimensions 3 and 1, respectively) so that the group transformations depend on four independent parameters, which are functions of the space-time coordinate x. The gauge invariance requires that there are four massless vector fields, corresponding to the four generators of the group. In order to generate the appropriate SSB mechanism, one need to introduce a four-component scalar field, with two neutral and two electrically charged components. After the SSB, the symmetry is reduced to

$$U(1)_{EM} \tag{19.31}$$

where we distinguish the final group $U(1)$ from that related to the initial Lagrangian (because it corresponds to a different generator); specifically, it is the gauge group of QED.

In the process of SSB, three out of the four scalar fields are absorbed by three vector fields that acquire mass, while a fourth vector field remains massless (it is the vector potential A_μ of QED that represents the photon). The remaining neutral scalar component represents what we call the *Higgs field* and its quanta are the *Higgs bosons*. The Higgs field acquires a vacuum expectation value (v.e.v.) different from zero that fills all the space and represents the ground state; it is its coupling to the other massless particles (quarks, leptons and vector bosons) that generates their masses.

After the SSB, the Lagrangian of the SM describes separately QED mediated by a massless boson (the photon) and the weak interaction part that includes three massive vector bosons, which are the famous $W^{\pm}$ and Z^0 particles. They were discovered at CERN in 1983 [6–8]. The experimental values of the masses of these bosons are

$$M_W = 80.385 \pm 0.015\,\mathrm{GeV/c^2}, \quad M_Z = 91.1876 \pm 0.0021\,\mathrm{GeV/c^2}, \tag{19.32}$$

in perfect agreement with the theoretical predictions.

On one hand, it was shown by 't Hooft and Veltman [15, 16] that the SM theory is renormalizable and, on the other hand, that all experiments have confirmed all its predictions.

Concluding this section, we would like to add a brief comment on the mass production by the Higgs field. The origin of the mass can be described in the following way, in the frame of the standard cosmological model. Few instants after the Big Bang, when the temperature of the Universe began to decrease, the Higgs field (that was permeating all the Universe) "crystallized" (in the sense that it acquired v.e.v. different from zero) and the electroweak symmetry was broken. Then the Higgs boson became massive and it produced the masses of the other particles. While photons remained massless, and propagated freely around the Universe, the weak vector bosons acquired mass and became very heavy; as a consequence, the range of the weak force became very small, and EM and weak interactions became very different from each others. At the same time, also quarks and leptons acquired their masses. One should assume that all the masses of the elementary particles are generated by the Higgs mechanism. However, one should distinguish between the *truly elementary particles*, like quarks and leptons, that are the fundamental constituents of matter, from the *composite particles*, like protons and neutrons, and the multitude of the so-called elementary particles that are all composite of quarks. As a matter of facts, one has to take into account that the mass generated by the Higgs mechanism is a small fraction of the masses of the composite particles. Let us consider, for example, the case of protons and neutrons. They are made of three quarks that are very light, but there is a lot of processes going on in their inside: production and absorption of gluons and quark-antiquark pairs. These processes produce a large amount of energy, and it is this energy that gives the main contribution to the masses of protons and neutrons.

19.5 The Discovery of the Higgs Boson

As already said in the introduction, the existence of the so-called Higgs boson was predicted, independently and almost simultaneously in 1964, by the groups: F. Englert and R. Brout (in August) [17], P.W. Higgs (in October) [18,19], G. Guralnik, C.R. Hagen and T.W.B. Kibble (in November) [20]. However, the name of Higgs boson was assigned to the new particle by Steven Weinberg and adopted up to now by everybody. Englert and Higgs have been awarded the 2013 Nobel in Physics for their theoretical predictions.

The discovery was announced at CERN on July 4, 2012; the speakers were Fabiola Gianotti, spokesperson of the ATLAS experiment [21], and by Joseph Incandela, spokesperson of the CMS experiment [22].

The discovery was done at the Large Hadron Collider (LHC) at CERN in Switzerland, but searches for the Higgs particle were conducted previously also at the Large Electron-Positron Collider (LEP) at CERN, and at the Tevatron, at Fermilab in the United States, as we briefly mention in the following.

LEP. The first extensive search for the Higgs boson was carried out at LEP, the Large Electron-Positron collider with a 27 km circumference and 209 GeV maximum energy. It provided stringent tests of the SM, and very accurate determination of its parameters. It found no conclusive evidence of the Higgs boson, and this implied that, if it existed, it would be heavier than $114\,\text{GeV}/c^2$. LEP was shut down and dismantled in 2000, to be replaced by LHC.

Tevatron. It is a circular proton-antiproton accelerator built at Fermiab, Batavia, Illinois. Its circumference is 6.3 km long; its maximun energy is $1.8\,\text{TeV} = 1.8 \times 10^3\,\text{GeV}/c^2$. It started operating in 1983 and it obtained lot of interesting results, in particular, the first observation of the *top*, the heaviest quark (mass about $175\,\text{GeV}/c^2$). It showed a small excess of events possibly indicating a signal in the mass range $115\text{–}140\,\text{GeV}/c^2$, but the result was not significant for the Higgs boson search. It ceased operations in 2011.

LHC. The Large Hadron Collider is at present the world largest and most powerful particle accelerator; it was built, in particular, for confirming or excluding the existence of the Higgs boson, but its aims go much further than this. It was installed in the place of LEP. Inside the tunnel, two beams of charged protons (or even heavy ions) travel close to the speed of light, in opposite directions in two separate beam pipes, before they are made to collide. They are guided around the ring by strong magnetic field maintained by superconducting electromagnets. The system is kept at very low temperature, close to the absolute zero. The maximum center-of-mass energy has been 7 TeV. LHC was shut down at the end of 2012 for upgrading; it will reopen at the beginning of 2015 with the energy of 14 TeV.

The Higgs boson was discovered by means of two independent very big detectors that gave their name to the two experiments cited above:

ATLAS (A Toroidal LHC Apparatus) and **CMS** (Compact Muon Solenoid). They have a cylindrical form along the direction of the proton beams; the collisions of the two beams occur in the central region. They consist of different layers of dense materials made of silicon (the particles traveling through produce tiny signals that are amplified and registered); they contain hadronic calorimeters that measure the energies of electrons and photons with great accuracy. The detectors are placed in strong magnetic fields which bend the trajectories of the charged particles; from the curvature one can determine their momenta. Both detectors are able to detect all particles, except neutrinos, and one can identify the nature of the particles created in the collisions. There is a huge number of collisions: automatic screening selects the events that have specific features. A very large amount of data have been collected, and advanced computing facilities have been necessary. The data have been distributed to all the nodes of the *GRID* (Global Research Information Database), which include 170 institutions in 36 countries.

In order to detect the Higgs boson, one need to know its production mechanism and its decay modes. In the following, we consider the properties that are expected for the Higgs boson within the SM: most of its properties are predicted by the model, but its mass is not determined and it is arbitrary.

On one hand, the coupling constants of the Higgs boson with the photon and the vector bosons (W and Z) are predicted by the SM. On the other hand, the theory does not predicts the values of its coupling constants with quarks and leptons, but we know that they are proportional to their masses. Therefore, with this information, one can determine the relative weights of the production processes of the Higgs boson, and of its decay modes.

Even if the value of the Higgs mass cannot be predicted within the SM, some estimates were obtained indirectly from radiative corrections of the SM parameters, due to Higgs loops in the higher order graphs. In fact, from the precision electroweak measurements the upper limit $M_H < 161\,\text{GeV}$ was obtained.

The value of the Higgs mass reported by the CERN experiment is

$$M_H = 125.9 \pm 0.4\,\text{GeV/c}^2 \tag{19.33}$$

The Higgs bosons can be produced at LHC by different mechanisms of which we quote here the main ones.

1. *Gluon fusion*. The most likely process is that two gluons, emerging from the two protons, combine in the collision producing a fermion loops, by which a Higgs boson is produced. Since the Higgs coupling is proportional to the fermion mass, the relevant process is that corresponding to a *top* or a *bottom*, that are the heaviest quarks. This kind of process is the dominant contribution for the Higgs production.
2. *Weak boson fusion*. A pair of quarks, emerging form the two colliding protons, can exchange a virtual W or Z boson, which in turn emits a Higgs boson. This process is the second most important for the production at LHC.
3. *Higgs Strahlung*. In a proton-proton collision a quark-antiquark pair can produce a W or a Z boson which, if it carries sufficient energy, can emit a Higgs boson. It would be the third largest production process.

The Higgs boson is very heavy with respect to the other particles; therefore it can decay in very many channels and its lifetime is extremely short. In fact, quantum mechanics says that, if it is possible for a particle to decay into a set of lighter particles, then it will do so. The SM predicts the relative branching ratios of the different decay modes of the Higgs boson, and a value of about 1.6×10^{-22} s is estimated for its mean lifetime.

Also for the decays, we quote the principal modes.

1. *Decays into a fermion-antifermion pair*. As a general rule, the Higgs is more likely to decay into heavy then light fermions. Then the most common decay is into a $b - \bar{b}$ quark pair; the decay is into $\tau - \bar{\tau}$ lepton pair is also relevant.
2. *Decays into a vector boson pair*. The Higgs boson can split into a pair of weak vector bosons. The most likely decay mode is into a pair of W bosons; the decays of the two W's into quarks are difficult to distinguish from the background, and the decays into leptons cannot fully be reconstructed, because two of them are neutrinos. A cleaner signal is given by the decay into a pair of Z-bosons, if each boson subsequently decays into a pair of charged leptons (electrons or muons).
3. *Decays into massless vector bosons*. Decays into massless vector bosons (i.e. gluons or photons) are also possible, but require an intermediate loop of virtual heavy quarks (top or bottom) or weak vector bosons. The decay into a photon pair is much rarer but it is relevant for experimental searches for the Higgs boson, because the energies and momenta of the two photons can be measured very precisely, giving an accurate reconstruction of the mass of the decaying particle.

As said above, LHC has been shut down for upgrading and it will reopen at the beginning of 2015. In the meanwhile an extensive analysis will be carried out to check if the new particle possesses all the properties predicted by the SM, such as spin, parity and coupling constants.

The analysis of the decays $H \to \gamma + \gamma$, $H \to W + W$ and $H \to Z + Z$ gives evidence for the spin 0 nature of the Higgs boson, with positive parity being strongly preferred [23]. Also the relative weights of the Higgs couplings to other particles are in agreement with the SM expectations [24].

Lot of new data will be obtained at the higher energy available (14 TeV) when LHC restarts. It will be possible to determine if the Higgs boson is the only specimen of scalar fundamental particle, or if it has some partners.

In fact, there are theoretical motivations that the SM has to be enlarged, with the implementation of a new kind of symmetry, specifically the *supersymmetry*, which is a symmetry between fermions and bosons. It implies the existence of partners of the fundamental particles: the supersymmetric partners of quark and leptons with spin zero, those of the the vector bosons γ, W and Z with spin $1/2$. Moreover, the minimal supersymmetric extension of the SM predicts the existence of five kinds of Higgs bosons (three neutral and two charged) and of their spin $1/2$ partners.

Finally, we are still faced with the problem of *dark matter*, of which we do not know the composition. We hope that the future experiments at the upgraded LHC will enlighten this mistery. Another theme will be the search for new particles related to the hypothesis of *extra dimensions*. In conclusion, we expect lot of exciting surprises from the future of LHC.

References

1. Costa G (1997) Unification of the fundamental interactions: problems and perspectives. In: Di Bartolo B, Kyrkos S (eds) Spectroscopy and dynamics of collective excitations in solids. NATO ASI series, vol 356. Springer, Boston, pp 581–597
2. Glashow SL (1961) Partial-symmetries of weak interactions. Nucl Phys 22:579–588
3. Weinberg S (1967) A model of leptons. Phys Rev Lett 19:1264–1266
4. Salam A (1968) Weak and electromagnetic interactions. Conf Proc C680519:367–377
5. Feynman RP, Gell-Mann M (1958) Theory of the Fermi interaction. Phys Rev 109:193–198
6. UA1 Collaboration (1983) Experimental observation of isolated large transverse energy electrons with associated missing energy at s=540 gev. Phys Lett B 122(1):103–116
7. UA2 Collaboration (1983) Observation of single isolated electrons of high transverse momentum in events with missing transverse energy at the {CERN} pp collider. Phys Lett B 122(5–6):476–485
8. UA2 Collaboration (1983) Evidence for z0e+e at the {CERN} pp collider. Phys Lett B 129(1–2):130–140
9. Peskin ME, Schroeder DV (1995) An introduction to quantum field theory. Addison-Wesley, Reading, pp 196–198
10. Anderson PW (1963) Plasmons, gauge invariance, and mass. Phys Rev 130:439–442
11. Nambu Y, Jona-Lasinio G (1961) Dynamical model of elementary particles based on an analogy with superconductivity. I. Phys Rev 122(1):345–358
12. Nambu Y (1960) Axial vector current conservation in weak interactions. Phys Rev Lett 4:380–382
13. Goldstone J (1961) Field theories with ≪ superconductor ≫ solutions. Il Nuovo Cimento 19(1):154–164
14. Costa G, Fogli G (2012) Symmetries and group theory in particle physics. Volume 823 of lecture notes in physics. Springer, Berlin/Heidelberg, pp 192–199
15. 't Hooft G (1971) Renormalizable Lagrangians for massive yang-mills fields. Nucl Phys B 35(1):167–188

16. 't Hooft G, Veltman M (1972) Regularization and renormalization of gauge fields. Nucl Phys B 44(1):189–213
17. Englert F, Brout R (1964) Broken symmetry and the mass of gauge vector mesons. Phys Rev Lett 13:321–323
18. Higgs PW (1964) Broken symmetries and the masses of gauge bosons. Phys Rev Lett 13:508–509
19. Higgs PW (1966) Spontaneous symmetry breakdown without massless bosons. Phys Rev 145:1156–1163
20. Guralnik GS, Hagen CR, Kibble TWB (1964) Global conservation laws and massless particles. Phys Rev Lett 13:585–587
21. ATLAS Collaboration (2012) Observation of a new particle in the search for the standard model Higgs boson with the {ATLAS} detector at the {LHC}. Phys Lett B 716(1):1–29
22. CMS Collaboration (2012) Observation of a new boson at a mass of 125 gev with the {CMS} experiment at the {LHC}. Phys Lett B 716(1):30–61
23. ATLAS Collaboration (2013) Evidence for the spin-0 nature of the Higgs boson using {ATLAS} data. Phys Lett B 726(1–3):120–144
24. ATLAS Collaboration (2013) Measurements of Higgs boson production and couplings in diboson final states with the {ATLAS} detector at the {LHC}. Phys Lett B 726(1–3):88–119

Part III
Short Seminars

20 Chemical and Strain Engineering of Functional Oxides

Ausrine Bartasyte, Valentina Plausinaitiene, Adulfas Abrutis, and Samuel Margueron

Abstract $LiNbO_3$ and $LiTaO_3$ are the two of the most important crystals, being the equivalent in the field of optics, nonlinear optics and optoelectronics to silicon in electronics. Thus, the studies about epitaxial ferroelectric $LiNbO_3$ and $LiTaO_3$ thin films are of great interest because of their potential application as miniaturized and integrated elements in acoustic delay lines, resonators, pyroelectric sensors, optical waveguides, etc. Although $LiNbO_3$ and $LiTaO_3$ films have been fabricated by different techniques, many electrical and electro-optical properties reported are not comparable for those of $LiNbO_3$ and $LiTaO_3$ single-crystals. The growth of high quality $LiNbO_3$/$LiTaO_3$ films is far from being a routine due to a difficulty to estimate the phase composition and Li stoichiometry. Thus, the deposited films are hardly reproducible, often contain undesirable phases (for example, $LiNb_3O_8$ or Li_3NbO_4) or are stoichiometrically inhomogeneous. This inhibits the applications of the materials in the thin film form. The indirect methods, used to estimate Li concentration in the single crystals, cannot be applied directly due to the presence of strain, size effects or other defects in the films, which influences also the structural, optical and other physical properties. Thus, the indirect method for estimation of nonstoichiometry of $LiNbO_3$ and $LiTaO_3$ films, based on Curie temperature and dampings of Raman modes and taking into account secondary effects, was developed. Raman spectroscopy, a fast, local and non-destructive probe, was applied for investigation and mapping of phase composition, Li stoichiometry and stresses in $LiNbO_3$ ($LiTaO_3$) films. The methodology of identification of different phases and

A. Bartasyte (✉)
Institute Jean Lamour (UMR 7189) CNRS – Lorraine University, Nancy, France
e-mail: ausrine.bartasyte@gmail.com

V. Plausinaitiene • A. Abrutis
Department of General and Inorganic Chemistry, University of Vilnius, Vilnius, Lithuania

S. Margueron
Laboratoire Matériaux Optiques, Photonique et Systèmes, EA 4423, Lorraine University and Supelec, Metz, France

B. Di Bartolo et al. (eds.), *Nano-Structures for Optics and Photonics*,
NATO Science for Peace and Security Series B: Physics and Biophysics,
DOI 10.1007/978-94-017-9133-5_20, © Springer Science+Business Media Dordrecht 2015

extraction of residual stress and precise Li stoichiometry values from single Raman spectrum of $LiNbO_3$ film was presented in details. Moreover, this approach might be used for study of phase composition, stress and nonstoichiometry of other oxide films. In this work, the thin films were deposited by pulsed injection metalorganic chemical vapour deposition – a method providing the digital control of the growth. The physical and structural properties of $LiTaO_3$ and $LiNbO_3$ crystals and thin films were tuned by changing their chemical composition, as well. For example, optically isotropic ferroelectric materials (for thermal sensing) (Glazer et al., Sensing physical parameters using birefringence. WO Patent App. PCT/GB2011/051,111, 2011) or piezoelectric materials with reduced temperature coefficient of frequency (for acoustic delays lines) (Bartasyte and Elmazria, Patent F12 52410, 2012) were obtained by chemical engineering of $LiTaO_3$. It was shown that strain engineering is a powerful tool for governing the film symmetry, texture, domain structure (Borodavka et al., J Appl Phys 113(18):187216, 2013), phase transitions (Bartasyte et al., Appl Phys Lett 93(24):242907, 2008) and thermal expansion (Bartasyte et al., Appl Phys Lett 101(12):122902, 2012). The possibility to change by several times the thermal expansion of $LiNbO_3$/$LiTaO_3$ thin films by applying biaxial strain opens new avenues for temperature compensated devices.

Keywords Ferroelectrics • Lead titanate • Lithium niobate and tantalate • Strain • Chemical engineering

21
Gold Photonic Crystals and Photonics Quasi-crystals for Reproducible Surface-Enhanced Raman Substrates

Rossella Capasso, M. Pannico, P. La Manna, P. Musto, M. Rippa, P. Mormile, and L. Petti

Abstract Since the initial discovery of surface-enhanced Raman scattering (SERS), an increased amount of work has been done on the research of substrates for highly efficient Raman scattering enhancement due to their extraordinary potential for trace analysis and biological tags (Jarvis et al., Anal Chem 76:5198–202, 2004). The optical properties of noble metals with nanostructures have attracted enormous attention because of their potential application in optical sensing (Chen et al., Biosens Bioelectron 22:926–32, 2007), biosensor (Liu and Lu, J Am Chem Soc 125:6642–6643, 2003) and cell diagnostics (Huang et al., Nano Lett 7:1591–1597, 2007). Recently, the plasmonic optical responses of metal nanoparticles, based on localized surface plasmon resonances (LSPR) and significant fluorescence enhancement in the visible and near IR region, have been intensively researched. Many groups have demonstrated that the plasmon resonance is closely related to the size and shape of metal nanoparticles and the dielectric properties of the surrounding medium (Huang et al. Adv Mater 21:4880–4910, 2009). The possibility of engineering complex metal nanoparticle arrays with distinctive plasmonic resonances extending across the entire visible spectrum can have a significant impact on the design and fabrication of novel nanodevices based on broadband plasmonic enhancement (Gopinath et al., Opt Express 17:3741, 2009). In the present work we studied artificial electromagnetic (EM) nanomaterials to develop innovative plasmonic nanobiosensors based on SERS and working in the visible frequency band. The last decade has been characterized by artificial EM materials, including photonic crystals (PCs) and photonic quasi-crystals (PQCs) (Matarazzo et al., J Opt 13:015602, 2011; Rippa et al., Nanoscale 5:331–336,

R. Capasso (✉) • M. Rippa • P. Mormile • L. Petti
Institute of Cybernetics "E. Caianello" of CNR, Via Campi Flegrei 34, 80078 Pozzuoli, Italy
e-mail: r.capasso@cib.na.cnr.it

M. Pannico • P. La Manna • P. Musto
Institute of Chemistry and Technology of Polymers of CNR, Via Campi Flegrei 34, 80078 Pozzuoli, Italy

B. Di Bartolo et al. (eds.), *Nano-Structures for Optics and Photonics*,
NATO Science for Peace and Security Series B: Physics and Biophysics,
DOI 10.1007/978-94-017-9133-5_21,

2013), making these very attractive given that there are new possibilities to control the EM field in innovative way. With the use of PCs and PQCs, it is possible to synthesize novel artificial structures characterized by selective EM responses, which, in turn, undergo significant frequency shifts, in presence of biological material. Gold PQCs, in contrast to period structures, give rise to photonic-plasmonic resonances which extend across the entire visible spectrum. Gold PQCs are proposed for the engineering of reproducible surface enhanced Raman scattering (SERS) substrates. Using a molecular monolayer of pMA (p-mercaptoaniline) as a Raman reporter, we show that higher values of SERS enhancement factors can be achieved in PQCs structures compared to their periodic counterparts. The resulting PQCs nanostructured films can serve as good SERS substrates, exhibit large electromagnetic field enhancement factors for pMA, and can readily be used in ultrasensitive, molecule-specific sensing utilizing vibrational signatures.

22
Mid-infrared Surface Plasmon Polariton Sensors Resonant with the Vibrational Modes of Phospholipid Layers

Fausto D'Apuzzo, O. Limaj, A. Di Gaspare, V. Giliberti, F. Domenici, S. Sennato, F. Bordi, S. Lupi, and M. Ortolani

Abstract In this work we employed micrometric subwavelength hole arrays on thin metal films as surface plasmon-based biochemical sensors. This allowed for determining, from a single mid-infrared measurement, both the thickness and the absorption spectrum of liposome adsorbed phospholipid monolayers and trilayers. Improved sensitivity is observed due to anti-crossing behavior when mid-infrared molecular vibrations frequency matches surface plasmon mode resonance. The

F. D'Apuzzo (✉)
Department of Physics, Sapienza University of Rome, Rome, Italy

Istituto Italiano di Tecnologia, Rome, Italy
e-mail: Fausto.Dapuzzo@iit.it

O. Limaj
Department of Physics, Sapienza University of Rome, Rome, Italy

INFN Laboratori Nazionali di Frascati, Rome, Italy

A. Di Gaspare
CNR – Institute of Photonics and nanotechnologies, Rome, Italy

V. Giliberti • F. Domenici
Department of Physics, Sapienza University of Rome, Rome, Italy

S. Sennato • F. Bordi
CNR – Istituto dei Processi Chimico-Fisici, Rome, Italy

Department of Physics, Sapienza University of Rome, Rome, Italy

S. Lupi
Department of Physics, Sapienza University of Rome, Rome, Italy

INFN Laboratori Nazionali di Frascati, Rome, Italy

CNR – Istituto Officina dei Materiali, Rome, Italy

M. Ortolani
Department of Physics, Sapienza University of Rome, Rome, Italy

CNR – Institute of Photonics and nanotechnologies, Rome, Italy

B. Di Bartolo et al. (eds.), *Nano-Structures for Optics and Photonics*, NATO Science for Peace and Security Series B: Physics and Biophysics, DOI 10.1007/978-94-017-9133-5_22,

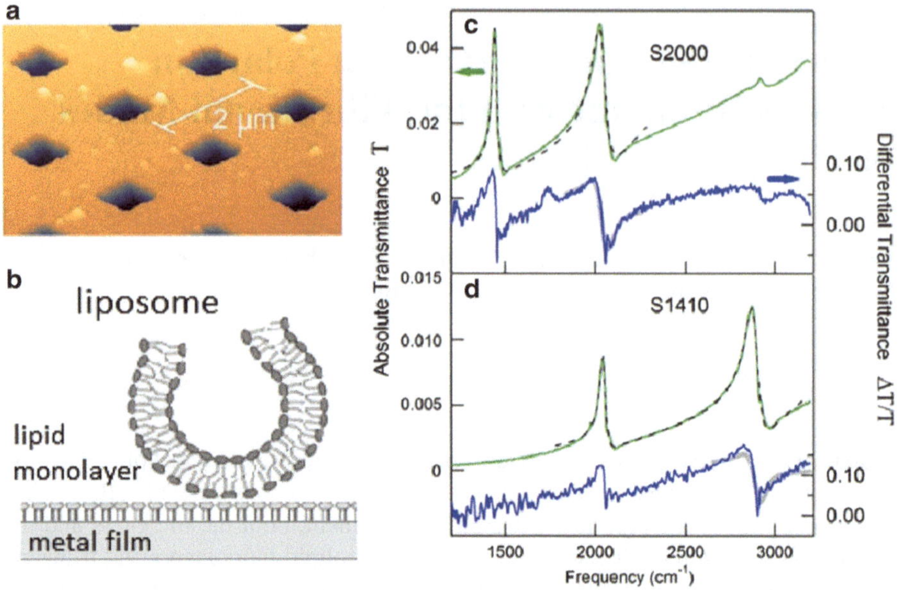

Fig. 22.1 AFM image of the nanohole square array with period 2 μm (**a**). Schematic description of liposome fusion on the Langmuir-Blodgett lipid monolayer on the sensor metallic surface (**b**). Transmittance spectra (left axis) and differential transmittance (right axis) calculated as the spectral difference before and after target deposition on the sensor, for the 2 μm (**c**) and the 1.41 μm period array (**d**)

development of optical sensors of few molecular layers is attracting considerable interest because they ensure non-contact interaction of the probing system with the target and portability for on-site biodiagnostics. Surface Plasmon Polaritons (SPP) propagating at the interface between a low-loss metal film and a dielectric medium have been exploited for increasing the sensitivity to few-molecule-thick layers, as the probing electric field is strongly confined by SPPs in the near-field region of the interface (Adato et al., P Natl Acad Sci U S A 106(46):19227–19232, 2009; Wu et al., Nat Mater 11(1):69–75, 2012; Limaj et al., Appl Phys Lett 98(9):091902, 2011). In the mid-infrared (IR) range (wavelengths λ between 2 and 10 μm) many classes of biomolecules naturally display a specific vibrational spectrum which could be used for a label-free identification of the target. In this work, we present the modeling, fabrication, and spectroscopic characterization of SPP sensors working in the mid-IR range. Here, in addition, we have set by design the SPP frequency at the vibrational mode of the target layer, in this case phospholipid mono- and multilayers representing simplified models of the cell plasma membrane.The sensor prototype was a hole array of square apertures with varying micrometric period in an Al film on Silicon substrate. The sensor was fabricated through electron beam lithography, but the ease of the design allows also for more cost effective nanoimprint techniques to be used. Finite-difference time-domain (FDTD) simulations have been performed to analyze the electric field spatial profile associated to SPP excitation (Limaj et al., Plasmonics 8:851–858, 2013). The phospholipid layers were prepared by first depositing a monolayer on the sensor surface with a Langmuir-Blodgett trough (estimated thickness was 2.6 nm). Then the sensor has been immersed in a solution of liposomes and left for incubation, so to allow the liposomes to adsorb on the surface and fuse with the lipid monolayer. The optical response of the sensors was investigated through Fourier Transform Infrared Spectroscopy transmittance (T) measurements. The SPP resonance profile has been fitted through a Fano lineshape and the SPP resonance frequency shift upon deposition of the phospholipid layers has been calculated. Within the same single measurement, it was possible to discriminate whether liposome fusion took place or not and identify the target fingerprints through the absorption spectrum (Fig. 22.1).

[illegible]

23
Structural and Spectroscopic Properties of Er^{3+}:$CdNb_2O_6$ Phosphors

Murat Erdem, S. Ghafouri, M.K. Ekmekçi, A. Mergen, and Gönül Özen

Abstract Rare earth (RE^{3+}) doped luminescence materials are particularly attractive in photonic applications due to their fluorescence arising from 5d→4f or 4f→4f transitions. High purity, compositionally uniform, single phase and uniform particle size powders are required for high resolution and high luminous efficiency in the photonic device developments. Moreover it should be pointed out that the selection of host lattices and active centers for such an optical material is important. So the host lattice of $CdNb_2O_6$ may be doped, because it shows thermal and, chemical stability. In this study, we present here the results of a detailed investigation of the spectroscopic properties of the erbium (Er^{3+}) doped cadmium niobate which were successfully prepared by the molten salt technique. It was observed that 0–6 mol % of Er_2O_3 addition produced single phase structure in XRD and SEM graphs. (Figs. 23.1 and 23.2). Luminescence spectra of the samples doped with Er^{3+} ion were conducted in the wavelength-range of 900–1700 nm at 300 K. Fig. 23.3 shows that the concentration quenching occurred in high Er^{3+} ions doped sample.

M. Erdem (✉)
Physics, Marmara University, Kadikoy, 34722, Istanbul, Turkey
e-mail: merdem@marmara.edu.tr

S. Ghafouri • G. Özen
Department of Physics, Istanbul Technical University, Maslak, 34469, Istanbul, Turkey

M.K. Ekmekçi
Chemistry, Marmara University, Kadikoy, 34722, Istanbul, Turkey

A. Mergen
Metalurgy & Material Engineering, Marmara University, Kadikoy, 34722, Istanbul, Turkey

B. Di Bartolo et al. (eds.), *Nano-Structures for Optics and Photonics*,
NATO Science for Peace and Security Series B: Physics and Biophysics,
DOI 10.1007/978-94-017-9133-5_23,

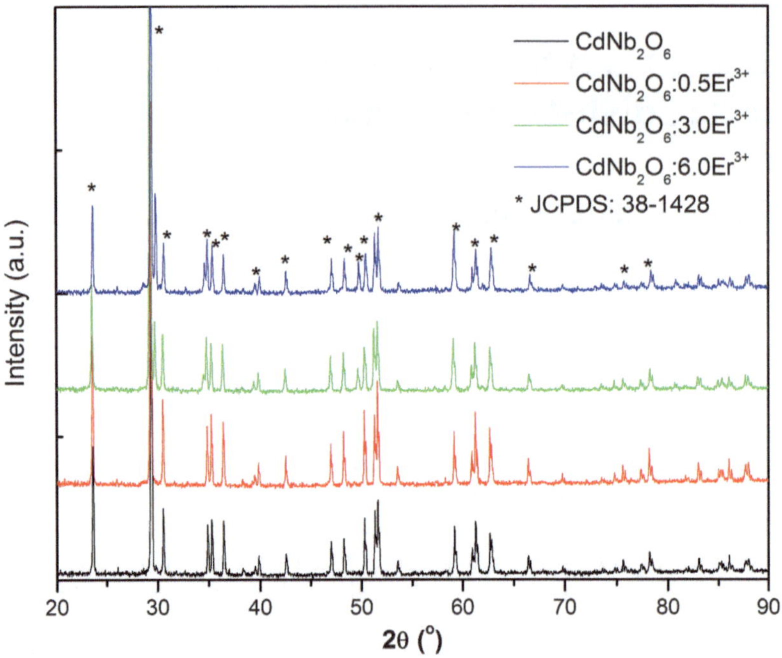

Fig. 23.1 Atomic force microscopy imaging of the cleaved surface of CdI_2 layered crystal (left – a growth helix on the fresh surface; right – pyramid-like structure grown in the process of in-air aging)

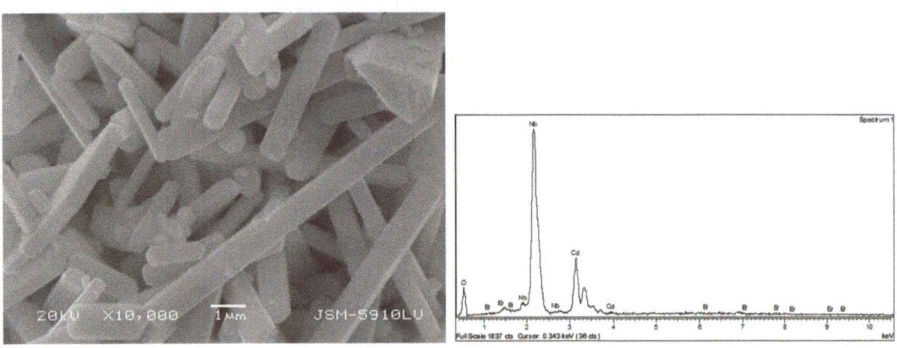

Fig. 23.2 The SEM graphs of $CdNb_2O_6$:$1.0Er^{3+}$ sample

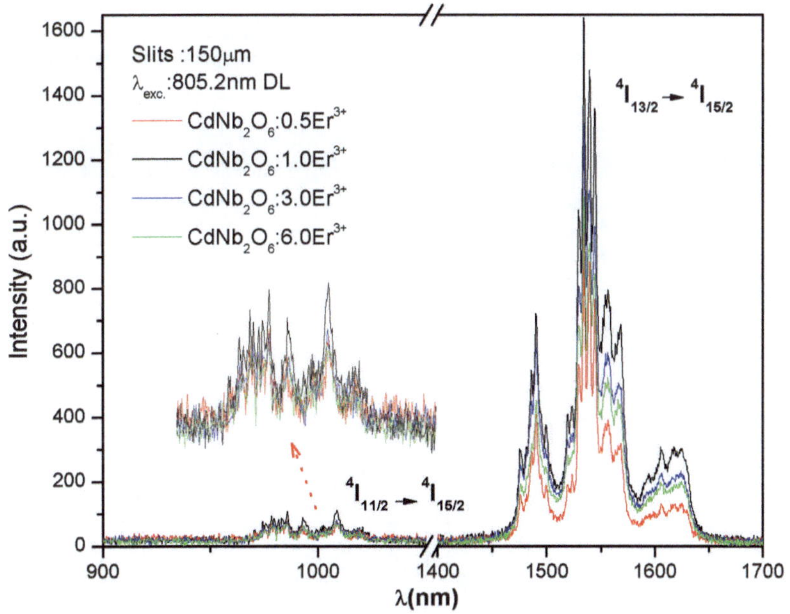

Fig. 23.3 The luminescence spectra of the samples at 300 K

24
Antibody Anchoring on QCM Gold Surfaces by UV Based Strategy

Riccardo Funari and Bartolomeo Della Ventura

Abstract Nowadays there is a strong interest in the development of measurement tools suitable for rapid and low cost analysis. Biosensors face this issue exploiting the intrinsic sensitivity and selectivity of a wide range of biomolecules. One of the main problems in biosensors development is the immobilization of the biological sensitive element onto the sensor surface. In this research topic we applied a novel strategy named Photonic Immobilization Technique (PIT) to anchor antibodies onto a gold electrode of a QCM with their antigen binding sites exposed to the environment (Della Ventura et al., Biomed Opt Express 2:3223–3231, 2011). The antibodies were activated by breaking the disulfide bridge in the triads Trp/Cys-Cys, which are typical of the immunoglobulins (Ioerger et al., Mol Immunol 36:373–386, 1999), through UV irradiation (Neves-Petersen et al., Protein Sci Publ Protein Soc 15:343–351, 2006). The new thiol groups so produced are exploited to anchor in a preferential way the proteins onto a gold surface. Even if the absorption mechanism is of the first order (one-photon absorption) the high UV power required by PIT needs some investigations concerning the most appropriate UV source. We compared femtosecond and nanosecond pulsed laser sources from different prospective, analyzing both their final effects on the biosensors and their effectiveness in protein activation by the Ellman's assay (Ellman, Arch Biochem Biophys 82:70–77, 1959). Moreover, conformational changes of the antibodies once irradiated were checked through steady state fluorescence (Garidel et al., Biotechnol J 3:1201–1211, 2008), on the other side AFM analysis of the gold surfaces in several irradiation conditions has also been performed. Our results show clear differences between the two laser sources, in particular the irradiation bynanosecond laser

R. Funari (✉)
Universitá degli Studi di Napoli "Federico II", via Cinthia, 26, 80126 Napoli, Italy
e-mail: funari.riccardo@gmail.com

B. Della Ventura
Universitá degli Studi di Roma "La Sapienza", p.le Aldo Moro, 5, 00185 Roma, Italy

B. Di Bartolo et al. (eds.), *Nano-Structures for Optics and Photonics*,
NATO Science for Peace and Security Series B: Physics and Biophysics,
DOI 10.1007/978-94-017-9133-5_24,

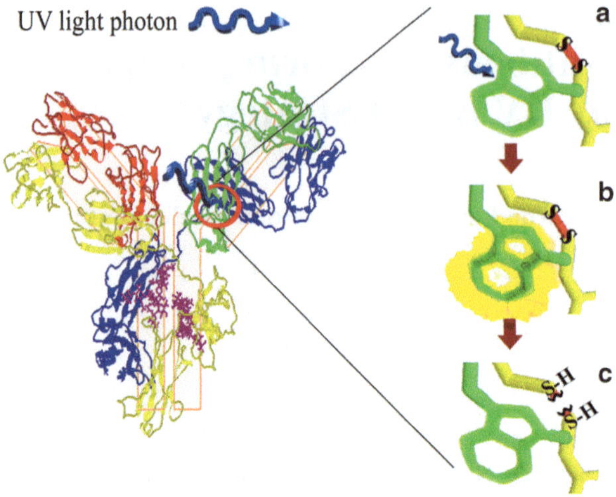

Fig. 24.1 UV activation of a typical IgG (1IGY). (**a**) The protein is irradiated. (**b**) One UV photon is absorbed by a tryptophan side-chain which transfers the energy to the near disulfide bridge. (**c**) The sulfur-sulfur bond opens (Funari et al., Anal Chem 85:6392–6397, 2013)

pulses seems to lead to a more rapid "photo-bleaching", as a consequence of the high fluence conveyed by these pulses, making femtosecond rather than nanosecond laser sources more suitable for PIT (Fig. 24.1).

Keywords Immunosensor • Quartz crystal microbalance • Antibody orientation • Parathion • Ultrashort UV pulses

25
Spectroscopic Properties of New Cubic Tungstates Doped with Eu^{3+} and Yb^{3+} Ions

Malgorzata Guzik, E.K. Bartosiewicz, M. Bieza, J. Iwanska, Yannick Guyot, E. Zych, and George Boulon

Abstract The motivation for this research was to find nanoscale tungstates with cubic structure, which then can be used to obtain the transparent ceramics. To our best knowledge any transparent ceramics of the rare-earth doped tungstates are not known and described in the literature, so obtaining them is therefore a great challenge. Two new classes of cubic compounds doped with Eu^{3+} and Yb^{3+} ions i.e. nano-and microcrystalline tungstates of chemical formula $Ba_3Y_2WO_9$ (Fig. 25.1) and microcrystalline $BaLaLiWO_6$ were synthesized by sol-gel method. For $Ba_3Y_2WO_9$ X-ray diffraction (XRD) analysis demonstrates single-phase nanopowders with high crystallite dispersion, and confirmed crystallization in cubic perovskite with the space group Fm-3m. However, for $BaLaLiWO_6$ only microcrystalline powder, despite the use of the same conditions of synthesis, confirmed by SEM method was obtained. The luminescence excitation and emission spectra in the ultraviolet-visible (UV) and NIR region at room and low temperature were used to investigate the optical properties of these phosphors. The investigations were completed by the site-selective excitation with luminescence and decay measurements. The comprehensive spectroscopic results was presented and discussed in depth.

M. Guzik (✉) • E.K. Bartosiewicz • M. Bieza • J. Iwanska • E. Zych
Faculty of Chemistry, Chemistry Department, University of Wroclaw, Joliot-Curie 14, PL-50-383 Wroclaw, Poland
e-mail: goguzik@poczta.fm

Y. Guyot • G. Boulon
Institute Light Matter, UMR5306 CNRS-University of Lyon 1, University of Lyon, 69622 Villeurbanne, France
e-mail: georges.boulon@univ-lyon1.fr

B. Di Bartolo et al. (eds.), *Nano-Structures for Optics and Photonics*, NATO Science for Peace and Security Series B: Physics and Biophysics,
DOI 10.1007/978-94-017-9133-5_25,

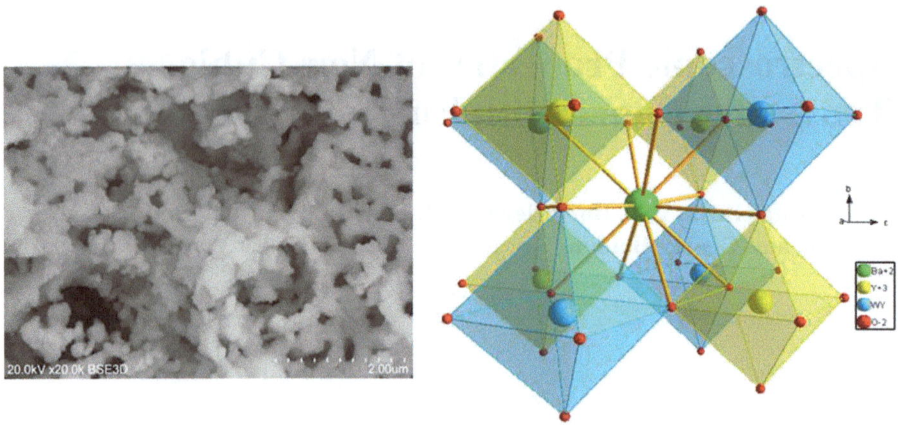

Fig. 25.1 SEM micrographs of $Ba_3Y_2WO_9$: Eu^{3+} ion and crystal structure of $Ba_3Y_2WO_9$

26
The Role of Surface Plasmon Resonance in Enhanced Transmission Through Metallic Gratings

Gergely Kajtár and P. Markoš

Abstract We investigate the physical origin of enhanced transmission features of one-dimensional (1D) metal gratings and identify two different surface plasmon resonances which influence the transmission of electromagnetic wave through a perforated metallic slab (Ebbesen et al., Nature 391:667–669, 1998). Theoretical analysis, similar to Roszkiewicz and Nasalski (J Phys B At Mol Opt Phys 46(2):025401, 2013), is confirmed by numerical simulations of transmission spectra of various metallic gratings. Our structure is shown in Fig. 26.1. The metallic 1D grating is periodic in the x-direction with period p and uniformly extended along the y-direction. Relative metallic permittivity $\varepsilon_m(f) = 1 - f_p^2/(f^2 + if\gamma)$ with $f_p =$ 2,147 THz and $\gamma = 5$ THz. We argue that enhanced transmission can be explained as an effect of two types of plasmon resonances (Figs. 26.1 and 26.2): *Lengthwise plasmon resonance – LPR* (symmetric and antisymmetric) propagating along the two interfaces of the metal slab (Fig. 26.1). This resonance appears as a sharp maximum (or two maxima, respectively) and a sharp minimum (Cao and Lalanne, Phys Rev Lett 88:057403, 2002). LPR is excited when the wave vector of the plasmon resonance coincides with $k \sin\varphi + m(2\pi/p)$ ($m = \pm 1, \pm 2 \ldots$). *Crosswise plasmon resonance – CPR* is excited in air gaps. The incident wave is guided through the grating if its frequency equals the CPR's resonant frequency. Consequently, the CPR appears as a Fabry-Perot spectral resonance in the transmission spectrum when its wavelength λ corresponds to $m(d/2)$. Both plasmonic resonances could be represented by surface waves propagating along *homogeneous* slabs. Using this approximation, dispersion relations could be calculated as in Markoš and Soukoulis (Wave propagation: from electrons to photonic crystals and left-handed materials (Google eBook). Princeton University Press, Princeton, 2008). Results are shown in Fig. 26.2. The transmission spectrum was calculated numerically by

---

G. Kajtár (✉) • P. Markoš
Department of Physics, INPE FEI STU, Ilkovičova 3, 812 19 Bratislava, Slovak Republic
e-mail: gergely.kajtar@stuba.sk

B. Di Bartolo et al. (eds.), *Nano-Structures for Optics and Photonics*, NATO Science for Peace and Security Series B: Physics and Biophysics,
DOI 10.1007/978-94-017-9133-5_26, © Springer Science+Business Media Dordrecht 2015

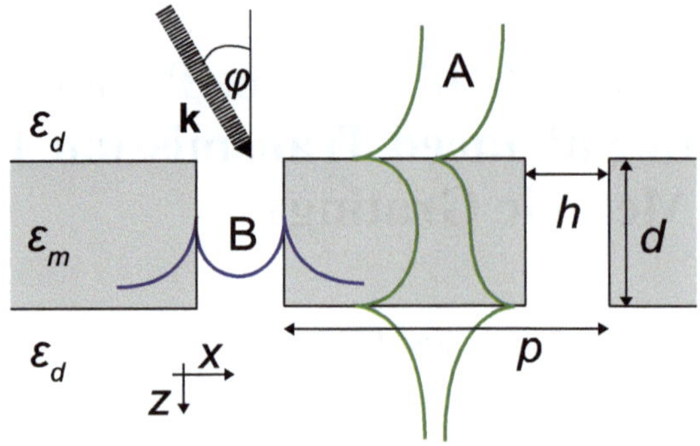

Fig. 26.1 Structure of the grating and two types of surface plasmons active in enhanced transmission: LPR symmetric and antisymmetric (*A*) and CPR (*B*). $p = 1{,}000$ nm, $h = 10$ nm, $\varepsilon_d = 1$. Incident wave is normal and TM polarised

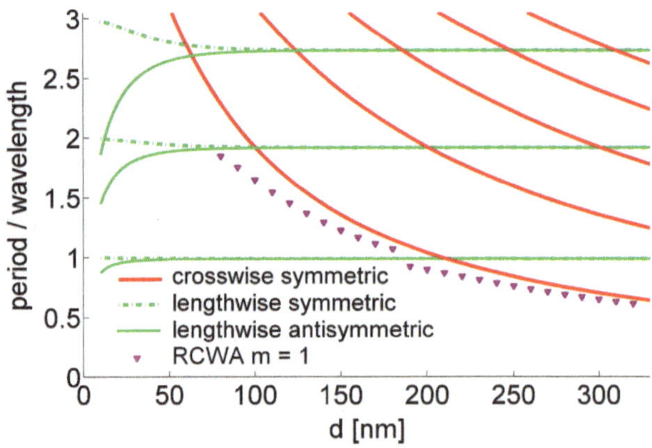

Fig. 26.2 Resonant frequencies of LPRs and CPRs as a function of grating thickness d. The first order CPR obtained by RCWA is plotted as *triangles*

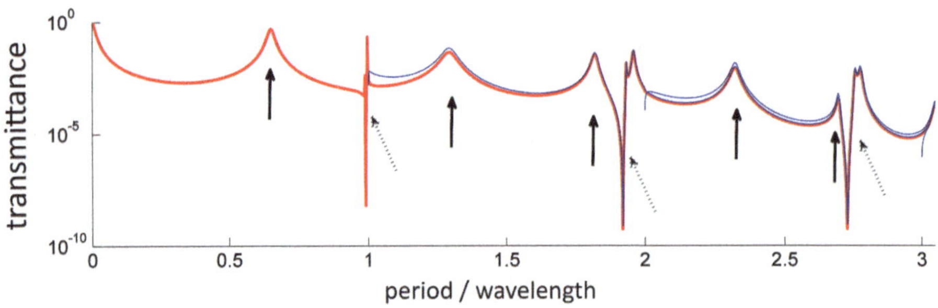

Fig. 26.3 Transmission spectrum (all diffracted orders) versus the normalised frequency of the electromagnetic wave for grating thickness $d = 300$ nm. *Arrows* indicate enhanced transmissions due to CPRs (*continuous*) and LPRs (*dashed arrow*)

our own numerical program based on the rigorous coupled-wave analysis – RCWA (Hench and John, Electron Trans Numer Anal (ETNA) 31:331–357, 2008). As shown in Fig. 26.3, two types of plasmon resonances are clearly observable and positions of resonances agree very well with predictions of the theory. In conclusion, we confirmed that surface plasmon resonances strongly influence the transmission of electromagnetic waves through 1D thin metallic gratings. Another resonances which might be excited in periodic metallic structures (antisymmetric CPR, guided waves) are discussed in Kajtár and Markoš (Opt Commun 313:382–387, 2014).

Keywords Enhanced transmission • Surface plasmon resonance • Metallic grating • RCWA

[illegible]

[illegible]

27
Experimental Observation of Self-Organized Nanostructures in Layered Crystals

Ivan Karbovnyk, I. Rovetskii, and I. Bolesta

Abstract Experimental evidence of nanostructures formation in layered crystals is discussed. Arrays of pyramid-like nanostructures formed in cadmium iodide crystals obtained either by vapor deposition or by using directional Bridgman-Stockbarger growth technique are directly observed using atomic force microscopy imaging and have influence on the optical spectra of the material. Nanostructured materials assembled from submicron-sized building blocks have a great potential for improving the performance of various electronic devices. Although modern nanoengineering allows manipulating the structure even at the atomic scale, there are also many self-organized nanosystems in nature (Rousset and Ortega, J Phys Condens Matter 18:null, 2006). Below, an example of such self-organized nanostructures will be addressed. Layered crystals exhibit specific properties due to their structure: atoms arrange in layers with ionic-covalent bonding, which are held together by relatively weak van der Waals forces. Specific defects, in particular, screw dislocations and related growth helices arising during crystal growth process can serve as nucleation centers for the formation of nanostructures in these materials.

Cadmium iodide, CdI_2, which belongs to a wider family of MX2 compounds, is often regarded as a model example of a layered structure (Singh et al., Surf Sci 422:L188–L191, 1999). Structural arrangement of CdI_2 is shared by a number of other materials, including iodides (PbI_2, MnI_2), bromides ($MgBr_2$, $FeBr_2$) and chalcogenides (TiS_2, $TiSe_2$). Therefore, pure and doped cadmium iodide crystals are good candidates to study the mechanisms of nanostructures formation. A typical growth helix site in CdI_2 crystal is shown in Fig. 27.1 on the left. It was established that in-air aging of the cleaved crystal surface stimulates the formation of pyramid-shaped structures (see Fig. 27.1, right). Further aging leads to accelerated growth of nanostructures and eventually to the effect of their self-organization into arrays. Time evolution of the nanostructures in nominally pure CdI_2 crystals is shown in

I. Karbovnyk (✉) • I. Rovetskii • I. Bolesta
Department of Electronics, Ivan Franko National University of Lviv, Lviv, Ukraine
e-mail: ivan_karbovnyck@yahoo.com

B. Di Bartolo et al. (eds.), *Nano-Structures for Optics and Photonics*,
NATO Science for Peace and Security Series B: Physics and Biophysics,
DOI 10.1007/978-94-017-9133-5_27, © Springer Science+Business Media Dordrecht 2015

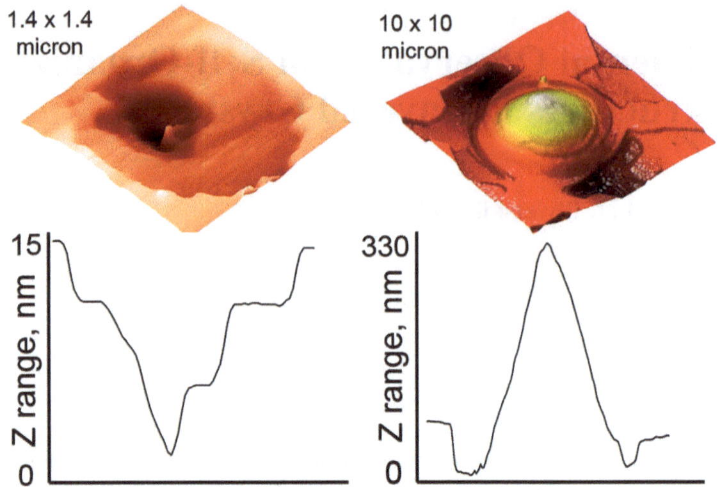

Fig. 27.1 Atomic force microscopy imaging of the cleaved surface of CdI_2 layered crystal (*left* – a growth helix on the fresh surface; *right* – pyramid-like structure grown in the process of in-air aging)

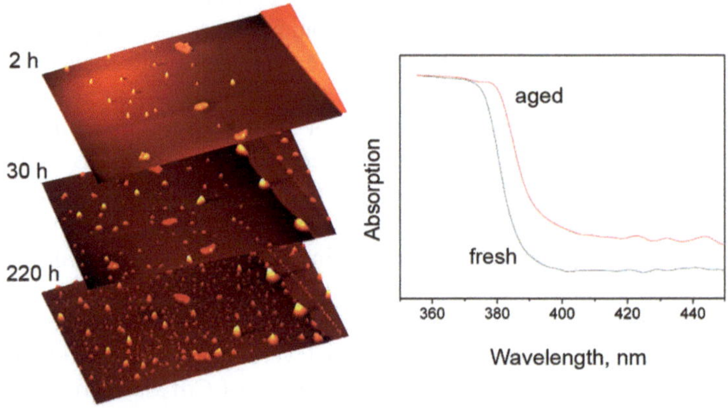

Fig. 27.2 Time evolution of nanostructures formed in CdI_2 observed by using atomic force microscope (*left*, each image represents 6 × 10 μm area) and changes in optical absorption (*right*) due to the formation of nanostructures

the left part of Fig. 27.2. Average height of individual structures increases from 25 nm (after 2 h of in-air aging) up to 100 nm (after 220 h). One should expect optical properties of the crystal to be influenced by such surface modifications (Bellucci et al., J Phys Condens Matter 19:395015, 2007). As confirmed by spectroscopic data (see Fig. 27.2, right) (Bellucci et al., J Phys Condens Matter 19:395015, 2007), optical absorption edge is slightly shifted to long wavelength region and overall background absorption in the transparency region increases. Rise of the background absorption is probably caused by the rough surface covered by light-scattering pyramid-like structures. Taking into account a wide size distribution of nanostructures it is not possible to verify if the absorption edge shift is dependent on their size. Concluding, atomic force microscopy is an efficient tool for tracking the evolution of nanostructures in solids and combined with optical spectroscopy can be applied for the comprehensive characterization of layered crystals at the nanoscale.

Keywords Nanostructures • Optical absorption • Atomic force microscopy

28
Nano-structuring for Molecular Motor Control

Mercy Lard, L. ten Siethoff, S. Kumar, M. Persson, G. te Kronnie, A. Månsson, and H. Linke

Abstract The interaction of self-propelled biological molecular-motors and cytoskeletal filaments holds relevance for a variety of applications such as biosensing, drug screening, diagnostics and biocomputation. The use of these systems for lab-on-a-chip biotechnology applications shows potential for replacement of microfluidic flow by active, molecular-motor driven transport of filaments. The ability to control, confine and detect motile objects in such a system is possible by development of nanostructured surfaces for on-chip applications and fundamental studies of molecular-motors. Here we describe the localized detection (Lard et al., Sci Rep 3:1092, 2013) and fast transport of actin filaments by myosin molecular-motors (Lard et al., Biosens Biolectron 48(0):145–152, 2013), inserted within nanostructures, as a method for biocomputation and molecular concentration. These results include extensive myosin driven concentration of actin filaments on a miniaturized detector, of relevance for use of molecular-motors in a diagnostics platform. Also, we discuss the local enhancement of the fluorescence signal of filaments, relevant for use in a biocomputation device where tracking of potentially thousands of motile objects is of primary significance.

M. Lard (✉) • H. Linke
The Nanometer Structure Consortium (nmC@LU), Division of Solid State Physics, Lund University, SE 221 00 Lund, Sweden
e-mail: mercy.lard@ftf.lth.se

L. ten Siethoff • M. Persson • A. Månsson
Department of Chemistry and Biomedical Sciences, Linnaeus University, SE-391 82 Kalmar, Sweden

S. Kumar
Department of Chemistry and Biomedical Sciences, Linnaeus University, SE-391 82 Kalmar, Sweden

Department of Biotechnology, Delhi Technological University, Delhi-110042, India

G. te Kronnie
Department of Women's and Children's Health, University of Padova, 35131 Padova, Italy

B. Di Bartolo et al. (eds.), *Nano-Structures for Optics and Photonics*,
NATO Science for Peace and Security Series B: Physics and Biophysics,
DOI 10.1007/978-94-017-9133-5_28,

[illegible] Women's and Children's Health, University of Padova, [illegible]

[illegible] Nano-Structures for Optics and Photonics, NATO Science for Peace and Security Series B: Physics and Biophysics,
DOI 10.1007/978-94-017-9133-5_28, © Springer Science+Business Media Dordrecht 2015

29 Fingerprint Imaging Enhancement by Deposition of Columnar Thin Films Nanostructures

Ismail Mekkaoui Alaoui

Abstract Imaging and visualization of latent fingerprints on smooth surfaces is not an easy task when the latent finger mark is on smooth and/or high reflecting surfaces. The surface where the fingerprint is located may present many physical and optical constraints, such as roughness, high background fluorescence which may be handled using time-resolved luminescence imaging (Mitchell and Menzel, Proc SPIE 1054:191–195, 1989; Alaoui et al., Forensic Sci Int 152:215–219, 2005). Optical techniques like oblique angle illumination and columnar-thin-film acquisition of fingerprint topology have been successfully used for visualization of fingermarks on smooth glass and mirror surfaces (Alaoui, Time-resolved luminescence imaging and applications. In: Byrnes (ed) Imaging for detection and identification. NATO security through science series. Springer, Netherlands, pp 243–248, 2007). We report preliminary results on the visualization of latent fingerprints on two smooth surfaces (glass and windows aluminum) using columnar thin films (CTF) nanostructures of CaF_2 followed by 1,2-indanedione fluorescence (Shaler, J Nanophotonics 5:051509, 2011). The preliminary results are promising. We have been able to visualize fingerprints on glass under dark field measurement after CTF deposition and observation under microscope (Dutta et al., Forensic Sci Int 228:32–37, 2013). Fingerprints on frame windows aluminum were able to be visualized after CTF deposition and 1,2-indanedione fluorescence. These results may open new options in fingerprint visualization and detection using columnar

I.M. Alaoui (✉)
Faculty of Sciences Semlalia, Physics Department, Cadi Ayyad University, BP 2390 Marrakech 40000, Morocco
e-mail: mekkaoui@uca.ma

B. Di Bartolo et al. (eds.), *Nano-Structures for Optics and Photonics*, NATO Science for Peace and Security Series B: Physics and Biophysics,
DOI 10.1007/978-94-017-9133-5_29,

thin films followed by fluorescent reagents. Our preliminary results show that this technique can be useful for latent fingerprints on nonporous surfaces. More work is necessary to optimize the CTF deposition process and the chemical post treatment (solution concentration, temperature, humidity, etc.).

Keywords Fingerprints • Imaging • Thin films • Nanostructure

30 Analytic Solution of the Rabi model

Alexander Moroz

Abstract The Rabi model (Rabi, Phys Rev 51:652–654, 1937) describes the simplest interaction between a cavity mode with a frequency ω and a two-level system with a resonance frequency ω_0. The model is characterized by the Hamiltonian (Rabi, Phys Rev 51:652–654, 1937; Schweber, Ann Phys 41:205–229, 1967)

$$\hat{H}_R = \hbar\omega \mathbb{1}\hat{a}^\dagger\hat{a} + \hbar g\sigma_1(\hat{a}^\dagger + \hat{a}) + \mu\sigma_3, \tag{30.1}$$

where $\hat{a}$ and $\hat{a}^\dagger$ are the conventional boson annihilation and creation operators satisfying commutation relation $[\hat{a}, \hat{a}^\dagger] = 1$, g is a coupling constant, $\mu = \hbar\omega_0/2$, $\mathbb{1}$ is the unit matrix, σ_j are the Pauli matrices in their standard representation, and we set the reduced Planck constant $\hbar = 1$. In the Bargmann space of entire functions (Bargmann, Commun Pure Appl Math 14:187–214, 1961), the eigenfunctions of the Rabi model can be determined in terms of an entire function

$$\varphi(z) = \sum_{n=0}^{\infty} \phi_n z^n. \tag{30.2}$$

In a series of articles (Moroz, Europhys Lett (EPL) 100:60010, 2012; Moroz, Ann Phys 338(0):319–340, 2013; Moroz, Ann Phys 340(1):252–266, 2014; Moroz, Quantum models with spectrum generated by the flows of polynomial zeros, p 9, 2014) we have shown that the expansion coefficients $\phi_n(\epsilon)$ are positive definite orthogonal polynomials of a discrete energy variable ϵ (Moroz, Europhys Lett (EPL) 100:60010, 2012; Moroz, Ann Phys 338(0):319–340, 2013; Moroz, Ann Phys 340(1):252–266, 2014; Moroz, Quantum models with spectrum generated by the flows of polynomial zeros, p 9, 2014). The weight function $\psi(\epsilon)$ in the orthogonality relations of the polynomials is an increasing step function. The spectrum of the Rabi

A. Moroz (✉)
Gitschiner Strasse, Berlin, Germany
e-mail: wavescattering@yahoo.com

B. Di Bartolo et al. (eds.), *Nano-Structures for Optics and Photonics*,
NATO Science for Peace and Security Series B: Physics and Biophysics,
DOI 10.1007/978-94-017-9133-5_30,

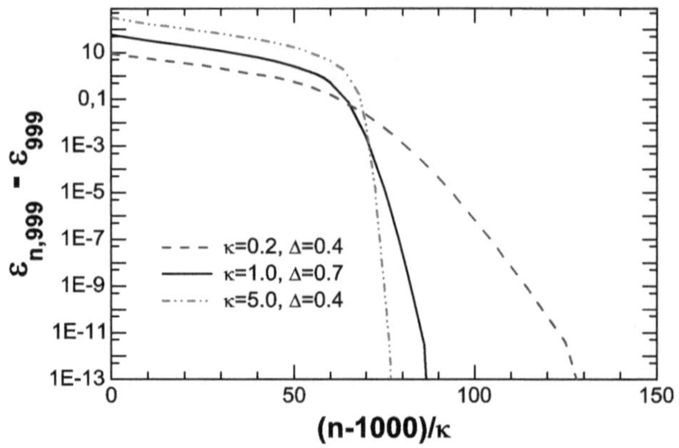

Fig. 30.1 Convergence of the the 1,000th zero $\epsilon_{n,999}$ of $\phi_n(\epsilon)$ toward the exact 1,000th eigenvalue ϵ_{999} = 998.907883759510, 997.950425260357 and 973.989087026621 of the Rabi model in the positive parity eigenspace for $(\kappa, \Delta) = (0.2, 0.4), (1, 0.7), (5, 0.4)$, respectively. The ground state energy ϵ_0 corresponds to the first zero of $\phi_n(\epsilon)$. Each decreasing sequence $\epsilon_{n,999}$ forms a discrete flow converging toward the corresponding exact eigenvalue down to machine precision

model corresponds to the points of discontinuity of $\psi(\epsilon)$. Numerically, the spectrum can be recovered as fixed points of monotonic flows of the zeros of orthogonal polynomials $\phi_n(\epsilon)$ in the limit $n \to \infty$. In practice only a *single* well-behaved $\phi_{n_0+n_t}(\epsilon)$ is required to determine any first n_0 eigenvalues of the Rabi model arranged in increasing order (cf. Fig. 30.1). For instance, $n_t \simeq 26$ for $n_0 = 1{,}000$ and $\kappa = 0.2$ if double precision is required. To reach unlimited precision, the limit $n \to \infty$ is required. For a comparison, Braak's approach (Braak, Phys Rev Lett 107:100401, 2011; Braak, Annalen der Physik 525(3):L23–L28, 2013) requires (i) to solve for an undetermined number of functions K_n having infinite number of poles on real axis, (ii) to assemble the functions K_n into a transcendental $G_\pm$, (iii) to solve for zeros of $G_\pm$. Not surprisingly, Braak's approach reaches its limits already at ca. 20 energy levels in double precision (Braak, Phys Rev Lett 107:100401, 2011; Braak, Annalen der Physik 525(3):L23–L28, 2013). An additional analytic continuation step could increase the number of calculable energy levels in Braak's approach only to around 100 levels (Braak, Phys Rev Lett 107:100401, 2011; Braak, Annalen der Physik 525(3):L23–L28, 2013).

31
Microscroll Invisibility Cloak

Philip Munoz and E. Mazur

Abstract We present a design for a self-assembled cylindrical invisibility cloak, based on strained-induced rolling. Here, a 2D slab metamaterial is patterned by electron beam lithography on a compressively strained InGaAs thin film. The metamaterial is composed of an array of periodically spaced silver nanorods embedded in a polymer background. The anisotropic effective permittivity of the slab is defined by on the aspect ratio and pitch of the nanorods, and is modeled by Effective Medium Theory and the Finite Difference Time Domain method (FDTD). The stained film is released from the substrate by wet etching, and the film strain relaxes, causing the combined metamaterial/thin-film to curl into a tight roll. The rolling radius is modeled by continuum strain theory and confirmed by experiment. Resulting microscroll has an anisotropic and radially-dependent effective permittivity, which is inherited from the original slab metamaterial. Depending on the nanorod spacing in the metamaterial layer, we can tune the radial dependence of the permittivity. This design can be used to realize a variety of transformation optical devices with cylindrical symmetry. In particular, we analyze the case of a TE cylindrical invisibility cloak with reduced parameters using FDTD.

P. Munoz (✉) • E. Mazur
School of Engineering and Applied Sciences, Harvard University, Cambridge, MA, USA
e-mail: pmunoz@seas.harvard.edu

B. Di Bartolo et al. (eds.), *Nano-Structures for Optics and Photonics*, NATO Science for Peace and Security Series B: Physics and Biophysics, DOI 10.1007/978-94-017-9133-5_31,

21
Microwave Invisibility Cloak

B. Di Bartolo et al. (eds.), *Nano-Structures for Optics and Photonics*,
NATO Science for Peace and Security Series B: Physics and Biophysics,
DOI 10.1007/978-94-017-9133-5_21,

32
The Identification of the Inorganic Pigments in the Cultural Heritage Objects Using Surface-Enhanced Raman Scattering

E. Shabunya-Klyachkovskaya, S. Vaschenko, V. Stankevich, and S. Gaponenko

Abstract The Ge/Si nanostructures with Au coating were used for surface enhanced Raman scattering by inorganic α-HgS microcrystals. The enhancement factor in the case of using these nanostructures like SERS-active substrates was higher than in the case of using Ag gel-film. We also compared Raman scattering efficiency in dependence on the laser excitation wavelength.

To date, surface enhanced Raman scattering (SERS) is known as one of the most sensitive technique for the identification of small quantities of analytes. Since organic molecules are much smaller than the nanoobjects used like SERS-active substrates, SERS is often used for the identification of the organic art pigments. To identify inorganic art pigments, micro-Raman spectroscopy is usually applied.

In our recent work (Klyachkovskaya et al., Plasmonics 6:413–418, 2011) we were able to get 100-fold enhancement of the Raman spectra of ultramarine microcrystals. In addition, it was shown that self-assembled Ge/Si nanostructures coated with a noble metal can be used like very perspective SERS-active substrates. In this paper, we use these substrates for surface enhanced Raman scattering by inorganic α-HgS microcrystals (vermilion). In addition we compare Raman scattering efficiency in dependence on the type of SERS-active substrate and the laser excitation wavelength.

So, two types of SERS-active substrate were used for the experiments. The first of them were the self-assembled Ge/Si nanostructures and the second one were the gel-films of the silver nanoparticles (10–25 nm in size). The Ge/Si nanostructures were grown by chemical vapor deposition (Stoica et al., Nanotechnology 18:455307, 2007). Silver sol has been prepared via Ag ions reduction with dextroglucose under heating (Fang et al., Chem. Phys. Lett. 401:271–275, 2005). The Raman measurements were performed in the backscattering configuration at room temperature.

E. Shabunya-Klyachkovskaya (✉) • S. Vaschenko • V. Stankevich • S. Gaponenko
Institute of Physics of NAS of Belarus, Nezavisimosti Ave. 68, 220072 Minsk, Belarus
e-mail: e.shabunya-klyachkovskaya@ifanbel.bas-net.by; s.gaponenko@ifanbel.bas-net.by

B. Di Bartolo et al. (eds.), *Nano-Structures for Optics and Photonics*, NATO Science for Peace and Security Series B: Physics and Biophysics, DOI 10.1007/978-94-017-9133-5_32,

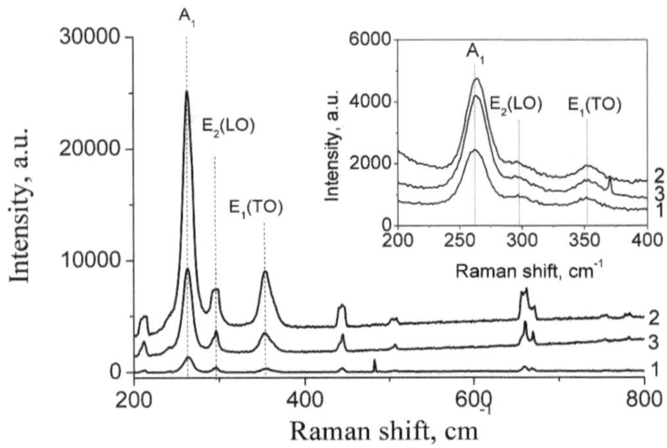

Fig. 32.1 Raman (1) and SERS spectra of vermilion, obtained using the Si/Ge – nanostrucres (2) and Ag gel film (3). The wavelength of excitation is 632.8 nm. In the inset: Raman and SERS spectra of vermilion excited with the Nd:LSB laser (531 nm)

The Nd:LSB laser (531 nm) as well as the He-Ne laser (632.8 nm) were used for Raman spectra excitation. The registration system consists of a spectrograph with a diffraction grating 1,200 grooves per mm (Solar TII S3901) and a CCD matrix, which is cooled with liquid nitrogen (Princeton instruments).

The bands at 262 cm^{-1} (secondary vibration of A_1), 296 cm^{-1} ($E_2(LO)$ optical phonon transverse vibrations) and 351 cm^{-1} ($E_1(TO)$ secondary longitudinal oscillations) occur in the Raman spectra. When Raman spectra are excited with the He-Ne laser (632.8 nm), the intensity increases in 15 fold in the case of using the Si/Ge nanostructures like the SERS-active substrate and in five fold in the case of using the silver gel-film as the substrate (see Fig. 32.1). In the case of using the Nd:LSB laser (531 nm) for Raman spectra excitation, the intensity increases only 1.9 and 1.7 fold, respectively (see the inset in the Fig. 32.1).

The dependence of the enhancement on the type of the SERS-active substrate may be related to their topography. The Ge islands on the Si substrate have a quadrangular pyramid shape, which leads to resonance with dipole plasma oscillations (Sajanlal et al., Nano Rev 2(0), 2011). In result, the local electric field at the top of the "pyramids" is essentially magnified. Additionally, the gold is the inert metal unlike the silver which is easily oxidized. As for the dependence of the Raman intensity on the excitation source, it's probably due to the fact that the He-Ne laser has a narrow contour of the excitation line and works in single mode unlike the Nd:LSB one, that has quite wide contour of the excitation line and excite Raman scattering over a wide range.

So, obtained results can be applied for the optimization SERS technique for the identification of the inorganic pigments in the Cultural heritage objects. The authors are grateful to Prof. D. Grutzmacher for providing the Ge/Si nanostructures for the experiments.

Keywords SERS • Vermilion • Inorganic pigments • Si-Ge nanostructures

33
Polymeric Microresonators for Biosensing Applications

Sarah Wiegele, Torsten Beck, Tobias Grossmann, Martin Mai, Timo Mappes, and Heinz Kalt

Abstract We report on whispering gallery mode (WGM) resonators consisting of poly(methyl methacrylate) (PMMA) which are promising candidates for label-free biosensing applications. The optical modes in these resonators experience a wavelength shift when the refractive index of the surrounding medium changes or when molecules attach to the resonator. Different methods can be used for functionalization of the resonator surface to obtain sensors which are sensitive to specific molecules. To obtain higher sensitivities coupled resonators can be produced. If the resonators are close enough the modes of the single cavities couple and supermodes extending over both cavities develop. Analogously to the hydrogen molecule a bonding and an anti-bonding optical mode is observed. The bonding mode shows a field enhancement in the gap which results in a stronger interaction of the mode with the surrounding medium.

For biosensing measurements we use goblet shaped resonators consisting of PMMA with Q-factors up to 1.3×10^7 (Grossmann et al., Appl Phys Lett 96:013303, 2010). For the surface modification one promising approach is spatially controlled photo-functionalization (Beck et al., Appl Phys Lett 102(12):121108, 2013). We use benzophenone-dPEG$_3$-biotin (BPB) to provide binding sites for the protein strepatvidin. Under UV-radiation ($\lambda = 325\,\text{nm}$) benzophenone binds covalently to the CH-groups of PMMA. After immersing the whole sample with BPB dissolved in ethanol single resonators can be irradiated with UV-light. Afterwards unbound BPB is removed by rinsing the sample with distilled water. To test the functionalization

S. Wiegele (✉) • T. Beck • T. Grossmann • M. Mai • H. Kalt
Institute of Applied Physics, Karlsruhe Institute of Technology, 76131 Karlsruhe, Germany
e-mail: sarah.wiegele@kit.edu

T. Mappes
Institute of Microstructure Technology, Karlsruhe Institute of Technology, 76128 Karlsruhe, Germany

Carl Zeiss AG, Corporate Research and Technology, Carl-Zeiss-Promenade 10, 07745 Jena, Germany

B. Di Bartolo et al. (eds.), *Nano-Structures for Optics and Photonics*, NATO Science for Peace and Security Series B: Physics and Biophysics, DOI 10.1007/978-94-017-9133-5_33, © Springer Science+Business Media Dordrecht 2015

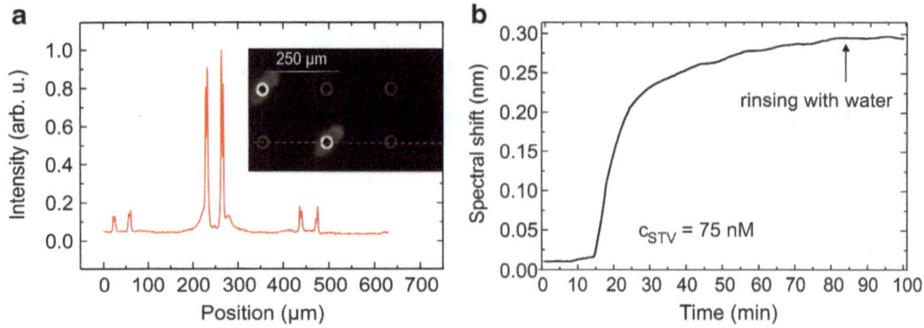

Fig. 33.1 (**a**) Fluorescence image of functionalized resonators flushed with fluorescently labeled streptavidin. (**b**) Shift of a resonant mode when streptavidin binds to the biotin

Fig. 33.2 The BRIS of coupled microdisks (*squares*) is approximately three times higher than the BRIS of a single microdisk (*triangles*)

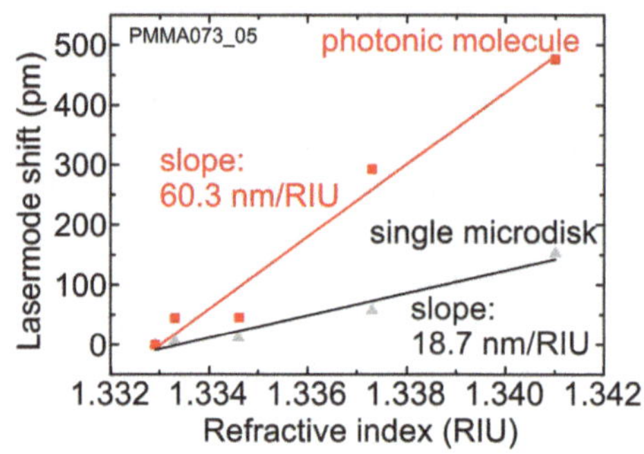

we rinse the sample with fluorescently labeled streptavidin and investigate the sample with a fluorescence microscope (Fig. 33.1a). It is clearly visible that the functionalized resonators show enhanced fluorescence. To demonstrate specific binding a functionalized resonator was placed in a chamber filled with water. A streptavidin solution was filled to the chamber carefully using a peristaltic pump. The spectral shift of a resonant mode was monitored over time and is shown in Fig. 33.1b.

To obtain higher sensitivities we made use of the field enhancement between two coupled resonators. We used lithografic fabrication to realize strongly coupled dye-doped polymeric disks. For coupling gaps smaller than 150 nm (which were obtained by a covering parylene layer) we were able to observe supermodes extending over both resonators (Grossmann et al., Light Sci Appl 2:e82, 2013). To analyze the overlap of these supermodes with the surrounding medium bulk refractive index sensitivity (BRIS) measurements were performed with coupled microdisks. Therefore the sample was incubated in solutions with different refractive indices and the spectral shift of the supermode was recorded. In Fig. 33.2 the shift of the supermodes is compared to the shift of a single disk resonator. The

BRIS of the coupled disks is approximately three times higher than the BRIS of a single resonator. This could indicate that the interaction of supermodes with the environment is higher than the interaction of normal modes. For this reason coupled microresonators are promising candidates for label-free biosensors with high sensitivity.

34
On the Possibility of Indicating Protein Conformational Changes via Energy Transfer Between Intrinsic Fluorophores

Nadezda G. Zhdanova, E.A. Shirshin, I.M. Panchishin, V.V. Fadeev, and A.A. Maskevich

Abstract The rate of intermolecular energy transfer (ET) between tyrosine and triptophan residues in human serum albumin (HSA) was estimated based on the information about its structure from X-ray crystallographic data provided in protein databank (PDB) as well as from fluorescent measurements. The possibility of indicating of protein molecule's conformational changes via intrinsic ET was shown by the analysis of PDB-structures of HSA and its complexes with different ligands. The manifestation of conformational changes of HSA in its fluorescence properties connected with the tyrosine-tryptophan ET was studied by means of steady-state spectroscopy and picosecond time-resolved technique.

Human serum albumin (HSA) is the most abundant protein of human blood plasma responsible for the transportation of different substances (e.g. fatty acids, drugs, metabolites). Binding of drugs to HSA and the dependency of the efficiency of this process on microenvironment is being extensively studied in literature. It is well-accepted that protein conformation, its structure and position of binding sites, is the basic property determining binding efficiency (Varshney et al., Chirality 22:77–87, 2010).

To investigate conformational changes in proteins different techniques can be applied. While the analysis of X-ray structure of HSA and its complexes with different ligands provide the information about conformational changes on a subnanometer scale, optical spectroscopy doesn't possess atomic resolution, but it is much more accessible and allows one to study structural changes in real time.

Among different optical techniques the method based on fluorescence (Forster) resonance energy transfer (FRET) (Lakowicz, Principles of fluorescence spectroscopy. Springer, New York, 2007) should be noted as for it allows one to

N.G. Zhdanova (✉) • E.A. Shirshin • I.M. Panchishin • V.V. Fadeev
Department of Physics, M.V. Lomonosov Moscow State University, Moscow, Russia
e-mail: nadezhda1989@gmail.com

A.A. Maskevich
Department of Physics, Y. Kupala State University of Grodno, Grodno, Belarus

B. Di Bartolo et al. (eds.), *Nano-Structures for Optics and Photonics*,
NATO Science for Peace and Security Series B: Physics and Biophysics,
DOI 10.1007/978-94-017-9133-5_34,

measure changes in relative position of pair of fluorophores. The investigation of conformational changes in protein in different environmental conditions via FRET techniques is one of the hot topics in modern biophysics.

A number of works make use of protein labeling with fluorophores that bind specifically to particular sites and can serve as energy acceptor (or donor, or both). It should be noted that binding of an external label can induce changes in protein structure as well as the investigated direct influence by chemical agents (Anand et al., J Phys Chem B 114:15839–15845, 2010) (e.g. surfactants, guanidine hydrochloride, urea) or alterations in temperature, pH, etc. At the same time it's well-known that many proteins contain the internal FRET pair: tyrosine (Tyr) and tryptophan (Trp) residues that represent a donor-acceptor FRET pair (Saito et al., Photochem Photobiol 33:289–295, 1981; Chiu and Bersohn, Biopolymers 16:277–288, 1977; Eisenhawer et al., Biochemistry 40:12321–12328, 2001). Despite that Tyr-Trp energy transfer is a well-known phenomenon in protein photophysics, it has not been thoroughly investigated due to experimental limitations. The recent advances in optical instrumental techniques and the accessibility of protein databank (PDB) allow one to reinvestigate the intramolecular FRET as a tool to detect the conformational changes in macromolecules.

In this paper the complex approach based on the analysis of PDB-structures and application of fluorescent techniques (steady-state and picosecond time-resolved methods) is developed for investigation the conformational changes in HSA and its complexes with different ligands. It was shown that Tyr-Trp energy transfer can be used as an effective indicator of structural changes in proteins.

Keywords Conformational changes • Human serum albumin (HSA) • Intrinsic energy transfer • Picosecond time-resolved fluorescence • Steady-state fluorescence

Part IV
Posters

35
Surface Plasmons and Strong Light-Matter Coupling in Metallic Nanoshells

Filippo Alpeggiani, S. D'Agostino, and L.C. Andreani

Abstract Light-matter interaction can be strongly enhanced by the confinement of the electric field due to the excitation of plasmonic modes at metal-dielectric interfaces (Gaponenko, Introduction to nanophotonics. Cambridge University Press, Cambridge/New York). In this work we give a theoretical formulation of the interaction between a dipole emitter and the localized surface plasmons of spherical metallic nanoparticles, particularly suitable for studying strong-coupling effects beyond perturbation theory, such as the appearance of vacuum Rabi splitting in the emission spectrum of the dipole (Savasta et al., ACS Nano 4:6369–6376, 2010; Van Vlack et al., Phys Rev B 85:075303, 2012). The threshold for entering the strong-coupling regime can be expressed with a formalism analogous to that of cavity-quantum-electrodynamics, based on a proper definition of the modal effective volume (Maier, Opt Express 14(5):1957, 2006; Agio, Nanoscale 4:692–706, 2012). In particular, we apply the results to the case of a spherical metallic nanoshell in the quasistatic approximation, deriving a spectrum of sphere-like and cavity-like surface plasmon excitations (Prodan and Nordlander, Chem Phys Lett 352:140–146, 2002) whose frequencies and effective volumes are calculated as a function of various geometrical parameters. We show that a vacuum Rabi splitting can occur in emission spectra for dipole oscillator strengths as small as a few units (Alpeggiani et al., Phys Rev B 86:035421, 2012), which can be easily achieved with organic molecules or quantum dots. The most favorable situation for strong coupling is when the dipole is located inside the nanoshell. Surprisingly, this dipole couples with sphere-like modes more strongly than with cavity-like ones, if the shell is thin enough. Our framework can be extended to a fully computational approach in order to investigate the strong-coupling regime of light-matter interaction in more complex geometries.

Keywords Surface plasmons • Strong coupling • Nanoshell

F. Alpeggiani (✉) • S. D'Agostino • L.C. Andreani
Department of Physics, University of Pavia, via Bassi 6, 27100 Pavia, Italy
e-mail: filippo.alpeggiani01@universitadipavia.it

B. Di Bartolo et al. (eds.), *Nano-Structures for Optics and Photonics*, NATO Science for Peace and Security Series B: Physics and Biophysics,
DOI 10.1007/978-94-017-9133-5_35, © Springer Science+Business Media Dordrecht 2015

B. Di Bartolo et al. (eds.), *Nano-Structures for Optics and Photonics*,
NATO Science for Peace and Security Series B: Physics and Biophysics,
DOI 10.1007/978-94-017-9133-5_25, © Springer Science+Business Media Dordrecht 2015

36
Photoluminescence and Electroluminescence Characterization of a-Si:H/c-Ci Interfaces

José Alvarez, J.P. Kleider, and S. Ibrahim

Abstract Devices based on silicon heterojunctions (HT) formed between hydrogenated amorphous silicon (a-Si:H) and crystalline silicon (c-Si) are very attractive for solar cell applications; efficiencies of 24.7 % have been reached (Terakawa et al., MRS Proc 1536:17–26, 2014), with relative low cost of fabrication. The interface of the a-Si:H/c-Si HT solar cell and its characterization remain critical issues in order to achieve optimal efficiencies. Luminescence spectroscopy techniques are powerful approaches that are nowadays currently used for the characterization of solar cells. More specifically, photoluminescence (PL) and electroluminescence (EL) are increasingly carried out under different intensity-frequency regimes to investigate interfaces, surface passivation and bulk properties at the different steps of the a Si:H/c Si HT solar cell fabrication. In the first part of this study, PL and EL measurements have been performed under different excitation regimes, namely steady-state and variable frequency small signal modulated conditions, and conducted on different HT configurations: a Si:H(p)/c-Si(n) and a-Si:H(p)/a Si:H(i)/c-Si(n). In modulated conditions, experimental results show that the light intensity dependence of the effective lifetime extracted by PL is in good agreement with the one determined from EL technique (Fig. 36.1). The effective lifetime estimated for the different HT configurations points out higher values for the HTs with an intrinsic layer. This confirms that passivation with an intrinsic layer improves solar cells performances. The second part of the study concerns a multiscalar PL analysis of c-Si through two complementary experimental setups. A confocal microscope has been used for the microscopic scale approach, and a non-confocal system for the macroscopic scale. The same excitation wavelength

J. Alvarez (✉) • J.P. Kleider • S. Ibrahim
LGEP, CNRS UMR 8507, Supelec, University of Paris-Sud, UPMC Univ Paris 06, 11 rue Joliot-Curie, Plateau de Moulon, F-91192 Gif-sur-Yvette Cedex, France
e-mail: Jose.Alvarez@lgep.supelec.fr

B. Di Bartolo et al. (eds.), *Nano-Structures for Optics and Photonics*, NATO Science for Peace and Security Series B: Physics and Biophysics,
DOI 10.1007/978-94-017-9133-5_36,

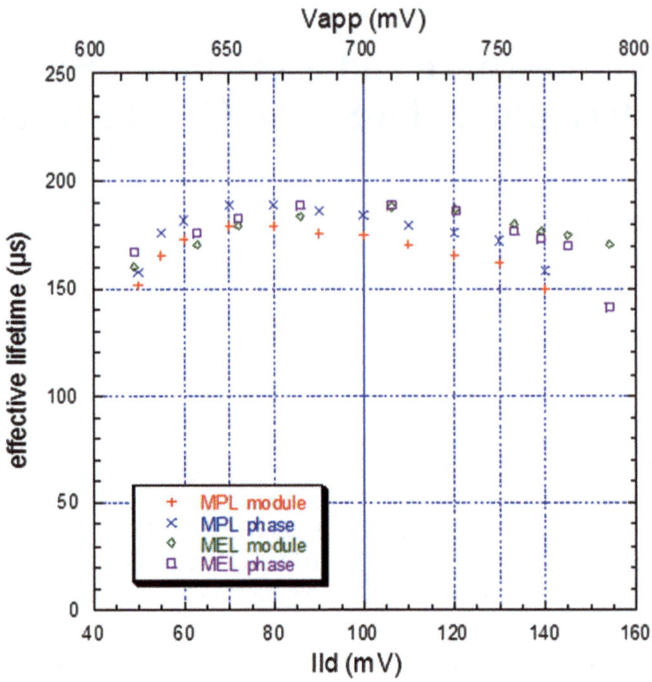

Fig. 36.1 Effective lifetime evolution versus the current of the laser diode (Ild) and the applied voltage (Vapp) in case of modulated PL and EL, respectively

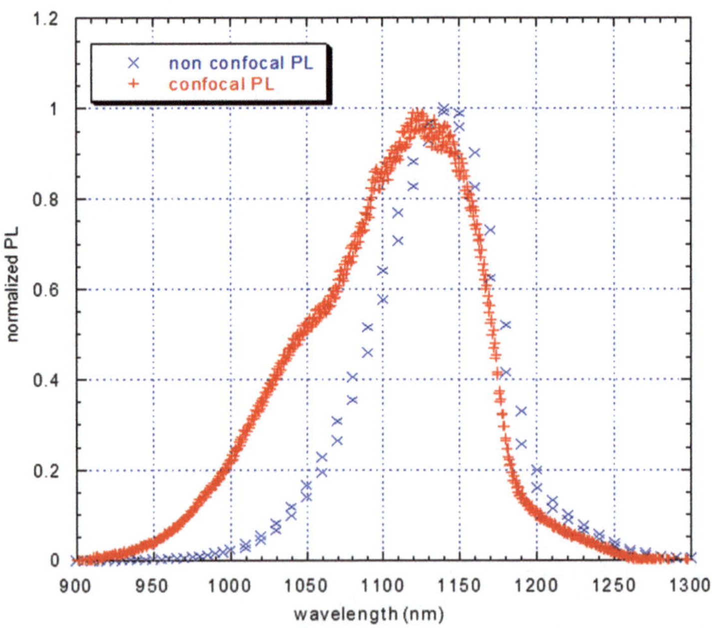

Fig. 36.2 PL spectra of c-Si obtained from a confocal and non-confocal system

(785 nm) was used in both experiments. Depending on the PL setup we observed different PL spectra behavior. At shorter wavelengths, a shoulder is evidenced for the confocal PL system, which does not exist for the non-confocal PL system (Fig. 36.2). The difference between both spectra is explained by the depth scan capabilities of the confocal system that allow one to discriminate between a volume element of photons emitted from the near-surface and the bulk. On the other hand, the non-confocal PL system integrates a PL signal from different depths where reabsorption has a high probability to take place for the more energetic photons; the relative weight of these high energy photons in the measured PL is thus reduced, making the short wavelength shoulder disappear. The following interpretation has been discussed and supported by numerical modeling.

37
Structural, Optoelectronic and Electrical Properties of GaAs Microcrystals Grown from (001) Si Nano-areas

José Alvarez, A. Jaffré, C. Renard, N. Cherkasin, T. Molière, L. Vincent, R. Hamouche, V. Yam, F. Fossard, J.P. Kleider, D. Mencaraglia, and D. Bouchier

Abstract An innovative approach is being investigated to develop III–V compounds on silicon (Si) substrates with the purpose to offer a technological alternative for the development of high efficiency solar cells ($\sim$30 %). Until now, germanium (Ge) substrate has been the privileged material for the development of III–V multijunctions (MJ) solar cells mainly dedicated to space applications. Ge offers several advantages, namely the lattice matching to Si and its use as a bottom cell in the MJ. However, the main drawback remains the cost of Ge substrates, which makes it inappropriate for terrestrial photovoltaic (PV) applications. New routes for high efficiency MJ solar cells are expected through the significant improvements of the selective area epitaxy (Li et al., J Appl Phys 103:106102, 2008; Deura et al., J Cryst Growth 310:4768–4771, 2008; Hsu et al., Appl Phys Lett 99:133115, 2011) allowing defect free III–V compounds to be grown on Si substrates patterned with dielectric films. In this work, Si nanoscale areas opened through a SiO_2 layer (<1 nm) formed on (001) Si have been used to grow GaAs microcrystals by chemical beam epitaxy (CBE) in the temperature range 550–600 °C (Renard et al., Appl Phys Lett 102:191915, 2013). Structural, optoelectronic and electrical properties of GaAs microcrystals have been analyzed at room temperature by micro-Raman, photoluminescence and conductive probe atomic force microscopy (CP-AFM). The fine structure of crystals (facet orientations, crystal defects) has also been

J. Alvarez (✉) • A. Jaffré • T. Molière • D. Mencaraglia • J.P. Kleider
LGEP, CNRS UMR 8507, Supelec, University of Paris-Sud, UPMC Univ Paris 06, 11 rue Joliot- Curie, Plateau de Moulon, F-91192 Gif-sur-Yvette Cedex, France
e-mail: Jose.Alvarez@lgep.supelec.fr

C. Renard • L. Vincent • R. Hamouche • V. Yam • F. Fossard • D. Bouchier
IEF, CNRS UMR 8622, University of Paris-Sud, 91405 Orsay, France

N. Cherkasin
CEMES, UPR CNRS 8011, 29 rue Jeanne Marvig, 31055 Toulouse, France

B. Di Bartolo et al. (eds.), *Nano-Structures for Optics and Photonics*,
NATO Science for Peace and Security Series B: Physics and Biophysics,
DOI 10.1007/978-94-017-9133-5_37,

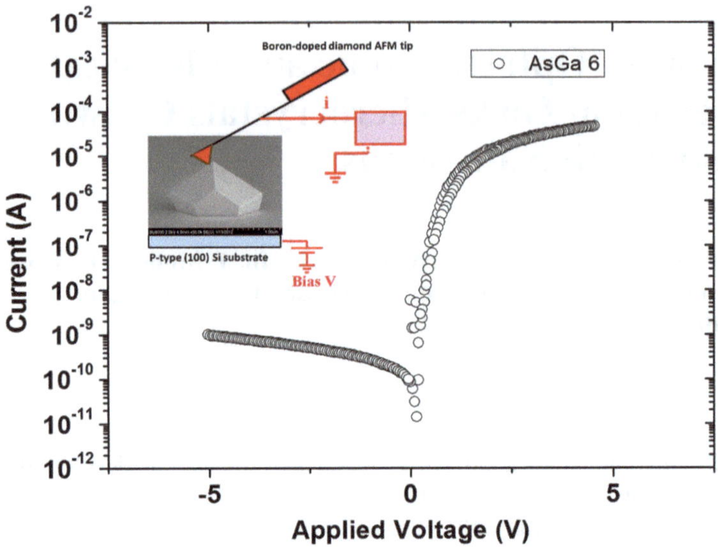

Fig. 37.1 Local I–V measurements performed on a GaAs microcrystal

investigated by transmission electron microscopy (TEM). Linear polarized Raman spectroscopy performed on multiple microcrystals shows exclusively the TO mode which is typically expected for (110) GaAs plane orientations and/or heavily n-type Si-doped GaAs (Zardo et al., Phys Rev B 80:245324, 2009). TEM confirms that all facets are {110}, but unintentionally Si doping cannot be excluded. Indeed, PL measure-ments point out a red shift for the microcrystals for which nucleation seeds were created by silane exposure. CP-AFM imaging of GaAs microcrystals performed at +1 and −1 V, respectively, points out a current rectification behavior confirmed by local I–V measure-ments (Fig. 37.1). These results can be interpreted as a sign of the presence of a p-n junction, which agrees well with the p-type doping of Si substrates used in this study (1–5 Ωcm) and the unintentionally n-type doping of GaAs microcrystals suggested by PL measurements (Pavesi and Henini, Microelectron J 28:717–726, 1997).

Keywords GaAs microcrystals • Si nano-areas

38
Plasmonic Sensors for Aromatic Hydrocarbon Detection

Laura Brigo, M. Cittadini, E. Gazzola, G. Zacco, F. Romanato, M. Guglielmi, A. Martucci, and G. Brusatin

Abstract The development of innovative materials for sensitive and selective gas sensing is a very relevant field for the current nanotechnology research. A strong effort is dedicated to the fabrication of low-cost and efficient nanoscale devices capable of a fast detection. Resistive electrical devices are the most adopted solutions for in-situ and real-time detection, but their main drawbacks are the low selectivity, response drift, electromagnetic noise dependence and need of contact measurements. Optical gas sensors allow to overcome such limits, and could moreover exhibit thermal and mechanical stability, operate at room temperature, and be integrated on-chip. Within this framework, plasmon-based optical devices are knowing an increasing development and diffusion. Herein plasmonic sensors for aromatic hydrocarbon detection are presented. These systems are based on aryl-bridged polysilsesquioxanes (aryl-PSQs), obtained either ① coupling such hybrid films with Au nanoparticles (NPs), aiming to the excitation of localized surface plasmon resonances (LSPRs), or ② depositing them onto metallic waveguiding layers, to form gratings supporting the propagation of surface plasmon polaritons (SPPs). Aryl-PSQs are sol-gel materials characterized by a native controlled porosity and other functionalities (Loy and Shea, Chem Rev 95:1431–1442, 1995; Dabrowski et al., Appl Surf Sci 253:5747–5751, 2007; Brigo et al., Nanotechnology 23:325302, 2012). Temperature programmed desorption investigations of xylene on phenyl-bridged (ph-PSQ) and diphenyl-bridged (diph-PSQ) PSQ films indicate a specific $\pi - \pi$ interaction between the organic component of the films and

L. Brigo (✉) • M. Cittadini • M. Guglielmi • A. Martucci • G. Brusatin
Industrial Engineering Department and INSTM, University of Padova, Padova, 35131, Italy
e-mail: laura.brigo@unipd.it

E. Gazzola • G. Zacco • F. Romanato
Physics and Astronomy Department, University of Padova, Padova, 35131, Italy

B. Di Bartolo et al. (eds.), *Nano-Structures for Optics and Photonics*,
NATO Science for Peace and Security Series B: Physics and Biophysics,
DOI 10.1007/978-94-017-9133-5_38,

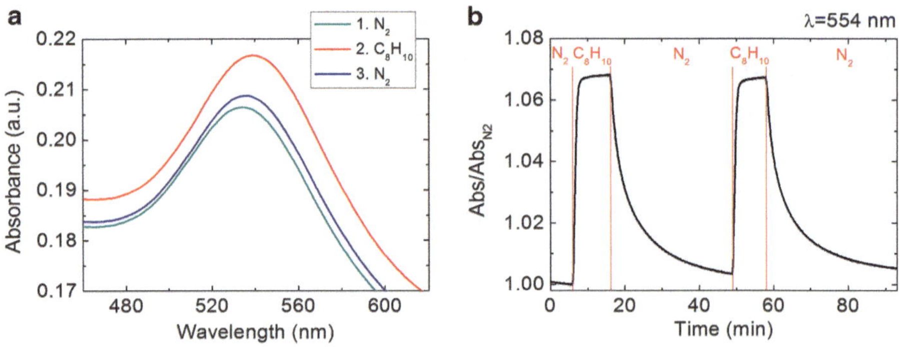

Fig. 38.1 LSPR absorption band and dynamic response of Au NP-based sensor in N_2 and 30 ppm xylene

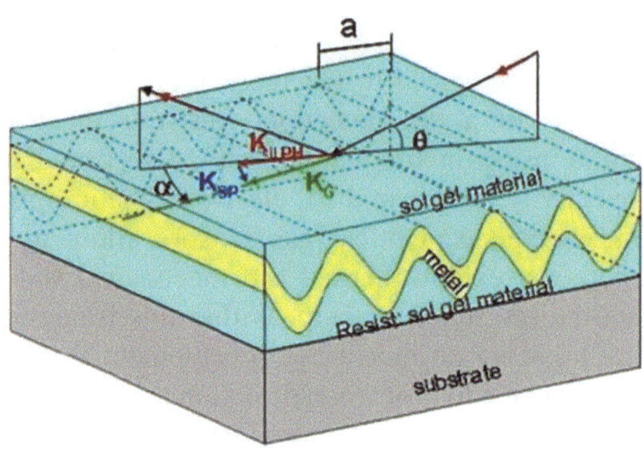

Fig. 38.2 Sketch of the multilayer grating for LR and SR SPP excitation

xylene molecules: the interaction energy is quantified in 38 ± 14 kJ/mol and 115 ± 13 kJ/mol, respectively (Brigo et al., J Mater Chem C 1:4252, 2013). For type ① sensors, a thin film of aryl-PSQ was deposited on a submonolayer of Au NPs coating a fused silica substrate. These sensors were tested monitoring the variation of the LSPRs under cycles of exposure to N_2 and to 30 ppm xylene in N_2. Figure 38.1a shows that in the presence of xylene molecules, the resonance undergoes an intensity increase and red-shift. An increase in refractive index of the dielectric, as a consequence of the benzene-xylene coupling, might determine the resonance shift to lower frequencies. The plot of Fig. 38.1b shows the dynamic response of a 150 nm thick sensor: response intensity of 0.068, response time of 1′ and recovery time of 20′. For type ②, sinusoidal surface plasmon gratings were embedded in ph-PSQs (Fig. 38.2). The sensor exhibits two SPP modes: the Long Range (LR) and the Short Range (SR) SPP. Figure 38.3 shows the reflectance upon exposure to 30 ppm xylene, and after regeneration. A 2.9 ± 0.9 nm red-shift of the LR and a 2.5 ± 1.3 nm red-shift of the SR were correlated to the interaction with the analyte. Collected data are compatible with theoretical predictions assuming a film

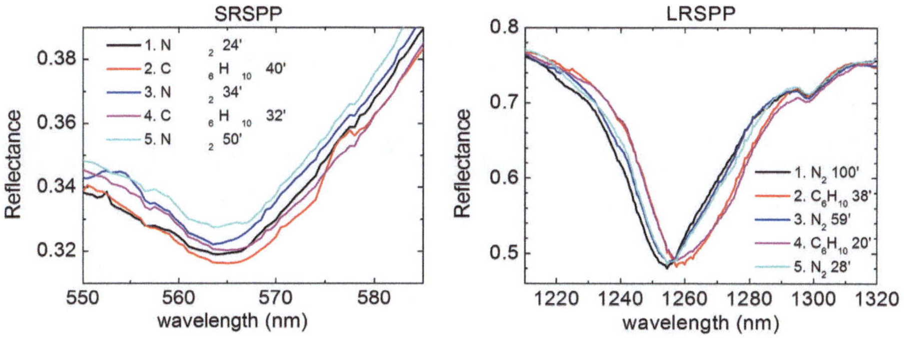

Fig. 38.3 Reflectance spectra measured in N_2 atmosphere and in 30 ppm xylene

Δn of 0.011 ± 0.005 (Brigo et al., Nanotechnology 24:155502, 2013). The reported sensors demonstrate a major performance in terms of sensitivity, ease in fabrication procedure, and promising dynamics.

Keywords Plasmonics • Gas sensing • Bridged polysilsesquioxanes • Sol-gel process

39
Optical Annealing of Black Silicon

Benjamin Franta, D. Pastor, M. Aziz, and E. Mazur

Abstract Silicon hyperdoped with chalcogens using femtosecond (fs) laser irradiation ("black silicon") shows strong broad-band absorption due to intermediate band formation and surface texturing; thus, this material has promise for advanced optoelectronic devices such as silicon-based infrared photodetectors and intermediate band solar cells. To reduce structural defects and form a rectifying junction after fs laser irradiation, however, annealing is necessary, which deactivates the optical effect of the dopant. Here, we investigated the use of laser-based optical annealing using nanosecond (ns) excimer pulses and found that it removes amorphous material resulting from the black silicon fabrication process while maintaining high above-bandgap and sub-bandgap absorption. We also studied the effect of ns optical annealing on the surface morphology of black silicon.

B. Franta (✉) • D. Pastor • M. Aziz • E. Mazur
School of Engineering and Applied Sciences, Harvard University, Cambridge, MA, USA
e-mail: bafranta@fas.harvard.edu

B. Di Bartolo et al. (eds.), *Nano-Structures for Optics and Photonics*,
NATO Science for Peace and Security Series B: Physics and Biophysics,
DOI 10.1007/978-94-017-9133-5_39,

40
Strongly Confined Gap Plasmon Modes in Graphene Sandwiches and Graphene-on-Silicon

Yan Francescato, V. Giannini, and S.A. Maier

Abstract Graphene has emerged as a radically new platform in nanotechnology and its tunable optical and electrical properties make it a material of choice for future nanocircuitry (Vakil and Engheta, Science 332(6035):1291–1294, 2011). A key element towards actual devices is the exploration of graphene nanostructures. For instance, graphene nanoribbons, readily fabricated by electron beam lithography, have been shown to be attractive waveguides for plasmons. These bound surface waves arising from the coupling between light and collective oscillations of the charge carriers exhibit indeed unusually strong confinement in graphene (Nikitin et al., Phys Rev B 84:161407, 2011; Christensen et al., ACS Nano 6(1):431–440, 2012). Our work focuses on the physics and the classification of plasmon waveguide modes in structures consisting of two infinitely long graphene ribbons vertically offset by a gap, a "sandwich" geometry. We find strongly hybridized plasmonic modes, some of which are tightly confined within the gap region, and therefore hold promise for nanodevices (Francescato et al., New J Phys 15(6):063020, 2013). In order to aid the understanding of the fundamental physics of the different classes of waveguide modes encountered, we introduce a convention for plotting the mode spectrum which allows to group the modes by shared characteristics. This representation is particularly useful when coupling occurs, because the mode density increases considerably. In this manner, and varying the critical parameters of width, gap and operation wavelength, different regimes, coupling mechanisms and mode families can be recognized. We confirm our findings by considering experimentally realizable systems with tunable graphene doping in a geometry where a single ribbon is placed on top of a highly doped silicon substrate via a dielectric spacer layer. Remarkably, we show that the new gap modes still survive in the latter case. More, we report on an unprecedented level of confinement of a terahertz wave of nearly 5 orders of magnitude. Because of their remarkable

---

Y. Francescato (✉) • V. Giannini • S.A. Maier
The Blackett Laboratory, Imperial College London, London SW7 2AZ, UK
e-mail: yan.francescato10@imperial.ac.uk

B. Di Bartolo et al. (eds.), *Nano-Structures for Optics and Photonics*, NATO Science for Peace and Security Series B: Physics and Biophysics,
DOI 10.1007/978-94-017-9133-5_40, © Springer Science+Business Media Dordrecht 2015

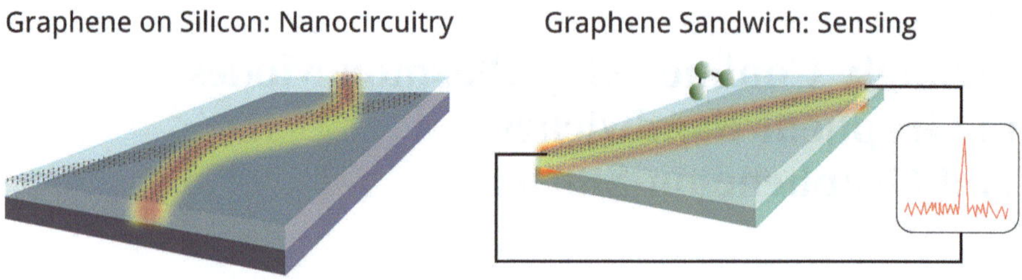

Fig. 40.1 *Left*: When graphene is placed on top of a conducting substrate separated by a thin dielectric spacer such as an oxide or an organic layer, extremely confined modes within this gap can transfer information across the device acting like building blocks for nanocircuits. *Right*: two graphene ribbons stacked around a dielectric spacer can support very sensitive modes right at their surface and therefore could detect electrically single adsorption event

field distributions and extreme confinement, the families of modes presented here could serve as the building blocks for both graphene-based integrated optics and ultrasensitive sensing modalities (Francescato et al., New J Phys 15(6):063020, 2013) (Fig. 40.1).

Keywords Strongly confined plasmon • Graphene ribbon sandwich • Graphene on silicon

41
Investigation of Polarized Light Emitting Diode

Tibor Gál, S. Örs, and P. Koppa

Abstract The significance of LED technology in the lighting industry is increasing abruptly. However many application (LCD backlight, vehicle lights) require polarized light sources. For these applications polarized light is produced by polar filters, which limits the efficiency to less than 50 %. The goal of this research was to develop a LED providing polarized light, so it is not necessary to use polar filters. The object of this work was to investigate oriented metallic nanoparticles to improve the efficiency and to manipulate the polarization of the outcoming light of a LED source. To calculate the optical properties of this problem it is necessary to solve the Maxwells equations with numerical methods. For this purpose the finite element method (FEM) was used. We observed that this interaction is strongly dependent on the polarization, distance, and orientation of the particles.

T. Gál (✉) • S. Örs • P. Koppa
Department of Atomic Physics, Budapest University of Technology and Economics, Budapest, Hungary
e-mail: gal.tiborpeter@gmail.com

B. Di Bartolo et al. (eds.), *Nano-Structures for Optics and Photonics*,
NATO Science for Peace and Security Series B: Physics and Biophysics,
DOI 10.1007/978-94-017-9133-5_41,

B. Di Bartolo et al. (eds.), Nano-Structures for Optics and Photonics,
NATO Science for Peace and Security Series B: Physics and Biophysics,
DOI 10.1007/978-94-017-9133-5_41, © Springer Science+Business Media Dordrecht 2015

42 Linking Optical Properties and Nanostructure of NiCrOx Cermet Nanocomposite for Solar Thermal Application

Lucie Gaouyat, Z. He, Jean-François Colomer, D. Schryvers, F. Mirabella, and Olivier Deparis

Abstract Conversion of solar energy into thermal energy helps reducing consumption of non-renewable energies. Cermets (ceramic-metal composites) are versatile materials suitable, amongst other applications, for solar selective absorbers. The presence of metallic Ni nanoparticles (NPs) in the dielectric matrix is a prerequisite for efficient solar selective absorption in NiCrOx cermets (Zhao and Wackelgard, Solar Energy Mater Solar Cells 90:1861–1874, 2006). By combining comprehensive chemical and structural analyses, but also electronic microscopy, we revealed the origin of the remarkable optical properties of this cermet material. A solar absorber multilayer stack on aluminium substrate using a sputtered NiCrOx layers allows us to achieve solar absorptance as high as $\alpha = 96.1\,\%$ while keeping thermal emissivity as low as $\epsilon = 2.2\,\%$ both values are comparable to best values recorded so far (Buhrman, Physics of solar selective surfaces. In: Ber (ed) Advances in solar energy. Springer, New York, pp 207–282, 1986). Determining the origin of sputtered NiCrOx cermets optical properties drives further optimization of solar absorbers.

L. Gaouyat (✉) • J.-F. Colomer • O. Deparis
Department of Physics, University of Namur, rue de Bruxelles, 61, B-5000, Namur, Belgium
e-mail: lucie.gaouyat@unamur.be

Z. He • D. Schryvers
Electron Microscopy for Materials Research (EMAT), University of Antwerp, Groenenborgerlaan 171, Antwerp B-2020, Belgium

F. Mirabella
CRM Group AC&CS, Boulevard de Colonster 57B, Liege B-4000, Belgium

B. Di Bartolo et al. (eds.), *Nano-Structures for Optics and Photonics*,
NATO Science for Peace and Security Series B: Physics and Biophysics,
DOI 10.1007/978-94-017-9133-5_42,

[illegible]

[illegible]

B. Di Bartolo et al. (eds.), *Nano-Structures for Optics and Photonics*,
NATO Science for Peace and Security Series B: Physics and Biophysics,
DOI 10.1007/978-94-017-9133-5_32, © Springer Science+Business Media Dordrecht 2015

43
Ultrafast Active Control of Plasmonic Resonances at THz Frequencies

Giorgos Georgiou, M.C. Schaafsma, and J.G. Rivas

Abstract Semiconductors are promising materials for THz plasmonics. They acquire metallic behavior when sufficient free carriers are present. Plasmonic structures fabricated out of these semiconductor materials can sustain localized surface plasmon polaritons (LSPPs). The plasmonic behavior of such structures is determined by both their geometry and dielectric properties. The carrier density – hence the plasmonic behavior – can be actively controlled for a given geometry by optical excitation of free carriers in the semiconductor.

We have demonstrated that by ultrafast optical pumping of free carriers in a random array of Si bowtie antennas, it is possible to activate LSPPs on picosecond timescales (Berrier et al. Opt Express 18:23226–23235, 2010). Bowtie antennas, Fig. 43.1, have the ability to resonantly enhance the field intensity at the gap separating the two triangular resonators, which allows the concentration of THz radiation beyond the diffraction limit. THz field intensity enhancements of several orders of magnitude in deep subwavelength volumes have been predicted (Giannini et al. Opt Express 18(3):2797–2807, 2010). These large local field enhancements have been already used for the enhanced detection of nanometric inorganic layers and biological films (Berrier et al. Opt Express 20(5):5052–5060, 2012; Berrier et al. Biomed Opt Express 3(11):2937–2949, 2012).

In this contribution we will show the first THz measurements of the enhanced resonant extinction of single semiconductor plasmonic antennas. This demonstration is achieved by placing the antenna at the output aperture of a conically tapered waveguide, which enhances the intensity of the incident THz field at the antenna position by a factor of 10, and suppresses the background radiation that otherwise is transmitted without being scattered by the antenna (Schaafsma et al. New J Phys 15(015006):1–14, 2013). This far-field investigation of single plasmonic antennas

G. Georgiou (✉) • M.C. Schaafsma • J.G. Rivas
Center for Nanophotonics, FOM Institute AMOLF, Science Park 104, 1098XG, Amsterdam, The Netherlands
e-mail: georgiou@amolf.nl

B. Di Bartolo et al. (eds.), *Nano-Structures for Optics and Photonics*, NATO Science for Peace and Security Series B: Physics and Biophysics, DOI 10.1007/978-94-017-9133-5_43,

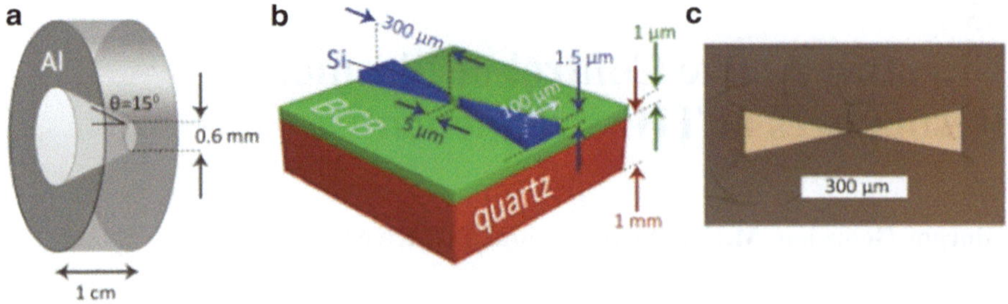

Fig. 43.1 Schematic representation of a conical waveguide (**a**) and a bowtie antenna (**b**). Optical microscope image of a bowtie antenna (**c**). The antenna is formed by two triangular monomers of Si, bonded onto a quartz substrate with BCB. The cracks in the microscope image are in the supporting BCB layer

may open the possibility of THz time-domain spectroscopy of single nanostructures by bridging together length scales as unequal as the THz wavelengths and the dimensions of nanostructures.

Keywords Terahertz THz • Plasmonics • Bow-tie antennas • Localized surface plasmon resonance • Semiconductors

44 An Eigenvector-Expansion Method for Localized Plasmon Modes: Application to Extinction and Electron Energy Loss Spectra of Isolated and Coupled Metallic Nanoparticles

Stéphane-Olivier Guillaume, F. Javier García de Abajo, and Luc Henrard

Abstract The dominant features in the UV/visible spectra of metallic nanoparticles are governed by collective oscillations of valence electrons. These resonances are called localized surface plasmon (LSP) and have positions, widths and intensities that depends strongly on the nanoparticle's geometry and environment but also on the nature of the probing excitation. When one is interested in coupled nanoparticles, the plasmon modes of these systems can be described with an hybridization scheme (Prodan et al. Science 302(5644):419–422, 2003). With help of this approach, modes can be classified in "bright" or "dark" mode according to whether they have a net dipole moment. In the last case, the mode cannot be excited with traditional optical excitation but one has to use a focused beam of fast electrons having electric field with low spatial extension that allows local stimulation (Nelayah et al. Nat Phys 3:348–353, 2007). Here, we are interested in the simulation of the extinction and electron energy loss (EEL) spectra of isolated and coupled metallic particles. To achieve this, we have developed a new approach based on the Discrete-Dipole Approximation (DDA) (Draine, Astrophys J 333:848–872, 1988; Geuquet and Henrard, Ultramicroscopy 110:1075–1080, 2010) that uses eigenvectors of the particle's propagator to make a truncated basis for physical quantities (Guillaume et al. Phys Rev B 88(245439), 2013). The response of the cluster of particles is then obtained by adding interaction terms which account for multiple scattering between the particles.

S.-O. Guillaume (✉) • L. Henrard
Department of Physics, University of Namur, rue de Bruxelles, 61, B-5000, Namur, Belgium
e-mail: stephane-olivier.guillaume@unamur.be

F.J. García de Abajo
ICFO–Institut de Ciencies Fotoniques, Mediterranean Technology Park, Av. Carl Friedrich Gauss 3, 08860 Castelldefels (Barcelona), Spain

B. Di Bartolo et al. (eds.), *Nano-Structures for Optics and Photonics*, NATO Science for Peace and Security Series B: Physics and Biophysics,
DOI 10.1007/978-94-017-9133-5_44,

Keywords Discrete-dipole approximation • Localized plasmon • Metallic nanoparticle • Coupling • Hybridization • Electron energy-loss spectroscopy • Optical extinction

45
Characterisations of New Nd^{3+}-Doped Scheelite-Type Molybdates for Laser Materials

Malgorzata Guzik, E. Tomaszewicz, Yannick Guyot, P. Tomaszewicz, J. Legendziewicz, and George Boulon

Abstract The necessity to develop new laser materials was our motivation to study a new family of molybdates activated by the Nd^{3+} ions, which crystallizes in the scheelite-type structure (the tetragonal symmetry, the space group $I4_{1/a}$). A series of $Cd_{1\text{-}3x}Nd_{2x-x}MoO_4$ solid solutions with concentration of Nd^{3+} ions from 0.1 to 66 mol % with respect Cd^{2+} ions have been successfully synthesized by a high-temperature annealing, using $CdMoO_4$ and $Nd_2(MoO_4)_3$ as the starting materials. Structural and spectroscopic characterizations of Nd^{3+}in substitution of Cd^{2+} were carried out. The Nd^{3+} ions in this matrix most probably occupy two non-equivalent symmetry sites (see Fig. 45.1). The substitution of Cd^{2+} by trivalent Nd^{3+} cations leads to the formation of cationic vacancies in a framework (which are denoted in the chemical formula as _) and its concentration depends essentially on the composition of initial $CdMoO_4/Nd_2(MoO_4)_3$ mixtures. The analysis of the morphology using SEM reveals high homogeneity of the spherical-shape products with the average grain size of about 10 μm. The optical analysis and the laser performance parameters suggest $Cd_{1\text{-}3x}Nd_{2x-x}MoO_4$ as a potential laser powder material: both the values of the absorption cross-section and the very strong emission of the $^4F_{3/2} \longrightarrow {}^4I_{11/2}$ laser channel of Nd^{3+} recorded under Xe

M. Guzik (✉) • J. Legendziewicz
Faculty of Chemistry, Chemistry Department, University of Wroclaw, Joliot-Curie 14, PL-50-383 Wroclaw, Poland
e-mail: goguzik@poczta.fm

E. Tomaszewicz
Department of Inorganic and Analytical Chemistry, West Pomeranian University of Technology, Al. Piastw 42, PL- 71-065 Szczecin, Poland

Y. Guyot • G. Boulon
Institute Light Matter, UMR5306 CNRS-University of Lyon 1, University of Lyon, 69622 Villeurbanne, France

P. Tomaszewicz
Institute of Low Temperature and Structure Research, PAS, Okolna 2,50-422 Wroclaw, Poland

B. Di Bartolo et al. (eds.), *Nano-Structures for Optics and Photonics*,
NATO Science for Peace and Security Series B: Physics and Biophysics,
DOI 10.1007/978-94-017-9133-5_45, © Springer Science+Business Media Dordrecht 2015

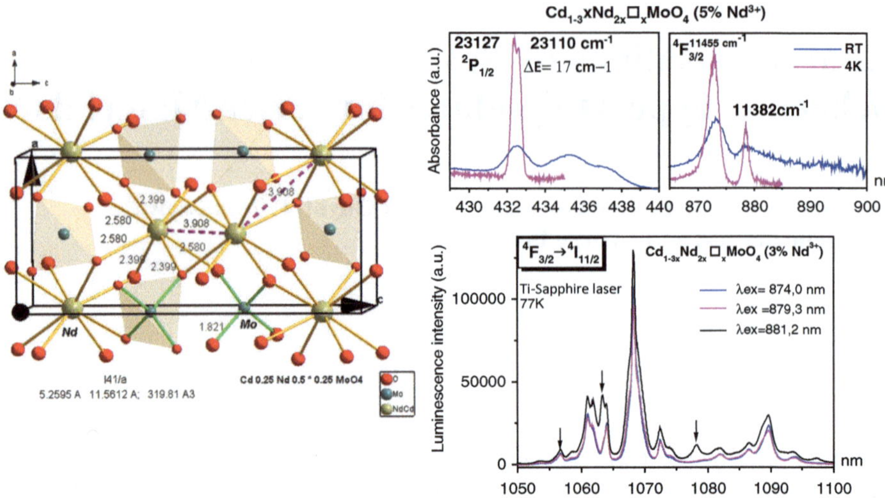

Fig. 45.1 (*left*) The structure of low-polymorphic modification of $Cd_{1\text{-}3x}Nd_{2x\ x}MoO_4$ solid solutions, (*top*) Selected transitions ($^4I_{9/2} \longrightarrow {}^2P_{1/2}$, and $^4I_{9/2} \longrightarrow {}^4F_{3/2}$) of the absorption spectra of $Cd_{1\text{-}3x}Nd_{2x-x}MoO_4$ at 298 and 4K, (*bottom*) Emission spectra of $Cd_{1\text{-}3x}Nd_{2x-x}MoO_4$ (3 % Nd^{3+}) under selective Ti-Shappire laser line at 77 K

lamp excitation or OPO laser pumping, as well as, the radiative lifetimes of IR luminescence are appropriate for potential applications of this phosphor as a solid-state laser. Our earlier studies have shown that the presence of the cationic vacancies in the framework may significantly improve the laser parameters (Guzik et al., J Mater Chem 22:14896, 2012).

46
Cluster Implantation and Deposition Apparatus: Design and Capabilities

Muhammad Hanif and Vladimir Popok

Abstract Aggregates of atoms and molecules (clusters) can be considered to be a distinct form of matter, a "bridge" between atoms on the one hand and bulk on the other. Cluster can be used as models for investigation of fundamental physical aspects of the transition from atomic to macro scale as well as for fabrication of nanostructured materials with unique range of optical, magnetic, mechanical and catalytic properties (Jena and Castleman, Proc Natl Acad Sci U S A 103:10560–9, 2006; Wegner et al. J Phys D Appl Phys 39:R439–R459, 2006; Toyoda and Yamada, IEEE Trans Plasma Sci 36:1471–1488, 2008; Popok and Campbell, Rev Adv Mater Sci 11:19–45, 2006). Clusters kinetic energy is an important parameter which defines the cluster-surface interaction regime. These regimes can vary from soft landing to implantation (see Fig. 46.1). Different approaches can be used for generation of clusters in vacuum offering a wide choice of clusters species and providing a possibility for control of cluster size and energy distribution (Popok and Campbell, Rev Adv Mater Sci 11:19–45, 2006; Popok et al. Surf Sci Rep 66:347–377, 2011). In the current report, a design and capabilities of a cluster implantation and deposition apparatus (CIDA) involving two different cluster sources are described. The experimental setup represents a modified and upgraded CIDA described in detail in (Popok et al. Rev Sci Instrum 73:4283, 2002). Schematic picture of the new apparatus is shown in Fig. 46.2. One of the sources is a new model of a pulsed cluster source (PuCluS-2) based on a high repetition rate Even-Lavie valve. This source can be used for the production of cluster ions from various gas phase precursors. The core part of the other part is commercial gas condensation nanocluster source (NC200U, Oxford Applied Research Ltd.). This source is intended for the production of metal clusters. Both sources are connected to CIDA where the size-selection can be carried out and cluster kinetic energy can

M. Hanif (✉) • V. Popok
Department of Physics and Nanotechnology, Aalborg University, Skjervej 4A, 9220 Aalborg, Denmark
e-mail: mh@nano.aau.dk; vp@nano.aau.dk

B. Di Bartolo et al. (eds.), *Nano-Structures for Optics and Photonics*,
NATO Science for Peace and Security Series B: Physics and Biophysics,
DOI 10.1007/978-94-017-9133-5_46,

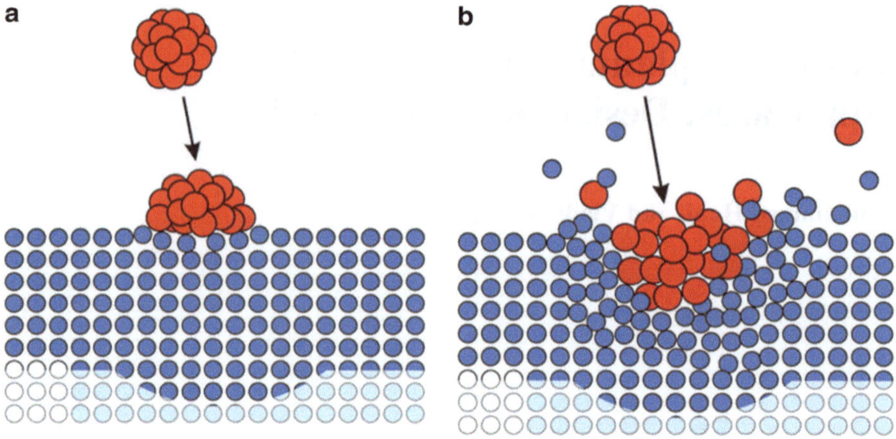

Fig. 46.1 Schematic pictures of (**a**) cluster deposition and (**b**) cluster implantation

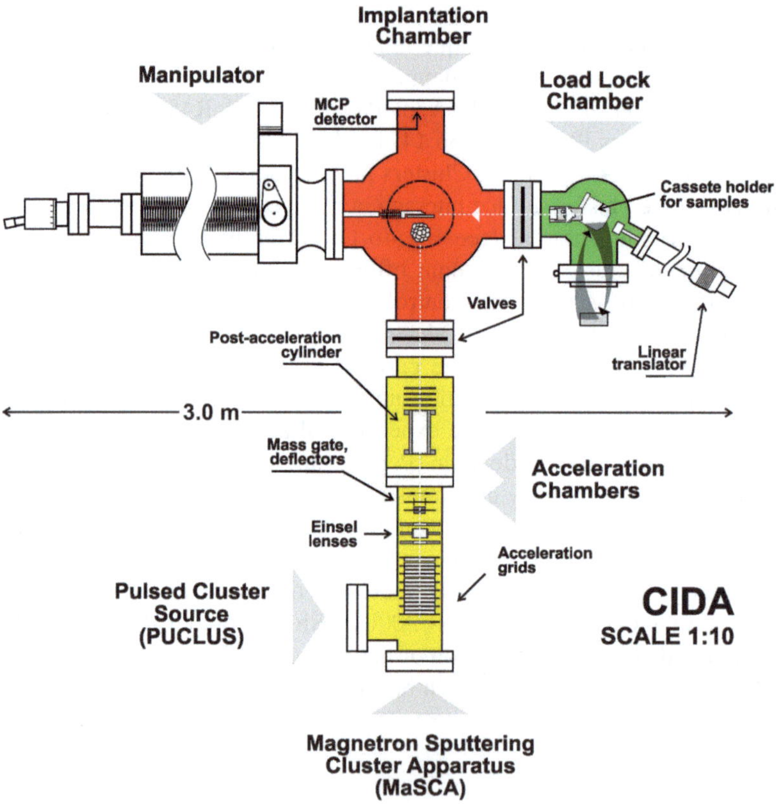

Fig. 46.2 Schematic drawing of CIDA containing a magnetron sputtering source and pulsed cluster source from gaseous precursors

be varied up to ca. 20 keV/cluster. The clusters can be deposited on or implanted in different substrates at ultra-high vacuum conditions. The clusters produced from gas precursors (Ar, N etc.) by PuCluS-2 can be used to study cluster ion implantation in order to develop contributions to the theory of cluster stopping in matter as well as for practical applications requiring ultra-shallow implantation and modification of surfaces on the nanoscale. Metal clusters from the magnetron cluster source are of interest for the production of optical sensors to detect specific biological objects (for instance, enzymes and proteins) as well as for the enhancement of photocatalytic reactions, for example, in hydrogen production. In both cases the phenomenon of localised surface plasmon resonance is utilized. Regime of metal cluster pinning can be used for processing of graphene.

Keywords Cluster beams • Cluster deposition • Cluster implantation • Surface plasmon resonance

47
Roles of Surface Pattern Morphology and Sunlight Incoherence on Solar Cell Optimization

Aline Herman, M. Sarrazin, and Olivier Deparis

Abstract In order to improve the light absorption efficiency of ultra-thin solar cells, front-side or/and back-side surface texturing are commonly used. This aim is to help coupling incident light into the active layer of the solar cell via light trapping. It has been shown that the morphology of the corrugation pattern has a strong influence on the solar cell efficiency (Herman et al. J Appl Phys 112(11):113107, 2012). However, most of the current optimizations consider that the cell is illuminated under coherent light. This hypothesis is not accurate since the sun is an incoherent source, with a coherence time around 3 fs (Hecht E, Optics, 4th edn. Addison-Wesley, Reading, 2001). Recently, we developed a rigorous theory allowing to take into account the effects of the temporal incoherence of light (Sarrazin et al. Opt Express 21(Suppl 4):A616–A630, 2013). Using this theory we showed that the photocurrent produced by a thin-film solar cell strongly depends on the coherence time of the incident light (Herman et al. New J Phys 16:013022, 2014). Therefore, optimal geometries found under coherent light should be reconsidered because they do not necessarily lead to maximum efficiency under incoherent light.

A. Herman (✉) • M. Sarrazin • O. Deparis
Department of Physics, University of Namur, rue de Bruxelles, 61, B-5000, Namur, Belgium
e-mail: aline.herman@unamur.be

B. Di Bartolo et al. (eds.), *Nano-Structures for Optics and Photonics*,
NATO Science for Peace and Security Series B: Physics and Biophysics,
DOI 10.1007/978-94-017-9133-5_47,

48
Processing and Characterization of Nanowire Arrays for Photodetectors

Vishal Jain, J. Wallentin, A. Nowzari, M. Heurlin, D. Asoli, M.T. Borgström, F. Capasso, L. Samuelson, and H. Pettersson

Abstract We present a fabrication scheme of contacting arrays of vertically standing nanowires (NW) for LEDs (Duan et al. Nature 409:66–69, 2001), photodetectors (Wang et al. Science (NY) 293:1455–1457, 2001) or solar cell applications (Wallentin et al. Science (NY) 339:1057–1060, 2013). Samples were prepared by depositing Au films using nano-imprint lithography (Må rtensson et al. Nano Lett 4:699–702, 2004) which are used as catalysts for NW growth in a low-pressure metal organic vapour phase epitaxy system where III-V precursors and dopant gases are flown at elevated temperatures which lead to the formation of NWs with different segments (Borgström et al. Nano Res 3:264–270, 2010). An insulating SiO_2 layer is then deposited and etched from the top segments of the NWs followed by sputtering of a transparent top conducting oxide and opening up $1 \times 1\,\text{mm}^2$ device areas through a UV lithography step and etching of the top contact from non-device areas. A second UV lithography step was subsequently carried out to open up smaller windows on the ITO squares for bond pad definition, followed by metallization and lift-off; and the substrate is used as back contact. We also report on the electrical and optical properties of near-infrared $p^+ - i - n^+$ photodetectors/solar cells based

V. Jain (✉) • H. Pettersson
Solid State Physics and the Nanometer Structure Consortium, Lund University, Box 118, SE-22100 Lund, Sweden

Laboratory of Mathematics, Physics and Electrical Engineering, Halmstad University, Box 823, SE-301 18 Halmstad, Sweden
e-mail: vishal.jain@ftf.lth.se

J. Wallentin • A. Nowzari • M. Heurlin • M.T. Borgström • L. Samuelson
Solid State Physics and the Nanometer Structure Consortium, Lund University, Box 118, SE-22100 Lund, Sweden

D. Asoli
Sol Voltaics AB, Ideon Science Park, Scheelevägen 17, SE-22370 Lund, Sweden

F. Capasso
School of Engineering and Applied Sciences, Harvard University, Cambridge, MA 02138, USA

B. Di Bartolo et al. (eds.), *Nano-Structures for Optics and Photonics*, NATO Science for Peace and Security Series B: Physics and Biophysics, DOI 10.1007/978-94-017-9133-5_48, © Springer Science+Business Media Dordrecht 2015

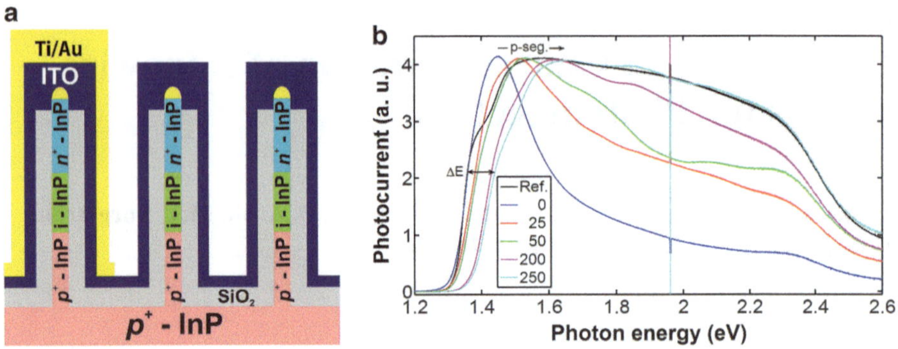

Fig. 48.1 (**a**) Schematic detector layout. (**b**) Spectrally resolved short-circuit photocurrent of samples with varying p-segment lengths measured at 300 K along with a planar InP mesa sample. The intensity has been normalized for comparison. http://link.springer.com/article/10.1007%2Fs12274-014-0422-2

on square millimeter ensembles of InP nanowires grown on InP substrates. The study includes a sample series where the p^+-segment length was varied between 0 and 250 nm, as well as solar cell samples with 9.3 % efficiency with similar design. The NWs have a complex modulated crystal structure of alternating wurtzite and zincblende segments, a polytypism that depends on dopant type. The electrical data for all samples display excellent rectifying behavior with an ideality factor of about 2 at 300 K. From spectrally resolved photocurrent measurements, we conclude that the photocurrent generation process depends strongly on the p^+-segment length. Without p^+-segment in the NWs, photogenerated carriers funneled from the substrate into the NWs contribute significantly to the photocurrent. Adding a p^+-segment shifts the depletion region up into the i-region of the NWs reducing the substrate contribution to photocurrent while strongly improving the collections of carriers generated in the NWs, in agreement with theoretical modeling (Fig. 48.1).

Keywords Nanophotonics • Nanowires • Infrared (IR) • Photodetectors • Solar cells

49
Structural Analysis of Ge-Ga-Se/S-CsCl Glasses and Glass-Ceramics for Application in Chalcogenide and Chalcohalide Photonics

Halyna Klym, A. Ingram, O. Shpotyuk, and L. Calvez

Abstract The current state-of-the-art of the modern photonics indicates on significant importance of glassy-like multifunctional media with improved exploitation proper-ties. For example, chalcogenide and chalcohalide glasses were considered attractive materials for infrared (IR) optics because of their excellent IR transmittance, low-cost raw material and relatively simple molding process (Yang Z et al. Mater Res Bull 43:3596–3600, 2008). The nearest atomic arrangement in a glass and glass-ceramics can be adequately studied with numerous experimental measuring techniques. However, the row of probes available to study atomic-deficient distribution is rather limited, especially at sub-nanometer scale. One of the best (if not unique) techniques capable to probe such finest free volumes is the positron annihilation lifetime (PAL) spectroscopy, the method grounded on physical phenomena of positron-electron interaction in a matter (Klym et al. physica status solidi (c) 4:715–718, 2007). In the present paper, we imply, for the first time, the PAL method to reveal changes in free-volume structure of $80GeSe_2$-$20Ga_2Se_3$ chalcogenide glasses (ChG) subjected to thermally-activated crystallization and $(80GeS_2\text{-}20Ga_2S_3)_{100-x}(CsCl)_x$, $0 < x < 15$, chalcohalide glass-ceramic composites. The PAL spectra were recorded with conventional system (ORTEC) of 230 ps resolution at the temperature T = 22°C and relative humidity RH = 35 %. The measured PAL spectra of ChG were processed with standard LT 9.0 computer

H. Klym (✉)
Lviv Polytechnic National University, 12 Bandera str., Lviv, 79013, Ukraine
e-mail: klymha@yahoo.com

A. Ingram
Opole University of Technology, 75 Ozimska str., Opole, 45370, Poland

O. Shpotyuk
Lviv Institute of Materials of SRC Carat, 202, Stryjska str., Lviv, 79031, Ukraine

L. Calvez
Equipe Verres et Céramiques, UMR-CNRS 6226, Institute des Sciences chimiques de Rennes, Université de Rennes 1, 35042 Rennes Cedex, France

B. Di Bartolo et al. (eds.), *Nano-Structures for Optics and Photonics*, NATO Science for Peace and Security Series B: Physics and Biophysics, DOI 10.1007/978-94-017-9133-5_49, © Springer Science+Business Media Dordrecht 2015

Table 49.1 Fitting parameters and positron trapping modes for PAL spectra of $80GeSe_2$-$20Ga_2Se_3$ and $80GeS_2$-$20Ga_2S_3$ glasses

	Fitting parameters				Positron trapping modes		
	τ_1	I_1	τ_2	I_2	τ_{av}	τ_b	κ_d
Sample	ns	a.u.	ns	a.u.	ns	ns	ns^{-1}
$80GeSe_2$-$20Ga_2Se_3$/base	0.214	0.69	0.439	0.31	0.283	0.253	0.74
$80GeSe_2$-$20Ga_2Se_3$/10 h	0.218	0.71	0.453	0.29	0.287	0.257	0.70
$80GeSe_2$-$20Ga_2Se_3$/25 h	0.216	0.65	0.426	0.35	0.288	0.260	0.79
$80GeSe_2$-$20Ga_2Se_3$/50 h	0.209	0.65	0.416	0.35	0.281	0.252	0.83
$80GeS_2$-$20Ga_2S_3$/base	0.234	0.67	0.481	0.33	0.32	0.28	0.73

program, the obtained curve being fitted by two components with τ_1, τ_2 lifetimes and I_1, I_2 intensities ($I_1 + I_2 = 1$). Therefore, the positron trapping modes in the studied ChG, e.g. average positron lifetimes τ_{av}, positron lifetime in defect-free bulk τ_b, and positron trapping rate in defects κ_d were calculated using a formalism of two-states trapping model (Klym et al. physica status solidi (c) 4:715–718, 2007). It is shown that with increase of annealing duration from $80GeSe_2$-$20Ga_2Se_3$ glass to samples annealed for 10 h, the lifetimes τ_2 increases and I_2 intensities decreases to 0.29 because of void expansion (Table 49.1). This trend correspondingly reduces the positron trapping rate in defect κ_d without significant changes in τ_{av} and τ_b lifetimes. With further annealing duration proceeding to 25 and 50 h, the I_2 intensity ceases to increase, while lifetime τ_2 appreciably decreases to 0.426 and 0.416 ns, respectively. These changes result in increase of positron trapping rate κ_d. During crystallization, the glass structure relaxes towards more thermodynamically favourable state (crystallization shrinkage or densification), eliminating the excess of free volume of neighbouring voids. It means that existing free volume voids either disappear or convert into a greater number of smaller ones. We can argue that crystallization in $80GeSe_2$-$20Ga_2Se_3$ glass induced by long-term annealing (25–50 h) is accompanied by void growth because their expansion and appearance of new tiny voids due to mismatch between glassy and crystalline regions. So the fragmentation of larger free-volume entities into smaller ones occurs. Such process is accompanied by essential decrease in τ_2 lifetime and correspond-ing increase in I_2 intensity in full agreement with above scenario. The described crystallization precedes opposite changes caused by nucleation in $80GeSe_2$-$20Ga_2Se_3$ glasses at lower annealing times (10 h). Main trends in $(80GeS_2\text{-}20Ga_2S_3)_{100-x}(CsCl)_x$, $0 < x < 15$, chalcohalide glass-ceramic composites are shown in τ_{av}.

50
Regularities in Holographic Formation of Periodic Structures in the System Polymer-Metal Nanoparticles

Lyudmyla M. Kokhtych, A.S. Kutsenko, and T.N. Smirnova

Abstract We represent a new method for distribution of metal nanoparticles (NP) in a polymer matrix, which enables to synthesize NP from the metal precursor preliminary distributed in this polymer matrix by using the light interference field (Smirnova et al. Nanotechnology 20(40):405301–405312, 2009). Nanostructures based on noble metal NP embedded into transparent matrix possess specific optical and electric properties. Their unique properties related mainly with excitation of local surface plasmons attract considerable interest both from the viewpoint of their fundamental investigations and practical applications (Kreibig and Vollmer, Optical properties of metal cluster. Springer, Berlin, 1995; Nicolais and Carotenuto, Metal-polymer nanocomposites. Wiley, New Jersey, 2005; Link and El-Sayed, J Phys Chem B 103(40):8410–8426, 1999). Exploration of periodically ordered metal microstructures has shown that properly structured medium can convert an incoming light wave to a surface plasmon-polariton wave and vice versa. This is a powerful basis for designing new optical diffractive elements with ultrahigh dispersion (Mikhailov et al. Phys Rev Lett 99(8):083901–083904, 2007). To realize the offered method, we developed photopolymerizable composition based on acrylic monomers and silver precursor ($AgNO_3$ solution in acetonitrile) that provides irreversible volume diffusion redistribution of the monomer component and metal precursor. We found an optimal composition of the operation solution and conditions for holographic recording that provide formation of volume structures with the average diameter of NP within the range 3–5 nm, low level of light scattering OL $< 3\,\%$ and long-term stability of parameters (no less than 5 years). Shown in this work is a high

L.M. Kokhtych (✉) • T.N. Smirnova
Institute of Physics, NAS of Ukraine, 03680 Kyiv, Ukraine
e-mail: kokhtych@iop.kiev.ua

A.S. Kutsenko
L.V. Pisarzhevsky Institute for Physical Chemistry, 03028 Kyiv, Ukraine

B. Di Bartolo et al. (eds.), *Nano-Structures for Optics and Photonics*,
NATO Science for Peace and Security Series B: Physics and Biophysics,
DOI 10.1007/978-94-017-9133-5_50,

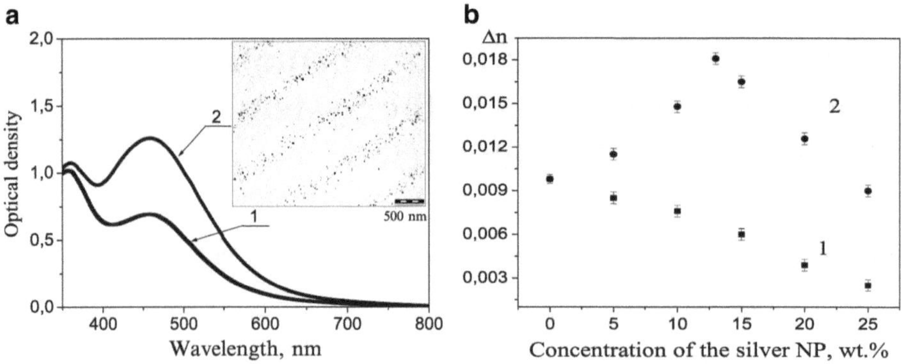

Fig. 50.1 (**a**) Local surface-plasmon band in the absorption spectrum of a grating (included photo contains TEM-image of the grating cross-section) : *1* – after holographic exposure, *2* – after thermal post-processing; (**b**) Δn as a function of the concentration of the silver NP precursor : *1* – after holographic exposure, *2* – after thermal post-processing

degree of the spatial modulation of NP localized inside the planes of a holographic grating corresponding to minimums of the interference pattern (included photo on the Fig. 50.1a). It was ascertained that the increase in the modulation amplitude of the refractive index at the level $\Delta n \cong 0{,}02$ takes place as a result of additional action of illumination as well as heat treatment, which leads to considerable enhancement of silver NP reduction (Fig. 50.1a, b). We have obtained stable 1D – 2D periodical "polymer-silver NP" structures with the thickness up to 50 μm, spatial period 0.3–3.5 μm, the diffraction efficiency of which at the wavelength 632.8 nm reaches 95 % for 1D structures and 80 % – for the 2D ones, respectively. The prepared periodic structures can be used as a light-controlled non-linear diffraction elements as well as resonators with a distributed feedback in waveguide lasers.

Keywords Polymer-metal nanoparticles system • Periodic structure • Surface plasmon

Acknowledgements This work was supported by the State Fund for Fundamental Researches joint projects "SFFR-BRFFR-2013", No. 54.01/014.

51
Modeling the Effect of Different Dimensions in High Contrast Grating Mirror

M. Aghaeipour, V. Ahmadi, and V. Qaradaghi

Abstract The reflectivity spectrum in Vertical-cavity surface-emitting lasers (VCSELs) is investigated by replacing High Contrast Grating (HCG) layer instead of distributed Bragg reflectors (DBR) as a top mirror. In addition, the scalability feature in the HCG layer is used to design a reflector for $\lambda = 1{,}300$ nm. High reflectivity (>99 %) mirrors in VCSELs are mandatory because of the short gain length. In conventional VCSELs, distributed Bragg reflectors (DBR) are used to compensate the weak gain. Recently, the Hasnain group presented High Contrast Grating (HCG) layer as a high reflectivity monolayer (Huang et al., IEEE J Sel Top Quantum Electron 15(5):1485–1499, 2009). Not only is the HCG layer considerably thinner than a conventional DBR, but it also provides a much broader reflectivity spectral width. Furthermore, the HCG layer is polarization sensitive so that using correct dimensions it can be made to reflect the TE-mode or the TM-mode. Here, we compare the reflectivity spectrum from a HCG monolayer with a DBR. In HCG, there are three important factors that determine the spectrum range that can be used for tuning: thickness (t), periodicity (Λ) and duty-cycle (η) (Fig. 51.1). The structure is shown in Fig. 51.1. Two structures are investigated in which the bottom mirror in both of them is the same and is composed of alternating layers of n-type AlAs and AlGaAs with quarter wavelength thickness and a 27.5-period. The top mirror in DBR case (conventional VCSEL) is a 19.5-period p-type AlAs and AlGaAs DBR mirror (Aghaeipour et al., Opt Quantum Electron 45:115–126, 2013). In the HCG case (HCG-VCSEL), a layer of AlGaAs/air HCG is used instead of p-DBRs. In the following, we compare reflectivity spectrum in conventional VCSEL and HCG-one for $\lambda = 850$ nm (Fig. 51.2). The reflectivity spectrum in HCG-VCSEL is much broader than that of the convention DBR one. In the next

M. Aghaeipour (✉)
Solid State Physics, Lund University, Lund, Sweden
e-mail: mahtab.aghaeipour@ftf.lth.se

V. Ahmadi • V. Qaradaghi
Department of Electrical and Computer Engineering, Tarbiat Modares University, Tehran, Iran

B. Di Bartolo et al. (eds.), *Nano-Structures for Optics and Photonics*, NATO Science for Peace and Security Series B: Physics and Biophysics,
DOI 10.1007/978-94-017-9133-5_51,

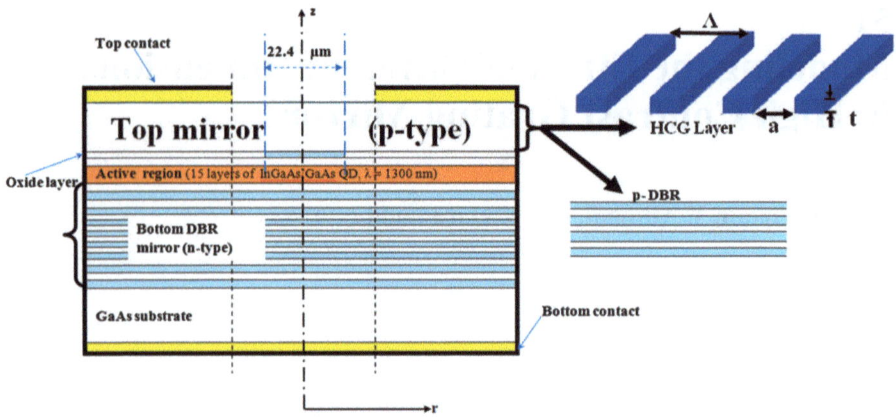

Fig. 51.1 Structure of VCSEL, active region contains 15 InGaAs/GaAs QD layers between 27.5 AlAs/AlGaAs layers for the n-DBR bottom mirror and top mirror including p-DBR or one HCG layer

Fig. 51.2 Comparing reflectivity spectrum in TE-mode between conventional VCSELs and HCG-VCSEL for $t = 140\,\text{nm}$, $a = 400\,\text{nm}$, $\Lambda = 620\,\text{nm}$

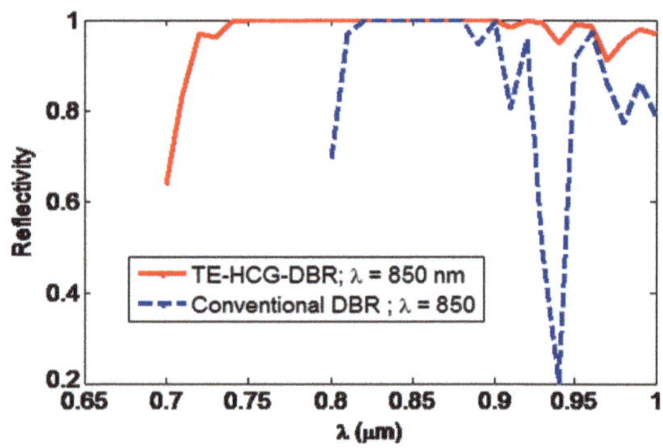

Fig. 51.3 Calculated reflectivity spectrum of HCG for $\lambda_c = 1{,}300\,\text{nm}$ with grating period (Λ) = 948 nm, thickness (t) = 214 nm, and duty cycle (η) = 35 %

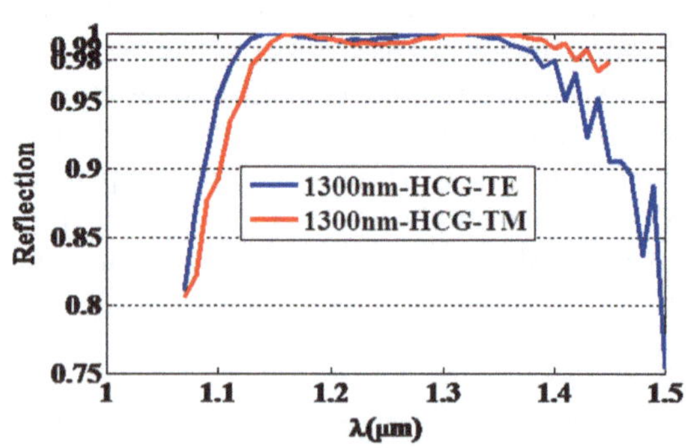

step, we change the dimensions of HCG and investigate the reflectivity spectrum at $\lambda = 1{,}300$ nm. One interesting point about HCG refers to its scalability (Huang et al., IEEE J Sel Top Quantum Electron 15(5):1485–1499, 2009). In the other word, using the same material we reach to a broader spectral range by changing the dimension of HCG layer. In Fig. 51.3, the three parameters (t, Λ and η) have been tuned for $\lambda = 1{,}300$ nm for TE-mode and TM-mode. In summary, a HCG layer is a good candidate for replacing DBR reflectors in VCSELs, on the one hand to reduce the dimension of the laser on the other hand to broaden the reflectivity spectrum. In addition to that, a HCG layer which is a 1D-Photonic Crystal plays an important role to control the output polarization.

52
InAs/GaAs Quantum Dots Covered by Graded GaAsSb Strain Reducing Layer

Jana Kubištová, M. Zíková, A. Hospodková, J. Pangrác, K. Kuldová, E. Hulicius, and J. Oswald

Abstract Two systems of QDs are mainly used to extend operating wavelength of the GaAs based optoelectronic devices towards telecommunication wavelengths 1.3 μm or 1.55 μm, In(Ga)As/GaAs and GaSb/GaAs QDs (Cui et al, Physica E 45:173–176, 2012). The most successful QD struc-tures combine InAs QDs with InGaAs or GaAsSb strain reducing layer (SRL) to red shift the QD emission wavelength (Zhang et al, Sci Rep 2:477, 2012; Liu et al, Appl Phys Lett 86(14):143108, 2005; Hospodková, Capping of InAs/GaAs quantum dots for gaas based lasers. In: Al-Ahmadi A (ed) Quantum dots: a variety of new applications. InTech, pp 27–46, 2012; Hospodková et al, J Crys Growth 370:303–306, 2013). The tasks of the SRL are to re-duce the strain in QDs, to conserve their height and shape and to prevent their dis-solution. SRL also changes the hole localization: depending on the amount of Sb in GaAsSb, the hole may be localized in the QD (type-I heterostructure, less than about 14 % of Sb) or outside the QD in the SRL (type-II, more than 14 % of Sb); electrons are in both cases inside the QD. In this work, the decrease of the ground state emission energy was studied while keeping type-I band alignment between InAs QDs and GaAsSb SRL. The InAs/GaAs QD structures were prepared by low-pressure Metalorganic Vapour Phase Epitaxy (LP MOVPE) in Stranski-Krastanow growth mode. All samples were covered by the GaAsSb SRL with different Sb content and gradation and were studied by photoluminescence (PL). The theoretical simulations of the QD structures were done and compared with the PL results.

J. Kubištová (✉)
Faculty of Nuclear Sciences and Physical Engineering, Czech Technical University in Prague, Technická 2, 166 27 Prague 6, Czech Republic

Institute of Physics of the Academy of Sciences of the Czech Republic, v. v. i., Cukrovarnická 10, 16200, Prague 6, Czech Republic
e-mail: kubistova@fzu.cz

M. Zíková • A. Hospodková • J. Pangrác • K. Kuldová • E. Hulicius • J. Oswald
Institute of Physics of the Academy of Sciences of the Czech Republic, v. v. i., Cukrovarnická 10, 16200, Prague 6, Czech Republic

B. Di Bartolo et al. (eds.), *Nano-Structures for Optics and Photonics*,
NATO Science for Peace and Security Series B: Physics and Biophysics,
DOI 10.1007/978-94-017-9133-5_52,

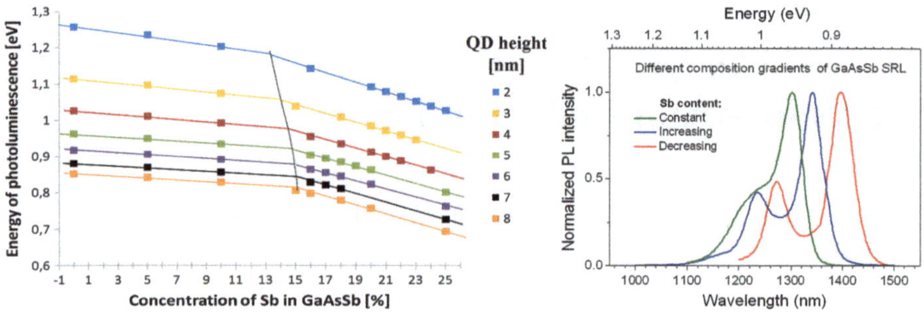

Fig. 52.1 The dependence of PL energy on the concentration of Sb in GaAsSb SRL. (*left*) In type-II re-gime, the decrease of the PL energy with growing Sb content is more pronounced. (*right*) The measured PL spectra of QD samples with different composition gradient of the SRL

As both theoretical and experimental results confirm, the type-I to type-II transi-tion with increasing Sb content in SRL depends on the size of QDs; for smaller QDs, less Sb is needed to obtain the type-II band alignment (13 % Sb in the SRL for 2 nm high QDs vs. 15 % Sb for 8 nm high QDs). The redshift of PL wavelength is remarkable with increasing Sb amount and is more pronounced for type-II heterostructures. Furthermore, the influence of Sb gradation in SRL on the QD properties was studied. Decreasing Sb concentration creates a triangular barrier in the valence band of the structure, which may split the ground state energy of holes. For low total Sb content, the hole ground state is localized in the QD and for high content, the ground state of holes is in the SRL, which means that the type-I to type-II transition may be tuned by the Sb content. If the concentration of Sb is increasing in the SRL, the hole ground state energy is the same in the QD and in SRL, which shifts the probability density of finding the hole towards the top of the QD (to the SRL). The QD structures with graded Sb content (with 5 nm thick SRL and total amount of Sb in SRL 10 %) were prepared and the PL was measured. The shortest PL wavelength was measured for samples with constant Sb content in the SRL, which means that both types of SRL gradation are efficient to extend the PL wavelength. The longest PL wavelength (1,391 nm) of type-I alignment was achieved for samples with decreasing Sb content, which is in accordance with the theoretical model for low Sb content (Fig. 52.1).

Keywords Quantum dots • Strain reducing layer • GaAsSb • InAs • GaAs

53
Compact Wavelength- and Pulse-Duration-Tunable Ultrafast Laser System for Coherent Raman Microscopy

Aleksandr A. Lanin, L.V. Doronina-Amitonova, Andrei B. Fedotov, and Aleksei M. Zheltikov

Abstract We demonstrated a three-dimensional nonlinear microscopy of a diamond film and brain tissue using temporally and spectrally shaped ultrashort laser pulses, delivered by a oscillator-only laser system, which integrates a photonic-crystal-fiber and a nonlinear crystals. After the demonstration of unique properties and excellent performance of photonic-crystal fibers (PCFs) as frequency converters and elements of advanced fiber lasers, waveguides of this class find growing applications as components of optical devices for a broad variety of applications (Knight, J Opt Soc Am B 24(8):1661, 2007). In optical imaging, PCFs have been advantageously integrated into the apparatus for coherent Raman microscopy (Andresen et al., Opt Lett 31(9):1328, 2006; Sidorov-Biryukov et al., Opt Lett 31(15):2323, 2006), radically simplifying the design of a multicolor optical source required for this method of imaging. Here, we demonstrate that soliton self-frequency shift (SSFS) in a PCF pumped by an increased-energy

A.A. Lanin (✉)
Physics Department, International Laser Center, M.V. Lomonosov Moscow State University, Moscow, 119992, Russia
e-mail: alexlanin@mail.ru

L.V. Doronina-Amitonova • A.B. Fedotov
Department of Neuroscience, Kurchatov Institute National Research Center, Moscow, Russia

Physics Department, International Laser Center, M.V. Lomonosov Moscow State University, Moscow, 119992, Russia

A.M. Zheltikov
Physics Department, International Laser Center, M.V. Lomonosov Moscow State University, Moscow, 119992, Russia

Department of Neuroscience, Kurchatov Institute National Research Center, Moscow, Russia

Department of Physics and Astronomy, Texas A&M University, College Station TX, 77843-4242, USA

B. Di Bartolo et al. (eds.), *Nano-Structures for Optics and Photonics*, NATO Science for Peace and Security Series B: Physics and Biophysics, DOI 10.1007/978-94-017-9133-5_53,

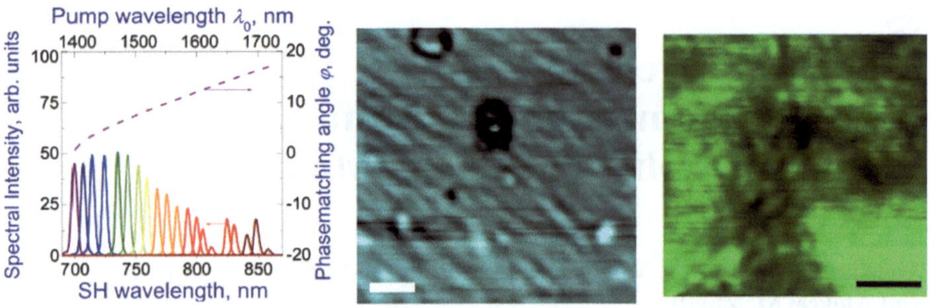

Fig. 53.1 (*left*) The spectra of the second-harmonic output of 20-mm-thick LBO crystal pumped by wavelength-shifted solitons from a 20-cm-long PCF. The *dashed line* shows the SHG phase-matching angle ϕ as a function of the soliton pump (the upper abscissa axis) and the second-harmonic (the lower abscissa axis) wavelength for LBO. (*center*) CARS image of a diamond film surface. The scale bar is 5 μm. (*right*) CARS images of a brain segment adjacent to a ventricle. The scale bar is 30 μm

output of a long-cavity mode-locked Cr: forsterite laser can be combined with second-harmonic generation (SHG) in a nonlinear crystal to provide a broad wavelength and pulse-width tunability of ultrashort light pulses in the range of wavelengths from 680 to 1,800 nm at the 20-MHz repetition rate. We show that LBO crystals can support phase matching for short-pulse SHG within the entire tunability range of the nanojoule-level PCF output, stretching from 1.35 to 1.80 μm (Fig. 53.1 left) (Lanin et al., Opt Lett 37:1508–1510, 2012). Thin LBO crystals are employed in our experiments to generate broadly tunable sub-100-fs second-harmonic pulses in the range of wavelengths from 680 to 900 nm. Thicker nonlinear crystals with properly adapted beam focusing are shown to provide up-to-40 %-efficiency frequency doubling of the PCF output with simultaneous spectral narrowing, yielding ultrashort light pulses with an average power up to 10 mW ideally suited for coherent Raman microspectroscopy. Simultaneous wavelength and pulse-width tunability provided by a combination of the SSFS and SHG technologies is especially helpful for nonlinear Raman spectroscopy and microscopy, where the wavelength tunability is needed for a selective excitation of Raman modes, while the pulse-width tunability is instrumental in reducing the coherent background, related to nonresonant four-wave mixing, in the overall coherent Raman response. In Fig. 53.1 we present typical images provided by coherent anti-Stokes Raman scattering (CARS) which were obtained for synthetic diamond film and a slice of brain (Doronina-Amitonova et al., Appl Phys Lett 99:231109, 2011). The pump field in these experiments had a fixed wavelength, $\lambda_p = 623$ nm, and was provided by the Cr: forsterite laser output frequency-doubled in a periodically poled lithium niobate (PPLN) waveguide. The Stokes field with a tunable wavelength λ_s was delivered by PCF-based SSFS combined with SHG as described above.

54
Random Lasing in ZnSe and CdSe Semiconductor Powders

Maksim S. Leanenia, E.V. Lutsenko, and G.P. Yablonskii

Abstract Random lasing determines the effect of light generating in highly disordered scattering gain media and was firstly theoretically predicted by V.S. Letokhov in 1967 (Letokhov, JETP Lett 5:212–215, 1967). The feedback in random lasers is based on randomly appearing gain loops between neighbouring active particles of scattering media. The materials which are considered to produce random lasing are $A^{II}B^{VI}$ semiconductor powders (Cao, Waves Random Media 13:R1–R39, 2003; Takahashi et al, Opt Lett 34:3923–3925, 2009) and vertically aligned nanowires (Chen at al, Excitonic properties and near-infrared coherent random lasing in vertically aligned CdSe nanowires. Adv Mater (Deerfield Beach, Fla.) 23:1404–1408, 2011) and laser dye solutions containing scattering nanoparticles (Cao, Waves Random Media 13:R1–R39, 2003). In the present work, random lasing in the visible spectra is obtained in ZnSe (475 nm) and CdSe (735 nm) wide bandgap semiconductor microcrystalline powders. A dramatic intensity increasing, spectrum shortening and a structure appearance were observed in ZnSe and CdSe powders with the crystalline size of about 1 μm in diameter after exceeding the threshold excitation intensity. The lasing is due to the emergence of feedback through the formation of random gain loops between the active particles of the scattering medium. Random lasing threshold in the micron sized ZnSe active powder particles is 620 kW/cm^2 that is lower than in work (Takahashi et al, Opt Lett 34:3923–3925, 2009). Random lasing in CdSe powder was obtained for the first time with the threshold excitation intensity 480 kW/cm^2. The key feature of random lasers is low cost of its production (no complex cavity is needed) and possibility to be deposited

M.S. Leanenia (✉) • E.V. Lutsenko • G.P. Yablonskii
Stepanov Institute of Physics of NAS of Belarus, Nezalezhnasti
Ave. 68, 220072, Minsk, Belarus
e-mail: max.leanenia@gmail.com

B. Di Bartolo et al. (eds.), *Nano-Structures for Optics and Photonics*,
NATO Science for Peace and Security Series B: Physics and Biophysics,
DOI 10.1007/978-94-017-9133-5_54,

on any type of surface. Described lasers can find their applications in visualization systems and lighting technology (Cao, Waves Random Media 13:R1–R39, 2003), in data transmission, in medicine as biosensors and in identification systems “friend or foe” (Dubois and Rochelle, Active cooperative tuned identification friend or foe (actiff). US Patent 5,966,227, 12 Oct 1999).

55
Design of an Ultra-broadband Super Absorbing Plasmonic Metamaterials

Michaël Lobet, Olivier Deparis, and Luc Henrard

Abstract In recent years, the need for thin films structures that absorb electromagnetic radiation over a broadband spectrum has been shown to be paramount in solar harvesting for photovoltaic cells, thermal emitters or infrared detectors systems operating in 1–3 μm and in 3–5 μm atmospheric windows used for both civilian and military applications. Recently, Cui et al. (Nano Lett 12:1443–7, 2012), Lobet et al. (Opt. Expr 22:12678–12690, 2014) presented saw-toothed anisotropic metamaterials acting as thin film broadband absorbers in the infrared regime using alternating gold and dielectric layers. The full absorption normalized width at half the maximum was $\Delta\omega/\omega = 86\,\%$ over the 3–5.5 μm wavelength range. In the present work, we design a metamaterial absorber (MMA) showing averaged absorptivity of $\langle \bar{A} \rangle = 93.3\,\%$, over a much broader spectrum, between 200 nm and 5.2 μm. This MMA is made of a periodical array of truncated pyramids with a squared base disposed in air (Fig. 55.1). The truncated pyramids deposited on a semi-infinite gold substrate consist of 20 layers a stratified medium with alternating Au (thickness $t_{Au} = 15\,\text{nm}$) and Ge (thickness $t_{Ge} = 35\,\text{nm}$) slices whose lateral side linearly reduces from $L = 600\,\text{nm}$ at the basis to $l = 150\,\text{nm}$ at the top of the pyramid. A stack of twenty layers is used in the present MMA resulting to a 1 μm high thin structure. The truncated pyramids are deposited on a 200 nm thick gold film. A SiO_2 matrix with a refractive index of $n_{SiO_2} = 1.44$, acts as semi-infinite substrate. The periodicity is set to 800 nm. Gold and germanium permittivities are described using tabulated data (Potter, Germanium (Ge). In. Palik (ed) Handbook of optical constants of solids. Academic, Orlando, 1985; Johnson and Christy, Phys Rev B 6:4370–4379, 1972). Numerical results were obtained with a rigorous coupled-wave analysis (RCWA) method, which is perfectly suited for

M. Lobet (✉) • O. Deparis • L. Henrard
Laboratoire de Physique du Solide, University of Namur, 5000 Namur, Belgium
e-mail: michael.lobet@unamur.be

B. Di Bartolo et al. (eds.), *Nano-Structures for Optics and Photonics*, NATO Science for Peace and Security Series B: Physics and Biophysics, DOI 10.1007/978-94-017-9133-5_55, © Springer Science+Business Media Dordrecht 2015

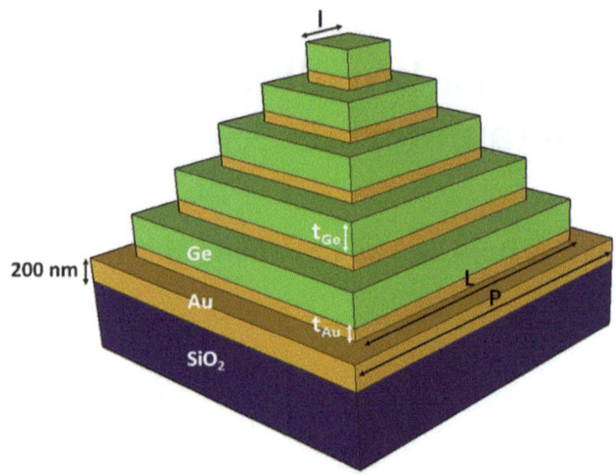

Fig. 55.1 Metamaterial absorber (MMA) made of truncated pyramids forming a periodic array of period P = 800 nm, deposited on a 200 nm *thick gold layer* on a SiO_2 matrix substrate. The pyramid consists of 20 layers with alternating Au (thickness t_{Au} = 15 nm) and Ge (thickness t_{Ge} = 35 nm) layers. The devices possess a total thickness t = 1 μm and the lateral size of the pyramid reduces from L = 600 nm to l = 150 nm

periodic systems. Since the gold layer is sufficiently thick, no transmission ($T = 0$) occurs, and the absorption (A) is directly deduced from reflectivity (R) calculations using $A = 1 - R$. We distinguish three parts in the absorption spectrum: a first part ranging from 200 nm to 2.0 μm (UV, visible, NIR, MIR) with an averaged absorption of $\langle \bar{A}_I \rangle = 92.1\,\%$, a second part encompassing wavelengths between 2.0 and 2.7 μm (MIR) showing a weaker absorption of $\langle \bar{A}_{II} \rangle = 80.1\,\%$ and, finally, a third part from 2.7 to 5.2 μm (MIR) with an absorption of $\langle A_{III} \rangle = 97.8\,\%$ (Lobet et al., Opt. Expr 22:12678–12690, 2014). The absorption mechanism is attributed to the excitation of stationary surface plasmon (SSP) at the interface between metal and dielectric layers. Each layer of the pyramid contributes to several narrow absorption peaks and, since SSP are shape and size dependent, the associations of different sizes of the layers sum up constructively, boosting absorption in a broadband manner. Modelling of the absorption of individual layers was conducted using a Discrete Dipole Approximation DDSCAT code (Draine and Flatau, J Opt Soc Am A 11:1491, 1994) and enabled to identify modes optically active in individual systems. Those modes correspond to excitation of localized surface plasmons modes (Lobet et al., Opt. Expr 22:12678–12690, 2014). We identified plasmonic modes in absorption spectra of square parallelepipeds of thickness $t_{Au} = 15$ nm and length $L = 150, 300, 450$, and 600 nm embedded in a Ge dielectric medium. Those peaks correspond to dipolar, sextupolar and octopolar modes, according to field map distributions. We show the mode resonance wavelength shift $\Delta\lambda$ with respect to the resonance of the smallest system ($L = 150$ nm) as a function of the aspect ratio $R = L/t$ which is consistent with the dispersion relation of surface plasmon polariton of infinite thin film of Au (Lobet et al., Opt. Expr 22:12678–12690,

2014). This proves that, when L increases, longer wavelengths are absorbed, which explains the broadband spectrum. The ultra-broadband characteristics of the MMA, extending from visible to mid-IR ranges, give huge versatility for the design of various devices.

Keywords Metamaterials • Plasmonics • Subwavelength structures • Nanostructures

56 High-Order Harmonic Generation and Plasmonics

Eleonora Lorek, E.W. Larsen, C.M. Heyl, P. Rudawski, M. Miranda, C. Guo, E. Mårsell, S. Carlström, C.L. Arnold, D. Paleček, D. Zigmantas, A. Mikkelsen, A. L'Huillier, and J. Mauritsson

Abstract Attosecond pulses allow for imaging of very fast processes, like electron dynamics. Stockman et al. suggested to use these pulses in connection with a Photoemission electron microscope (PEEM) to study the ultrafast dynamics of plasmons (Stockman et al. Nat Photonics 1:539–544, 2007). For efficient plasmon studies, the repetition rate of the attosecond pulses used needs to be higher than a few kHz (Mikkelsen et al. Rev Sci Instrum 80:123703, 2009). Attosecond pulses are produced in a process called high-order harmonic generation (HHG) (Paul et al. Science 292(5522):1689–1692, 2001; Ferray et al. J Phys B At Mol Opt Phys 21:L31–L35, 1988). In HHG, a strong laser field allows an electron to tunnel out, get accelerated and recombine with a high kinetic energy resulting in extreme ultraviolet attosecond pulses. The large intensity needed to drive the process normally limits the repetition rate of the laser to a few kHz. Using a tight focusing scheme (Heyl et al. Phys Rev Lett 107:033903, 2011; Vernaleken et al. Opt Lett 36:3428–3430, 2011), we, however, generate harmonics at a repetition rate of 200 kHz, both with a commercial turn-key laser and with an advanced laser system. Suitable nanostructures for a strong field enhancement are produced in-house and the field enhancement is studied with PEEM in a non-time resolved manner. With high-order harmonics produced at a high repetition rate, we hope to be able to follow also the ultrafast dynamics of plasmons in these structures (Mårsell et al. Ann der Phys 525:162–170, 2013).

Keywords High-order harmonic generation • Nanostructures • Plasmonics

---

E. Lorek (✉) • E.W. Larsen • C.M. Heyl • P. Rudawski • M. Miranda • C. Guo • E. Mårsell
S. Carlström • C.L. Arnold • A. Mikkelsen • A. L'Huillier • J. Mauritsson
Department of Physics, Lund University, Box 118, 221 00 Lund, Sweden
e-mail: eleonora.lorek@fysik.lth.se

D. Paleček • D. Zigmantas
Department of Chemistry, Lund University, Box 118, 221 00 Lund, Sweden

B. Di Bartolo et al. (eds.), *Nano-Structures for Optics and Photonics*,
NATO Science for Peace and Security Series B: Physics and Biophysics,
DOI 10.1007/978-94-017-9133-5_56,

NATO Science for Peace and Security Series B: Physics and Biophysics,
DOI 10.1007/978-94-017-9132-8_56, © Springer Science+Business Media Dordrecht 201

57
Low Dispersion Propagation of Broad-Band THz Pulses in a Two-Wire Waveguide

Anna Mazhorova, M.K. Mridha, M. Daneau, M. Clerici, M. Peccianti, P.-L. Lavertu, X. Ropagnol, F. Vidal, and R. Morandotti

Abstract Terahertz (THz) waveguides are the subject of an intensive research aimed at transferring THz radiation over long distances while preserving the pulse's amplitude and phase. Standard, optical guided waves approach are not suitable since most of the THz generation techniques deliver extremely large bandwidth pulses, corresponding to single-cycle electric wave packets, which will strongly disperse in dielectric waveguides. This issue is even more important considering that one of the main applications of THz radiation is time-domain spectroscopy, where the temporal coherence of the probing pulse is a must. On the other hand, dispersion-less waveguides have been known at longer wavelengths, e.g. microwave and radio-signals, for several years (Collin, Foundations for microwave engineering. Wiley, New York, 2001). As previously proposed also by Mbonye and coworkers for gigahertz signals (Mbonye et al. Appl Phys Lett 95(23):233506, 2009), we have investigated a two-wire configuration properly scaled for the transmission of THz signals. Here, we demonstrate the characterization of a two-wire waveguide for the dispersion-less delivery of broadband THz pulses, due to a quasi-TEM mode excitation (Tannouri et al. Chin Opt Lett 9(11):110013, 2011). We present our experimental results indicating that low dispersion could be achieved for a broad THz bandwidth over 20 cm of propagation by employing Copper wires separated

---

A. Mazhorova (✉) • M.K. Mridha • P.-L. Lavertu • X. Ropagnol • F. Vidal • R. Morandotti
INRS-EMT, 1650 Blvd. Lionel-Boulet, Varennes, QC J3X 1S2, Canada
e-mail: mazhorova@emt.inrs.ca

M. Daneau
INRS-EMT, 1650 Blvd. Lionel-Boulet, Varennes, QC J3X 1S2, Canada

University of Ottawa, Ottawa, ON K1N 6N5, Canada

M. Clerici
INRS-EMT, 1650 Blvd. Lionel-Boulet, Varennes, QC J3X 1S2, Canada

School of Engineering and Physical Sciences, Edinburgh EH14 4AS, UK

M. Peccianti
Institute for Complex Systems-CNR, Via dei Taurini 19, 00185 Rome, Italy

B. Di Bartolo et al. (eds.), *Nano-Structures for Optics and Photonics*,
NATO Science for Peace and Security Series B: Physics and Biophysics,
DOI 10.1007/978-94-017-9133-5_57,

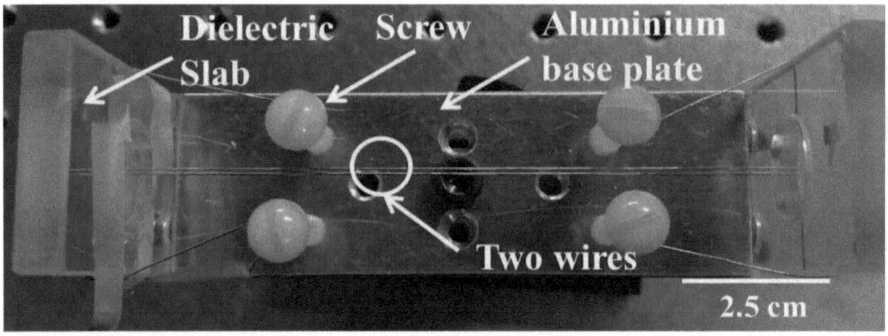

Fig. 57.1 10 cm long two-wire waveguide. The wires are held under tension by wrapping around plastic screws. An Aluminum base plate supports the two wire waveguide design

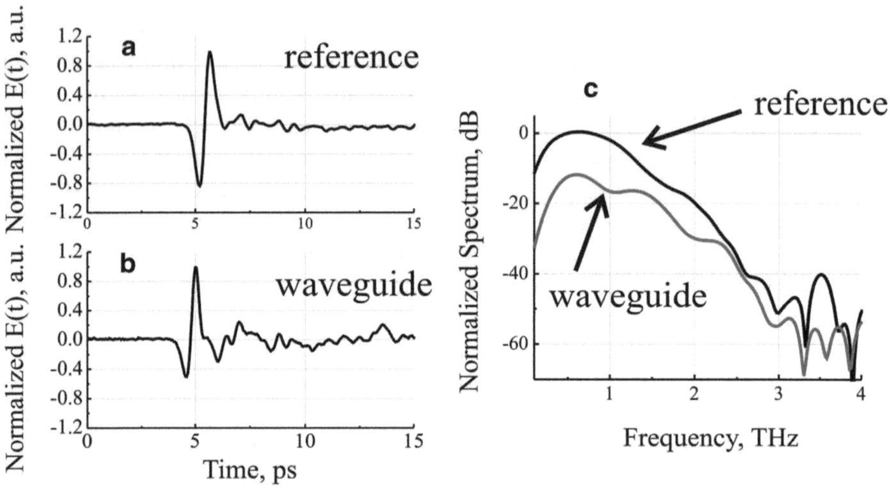

Fig. 57.2 Measured normalized THz pulse for both (**a**) the reference signal and (**b**) the transmitted one over the 20 cm long waveguide; (**c**) normalized amplitude spectra of the reference and the waveguide signal

by 300 μm gap in the air. The waveguide is supported by an Aluminum base plate which acts as its backbone. Two dielectric slabs are then attached to either ends of the base plate. The dielectric slabs have a 800 μm diameter hole drilled through their centres (Fig. 57.1). As shown in the figure, the two copper wires of 250 μm diameter each pass through the centre of these holes on either side of the base plate. As the diameter of the hole is 800 μm, a space of 300 μm is left between the two wires. In order to maintain a uniform separation between the wires, they are kept under tension. Tension is applied by wrapping the four free ends of the wires around screws. A THz TDS setup was used to test the waveguide (80 MHz rep. rate, 125 fs, 800 nm). The broadband THz radiation is generated by an 80 μm gap-size photo-conductive antenna. One of the major issues related to the investigated two-wire configuration is the efficient coupling between the quasi-TEM00 mode of the waveguide and the generated light. We addressed this issue by placing the emitter in close proximity to the input of the waveguide. Figure 57.2

shows the measured THz signal (normalized) for both (a) the signal injected into the waveguide and (b) the one transmitted by the two-wire waveguide and measured via electro-optical sampling. In Fig. 57.2c the normalized amplitude spectra are shown for both the input and the transmitted pulse. The recorded data clearly demonstrate that the single-cycle character of the input pulse is preserved after 20 cm of propagation, with negligible modulations in the power spectral density, on a more than 2 THz bandwidth. Remarkably, the two-wire waveguide supports quasi-TEM mode confined on very small area in between the wires, opening intriguing perspectives e.g. for high-spatial resolution imaging.

Keywords Spectroscopy • Terahertz • Fiber optics • Optical communications • Optoelectronics • Plasmonics • Waveguides

58
Femtosecond Laser Micromachining of a-Si:H

Michael Moebius and E. Mazur

Abstract Femtosecond laser micromachining has been used to write bulk waveguides and photonic devices in glasses, polymers, and crystalline silicon (Gattass and Mazur, Nat Photonics 2:219–225, 2008). Refractive index changes in these materials tend to be less than a percent, which sets limitations on applications due to low light confinement. Hydrogenated amorphous silicon (a-Si:H) presents a unique, versatile material platform with incredible potential. Variations in the hydrogen content can produce refractive index changes as high as 40–80 % (Fortmann et al. Thin Solid Films 395:142–146, 2001). There is potential for integration on a silicon platform. We use femtosecond laser micromachining to locally reduce the hydrogen content of the material, with the goal of increasing the refractive index. We will use this method to directly write three-dimensional (3D) photonic devices in a-Si:H. This novel, simple method for photonic device fabrication in a-Si:H will facilitate many applications and device integration. Femtosecond laser processing enables 3D fabrication through nonlinear interactions. An ultrafast pulsed laser is focused inside the bulk of a transparent material. Due to tight focusing, material modifications only occur at the focal point of the laser. Complex patterns can by written by translating the sample with respect to the laser focus using x-, y- and z-translation. Intensity dependence of nonlinear processes enables fabrication of features with dimensions below the diffraction limit of light through careful selection of laser exposure. 3D fabrication allows greater versatility in devices and can increase device density, which is critical for modern optics and electronics where space is at a premium. Waveguide fabrication through variations in hydrogen content has been demonstrated in 2D using traditional micro- and nano-fabrication techniques (Fortmann et al. Thin Solid Films 430:278–282, 2003; 501:350–353, 2006). This process requires many steps, including ion implantation to locally control hydrogen content. The waveguides consist of regions with low

M. Moebius (✉) • E. Mazur
School of Engineering and Applied Sciences, Harvard University, Cambridge, MA, USA
e-mail: mmoebius@fas.harvard.edu

B. Di Bartolo et al. (eds.), *Nano-Structures for Optics and Photonics*, NATO Science for Peace and Security Series B: Physics and Biophysics, DOI 10.1007/978-94-017-9133-5_58, © Springer Science+Business Media Dordrecht 2015

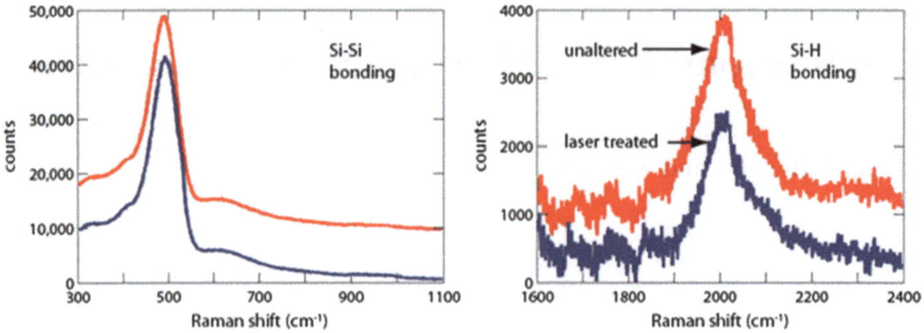

Fig. 58.1 Raman spectroscopy measurements of unaltered and laser treated a-Si:H. Peak intensity ratio of Si-H bonding over Si-Si bonding is reduced by 30 % during laser processing

hydrogen content. Fabrication of 3D photonic devices would require stacking many 2D layers. We are developing femtosecond laser processing to directly write complex 3D patterns in a single-step process, introducing greater versatility and processing speed. We have succeeded in lowering the hydrogen concentration in 2D patterns within a 1-μm film of a-Si:H using a near infrared (1,050 nm) femtosecond laser. Contrast is visible between unaltered and laser processed material under optical microscopy, suggesting index changes due to a reduction in hydrogen content. Contrast increase with laser fluence or a greater number of incident laser pulses. AFM measurements verify that regions with reduced hydrogen content do not exhibit changes in surface topography. We used Raman spectroscopy to verify the reduction in hydrogen content by examining the intensity ratio of Si-H bonding peaks (2,000 and 2,100 cm^{-1}) to Si-Si bonding around 480 cm^{-1} (Fig. 58.1). A reduction in the Si-H intensity corresponds to reduced hydrogen content (Brodsky et al. Phys Rev B 16:3556–3571, 1977). These initial results provide a proof of concept and valuable knowledge for the fabrication of devices.

Keywords Direct laser writing • Integrated optics • Amorphous silicon

59
Low-Level Monoclinic Content Detection in Zirconia Implants Using Raman Spectroscopy

Marienette Morales Vega, A. Bonifacio, V. Lughi, and V. Sergo

Abstract Stabilized zirconia is a ceramic material known for its superior mechanical properties and biocompatibility, and has been widely employed as a biomaterial (Chevalier, Biomaterials 27(4):535–543, 2006) – most recently including dental and spinal implants (Lughi and Sergo, Dent Mater Off Publ Acad Dent Mater 26(8):807–820, 2010). However, the properties of zirconia ceramic may be reduced when it contacts humid environments. Low-Temperature Degradation (LTD) – associated to a phase transformation from tetragonal to monoclinic, and thus to a loss of mechanical properties (Lughi and Sergo, Dent Mater Off Publ Acad Dent Mater 26(8):807–820, 2010; Chevalier et al. Annu Rev Mater Res 37(1):1–32, 2007) – makes detection of low levels of monoclinic a critical aspect in developing and engineering zirconia-based biomaterials. In this work, detection of extremely little amounts of monoclinic phase is demonstrated on implant prototypes by combining Raman spectroscopy and multivariate analysis. Implant prototypes made of Yttria-Stabilized Tetragonal Zirconia were subjected to pin-on-disk tests, leaving distinct wear marks on the samples surface. Arrays of Raman spectra were collected in areas across the pristine and the worn surface, using a 785 nm laser with a 50× magnification objective, 0.75 NA. The spectral array was treated by means of standard univariate analysis and by multivariate analysis based on Principal Component Analysis (PCA). The results of the multivariate analysis are represented as maps. Figure 59.1 shows the scores corresponding to the loadings of the relevant component, *i.e.* the one that most emphasize the presence of monoclinic phase. Note that individual Raman spectrum would suggest that the sample is purely tetragonal. However, PCA results will reveal presence of traces of monoclinic phase of less than 1 vol% (Fig. 59.2). Detection of very low amounts of monoclinic phase, indicating

M.M. Vega (✉) • A. Bonifacio • V. Lughi • V. Sergo
Center of Excellence for Nanostructured Materials, Graduate School of Nanotechnology, University of Trieste, Trieste, Italy
e-mail: marienettemorales.vega@phd.units.it

B. Di Bartolo et al. (eds.), *Nano-Structures for Optics and Photonics*, NATO Science for Peace and Security Series B: Physics and Biophysics,
DOI 10.1007/978-94-017-9133-5_59, © Springer Science+Business Media Dordrecht 2015

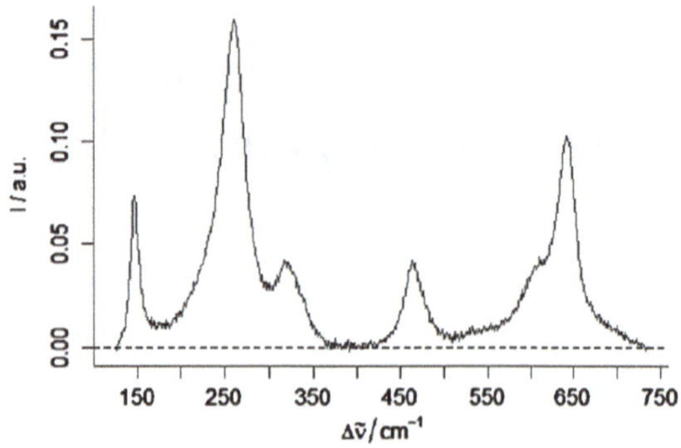

Fig. 59.1 Single spectrum from the worn area; the main characteristic peaks from the monoclinic phase (182 and 191 cm^{-1}) are not detected

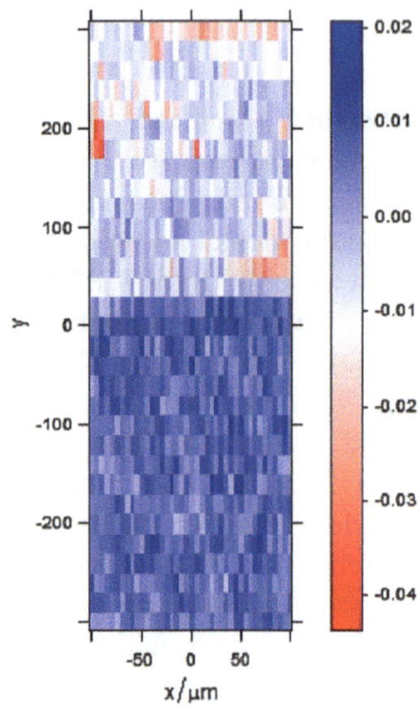

Fig. 59.2 Map of the sample across the area subjected to the pin-on-disk test. Color scale represents the scores of the component that most emphasize the presence of monoclinic (*blue*: fully tetragonal; *red*: about 1 % monoclinic)

incipient t-m transformation, was made possible by using PCA on Raman spectral arrays. This is a promising approach for early detection of LTD, a key aspect in the development of zirconia-based biomaterials.

Acknowledgements This work is part of the LONGLIFE project that has received funding from the European Community's Seventh Framework Programme (FP7/2007-2013) under grant agreement no280741. Dr Falk Bernsmann is kindly acknowledged for preparing the samples.

60
Modeling Additive Color Effect in Natural Photonic Polycrystals Using the Layer Homogenization Method: The Case of the Diamond Weevil

Sébastien Mouchet, Jean-Pol Vigneron, Jean-François Colomer, and Olivier Deparis

Abstract The original method we propose allows to compute the spectral reflectance of photonic crystals whose unit cell is composed of form-birefringence anisotropic elements such as cylinders or parallelepipeds. This method relies on the layer homogenization of the photonic structure. It is especially useful for the calculation of reflectance according to different crystal orientations. The coloration due to an additive color effect in the photonic polycrystal found on the diamond weevil, *Entimus imperialis*, was investigated. Simulating the reflectance of photonic polycrystals often turns out to be necessary in the study of such structures as well as in the design and production of bioinspired devices (Vigneron and Lousse, Proc SPIE 6128:61281G, 2006; Deparis and Vigneron, Mater Sci Eng B-Adv 169(1–3):12–15, 2010). In this context, the computation of the reflectance of photonic crystals (PCs) displaying form-birefringence anisotropic elements in the unit cell (e.g., cylinders, parallelepipeds...) turns out to be cumbersome, particularly when the reflectance is calculated for different crystal orientations as in the case of polycrystals. The method proposed here solves this problem in the particular case of a PC with a periodicity that is such that there is only specular reflection and no higher-order diffraction (Mouchet et al. Opt Express 21(11):13228–13240, 2013). In this method, the structure with a particular crystal orientation is sliced into layers and the periodic dielectric function $\varepsilon\left(\vec{r}\right)$ is homogenized within each layer (Fig. 60.1a). Using this Layer Homogenization (LH) method, the reflectance of one single domain of the polycrystal can be computed in an arbitrary orientation thanks to a standard thin film solver. The reflectance due to the additive color effect created by the disorder in the crystal domain orientation of the polycrystal is modeled by averaging reflectance spectra computed for several incidence angles and crystal domain orientations. Our method

S. Mouchet (✉) • J.-P. Vigneron • J.-F. Colomer • O. Deparis
Department of Physics, University of Namur, Rue de Bruxelles, 61, B-5000 Namur, Belgium
e-mail: sebastien.mouchet@unamur.be

B. Di Bartolo et al. (eds.), *Nano-Structures for Optics and Photonics*,
NATO Science for Peace and Security Series B: Physics and Biophysics,
DOI 10.1007/978-94-017-9133-5_60,

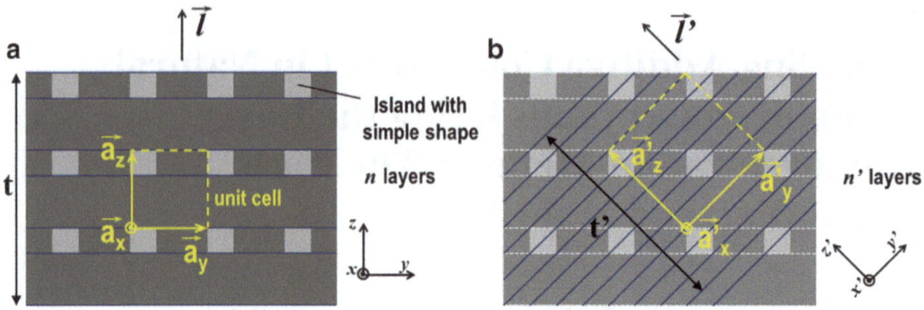

Fig. 60.1 Layer Homogenization (LH) method: (**a**) the PC is defined for one crystal domain orientation $\vec{l}$ as a stratified medium with two-dimensional lattices of islands having simple shapes (e.g., cylinders). The unit cell is described by $\vec{a}_x$, $\vec{a}_y$ and $\vec{a}_z$ vectors. (**b**) The photonic crystal is redescribed by a laterally homogeneous stratified medium with an orientation $\vec{l}'$ that is different from $\vec{l}$. A new crystal unit cell is defined by a new set of translation vectors $\vec{a}'_x$, $\vec{a}'_y$ and $\vec{a}'_z$

Fig. 60.2 (**a**) *Entimus imperialis* is a remarkable example of additive color effect. A dull green coloration is perceived by the human eye. (**b**) Its cuticle is however covered by a photonic polycrystal made of different colorful domains

was applied to the case of a natural photonic polycrystal found on the cuticle of the diamond weevil *Entimus imperialis* (Fig. 60.2a). Its coloration is due to an additive color effect created by PC domains with various orientations (Fig. 60.2b) (Deparis and Vigneron, Mater Sci Eng B-Adv 169(1–3):12–15, 2010; Mouchet et al. Proc SPIE 8480:848003, 2012): a single PC domain gives rise to a gleaming iridescent color (from blue to orange) but the disorder in the crystal domain orientation results in a non-iridescent dull color (Fig. 60.2a). Investigating such a structure is relevant in the development of bioinspired applications such as biomimetic devices e.g., gas, temperature or pH sensors (Van Opdenbosch et al. Photon Nanostruct 10(4):516–522, 2012).

Keywords Photonic crystals • Photonic bandgap materials • Natural photonic crystals • Structural color • Computational electromagnetic methods • Beetle • Nanoarchitecture • Bioinspiration • Biomimetics

61 Transformation of Optical Properties for Gamma-Irradiated Lithium Fluoride Crystals Under Transition from Bulk to Nanocrystals

V.S. Kalinov, Aleksei N. Novikov, and A.P. Stupak

Abstract Radiation defects (color centers (CC)) substantially change optical characteristics of crystals. Lithium fluoride (LiF) crystals with radiation-induced CC are widely used and studied. LiF crystal is a model crystal to study radiation effects on solids. There are still many unexplained questions regarding radiation defects formation in nanocrystals: defects aggregation processes, possibilities of formation of defects types which are not observed in bulk crystals. In this work, the results of studying optical properties for gamma-irradiated lithium fluoride crystals under transition from bulk to nanocrystals are presented. Crystal samples for the research were prepared in the form of plates with the sizes $(0.7-1.5)\times(5-7)\times(7-10)\,\text{mm}^3$ by cutting out samples from nominally pure LiF single crystal along the plane {100}. Mechanical fragmentation of the same crystals was used for nanocrystals production. The nanocrystals sizes were in the hundreds of nanometers range ($d < 1\,\mu\text{m}$). Nanocrystals were pressed in pellets for convenience purposes. Defects were created by gamma quanta irradiation from a ^{60}Co source. The samples were irradiated at liquid nitrogen temperature (LNT) $T_{irr} = 77\,\text{K}$. After irradiation at LNT and the extraction of samples from liquid nitrogen, their spectra of a photoluminescence (PL) were measured at the ambient temperature. Four bands are observed in photoluminescence excitation spectra on registration wavelength $\lambda_{det} = 680\,\text{nm}$ in the pellets (Fig. 61.1, curve 1). One of them with the maximum at $\lambda \approx 445\,\text{nm}$ belongs to centers F_2. Intensities of three other bands with maxima at 393, 497 and 563 nm are changed in identical ways at various influences on samples. This experimental fact allows one to conclude that these bands belong to the same type of the centers formed in nanocrystals. These bands are completely absent in the spectra of the bulk crystal irradiated at LNT (Fig. 61.1, curve 2). Excitation of the pellets irradiated at LNT by radiation

V.S. Kalinov • A.N. Novikov (✉) • A.P. Stupak
B.I. Stepanov Institute of Physics, National Academy of Sciences of Belarus, 68 Nezavisimosti Ave., Minsk 22072, Belarus
e-mail: novalexey@mail.ru

B. Di Bartolo et al. (eds.), *Nano-Structures for Optics and Photonics*, NATO Science for Peace and Security Series B: Physics and Biophysics, DOI 10.1007/978-94-017-9133-5_61,

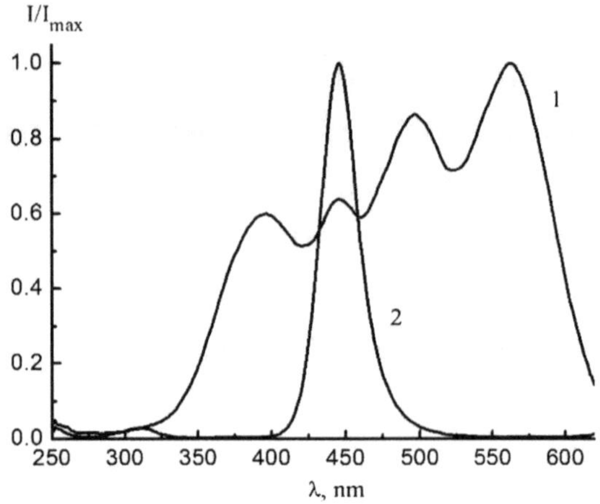

Fig. 61.1 Luminescence excitation spectra at registration wavelength $\lambda_{det} = 680$ nm for a pellet (*1*), a bulk crystal (*2*) irradiated at LNT. Spectra are registered after the termination of post-radiation CCs aggregation processes and normalized to their maxima

with wavelength $\lambda_{exc} = 570$ nm leads to the luminescence, whose spectrum very strongly overlaps the CCs F_2 PL spectrum. The results of additional experiments (ultra-violet irradiation of the samples and heating to 200–250 °C) give the basis to conclude that this luminescence is caused by the presence of new CC, which were not observed in LiF crystals before.

62
Characterization of Polyethylene Glycol Self-Assembled Monolayers by Means of Sum-Frequency Generation Spectroscopy for Biosensor Applications

Nadia Peerboom, F. Cecchet, Y. Caudano, J. Moreau, S. Wautier, J. Marchand-Brynaert, M. Henry, S. Demoustier-Champagne, and D. Lis

Abstract Protein biochips are miniaturized biological sensors intended to analyze and characterize biomolecule interactions with high throughput. An important issue when developing such biochips is substrate passivation. The support is rendered inert in order to avoid non-specific adsorption. Strategic control of the non-specific protein adsorption can be achieved by creating a resistant self-assembled monolayer (SAM) based on polyethylene glycol (PEG). The degree of resistance depends on the PEG surface density, i.e. the number of PEG units the molecule contains. Infrared-visible sum-frequency generation (SFG) spectroscopy (Lambert et al., Appl Spectrosc Rev 40:103–145, 2005) is used to in-vestigate the vibrational fingerprint of a PEG self-assembled monolayer adsorbed on a flat platinum surface, in the 2,750–3,050 cm^{-1} frequency range. The objec-tive is to characterize the SFG baseline of the biosensor that will be further developed by mixing the PEG antifouling layer with bioactive host molecules. Nanostructures will then be implemented on the substrate in order to enhance the SFG signal through localized surface plasmon resonances (Lis et al., Adv Opt Mater 1:244–255, 2013). The ultimate goal will be to detect the SFG signature of the antigen/antibody recognition process

N. Peerboom (✉) • F. Cecchet • Y. Caudano • D. Lis
Research Centre in Physics of Matter and Radiation (PMR), University of Namur, Rue de Bruxelles 61, B-5000 Namur, Belgium
e-mail: nadia.peerboom@unamur.be

J. Moreau • S. Wautier • J. Marchand-Brynaert
Laboratoire de Chimie Organique et Médicale, Institute of Condensed Matter and Nanosciences (IMCN), Université Catholique de Louvain (UCL), Bâtiment Lavoisier, Place Louis Pasteur L4.01.02, B-1348 Louvain-la-Neuve, Belgium

S. Demoustier-Champagne
Bio and Soft Matter (BSMA), Institute of Condensed Matter and Nanosciences (IMCN), Université Catholique de Louvain (UCL), Place Croix du Sud 1/box 4, B-1348 Louvain-la-Neuve, Belgium

M. Henry

B. Di Bartolo et al. (eds.), *Nano-Structures for Optics and Photonics*, NATO Science for Peace and Security Series B: Physics and Biophysics, DOI 10.1007/978-94-017-9133-5_62,

at the interface of the above biosensing layer. The general chemical formula of the molecule investigated is $C_3H_5S_2{-}(CH_2)_4{-}C{=}O{-}NH{-}(CH_2{-}CH_2{-}O)_n{-}CH_3$. The molecule studied here holds 7 polyethylene glycol units (n = 7). Former studies carried out by quartz crystal microbalance (QCM) and electrochemistry showed that molecules containing 7 PEG chains have the best antifouling properties. The SFG spectra present several vibrational modes (Sartenaer et al., Biosens Bioelectron 22:2179–2183, 2007; Cimatu et al., J Phys Chem C 112:14529–14537, 2008; Even et al., Macromolecules 39:9396–9401, 2006). The most significant contributions come from the CH_2, $O{-}CH_2$ and $O{-}CH_3$ modes.

63 InAs Quantum Dots Covered by GaAsSb Strain Reducing Layer

Martin Pokorný, M. Kozák, F. Trojánek, J. Pangrác, and A. Hospodková

Abstract This work is focused on examining photoluminescent properties of InAs quantum dots (QDs) on GaAs substrate covered by $GaAs_{1-x}Sb_x$ strain reducing capping layer (SRL) prepared by Stranski-Krastanow method. We measured luminescence decay time of two samples with different concentration of Sb in this layer. We investigated the influence of temperature, as well as the intensity and wavelength of the excitation pulse. We also compared the properties of the samples after excitation by 760 nm pulse and 850 nm pulse – the former one is energetically above the substrate band gap; in the second case we excited only the QDs and the wetting layer (WL). We consequently derived recombination and relaxation processes occurring inside InAs QDs and also the transport of charge carriers from the substrate and the WL into QDs.

Keywords Quantum dots • Upconversion • Luminescence

M. Pokorný (✉) • M. Kozák • F. Trojánek
Faculty of Mathematics and Physics, Charles University in Prague, Ke Karlovu 5, 120 00, Praha 2, Czech Republic
e-mail: martin@martin-pokorny.cz

J. Pangrác • A. Hospodková
Institute of Physics of the Academy of Sciences of the Czech Republic, v. v. i., Cukrovarnická 10, 16200, Prague 6, Czech Republic

B. Di Bartolo et al. (eds.), *Nano-Structures for Optics and Photonics*, NATO Science for Peace and Security Series B: Physics and Biophysics,
DOI 10.1007/978-94-017-9133-5_63,

B. Di Bartolo et al. (eds.), *Nano-Structures for Optics and Photonics*,
NATO Science for Peace and Security Series B: Physics and Biophysics,
DOI 10.1007/978-94-017-9133-5_53, © Springer Science+Business Media Dordrecht 2015

64
A Prediction Made When Playing with NeoCube® Is Confirmed by Impurity Spectroscopy

Indrek Renge

Abstract Earlier, we have used Lennard-Jones (L-J) model of glassy state to describe the inhomogeneous band shape of impurity spectra and its transformation under compression. Color effects on spectral holes burned at different frequencies could be rationalized for temperature and pressure changes (Lin et al. Chem Phys Lett 576:15–20, 2013; Renge, J Chem Phys 127:034504, 2007). Qualitatively, observation of a heap of balls allows one to draw important conclusions about the behavior of glassy state. For example, tight packed lattices, hexagonal or cubic, are not formed, because of friction, so that glasses contain ∼10 % of free space. The coordination number of a sphere is expected to be 8–11, instead of 12. Huge amount of potential energy must be locked in glassy matter. On the other hand, pair wise distances remain the same as in the "crystal". Structurally, glass is not a frozen liquid. Moreover, glass formation cannot be regarded as a phase transition, since no phase boundary is created. Below glass temperature T_g the Boltzmann thermometer fails, and so does (equilibrium) thermodynamics (Lin et al. Chem Phys Lett 576:15–20, 2013; Renge, J Chem Phys 127:034504, 2007). The visualization of salient features of the glassy state is possible, notwithstanding the fact that isotropic, dispersive attraction of L-J 6–12 potential is replaced simply by gravitation. Depending on the size, nanoparticles are expected to show intermediate behavior between molecules and macroscopic bodies. "Crystallization" of uniform nanoparticles as a function of interaction type is a non-trivial phenomenon, and is of considerable interest. Collections of Nd magnets, typically 216 spherical beads of diameter 5 mm have recently been commercialized as NeoCube®. Even without computer modeling, it becomes obvious that simple cubic lattice of magnetic dipoles is metastable and collapses easily into disorder. Also, close packed face centered

I. Renge (✉)
Institute of Physics, University of Tartu, 142 Riia Street, EE51014 Tartu, Estonia
e-mail: indrek.renge@ut.ee

B. Di Bartolo et al. (eds.), *Nano-Structures for Optics and Photonics*,
NATO Science for Peace and Security Series B: Physics and Biophysics,
DOI 10.1007/978-94-017-9133-5_64,

Table 64.1 Separation of optical and static permittivity effect on solvent shift $\nu - \nu_0$, according to the equation $p(n^2-1)/(n^2+2)+e(\varepsilon'-1)/(\varepsilon'+2)$ (n, refractive index, ε' corrected dielectric constant), data from (Meyer and Mialocq, Opt Commun 64:264–268, 1987; Arzhantsev et al. J Phys Chem A 110:3454–3470, 2006)

Dye	A/Fl	$p(\text{cm}^{-1})$	$e(\text{cm}^{-1})$	N	r^2
Coum 153	A	−6,700 ± 1,200	−1,600 ± 100	14	0.981
	F	−3,400 ± 3,100	−3,700 ± 200	9	0.975
DMA-CS	A	−7,900 ± 1,700	−800 ± 100	9	0.948
	F	−4,300 ± 3,000	−4,500 ± 200	9	0.993
DCM	A	−11,200 ± 1,000	−900 ± 50	12	0.984
	F	−6,500 ± 1,300	−3,100 ± 100	8	0.998

cubic or hexagonal structures appear difficult to obtain. Instead, strings, loops, voids, and helixes tend to form. Such behavior of dipolar hard spheres is well known from Monte Carlo simulations of polar liquids, liquid crystals, polymers, Janus colloids, ferrofluids, nanoparticles, self-assembly of photonic materials, etc. (Weis and Levesque, Phys Rev Lett 71:2729–2732, 1993; Groh and Dietrich, Phys Rev E 63:021203, 2001). We wish to establish a closer connection of this active field of statistical physics with impurity spectroscopy. More realistic description of van der Waals glasses should be possible with Stockmayer potential comprising both the L-J and dipolar components. In a demonstration experiment, solvatochromism of a polar dye was considered. A greatly enhanced dipole moment in the excited state should increase average intermolecular distance. As a result, the matrix density dependence of absorption and fluorescence band shifts will be different. Indeed, the refractive index dependence of emission band positions is less steep than that of absorption for widely used laser dyes DCM, Nile Red, and Coumarins. In accordance with the Lorenz-Lorentz and London formulas, the solvent refractive index dependent solvent shifts, owing to dispersion and induction, have been assumed constant in literature for both absorption and fluorescence maxima. By contrary, it follows from Table 64.1 that emission bands are much less sensitive. Considerably higher dipole moment in the excited state causes an increase of effective distance between the solute and solvent molecules r, diminishing the absolute value of coefficient p. The dependence on ε' is affected both by the change of r, and Onsager's reaction field that is stronger in the zwitter-ionic excited state.

Keywords Dipolar particle packing • Liquid structure • Fluorescence

65
Maximizing Intensity in TiO_2 Waveguides for Nonlinear Optics

Orad Reshef, C.C. Evans, S. Griesse-Nascimento, J.D.B. Bradley, and E. Mazur

Abstract Titanium dioxide (TiO_2) represents an attractive candidate for nonlinear optical devices due to its large refractive index and large Kerr nonlinearity. These properties can strongly enhance confinement and nonlinear interactions. TiO_2 also possesses high transparency, exhibiting no linear absorption within the entire visible spectrum and no two-photon absorption at wavelengths above 800 nm. Considering these qualities, TiO_2 is capable of outperforming most other widely transparent materials, such as fused silica, silicon nitride, and diamond. Using electron beam lithography and a liftoff procedure followed by reactive ion etching, we structure both amorphous TiO_2 as well as polycrystalline anatase thin films to create photonic devices that exploit this material's properties in order to study nonlinear optics (Bradley et al. Opt Express 20:23821–23831, 2012; Evans et al. Opt Express, submitted) Nonlinear optics benefit from prolonged interactions, necessitating large intensities along extended waveguide lengths. For this reason, waveguide losses need to be minimized. We study the effects of mask materials and annealing procedures on waveguide propagation losses. We also study a variety of taper structures and optimize the insertion losses of these waveguides. For ultrafast pulses, dispersion becomes an important parameter since strong dispersion can elongate a pulse and lower the large peak intensities needed for nonlinear optics. Within nano-scale structures, this parameter can be tailored without difficulty by changing the waveguide geometry. We present a finite-element analysis that demonstrates the geometries necessary to obtain negligible or anomalous dispersion and thus maintain large pulse intensities. These techniques can readily be applied to other novel photonic material platforms.

O. Reshef (✉) • C.C. Evans • S. Griesse-Nascimento • J.D.B. Bradley • E. Mazur
School of Engineering and Applied Sciences, Harvard University, Cambridge, MA, USA
e-mail: orad@seas.harvard.edu

B. Di Bartolo et al. (eds.), *Nano-Structures for Optics and Photonics*, NATO Science for Peace and Security Series B: Physics and Biophysics, DOI 10.1007/978-94-017-9133-5_65, © Springer Science+Business Media Dordrecht 2015

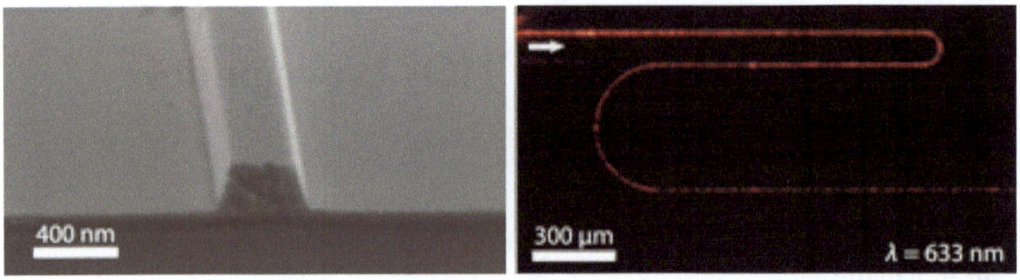

Fig. 65.1

66
An Improved Method for T-Matrix Calculations of Light Scattering by Spheroidal Particles

Walter R.C. Somerville, B. Auguié, and E.C. Le Ru

Abstract The modeling of light scattering by particles has a long history, starting with the case of spheres (Mishchenko and Travis LD, Bull Am Meteorol Soc 89:1853–1861, 2008). Spheres however provide a constrained shape to model particles, and the extension to spheroids allows the modeling of many more systems. However, the modeling of light scattering by spheroids remains a more challenging problem than for spheres. One common semi-analytic technique to model spheroids is the Extended Boundary-Condition Method (EBCM) to compute the Transition-matrix (T-matrix) (Waterman, Proc IEEE 53:805–812, 1965), which may be thought of as an extension of Lorenz-Mie theory to particles of arbitrary shape. The T-matrix method allows the calculation of near-and far-field optical properties, as well as the easy calculation of orientation-averaged scattering properties, as well as considering scattering by collections of particles. It is thus an attractive method for the modeling of light scattering in plasmonics and other fields (Boyack and Le Ru, Phys Chem Chem Phys 11:7398–7405, 2009). However, in the case of spheroids, there are problems in most implementations of the EBCM that limit the parameter space that the method is useful for. The EBCM consists of populating a matrix with surface integrals over the surface of the particle being modeled, followed by the inversion of the system, to obtain the T-matrix. It has long been known that this method suffers numerical problems in the case of large or high aspect-ratio particles. Until recently, the solution to these problems was to use extended-precision arithmetic to extend the range of applicability (Mishchenko and Travis, Opt Commun 109(1–2):16–21, 1994). Recently, we demonstrated that the cause of these problems (Somerville et al., J Quant Spectrosc Rad Transf 113(7):524–535, 2012), for the special case of spheroids, is the cancellation of leading terms in the integral, which integrate

W.R.C. Somerville (✉) • B. Auguié • E.C. Le Ru
The MacDiarmid Institute for Advanced Materials and Nanotechnology, School of Chemical and Physical Sciences, Victoria University of Wellington, PO Box 600, Wellington 6140, New Zealand
e-mail: walter.somerville@vuw.ac.nz

B. Di Bartolo et al. (eds.), *Nano-Structures for Optics and Photonics*, NATO Science for Peace and Security Series B: Physics and Biophysics, DOI 10.1007/978-94-017-9133-5_66,

identically to zero but dominate the integrand. This leads to a loss of precision, which in some cases is catastrophic. We present here an implementation of the EBCM which formulates the integrals without the problematic terms, which enables the efficient modeling of light scattering of particles which were previously difficult to model (Somerville et al., J Quant Spectrosc Rad Transf 123:153–168, 2013). Examples of interest to plasmonics are provided of simulations using the improved method, in cases where traditional EBCM codes would fail. This allows the easy modeling of previously inaccessible systems.

67
Automated Algorithms for Multilayer Thin Film Filter Design Using Metamaterials

Timothy Threadgold, John W. Bowen, and G. Hawkins

Abstract Currently, infrared filters for astronomical telescopes and satellite radiometers are based on multilayer thin film stacks of alternating high and low refractive index materials. However, the choice of suitable layer materials is limited and this places limitations on the filter performance that can be achieved. With the introduction of metamaterials into the design of multilayer thin film filters, it is now possible to include layers with arbitrary refractive indices to ensure an optimal transmission profile. However, the introduction of more variables into the design process has made many traditional iterative design methods near obsolete. Evolutionary algorithms are extremely effective in solving problems with large solution spaces such as the design of multilayer thin film filters. By using an altered differential evolutionary algorithm (Pantoja et al., Design of thin film filters using differential evolution optimization technique. In: 2009 SBMO/IEEE MTT-S international microwave and optoelectronics conference (IMOC), pp 447–452, 2009), multilayer thin-film filters which have the desired transmission profile can be designed with minimal human input. The algorithm was adapted to design layers which have arbitrary refractive indices, as opposed to being designed using only traditional materials. Once the design of a multilayer thin film filter is completed by a evolutionary algorithm, each layer can be designed as a metamaterial with the required refractive index – in this case a bulk material of known refractive index with sub-wavelength structures (Wada et al., Appl Phys Express 3(10):102503, 2010). The period and size of the sub-wavelength structures for each layer were designed using Bruggeman's effective material approximation to give the required refractive index and depth for that layer.

Keywords Genetic algorithm • Metamaterials • Multilayer filters • Optimisation • Filter design

T. Threadgold (✉) • J.W. Bowen • G. Hawkins
School of Systems Engineering, University of Reading, Whiteknights, Reading, RG6 6AY, UK
e-mail: t.m.threadgold@pgr.reading.ac.uk; j.bowen@reading.ac.uk

B. Di Bartolo et al. (eds.), *Nano-Structures for Optics and Photonics*, NATO Science for Peace and Security Series B: Physics and Biophysics, DOI 10.1007/978-94-017-9133-5_67, © Springer Science+Business Media Dordrecht 2015

68
Surface Charge Effects on the Interaction Between a Solid-Supported Model Lipid Membrane and AuNPs Studied by SFG Spectroscopy

Xavier Toledo-Fuentes, D. Lis, and F. Cecchet

Abstract In this study, we use the selectivity and sensibility of sum-frequency generation spectroscopy (SFG) to interfacial media (Chen and Chen, Biochimica et biophysica acta 1758:1257–1273, 2006) in order to characterize the interactions, and their effects, between a cell membrane model, here made of a li-pid bilayer, and gold nanoparticles. With the aim of giving to this study a sound basis, we did first realize separately the SFG characterization of the membrane model and the gold nanoparticles. Here, the membrane model is a bilayer made of phosphatidylcholine – one of the major components of living membranes – adsorbed on a CaF_2 or SiO_2 prism. The gold nanoparticles have a diameter of 6, 12 or 38 nm and a negative surface state of charge. We highlighted that as long as it stays in water, the membrane model is SFG inactive, due to the globally centrosymmetric nature of the lipid bilayer and to the SFG selection rule, which forbids the emergence of a SFG signal from centrosymmetric media. On the other hand, once it has been exposed to air, the bilayer gives a SFG signal. This fact is explained by the folding (flip-flop) of a part of the bilayer on itself at the contact with the air, due to its hydrophobic nature. Concerning the nanoparticles, they generate a SFG signal coming from the surfactant giving them their state of charge while preventing them binding together. That signal has been measured at the interface with a CaF_2 prism. On the other hand, in the case of a SiO_2 prism, no signal coming from the nanoparticles has been detected. The same results have been observed when the lipid bilayer was present on the prisms. This is explained by the fact that at neutral pH, the SiO_2 has a distinctly negative state of charge (Yeganeh et al. Phys Rev Lett 83:1179–1182, 1999), repelling so the nanoparticles beyond the SFG field of detection at the interface. As for CaF_2, it has a positive state of charge at neutral pH (Schrodle et al. J Phys Chem C 111:10088–10094, 2007), which leads to an

X. Toledo-Fuentes (✉) • D. Lis • F. Cecchet
Laboratoire Lasers et Spectroscopies – Centre PMR – NAmur Research Institute for LIfe Sciences, University of Namur, Rue de Bruxelles 61, B-5000 Namur, Belgium
e-mail: xavier.toledo@unamur.be

B. Di Bartolo et al. (eds.), *Nano-Structures for Optics and Photonics*, NATO Science for Peace and Security Series B: Physics and Biophysics, DOI 10.1007/978-94-017-9133-5_68,

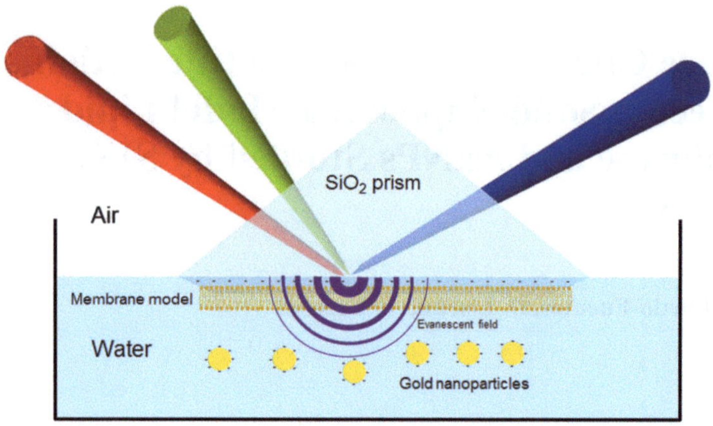

Fig. 68.1 In this situation, the nanoparticles, the bilayer and the prism surface are all negatively charged, so the NPs stays out of the beams evanescent field range and are undetected

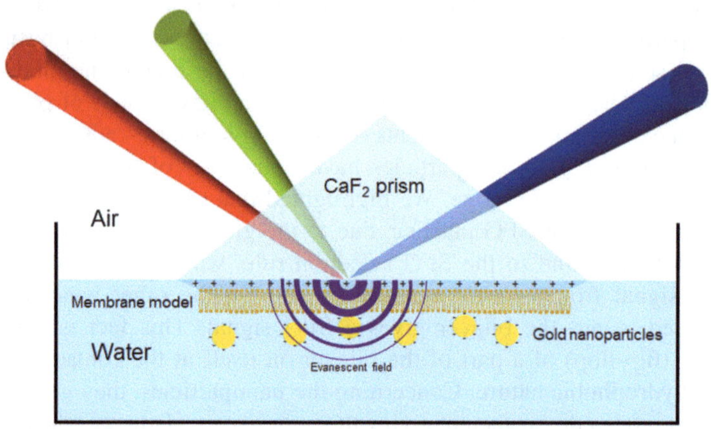

Fig. 68.2 Here, the prism surface is now positively charged and the NPs are closer to it, lying in the evanescent field range and are so detected

attractive interaction with the negatively charged nanoparticles and so allows their detection. Furthermore, since it has been shown (Lin et al. ACS Nano 4:5421–5429, 2010) that in pH-neutral water, phosphatidyl-choline bilayers have a negative state of charge, the positive nanoparticles are mostly expected to interact with these bilayers. To sum up, we have experimentally confirmed that negative nanoparticles weakly interact with phosphatidylcholine bilayer, even with a substrate taking on a positive state of charge in contact with water. On the basis of these results, the next step will be to confirm the interaction of positive nanoparticles with the bilayer, knowing that this interaction should be even more important if the substrate takes on a negative state of charge in water (Figs. 68.1 and 68.2).

69
Photoluminescence of Colour Centres in Thermally-Evaporated LiF Films for Radiation Imaging Detectors

Maria Aurora Vincenti, F. Bonfigli, G. Messina, R.M. Montereali, A. Rufoloni, and E. Nichelatti

Abstract Lithium fluoride (LiF) is a well-known dosimeter material in pure (McLaughlin et al. Nucl Instrum Methods 175:17–18, 1980) and doped (Lakshmanan et al. Phys Status Solidi (a) 153:265–273, 1996) form. Radiation detectors based on microcrystalline dispersion of LiF in a polymeric matrix have been introduced for gamma and electron high-dose dosimetry (Kovacs et al. Radiat Phys Chem 57:691–695, 2000). In recent years the area of growth and characterisation of LiF thin films has seen a considerable expansion. Polycrystalline LiF films grown by thermal evaporation were proposed and tested as nuclear sensors for neutrons (Cosset et al. Thin Solid Films 303:191–195, 1997) and for gamma dosimetry (Montecchi et al. Point defects in lithium fluoride films induced by gamma irradiation, ch. 116. In: Proceedings of the 7th international conference on advanced technology and particle physics, Como, pp 819–825, 2002). In the last years many efforts have been devoted to the development of novel imaging detectors for extreme-ultraviolet radiation and soft X-rays (Baldacchini et al. J Nanosci Nanotechnol 3:483–486, 2003), hard X-rays (Almaviva et al. Appl Phys Lett 89(5):054102, 2006), as well as low (Baldacchini et al. J Phys Condens Matter 10:857–867, 1998) and high energy electrons. Such solid-state detectors are based on the optical reading of photoluminescence (PL) from stable, visible-emitting colour centres (CCs), produced by irradiation with ionising radiations.

M.A. Vincenti (✉) • F. Bonfigli • G. Messina • R.M. Montereali
ENEA C.R. Frascati, UTAPRAD-MNF, Via E. Fermi 45, 00044 Frascati, Rome, Italy
e-mail: maria_aurora.vincenti@enea.it

A. Rufoloni
ENEA C.R. Frascati, UTFUS-COND, Via E. Fermi 45, 00044 Frascati, Rome, Italy

E. Nichelatti
ENEA C.R. Casaccia, UTTMAT-OTT, Via Anguillarese 301, 00123 S. Maria di Galeria, Rome, Italy

B. Di Bartolo et al. (eds.), *Nano-Structures for Optics and Photonics*, NATO Science for Peace and Security Series B: Physics and Biophysics, DOI 10.1007/978-94-017-9133-5_69, © Springer Science+Business Media Dordrecht 2015

These aggregate electronic defects are F_2 and F_3^+ CCs (two electrons bound to two and three close anion vacancies, respectively), which possess almost overlapping absorption bands, peaked at about 450 nm, called M band (Nahum, Phys Rev 157:817–830, 1967). By optical pumping in this spectral region, F_2 and F_3^+ CCs emit broad PL bands peaked at 678 and 541 nm, respectively. LiF films of different thickness were grown by thermal evaporation on different substrates, such as glass and silica, as well as plastic ones (Di Lazzaro et al. Extreme ultraviolet marking system for anti-counterfeiting tags with adjustable security level. In: Allakhverdiev KR (ed) XIX international symposium on high-power laser systems and applications, Istanbul. International Society for Optics and Photonics, p 86770T, 2013). Even in LiF thin films grown on transparent substrates, the direct use of optical absorption spectra to individuate the presence of different kinds of point defects is often precluded by the presence of interference fringes due to the refractive index difference between film and substrate (Nichelatti et al. J Non-Crystalline Solids 322:117–121, 2003). PL measurements are more sensitive and a very effective investigation method, called Combined Excitation-Emission Spectroscopy (Gill et al. Appl Phys Lett 64(19):2483, 1994), can be applied. In this work we present the basic characteristic of this technique and the first results of the investigation of polycrystalline LiF films grown by thermal evaporation on glass substrates and coloured by high-energy electrons in air.

Keywords Lithium fluoride • Thin films • Colour centres • Photoluminescence • Combined excitation-emission spectroscopy

70
P3HT:PCBM Based Organic Solar Cells: Structure Optimization and Improving External Quantum Efficiency by Plasmonic Nanoparticles Incorporation

Andrej Vojtko, M. Benkovicova, Y. Halahovets, M. Jergel, M. Kotlar, M. Kaiser, P. Siffalovic, V. Nadazdy, and E. Majkova

Abstract We optimized morphology of the active layer of organic solar cells based on poly(3-hexylthiophene-2,5-diyl) (P3HT) and [6,6]-phenyl-C61-butyric acid methyl ester (PCBM) bulk heterojunction. The optimization process includes a 20 min solvent annealing followed by a 5 min thermal annealing at 110 °C. After such a procedure a ∼3 % power conversion efficiency (PCE) was achieved. Adding of gold plasmonic nanoparticles (0.1 wt %) resulted in external quantum efficiency improvement.

Figure 70.1 depicts surface of the P3HT:PCBM active layer before and after annealing processes. The corresponding power spectral density functions reveal an increase of the typical percolation length from 21 to 35 nm with much more homogeneity and better evolved interfaces in the latter case. The reflectivity of the solar cell without a hole transport layer (HTL) calculated by using the ellipsometrically obtained optical constants (Fig. 70.2) shows a close correlation with the experimentally measured PCEs (Fig. 70.3a). Optical optimization of the thicknesses of the active layer with HTL is shown in Fig. 70.3b. Different concentrations of 6–7 nm plasmonic gold nanoparticles were incorporated into the active layer. After incorporation of 0.1 wt% of nanoparticles, an increase of the external quantum

A. Vojtko (✉) • M. Benkovicova • Y. Halahovets • M. Jergel • M. Kaiser • P. Siffalovic
V. Nadazdy • E. Majkova
Department of Multilayers and Nanostructures, Institute of Physics, Slovak Academy of Sciences, Dubravska cesta 9, 845 11 Bratislava, Slovak Republic
e-mail: andrej.vojtko@savba.sk

M. Kotlar
Department of Multilayers and Nanostructures, Institute of Physics, Slovak Academy of Sciences, Dubravska cesta 9, 845 11 Bratislava, Slovak Republic

Institute of Electronics and Photonics, Slovak University of Technology, Ilkovicova 3, 812 19 Bratislava, Slovak Republic

B. Di Bartolo et al. (eds.), *Nano-Structures for Optics and Photonics*, NATO Science for Peace and Security Series B: Physics and Biophysics,
DOI 10.1007/978-94-017-9133-5_70,

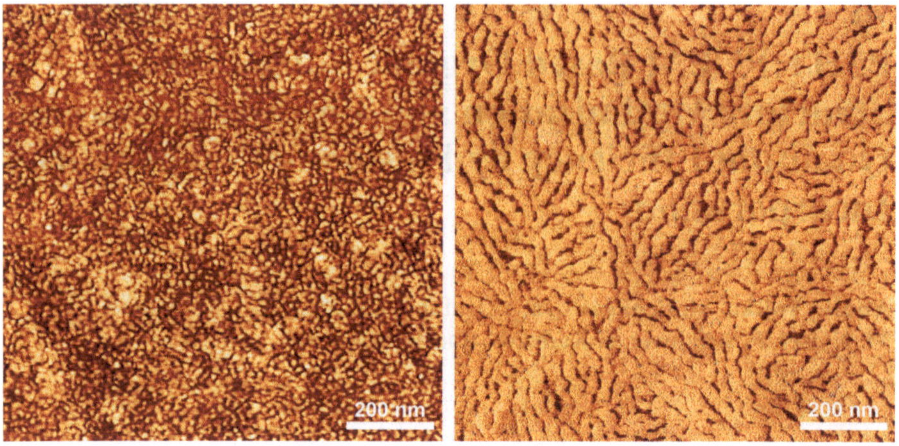

Fig. 70.1 AFM images of P3HT:PCBM before (*left*) and after (*right*) annealing processes

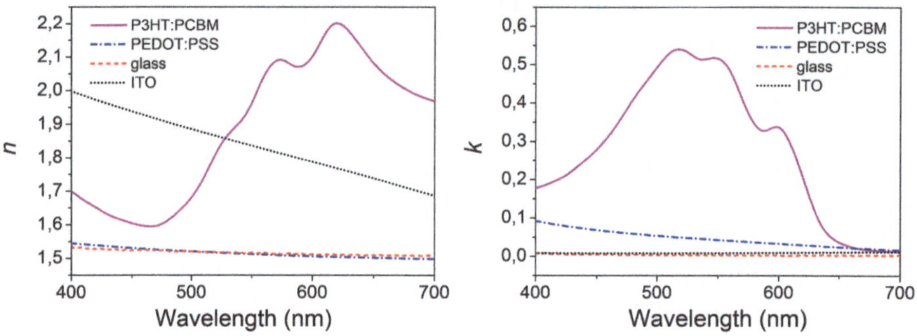

Fig. 70.2 Optical constants determined by spectroscopic ellipsometry

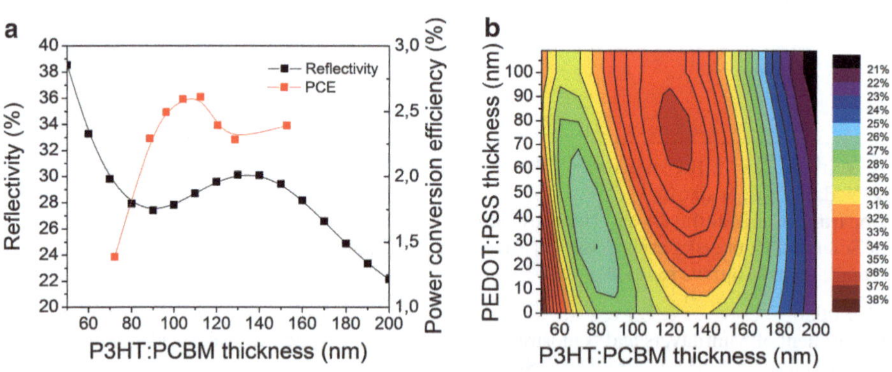

Fig. 70.3 (**a**) Comparison of the reflectivity and PCE of the solar cell without HTL. (**b**) Reflectivity map of the solar cell with HTL

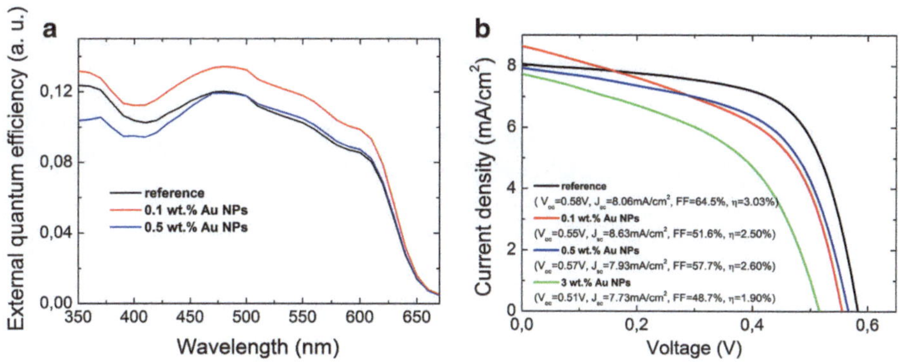

Fig. 70.4 (**a**) External quantum efficiency of the reference P3HT:PCBM blend and the blend with incorporated 6–7 nm gold plasmonic nanoparticles. (**b**) I-V curves of the solar cells with different concentrations of the incorporated nanoparticles

efficiency was observed (Fig. 70.4a). In the following, the carrier collection has to be also optimized in order to enhance the fill factor and PCE of the solar cell shown in Fig. 70.4b.

Keywords P3HT:PCBM • Optimization • Reflectivity • Optical constants • Gold plasmonic nanoparticles

71 Optical Properties Nanocomposite Composed of Ag Nanoparticles Embedded in a DLC Film

Iryna Yaremchuk, A. Tamulevičienė, T. Tamulevičius, K. Šlapikas, M. Andrulevicius, and S. Tamulevičius

Abstract The results of theoretical and experimental investigation of the optical absorption spectra of thin films of nanocomposite materials composed of silver nanoparticles embedded in a diamond like carbon (DLC) matrix are presented. The optical characteristics of DLC-Ag nanocomposite are considered within the framework of the effective medium approximation. Fundamental properties of nanostructured materials are currently excessively studied because of their potential application in numerous fields such as electronic devices, opto-electronics, optics, tribology, biotechnology, human medicine and others. The noble metal nanoparticles embedded in a matrix such as amorphous carbon have been studied intensively (Montiel-González Z et al. Thin Solid Films 519:5924–5932, 2011), since such type of nanocomposites are expected to exhibit interesting optical property so called surface plasmon resonance and are promising materials for developing the elemental base of different physics devices. Amorphous hydrocarbon coatings, often called diamond-like carbon (DLC) coatings are nowadays used in many applications due to low friction coefficient and high hardness and also corrosion resistance, chemical inertness, high electrical resistivity, infrared-transparency, high refractive index (Robertson, Mater Sci Eng R Rep 37:129–281, 2002). In this work, the optical absorption of DLC-Ag films was experimentally and theoretically investigated. The experimental absorption spectra of DLC based silver nanocomposites deposited by reactive magnetron sputtering on fused silica were compared with the simulated

I. Yaremchuk (✉)
Department of Photonics, Lviv Polytechnic National University, S. Bandera Str. 12, Lviv 79013, Ukraine
e-mail: yaremchuk@polynet.lviv.ua

A. Tamulevičienė • T. Tamulevičius • K. Šlapikas • M. Andrulevicius, • S. Tamulevičius
Institute of Materials Science, Kaunas University of Technology, Savanoriu Ave. 271, Kaunas LT-50131, Lithuania

B. Di Bartolo et al. (eds.), *Nano-Structures for Optics and Photonics*, NATO Science for Peace and Security Series B: Physics and Biophysics, DOI 10.1007/978-94-017-9133-5_71, © Springer Science+Business Media Dordrecht 2015

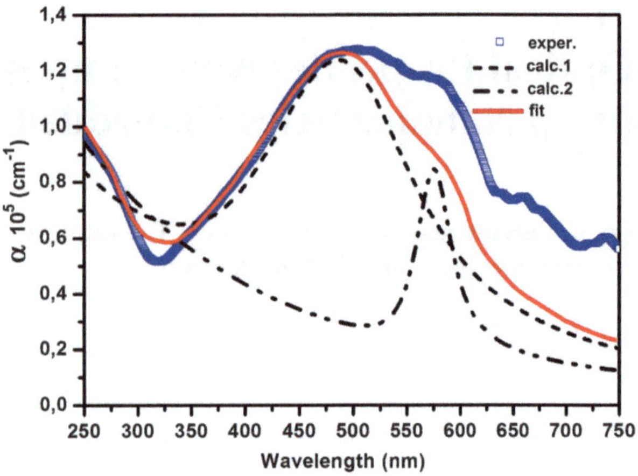

Fig. 71.1 Experimental and calculated optical absorption spectra of DLC-Ag thin film

ones. Role of the host matrix and silver nano-particles on the tunable effective refractive index of the films was analyzed. Using Mie theory it was researched that absorption of the nanocomposite thin films is sensitive to both refractive index and extinction coefficient parameters of the DLC host and in the evaluation of the absorption of the film it is necessary to take into account the spectral dependence of the complex refractive index of DLC. Reduction in size of the metal particles results in blue-shift of the surface plasmon resonance peak but spectral position of the absorption peak is more sensitive to the dielectric constant of DLC, than to the radius of nanoparticles. The absorption peaks related to the surface plasmon resonance are red-shifted with the increase in Ag volume fraction and this shift may be attributed to the increased electromagnetic interaction between the particles and change in the effective permittivity of the DLC-Ag nanocomposite (Yaremchuk et al. Physica Status Solidi (a) 211:329–335, 2014). In Fig. 71.1 experimental and calculated optical absorption spectra of DLC-Ag thin film with 20 % Ag concentration are depicted. The optical absorption peak due to the surface plasmon resonance of Ag particles centered at around 530 nm was observed. Strong absorption peak centered at around 480 nm we address to the nanoparticles with a radius of 8.5 nm (calculation 1, see Fig. 71.1) and peak at around 580 nm – to the nanoparticles with a radius of 30 nm (calculation 2, Fig. 71.1). The approximation curve consists of these two peaks and provides reasonable fit of the experimental curve. Application of the extended Maxwell–Garnett effective theory provides good fit of the experimental optical absorption spectra of the DLC-Ag films with the silver nanoparticle concentration up to 20 %. The difference between the Maxwell-Garnett theory prediction and the experimental results is that a surface plasmon resonance is broader than the computed one as a result of wide distribution of the particle size consequently resulting in the broadening of the experimental plasmon resonance peak.

Keywords Nanocomposites • Nanoparticles • Effective theory • DLC-Ag

72
List of Participants

1. Prof. Ismail Mekkaoui Alaoui
   Professor of Physics
   Faculty of Sciences Semlalia, Cadi Ayyad University
   BP 2389 Marrakech 40000, MOROCCO
   Tel: 212-6-68046017; Fax: 212-5-24437410
   e-mail: mekkaoui@uca.ma

2. Filippo Alpeggiani
   Department of Physics
   University of Pavia
   Via Bassi 6 27100, Pavia, ITALY
   filippo.alpeggiani01@ateneopv.it

3. José Alvarez,
   CNRS Researcher
   Laboratoire de Genie Electrique de Paris (LGEP)
   CNRS UMR 8507 ; Supelec ; Univ. Paris Sud ; Univ. Paris 6
   11 rue Joliot Curie,
   Plateau de Moulon,
   91192, Gif-sur-Yvette, FRANCE
   Phone : 33 (0)1 69 85 16 43; Fax : 33 (0)1 69 41 83 18
   jose.alvarez@lgep.supelec.fr

4. Dr. Nicola Armaroli
   ISOF - CNR Building 4 (ex FRAE)
   Via Gobetti, 101 - 40129 Bologna, ITALY
   Tel.: +39 051 639 9820; Fax: +39 051 639 9844
   nicola.armaroli@isof.cnr.it

B. Di Bartolo et al. (eds.), *Nano-Structures for Optics and Photonics*,
NATO Science for Peace and Security Series B: Physics and Biophysics,
DOI 10.1007/978-94-017-9133-5_72,

5. Prof. Steve Arnold
Polytechnic Institute of NYU - 6 Metroteck Center
Brooklyn, NY 11201, USA
Tel. (1) 718 260-3899; 917 568-6549 (cell)
arnold@photon.poly.edu; sarnold935@aol.com

6. Prof. Rolindes Balda
Departamento de Fisica Applicada I- Escuela Tecnica Superior de
Ingenieros Industriales Y de Telecomunicacion-Universidad del Pais Vasco
Alda, Urquio, S/N49013 Bilbao, SPAIN
Tel.: (34)-946 01 4051 ; Fax.: (34) 94 601 41 78
rolindes.balda@ehu.es

7. Prof. Ausrine Bartasyte
P2M, Institut Jean Lamour, Nancy Université
Boulevard des Aiguillettes
B.P. 239
54506 Vandoeuvre-les-Nancy cedex, FRANCE
Tel. 03 83 68 49 35; Fax. 03 83 68 49 33
ausrine.bartasyte@gmail.com

8. Dr. Romeo Beccherelli
Senior Researcher
CNR - Istituto per la Microelettronica e Microsistemi
Area della Ricerca di Roma 2 - Tor Vergata
Via del Fosso del Cavaliere, 100
00133 Roma, ITALY
Tel +39 0649934538
Fax +39 0645488066
Mobile +39 3470054613
Skype: romeo.beccherelli
romeo.beccherelli@artov.imm.cnr.it

9. Muriel Bechu
Karlsruher Institut für Technologie
Institut für Angewandte Physik
Wolfgang- Gaede-Strasse 1
76131 Karlsruhe, GERMANY
muriel.bechu@kit.edu

10. Sean Blakley
Department of Physics
Texas A& M University
College Station, TX 77843, USA
s.blakley86@gmail.com

11. Prof. Georges Boulon
Université Claude Bernard- Lyon I
Laboratoire de Physico-Chimie des Materiaux Luminescents, Bât. 205
43, Boulevard du 11 Novembre 1918
F-69622 Villeurbanne Cedex, FRANCE
Tel. (33) 7 244-8271; FAX (33) 7 243-1130
BOULON GEORGES georges.boulon@univ-lyon1.fr

12. Dr. John W. Bowen
The University of Reading - Terahertz Laboratory
Reading RG6 6AH, UK
Tel : (44)(0)118 378 6702; FAX: (44) (0)118 378 8220
cybjb@cyber.reading.ac.uk

13. Dr. Laura Brigo
University of Padova
Industrial Engineering Department
Via Marzolo 9, I-35131 Padova, ITALY
Phone: +39.049.827.5023
Fax: +39.049.827.5505; Mobile: +39.347.4225329
laura.brigo@unipd.it

14. Giuseppe Cacciato
Dipartimento di Fisica e Astronomia
Università di Catania
Catania I 95123 ITALY
giuseppe.cacciato@ct.infn.it

15. Rossella Capasso
PhD student in "Novel Technology for Materials, Sensors and Imaging"
University of Naples "Federico II"
Institute of Cybernetics, CNR
Via Campi Flegrei, 34 - Comprensorio Olivetti
80078 Pozzuoli (Napoli), ITALY
r.capasso@cib.na.cnr.it

16. Dr. Maura Cesaria
University of Salento-Department of Physics
Via Arnesano, 73100 Lecce (ITALY)
ph: +39 0832 297555
maura.cesaria@le.infn.it

17. Prof. John Collins
Department of Physics
Wheaton College

Norton, MA 02766, USA
Tel.: (1) 508 286-3977; FAX (1) 508 286-3977
jcollins@wheatonma.edu

18. Prof. Giovanni Costa
Università degli Studi di Padova
Padova, ITALY
giovanni.costa@pd.infn.it

19. Fausto D'Apuzzo
Department of Physics
University "La Sapienza"
Roma, ITALY
Fausto.Dapuzzo@iit.it

20. Prof. Baldassare Di Bartolo (Co-Director of the Institute)
Department of Physics
Boston College
Chestnut Hill, MA 02167, USA
Tel. (1) 617 552-3601; home (1) 781 483-3993
FAX (1) 617 552-8478
dibartob@bc.edu

21. Prof. Enzo Di Fabrizio
Istituto Italiano di Tecnologia
Via Morego, 30
16163 Genova, ITALY
Tel.: 39 040 3757713
Enzo.Difabrizio@iit.it; Enzo.DiFabrizio@kaust.edu.sa

22. Dr. Murat Erdem
Marmara University
Science and Letter Faculty, Physics Department
C-123 Goztepe Kampusu, Kadikoy
34777, Istanbul, TURKEY
merdemmerdem@marmara.edu.tr

23. Markus Erola
Department of Chemistry
University of eastern Finland
P. O. Box 111
FI-80101 Joensuu, FINLAND
markus.erola@uef.fi

24. Prof. Joaquin Fernandez
Departamento de Fisica Applicada I- Escuela Tecnica Superior de Ingenieros Industriales Y de Telecomunicacion-Universidad del Pais Vasco
Alda Urquio, S/N 49013 Bilbao, SPAIN
Tel.: (34)-94 601 40 52; Fax.: (34) 94 601 41 78
joaquin.fernandez@ehu.es

25. Yan Francescato
Department of Physics (EXSS)
Imperial College London
Prince Consort Road
London, SW7 2BZ, UK
yan.francescato10@imperial.ac.uk

26. Ben Franta
Mazur Group
School of Engineering and Applied Sciences
Harvard University
Cambridge, MA 02138, USA
Tel: 704 502-6156
bafranta@fas.harvard.edu

27. Riccardo Funari
Dipartimento di Fisica
Università di Napoli "Federico II"
Napoli, ITALY
riccardo.funari@unina.it

28. Dr. Zeno Gaburro
Università degli Studi di Trento
Trento, ITALY
gaburro@science.unitn.it

29. Tibor Gál
Department of Atomic Physics
Technical Univresrity of Budapest
Budafoki út 8
H-111 Budapest, HUNGARY
gal.tiborpeter@gmail.com

30. Lucie Gaouyat
Department of Physics
Université de Namur ASBL
Rue de Bruxelles 61
5000 Namur, BELGIUM
lucie.gaouyat@unamur.be

31. Prof. Sergei Gaponenko
National Academy of Sciences
Institute of Physics
Minsk, BELARUS
nanoscience@tut.by

32. Julien Gateau
Faculté des Sciences
Biophotonics Group
Gèneve, SWITZERLAND
Tel.: (41) 22 37965 94
Julien.Gateau@unige.ch

33. Giorgos Georgiou
PhD student
FOM Institute AMOLF
Science Park 104
1098 XG Amsterdam, THE NETHERLANDS
georgiou@amolf.nl

34. Stéphane-Olivier Guillaume
Department of Physics
Université de Namur ASBL
Rue de Bruxelles 61
5000 Namur, BELGIUM
stephane-oivier.guillaume@unamur.be

35. Dr. Małgorzata Guzik
Faculty of Chemistry
University of Wrocław
F. Joliot-Curie 14
50-383 Wrocław, POLAND
Phone: +48 71 375 7373; Fax:+48 71 328 2348
goguzik@poczta.fm

36. Muhammad Hanif
Ph.D. Stipendiat
Institut for Fysik og Nanoteknologi
Aalborg Universitet
Skjernvej 4A
Rom: 5.215
9220-Aalborg Øst, DENMARK
mh@nano.aau.dk

37. Aline Herman
Department of Physics
Université de Namur ASBL
Rue de Bruxelles 61
5000 Namur, BELGIUM
aline.herman@unamur.be

38. Vishal Jain
Solid State Physics Department
Lund University
Lund, SWEDEN
vishal.jain@ftf.lth.se

39. Gergely Kajtár
Department of Physics,
Faculty of Electrical Engineering and Information Technology,
Slovak University of Technology in Bratislava
Ilkovičova 3, 812 19 Bratislava, SLOVAK REPUBLIC
Greg gregmaster777@gmail.com

40. Ivan Karbovnyk
Department of Electronics
Ivan Franko National University of Lviv
107 Tarnavskogo str
Lviv, UKRAINE
ivan_karbovnyck@yahoo.com

41. Prof. Halyna Klym
Institute of Computer Technology
Lviv Polytechnic National University
12 Bandera St.
Lviv., 79013, UKRAINE
klymha@yahoo.com

42. Lyudmyla Kokhtych,
Junior Researcher Scientist of
Coherent and Quantum Dept.,
Institute of Physics of NAS of Ukraine
Kyiv, UKRAINE
kokhtych@iop.kiev.ua

43. Mahtab Aghaeipour Kolyani
Phd student
Solid State Physics Department
Lund University

Lund, SWEDEN
mahtab.aghaeipour_kolyani@ftf.lth.se

44. Jana Kubistova
Faculty of Nuclear Sciences and Physical Engineering
Czech Technical University
Na Slovance 199/2
182 21 Praha 8, CZECH REPUBLIC
kubistova@fzu.cz

45. Aleksandr Lanin
Physics Department of the Moscow State University and the Russian Quantum Center.
Moscow, RUSSIA
alexlanin@mail.ru

46. Mercy Lard
PhD student
Lund University
Div. of Solid State Physics,
Lund, SWEDEN
mercy.lard@ftf.lth.se

47. Maxim Leanenia
Laboratory of Semiconductor Physics and Techniques
B. I. Stepanov Institute of Physics
National Academy of Sciences of Belarus
Minsk, BELARUS
max.leanenia@gmail.com

48. Prof. Uli Lemmer
Lichttechnisches Institut
Universitaet Karlsruhe (TH)
76131 Karlsruhe, GERMANY
Tel.: +49-721-608-2530; FAX:+49-721-608-2590
Mobile: +49-175-202-7612
Secretary: +49-721-608-2531
uli.lemmer@kit.edu

49. Haichun Liu
PhD Student
Atomic Physics Division
Department of Physics
Lund University
P.O. Box 118

SE-221 00 Lund, SWEDEN
Phone: +46 (0)46 222 7471; Fax: +46 (0)46 222 425
haichun.liu@fysik.lth.se

50. Michael Lobet
Department of Physics
Université de Namur ASBL
Rue de Bruxelles 61
5000 Namur, BELGIUM
michael.lobet@unamur.be

51. Ardi Loot
Institute of Physics
University of Tartu,
Tartu, ESTONIA
ardi.loot@ut.ee

52. Eleonora Lorek
PhD student
Atomic Physics
Lund University, Lund, SWEDEN
eleonora.lorek@fysik.lth.se

53. Samuel Margueron
Laboratoire Matériaux Optiques, Photonique et Systémes (LMOPS)
Université de Lorraine et Supélec
2 Rue E.Belin, F-57070 Metz Cedex FRANCE
tel: (33) (0)3 87 37 85 49/ fax: (0)3 87 37 85 59
samuel.margueron@supelec.fr

54. Dr. Anna Mazhorova
Ultrafast Optical Processing Group
INRS-EMT Università du Québec
1650 Blvd. Lionel Boulet, Varennes,
Québec J3X 1S2, Canada
anna.mazhorova@gmail.com

55. Prof. Eric Mazur
School of Engineering and Applied Sciences; Harvard University
Cambridge, MA 02138, USA
Tel. (1) 617 495-8729; FAX (1) 617 495-1229 or 9837
mazur@physics.harvard.edu

56. Michael Moebius
Mazur Group

School of Engineering and Applied Science
Harvard University
Cambridge, MA 02138, USA
mmoebius@fas.harvard.edu

57. Marienette Morales Vega
PhD Student
Center of Excellence for Nanostructured Materials (CENMAT)
Graduate School of Nanotechnology
University of Trieste
Via Valerio, 6a
34127 Trieste, ITALY
MARIENETTEMORALES.VEGA@phd.units.it

58. Alexander Moroz
Gitschiner Strasse 103
D-10969 Berlin, GERMANY
wavescattering@yahoo.com

59. Sébastien Mouchet
Department of Physics
Université de Namur ASBL
Rue de Bruxelles 61
5000 Namur, BELGIUM
sebastien.mouchet@unamur.be

60. Philip Munoz
Mazur Group
School of Engineering and Applied Sciences
Harvard University
Cambridge, MA 02138, USA
pmunoz@fas.harvard.edu

61. Aliaksei Novikau
Voitovich Research Group
B. I. Stepanov Institute of Physics
National Academy of Sciences
Minsk, BELARUS
novalexey@mail.ru

62. Prof. Lukas Novotny
ETH Zurich, Photonics Laboratory
8093 Zurich, SWITZERLAND
lnovotny@ethz.ch

63. Prof. Willie Padilla
Boston College, Department of Physics
140 Commonwealth Ave.
Chestnut Hill, MA 02467
617 552 0667 PHONE
617 552 8478 FAX
www2.bc.edu/~padillaw
willie.padilla@bc.edu

64. Nadia Peerboom
Department of Physics
Université de Namur ASBL
Rue de Bruxelles 61
5000 Namur, BELGIUM
nadia.peerboom@unamur.be

65. Dr. Wolfram Pernice
Karlsruhe Institute of Technology
D-76128 Karlsruhe, GERMANY
wolfram.pernice@googlemail.com

66. Prof. Yves-Alain Peter
Microphotonics Laboratory
Department of Engineering Physics
Ecole Polytechnique de Montréal
P.O. Box 6079, Station Centre-Ville
Montréal, Qué
bec H3C 3A7, CANADA
Phone : + 1 514 340 4711 ext. 3100
Fax : + 1 514 340 5195
Cell : + 1 514 465 7636
yves-alain.peter@polymtl.ca
http://www.polymtl.ca/pomp/en

67. Kasey Phillips
Mazur Group
School of Engineering and Applied Sciences
Harvard University
Cambridge, MA 02138, USA
Tel: 704 502-6156
kasey@seas.harvard.edu

68. Martin Pokorný
Graduate student at Department of Chemical Physics and Optics
Faculty of Mathematics and Physics

Charles University
Prague, CZECH REPUBLIC
martin@martin-pokorny.cz

69. Prof. Dr. Markus Pollnau
Chair, Integrated Optical MicroSystems Group
MESA+ Institute for Nanotechnology
University of Twente
P.O. Box 217, 7500 AE Enschede, THE NETHERLANDS
Building Carre, Room CR 2.429
Phone: +31-53-4891037
Fax: +31-53-4893996
E-mail: m.pollnau@utwente.nl
URL: http://www.utwente.nl/ewi/ioms/
Private: +49-5921-7265688

70. Nini Elisabeth Abildgaard Reeler
Nano-science Center
Department of Chemistry
University of Copenhagen
Copenhagen, DENMARK
nini@chem.ku.dk

71. Dr. Indrek Renge
Institute of Physics
University of Tartu
142 Riia Street
EE51014 Tartu, ESTONIA
renge@fi.tartu.ee

72. Orad Reshef
Mazur Group
School of Engineering and Applied Sciences
Harvard University
Cambridge, MA 02138, USA
orad@seas.harvard.edu

73. Taavi Repän
Institute of Physics
Laboratory of Laser Spectroscopy
University of Tartu
Tartu, ESTONIA
taavi.repan@ut.ee

74. Regina Schmitt
PhD student
Division of Solid State Physics
Lund University
Lund, SWEDEN
regina.schmitt@ftf.lth.se

75. Anton Sergeyev
Institut for Applied Physics
Albert-Einstein-Strasse 15
07745 Jena, GERMANY
anton.sergeyev@uni-jena.de

76. Elena Shabunya-Klyachkovskaya
National Academy of Science of Belarus
Institute of Physics, Laboratory of Nano-optics
Nezavisimosti Ave., 68
220072, Minsk, BELARUS
e.sh-k@ifanbel.bas-net.by
elena_kl@mail.ru

77. Qiang Shi
Karlsruhe Institut für Technologie
Institut für Angewandte Physik
Wolfgang- Gaede-Strasse 1
76131 Karlsruhe, GERMANY
qiang.shi@kit.edu

78. Luciano Silvestri
Department of Physics
Boston College
Chestnut Hill, MA 02467
Tel 617 840-2373
silveslu@bc.edu

79. Jiri Slaby
Biosensors Research Group
Institute of Photonics and Electronics ASCR, v.v.i.
Academy of Sciences of the Czech Republic
Chaberska 57
182 51 Prague 8, CZECH REPUBLIC
slaby@ufe.cz

80. Evegeny Smirnov
Doctoral Assistant

Laboratory of Physical and Analytical Electrochemistry (LEPA)
Ecole Polytechnique Federale de Lausanne (EPFL)
Lausanne, SWITZERLAND
Phone: +41 21 69 33163
evgeny.smirnov@epfl.ch

81. Walter Somerville
School of Chemical and Physical Sciences
Victoria University of Wellington
Wellington 6140, NEW ZEALAND
Walter.Somerville@vuw.ac.nz

82. Prof. Costas Soukoulis
IESL-FORTH
Heraklion, Crete, GREECE
Tel.: +302810391303; Fax: +30281039130
soukouli@iesl.forth.gr
soukoulis@materials.uoc.gr
soukoulis@ameslab.gov

83. Professor Emeritus Shammai Speiser
The Schulich Faculty of Chemistry
Technion-Israel Institute of Technology
Haifa 32000 ISRAEL
Phone: 972 4 829 3735; mobile: 972 52 3235369;
Fax: 972 4 829 5703
email: speiser@technion.ac.il
URL: http://schulich.technion.ac.il/faculty_member.php?id=30

84. Prof. Mark I. Stockman
29 Peachtree Center Avenue
Department of Physics and Astronomy
Georgia State University
Atlanta, GA 30302, USA
Tel.: (1) 678 457-4739
mstockman@gsu.edu

85. Timothy Threadgold
School of Systems Engineering
University of Reading
Reading RG6 6AH, UK
timothy.threadgold@gmail.com

86. Dr. Alessio Tierno
ISMN-CNR

Via P. Gobetti 101
40129 Bologna, ITALY
Tel.: + 39 0516398544
mobile : +39 3388953429
alessiotierno@gmail.com

87. Xavier Toledo-Fuentes
Department of Physics
Université de Namur ASBL
Rue de Bruxelles 61
5000 Namur, BELGIUM
xavier.toledo@unamur.be

88. Mario Tonin
Institut de Physiques de la Matière Condenée
Laboratory of Quantum OptoElectronics
CH-1015 Lausanna, SWITZERLAND
mario.tonin@epfl.ch

89. Neimantas Vainorius
Solid State Physics
Lund University
Lund, SWEDEN
neimantas.vainorius@ftf.lth.se

90. Maria Aurora Vincenti, PhD
ENEA C.R. Frascati
UTAPRAD-MNF
Via Enrico Fermi, 45
00044 Frascati (RM)
maria_aurora.vincenti@enea.it
Tel. +39 06 9400 5668
Fax +39 06 9400 5334
web page http://www.frascati.enea.it/UTAPRAD/statosolido_film.htm

91. Academician Aleksandr P. Voitovich (Co-Director of the Institute)
National Academy of Sciences of Belarus
66 Skaryna Av.
Minsk 220072 BELARUS
Tel.: (375) 17 284-1732; FAX: (375) 17 2840879
voitovich@imaph.bas-net.by; avoitovich@gmail.com

92. Andrej Vojtko
Department of Multilayers and Nanostructures
Institute of Physics

Slovak Academy of Sciences
Dúbravaská cesta 9
84511 Bratislava, SLOVAK REPUBLIC
andrej.vojtko@savba.sk

93. Prof. Martin Wegener
Karlsrhue Institute of Technology
D-76128 Karlsruhe, GERMANY
Tel. (49) 721 608-3400; FAX (49) 721 607-59
martin.wegener@kit.edu

94. Sarah Wiegele
Karlsruher Institut für Technologie (KIT)
Institut für Angewandte Physik
Centrum für funktionelle Nanostrukturen
Wolfgang-Gaede-Straße 1
Geb. 30.23 (Physikhochhaus)
76131 Karlsruhe, GERMANY
Sarah.Wiegele@kit.edu

95. Prof. Jean-Pierre Wolf
Faculté des Sciences
University of Geneva
Biophotonics Group
Gèneve, SWITZERLAND
Tel.: (41) 22 37965 94
jean-pierre.wolf@unige.ch

96. Iryna Yaremchuk
Deparment of Photonics
Lviv Polytechnic National University
12, Bandera str.,
UA-79013, Lviv, UKRAINE
yaremchuk@polynet.lviv.ua

97. Nadezda Zhdanova
PhD student
M.V. Lomonosov Moscow State University
Faculty of Physics
Quantum Electronics Division
Moscow, RUSSIA
nadezhda1989@gmail.com

98. Prof. Aleksei Zheltikov
Moscow State University

Moscow, RUSSIA
zheltikov@phys.msu.ru

99. Dr. Dimitrios Zografopoulos
Researcher, CNR Istituto per la Microelettronica e Microsistemi
Area della Ricerca di Roma 2 Tor Vergata
Via del Fosso del Cavaliere 100, 00133, Roma
Tel: +39 0649934579; Fax: +39 0645488066
Mobile: +39 3421870126
dimitrios.zografopoulos@artov.imm.cnr.it

Index

B. Di Bartolo et al. (eds.), *Nano-Structures for Optics and Photonics*,
NATO Science for Peace and Security Series B: Physics and Biophysics,
DOI 10.1007/978-94-017-9133-5,

MIX
Papier aus verantwortungsvollen Quellen
Paper from responsible sources
FSC® C105338

If you have any concerns about our products,
you can contact us on
ProductSafety@springernature.com

In case Publisher is established outside the EU,
the EU authorized representative is:
Springer Nature Customer Service Center GmbH
Europaplatz 3, 69115 Heidelberg, Germany

Printed by Libri Plureos GmbH
in Hamburg, Germany